Edited by
John Villadsen

Fundamental Bioengineering

Related Titles

Kadic, E., Heindel, T.J.

An Introduction to Bioreactor Hydrodynamics and Gas-Liquid Mass Transfer

2014

Print ISBN: 978-1-118-10401-9; also available in electronic formats

Lutz, S., Bornscheuer, U.T. (eds.)

Protein Engineering Handbook

Volume 3

2013

Print ISBN: 978-3-527-33123-9; also available in electronic formats

Buchholz, K., Kasche, V., Bornscheuer, U.T.

Biocatalysts and Enzyme Technology

2 Edition

2012

Print ISBN: 978-3-527-32989-2; also available in electronic formats

Mandenius, C., Björkman, M.

Biomechatronic Design in Biotechnology

A Methodology for Development of Biotechnological Products

2011

Print ISBN: 978-0-470-57334-1; also available in electronic formats

Mosier, N.S., Ladisch, M.R.

Modern Biotechnology

Connecting Innovations in Microbiology and Biochemistry to Engineering Fundamentals

2009

Print ISBN: 978-0-470-11485-8; also available in electronic formats

Planned Volumes of the "Advanced Biotechnology" Series:

Applied Bioengineering

T. Yoshida (Osaka University, Japan)

Emerging Areas in Bioengineering

H. N. Chang (KAIST, Korea)

Micro- and Nanosystems for Biotechnology

C. Love (MIT, USA)

Industrial Biotechnology

C. Wittmann & J. Liao (Saarland University, Germany & UCLA, USA)

Synthetic Biology

S. Panke & C. Smolke (ETH Zürich, Switzerland & Stanford University, USA)

Systems Biology

J. Nielsen & S. Hohmann (Chalmers University, Sweden & Gothenburg University, Sweden)

Edited by John Villadsen

Fundamental Bioengineering

WILEY-VCH
Verlag GmbH & Co. KGaA

Volume Editor

John Villadsen
Technical University of Denmark
Department of Chemical and Biochemical Engineering
Søltofts Plads
Building 229
2800 Kgs. Lyngby
Denmark

Series Editors

Sang Yup Lee
KAIST
Department of Chemical & Biomolecular Engineering
373-1; Guseong-Dong
291 Daehak-ro,Yuseong-gu
305-701 Daejon
South Korea

Jens Nielsen
Chalmers University
Department of Chemical and Biological Engineering
Kemivägen 10
412 96 Göteborg
Sweden

Gregory Stephanopoulos
Massachusetts Institute of Technology
Department of Chemical Engineering
77 Massachusetts Avenue
Cambridge, MA 02139
USA

Library of Congress Card No.: applied for

British Library Cataloguing-in-Publication Data
A catalogue record for this book is available from the British Library.

Bibliographic information published by the Deutsche Nationalbibliothek
The Deutsche Nationalbibliothek lists this publication in the Deutsche Nationalbibliografie; detailed bibliographic data are available on the Internet at <http://dnb.d-nb.de>.

ISSN: 2365-3035
Print ISBN: 978-3-527-33674-6
ePDF ISBN: 978-3-527-69746-5
ePub ISBN: 978-3-527-69745-8
Mobi ISBN: 978-3-527-69747-2
oBook ISBN: 978-3-527-69744-1

Cover Design Adam Design, Weinheim
Typesetting Thomson Digital, Noida
Printing and Binding
CPI Group (UK) Ltd, Croydon, CR0 4YY

C9783527336746_130326

The manufacturer's authorized representative according to the EU General Product Safety Regulation is Wiley-VCH GmbH, Boschstr. 12, 69469 Weinheim, Germany, e-mail: Product_Safety@wiley.com.

Contents

List of Contributors

Basti Bergdahl
Danmarks Tekniske Universitet
Novo Nordisk Foundation Center
for Biosustainability
Kogle Allé 6
2970 Hørsholm
Denmark

Jochen Förster
Danmarks Tekniske Universitet
Novo Nordisk Foundation Center
for Biosustainability
Kogle Allé 6
2970 Hørsholm
Denmark

Thomas Grotkjær
Novozymes A/S
Biomass Conversion
Business Development
Novo Allé
2880 Bagsvaerd
Copenhagen
Denmark

Markus Herrgård
Danmarks Tekniske Universitet
Novo Nordisk Foundation Center
for Biosustainability
Kogle Allé 6
2970 Hørsholm
Denmark

Jakob Kjøbsted Huusom
Technical University of Denmark
Department of Chemical and
Biochemical Engineering
Anker Engelunds Vej 1
Bygning 101A
2800 Kgs Lyngby
Denmark

Daniel Machado
Universidade do Minho
Centro de Engenharia Biológica
4710-057 Braga
Portugal

Sunil Nath
Indian Institute of Technology,
Delhi
Department of Biochemical
Engineering and Biotechnology
Hauz Khas
New Delhi 110016
India

Sebastian Niedenführ
Forschungszentrum Jülich GmbH
Institute of Bio- and Geosciences
IBG-1: Biotechnology
52425 Jülich
Germany

Katharina Nöh
Forschungszentrum Jülich GmbH
Institute of Bio- and Geosciences
IBG-1: Biotechnology
52425 Jülich
Germany

Henk Noorman
DSM Biotechnology Center
A. Fleminglaan 1
2613 CA Delft
The Netherlands

Matthias Reuss
University of Stuttgart
Stuttgart Research Center
Systems Biology
Nobelstr. 15
70569 Stuttgart
Germany

Nikolaus Sonnenschein
Danmarks Tekniske Universitet
Novo Nordisk Foundation Center
for Biosustainability
Kogle Allé 6
2970 Hørsholm
Denmark

Bernhard Sonnleitner
Zurich University of
Applied Sciences (ZHAW)
Institute for Chemistry and
Biological Chemistry (ICBC)
Biochemical Engineering
Einsiedlerstrasse 29
8820 Waedenswil
Switzerland

John Villadsen
Technical University of Denmark
(DTU)
Department of Chemical and
Biochemical Engineering
Building 229
2800 Kgs Lyngby
Denmark

Wolfgang Wiechert
Forschungszentrum Jülich GmbH
Institute of Bio- and Geosciences
IBG-1: Biotechnology
52425 Jülich
Germany

About the Series Editors

Sang Yup Lee is Distinguished Professor at the Department of Chemical and Biomolecular Engineering at the Korea Advanced Institute of Science and Technology. At present, Prof. Lee is the Director of the Center for Systems and Synthetic Biotechnology, Director of the BioProcess Engineering Research Center, and Director of the Bioinformatics Research Center. He has published more than 500 journal papers, 64 books, and book chapters, and has more than 580 patents (either registered or applied) to his credit. He has received numerous awards, including the National Order of Merit, the Merck Metabolic Engineering Award, the ACS Marvin Johnson Award, Charles Thom Award, Amgen Biochemical Engineering Award, Elmer Gaden Award, POSCO TJ Park Prize, and HoAm Prize. He is Fellow of American Association for the Advancement of Science, the American Academy of Microbiology, American Institute of Chemical Engineers, Society for Industrial Microbiology and Biotechnology, American Institute of Medical and Biological Engineering, the World Academy of Science, the Korean Academy of Science and Technology, and the National Academy of Engineering of Korea. He is also Foreign Member of National Academy of Engineering, USA. In addition, he is honorary professor of the University of Queensland (Australia), honorary professor of the Chinese Academy of Sciences, honorary professor of Wuhan University (China), honorary professor of Hubei University of Technology (China), honorary professor of Beijing University of Chemical Technology (China), and advisory professor of the Shanghai Jiaotong University (China). Apart from his academic associations, Prof. Lee is the editor-in-chief of the *Biotechnology Journal* and is also contributing to numerous other journals as associate editor and board member. Prof. Lee is serving as a member of Presidential Advisory Committee on Science and Technology (South Korea).

Jens Nielsen is Professor and Director to Chalmers University of Technology (Sweden) since 2008. He obtained an MSc degree in chemical engineering and a PhD degree (1989) in biochemical engineering from the Technical University of Denmark (DTU) and after that established his independent research group and was appointed full professor there in 1998. He was Fulbright visiting professor at MIT in 1995–1996. At DTU, he founded and directed the Center for Microbial Biotechnology. Prof. Nielsen has published more than 350 research papers and coauthored more than 40 books, and he is inventor of more than 50 patents. He has founded several companies that have raised more than 20 million in venture capital. He has received numerous Danish and international awards and is member of the Academy of Technical Sciences (Denmark), the National Academy of Engineering (USA), the Royal Danish Academy of Science and Letters, the American Institute for Medical and Biological Engineering and the Royal Swedish Academy of Engineering Sciences.

Gregory Stephanopoulos is the W.H. Dow Professor of Chemical Engineering at the Massachusetts Institute of Technology (MIT, USA) and Director of the MIT Metabolic Engineering Laboratory. He is also Instructor of Bioengineering at Harvard Medical School (since 1997). He received his BS degree from the National Technical University of Athens and his PhD from the University of Minnesota (USA). He has coauthored about 400 research papers and 50 patents, along with the first textbook on metabolic engineering. He has been recognized by numerous awards from the American Institute of Chemical Engineers (AIChE) (Wilhelm, Walker and Founders awards), American Chemical Society (ACS), Society of Industrial Microbiology (SIM), BIO (Washington Carver Award), the John Fritz Medal of the American Association of Engineering Societies, and others. In 2003, he was elected member of the National Academy of Engineering (USA) and in 2014 President of AIChE.

1
Introduction and Overview

John Villadsen

Bioengineering is a relatively new addition to a long list of terms starting with "bio." It is broadly defined as "the application of engineering principles to biological systems." Bioengineering can include elements of chemical, electrical and mechanical engineering, computer science, materials, chemistry and biology. The systems that are analyzed range from cell cultures and enzymes applied in the bio-industry and in bioremediation to prosthetics, construction of models for organs such as liver, drug-delivery systems and numerous other subjects in biomedical engineering, all requiring an understanding of transport phenomena (mass, heat, and momentum transfer) and kinetics, combined in often large mathematical models. Besides a working knowledge of these core chemical engineering disciplines, a successful study of a problem in bioengineering requires an insight into the core disciplines of biology and biochemistry, specifically in human physiology when the goal is, for example, to construct a new cancer drug delivery system.

In this volume, coauthored by nine scientists, mostly working in academic institutions or in the bio-industry, the focus is on application of bioengineering in the emerging "white biotechnology" industry. The design of bioremediation systems closely follows the principles of analysis and design of industrial bioprocesses. This text will also prove valuable for environmental engineers. The biomedical applications of the text are, however, also quite obvious. Thus, the important but complex application of mesenchymal stem cells to treat osteoporosis is based on an optimal growth strategy of the cell culture on a scaffold at the right liquid flow with the right oxygen and nutrient availability. Here, kinetics and transport phenomena are coupled to basic biology and biochemistry, and design of the system is based on a complex model for the interaction between scaffold, cells, and nutrients.

In Chapters 5, 6, and 8, the reader will find self-contained accounts of the tools that together make it possible to understand the behavior of cell cultures and enzymatically catalyzed reactions: The interaction of metabolic network reactions in steady state and during transients, analyzed by mathematical models and solved by state-of-the-art computer software.

Fundamental Bioengineering, First Edition. Edited by John Villadsen.

In Chapter 16, a structural framework for successful scale-up of bioreactions from laboratory scale to large industrial scale is presented. In Chapter 17, the sequence of management decisions that may lead to new business ventures in the bio-industry is discussed.

The analyses of cultures on the level of the cell are authored by three leading European scientists. Each author gives - as far as possible - a complete account of his subject, illustrated with examples and with sufficient detail to give readers, both in industry and in graduate classes at universities, a fair chance to understand and *utilize* the very powerful analytical tools presented in the three chapters.

The two Chapters 16 and 17 on large-scale bioreactors and on the business opportunities in the bio-industry are written by leading experts from two major bio-industrial companies, DSM in the Netherlands and Novozymes in Denmark. These chapters could serve as guidelines for prospective business ventures in the industry.

Although the focus of this book is on the bioreactor, Chapters 12 and 13 cover further processing of the effluent from the bioreactor. The author, a distinguished Indian bioscientist, gives a short introduction to the subject of downstream processing. Also, a survey of measuring, monitoring, and control of bioreactions is included. In Chapter 14, a leading expert on chemical analysis to capture key fermentation variables and on using the experimental data in analysis of fermentation broths gives an easy-to-read but largely complete survey of the subject. In Chapter 15, a young expert in control of chemical processes, discusses control problems in bioreactors, specifically addressing the challenges of bio-system control.

Finally, the content of the book is tied together by seven chapters (2, 3, 4, 7, 9, 10, and 11) written by the editor of this work. These chapters introduce a common nomenclature for the whole book, with introductory material on stoichiometry, kinetics, thermodynamics, and design of ideal and real bioreactors. It is hoped that the introductory chapters, illustrated with many simple examples, will make it easier to read the advanced chapters, especially since there are frequent cross references between introductory and advanced chapters.

Part One
Fundamentals of Bioengineering

2
Experimentally Determined Rates of Bio-Reactions

John Villadsen

Summary

Rates of bioreactions are introduced as measured terms in steady-state mass balances for a continuous stirred tank reactor (CSTR). Both mass balances and the reaction rates have the same form for enzymatically catalyzed reactions and for reactions with living cells. In cell reactions, the rate of biomass formation is included through a separate mass balance. Reactants absorbed in the liquid phase from a gas phase are treated separately, and it is shown how experimental errors can lead to errors in the calculated rates. The black-box model for a cell-reaction stoichiometry is introduced and the yield coefficients are defined. Finally, different methods of controlling the CSTR at steady state are discussed.

2.0
Introduction

The rate of an enzymatically catalyzed bioreaction, or of a reaction that involves living cells (microbial, animal, or plant cells), can be determined experimentally in a *bioreactor.* The bioreactors used in academic research or in an industrial R&D department to obtain reaction rates are normally glass vessels of 0.5–5 l working volume V. A typical laboratory reactor is shown in Figure 2.1. It is well stirred either by an internal mechanical stirrer (a hydrofoil or a turbine) or by a magnetic stirrer, operated from the outside of the reactor. In all cases, the mixing of liquid feed into the medium volume V is supposed to be good enough to ensure that there is no spatial variation of substrates or products in the reactor. Batch operation or continuous operation of the reactor is typically used, and the assumption of perfect mixing in the medium volume V will ensure that the concentrations S_i of substrates and P_i of products are the same at any point in the reactor. If the continuous stirred tank reactor (CSTR) is operated in *steady state,* there is no accumulation of either products or substrates. The liquid flow v_f and the feed concentrations of substrates $s_{f,i}$ are kept constant in time. Now the medium volume V and the concentrations of substrates and products both in the reactor $[s_i, p_j]$ and in the effluent concentrations $[s_{e,i}, p_{e,i}]$ are constant in time. In the batch reactor, one starts with a high concentration of

Fundamental Bioengineering, First Edition. Edited by John Villadsen.

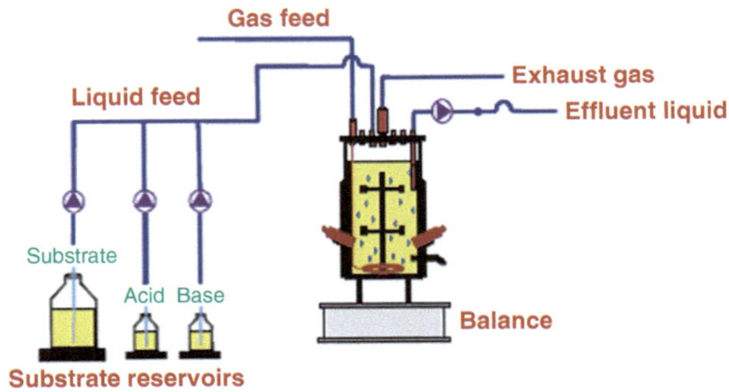

Figure 2.1 CSTR with substrate feed of liquid medium and of gas through a sparger. A separate feed of acid/alkali for pH control is used. In pH auxostats (see below) the feed of alkali is used as control variable.

substrates s_i^0 (+ a small amount of biomass for a fermentation), and S_i are converted to P_j over time. The volume of medium V in the reactor is constant in time.

A gas-phase substrate is introduced to the liquid through a sparger. It is absorbed in the liquid and is consumed by the reaction. Gaseous products are transferred to the gas phase by desorption from the liquid.

Then, mass balances for substrates and products are set up. These mass balances define the reaction rates, and solution of the mass balances allows the rates to be calculated based on measured concentrations $[s_i, p_i]$. To illustrate the procedure we shall use mass balances derived for a steady state CSTR, and the equipment is shown in Figures 2.1 and 2.2.

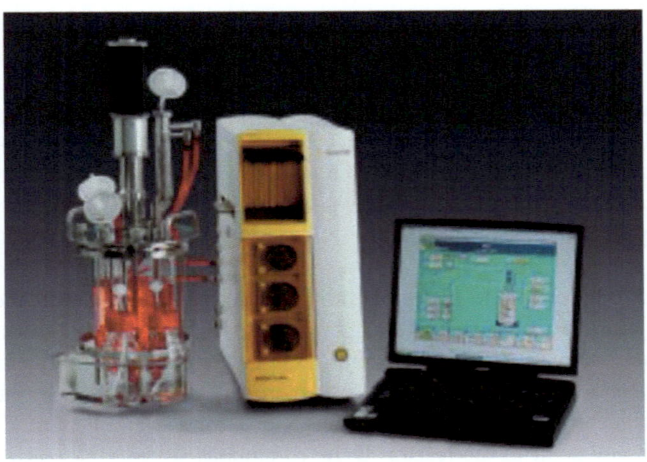

Figure 2.2 A commercial laboratory reactor (Biostat® A plus from Sartorious) for medium volume between 0.4 and 5 l. The reactor is supported by packages for either microbial cell or animal/plant cell cultures. The microbial package includes two 6-blade Rushton turbines (see Chapter 11), two gas-inlets, ports for inoculation, automatic and manual samplers, and temperature control via a heating blanket and cooling finger. Control of pH, T, dissolved oxygen (DO), stirrer speed, air flow rate, and foam control.

2.1
Mass Balances for a CSTR Operating at Steady State

The steady-state mass balances for a continuous reactor with a working volume V, for example, in m^3 liquid medium, in which an enzymatically catalyzed reaction occurs, is given by Eq. (2.1).

$$q_{s,i}\,V + v\,(s_{f,i} - s_{e,i}) = 0,\; q_{p,i}\,V + v\,(p_{f,i} - p_{e,i})\text{ or } q_{s,i} + D\,(s_{f,i} - s_{e,i}) = 0,\; q_{p,j} + D\,(p_{f,j} - p_{e,j}) = 0. \tag{2.1}$$

In Eq. (2.1), v is the liquid flow through the reactor in, for example, m^3 liquid h^{-1}. v/V is defined as the dilution rate D (h^{-1}). s_i is the concentration of the ith substrate in the reactor, $s_{e,i}$ the concentration in the liquid effluent, and $s_{f,i}$ the concentration in the feed. For the steady-state continuous reactor (see also Section 10.1), $s_i = s_{e,i}$. The same nomenclature is used for the product concentrations.

For each substrate, the *volumetric rate of production* $q_{s,i}$ of S_i, for example, in units of g S_i or C-mol S_i l^{-1} h^{-1}, is multiplied by the reactor volume V to give the production rate of S_i (e.g., in g S_i h^{-1}). To the production term is added $vs_{f,I}$, the amount of S_i introduced through the feed and subtracted $vs_{e,i}$, the substrate that leaves the reactor. The sum of the three terms of the steady-state mass balance is zero. The mass balance for product P_j contains the same terms.

$q_{s,i}$ ($i = 1, 2, \ldots, N$) and $q_{p,j}$ ($j = 1, 2, \ldots, M$) are always defined as (volumetric) rates of *production*. Hence $q_{s,i}$ are always negative and $-q_{s,i}$ is, therefore, *the volumetric rate of consumption* of S_i.

As is the case for any catalyzed reaction, the rate can be defined either per volume reactor (q) or per unit of catalyst (r), for example, per unit mass of catalyst added to each l of reactor. This second definition defines the *specific* reaction rates $r_{s,i}$ and $r_{p,j}$, which are obtained from $q_{s,i}$ and $q_{p,j}$ by division with e, the concentration of enzyme E, for example, in units of g E l^{-1}. The specific rates define the *activity* of the enzyme E to convert S_i to P_j. These definitions are further discussed in Section 7.1.

In reactions with living cells, the cell mass catalyzes the conversion of S_i to P_j, but the *substrate is also used to form more biomass X* – the reaction is autocatalytic. Hence biomass is also a product, and similar to Eq. (2.1) one obtains:

$$-q_{s,i} = D\,(s_{f,i} - s_i);\; q_{p,j} = D\,(p_j - p_{j,f})\text{ and } q_x = D\,(x - x_f);\; D = v/V. \tag{2.2}$$

$$-r_{s,i} = D\,(s_{f,i} - s_i)/x;\; r_{p,j} = D\,(p_j - p_{j,f})/x\text{ and } r_x = D\,(x - x_f)/x;\; D = v/V. \tag{2.3}$$

The unit of r_i could be g S_i produced/g biomass/h (i.e., r_{si} is negative), g P_i (g X h)$^{-1}$, g X (g X h)$^{-1}$.

r_x is defined as *the specific growth rate* of the culture, and in most biotechnology literature r_x is called μ. We shall only use this latter symbol when its meaning is obvious. It is seen that μ is the *ability* (or *activity*) of the biomass in the reactor to make more biomass. An active culture can make much biomass per g biomass present per hour – a resting culture has a low value of $r_x = \mu$.

Since some substrates (e.g., O_2) are captured from the gas phase and some products (e.g., CO_2) are released to the gas phase, one needs to add an extra term in Eqs. (2.1) and (2.2) for these reaction species. This term is $q^{\mathrm{T}}_{\mathrm{sk}}$ or $q^{\mathrm{T}}_{\mathrm{pm}}$ (e.g., in moles of O_2 or CO_2 transferred $(\mathrm{l\ h})^{-1}$).

$$(q_{\mathrm{sk}} + q^{\mathrm{T}}_{\mathrm{sk}}) = D(s_{\mathrm{kf}} - s_{\mathrm{k}}) \text{ and } (q_{\mathrm{pm}} + q^{\mathrm{T}}_{\mathrm{pm}}) = D(p_{\mathrm{m}} - p_{\mathrm{m,f}}). \tag{2.4}$$

In Eq. (2.4), the term $q^{\mathrm{T}}_{\mathrm{sk}}$ is positive (there is an influx of O_2 to the liquid), while $q^{\mathrm{T}}_{\mathrm{pm}}$ is negative.

The amount of reaction species k and m transferred between the gas and the liquid phase can be determined from a mass balance *on the gas phase.*

$$q^{\mathrm{T}}_{\mathrm{sk}}\, V = \frac{1}{RT}(v_{\mathrm{g,f}}\, \pi_{\mathrm{k,f}} - v_{\mathrm{g}}\, \pi_{\mathrm{k}}), \qquad q^{\mathrm{T}}_{\mathrm{pm}}\, V = \frac{1}{RT}(v_{\mathrm{g,f}}\, \pi_{\mathrm{m,f}} - v_{\mathrm{g}}\, \pi_{\mathrm{m}}). \tag{2.5}$$

Here, $(v_{\mathrm{g,f}}, v_{\mathrm{g}})$ are the volumetric gas flows $(\mathrm{l\ h}^{-1})$ in and out of the reactor, and $(\pi_{\mathrm{k,f}}, \pi_{\mathrm{k}})$ and $(\pi_{\mathrm{m,f}}, \pi_{\mathrm{m}})$ are the *partial pressures* of the substrate k and of the product m in the inlet and in the outlet from the reactor. $\pi_{\mathrm{k,f}} = y_{\mathrm{k}}\, P$, where y_{k} is the volume fraction of S_{k} in the gas (e.g., 0.21 mol O_2 $(\mathrm{mol}\ v_{\mathrm{gf}})^{-1}$ for O_2 in the inlet when the reactor is sparged with air) and P is the *total pressure.* $v_{\mathrm{g,f}}\, \pi_{\mathrm{k,f}}/RT$ = mol O_2 fed to the reactor per hour, and $(1/RT)(v_{\mathrm{g,f}}\, \pi_{\mathrm{k,f}} - v_{\mathrm{g}}\, \pi_{\mathrm{k}})$ is the moles of O_2 transported to the liquid phase. With π in atm and T in K, the value of the gas constant R is 0.082051 atm $(\mathrm{mol\ K})^{-1}$.

The mass transfer can also be calculated by Eq. (2.6). This is an empirical relation between a driving force $(s^*_{\mathrm{k}} - s_{\mathrm{k}})$ and $q^{\mathrm{T}}_{\mathrm{sk}}$, the volumetric mass transfer. s_{k} is the liquid phase concentration of substrate S_{k} and s^*_{k} is the *saturation* concentration of S_{k} in equilibrium with a gas phase with (approximately) the partial pressure $(\pi_{\mathrm{k,f}} + \pi_{\mathrm{k}})/2$ – or a more complicated expression for the *average* gas-phase partial pressure as explained in Chapter 11. $k_{\mathrm{l}}a$ is *the volumetric mass transfer coefficient,* a first-order rate constant (unit e.g., h^{-1}) for the mass transfer process. The mass transfer coefficient depends on the *power consumption* (unit, for example, W) to mix the liquid phase.

$$q^{\mathrm{T}}_{\mathrm{sk}} = k_{\mathrm{l}}a(s^*_{\mathrm{k}} - s_{\mathrm{k}}). \tag{2.6}$$

Combination of Eqs. (2.5) and (2.6) is one way of experimentally determining $k_{\mathrm{l}}a$ by simultaneous measurement of $[s_{\mathrm{k}}, \pi_{\mathrm{kf}}, \pi_{\mathrm{k}}]$ and $[v_{\mathrm{gf}}, v_{\mathrm{g}}]$ at different stirring intensities of the liquid. Once an empirical relation has been set up between the power input and the rate constant $k_{\mathrm{l}}a$, the relation can be used under similar conditions to predict the mass transfer $q^{\mathrm{T}}_{\mathrm{sk}}$ for any power input.

When the rate of formation of *one* of the reaction species has been determined, then, for a *single* enzymatic reaction or fermentation process, the rates of formation for all other reaction species can be calculated via the *yield coefficients* Y_{ij}. Y_{ij} is defined as the rate of formation of component j relative to the formation of another component i. $Y_{ij} = |r_j/r_i| = |q_j/q_i|$. The symbol || is used to ensure that Y_{ij} is positive also when the numerator and denominator have different signs.

Based on the yield coefficient, any rate q_j *different* from the *key rate* q_i can be found as follows:

$$q_j = Y_{ij}\, q_i, \quad \text{and } r_j = q_j/e, \quad \text{or } q_j/x;\ Y_{ij} = \left|\frac{c_{j,\mathrm{f}} - c_j}{c_{i,\mathrm{f}} - c_i}\right|. \tag{2.7}$$

It must be emphasized that Eq. (2.7) is *only* true if the yield coefficients are constant in the range of investigation of the reaction. This is the case if the reaction is described by a single stoichiometric equation for all investigated values of the dilution rate D.

Example 2.1 Calculation of rates and stoichiometry based on the key reactant

Catalase breaks down hydrogen peroxide by the stoichiometric equation $H_2O_2 \rightarrow H_2O + ½O_2$. This is the *only* overall stoichiometric equation by which O_2 is formed from H_2O_2. *The kinetic mechanism* of the reaction is, however, not at all revealed from the overall stoichiometry. Similarly, lactobionic acid (lba) is produced from lactose in the reaction: lactose + O_2 + $H_2O \rightarrow$ lba + H_2O_2 catalyzed by a carbohydrate oxidase, a reaction discussed further in Example 7.4. In the first reaction, $Y_{H_2O_2,O_2} = 1/2$ mol/mol, and in the second $Y_{\text{lactose},O_2} = Y_{\text{lactose},H_2O_2} = Y_{\text{lactose,lab}} = 1$ mol/mol. The production of lactobionic acid is $q_{\text{lba}} = Y_{\text{lac,lba}}\,(-q_{\text{lactose}})$ or $D\,(c_{\text{lba}} - c_{\text{lba,f}}) = Y_{\text{lac,lba}}\,(c_{\text{lactose,f}} - c_{\text{lactose}})$. The last relation shows that the effluent concentration of lactobionic acid (and of H_2O_2) can be calculated from $[c_{\text{lactose,f}}, c_{\text{lactose}}]$.

In enzymatically catalyzed reactions, the yield coefficients Y_{ij} are true stoichiometric coefficients, exact numbers just as in conversion of N_2 and H_2 to NH_3 by the stoichiometry $N_2 + 3\,H_2 \rightarrow 2\,NH_3$. This is also the case for reactions involving living cells if the same stoichiometry holds for all operational conditions, for example, for all permissible D values between 0 and $D_{\max}$. At $D_{\max} > v_{\max}/V$, the flow through the reactor is larger than the maximum production rate of cells. The culture washes out and steady-state operation cannot be maintained. In the stoichiometry of fermentation reactions, the coefficients are empirical numbers that can be interpreted when, as in Chapter 4, a good metabolic model is constructed for the reaction. Thus, in anaerobic fermentation with the yeast *Saccharomyces cerevisiae*, the substrate, glucose, is converted to ethanol, glycerol, CO_2, and more biomass. When the elementary composition of *S. cerevisiae* is given by $X = CH_{1.74}O_{0.6}N_{0.12}$, one obtains the following equation [1]:

$$\begin{aligned} CH_2O + 0.01658\,NH_3 \rightarrow\ & 0.1381\,X + 0.5160\,CH_3O_{0.5} + 0.2759\,CO_2 \\ & + 0.070\,CH_{8/3}O + 0.037\,H_2O. \end{aligned} \tag{1}$$

In Eq. (1) all carbon-containing compounds are normalized to contain 1 C-mol carbon. This leads to somewhat unusual formulas for ethanol

($CH_3O_{0.5}$), glycerol ($CH_{8/3}O$), and glucose (CH_2O), but the nomenclature is useful in quantitative work since the basis is always 12 g carbon, whereas molecules such as $C_6H_{12}O_6$ break down to smaller molecules (such as trioses in the glycolysis pathway (see Figure 4.2) or combine to larger molecules (such as Butyryl-CoA by condensation of two molecules of Acetyl-CoA in the solvent producing pathways of *Clostridium acetobutylicum* (see Figure 4.14). The advantage of writing the fermentation stoichiometry on a C-mol basis becomes apparent in later chapters. Here, we are content to notice that the closing of the carbon balance is immediately obvious when the yield coefficients on both sides of Eq. (1) are compared: $1 = \Sigma Y_{sj}$ ($j = 1$–4). $Y_{sn} = 0.01658$ can be obtained from Y_{sx} since for each carbon in the biomass 0.12 mol N is consumed and there are no other sinks for N in the stoichiometry, that is, $Y_{xn} = 0.12$, $Y_{sn} = 0.12 \cdot 0.1381 = 0.01658$. $Y_{sw} = 0.037$ can be obtained from either an H or an O balance on the stoichiometry.

The rate of production of, for example, ethanol is related to the rate of consumption of glucose by $q_e = 0.5160\ (-q_s)$ or $c_e - c_{ef} = 0.5160\ (23/30)\ (s_f - s)\ \text{g l}^{-1}$. Thus, if the feed contains 25 and the effluent 1.378 g glucose l^{-1}, then $c_e - c_{ef} = 9.34$ g/l. Unless the reactor considered receives its feed from a preceding reactor in a sequence of CSTRs, the feed is sterile and contains no products, $[x_f, p_{jf}] = [0,0]$. In that case, $[x, c_e = e, c_{glycerol} = g] = [0.1370 \cdot 25.02 \cdot 23.622/30 = 2.70, 9.34, 1.67]$ g/l. In this calculation, the formula weights of [S, X, E, G] = [30, 25.02, 23, 30.33] g $(\text{C-mol})^{-1}$ have been used to find $Y_{s,pj}$ in g g^{-1} from the stoichiometry given in Eq. (1).

Since the stoichiometry (1) is supposed to hold for all acceptable D values, one can finally calculate the reactor volume needed for a given production rate. Let the desired production rate of ethanol be 900 kg h^{-1} in an industrial reactor design based on the stoichiometry (1). Then $v = 900/9.34 = 96.4\ \text{m}^3\ \text{h}^{-1}$. If the process is carried out at $D = 0.32\ \text{h}^{-1}$, one needs a reactor of working volume $V = 96.4/0.32 = 301\ \text{m}^3$.

This last example illustrates how the so-called "black box" stoichiometric Eq. (1) can be set up based on steady-state experiments in a CSTR. To find the yield coefficients, one just needs the effluent concentrations $[s, x, e, g] = [23.622, 2.70, 9.34, 1.67]$ calculated above. The yield coefficient for CO_2 can be found using the carbon balance, even when the CO_2 production is not measured. In fact, it will be demonstrated in Example 4.5 that with a reliable metabolic model in hand only *one* production rate, for example, q_x (or r_x), needs to be measured to find *all* the yield coefficients on the right-hand side of Eq. (1) when the substrate consumption rate is normalized to 1. Nevertheless, it is highly advisable to include all available measured data in the vector of experimental results since, as shown in Chapter 4, experimental errors can significantly distort the calculated yield coefficients if only the minimum amount of data (in the example above the effluent glucose and biomass concentrations) are used. Although the experimental work of [1] on which Eq. (1) is based is done with the highest possible precision obtainable in a state-of-the-art laboratory bioreactor, the C balance (the author had also measured q_{CO_2}) fits to "only" 0.995. Today, one must demand that experimental data obtained in a good laboratory reactor are accurate enough to make the carbon balance close to 0.985–0.99; otherwise, the data are suspicious. The use of the carbon and a redox balance to be introduced in Chapter 3 is *either* to calculate missing yield coefficients *or* to check for inconsistencies in the experimental data. Another use of these two *fundamental* balances is illustrated in Chapter 4. There the goal is to find parameters in metabolic models, especially empirical yield coefficients in the metabolic reaction that describes biomass formation.

Example 2.2 Calculation of rates for reaction species that are both in the gas phase and in the liquid phase

It appears to be fairly simple to apply Eqs. (2.4) and (2.5) to determine the transfer rate q_i^T and thereafter the rate of production q_i of reaction species that are present both in the liquid phase and in the gas phase. Still the rates can be erroneously calculated, and in this example it will be shown how the measured data must be treated to obtain the correct rate values.

First the rates of transfer of oxygen and carbon dioxide $[q_o^T, q_c^T]$ as determined from continuous readings of the headspace molar fractions will be considered. All modern bioreactors are equipped with sufficiently good monitoring equipment for these variables, and readings are considered to be accurate to within a few percent relative. Hence, they are considered as keys to an efficient control of the reactor. For example, in aerobic cultivations *with glucose as C substrate*, the *respiratory quotient* RQ $q_c/(-q_o)$ is in the range 1.04–1.06 when no metabolic products are formed, except biomass and CO_2.

Consider the production of biomass by aerobic cultivation of the bacterium *Methylococcus capsulatus* on methanol. *M. capsulatus* can grow on both methane and methanol, and the product is sold as "single-cell protein (SCP)" as feed to a range of animals both in husbandry and in aquacultures. In this process, a valuable product (~US$ 1600 per ton) is obtained from raw materials that are available in huge quantities and with a price in the range of US$ 200 per ton based on shale gas.

A 10 l well-stirred continuous bioreactor is sparged with air (21% O_2, 79% N_2 + Ar, and 0.04% CO_2) at a feed rate of $v_{g,f} = 10\ \text{l min}^{-1}$ (1 volume gas per volume medium per minute = 1 vvm). The total pressure is 1 atm, and the temperature of the air is equal to the cultivation temperature, $T = 45\ °C$. Both the inlet and the effluent gas are free of water. In the effluent, one measures the mole fractions $[y_o, y_c] = [0.180, 0.0174]$.

The flow of air from the reactor v_g (45 °C, 1 atm) is calculated in the following way: 100 l of $v_{g,f}$ contains 79 l N_2 + Ar, while 100 l of v_g contains 100 − 18 − 1.74 l = 80.3 l of the inert gases N_2 + Ar. Since the content of inert gases has not changed, v_g must be smaller than $v_{g,f}$, and $v_g = 79/80.3\ v_{g,f} = 0.9838\ v_{g,f}$. The 10 l reactor has received $600 \cdot (0.21 - 0.18 \cdot 0.9838) = 19.75$ l O_2 h^{-1} or $(19.75 \cdot 1$ (atm)/0.08206/318.2 (K))/10 (l) = 0.0756 mol O_2 (l h)$^{-1} = q_o^T$. $(-q_c^T) = 600 \cdot (0.0174 \cdot 0.9838 - 0.0004) \cdot 0.00383 = 0.0385$ mol CO_2 (l reactor h)$^{-1}$.

The liquid-phase concentration of oxygen is very small, both in the feed and in the effluent. From Eq. (9.2) at 45 °C and with pure O_2 in the gas phase ($\pi_{O_2} = 1$ atm), one obtains $s_o^* = 0.98 \cdot 10^{-3}$ mol (l reactor)$^{-1}$. When the gas phase is air with $\pi_{O_2} = 0.21$ atm, the saturation concentration is $s_o^* = 0.21 \cdot 0.98 \cdot 10^{-3} = 0.206 \cdot 10^{-3}$ mol 1 (reactor)$^{-1}$. The effluent concentration of O_2 is even smaller, perhaps 10% of s_o^*. Hence, in Eq. (2.4), the term $D(s_{o,f} - s_o)$ is several orders of magnitude smaller than q_o^T, and the rate of oxygen consumption $(-q_o)$ is very close to the experimentally determined $q_o^T = 0.0756$ mol O_2 (l reactor h)$^{-1}$. Similarly, for steady-state cultivation, the total concentration of CO_2 (as dissolved CO_2, HCO_3^-, and CO_3^{-2}) is virtually the same in the inlet and outlet, and $q_c \sim (-q_c^T) = 0.385$ mol CO_2 (l reactor h)$^{-1}$.

The accuracy of, especially, the calculation of q_o^T increases when the difference between the mole fraction in the feed gas and the exhaust gas increases. Thus, with $v_{g,f} = 300$ l h^{-1} (0.5 vvm), one obtains the same $q_o^T = 0.0756$ mol O_2 (l reactor h)$^{-1}$ at an exhaust O_2 mole fraction $y_o = 0.1465$. This clearly gives a higher accuracy in the determination of q_o^T, but one must not decrease $v_{g,f}$ too much, since the mass transfer coefficient $k_l a$ decreases with $v_{g,f}$, and it will eventually become difficult to transfer the required q_o^T. Also, at a low $v_{g,f}$ value, the oxygen in the gas bubbles can become severely diluted with a gaseous product such as CO_2, and especially if the inert content of $v_{g,f}$ is small as is the case when enriched air is used, the partial pressure of O_2 in the bubbles decreases when the bubbles pass through the liquid phase.

Serious errors in the determination of q_i^T can result if the content of water in $v_{g,f}$ differs from that of v_g. Thus, when $v_{g,f}$ is bone-dry and v_g is not adequately dried, q_o^T is severely overestimated if the water evaporated into the gas is not taken into account. At 45 °C and 1 bar, the vapor pressure of H_2O over liquid water is 71.9 mm Hg, that is, the equilibrium mole fraction of H_2O in the exhaust gas is 71.9/760 = 0.0946 bar. In the example given above with $v_{g,f} = 600$ l h^{-1} bone-dry air and $q_o^T = 0.0756$ mol O_2 (l reactor h)$^{-1}$, the exhaust (wet) air would be $v_g = 600 - 19.75 + 10.03 + 0.0946\ v_g \rightarrow v_g = 652$ l h^{-1}. The mole fraction of O_2 would be $y_o = (0.21 \cdot 600 - 19.75)/652 = 0.1630$. Thus, if one *measures* v_g to be 652 l h^{-1}, the measured value of $y_o = 0.1630$ in the exhaust gas correctly predicts $q_o^T = 0.0756$ mol O_2 (l reactor h)$^{-1}$. If, however, one erroneously assumes that $v_g = 0.9838\ v_{g,f} = 590$ l h^{-1} as is the case when the exhaust air is bone-dry, then the transferred volume of O_2 would be calculated to $600 \cdot (0.21 - 0.9838 \cdot 0.1630) = 29.8$ l h^{-1}, and q_o^T would be overestimated by a factor 29.8/19.75 = 1.5.

Another kind of error in the calculation of rates occurs if one of the products formed in the liquid phase is partly stripped to the gas phase.

This happens in aerobic cultivations where O_2 is supplied by sparging the medium with air (or air enriched with O_2). But it also happens in anaerobic cultivations when fully anaerobic conditions are ensured by sparging the medium with an inert, for example, N_2.

Let the mole fraction of the product P in the liquid phase be $x_p = (p/55.5)$ where p is the concentration of P in the medium (mol l^{-1}) and $55.5 = 1000/18 =$ moles H_2O/l H_2O. The equilibrium mole fraction in the gas above the medium $y_p = \gamma_p\, x_p\, \pi_p$, where γ_p is the *activity coefficient* of P in the liquid phase and π_p is the partial pressure of P in the gas phase = vapor pressure at medium temperature T/total pressure. The amount of P removed to the gas phase is $v_{g,f}\, y_p$ when $v_g = v_{g,f}$, a reasonable assumption since y_p is usually very small. For $x_p \to 0$, the situation in most bioreactions, $\gamma \to \gamma_{\text{inf dilution}}$, which can be found in many tables or calculated by group contribution methods such as UNIFAC. Thus, for the three fermentation products, ethanol, acetaldehyde, and 1-butanol, $\gamma_{\text{inf dilution}} = [6.61, 7.5, 49.4]$ at 30 °C. Even a small x_p is magnified to an appreciable y_p. Thus, in a 2 wt% 1-butanol aqueous solution $x_p = 2/(74.12 \cdot 55.5) = 4.86 \cdot 10^{-4}$. The vapor pressure of 1-butanol is 10 mm Hg at 30°, and $y_p = 49.4 \cdot 4.86\ 10^{-4} \cdot 10/760 = 3.16\ 10^{-4}$. With an inert gas flow of 600 l h^{-1} to a 10 l reactor, the loss of 1-butanol to the gas phase is $7.62\,10^{-4}$ mol $(l\ h)^{-1}$. At $D = 0.2\,h^{-1}$, the measured butanol rate of production is $= 0.2 \cdot 2/74.12 = 5.4\ 10^{-3}$ mol $(l\ h)^{-1}$, while the true production rate is $q_p = 6.16\,10^{-3}$ mol $(l\ h)^{-1}$.

Acetaldehyde with a vapor pressure of 1076 mm Hg at 30 °C will almost quantitatively be stripped to the gas phase, and the total acetaldehyde production is a factor 7–8 higher than that calculated on the basis of the liquid effluent acetaldehyde concentration.

2.2 Operation of the Steady-State CSTR

Examples 2.1 and 2.2 have shown how production rates q_i are calculated for a steady-state continuous stirred tank reactor. In the following the typical laboratory CSTR is described and various ways of obtaining steady-state cultivation conditions are discussed. Figures 2.1 and 2.2 are, respectively, a sketch of a CSTR and a photograph of a state-of-the-art laboratory bioreactor Biostat® A plus from Sartorious.

Figure 2.1 shows the glass vessel mounted on a balance that continuously records the amount of liquid and biomass in the vessel. An accuracy of a few milligrams in the reading of the weight can be expected. The liquid feed comes from several reservoirs. The main substrate reservoir contains the carbon source, glucose, sucrose, glycerol, methanol, and ethanol-whichever carbon source is to be studied. To keep the pH constant, there will always be reservoirs with acid (H_2SO_4, HCl, etc.) and with base (NaOH, KOH, NH_4OH, etc.). The flow of alkali and acid is controlled individually based on continuous pH measurement in the vessel, while the carbon substrate feed concentration and the concentration of minerals and trace elements is kept constant through each steady-state experiment, hence the size of the reservoir should preferably be large enough to allow each steady state to be studied with only one batch of premixed substrate solution. In order to avoid dilution of the substrate by titration to a fixed pH, it is advantageous to use concentrated acid and base. Here, a common feed line of liquid substrate as shown in Figure 2.1 can be advantageous in order to avoid transient upsets of pH in the mixing of droplets of strong base/acid directly into the vessel. The feed should preferably be in the middle of the reactor, close to the agitator in order to avoid C substrate gradients. In the so-called pH auxostat (see below), the feed of base to keep pH constant when H^+ is liberated by uptake and metabolism of substrate must, however, be kept separate from the feed of carbon source. Different gas-sparging systems and mixers are used, often of the so-called Ruston turbine design (Chapter 11), which provide for a considerable power input to give a high mass transfer coefficient $k_l a$, but in stress-sensitive cultures (mammalian cells or plant cells) gentler mixers are used. Small reactors can be stirred by a magnetic stirrer. Concentrations of liquid- and gas-phase effluents are measured online (CO_2 and O_2), at-line, or offline as discussed further in Chapter 14.

All high-quality laboratory reactors (the so-called "ideal bioreactors") are equipped with a number of monitors for state variables, and some of these are used to control the process. Industrial reactors have much fewer measurements. Typically, temperature T, pressure P, agitator speed, RQ, the ratio between CO_2 and O_2 concentration in the gas-phase effluent, and the O_2 tension in the liquid phase are continuously recorded. A few liquid-phase variables, s and e, are measured at-line. All measurements are stored in the process computer as shown on Figure 2.2.

Steady-state operation of a bioreactor is characterized by time-independent values of all culture variables. The feed concentration s_o of the C substrate and the concentration of all minerals, including $H_2PO_4^-$, Cl^-, SO_4^{-2}, Na^+, K^+, Cu^+, and Mn^{+2}, hormones, growth essential compounds, and so on must be constant in time. The dilution rate $D = v/V$ is a key variable, as well as pH, T, and dissolved oxygen (DO) concentration s_{O2} that is often measured relative to the saturation concentration $s^*_{O_2}$, namely, $\mathrm{DO} = s_{O_2}/s^*_{O_2}$. If in a well-stirred reactor, these essential variables are kept constant by accurate measurements, and with good control algorithms to process the measurements, one can expect all concentrations of substrates and products in the effluents to be virtually time-independent and equal to their concentrations in the reactor medium.

In a few but exhaustively studied cases (see, for example, [2,3]), it is virtually impossible to attain steady-state operation of the continuous stirred tank reactor. This is the case when aerobic yeast fermentation is carried out in well-mixed "ideal" laboratory bioreactors at a dilution rate close to the so-called critical dilution rate, D_{crit}. Here, the metabolism suddenly changes from purely respiratory growth with biomass and CO_2 as the only products to fermentative (respiro-fermentative) growth in which also ethanol and glycerol are produced. The kinetics of the process around D_{crit} is highly nonlinear, and transients between even closely spaced D values are extraordinarily long and follow nearly unpredictable courses. One may even obtain long lasting and apparently stable free oscillations in the outlet concentrations of x, CO_2, ethanol, and so on from an uncontrolled continuous reactor where the dilution rate and all input substrate concentrations are meticulously kept constant. It is worth noting that in industrial bioreactors, where the mixing is less perfect and synchronization of the culture does not happen, these oscillations are never observed. This is, indeed, fortunate for the Novo Nordisk company that uses a genetically engineered yeast to produce an expensive, growth-associated product, human insulin, in large ($\approx 75\,m^3$) stirred tanks operated in steady state and close to D_{crit} to obtain a high volumetric productivity q_p of the desired hormone product. Even slight variations in the effluent concentrations are intolerable for a product that has to pass strict FDA quality tests.

Except in these "pathological" situations, it is easy to study the steady state of a bioreaction in a state-of-the-art bioreactor equipped with a number of control loops and serviced by a computer that automatically steps through a range of D values. At each D value, steady state is reached when the culture is in a "metabolically steady state," a term to be discussed further in Chapter 7. Basically, metabolic steady state is reached when the catabolic ("energy producing") pathways and the anabolic ("biomass-building and energy-consuming") pathways run at the same rate. When a selected D value is close to one at which steady state has been reached, it does not take long for the cell reactions to harmonize, and the new steady state is reached after a time $t \leq D^{-1}$. This is the basis for operation of the *accelerostat* where D is increased (or decreased) in small steps, and a series of steady states is obtained after a reasonably small experimental time. If the new D is far from the preceding D value, D_0, especially if D_0 is very small, it takes a very long time, perhaps $t = 5\,D_0^{-1}$, to reach a new steady state. Close to D values where the metabolism of the cell changes drastically (as at D_{crit} in aerobic yeast cultivation), the transient can last even much longer.

As illustrated below, it is, however, important to choose the right variable to be controlled, since the choice depends on the range of dilution rates to be studied. The example of Figure 2.3 illustrates how different control strategies are chosen.

For low dilution rate (until about $D = 0.2–0.22\,h^{-1}$), the effluent biomass concentration x in Figure 2.3a is almost constant. The specific rates of CO_2 production and of O_2 consumption increase linearly with D. A good control of steady state is obtained by continuous measurement of the medium weight $W(t)$ and controlling v to give a specific D value $= v\rho_{medium}/W$. The biomass concentration is nearly constant and the measurement of x cannot be used to obtain a desired steady state (D). The ethanol concentration is nearly zero, as is the glucose concentration. With modern, very sensitive measurement methods for glucose, one might use s as control variable, but it is better to use $W(t)$. In this mode of control, the reactor is called a *chemostat.* At the other end (Figure 2.3b), where washout is reached and s approaches $s_o = 20\,g\,l^{-1}$, measurement of $W(t)$ is not sensitive enough to secure a given steady-state value of D, but both $x(t)$ and $s(t)$ can be used. In the *auxostat,* one is even able to control the reactor at exactly the washout dilution rate, and this can be used to find the maximum specific growth rate r_x for a series of different strains [6]. When $x(t)$ is used as control variable

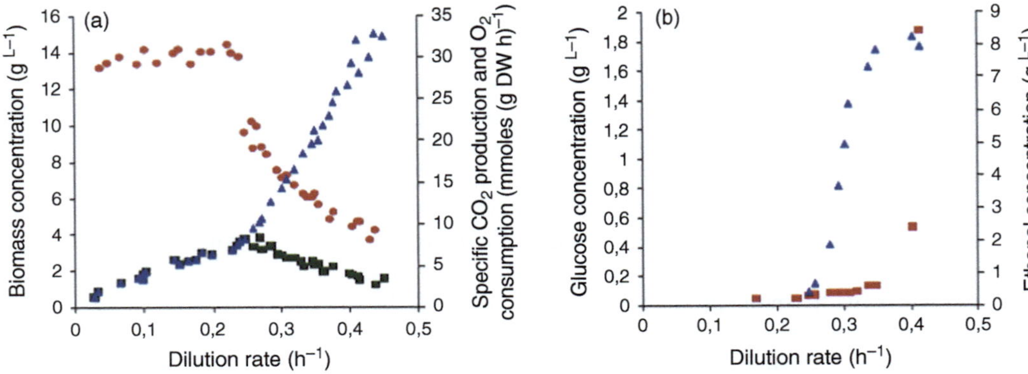

Figure 2.3 Steady-state aerobic cultivation of *Saccharomyces cerevisiae*. (a) Biomass concentration x(g l^{-1}) (solid circles), specific CO_2 production rate (mol CO_2(g X h)$^{-1}$, solid triangles), specific O_2 consumption rate (mol O_2(g X h)$^{-1}$, filled squares). (b) Effluent glucose concentration s (g l^{-1}, solid squares) and ethanol concentration e (gl^{-1}, filled triangles). Feed concentration of glucose 20 gl^{-1} with no biomass or ethanol in the feed. (Data from Ref. [4,5].)

(via a crude model for the kinetics of the reaction), the reactor is said to be operated as a *turbidostat*. Close to $D = 0.25\,h^{-1}$ where the effluent ethanol concentration e changes dramatically, one may use the liquid-phase ethanol concentration (or less accurately the gas-phase ethanol concentration) to control the reactor at a desired steady-state value of D. This is called a *productostat* [7], and again a crude model for the kinetics is used to translate the measured value of ethanol concentration to the desired value of D. The models for different control systems are further discussed in Chapter 15.

References

1 Duboc, P., von Stockar, U., and Villadsen, J. (1998) Simple generic model for dynamic experiments with *Saccharomyces cerevisiae* in continuous culture. *Biotechnol. Bioeng.*, **60**, 180–189.

2 Keulers, M., Satroutdinov, A.D., Suszuki, T., and Kuriyama, H. (1996a) Synchronization affector of autonomous short period sustained oscillation of *Saccharomyces cerevisiae*. *Yeast*, **12**, 673–682.

3 Keulers, M., Suzuki, T., Satroutdinov, A.D., and Kuriyama, H. (1996b) Autonomous metabolic oscillation in continuous culture of *Saccharomyces cerevisiae* grown on ethanol. *FEMS Microbiol. Lett.*, **142**, 253–258.

4 von Meyenburg, K.H. (1969a) Katabolit Repression und der Sprossungszyklus von *Saccharomyces cerevisiae.* Diss. ETH, Zürich.

5 von Meyenburg, H.K. (1969b) Energetics of the budding cycle of *Saccharomyces cerevisiae* during glucose limited aerobic growth. *Arch. Mikrobiol.*, **66**, 289–303.

6 Larsson, G., Enfors, S.O., and Pham, H. (1990) The pH auxostat as a tool for studying microbial dynamics in continuous culture. *Biotechnol. Bioeng.*, **36**, 224–232.

7 Lei, F., Olsson, L., and Jørgensen, S.B. (2003) Experimental investigations of multiple steady states in aerobic continuous cultivations of *Saccharomyces cerevisiae. Biotechnol. Bioeng.*, **82**, 766–777.

3
Redox Balances and Consistency Check of Experiments

John Villadsen

Summary

A black-box stoichiometry for a single bioreaction is introduced using element balances to calculate some of the yield coefficients. Using a redox balance on the stoichiometry is shown to reduce the computational work. In a number of examples, redox balances are used to check experimentally determined yield coefficients. The concept of a theoretical maximum yield of a bioreaction is defined and used on several industrially important processes. A given experimental stoichiometry is split into several elementary stoichiometries with the purpose of finding potential process improvements. Finally, redox balancing is used to analyze typical bio-remediation processes.

3.1
Black-Box Stoichiometry Obtained in a CSTR Operated at Steady State

The stoichiometry (1) of Example 2.1 and Eq. (3.1) quantitatively describes how much biomass, ethanol, CO_2, and glycerol is produced by conversion of one C-mol glucose *at a given steady-state operation condition*, that is, at given values of temperature, pH, and dilution rate $D = v/V$.

$$\begin{aligned} CH_2O + 0.01658\,NH_3 \rightarrow\ & 0.1381CH_{1.74}O_{0.6}N_{0.12} + 0.5160\,CH_3O_{0.5} \\ & + 0.2759\,CO_2 + 0.070\,CH_{8/3}O + 0.037\,H_2O. \end{aligned} \tag{3.1}$$

In the particular Example 2.1 on the anaerobic cultivation of the yeast *Saccharomyces cerevisiae* at $T = 30\,°C$ and $pH = 6.5$, the stoichiometry appears to hold quite well for dilution rates between 0.05 and $0.42\,h^{-1}$ which is close to the washout dilution rate. In most other cases, a stoichiometry like (3.1) is only a

Fundamental Bioengineering, First Edition. Edited by John Villadsen.

snapshot of the reaction *at a given value of D*. Even when temperature and pH are kept constant throughout the series of experiments with different D values, the overall stoichiometry for all operating conditions must be patched together from a series of these snapshots. Experiments carried out in a steady-state CSTR are time-consuming since, as described in Section 2.2, one has to wait quite some time before steady state (i.e., *metabolic steady state* of the culture) has been reached. The accuracy in the steady-state effluent concentration that can be obtained in a state-of-the-art laboratory bioreactor is in the range of mg l^{-1} but only if true steady state has been reached. In a narrow D range, for example, $0.45 < D < 0.55\,h^{-1}$, steady state values can be obtained in an accellerostat with about D_{av}^{-1} between measurements, that is, 2 h for each experiment. Animal cells grow much slower, and a series of six steady-state experiments carried out in D-steps of $0.004\,h^{-1}$ between $D = 0.02$ and $0.04\,h^{-1}$ would require a total experimental time in the order of $6 \cdot (5 \cdot 0.03^{-1}) = 1000\,h$ to complete when an ordinary chemostat is used compared to about 200 h in a well-controlled accelerostat. This long experimental time puts exceedingly hard demands on the reliability of the equipment – substrate bottles must be changed without changing *any* variable in the feed, infection of the culture through air-filters that could get soaked must be avoided, and any breakdown of temperature, pH or aeration control is likely to ruin the experiment. In an attempt to speed up the investigation, other approaches are commonly used, especially in the industry: batch cultivation or fed-batch culture is used instead of steady-state continuous culture. These modes of operation are investigated further in Chapters 10 and 11, and although they are frequently used in *production*, they *do not give nearly as much information on the stoichiometry of the reaction as does steady-state experiments*. Batch experiments can usually be carried out in a matter of a few hours, but they usually give only information of the maximum growth rate, and the transient nature of the experiment makes it very difficult to obtain the correct stoichiometry, except at the maximum specific growth rate. Fed-batch experiments are also carried out under continuously changing operational conditions, and one cannot obtain information on the influence of perhaps important additives. Thus, in order to find true stoichiometries and rate data that can be extrapolated to other process conditions, one needs to do steady-state CSTR experiments despite the difficulties associated with this type of operation. As a final note, none of the above experimental difficulties are encountered in enzymatic reactions which change from one steady state to the next almost immediately after the experimental conditions have changed.

In Eq. (3.1), there is one carbon source (glucose), and the stoichiometry is normalized to describe the conversion of one C-mol glucose to four carbon-containing compounds, each normalized to one C-mol. Water is formed as a by-product. There are four elemental balances for C, H, O, and N. With the normalization of the glucose consumption rate to one C-mole per unit time, and with the rates of formation or consumption of the other carbon-containing compounds also based on one C atom, the four elemental balances can be written as follows:

$$\begin{array}{ll} \text{C balance:} & Y_{sx} + Y_{se} + Y_{sg} + Y_{sc} = 1 \\ \text{H balance:} & 1.74\,Y_{sx} + 3Y_{se} + 8/3\,Y_{sg} - 3\,Y_{sn} + 2\,Y_{sw} = 2 \\ \text{O balance:} & 0.6\,Y_{sx} + 1/2\,Y_{se} + Y_{sg} + 2\,Y_{sc} + Y_{sw} = 1 \\ \text{N balance:} & 0.12\,Y_{sx} - Y_{sn} = 0. \end{array} \tag{3.2}$$

There are six yield coefficients to be determined, and there are four constraints, the four linear equations in Eq. (3.2). When two rates are measured, the remaining four rates can be calculated by solutions of the linear equations. The pair $[Y_{sx}, Y_{sn}]$ cannot be chosen since the two yield coefficients are linearly dependent when all N consumed ends up in the biomass.

The measurement of Y_{sn} can be used only to check if the experimental value obtained for Y_{sx} is correct, or one may find Y_{sx} on the basis of measurement of the ammonia consumption. This could be an advantage if the medium is dark colored, and a measurement of the biomass production by light absorption is difficult. Another independent way to determine Y_{sx} in Eq. (3.1) where no protonated compounds is formed is to measure the amount of base used to titrate the medium to a constant pH. The reason is that whereas ammonia is protonated at pH 6.5 and is imported to the cell as NH_4^+, nitrogen is incorporated into the biomass as NH_3, and the proton must be transported back into the medium to maintain a constant pH (around 7) inside the cell.

Y_{sw} should not be one of the experimentally determined yield coefficients since the amount of water formed in most bioreactions is too small to be correctly measured. Thus in Eq. (3.1) with a feed of $20\,g\,l^{-1}$ glucose the amount of H_2O produced at total consumption of the glucose feed is only $0.037 \cdot 20/30 \cdot 18\,g\,l^{-1} = 0.44\,g\,l^{-1}$, which will never be recognized considering that the medium contains about $1000\,g\,H_2O\,l^{-1}$.

If the relation $Y_{sn} = 0.12\,Y_{sx}$ is inserted in the second equation of Eq. (3.2) and two out of the six combinations $[(Y_{sx}, Y_{se}), (Y_{sx}, Y_{sg}), (Y_{sx}, Y_{sc}), (Y_{se}, Y_{sg}), (Y_{se}, Y_{sc})$, and $(Y_{sg}, Y_{sc})]$ are chosen as measured quantities, the three remaining yield coefficients can be determined by solution of the three equations. For example, when (Y_{sx}, Y_{se}) are the measured quantities:

$$\begin{pmatrix} 1 & 1 & 0 \\ 8/3 & 0 & 2 \\ 1 & 2 & 1 \end{pmatrix} \begin{pmatrix} Y_{sg} \\ Y_{sc} \\ Y_{sw} \end{pmatrix} = \begin{pmatrix} 1 - Y_{sx} - Y_{se} \\ 2 - 1.38\,Y_{sx} - 3Y_{se} \\ 1 - 0.6Y_{sx} - 0.5Y_{se} \end{pmatrix} = \begin{pmatrix} 0.3459 \\ 0.2614 \\ 0.6591 \end{pmatrix}. \tag{3.3}$$

The solution is $[Y_{sg}, Y_{sc}, Y_{sw}] = [0.0700, 0.2759, 0.0373]$, and $Y_{sn} = 0.01657$.

Here, the accuracy of the calculated yield coefficients is excellent, but as will be discussed in Section 4.9, the accuracy depends on the choice of measured rates, which influences the magnification of experimental errors in the calculated yields.

The algebra that accompanies the determination of nonmeasured yield coefficients (or rates) is modest in the simple example Eq. (3.1). In a more complex stoichiometry with more substrates and products such as Eq. (3.4), the calculation of nonmeasured quantities increases dramatically and larger errors are to be expected.

$$\begin{aligned}&CH_{a1}O_{b1}N_{c1}(S_1) + \sum_2^{N-1} Y_{s_1,s_j}\, S_j + Y_{s_1,n}\, NH_3 + Y_{s_1,O_2} O_2 \rightarrow \\ &Y_{s_1x}X + Y_{s_1,c}CO_2 + \sum_1^{M-2} Y_{s_1,p_j}\, P_j + Y_{s_1,w}H_2O.\end{aligned} \tag{3.4}$$

Again, it is advantageous to normalize the stoichiometry to one C-mol of S_1, one of the N carbon-containing substrates, and to use a basis of 1 C-mol also for all other substrates and carbon-containing products. In Eq. (3.4) where other N-containing compounds besides X are produced, there is, of course, no proportionality between the biomass yield coefficient and $Y_{s_1,n}$.

In the approach outlined above, one has invoked the preservation of elements (C, H, O, N) in the chemical reaction. Conservation of electrical charge is another obvious condition. Thus, if in Eqs. (3.1) and (3.4) the ammonia feed is NH_4^+, a term $Y_{s_1,n}H^+$ should be added as a product. The N-containing substrates in Eq. (3.4) could be amino acids, and these are also dissociated. The production of protons is now also dependent on these N sources, and a proper yield coefficient Y_{s_1,H^+} can be calculated. In general, uncharged species will be used in stoichiometries of this text, but still the rate of biomass formation or $Y_{s_1,x}$ cannot be found by titration in a more complex stoichiometry such as Eq. (3.4).

3.2
Calculation of Stoichiometric Coefficients by Means of a Redox Balance

Another principle of preservation is that of *redox* in a stand-alone black-box stoichiometric equation, an equation that relates substrates fed to the reactor and products retrieved from the reactor. Balancing of redox will be used in the remainder of this chapter as a simpler method to calculate stoichiometric coefficients. In Chapter 4 where specific pathways from substrate to specified products are used to analyze metabolic networks, the redox imbalance between the substrate and the products will be added to the stoichiometry as "redox packages." Each redox package will be shown to be equivalent to the redox power of one molecule of H_2. When the products are more oxidized than the substrate(s), one must add redox packages on the product side of the stoichiometry, and vice versa.

Redox balances are introduced in the following way:

1) The unit of redox is defined as one atom of hydrogen. Hence, H_2 contains two redox units.
2) For each element, *a redox-neutral compound* is defined. We choose H_2O for oxygen, CO_2 for carbon, NH_3 for nitrogen, and the acids H_2SO_4 and H_3PO_4 for sulfur and phosphorous, respectively.
3) With these definitions of redox-neutral compounds and with the defined unit for redox, one obtains that the elements [O, C, N, S, P] have redox levels [−2, 4, −3, 6, 5]. Any other set of redox-neutral compounds could

have been chosen. This would have no influence on redox balancing of the stoichiometry since all elements are preserved in the reaction. The choice of [H_2O, CO_2, NH_3] for the elements [O, C, N] is convenient since usually these compounds occur in the stoichiometry for bioreactions.

4) After obtaining the redox level of the commonly occurring elements, one can now calculate the redox level of any compound in a bioreaction. Thus, glucose $C_6H_{12}O_6$ contains 24 redox units or 4 units per C-atom. Lactic acid ($CH_3CHOHCOOH$) contains 12 redox units or 4 per C-atom. Lactic acid is, in fact, one possible end product of glycolysis (see Chapter 4) in which glucose is converted to two molecules of lactic acid. Ethanol ($C_2H_5OH = (CH_3O_{0.5})_2$ contains six redox units per C-mol. The degree of reduction of a compound is denoted by κ_i, for example, κ_x for the biomass. If the stoichiometry contains charged compounds, a positive charge reduces the degree of reduction of the compound by one, a negative charge increases κ_i by one unit. Hence, $\kappa_{NH_4^+} = \kappa_{NH_3} = 0$ and $\kappa_{H_2S} = \kappa_{HS-} = \kappa_{S-2} = 8$. Both H^+ and OH^- are redox neutral.
5) In every stand-alone stoichiometry, the redox content of the products must equal that of the substrates. Thus for Eq. (3.1),

$$CH_2O : 4, NH_3 : 0, X = CH_{1.74}O_{0.6}N_{0.12} : 4.18, CH_3O_{0.5} : 6, CH_{8/3}O : 14/3, \\ CO_2 : 0, H_2O : 0, \text{ and } 4 + 0.01658 \cdot 0 = 0.1381 \cdot 4.18 + 0.5160 \cdot 6 \\ + 0.0700 \cdot 14/3 + 0.2759 \cdot 0 + 0.037 \cdot 0 = 4.$$

This means that the redox balance in Eq. (3.1) closes. Together with the observation that the carbon balance closes, this is a very solid indication that the experimental investigation that leads to the stoichiometry Eq. (3.1) is accurately performed. If only the carbon balance closes, the products included in the balance may not have been the correct ones. For example, if the measurement of CO_2 production had been wrong, this might have been compensated in the carbon balance by inaccurate measurements of the other C-containing compounds in the effluent.

6) In a pathway that does not stand alone but where redox has to be delivered from another pathway, one needs to balance the pathway by adding redox equivalents on one side of the stoichiometry. Thus, in the reduction of glucose to glycerol and in the oxidation of glucose to CO_2:

$$CH_2O + 2/3 \text{ redox units } (= 1/3\text{“}H_2\text{”}) \rightarrow CH_{8/3}O. \tag{3.5}$$

$$CH_2O(+H_2O) \rightarrow CO_2 + 2\text{“}H_2\text{”}. \tag{3.6}$$

In the analysis of metabolic networks in Chapter 4 the redox equivalents H or “H_2” will be identified as cofactors in bioreactions which transport redox from one part of the metabolic network to another.

A stand-alone reaction can be constructed from Eqs. (3.5) and (3.6) by balancing the redox units:

$$7\,CH_2O + H_2O \rightarrow 6\,CH_{8/3}O + CO_2. \tag{3.7}$$

7) Returning to the stoichiometry Eq. (3.1), we find that the computational work needed when the yield coefficients are determined by a combination of the carbon balance and the redox balance is significantly smaller than when the four elemental balances were used to determine nonmeasured yield coefficients.

$$\begin{aligned}\text{Stoichiometry}: \ & CH_2O + 0.12\,Y_{sx}\,NH_3 \\ & \rightarrow Y_{sx}\,CH_{1.74}O_{0.6}N_{0.12} + Y_{se}\,CH_3O_{0.5} + Y_{sc}\,CO_2 + Y_{sg}\,CH_{8/3}O \\ & \quad + Y_{sw}\,H_2O.\end{aligned}$$

$$\text{Carbon Balance}: 1 = Y_{sx} + Y_{se} + Y_{sc} + Y_{sg}. \tag{3.8}$$

$$\text{Redox Balance}: 4 = 4.18\,Y_{sx} + 6\,Y_{se} + 14/3\,Y_{sg}. \tag{3.9}$$

Again, we need two measured yield coefficients to find the remaining two yield coefficients of carbon-containing compounds. But the problem has reduced to a 2×2 linear algebraic problem from the 3×3 linear algebraic problem of Eq. (3.3). With Y_{sx} and Y_{se} as the measured quantities one obtains

$$\begin{aligned}& Y_{sg} = 6/7 - 0.8957\,Y_{sx} - 9/7\,Y_{se}, \text{ and thereafter} \\ & \quad Y_{sc} = 1/7 - 0.1043\,Y_{sx} + 2/7\,Y_{se}.\end{aligned} \tag{3.10}$$

Inserting the values of $[Y_{sx}, Y_{se}]$ from Eq. (3.1) one obtains $[Y_{sc}, Y_{sg}] =$ [0.2759, 0.0700], and to obtain this (correct) result the amount of algebra involved is trivial compared to the solution of Eq. (3.3).

The only yield coefficient missing is that of H_2O. This can be obtained from either an O-balance or an H balance on Eq. (3.1) where the known yield coefficients for carbon-containing products are inserted. If the result for Y_{sw} is the same for both the O-balance and the H balance, the calculation of the important yield coefficients has been done correctly. In itself Y_{sw} has little importance. What was, in fact, done was to combine the elemental balances for O and for H into one balance, the redox balance, and this reduced the dimension of the algebraic problem to 2×2.

8) The concept of a redox balance applied to the stoichiometry has wider implications than just cutting down the computational work to find missing yield coefficients. In Chapter 9 where the thermodynamics of bioreactions is treated, it will be shown that the metabolic heat $(-\Delta H_R) = Q_R$ produced by converting 1 C-mol substrate to products in a given stoichiometry such as Eq. (3.1) can be calculated very simply, and fairly accurate, from the yield coefficients and using a redox balance like Eq. (3.9). Even the free energy $(-\Delta G_R)$ change that accompanies the reaction can often be calculated with reasonable accuracy.

3.3
Applications of the Redox Balance

The foremost applications of the redox balance are (1) to assist in calculation of nonmeasured stoichiometric coefficients and (2) to check whether the experimental data are consistent.

Consider the experimental data for aerobic yeast cultivation in Figure 2.3. The experiments were performed at pH 5.5 in a mechanically stirred bioreactor with working volume 2.7 l, sparged with air at a flow rate of 2.5 vvm ($=2.5 \cdot 2.7(\text{l})\ \text{min}^{-1} = 405\ \text{l h}^{-1}$). For $D \leq 0.25$, the biomass concentration in the effluent is approximately constant, $14\ \text{g l}^{-1}$. At $D \leq 0.25\ \text{h}^{-1}$, there is virtually no glucose in the effluent, and since $s_o = 28\ \text{g l}^{-1}$, $Y_{sx} = 14/28 = 0.5\ \text{g g}^{-1} = (14/25.17) \cdot (28/30) = 0.5959$ C-mol X/C-mol glucose for $X = CH_{1.83}O_{0.56}N_{0.17}$.

Since no other products but biomass and CO_2 are produced for $D < 0.25\ \text{h}^{-1}$, the stoichiometry is

$$CH_2O\ (+0.1013\ NH_3) + 0.3743\ O_2 \rightarrow 0.5959\ X + 0.4041\ CO_2\ (+0.6067\ H_2O). \tag{3.11}$$

Note: The biomass concentration was obtained by filtration of a given volume of (homogeneous) medium, drying the filter paper + biomass and weighing. Dry mass from yeast fermentation usually has 6–8 wt% ash (residue of minerals) after burning. Neither of the references von Meyenburg [1,2] contains information on the ash content of the weighed biomass sample, a typical omission in published papers. Since one presumes that no trivial error in the biomass concentration has been made, it is assumed that the biomass was on an ash-free basis.

The value of Y_{so} is found from a redox balance. The redox content of the substrates is 4–4 Y_{so} since each O_2 has a negative redox content of four units. The redox content of the products is $Y_{sx}\kappa_x = 0.5959 \cdot 4.20 = 2.503$ units. Hence, $Y_{so} = (4 - 2.503)/4 = 0.3743$ mol O_2 C-mol glucose^{-1}. The yield coefficient of $NH_3 = 0.17\ Y_{sx} = 0.1013$, and Y_{sw} is found from an O- or an H balance.

Figure 2.3 a also contains experimental data for the specific rates r_c and $(-r_o)$. Assuming that r_c and $(-r_o)$ are both proportional to D up to $D = 0.25\ \text{h}^{-1}$, one obtains $r_c = 6.9$ (mmol CO_2 h^{-1} g biomass $^{-1}$)/0.25 h^{-1} = 0.694 mol CO_2 C-mol X^{-1}. This value is close to $Y_{xc} = 0.4041/0.5959 = 0.678$ obtained from Eq. (3.11), and considering the difficulty of obtaining the correct slope of the straight line from Figure 2.3a, it is concluded that the measurement of r_c gives an independent confirmation of the fully aerobic stoichiometry Eq. (3.11). On Figure 2.3a, the measurements of $(-r_o)$ are not distinguishable from those of r_c. That this is not a misprint is confirmed by the tables in [2]. From Eq. (3.11), it is seen that $r_c/(-r_o)$ should be $0.4041/0.3743 = 1.08$. From a number of experimental studies, it is known that the *respiratory quotient* $\text{RQ} \equiv Y_{oc} = r_c/(-r_o)$ is in the range of 1.05–1.08 for fully respiratory growth of biomass, and it is likely that Figure 2.3 is in error when it presents r_c as more or less the same as $(-r_o)$. In process control of aerobic fermentations, the value of RQ is sometimes the only physiological quantity measured. When RQ is around 1.06, the process runs with no redox containing products besides biomass. When glucose is oxidized to CO_2 by molecular oxygen, the value of RQ is of course exactly 1.

From the calculations given above, it follows that measurement of more variables than are necessary to fix the stoichiometry is required to check the consistency of the experimental data. Needless to say, the checking of experimental data should be done on a regular basis, and not at the end of a long experimental investigation that can turn up to be of no value for scientific studies or for process design due to some systematic error that cannot be accounted for once the experimental study is finished.

On Figure 2.3b data for ethanol production are shown. No ethanol is produced for D less than $\sim 0.25\,h^{-1}$, and in a short interval ($\sim 0.25 < D < \sim 0.37$), the effluent concentration of ethanol increases to about $8\,g\,l^{-1}$. This is due to a dramatic shift in metabolism of the yeast from purely respiratory to oxido-fermentatitive. The biomass production rate decreases abruptly at $D \sim 0.25\,h^{-1}$, while the specific rate of CO_2 formation increases and $(-r_o)$ stays constant.

Note: In Figure 2.3b, it appears that r_o decreases for $D > 0.25\,h^{-1}$. This is an experimental error, corrected in later papers by von Meyenburg *et al.*

An analysis of the experimental data in Figure 2.3b shows that some ethanol is stripped to the gas phase due to the very high air sparging rate ($405\,l\,h^{-1}$) used by von Meyenburg. This is an experimental error of the type discussed in Example 2.2. Also it appears [3] that a rapid production of acetaldehyde around $D = 0.25\,h^{-1}$ has not been recognized in the experimental study. Such errors can make the study of the stoichiometry of a bioreaction very difficult, and an attempt to find the kinetics of the overall reaction will surely fail.

Example 3.1 A consistency test reveals a typical systematic error in an experimental study

In a strictly anaerobic cultivation of *S. cerevisiae*, the reactor with working volume 1 l is sparged with N_2 (to keep out O_2) at a rate of $v_g = 15\,l\,h^{-1}$ (1 atm, 30 °C). The inlet gas contains no CO_2, but the dry effluent gas contains 2.88 vol % CO_2. The inlet and outlet gas flows are assumed to be the same. The liquid feed of $v = 0.1\,l\,h^{-1}$ contains $23\,g\,l^{-1}$ glucose and $1\,g\,l^{-1}$ NH_3, but no products. The cultivation temperature is 30 °C. In the liquid effluent, one measures in $g\,l^{-1}$: glucose (s) = 0.048, NH_3 (n) = 0.728, biomass (x) = 2.36, ethanol (e) = 8.85, and glycerol (g) = 2.39. The biomass composition is $X = CH_{1.76}O_{0.56}N_{0.17}$.

1) Set up a C balance for the data and check if the C balance closes. Find $Y_{s,pi}$ for a possible missing product.
2) Give an educated guess of the identity of the missing product. What could have gone wrong in the study?

Carbon balance: The dilution rate $D = 0.1/10 = 0.1\,h^{-1}$, $M_x = 25.10$ g C-mol X^{-1}, and $\kappa_x = 4.13$.

$q_e = De = 0.1 \cdot 8.85/23 = 0.03\,848$ C-mol ethanol $(\mathrm{l\ h})^{-1}$, $q_g = 0.1 \cdot 2.39/30.667 = 0.00\,779$ C-mol glycerol $(\mathrm{l\ h})^{-1}$ $q_x = 0.1 \cdot 2.36/25.10 = 0.009402$ C-mol X $(\mathrm{l\ h})^{-1}$, $q_{CO2} = 0.0288 \cdot 15 = 0.432$ l (l reactor h)$^{-1}$ $= 0.432 \cdot (273/303)/22.4 = 0.01738$ mol (l reactor h)$^{-1}$, $(-q_s) = 0.1 \cdot (23\text{–}0.048)/30 = 0.07651$ C-mol glucose $(\mathrm{l\ h})^{-1}$.

Per C-mol glucose one obtains $\Sigma(Y_{s,p}) = 0.5029 + 0.1018 + 0.1228 + 0.2271 = 0.9546$, which is 0.0454 less than the one C-mol carbon feed. Since the carbon balance ought to close at least to 0.99, some product appears to be missing. One may learn more about the missing product by setting up a redox balance:

$4 = \Sigma(\kappa_i\ Y_{s,pi}) = 3.99 + 0.0454\ \kappa_p$. Hence the unknown product has a degree of reduction very close to 0. There is every reason to believe that the "missing compound" is CO_2.

The photoacoustic CO_2 analysis cell has a guaranteed accuracy of about 2% relative, and the error is probably due to an erroneous measurement of the gas flow. In an analysis of the experimental setup, it was found that v_g was measured with a flow meter that had a maximum range of 2 l min^{-1} (120 l h^{-1}). When a flow meter of a more suitable range of 0.5 l min^{-1} was used, it was found that the v_g was in reality 18 l h^{-1} and the C balance closed.

Finally, it may be added that the measurement of NH_3 consumption gives a check on q_x.

$(-q_n) = 0.1 \cdot (1 - 0.728)/17 = 0.0016$ mol N $(\mathrm{l\ h})^{-1}$. Hence, with 0.17 N/C in X one obtains $q_x = 0.0016/0.17 = 0.09412$ C-mol X $(\mathrm{l\ h})^{-1}$ since the nitrogen is absorbed only in the biomass.

Example 3.2 Oxalic acid as a by-product in the production of citric acid

Citric acid is an enormously important product in the bio-industry. It is produced in an annual amount of about 1.8 million tons in huge bubble column reactors of volume in excess of 350–400 m^3. The reactor is operated in fed-batch mode (see Section 10.5), and O_2 is a substrate that is fed through spargers. The upward flow of the gas bubbles induces a slow circulation motion to the gas + liquid phase and thereby provides sufficient mixing.

The carbon substrate is glucose, and as nitrogen substrate NH_4NO_3 is used, pH is controlled at 2.5.

The filamentous fungus *Aspergillus niger* is an excellent producer of citric acid, and it is used in almost all industrial citric acid processes. Citric acid is a metabolite extracted from the TCA cycle (Figure 4.6) that as discussed in Section 4.3 converts pyruvic acid ($CH_3COCOOH$) to CO_2 + a large number of metabolic products and building blocks for cell synthesis. For the purpose of the present example, it is enough to state that citric acid is formed by condensation of an energetically active form of acetic acid = $C_2H_4O_2$ and the dicarboxylic acid–oxaloacetic acid COOH-CH_2-CO-COOH = $C_4H_4O_5$ to the six-carbon tricarboxylic acid COOH-CH_2-(CH COOH)-CHOH-COOH = $C_6H_8O_7$. The two substrates are produced from one glucose molecule that contains 24 redox units, while the product contains 18 redox units. In the absence of other metabolic reactions, which could consume redox, the six excess redox units must be oxidized by oxygen:

$$C_6H_{12}O_6 + 3/2\,O_2 \rightarrow C_2H_4O_2 + C_4H_4O_5 + 2H_2O \rightarrow C_6H_8O_7 + 3\,H_2O. \quad (1)$$

When biomass formation is added to the stoichiometry, less oxygen is used since the degree of reduction of biomass is higher (typically 4.2 per C-mol) compared to citric acid (3 units per C-mol). To reduce the O_2 consumption, a highly oxidized N source (ammonium nitrate NH_4NO_3 with $\kappa_n = -8$) is used.

On a C-mol basis, the stoichiometry (with 0.1 CO_2 produced per C-mol X as explained further in Chapter 4), Eq. (4.26) becomes

$$CH_2O + Y_{so}\,O_2 + Y_{sn}\,NH_4NO_3 \rightarrow Y_{sx}\,X\,(CH_{1.8}O_{0.5}N_{0.2}) + 0.1\,Y_{sx}\,CO_2 + Y_{sp}CH_{8/6}O_{7/6} \quad (2)$$

Assume that the yield of citric acid is $Y_{sp} = 0.9$ g/g glucose $= 0.9 \cdot 30/32 = 0.8438$ C-mol citric acid (C-mol glucose)$^{-1}$. A product yield of close to 1 is obtainable in an up-to-date citric acid process, and the high yield is necessary to cover the capital and production costs, since there is a relatively small difference between the prices of glucose and citric acid.

From a nitrogen balance, $Y_{sn} = ½ \cdot 0.2 \cdot Y_{sx}$, and Y_{so} is calculated by applying a redox balance to Eq. (1):

$$Y_{so} = \tfrac{1}{4}(4 - 0.1 \cdot (1 - 0.8438) \cdot 0.9 \cdot 8 - (1 - 0.8438) \cdot 0.9 \cdot 4.20 - 0.8438 \cdot 3) = 0.1914.$$

$$CH_2O + 0.1914\,O_2 + 0.01406\,NH_4NO_3 \rightarrow 0.1406\,X + 0.8438\,CH_{8/6}O_{7/6} + 0.0156\,CO_2 + 0.3391\,H_2O. \quad (3)$$

Note that $Y_{so} = 0.1914$, whereas it would be one-fourth when no biomass is formed as in Eq. (1).

In an upset of the pH control, the pH of the cultivation increases to 4.5–5 for an extended period of time. The biomass yield is a little lower, 0.1333 C-mol X (C-mol glucose)$^{-1}$, but the yield coefficient Y_{so} remains almost the same ~0.19. The yield of citric acid does, however, decrease to 0.80 C-mol (C-mol glucose)$^{-1}$.

Other products have been formed since the C balance must still close, but apparently the new products must contain almost the same redox as citric acid, since Y_{so} hardly changes. A situation such as this can be analyzed only when more is known about the possible reactions in the TCA cycle. What happens is actually that an enzyme, oxaloacetate hydrolase, becomes activated at a higher pH value of 4.5–5, and now a side reaction Eq. (4) in which oxaloacetate is converted to acetic acid and oxalic acid $(COOH)_2$ competes with the main reaction Eq. (1) by which oxaloacetate and acetic acid condense to form citric acid:

$$C_6H_{12}O_6 + O_2 + 2\,CO_2 \rightarrow 2(COOH)_2 + 2\,CH_3COOH. \tag{4}$$

Multiply Eq. (4) by 1/6 to normalize to one C-mole:

$$CH_2O + 1/6\,O_2 + 1/3\,CO_2 \rightarrow 2/3\,(CO_2H + CH_2O\,(\text{acetic acid, HAc})). \tag{5}$$

Add the stoichiometry Eq. (5) multiplied by α to Eq. (3). From $(1+\alpha)$, C-mol glucose 0.80 C-mol citric acid is produced.

Hence, $0.8438 = 0.8 \cdot (1+\alpha)$ or $\alpha = 0.05\,475$. The total stoichiometry, normalized to one C-mol glucose, is as follows:

$$CH_2O + 0.1901\,O_2 + 0.01333\,NH_4NO_3 + 0.00265\,CO_2 \rightarrow \\ 0.1333\,X + 0.8\,CH_{8/6}O_{7/6} + 0.0346 \cdot (CO_2H + CH_2O(HAc)) + 0.322\,H_2O. \tag{6}$$

Due to the small net uptake of CO_2 in Eq. (5), the sum of the yield coefficients of C-containing compounds on the right-hand side of Eq. (6) is slightly greater than 1. The sum of yield coefficients for citric acid and the two by-products is 0.8692, mainly because compared to Eq. (3) 0.825 0.01825 CO_2 has been taken up and also due to the slightly lower biomass yield.

One must certainly avoid formation of the poisonous oxalic acid by the side reaction Eq. (4). A strict pH control is a possibility since the enzyme oxaloacetate hydrolase is completely inactive at pH = 2.5, but a safer way is to use a strain where the enzyme has been knocked out by genetic engineering. An *A. niger* strain with a single, well-defined gene deletion can be used without violating the food-related codes for gene manipulation. Pedersen *et al.* [4] demonstrated that the same citric acid yield and the same activity of the cell $(r_x = \mu) = 0.20\,h^{-1}$ as the parent strain could be obtained in a "safe" knockout strain.

The citric acid example demonstrates that in more complex cell reactions the construction of black-box models cannot be done without the knowledge of metabolic pathways, the subject of Chapter 4. Missing citric acid with a degree of reduction $\kappa_p = 3$ could be interpreted as an unlikely production of a number of other TCA cycle acids with a lower degree of reduction (see Figure 4.6), while the occurrence of a mixture of oxalic acid and acetic acid would not be anticipated without knowing how citric acid is formed.

3.4
Composition of the Biomass *X*

In the examples of the previous sections, biomass of different compositions $CH_aO_bN_cS_dP_e$ has been used. With the redox level of the elements defined as $[\kappa_C, \kappa_H, \kappa_O, \kappa_N, \kappa_S, \kappa_P] = [4, 1, -2, -3, 6, 5]$, the degree of reduction of X is $\kappa_x = 4 + a - 2b - 3c + 6d + 5e$. Healthy biomass is cultivated on a feed that contains all the above elements and cations (Na^+, K^+, Ca^{+2}, Mg^{+2}, $Fe^{+2 \text{ or } 3}$, Cu^{+2}) in millimolar quantities and other cations, for example, Mn^{+2}, in micromolar quantities, and "growth enhancers" such as certain hormones. The degree of reduction κ_x of biomass varies between microorganisms, but usually it stays within the interval [4.13, 4.22]. The composition of the biomass and also κ_x varies much more with growth conditions than it does between different organisms. At low growth rate, the amount of cell walls/g DW (dry weight of cell) is higher than at high growth rate where more cell machinery is present. Since the cell wall material is mainly carbohydrate with a degree of reduction close to 4, and the cell machinery is mainly proteins with a degree of reduction closer to 4.20, the degree of reduction for a continuous culture grown at low D is somewhat lower than that of a culture grown at high D. The change of κ_x between a culture grown on a feed with sufficient concentrations of all nutrients and that of a culture where an essential nutrient is missing is, however, much more pronounced. Thus, yeast cells grown at severe N limitation will accumulate glucose as carbohydrate storage compounds, glycogen ($CH_{1.67}O_{0.83}$) and trehalose ($CH_{1.78}O_{0.89}$). In a transient chemostat experiment where the feed contains no NH_3 after $t = 0$, and the production of active (reproducing) cells stops at $t = 0$, the glycogen content of the biomass increases to 25 wt%, and the trehalose content increases to 6 wt% after 20 h. If "normal" biomass has the *standard composition* $CH_{1.8}O_{0.5}N_{0.2}$ ($\kappa_x = 4.20$), the composition of the biomass (which after 20 h contains 0.921 mol glycogen, 0.214 mol trehalose, and 2.805 C-mol "standard biomass" per 100 g X) has changed to $CH_{1.768}O_{0.598}N_{0.142}$ ($\kappa_x = 4.14$) after 20 h cultivation without N supply.

The biomass composition can be determined by *elemental analysis*. A certain amount of biomass is collected on a dry and preweighed filter. The biomass and filter are dried at 120 °C in a laboratory oven for a given time (e.g., 4 h or longer) and transferred to an exsiccator to cool to room temperature. After subtracting the weight of the filter paper, the dry weight (DW) of the sample is determined. The amount of C, H, and N in the sample is determined in the gas phase after incineration. The residue is ash.

The ash content must of course be subtracted from the DW sample weight. Still inaccuracies occur if the sample is not dried to perfection, and some residual water is left in the incinerated sample.

Example 3.3 Determination of biomass composition and yield coefficient of biomass on substrate

The bacterium *Mycobacterium vaccae* can grow aerobically on ethane as sole carbon and energy source and with NH_3 as nitrogen source. In an experiment the yield of biomass is found to be 23.7 g DW (mol ethane)$^{-1}$. Except for small amounts of S and P, an elemental analysis of dry cell mass gives C: 47.4, H: 7.43, and N: 8.30 wt%. The ash content is 4.0 wt%.

The oxygen content of the biomass (it is found in the ash) must be determined from a total mass balance.

A 100 g dried biomass contains 47.4 g C, 7.43 g H, 8.30 g N, and 4 g "ash," and consequently 32.87 g O. On an ash-free basis, one obtains $47.4 \cdot 0.96/12 = 3.792$ mol C, 7.133 mol H, 0.569 mol N, and 1.972 mol O.

Hence the composition of the ash-free biomass is $CH_{1.88}O_{0.520}N_{0.150}$, $M_X = 24.3$ g (C-mol)$^{-1}$, and $\kappa_X = 4.39$.

The yield coefficient $Y_{sx} = 0.96 \cdot (23.7/24.3) = 0.9362$ C-mol X (mol C_2H_6)$^{-1}$ = 0.4681 C-mol (C-mol)$^{-1}$

$$(C_2H_6)/2 + 1.236\,O_2 + 0.0702\,NH_3 \rightarrow 0.4681 CH_{1.88}O_{0.52}N_{0.15} + 0.5319\,CO_2 + 1.165\,H_2O. \quad (1)$$

Note that the O_2 consumption is very high ($RQ = Y_{oc} = 0.43$), a typical feature of single-cell protein processes with hydrocarbons as C substrate.

"Normal" biomass (for which we shall often use a standard formula $X = CH_{1.8}O_{0.5}N_{0.2}$) has a composition of macromolecules as seen in Table 3.1.

The composition of the biomass can be based on the more or less known weight fractions of the seven main groups of biomass constituents listed in Table 3.1. The so-called genome-scale metabolic models to be discussed in Chapter 6 rely on the absorption of precursor metabolites and cellular building blocks into the different types of macromolecules in the table. Naturally, an

Table 3.1 Average composition of a "healthy," ash-free *Saccharomyces cerevisiae*.

Macromolecule (C-mol)$^{-1}$	Elemental composition	Weight %	Mole weight g
Protein	$CH_{1.58}O_{0.31}N_{0.27}S_{0.004}$	57	22.45
RNA	$CH_{1.25}O_{0.75}N_{0.38}P_{0.11}$	16	34.0
DNA	$CH_{1.15}O_{0.62}N_{0.39}P_{0.10}$	3	31.6
Carbohydrates	$CH_{1.67}O_{0.83}$	10	27.0
Phospholipids	$CH_{1.91}O_{0.23}N_{0.02}P_{0.02}$	10.8	18.5
Neutral fat	$CH_{1.84}O_{0.12}$	2.5	15.8
Pool of cellular metabolites	$CH_{1.8}O_{0.8}N_{0.2}S_{0.01}$	0.7	29.7

Note: The RNA pool consists of ribosomal RNA (rRNA), transfer RNA (tRNA), and messenger RNA (mRNA). Whereas rRNA and tRNA are very stable, mRNA has a half-life of less than a minute. It serves as a template for protein synthesis, and it is synthesized very fast in a rapidly growing cell. Only ~1 wt% of the cell DW is mRNA. The macromolecular composition of other microorganisms is similar to that of *S. cerevisiae*. The "pool of cellular metabolites" consists of small molecules, precursor metabolites, and building blocks for macromolecular compounds that make up the biomass. These compounds are discussed further in Table 4.2.

uncertainty of the genome-scale metabolic models is associated with the variation in cellular composition of the macromolecules that is seen in cells growing at different external conditions.

On the basis of Table 3.1, the elementary composition of a "healthy" yeast cell can be calculated:

$$X = CH_{1.596}O_{0.396}N_{0.216}S_{0.0024}P_{0.017}. \tag{3.12}$$

There are small differences in the cellular composition determined from elemental analysis and from dissection of the macromolecular composition of the cell, but the two methods are probably equally good according to the very detailed statistical analysis of Lange and Heijnen [5]. One should, of course, not hesitate to use a trustworthy experimentally determined biomass composition.

Note: A 100 g of biomass contains 57 g protein = 57/22.45 = 2.539 C-mol protein. Repeating for the other six components one finds that 100 g biomass contains 4.241 C-mol, that is, 4.241 carbon atoms. The O content of the 4.241 C-mol is $0.31 \cdot 2.539 + 0.75 \cdot 0.471 + \ldots = 1.678$ O atoms or 1.678/4.241 O = 0.396 O per C atom. The same calculation is done for other elements in X.

3.5 Combination of Black-Box Models

In Example 3.2, a (semiquantitative) model for the influence of a side reaction on the yield of citric acid was derived by a combination of two black-box stoichiometries. Basically, all that is needed is to construct and combine stoichiometries that satisfy the requirements of closing the carbon and redox balances. In Example 3.2, one also needed insight into the metabolic pathways of *A. niger* function, and the possible activation of oxaloacetate hydrolase should be known, but often quite obvious combinations of stoichiometries will suffice.

Production of a desired metabolite is just one possible fate of the substrate. If some of the enzymes in the pathway to the desired metabolite have insufficient capacity, the carbon may overflow into other pathways to produce undesirable end products. If the culture is stressed, for example, at a pH lower than the optimal, substantial carbon is used to keep the culture alive. This is *maintenance substrate consumption*, a subject that is further treated in Chapter 7. A general black-box model for substrate consumption is as follows:

$$(-r_s) = Y_{xs}^{true} r_x + m_s. \tag{3.13}$$

Here, r_x is the specific growth rate of the culture (defined in Eq. (2.3)), Y_{xs}^{true} is the amount of substrate needed to produce a certain amount of biomass *by the stoichiometry of the main reaction*, and m_s is the amount of substrate needed to keep a given biomass alive (in units of, for example, (g substrate $(\text{g biomass} \cdot \text{h})^{-1}$)).

The substrate spent in the maintenance reaction will end up as the final product of the catabolic pathway that produces free energy to sustain the existing biomass. Thus, glucose can end up as lactic acid, ethanol, and other metabolic products, or CO_2 can be the end product. In any case, the amount of product, measured in C-mol product (C-mol glucose substrate)$^{-1}$, is known from metabolic pathways. From the metabolic pathways discussed in Chapter 4, it follows that m_p/m_s is, respectively, 1, 2/3, and 1 when the final product is lactic acid, ethanol, and CO_2 by respiration.

Example 3.4 Combination of several stoichiometries to give an experimentally determined stoichiometry

Propane-1,3 diol (CH_2OH-CH_2-CH_2OH,"PDO") is condensed with terephthalic acid (benzene-1,4 dicarboxylic acid) to give the polyester Sorona®, an important new product from the DuPont company. The alcohol is made from glucose in a fermentation process, which is both cheaper and far more environment-friendly than a corresponding chemical route. The organism is a highly engineered *Escherichia coli* into which the pathway from glucose to glycerol has been inserted from yeast, and the two-step pathway from glycerol to PDO (see Figure 4.2) is inserted from a pathogen, *Klebsiella pneumonica*. The process is aerobic, but one tries to keep the oxygen uptake low since much redox is used to reduce glucose ($\kappa_s = 4$) to PDO ($\kappa_p = 16/3$), and both biomass and CO_2 produced in respiration, a competing sink for redox, are waste products. An "ideal" redox-neutral process in which $Y_{so} = 0$, and with no products except PDO and CO_2, is

$$CH_2O \rightarrow 0.75\,CH_{8/3}O_{2/3} + 0.25\,CO_2. \tag{1}$$

In Eq. (1), the redox content in one C-mole glucose is transferred to that of 0.75 C-mole PDO. It is, however, improbable that the ideal yield coefficient 0.75 can be obtained. Some biomass must be produced, and some glucose has to be converted to CO_2 by respiration to provide sufficient free energy to drive the process. The *theoretically maximum yield* coefficient 0.75 derived from Eq. (1) is, however, an important quantity. In the planning of any bioprocess, Y_{sp}^{max} is a fundamental economic parameter: If the relative cost of product to substrate P_{sp} is not high enough to make $P_{sp} \cdot Y_{sp}$ substantially larger than 1, the process has no economic potential.

Let us assume that the measured stoichiometry in a pilot plant fermentation of one of the strains developed (by Genencor) in the long development process toward "the golden bug" is given by Eq. (2):

$$\begin{aligned} CH_2O + 0.15\,O_2 + 0.020\,NH_3 \rightarrow\ & 0.120\,CH_{1.83}O_{0.56}N_{0.17} \\ & + 0.543\,CH_{8/3}O_{2/3} \\ & + 0.337\,CO_2(+0.197\,H_2O). \end{aligned} \tag{2}$$

It might be of considerable help in the further genetic work toward producing the golden bug to estimate how much of the carbon in Eq. (2) goes to biomass formation that is not accompanied by product formation due to limitations in the pathway leading to PDO and how much carbon is used to maintain the biomass present in a steady-state continuous reactor.

The three stoichiometries to be considered are as follows:

$$CH_2O \rightarrow 0.75\, CH_{8/3}O_{2/3} + 0.25\, CO_2. \tag{3}$$

$$\begin{aligned} &CH_2O\,(+0.1013\, NH_3) + 0.3743\, O_2 \\ &\quad \rightarrow 0.5959\, X + 0.4041\, CO_2 + 0.6067\, H_2O\,(3) = \text{Eq. } (3.11). \end{aligned}$$

$$CH_2O + O_2 \rightarrow CO_2 + H_2O. \tag{4}$$

For simplicity, the biomass production stoichiometry Eq. (3) has been chosen to be the same as Eq. (3.11). In Chapter 4 that also includes a free energy balance, one will obtain a better understanding of the amount of O_2 needed in Eq. (3) and consequently the stoichiometry of Eq. (3) will be seen to change with the efficiency of the respiratory process. For the present, it is enough to state that the stoichiometry Eq. (3) is derived from well-documented experimental studies of purely respiratory growth.

The first reaction is defined to have the rate 1, the two others α and β, respectively. The total glucose consumption is $1+\alpha+\beta$ and the observed yield coefficients of PDO and biomass X are given by Eq. (5):

$$(1+\alpha+\beta)^{-1} \cdot 0.75 = 0.543;\; 0.5959\,\alpha\,(1+\alpha+\beta)^{-1} = 0.120. \tag{5}$$

The two equations are solved to obtain $1+\alpha+\beta=1.3812$ and $\alpha=0.2781$. Hence, $\beta=0.1031$.

When these parameters are inserted into Eq. (6), one obtains the experimentally derived stoichiometry Eq. (3.2) after normalizing.

$$\begin{aligned} &1.381 CH_2O + 0.101\,\alpha\, NH_3 + (0.374\,\alpha + \beta)\, O_2 \\ &\quad \rightarrow 0.596\,\alpha\, X + 0.75\, \text{PDO} + (0.25 + 0.404\,\alpha + \beta)\, CO_2 \\ &\qquad + (0.607\,\alpha + \beta)\, H_2O. \end{aligned} \tag{6}$$

The example demonstrates that any experimental stoichiometry like Eq. (2) can be broken into a number of "standard" stoichiometries. In the present case, biomass is produced at a C rate $0.596\,\alpha = 0.\,1657$ C for each 0.75 C of 1,3 PDO produced. Maintenance of the biomass requires an additional 10.3% of the total glucose used. One concludes that the pathways to PDO (the last two steps from glycerol to PDO) have to be opened up, perhaps by increasing the level of the two enzymes that catalyze the two steps. Also, a significant (but not an alarming) amount of glucose is used to maintain a healthy culture, essentially by producing more free energy by conversion of glucose to CO_2.

The PDO fermentation process is a landmark in the progress of "*white biotechnology*," the use of sustainable processes to produce a range of bulk chemicals. The progress from engineering a suitable microorganism and optimizing it by laboratory experiments, pilot-plant fermentations, and finally to industrial production from 2004 (125 000 ton/year with a sales value of US$ 157 million in 2012) by DuPont and Tate & Lyle in Loudon, Tennessee, is described in publications from DuPont and in many scientific papers. Reference [6] is an account by key researchers from DuPont of the metabolic engineering work that preceded production and Ref. [7] reports on further work on the process. In 2013, industrial production of another important diol, butane 1,4 diol (BDO), was started by Genomatica in cooperation with DuPont and Tate & Lyle at their Loudon plant.

Example 3.5 Constructing a stoichiometry that changes continuously with the operating conditions

The overall black-box stoichiometry may change both with the dilution rate and with other environmental conditions. This is illustrated by anaerobic yeast fermentation carried out at different values of pH:

$$\begin{aligned} CH_2O + 0.01658\,NH_3 \rightarrow\ & 0.1381\,X + 0.5160\,CH_3O_{0.5} + 0.2759\,CO_2 \\ & + 0.070\,CH_{8/3}O + 0.037\,H_2O \\ & = \text{Eq. (3.1)}. \end{aligned} \tag{1}$$

$$CH_2O \rightarrow 2/3\,CH_2O + 1/3\,CO_2. \tag{2}$$

The yield coefficient Y_{xs}^{true} of the main reaction Eq. (1) is 1/0.1381 = 7.241, and $X = CH_{1.74}O_{0.6}N_{0.12}$. The maintenance reaction Eq. (2) has ethanol (E) and CO_2 as the end products; $m_{p,ethanol}/m_s = 2/3$ C-mol ethanol/C-mol glucose and $m_{p,CO_2}/m_s = 1/3$ mol CO_2/C-mol glucose. The maintenance sugar consumption, m_s, rapidly increases when the pH of the fermentation decreases to less than 5.

$$\begin{aligned} &\text{Write (1) as } Y_{xs}^{true}\,CH_2O + 0.12\,NH_3 \\ &\quad \rightarrow X + Y_{x,ethanol}^{true}E + Y_{x,CO_2}^{true}CO_2 + Y_{x,glycerol}^{true}G + Y_{x,H_2O}H_2O. \end{aligned} \tag{3}$$

The rate of the reaction Eq. (3) in a steady-state CSTR with $x_f = 0$ is $r_x = D$ (see Eq. (2.3)). Reaction (2) has the rate $(-r_s) = m_s$. Hence, by adding Eqs. (2) and (3), one obtains the total glucose consumption and ethanol production:

$$(-r_s)^{total} = Y_{xs}^{true}D + m_s, \text{ and } r_p = Y_{x,ethanol}^{true}D + m_p.$$

The overall stoichiometry (without H_2O) is as follows:

$$\begin{aligned} &CH_2O + 0.12D/(DY_{xs}^{true} + m_s)NH_3 \\ &\quad \rightarrow D/(Y_{xs}^{true}D + m_s)X + (DY_{x,e}^{true} + 2/3m_s)/(Y_{xs}^{true}D + m_s)\,E \\ &\quad + DY_{xg}^{true}/(Y_{xs}^{true}D + m_s)\,G \\ &\quad + (DY_{x,e}^{true} + 1/3m_s)/(Y_{xs}^{true}D + m_s)\,CO_2. \end{aligned} \tag{4}$$

From the overall stoichiometry, one obtains the *observed yield coefficients* Y_{xs}^{obs} and $[Y_{se}, Y_{sg}, Y_{sc}]^{obs}$. The values of $[Y_{sx}, Y_{s,\,pi}]^{true}$ and of $[m_s, m_{pi}]$ are obtained from linear plots of $(-r_s)$ and r_{pi} versus dilution rate D as further discussed in Example 7.2, Figure 7.6. It follows from Eq. (4) that Y_{sx}^{obs} and Y_{sg}^{obs} decrease to zero when $D \rightarrow 0$, while $[Y_{se}, Y_{sc}]^{obs} \rightarrow [2/3, 1/3]$. Hence, maintenance is a good way of increasing the yield of ethanol on glucose, but certainly not the productivity of ethanol that, of course, decreases to 0 when $D \rightarrow 0$.

It has been shown in Section 3.5 that almost any experimentally obtained stoichiometry can be patched together from a few basic stoichiometries, for example, a stoichiometry obtained at high D and a suitable pH (where maintenance is not important) and a separate stoichiometry representing either maintenance or the occurrence of a side reaction as illustrated in Example 3.2.

The tools of Chapter 4 will allow a much better analysis of the physiology of the fermentation process by including the immense fund of data harbored in metabolic pathways.

3.6 Application of Carbon and Redox Balances in Bio-Remediation Processes

The examples given until now have been concerned with production of biochemicals by fermentation processes. In bio-remediation, the substrates are broken down to small molecules of CO_2 and H_2O, and (in anaerobic processes) some of the capital costs of the installation may be recovered, for example, biogas and CH_4.

Environmental engineering has a number of challenges that are not found in the bio-industry: The substrate composition and feed rate fluctuates with time, the biodegradation of pollutants is mediated by consortia of microorganisms, most of which are not sequenced, and the composition of the microbial consortium changes with the composition of the feed. Still, a number of fundamental calculations in bio-remediation can be done by the methods discussed in this chapter, and likewise the design of reactors for bio-remediation follows the general principles of Chapters 10 and 11. Thus, cross-talk between environmental engineers and the production industry can be beneficial for both parties: It will be of considerable advantage to the bio-remediation industry if some of the general calculation methods proposed in texts like this book are adopted. For practitioners of industrial biotechnology, environmental science offers very promising microorganisms and processes if they can be adapted to the environment of an industrial reactor.

The use of a redox balance in the analysis of bioremediation processes is an excellent tool that gives insight into the nature of the process and supports quantitative studies.

Thus, anaerobic degradation of cellulose to the end products CH_4 and CO_2 is accomplished in a series of redox-neutral reactions in which the redox present in the substrate is distributed in the products:

$$\begin{aligned}\text{Cellulose } \frac{1}{n}(CH_2O)_n \rightarrow\ & Y_{s,HAc}CH_2O\,(HAc) + (1 - Y_{s,HAc})\,CO_2 \\ & + (2 - 2\,Y_{s,HAc})\,H_2 + (Y_{s,HAc} - 1)\,H_2O.\end{aligned} \quad (3.14)$$

$$\begin{aligned}\text{Cellulose } \frac{1}{n}(CH_2O)_n \rightarrow\ & Y_{s,BAc}\frac{1}{4}C_3H_7COOH + (1 - Y_{s,BAc})\,CO_2 \\ & + (2 - 2.5\,Y_{s,BAc})\,H_2 + (1.5\,Y_{s,BAc} - 1)\,H_2O.\end{aligned} \quad (3.15)$$

In Eqs. (3.14) and (3.15), *fatty acids*, for example, acetic acid and butyric acid (HBu), are formed together with CO_2 and H_2. Acetic acid is decomposed to CH_4 and CO_2 as shown in Eq. (3.16), and it would be very desirable if butyric acid could also be decomposed by the second reaction in Eq. (3.16). Example 9.4 gives a more detailed description of the processes that actually occur.

$$\begin{aligned}CH_2O\,(HAc) \rightarrow \frac{1}{2}CH_4 + \frac{1}{2}CO_2,\ \text{and}\ \frac{1}{4}\ C_3H_7COOH + \frac{1}{4}H_2O \\ \rightarrow \frac{5}{8}CH_4 + \frac{3}{8}CO_2.\end{aligned} \quad (3.16)$$

In bio-remediation biological oxygen demand (BOD) and chemical oxygen demand (COD) are used as units of redox content of a substrate. BOD is the most important. It is defined as the amount of O_2 needed to oxidize all the reduced matter (solids or solutes) in, for example, a sample of wastewater to redox neutral compounds with the help of a consortium of microorganisms, either naturally occurring in the sample or added to it. Elsewhere in this series of books on biotechnology, there will be more detailed descriptions of the way BOD is measured in different countries, and also COD and total organic carbon (TOC) will be discussed. In the present context, it is enough to state that BOD is measured in units of, for example, mg O_2/l wastewater or kg O_2/ton of wastewater. Thus, a solution containing 0.2 g/l cellulose (mostly as colloids) has a BOD of 32 000 mg $O_2 \cdot (0.2/30) = 213$ mg l^{-1}, since the cellulose is degraded to redox neutral products by

$$\frac{1}{n}(CH_2O)_n + O_2 \rightarrow CO_2 + H_2O. \quad (3.17)$$

Typical raw sewage has a BOD value of about 300 mg l^{-1}.

Although BOD is measured in units of O_2 needed to decompose the organic material to redox-neutral final products, the process considered is not necessarily an aerobic process such as Eq. (3.17). In anaerobic processes, BOD is measured as the theoretical amount of O_2 which *would* be needed to convert the reducible material in the feed to redox-neutral compounds, although the actual products may well contain redox. Here, balancing of the redox between substrate and products is used to calculate the reaction stoichiometry as described in Example 3.6.

Example 3.6 Designing an anaerobic wastewater treatment unit for slaughterhouse waste

The wastewater from an animal processing plant is sent to an anaerobic digester. The major pollutant is a mixture of animal fats that will be represented as glycerol-tripalmitate (tripalmitin), $C_{51}H_{98}O_6$. The digester is a stirred tank reactor that operates in steady state. Laboratory measurements show that the feed stream has a BOD of 3180 kg per day of operation. There are three product streams: (1) a gaseous mixture of CH_4 and CO_2. (2) A liquid effluent that contains no fat. This effluent stream is measured to have a BOD of 330 kg/day, and the main component is characterized as acetic acid, HAc. (3) A solid effluent of biomass, 160 kg/day with an ash content of 7 wt% ($X = CH_{1.8}O_{0.5}N_{0.2}$). As the first step in the design of the digester, the stoichiometry of the process is to be determined.

A) First the feed rate of the tripalmitin solution is determined from the BOD of the feed.

   The degree of reduction of palmitin is $\kappa_s = (1/51)\ (4 \cdot 51 + 1 \cdot 98 - 2 \cdot 6) = 5.686\ (\text{C-mol})^{-1}$. Since all the fat is digested, this amount of redox has to be distributed between the products, CH_4, biomass, HAc and CO_2 (which, of course, is redox neutral). Since the total redox content of the feed corresponds to $3180/32 = 99.375$ kmol O_2 per day needed to oxidize all the redox in the feed, the feed stream is calculated from $M \cdot 5.686 = 99.375 \cdot 4$ or $M = 69.91$ kC-mol tripalmitin/day.

B) The liquid effluent (HAc) is calculated from $M_{HAc} \cdot 4 = (330/32) \cdot 4$ or $M_{HAc} = 10.31$ kC-mol HAc/day. The solids stream is calculated from $M_X = 160 \cdot 0.93/24.6 = 6.05$ kC-mol biomass/day. Thus, the total production of CO_2 and $CH_4 = M_{CO_2} + M_{CH_4} = 69.91 - 10.31 - 6.05 = 53.55$ kC-mol/day.

C) A total redox balance states how the redox content of the feed is distributed in the products: $69.91 \cdot 5.686 = 10.31 \cdot 4 + 6.05 \cdot 4.20 + M_{CH_4} \cdot 8 + M_{CO_2} \cdot 0 \rightarrow M_{CH_4} = 41.36\ \text{kC-mol}\ CH_4/\text{day}$. Hence $M_{CO2} = 12.19$ kC-mol CO_2/day.

The stoichiometry of the black-box model for the digestion of the fat is as follows:

$$\begin{aligned}(1/51)\, C_{51}H_{98}O_6 + 0.01731\, NH_3 \rightarrow\ & 0.0865\, X + 0.1744\, CO_2 \\ & + 0.1475\, \text{HAc} \\ & + 0.5916\, CH_4\ (+\text{water}). \qquad (1)\end{aligned}$$

From an economic point of view, the most important result is that the biogas production M_{CH4} (which to a certain extent can cover the cost of the wastewater treatment process) has been extracted by simple carbon and redox balances.

Oxygen is not the only redox sink (or electron acceptor), as a large number of anions or cations can have a similar role. SO_4^{-2}, NO_3^{-1}, $S_2O_3^{-2}$, and Fe^{+3} are typical examples. The sulfate in stagnant oxygen-depleted water is reduced to sulfide with organic waste material as carbon and energy source:

$$2\frac{1}{n}\,(CH_2O)_n + SO_4^{-2} \rightarrow H_2S + 2\,HCO_3^{-1}. \tag{3.18}$$

Typical sulfur-reducing bacteria such as *Desulfobacter* run the sulfur-reducing process Eq. (3.18) in parallel with their growth on the carbon source, using a little of the H_2S produced to build up more biomass. The free energy needed to sustain Eq. (3.18) is obtained by oxidation of the carbon source to CO_2.

Eq. (3.18) is a typical *dissimilatory* redox process. Such processes are utilized in many wastewater cleaning processes (which in contrast to Eq. (3.18) produce environment-friendly end products). Thus, both nitrate and ammonia can be removed by the following process:

$$\begin{aligned} CH_2O + 0.2\,Y_{sx}\,NH_3 + 2\,\alpha\,HNO_3 \rightarrow\; & Y_{sx}CH_{1.8}O_{0.5}N_{0.2} + \alpha\,N_2 \\ & + (1 - Y_{sx})\,CO_2 \\ & + (1.4 - 1.02\,Y_{sx})\,H_2O. \end{aligned} \tag{3.19}$$

A redox balance yields:

$$4 + 2\,\alpha \cdot (-8) = 4.20\,Y_{sx} + \alpha \cdot (-6). \tag{3.20}$$

$$\alpha = \frac{4 - 4.20Y_{sx}}{10}; \quad Y_{sw} = 1.4 - 1.02Y_{sx}. \tag{3.21}$$

For $Y_{sx} < 0.952$ (when $Y_{sx} = 0.952$, all the redox from the glucose is used to produce biomass), some nitrate can be reduced to N_2. The highest possible value $\alpha = 0.4$ for nitrate removal is obtained when $Y_{sx} = 0$, but this limit is not attainable since some of the carbon source must be used to produce biomass in order to maintain the process.

The stoichiometry Eq. (3.19) is, of course, not correct at the approximately neutral pH of the wastewater treatment process. In actual fact, the stoichiometry should be:

$$\begin{aligned} & CH_2O + 0.2Y_{sx}NH_4^+ + (8 - 8.4Y_{sx})/10\,NO_3^{-1} \\ & \rightarrow Y_{sx}X + (4 - 4.2\,Y_{sx})/10\,N_2 + (1 - Y_{sx})\,HCO_3^{-1} \\ & +(0.04\,Y_{sx} + 0.2)H^+ + (0.4 - 0.02\,Y_{sx})\,H_2O. \end{aligned} \tag{3.22}$$

Note, the main result, namely, Eq. (3.21) is obtained for the nitrate uptake that runs parallel to biomass production with concomitant uptake of ammonium ions. If one insists on using the complicated formulation Eq. (3.22,) one should first make the carbon and redox balances close. Next, the charge balance is closed and the coefficient to H^+ is obtained. Finally, Y_{sw} is calculated, either from the oxygen or from the hydrogen balance. Also note that the degree of reduction of NO_3^{-1}, NH_4^+, and HCO_3^{-1} is the same as that of HNO_3, NH_3, and CO_2 as stated earlier in point (4) of Section 3.2.

In Eq. (3.19) or (3.22), nitrate is removed at the expense of a cheap carbon source that could be HAc and degraded cellulose which are present in most wastewaters. Some denitrifiers, for example, *Pseudomonas denitrificans*, reduce nitrate all the way to NH_4^+ that is taken up by the organism using an *assimilatory* nitrate reduction process.

Other naturally occurring microorganisms oxidize NH_3 and H_2S to NO_3^{-1} and SO_4^{-2}, respectively. Here, O_2 is a substrate, and both these processes and the aerobic biomass formation are energetically favored. In the removal of both NH_4^+ and NO_3^{-1} from wastewater, an aerobic bioprocess is coupled in series with an anaerobic process. In the aerobic process, all NH_4^+ is oxidized to NO_3^{-1}, while in the anaerobic process the NO_3^{-1} is reduced to N_2 by Eq. (3.19) or (3.22).

References

References to the new (2013) industrial process for production of 1,4 butane-diol are given in the home page of the Genomatica company. The underlying patent (US patent 8067214) in which the diol is produced from the TCA cycle intermediate succinyl-CoA via succinate semi-aldehyde (COOH-$(CH_2)_2$-CHO) is only one of the several recent patents concerning bio-production of this important chemical based on glucose D-xylose (CH_2OH-$(CHOH)_3CHO$) and other sugars. The maximum theoretical yield of 1,4 butane diol (8/15 C-mol (C-mol glucose)$^{-1}$) is calculated in Example 4.3.

1 von Meyenburg, K.H. (1969a) Katabolit Repression und der Sprossungszyklus von *Saccharomyces cerevisiae.* Diss. ETH, Zürich.

2 von Meyenburg, H.K. (1969b) Energetics of the budding cycle of *Saccharomyces cerevisiae* during glucose limited aerobic growth. *Arch. Mikrobiol.*, **66**, 289–303.

3 Villadsen, J. and Jørgensen, S.B. (2014) Reflections on the aerobic fermentation stoichiometry of Crabtree positive yeasts. *Biotechnol. Bioeng.*, **111**, 632–637.

4 Pedersen, H., Christensen, B., Hjort, C., and Nielsen, J. (2000) Construction and characterization of an oxalic acid nonproducing strain of *Aspergillus niger. Metabol. Eng.*, **2**, 34–41.

5 Lange, H.C. and Heijnen, J.J. (2001) Statistical reconciliation of the elemental and molecular biomass composition of *Saccharomyces cerevisiae. Biotechnol. Bioeng.*, **75**, 334–344.

6 Nakamura, C.N. and Whited, G.M. (2003) Metabolic engineering for the microbial production of 1, 3 propane-diol. *Curr. Opin. Biotechnol.*, **14**, 454–459.

7 Saxena, R.K., Anand, P., Saran, S., and Isar, J. (2009) Microbial production of 1,3 propane-diol. Recent developments and emerging opportunities. *Biotechnol. Adv.*, **27**, 897–913.

4
Primary Metabolic Pathways and Metabolic Flux Analysis

John Villadsen

Summary

The objective of this chapter is to introduce the reader to metabolic flux analysis (MFA) – or more precisely to stoichiometry-based MFA. The first six sections of this chapter introduce the catabolic pathways in which free energy is produced to drive the cell-building anabolic pathways. An overview of these, so-called primary pathways will provide the reader who is primarily trained in the engineering sciences with at least a preliminary introduction to biochemistry and will also show how carbon is drained off the catabolic pathways to provide precursors for cell mass building and sometimes for important industrial products. Chapters 4.7 to 4.9 describe the quantitative basis for MFA. To analyze the flow of carbon in the metabolic network of cells is today a key discipline of bioengineering. Computer-based software for this purpose is freely available, and it is based on the theory of Chapters 4.8 and 4.9. In Chapter 5, this theory is extended to calculate the flow of labeled substrate in the network, and with the algorithms discussed in Chapter 5 a comprehensive analysis of the flow through complex networks is possible. As an introduction to MFA in "real" systems, the methods of Chapter 3 will, however, in Chapter 4.7 be treated systematically by the so-called substrate-to-product MFA technique. It is hoped that this "soft" introduction to a complex, but important discipline of bioengineering will help biologists and biochemists to appreciate the concepts of MFA and its enormous value both for academic and for industrial research.

4.0
Introduction

The examples used in Chapter 3 to illustrate how a stoichiometry can be set up were (mostly) discussed without any reference to the biochemistry that determines how carbon is distributed in the metabolic network and finally excreted in metabolic products and biomass. All that was needed was balances for the elements – often only the carbon balance and the redox balance. Apart from the

Fundamental Bioengineering, First Edition. Edited by John Villadsen.

fact that a cell can produce a given product only if a pathway from substrate to the product exists in the organism, this crude procedure ignores the information concerning free energy changes between substrate and product. It was shown in Eqs. (3.5) and (3.6) how two parallel reactions from glucose to products can be combined to one stand-alone reaction stoichiometry (3.7) by balancing the production and consumption of redox units, preliminarily called "H." In this chapter the true nature of the "redox packages" that are passed from one reaction to another will be revealed. Just as is the case for redox balancing of reactions, a reaction with high affinity (large positive value of $-\Delta G_R$) can pass free energy to other reactions. In contrast to reactions between organic or inorganic compounds, the living cells can harness part of the free energy change as "energy packages" that can be donated to other reactions and thereby give them an overall positive affinity. When an energy balance is used together with a redox balance, another constraint between the yield coefficients is introduced. This can be used to find the product yields with fewer measurements than were needed in Chapter 3.

In Sections 4.0–4.5 the primary pathways in cell metabolism will be discussed. These pathways have a dual function: (1) They primarily deliver metabolic energy to drive the synthesis of macromolecules, first of all proteins. These cell-building reactions have a negative ($-\Delta G_R$). (2) Next, they provide precursors that are further combined into building blocks for macromolecules.

The primary pathways have been studied for more than a century. The chemistry and the free energy release for each step in the primary pathways from substrate to end product is known exactly, whereas the energy demand for synthesis of the macromolecules in new cells is much more uncertain. First of all, the energy needed to build cells in *anabolism* depends on the operating conditions that do not influence the energy release per carbon molecule from the *catabolic* reactions.

The primary pathways produce almost all the low molecular compounds that are now considered as substitutes for the same products derived from oil-based substrates: alcohols, aldehydes, ketones, and carboxylic acids. Bio-industrial production of amino acids, antibiotics, precursors for synthetic rubber (e.g., isoprene, 2-methyl-1,3 butadiene) would normally be classified as the result of *secondary* metabolic pathways that are distinct from the energy-producing primary pathways. Also here the energy requirement can be exactly predicted from charts of metabolic pathways. When the final product is, for example, a heterologous protein, sometimes associated with biomass growth but often produced in a nongrowth stationary phase, the energy requirement must be found by more or less empirical rules.

The primary pathways to be examined in the following are: (1) glycolysis, primarily by the EMP pathway, but other glycolytic pathways will also be mentioned; (2) fermentative pathways in which the redox generated in the glycolytic reactions is consumed; (3) reactions in the TCA cycle (TCA = tricarboxylic acid), which produce biomass precursors and also redox power to be used for energy production in oxidative phosphorylation; and finally (4) reactions in the PP pathway (PP = pentose phosphate). In the PP pathway, precursors for synthesis of genetic material and secondary metabolic products are produced. Finally, NADPH, the redox carrier in most anabolic reactions, is an important product from the PP pathway.

Figure 4.1a (which has to be viewed on a 1.2 m × 0.9 m poster to read details) presents the entire metabolic network of cells. Pathways are shown that produce metabolic products, and also pathways where carbon is siphoned off to produce precursors and building blocks for macromolecules. Individual metabolic reactions are shown as being reversible, and enzymes that catalyze the reaction in either direction are stated. Black arrows signify general biochemical pathways, blue arrows are used for animal cells, green for plant cells, and red for unicellular organisms. Reaction species are shown in black, enzymes in blue, and cofactors in red. Activators (e.g., metal ions) are shown next to the reaction arrow with either a (+) or a (−) sign to show activation or deactivation.

In this chapter, only reactions in which the carbon source is either degraded to CO_2 or to primary metabolic products or used to produce biomass will be treated in detail. When the carbon flow is reversed, proteins, lipids, and other large molecules from biomass are degraded, serving as carbon and energy source for the synthesis of precursors to form new biomass. This is *gluconeogenesis* in contrast to *glycolysis*. The final product of gluconeogenesis is an activated glucose (glucose 6-phosphate, G6P) that serves to create genetic material to build up new cells. Gluconeogenesis is of fundamental importance in *nutrition*, but the main focus of this chapter is, as stated above, *production* of biomolecules either from (basically) sugar or from degraded carbohydrate polymers.

The chart in Figure 4.1a consists of 120 subsections denoted by coordinates A–L and 1–10. Each of the 120 subsections can be picked out for further study in the electronic version of the ExPASy chart. Figure 4.1b is a collation of four of these subsections and shows (hopefully in a readable format) the reactions associated with the TCA cycle (see Figure 4.6).

A large number of reactions feed carbon from various sources into the TCA cycle. They are the *anaplerotic reactions* (*plerotikos* in Greek means "to fill" and *ana* means "up"). These reactions are very important in the TCA cycle where in aerobic reactions carbon can be degraded to CO_2 and redox equivalents for use in energy production by *respiration*. But the TCA cycle also produces a number of precursors for cell constituents (proteins), and the metabolites thus extracted are replenished by anaplerosis. The drain of TCA cycle metabolites in *cataplerotic* reactions is especially critical in commercial amino acid production and in the production of carboxylic acids, for example, citric acid and succinic acid.

In practical work, the information on the chart is retrieved in digital form. All reactions in a given pathway are listed together with the enzymes catalyzing $X \rightarrow Y$ and $Y \rightarrow X$. The redox units involved in the reaction can, of course, be calculated from the redox levels of X and Y, but the name of the redox carrier, the compound acting as cofactor in the reaction, is stipulated on the chart. Likewise, the free energy package that is donated to some reactions in the thermodynamically challenged direction is named. The redox- and energy-containing

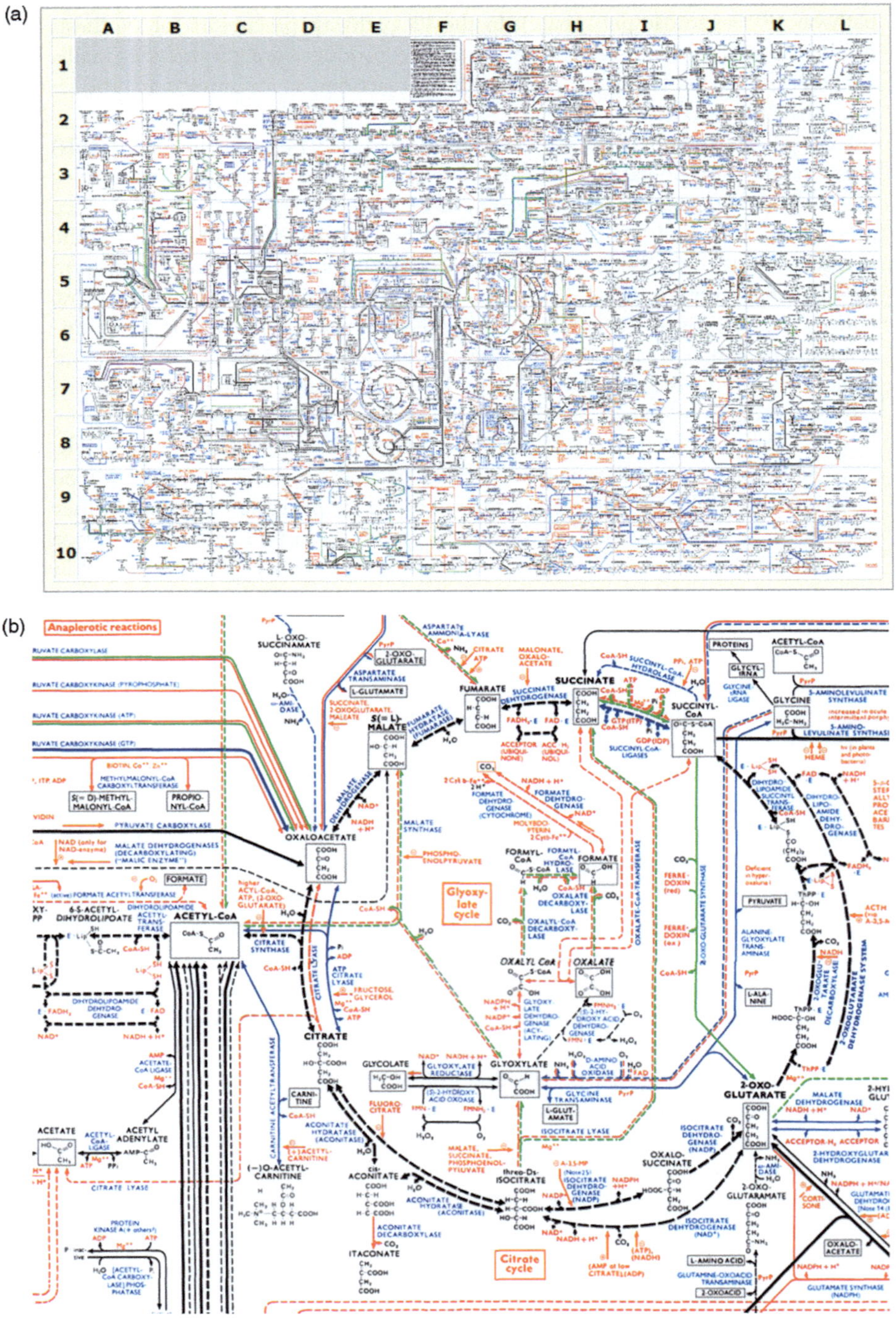

Figure 4.1 (a) Metabolic pathway chart from ExPASy Proteomics server and (b) an enlarged portion from the center of (a) (F5, F6, G6, G5) and showing reactions relevant for the TCA cycle that is shown schematically in Figure 4.6.

cofactors are discussed in the following, more detailed description of the primary pathways.

4.1 Glycolysis

In glycolysis, Figure 4.2, a glucose molecule is transported from the medium through the cell membrane by one of the many glucose transport carriers imbedded in the membrane. Inside the cell, the glucose molecule is energized by reaction with an energy package (ATP) to G6P. G6P is isomerized to F6P and is

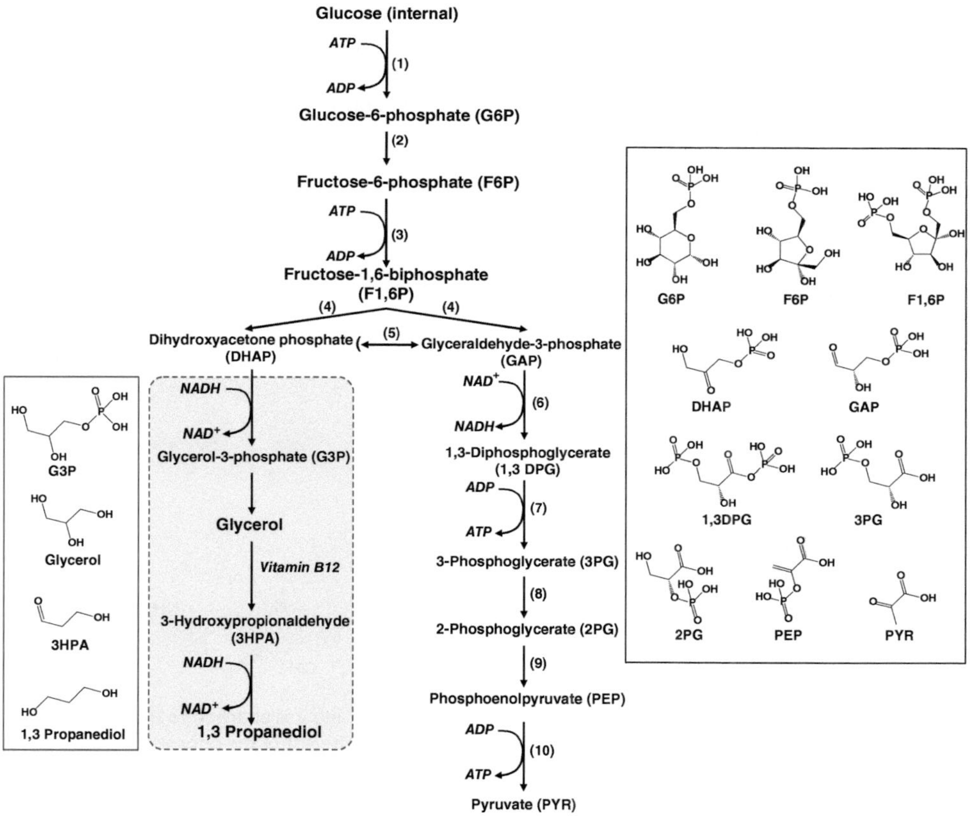

Figure 4.2 Schematic representation of the Embden–Meyerhof–Parnas glycolytic pathway: glucose → pyruvate. Also shown (shaded) is a reductive pathway starting at DHAP and leading to a number of valuable three-carbon compounds. In bacteria with a PTS transport system, glucose is directly converted to G6P on passing the cell membrane. The ATP is taken from reaction (10) and no intracellular free glucose exists. The thermodynamics of the glycolysis pathway is discussed in Chapter 9.

further energized to F1,6P by reaction with a second ATP. The internal energy of F1,6P is so high that it breaks down into two triosephosphates, dihydroxy-acetone phosphate (DHAP) and glyceraldehyde phosphate (GAP).

The two triose-phosphates are in equilibrium. If the end product of glycolysis, pyruvate (PYR), is withdrawn rapidly from the pathway, which is the case when glycolysis is (mainly) used for energy production, then all carbon is drawn from DHAP to GAP. The reactions in the shaded box show how carbon can also be channeled into glycerol and further to a number of other primary pathway metabolites, for example, propane 1,3 diol (PDO), as discussed in Example 3.5.

The enzymes are (1) hexokinase, (2) phosphohexose isomerase, (3) phosphofructokinase, (4) aldolase, (5) triosephosphate isomerase, (6) 3-phosphoglyceraldehyde dehydrogenase, (7) 3-phosphoglycerate kinase, (8) phosphoglycero mutase, (9) enolase, and (10) pyruvate kinase.

The primary function of glycolysis is to provide free energy for cell processes, especially for synthesis of macromolecules, but it also supplies many biomass precursors (see Table 4.2). The energy is provided in reactions (7) and (10) as "energy packages," adenosine triphosphate, abbreviated to ATP. The structural formula for ATP is shown in Figure 4.3a.

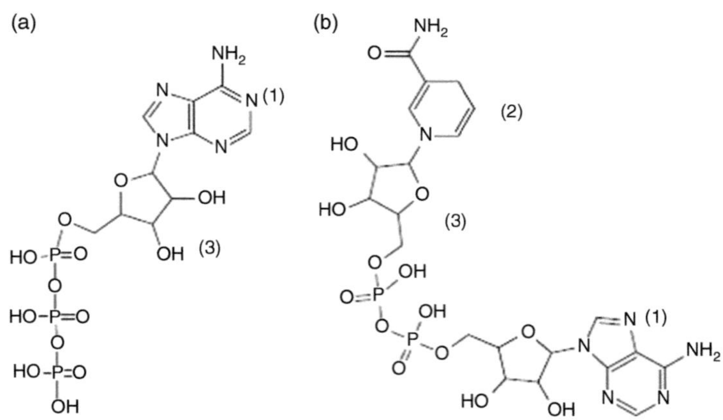

Figure 4.3 Structural formula of ATP (a) and of NADH (b). (a) The upper part of ATP is six-amino purine = *adenine*. This is attached to the 1′carbon of the C5 sugar *ribose*. The sugar is phosphorylated at the 5′carbon. When ATP delivers an energy package, it is hydrolyzed to ADP, adenosine diphosphate, which can be further hydrolyzed to AMP, adenosine monophosphate. (b) The structure of NADH is very similar to that of ATP. The adenine is recognized to the right in the picture, and it is attached to the 1′carbon in ribose. The last of the phosphate groups is exchanged for another nucleotide, *nicotine amide*. When NADH is oxidized to NAD^+, the 1′nitrogen in the pyridine ring of nicotine amide becomes positively charged, while at the 4′carbon one H is removed, resulting in a net loss of two redox units ("H_2"). The net reaction when redox is transferred is as follows: $R + NADH + H^+ \rightarrow RH_2 + NAD^+$.

At the normal pH 7 of cell reactions, all four acid groups (OH) in ATP are dissociated, and hydrolysis of ATP to ADP is given by the reaction:

$$ATP^{-4} + H_2O \rightarrow ADP^{-3} + H_2\,PO_4{}^{-1}(= \text{“P”}) + (-\Delta G^{\circ}{\sim}31-35\,\text{kJ}). \qquad (4.1)$$

$$\begin{aligned} \text{Glucose} + ATP &\rightarrow G6P \leftrightarrow F6P \text{ and } F6P + ATP \rightarrow F1,6P \\ &\rightarrow GAP + DHAP. \end{aligned} \qquad (4.2)$$

The free energy in GAP (and DHAP) is released in reactions (7) and (10) of Figure 4.2 and used to energize other reactions. It is not possible to convert glucose directly to the two trioses – glucose must first be energized.

The value for $-\Delta G^{\circ}$ shown in Eq. (4.1) corresponds to *standard reaction conditions.* As discussed in Chapter 9, the free energy released by hydrolysis of one ATP varies considerably with the concentration of the reactants that in the standard state is supposed to be 1 M for ATP^{-4}, ADP^{-3}, and $H_2PO_4{}^{-1}$, but 10^{-7} M for H^+. The range of ΔG° reflects the uncertainties in literature values.

In Figure 4.2, reactions from glyceraldehyde 3-phosphate (GAP) to pyruvate produce two ATP per triose (reactions (7) and (10)) and one NADH. Since DHAP is supposed to be transferred to GAP by the reversible reaction (5) when all the carbon ends up as PYR, the total ATP yield per glucose molecule is 4. To activate glucose to F1,6P, two ATPs have been used. Hence, the net-yield is two ATPs and two NADHs per glucose molecule. Since production of 1 C-mol of biomass from one C-mol glucose by an *anaerobic* process requires about two ATPs (see Section 4.5, Eq. (4.23)), one cannot expect a biomass yield > 1/6 C-mol X (C-mol glucose)$^{-1}$. A higher biomass yield requires an *aerobic* process. Much more ATP is produced per glucose molecule in *oxidative phosphorylation* (see Section 4.4) than by *substrate-level phosphorylation* in glycolysis. Glycolysis by the EMP pathway is, however, the most effective of all ATP-generating pathways when glucose is degraded to pyruvate.

Another process to convert glucose to pyruvate is the *Entner–Doudoroff* (ED) pathway. G6P is oxidized to 6- phosphogluconate (COO^{-1}- $(CHOH)_4$-CH_2-O-$HPO_4{}^{-2}$) by reaction (11) in Figure 4.7. This is next converted to KDPG (2-keto-3-deoxy-6-phosphogluconate = COO^{-1}-CO-CH_2-$(CHOH)_2$-CH_2-O-$HPO_4{}^{-2}$). As is the case for F1,6P, an aldolase splits the 6-carbon KDPG into two 3-carbon molecules, but here the products are pyruvate and GAP. Only the path from GAP to PYR yields two ATPs, but the ATP spent to activate F6P to F1,6P is saved. The net result is that two PYRs are produced, but only one ATP per glucose, that is, *half the ATP yield* of glycolysis by the *Embden–Meyerhof–Parnas* pathway (Figure 4.2). The ED pathway is probably of ancient origin and it functions in bacteria that lack phosphofructokinase, the enzyme in reaction (3) of Figure 4.2. In the bacterium genus *Zymomonas* glycolysis proceeds via the Entner–Doudoroff pathway, and very little biomass is produced in an anaerobic fermentation process. This is utilized to produce ethanol at close to the theoretical yield ⅔ C-mol (C-mol glucose)$^{-1}$ by *Zymomonas mobilis.* The downside of ethanol production by means of bacteria is that they cannot nearly sustain the same ethanol concentration as typical yeast strains (e.g., *Saccharomyces cerevisiae*). In state-of-the-art production of ethanol from corn syrup or lignocellulosic biomass *S. cerevisiae* is able to function also at an ethanol concentration of more than 20 wt%, while bacteria work only up to an ethanol concentration of 5–6 wt%. With yeast as production strain, the all-important energy input for up-concentration of the aqueous ethanol solution to nearly 100% ethanol is smaller by a factor of 4–6.

Reaction (6) of Figure 4.2 results in the oxidation of an aldehyde, glyceraldehyde, to an acid, glyceric acid. Two redox units are obtained per pyruvate. This corresponds to the total loss of 4 redox units when glucose that contains 24 redox units is converted to 2-pyruvic acid (= 2-CH_3-CO-COOH) that contains 20 redox units. The redox carrier is NADH, the organic molecule shown in Figure 4.3b.

NADH is a typical redox carrier in catabolic reactions, but there are many other cofactors with a similar function, for example, flavin adenine dinucleotide ($FADH_2$) and pyrroloquinone (PQQ). In eukaryotes (e.g., yeast), $FADH_2$ carries high-energy electrons in oxidative phosphorylation (Section 4.5 and Chapter 9). The oxidized form FAD (also named $FADH^+$) is tightly bound to the second protein complex in the oxidative phosphorylation chain. FAD is also part of the enzyme complex succinate dehydrogenase, the catalyst for oxidation of succinate to fumarate in the TCA cycle. PQQ is used in the catabolism of *methylotrophs* (bacteria that use methane or methanol as carbon source), for example, in the oxidation of methane to methanol.

In glycolysis by the Entner–Doudoroff pathway, one obtains one NADPH and one NADH rather than the two NADH, the result from the Emden–Meyerhof pathway. The redox cofactor NADPH (which has a structure very similar to that of NADH) is produced in catabolic pathways, but consumed in anabolic pathways. In contrast to NADH, it cannot be oxidized in respiration, and for this reason it does not contribute to the very large free energy production in aerobic processes.

The redox and the energy level in cells, the catabolic redox charge R_c, the anabolic redox charge R_a, and the energy charge R_e are experimentally found to be controlled within rather strict limits:

$$R_c = \frac{\text{NADH}}{\text{NADH} + \text{NAD}^+}; \quad R_a = \frac{\text{NADPH}}{\text{NADPH} + \text{NADP}^+};$$

$$0.03 < R_c < 0.07; \; 0.4 < R_a < 0.6. \tag{4.3}$$

$$R_e = \frac{\text{ATP} + \text{ADP}/2}{\text{ATP} + \text{ADP} + \text{AMP}}; \quad 0.85 < R_e < 0.95. \tag{4.4}$$

The two very similar cofactors NADH and NADPH can be converted to each other by the following reaction:

$$\text{NADH} + \text{NADP}^+ \leftrightarrow \text{NADPH} + \text{NAD}^+ \tag{4.5}$$

The reversible reaction (4.5) is catalyzed by NADPH:NAD-transhydrogenase, an enzyme present in a number of bacteria. When the values for R_c and R_a for normally functioning cells are inserted in the mass action fraction K for Eq. (4.5) it will, however, follow that reaction (4.5) is shifted far to the left, that is, toward

NADH and NADP$^+$. From Eq. (9.10), and with a standard free energy ($-\Delta G_R^\circ$) of reaction (4.5) approximately zero, ($-\Delta G_R$) in the direction from left to right is as follows:

$$(-\Delta G_R) = (-\Delta G_R^\circ) - RT \ln \frac{[\text{NADPH}][\text{NAD}^+]}{[\text{NADH}][\text{NADP}^+]}$$

$$\approx 0 - RT \ln\left[\left(\frac{1-R_c}{R_c}\right)\cdot\left(\frac{R_a}{1-R_a}\right)\right] \approx -RT \ln 19. \qquad (4.6)$$

Consequently, there is no chance that cloning of NADPH:NAD-transhydrogenase into yeast will convert NADH to NADPH that as seen in Example 4.5 would diminish the undesired by-product glycerol in production of bio-ethanol.

The production of redox units in a pathway can always be calculated as shown in Chapter 3 by comparing the redox levels of the substrate and the product of the pathway. The nature of the redox carrier produced or consumed (e.g., NADH or NADPH) cannot be decided without reference to the biological pathway charts. In contrast, it is impossible to calculate the ATP yield of a pathway without reference to the biology of the metabolic reactions in the pathway.

Note: Nissen *et al.* [1] measured the ratios NADH/NAD$^+$ (0.15 ± 0.01) and NADPH/NADP$^+$ (5.10 ± 0.55) in two yeast strains, which could not express NADPH:NAD transhydrogenase, and in one strain with an operative NADPH:NAD transhydrogenase (0.17 ± 0.01 and 2.96, respectively). These quite accurate measurements gave $R_c = 0.13$ and $R_a = 0.84$ for the strain that did not express the enzyme and [0.145, 0.75] for the "active" strain. The mass action fraction is even higher than 19 as in Eq. (4.6) and consequently even more NADH is obtained when reaction (4.5) is released by insertion of the gene for the enzyme. Presumably, the ratio NADH/NAD$^+$ is tightly controlled by other mechanisms and will not reach its equilibrium value by Eq. (4.6).

Whether an enzymatically catalyzed network reaction produces ATP or not will first of all depend on the free energy change of the reaction. Thus, hydrolysis of glycerol-3 phosphate to glycerol is not accompanied by production of an ATP, although the reverse reaction requires one ATP. The free energy change that accompanies hydrolysis of glycerol-3 phosphate is only about 13 kJ, not enough to produce one ATP (~35 kJ). In contrast, the reverse reaction needs one ATP.

In the metabolic charts, enzymes that produce ATP as the second product are named *kinases*, while enzymes that deliver the released energy, not as free energy but as heat ($-\Delta H = Q$), are named *hydrolases*. The metabolic heat produced is, of course, also needed by the microorganisms to sustain their optimal operating temperature (e.g., 37 °C for *Escherichia coli* and about 30 °C for yeast).

4.2 Fermentative Metabolism: Regenerating the NAD$^+$ Lost in Glycolysis

The purpose of glycolysis is to produce sufficient ATP to synthesize biomass in an anaerobic process where ATP is not produced in any other pathway.

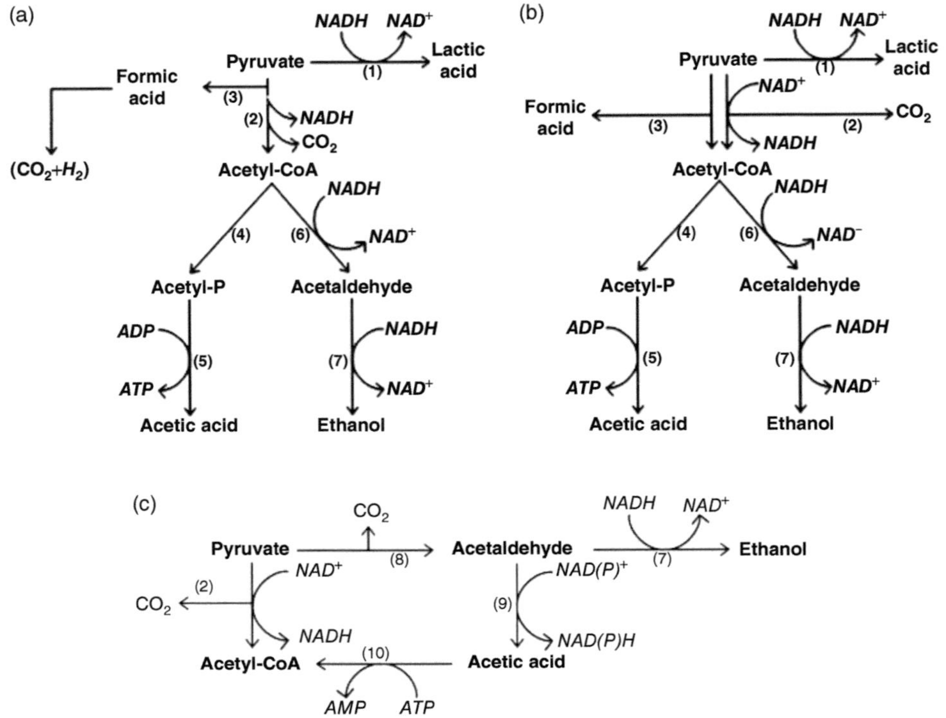

Figure 4.4 Fermentative pathways in *E. coli* (a), lactic bacteria (b), and in yeast (c). The enzymes in (a) and (b) are (1) lactic acid dehydrogenase, (2) pyruvate dehydrogenase, (3) pyruvate-formate lyase (also called formate acetyltransferase), (4) phosphate acetyl transferase, (5) acetate kinase, (6) acetaldehyde dehydrogenase, and (7) alcohol dehydrogenase (ADH). In *E. coli* (and in some other bacteria), formic acid can be converted to $H_2 + CO_2$. In yeast, the pyruvate (a metabolite in cytosol) can be converted to acetyl-CoA (AcCoA) in the mitochondria by (2), pyruvate dehydrogenase, and the AcCoA is further degraded in the TCA cycle. It can also be converted by (8) to acetaldehyde, which in (7) is further reduced by ADH to ethanol. This is then the final product of the fermentative metabolism. Acetaldehyde can also be oxidized to acetic acid (HAc) by (9) acetaldehyde dehydrogenase. The HAc is energized to (cytosolic) AcCoA by (10), acetyl-CoA synthase, and the cytosolic AcCoA can (1) be used in the synthesis of lipids or (2) be imported to the TCA cycle as mitochondrial AcCoA. Also in aerobic metabolism, a large part of the flux passes through the loop (8), (9), and (10) and ends up as mitochondrial AcCoA. No ethanol is formed, unless the flux of glucose exceeds a certain limiting value, and ethanol starts to be produced (oxido-fermentative metabolism of yeast).

2 NADH will also be produced. NAD^+ can be regenerated by sending carbon from DHAP toward glycerol or more reduced products such as PDO, but the typical solution is to convert pyruvate to more reduced products in the *fermentative pathways* shown in Figure 4.4. The final products are either lactic acid or a mixture of formic acid, acetic acid (HAc), and ethanol in bacteria (Figure 4.4a

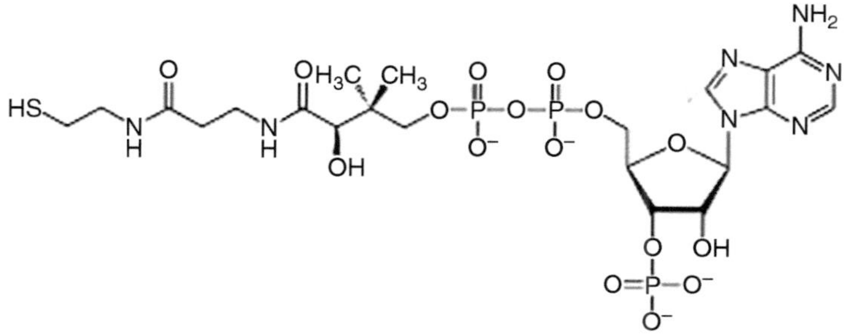

Figure 4.5 Coenzyme CoA. Note that the right-hand part of the compound is identical to ATP. When CoA reacts with a carboxylic acid, a thio-ester is formed, for example, AcCoA, BuCoA, or SA-CoA (succinyl-CoA in Figure 4.6).

and b) or ethanol alone (Figure 4.4c). Thus, *fermentative metabolism* provides a solution for regeneration of NAD^+, and furthermore the final metabolic products, ethanol and lactic acid, are important fermentation products.

The fermentative metabolism for two bacteria, *E. coli* and a typical lactic bacteria, *Lactococcus lactis*, are shown in Figure 4.4a and b, respectively. The fermentative pathways of the eukaryote *S. cerevisiae* is shown in Figure 4.4c.

The cofactor CoA is (like ATP and NADH) a large organic molecule (Figure 4.5) that in the end has a thiol (sulfur–alcohol) group. AcCoA is consequently a thio-ester of CoA and HAc and can be considered as an energized form of HAc. AcCoA can release its extra free energy as ATP by hydrolysis to HAc and CoA (reaction (5), via Acetyl-P, reaction (4)) or provide free energy for other reactions, for example, to form citric acid in the TCA cycle by combination with oxaloacetate.

In two almost identical pathways, Figure 4.4a, and b, the NADH is reconverted to NAD^+ for further use in glycolysis. In lactic bacteria fermentation, the end product is lactic acid, unless the sugar flux through glycolysis is very low, that is, when the glucose concentration in the medium is below about $15\,\mathrm{mg\,l^{-1}}$ (15 ppm). For concentrations above $\sim 15\,\mathrm{mg\,l^{-1}}$, glucose is converted to lactic acid (HLac) by the stand-alone stoichiometry (4.7) and for lower glucose concentrations, a mixture of HLac, HCOOH, HAc, and ethanol results as shown in the stand-alone stoichiometry (4.8):

$$Homo\text{-}lactic\,fermentation: \quad C_6H_{12}O_6 \rightarrow 2\,C_3H_6O_3 + 2\,\mathrm{ATP} \tag{4.7}$$

Hetero-lactic fermentation (*"mixed acid fermentation"*) using PFL (Pyruvate Formate Lyase), the enzyme of the reaction (3) on Figure 4a or 4b:

$$\begin{aligned} C_6H_{12}O_6 \rightarrow\ & 2\alpha\, C_3H_6O_3 + (1-\alpha)[CH_3COOH + C_2H_5OH + 2\,HCOOH] \\ & + 2\alpha\,\mathrm{ATP} + 3(1-\alpha)\mathrm{ATP} \end{aligned} \tag{4.8}$$

In Eq. (4.8), both the carbon and the redox balance are seen to be satisfied. When $\alpha = 1$, 2-ATP is produced per glucose. For $\alpha = 0$, the ATP yield on glucose is three, since one extra ATP is obtained per PYR in the mixed fermentation.

As will be seen in Section 4.6, the overall *anaerobic* conversion of glucose to mixed acids via the pyruvate dehydrogenase pathway (2) cannot be written as a stand-alone anaerobic stoichiometry since there will be an uncompensated production of NADH when CO_2 is a product rather than HCOOH. In the presence of strong sinks for NADH such as 1,3 propane diol in Figure 4.2, or in the presence of oxygen, a stand alone stoichiometry with HAc and CO_2 as the only byproducts can, however, be set up. Reoxidation of NADH by the reaction $NADH+½O_2 \rightarrow NAD^+$) is catalyzed by the enzyme NADH oxidase. Now, the ATP yield on glucose is four, that is, a doubling of the yield compared to homolactic fermentation.

With a little O_2 present, HAc, CO_2, and HLac can be the only products. The ATP yield is $2\alpha + 2\ (2 - \alpha)$. This is used to increase the biomass yield in the dairy industry.

In anaerobic yeast fermentation, glucose can be converted to ethanol by the stand-alone stoichiometry:

$$C_6H_{12}O_6 \rightarrow 2\,C_2H_5OH + 2\,CO_2 + 2\,ATP. \tag{4.9}$$

By respiratory fermentation (Section 4.5), the yield of metabolic products such as ethanol, HLac, or HAc is much smaller than given by Eq. (4.8) or (4.9), and CO_2 is the typical end product of catabolism as seen in Section 4.2.

In a real fermentation process, whether anaerobic or aerobic, all carbon cannot be sent to primary metabolic products. Some carbon is drained off both from the glycolysis pathway and from the other primary pathways to make a long list of precursors for biomass synthesis. Also when other metabolic products are drained away from the EMP pathway (e.g., propane 1,3 diol, amino acids, penicillin, etc.), these compete with the energy creating pathways. Table 4.2 shows where carbon is drained from the EMP pathway and the other primary pathways to make biomass precursors.

4.3
The TCA Cycle: Conversion of Pyruvate to NADH + $FADH_2$, to Precursors or Metabolic Products

The TCA cycle is fed by the end product of glycolysis, pyruvate. One pyruvate molecule (PYR) is carboxylated to oxaloacetate in reaction (10) of Figure 4.6 by expending one ATP (or the carboxylation starts at PEP and no ATP is used). The other pyruvate is converted to AcCoA by reaction (1). The two compounds react to form citric acid, a 6-carbon, tricarboxylic acid (TCA). Following the cycle in the direction of the arrows of Figure 4.6, one ends up with oxaloacetate that reacts with another pyruvate to start a new cycle. If no metabolites are withdrawn from the cycle, the net result is that for each turn of the cycle in the direction of the arrows one pyruvate is degraded to three CO_2. The 10 redox

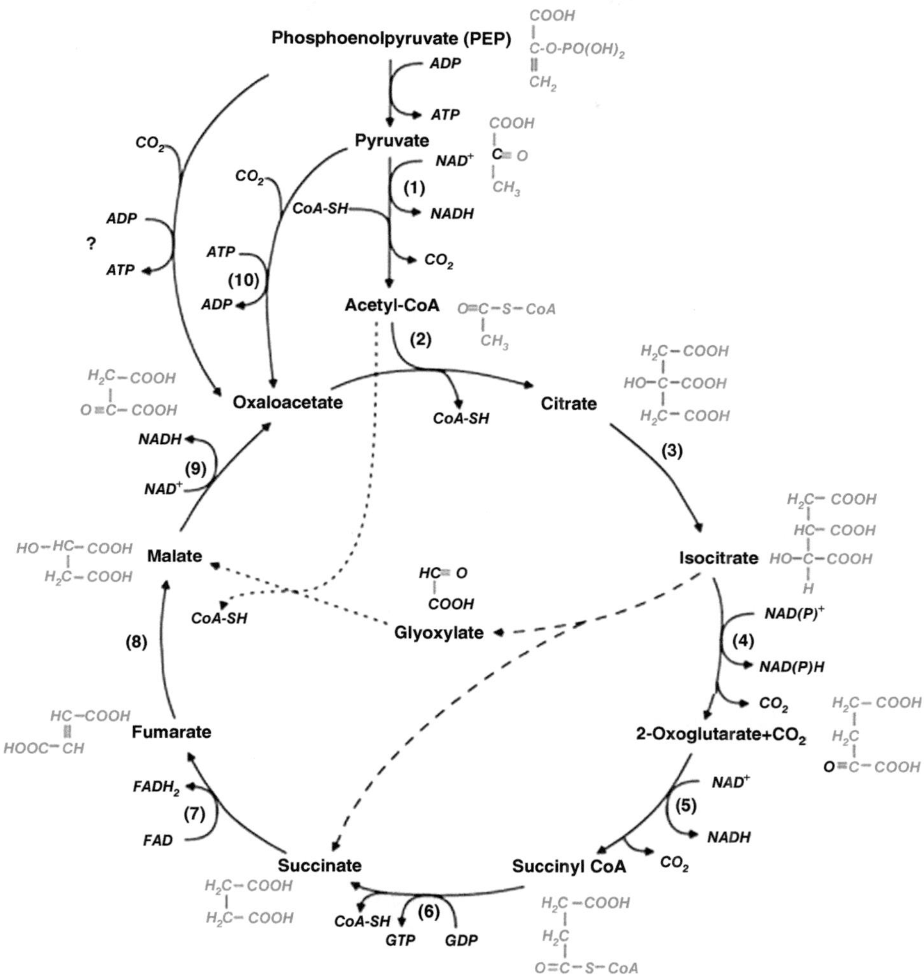

Figure 4.6 The TCA cycle (citric acid cycle or Krebs cycle) with two entry points, by carboxylation of PEP or PYR, or from PYR after dehydrogenation and release of one carbon as CO_2. Also shown is the glyoxylate shunt, the splitting of citric acid (CA) to glyoxylate (GOX), and succinic acid followed by reaction of GOX with another AcCoA to form malic acid (MA), which is reduced to succinic acid. The clockwise direction of the arrows on the figure corresponds to degradation of one PYR to three CO_2 with concomitant production of ten redox units (four NADH and one $FADH_2$). The enzymes are (1) pyruvate dehydrogenase, (2) citrate synthase, (3) aconitase, (4) isocitrate dehydrogenase, (5) 2-oxoglutarate dehydrogenase, (6) succinate thiokinase, (7) succinate dehydrogenase, (8) fumarase, (9) malate dehydrogenase, and (10 a and b) pyruvate and PEP carboxylase. There are two forms of isocitrate dehydrogenase, one (located in the mitochondria) uses NAD^+ and the other (located in the cytosol) uses $NADP^+$ as cofactor. The total degradation of PYR by Eq. (4.10) is written for the first cofactor and describes the situation in, for example, yeast. The other is used in bacteria (they have no mitochondria) where the TCA cycle operates in the cytosol. The NADPH released in bacteria by (4) adds to the pool of NADPH used, for example, for amino acid synthesis. The OAA production in the reaction PEP + CO_2 → OAA may or may not be accompanied by production of one ATP. The direction of carbon flow (shown by the arrows on the figure) can be reversed for many of the reactions.

units from $CH_3COCOOH$ are stored in four NADH and one $FADH_2$:

$$CH_3COCOOH \rightarrow 3\,CO_2 + 4\,NADH + FADH_2 + GTP. \tag{4.10}$$

GTP is an energy carrier with a structure exactly like that of ATP, except that another purine derivative, guanine (G), is used instead of adenine (A). Its energy content is the same as that of ATP.

In anaerobic processes, it would be fatal if the TCA cycle ran in the direction of the arrows in Figure 4.6 where 10 redox units are accumulated per pyruvate, besides the 1 NADH that accompanies the formation of one pyruvate molecule from glucose. In aerobic processes, the NADH and $FADH_2$ can, however, be used in respiration to form large quantities of ATP as described in Section 4.5 or NADH may be donated to other reactions that need an input of redox, for example, the reaction from glucose to 1,3 propane diol shown in Figure 4.2 with a net consumption of 2 NADH per molecule of propane diol produced.

In anaerobic processes, the TCA cycle is divided into a reductive branch from oxaloacetate to succinate and an oxidative branch from citrate to succinate, and as shown in Eq. (4.11)

$$2\,PYR + 2\,CO_2 \rightarrow 2\,OAA;\, PYR \rightarrow AcCoA + CO_2 + NADH;$$

$$(OAA = COOH\text{-}CH_2\text{-}CO\text{-}COOH).$$

$$(1): OAA + AcCoA \rightarrow SA + 2\,CO_2 + 2\,NADH;$$
$$(2): OAA + NADH + FADH_2 \rightarrow SA;\, (SA = (CH_2COOH)_2).$$

Total reaction (sum of reactions in the oxidative (1) and reductive (2) branch:

$$3\,PYR + FADH_2 \rightarrow 2\,SA + CO_2 + 2NADH. \tag{4.11}$$

In Eq. (4.11) NADH and $FADH_2$ both represent "H_2" = 2 redox units.

The glyoxylate shunt is an alternative path to SA, which also operates under anaerobic conditions:

$$4\,PYR \rightarrow 2\,OAA + 2\,AcCoA + 2\,NADH;\, OAA + NADH + FADH_2 \rightarrow SA.$$

$$\begin{aligned} OAA + AcCoA &\rightarrow \text{Citric acid} \\ &\rightarrow GOX(= CHO\text{-}COOH) + SA;\, GOX + AcCoA + FADH_2 \\ &\rightarrow SA. \end{aligned}$$

The net reaction is as follows:

$$4\,CH_3COCOOH + 2\,FADH_2 \rightarrow 3(CH_2COOH)_2 + NADH. \tag{4.12}$$

In Example 4.3, it will be shown how a stand-alone stoichiometry from glucose to succinic acid can be set up by a combination of the pathways used to derive Eqs. (4.11) and (4.12).

The TCA cycle is a major source for many of the metabolites produced in fermentation. The important metabolite succinic acid (SA) is produced as described in Examples 4.3 and 4.6, and most importantly, the TCA cycle is used for the production of a large number of amino acids.

Glutamic acid ("glutamate") is by far the industrially most important amino acid. It is obtained by amination of 2-oxo glutaric acid:

$$\underset{\text{2-oxo glutaric acid}}{\text{COOH-(CH}_2)_2\text{-CO-COOH}} + \text{NH}_3 + \text{NAD(P)H} \rightarrow \underset{\text{glutamic acid}}{\text{COOH-(CH}_2)_2 - \text{HC NH}_2\text{-COOH}}\,. \tag{4.13}$$

The annual production of glutamic acid and of its sodium salt, monosodium glutamate (MSG) = COOH-(CH$_2$)$_2$ – HC NH$_2$-COONa exceeds 1.5 million tons. Glutamic acid is used in sauces and in cheese, while MSG is a general flavor enhancer in the food industry. The nitrogen required in Eq. (4.13) is often present in the medium as NH_4^+, which is freely imported to the cell, but since nitrogen is fixed as NH_3 in the biomass, one ATP has to be spent when the proton is transported back to the medium. The affinity for NH_3 of the enzyme glutamate dehydrogenase (GDH) used in Eq. (4.13) is low, which means that the reaction takes place only when the NH_4^+ concentration is high in the medium. A second synthesis route for glutamate is via its amide, glutamine.

$$\text{Glutamate} + \text{NH}_3 \rightarrow \text{Glutamine(COOH-(CH}_2)_2\text{-HCNH}_2\text{-CONH}_2). \tag{4.14}$$

$$\text{Glutamine} + \text{2-oxoglutarate} + \text{NADPH} + \text{ATP} \rightarrow 2\,\text{Glutamate}. \tag{4.15}$$

In the so-called GC-GOGAT pathway Eqs. (4.14) and (4.15), the net result is also that one glutamate is formed from one 2-oxo-glutarate, but the process that requires an extra ATP is operative when $[NH_4^+]$ is low and the GDH path (4.13) becomes thermodynamically limited.

Besides being an important commercial product, glutamic acid is also used to synthesize a large number of other amino acids catalyzed by *amino-transferases*. Thus, the amino acid aspartate is produced from glutamate using the enzyme aspartate amino transferase:

$$\text{Oxaloacetate(OAA)} + \text{glutamate} \rightarrow \text{Aspartate(COOH-CH}_2\text{-HCNH}_2\text{-COOH)} + \text{2-oxo glutarate}. \tag{4.16}$$

By further reactions, the commercially very important lysine (annual production ~1 million tons, used to improve lysine-deficient vegetable-based protein in husbandry), is produced in an eight-step pathway:

$$\text{Asparatate} + \text{pyruvate} + 3\,\text{NADPH} + \text{NH}_3 \rightarrow \text{Lysine(COOH-HCNH}_2\text{-(CH}_2)_3\text{-H}_2\text{CNH}_2) + \text{CO}_2. \tag{4.17}$$

Since the transamination reaction (4.16) regenerates the two oxoglutarate, no carbon is lost except at the carbon split off at the end of the eight-step pathway to give the 6-carbon amino acid from four carbon in aspartate and three carbon in pyruvate. Also, in the pathway from glucose to oxaloacetate, no carbon is lost since CO_2 lost when pyruvate is converted to AcCoA is gained by carboxylation of another pyruvate to form OAA.

Thus, the reactions to produce lysine from glucose when NADPH is used in Eq. (4.13) are as follows:

$$\begin{aligned}&\text{Glucose} + CO_2 \rightarrow \text{PYR} + \text{OAA} + 2\,\text{NADH} + 2\,\text{ATP}; \text{OAA} + \text{Glutamate}\\&\rightarrow \text{Asparate} + \text{2-oxoglutarate}\\&\text{2-oxoglutarate} + \text{NADPH} \rightarrow \text{Glutamate}\\&\text{Aspartate} + \text{PYR} + 3\,\text{NADPH} + \text{ATP} \rightarrow \text{Lysine} + CO_2\\&\text{Net reaction}: \text{Glucose} + 4\,\text{NADPH} \rightarrow \text{Lysine} + 2\,\text{NADH} + \text{ATP}.\end{aligned} \tag{4.18}$$

In the set of reactions that lead to Eq. (4.18), glutamate is regenerated from 2-oxoglutarate at the expense of NADPH, and the OAA needed to make aspartate comes from glucose. PYR, the other part of glucose, is used to synthesize lysine from aspartate. CO_2 spent by carboxylation of PYR is regenerated. Hence, both the carbon balance and the redox balance appear to close, and the carbon yield of lysine should be 1. In the production organisms for lysine, NADH cannot be converted to NADPH and is respired to CO_2 in the aerobic process. The 4 NADPH have to be generated by Eq. (4.19) from 1/3 $C_6H_{12}O_6$ in the first two reactions (11) and (12) of the PP pathway, Figure 4.7, and the theoretical yield of lysine is only $(1/(4/3)) = 3/4$.

$$\frac{1}{3}C_6H_{12}O_6 + \frac{1}{3}\text{ATP} \rightarrow \frac{1}{3}\text{G6P} \rightarrow 4\,\text{NADPH} + 2\,CO_2. \tag{4.19}$$

A higher yield can be obtained only if some of the 2 NADH produced in glycolysis can be exchanged with NADPH. This could happen either if an NADH requiring iso-enzyme could be used instead of the NADPH requiring enzyme in regeneration of glutamate from 2-oxoglutarate by Eq. (4.13) or if any of the 3 NADPH requiring enzymes in the pathway from aspartate to lysine could be exchanged with an NADH requiring enzyme. Naturally, such reconstructions of the commercially used bacterium *Corynebacterium glutamicum* are obvious targets for gene modification of the organism. If in a genetically improved version of the production organism one could use NADH instead of one or more of the 3 NADPH in the lysine pathway, the carbon yield would increase. Incorporating the appropriate gene from another organism into *C. glutamicum* would increase the yield to $(1/(5/4)) = 0.80$ if just one of the NADPH in the pathway from aspartate to lysine could be replaced by NADH. The remaining NADH is converted to NAD^+ and about 2 ATP in oxidative phosphorylation, and this would provide sufficient free energy for some biomass growth to be sustained together with a larger production of lysine per C-mol carbon in glucose.

These issues are discussed in Marx *et al.* [2]. Takeno *et al.* [3] proposed to substitute the NAD^+-dependent GAPDH (the enzyme in reaction (6) in the EMP pathway, Figure 4.2) with a nonphosphorylating NADP-dependent enzyme (GAPN) from *Streptococcus mutans*, and the conversion (6) of GAP to 1,3 DPG

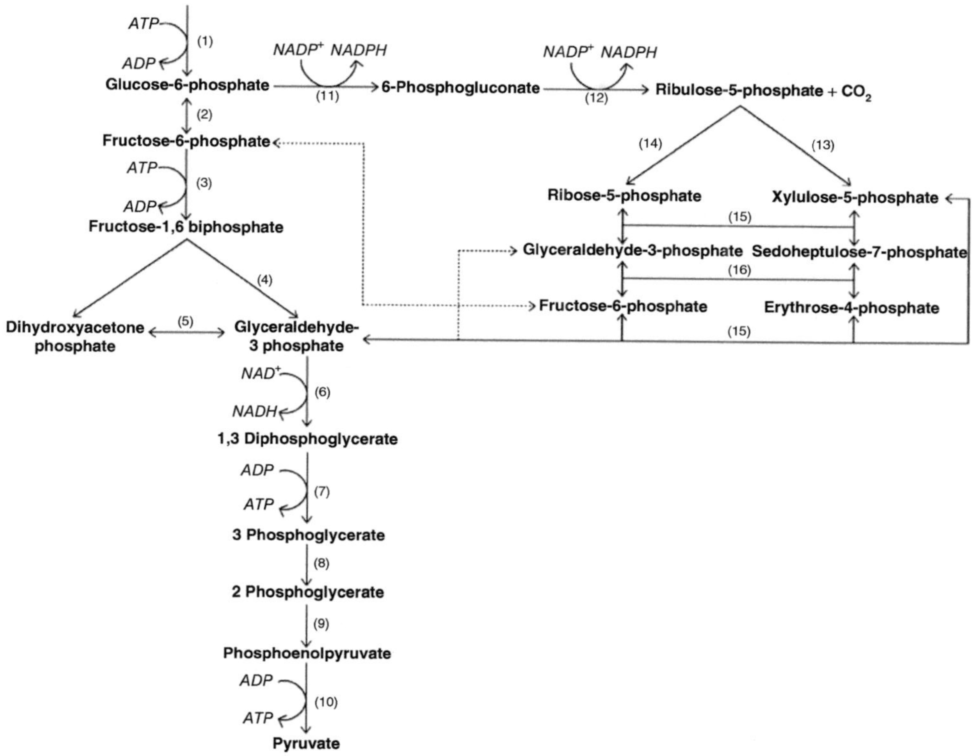

Figure 4.7 The pentose phosphate pathway (PP pathway) is used to produce anabolic redox power, genetic material, and substrate (E4P) for the production of aromatic amino acids. The enzymes are (11) G6P dehydrogenase, (12) 6-phosphogluconate dehydrogenase, (13) ribulose-phosphate 3-epimerase, (14) ribose-phosphate isomerase, (15) transketolase, and (16) transaldolase. Transketolase catalyzes two reactions, interconversion of R5P and Xu5P to GAP and S7P and interconversion of Xu5P and E4P to GAP and F6P, respectively. All carbon that is not used in the PP pathway is returned to the EMP pathway as GAP and F6P.

is done with the generation of NADPH, and now no NADH, but 2 NADPH result from glycolysis.

The production of the commercial bulk chemicals succinic acid and lysine in the TCA cycle is, of course, not part of the "natural" cell-building purpose of the consumption of carbon substrate in the TCA cycle. More than 50 wt% of the biomass is proteins, and the amino acids in proteins are for a large part derived from the TCA cycle. Thus, the commercial strains for production of, for example, SA are stressed since they carry the burden of extra genes for the production of the desired chemical.

Thus, glutamate is the starting point for the synthesis of the amino acids glutamine, proline, and arginine. Oxaloacetate provides the carbon for aspartate, the starting point for lysine, threonine and isoleucine, methionine and asparagine.

Table 4.1 The 20 physiologically important (L-) amino acids and their net chemical formula.

Essential	Nonessential
Arginine $C_6H_{14}N_4O_2$	Alanine $C_3H_7NO_2$
Histidine $C_6H_9N_3O_2$	Aspargine $C_4H_8N_2O_3$
Isoleucine $C_6H_{13}NO_2$	Aspartate $C_4H_7NO_4$
Leucine $C_6H_{13}NO_2$	Cysteine $C_3H_7NO_2S$
Lysine $C_6H_{14}N_2O_2$	Glutamate $C_5H_9NO_4$
Methionine $C_5H_{11}NO_2S$	Glutamine $C_5H_{10}N_2O_3$
Phenylalanine $C_9H_{11}NO_2$	Glycine $C_2H_5NO_2$
Threonine $C_4H_9NO_3$	Proline $C_5H_9NO_2$
Tryptophan $C_{11}H_{12}N_2O_2$	Serine $C_3H_7NO_3$
Valine $C_5H_{11}NO_2$	Tyrosine $C_9H_{11}NO_3$

Pyruvate is the progenitor of alanine, valine, and leucine. PEP and erythrose 4-phosphate (a metabolite from the PP pathway, Figure 4.7) are progenitors of the aromatic amino acids phenylalanine, tryptophan, and tyrosine. 3-Phosphoglycerate (3 PG), an intermediate in the glycolysis pathway, is the starting point for synthesis of serine, and from serine to glycine and cysteine. Ribose-5 phosphate, an intermediate in the PP pathway, leads to histidine.

All amino acids contain C, H, N, and O. Cystein and methionine also contain sulfur. Phenyl alanine and tyrosine have a benzene ring, tryptophan an indol ring, and histidine an imidazol ring. All four aromatic amino acids contain serine (2 amino-3 hydroxy propionic acid) in the 1′ position of the ring. The aliphatic amino acids are all derived from TCA cycle or glycolysis intermediates. The left-hand column of Table 4.1 contains the so-called *essential* amino acids that mammals (including humans) cannot naturally synthesize in their cells and that, therefore, must be supplied in the diet. Histidine *is* actually produced in the body of grown-ups, but not by infants, and arginine is produced in too small quantities. Hence, they are classified as essential amino acids. Milk protein contains all essential amino acids and is consequently an ideal diet for humans.

4.4
NADPH and Biomass Precursors Produced in the PP Pathway

The pentose phosphate pathway, Figure 4.7, can be conceived as a shunt on the glycolysis with one entry point at G6P and reentry at glyceraldehyde 3-phosphate (GAP) and at F6P. All reactions in the PP pathway are tightly regulated, and the pathway receives only a net carbon flux from glycolysis when need arises.

The main carbon drains from the PP pathway are as follows:

1) Pentoses to synthesize genetic material. Ribulose-5-phosphate (R5P) is converted to xylulose-5-phosphate (Xu5P). The 10 carbon in R5P and

Xu5P are reshuffled to form sedoheptulose 7-phosphate (S7P), a 7-carbon sugar, and GAP, the 3-carbon compound. Again the 10 carbon may be reshuffled, now to form F6P and a 4-carbon sugar erythrose-4P (E4P).

2) R5P is converted in a long pathway and with uptake of amino acids to adenine and guanine, two of the compounds that together with pentoses form genetic material, RNA and DNA, by polymerization reactions. E4P reacts with PEP to form aromatic amino acids. Phenylalanine is used in the production of artificial sweeteners and tyrosine is the starting point for the production of pharmaceutical L-DOPA. Hence, the path leading to E4P is often engineered to provide sufficient E4P for industrial processes.
3) Finally, the PP may stop after 2 NADPH are produced, and one out of the six carbon in G6P is lost to CO_2. The rest of the carbon is returned to glycolysis at F6P or at GAP without harvesting any of the PP pathway products. By the reversible reaction F6P ↔ G6P, the remaining five carbon in the original G6P can be cycled back to the PP pathway. When all carbon is returned to F6P, the total reaction is as follows:

$$\mathrm{G6P} \rightarrow 12\,\mathrm{NADPH} + 6\,\mathrm{CO_2}. \qquad (4.20)$$

Since the generation of both 12 NADPH and 12 NADH requires that one glucose molecule is converted to 6 CO_2, the use of NADH as a common denominator for both cofactors (and for $FADH_2$) will be used in Section 4.7 to set up stoichiometric models by the use of metabolic flux analysis (MFA). This simplifies the algebra, and the loss of glucose is the same whichever cofactor is synthesized.

4.5 Oxidative Phosphorylation: Production of ATP from NADH ($FADH_2$) in Aerobic Fermentation

Obligate anaerobes (e.g., methanogens, sulfate reducing bacteria, and soil bacteria) can only grow in the absence of O_2.

Facultative anaerobes (e.g., *E. coli*) can grow both with O_2 and without O_2. *E. coli* can actively use O_2 in *respiration*. Lactic bacteria are deficient in the electron transfer chain, an essential part of the respiratory mechanism. Hence, they are unable to gain more energy by respiration, but they are *aerotolerant*. They are not harmed by O_2 and may use O_2 to oxidize NADH, thereby gaining extra ATP, since HAc (+NADH) can be the only products in the anaerobic mixed acid fermentation pathway. Many yeasts (e.g., *S. cerevisiae*) and a few filamentous fungi are facultative anaerobes. Typically, certain growth enhancers, for example, sterols are added to an anaerobic *S. cerevisiae* cultivation.

Animal cells and some bacteria (e.g., *Mycobacterium tuberculosis*) are *obligate aerobes*. They die without a supply of O_2. The important filamentous fungi *Aspergillus oryzae* and *A. niger* are probably obligate aerobes.

Facultative anaerobes and aerobes create energy both by *substrate-level phosphorylation* in glycolysis and by *respiration*, the common name for *oxidative phosphorylation*. Obligate anaerobes can gain energy only by *substrate-level phosphorylation*, by reducing anions such as NO_3^{-1}, SO_4^{-2}, and cations such as Fe^{+3} (to Fe^{+2}), or in the case of methanogens by degrading carbohydrates to fatty acids and further to methane and CO_2.

Respiration is, however, a much more efficient method of harvesting free energy by degradation of carbohydrates, and it has given a vast evolutionary advantage both to facultative anaerobes and to obligate aerobes (unless they are deprived of O_2 as in drowning!).

Aerobic degradation of glucose (or carbohydrates in general) follows

$$\begin{aligned} C_6H_{12}O_6 \rightarrow 6\,CO_2 + 2\,NADH + (8\,NADH + 2\,FADH_2) + 2\,ATP \\ +GTP, \text{ and thereafter}: \end{aligned}$$

$$\begin{aligned} NADH + \tfrac{1}{2}\,O_2 \rightarrow (P/O)_{NADH}\,ATP, \text{and } FADH_2 + \tfrac{1}{2}\,O_2 \\ \rightarrow (P/O)_{FADH2}\,ATP \text{ by respiration} \end{aligned} \tag{4.21}$$

P/O is a stoichiometric coefficient in Eq. (4.21) and it determines how much ATP is generated per NADH or $FADH_2$ that is oxidized to, respectively, NAD^+ and FAD.

In Chapter 9, it is shown that oxidation of NADH to NAD^+ releases approximately 220 kJ/mol free energy, and it looks as if seven ATP (each containing 31–35 kJ free energy at standard conditions) could be generated per mole NADH. This high energy release, however, cannot be harvested from the oxidative phosphorylation mechanism discussed in Chapter 9. In studies of yeast fermentation, one finds a value of P/O in the range of 1.2–1.5 ATP per NADH or $FADH_2$ that is oxidized in oxidative phosphorylation [4]. Still, considering that 12 redox cofactors are produced in glycolysis + TCA cycle, the outcome is that a huge extra ATP production is obtained by oxidative phosphorylation compared to substrate-level phosphorylation where most of the free energy in the substrate is retrieved in the metabolic products such as ethanol or lactic acid, and the ATP yield is only two ATP/$C_6H_{12}O_6$.

O_2 is potentially dangerous for living cells (it can promote damage to the DNA), and when a facultative anaerobic organism chooses to grow with O_2 as substrate, a large number of protective agents are activated to prevent the formation of potentially lethal by-products of aerobic growth. The agents, for example, *catalase* and *manganese-superoxide dismutase* are induced and gradually synthesized. In yeast fermentation, a critical level of carbon flow toward the TCA cycle and respiration is reached. When the glucose flux exceeds the critical value, the extra substrate is directed toward ethanol as discussed in Figure 4.4c. This is known as the *Crabtree effect*. In the examples of Section 3.3, which were based on Figure 2.3, the critical specific growth rate $r_x = D_{crit}$ was found to be $\approx 0.25\,h^{-1}$, corresponding to $(-r_s) = Y_{xs} \cdot 0.25 = 0.42$ C-mol glucose consumed $(\text{C-mol X})^{-1}\,h^{-1}$ by the stoichiometry (3.11). The exact cause for the occurrence of the Crabtree effect is not known, and it is ascribed to "some limitation in the TCA cycle or in respiration."

It may be that the specific rate of oxygen consumption $(-r_o)$ becomes so high that the oxygen-induced damage to the mitochondria would become too marked for $(-r_s) > (-r_s)_{crit}$; otherwise, the production rate of ATP by respiration could exceed a value where either the transport of ADP to the active sites in the mitochondria membrane (Figure 9.2) or the flow of ATP back to the cell becomes a limiting factor.

When the rate of oxygen consumption becomes *too low*, only part of the glucose consumption occurs by oxidative phosphorylation and some metabolic products are produced, for example, ethanol in yeast fermentation. This is the *Pasteur effect* that is observed at *low* oxygen availability.

4.6
Summary of the Biochemistry of Primary Metabolic Pathways

The primary metabolic pathways described in Section 4.1–4.5 have two competing purposes: To provide free energy for synthesis of the macromolecular biomass components that are shown in Table 3.1 and to synthesize the biomass precursors and building blocks shown in Table 4.2.

The synthesis of biomass requires very large amounts of free energy (ATP), for example, about 29 mmol ATP (g cell mass)$^{-1}$ for the 57 wt% of biomass that is protein. Synthesis of genetic material (RNA and DNA) requires another 8 mmol (g cell mass)$^{-1}$ and lipids 3 mmol (g cell mass)$^{-1}$ out of a total of about 41 mmol ATP (g cell mass)$^{-1}$. Since degradation of glucose in catabolic reactions produces at most 2 ATP (mol $C_6H_{12}O_6$)$^{-1}$, the amount of glucose needed to sustain a specific growth rate of $r_x = \mu = 0.4\,h^{-1}$ (a typical value for the yeast *S. cerevisiae*) is $0.041/2 \cdot 0.4 \cdot 180 = 1.46$ g glucose (g biomass h)$^{-1}$ under anaerobic conditions. Under aerobic conditions, the ATP yield on glucose is much higher, about 20–25 mol ATP ($C_6H_{12}O_6$)$^{-1}$, but the ability of the cell machinery to harness this large energy production is smaller in aerobic than in anaerobic growth. Cell maintenance, i.e., the reconstitution of fresh cell mass from inevitable deterioration of living cells to cell debris requires an extra glucose consumption as discussed in Eq. (3.13) and Example 3.6. In aerobic growth much ATP is wasted in the so-called futile cycles where ATP is not produced in the forward direction but consumed in the reverse direction. The hydrolysis of glycerol 3-phosphate to glycerol by a hydrolase and the regeneration of the glycerol 3-phosphate from glycerol by a kinase that requires one ATP (Figure 4.2, shaded box) is an example, but much larger futile cycles exist. If the energy production in catabolic reactions does not exactly match the energy consumption in anabolic reactions, the excess ATP is "wasted" as metabolic heat production. This is the case in transients if the sugar availability is suddenly increased (e.g., by increasing the dilution rate D in a CSTR reactor). The generation of cell machinery to enable faster growth at the new and higher D value lags

Table 4.2 Precursor metabolites come from glycolysis and from the TCA cycle. G6P, PYR, AcCoA, 2-OG, and OAA are the most important precursors used to form building blocks for storage carbohydrates (glycogen ($CH_{1.67}O_{0.83}$), trehalose ($CH_{1.78}O_{0.89}$), and polyhydroxy acids (PHA) such as polyhydroxybutyrate (a commercial polymer)), amino acids, lipids, isoprenoids (a large and diverse group of compounds with the isoprene structure 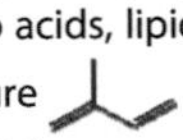 as backbone), and finally for genetic material. UDP-*N*-acetyl-glucosamine is one of the two precursors (the other is UDP glucuronic acid from G6P) that is used to produce the polysaccharide hyaluronic acid, an important product in the cosmetic industry and for regenerative surgery in the clinic.

Precursor metabolite	Building blocks	Amount required mmol (g cell mass)$^{-1}$
Glucose 6-phosphate (G6P)	Storage carbohydrates (glycogen, trehalose, PHA), R5P, NADPH	1.4
Fructose 6-phosphate (F6P)	UDP-*N*-acetyl glucosamine	0.07
Ribose 5-phosphate (R5P)	Histidine, tryptophan, nucleotides	0.9
Erythrose 4-phosphate	Phenylalanine, tryptophan, tyrosine	0.4
Glyceraldehyde 3-phosphate (GAP) → glycerol 3P (G3P)	Backbone of phospholipids	0.2
3-phosphoglycerate (3GP)	Serine, cysteine, glycine, choline purine bases (→ nucleotides)	1.5
Phosphoenolpyruvate (PEP)	Phenylalanine, tryptophan, tyrosine	0.5
Pyruvate (PYR)	Alanine, valine, leucine, isoleucine, lysine (and AcCoA)	2.8
Acetyl-CoA (AcCoA)	Lipids, isoprenoids, PHA	3.8
2-Oxo glutarate (2OG)	Glutamate → other amino acids	1.1
Oxaloacetate (OAA)	→ Many amino acids, nucleotides	1.8
Succinyl-CoA	(with glycine) → porphyrin ring in hemes, degradation of amino acids	Small amount

behind the production of more catabolic machinery by several hours, and "futile" sugar consumption is observed (see examples in Chapter 7). Sometimes, this is advantageous, for example, when the desired product is a metabolic product, such as ethanol. Here, the nonsteady-state yield of ethanol on glucose can be higher than the steady-state yield, for example, 0.516 C-mol (C-mol glucose)$^{-1}$ in Example 2.1.

The synthesis of biomass precursors competes with the energy-generating catabolic pathways, and considerable amounts of the carbon source are withdrawn not only to produce the precursors and building blocks of Table 4.2 but also to produce the anabolic redox carrier NADPH by reduction of NADPH$^+$. The two major drains of NADPH, mainly produced by oxidation of G6P in reactions (11) and (12) of the PP shunt, is to provide redox power for amino acid production (→ proteins) and to synthesize lipids from AcCoA by a series of reactions where CO_2 is incorporated to build up the long-chain fatty acids. About 12 mmol and 5 mmol NAPDH (g cell mass)$^{-1}$, respectively, is used for these two cell building purposes, corresponding to a glucose consumption of 0.26 g glucose (g cell mass)$^{-1}$. This number refers to the so-called *minimal medium* that consists of glucose supplemented with minerals, growth factors, hormones, and so on according to a recipe. In a *rich medium*, minerals, amino acids, and other building blocks are supplied together with sugar as yeast extract, casein peptone, and corn steep liquor. The rich (or "complex") media are used in most industrial bioprocesses, but for physiological studies they are not desirable since the medium composition is not exactly known and it varies from batch to batch.

The total amount of NADPH and precursor metabolites derived from glucose is about 35 mmol (g cell mass)$^{-1}$, approximately 85% of the glucose requirement for energy purposes. Clearly, the amount of precursors needed when the ultimate goal is to produce one of the final metabolic products, lysine, isoprene, PHB (poly-hydroxy-butyrate), citric acid, succinic acid, or hyaluronic acid far exceeds the amounts calculated for "normal" cell function. Complicated genetic engineering techniques are needed to avoid that these industrial processes fail due to insufficient energy production or by the diversion of too much carbon from synthesis of key components of the cell.

Quantitative data (for *E. coli*) in Table 4.2 are modified from Lengeler *et al.* [5] (p. 115). Data for yeast (*S. cerevisiae*) are available from several reviews. The most comprehensive is [6] (see also Chapter 6).

4.7 Metabolic Flux Analysis Discussed in Terms of Substrate to Product Pathways

The primary objective of metabolic flux analysis is to calculate the distribution of carbon in the easily measured product streams excreted to the medium. This is also the goal of the black-box models discussed in several examples of Chapter 3, for example, Examples 3.5 and 3.6. The tools used in Chapter 3 are carbon and redox balances that provide two constraints on the yield coefficients.

In Sections 4.6 and 4.7, a small part of the enormous fund of knowledge stored in metabolic pathway information is used, for example, to identify missing metabolites – the production of oxalic acid and acetic acid by a side reaction in the TCA cycle (Example 3.2) is a typical case. What is, however, first of all possible when the flow of carbon is traced through the metabolic network is to set up a *systematic* procedure for calculating the net rate of production of metabolites. In this procedure, an energy (ATP-) balance can be included along with a carbon and a redox balance. In the end, the analysis of a metabolic network can be used to construct an overall, black-box stoichiometry that includes the experimentally observed products. But now the influence of all the reactions that combine to give the black-box stoichiometry is systematically studied. The ad-hoc methods of Chapter 3 are extended to become what can reasonably be called a scientific approach to cell reactions. This is both an important tool for research in the physiology of microorganisms and a useful tool for engineering of the metabolic network to improve an industrial process.

With the information provided in Sections 4.0–4.5, part of the metabolic network has become available for a preliminary study of MFA. In the remainder of Chapter 4, the study of what will be called "stoichiometry-based MFA" will serve as an introduction to the stringent, "professional" analysis of metabolic networks in Chapter 5, where many more rate data are included through the use of ^{13}C-labeled substrates, and further in Chapter 6, where "total" cell models are discussed in genome-scale metabolic models.

Two representations of stoichiometry-based MFA will be discussed. In Section 4.7, all the reactions that occur from one metabolite pool to the next in a pathway that leads from a given substrate S to a given end product P_i will be combined to give just one carbon flux V_i from S to the product P_i that is excreted to the medium and measured there. This will be called the "substrate to product flux method" or "SP flux method," and the flux from S to P_i is called an "SP flux."

In Section 4.8, the internal fluxes v_j between each metabolite pool in the network are used to analyze the network. Since the number of internal fluxes v_j is bound to be larger than the SP fluxes V_i, the dimensionality of the analytical problem is larger in the internal flux method. One may argue that the SP flux method with its "black-box" representation of large parts of the metabolic network is particularly attractive for engineers, but for any newcomer to MFA the modest calculations of the SP flux method compared to the *internal flux method* will be attractive. Therefore, the SP flux method is explained first, and it is illustrated by a number of easy-to-understand examples. It must, however, be emphasized that for error-free data, the SP method and the internal flux method give exactly the same values for calculated rates of production r_{P_i}. More important, when proceeding to the analysis of more complex networks as in Chapter 5, the methodology of the internal flux method carries over unchanged and it is backed up by easy-to-use computer software, although the user may not quite understand how the software works.

The metabolic network model is based on a number of *nodes* N_i, $i = 1, 2 \ldots,$ M where each metabolic pool is represented by a node. Schematically, the network is represented as a *graph*. In the *internal flux method*, the nodes are connected by the *internal fluxes* v_j. In the *SP flux method*, the substrate and each of a set of *key products* P_i are connected by an *SP flux* pathway V_i. This is illustrated in Figure 4.8.

In the SP flux method, the main subject of Section 4.7, the total number V_{total} of SP pathway fluxes is given by Eq. (4.22). For each of the M nodes N_i one counts the number *out* of internal fluxes v_j that leave and subtracts the number *in* of internal fluxes v_k that enter the node. For a *diverging* node *out* > *in*, for a *converging* node *out* < *in*, and for a *neutral* node *out* = *in*

$$V_{total} = \sum_1^M (\text{out} - \text{in})_i + 1. \tag{4.22}$$

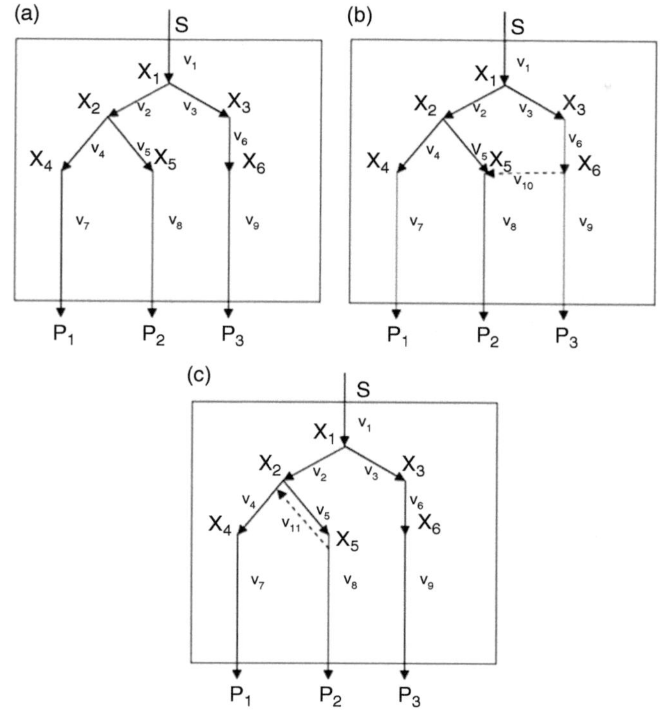

Figure 4.8 Graphs of simple metabolic networks. (b) and (c) are variants of the basic network (a). X_i = metabolite pools. In (a), the node points of the network are X_1 and X_2. Extra nodes are found in (b) and (c) as defined in the text. There are three SP- pathways V_i from substrate S to excreted key products P_1, P_2, and P_3 as defined in Eq. (4.22). The flux v_{10} through path X_6–X_5 in (b) or the individual fluxes v_5 and v_{11} between X_2 and X_5 in (c) can be determined only when a labeled substrate is used (see text).

Each SP pathway ends at a *key product* P_i, and by suitable normalization of the stoichiometry of the pathway, the specific rate of production r_{Pi} equals the SP flux V_i through the pathway. A key product P_i is defined as a product that is *only* produced in V_i. There may be many more metabolic products from the metabolic network than V_{total}, but the specific rate of production of these other products (e.g., CO_2 and the substrate S) can be found by linear combination of the rates of formation of the key products. Not all the key products may be *independent* products, since one may add constraints between r_{Pi} based on a total redox balance and also on an energy balance for the whole network. The constraints express that redox and ATP produced in some SP pathways is consumed in other SP pathways.

The stoichiometry for each SP pathway is consequently written with both redox and ATP included. The carbon balance is closed for each individual SP pathway and consequently also for the whole network. Only NADH and ATP are used to represent cofactors for redox and energy transfer between SP pathways (an assumption that is not used in more detailed MFA), and consequently there will be *at most* two constraints on the production rates of the V_{total} key products P_i.

When biomass X is a key product, a *separate* SP pathway from S to X is included. When *several* different substrates (e.g., glucose and O_2) are used, the distribution of each substrate to different P_i is shown in the graph. The reason why a separate SP pathway to biomass is included is that although carbon is drained off in many pathways in the metabolic network to form an *unstructured* biomass X, the contribution from pathways that lead to, for example, proteins and lipids is not distinguished, and the formation of X is represented in a simple way on the graph. Hence, one obtains the following:

$$P_{\text{total}} = V_{\text{total}},\ \text{where}\ V_{\text{total}} = \sum_1^M (\text{out} - \text{in})_i + 1\ \text{when no biomass is included.}$$

$$P_{\text{total}} = V_{\text{total}} = \sum_1^M (\text{out} - \text{in})_i + 2\ \text{when (unstructured) biomass X is a product.} \tag{4.23}$$

$$(-r_s)_{\text{total}} = (-r_s) + (-r_s)_X. \tag{4.24}$$

In Eq. (4.24), the specific rate of substrate consumption required to feed the metabolic network is $(-r_s)$, while the substrate consumption to support (unstructured) biomass formation is $(-r_s)_X$.

The M nodes also define the number of internal fluxes v_i in the network:

$$v_{\text{total}} = \sum_1^M \text{out}_i + 1. \tag{4.25}$$

In Eq. (4.25), nodes where out = in = 1 *are not included* – they define places in the network with no topological influence on the carbon flow – the two individual fluxes entering and leaving a *neutral* node are the same.

In Figure 4.8a, there are nodes at X_1–X_6. X_1 and X_2 are *diverging* nodes, while nodes X_3–X_6 are uninteresting neutral nodes. $V_{\text{total}} = 2\ (2-1) + 1 = 3 = P_{\text{total}}$. $v_{\text{non-trivial}} = 2 \cdot 2 + 1 = 5$, whereas each of v_6–v_9 are equal to one of the five remaining internal fluxes v_1–v_5. In Figure 4.8b, there are diverging nodes at X_1, X_2, and X_6 (each with out – in = 1), and one converging node X_5 with out – in = –1. X_3 and X_4 are neutral nodes. $V_{\text{total}} = P_{\text{total}} = 3\ (2-1) - 1 + 1 = 3$. The nodes at X_1, X_2, and X_6 have out = 2, and the node at X_5 has out = 1. Hence $v_{\text{non-trivial}} = (3 \cdot 2 + 1) + 1 = 8$. The non-trivial internal fluxes are v_1–v_5, v_8, v_9, and v_{10}, while $v_6 = v_3$ and $v_7 = v_4$.

In Figure 4.8c, the reversible reaction between X_2 and X_5 gives rise to two diverging nodes at X_1 and X_5 and one node at X_2 where out = in = 2.

$V_{total} = P_{total} = 2 \cdot (2-1) + 1 \cdot (2-2) + 1 = 3$. $v_{total} = 3 \cdot 2 + 1 = 7$, namely, v_1–v_5, v_8 and v_{11} at nodes X_1, X_2, and X_5, each having out = 2. $v_6 = v_9 = v_3$, and $v_7 = v_4$.

When the network is described using the internal fluxes v_i, there are internal mass balances to be considered. At each node, the flow of carbon into the node must equal the flow of carbon out. In Figure 4.8a, there are two internal mass balances: $v_1 = v_2 + v_3$ and $v_2 = v_4 + v_5$. In Figure 4.8b, four mass balances: $v_1 = v_2 + v_3$, $v_2 = v_4 + v_5$, and $v_3 = v_{10} + v_9$ and $v_8 = v_5 + v_{10}$. In Figure 4.8c, there are three mass balances: $v_1 = v_2 + v_3$, $v_5 = v_8 + v_{11}$, and $v_2 + v_{11} = v_4 + v_5$. When the mass balances on the internal fluxes are applied, the number of independent internal fluxes is found to be 3 in Figure 3.8a, and 4 in both Figures 4.8.b and 4.8.c. In all three cases the number of key-products Pi is 3.

Balancing of the redox and the ATP balances will give a further two constraints. If both balances are included (the ATP balance can be used only if production of biomass or another product that consumes ATP is included) V_{total} and $v_{independent}$ will be reduced by 2.

When the internal mass balances are taken in account, there will be three independent internal fluxes v_i in Figure 4.8a, and this is the same as the number of SP fluxes V_i. The two internal mass balances in the internal flux method just give rise to more calculation work and possibly (as discussed in Section 4.9) to a smaller accuracy of the calculated rates.

When redox and ATP balances are not included, one ends up with four independent internal fluxes in Figure 4.8b and c, while the number P_{total} of key products P_i remains at three. Here there is a difference between the two approaches. If there are only three distinguishable key metabolic products $P_{total} = 3$ in the three figures, then one of the internal fluxes cannot be determined. Only when labeled substrate S is used and the fraction of labeled substrate can be measured in product P_2 and P_3 together with the total amounts P_2 and P_3 will it be possible to find v_{10} in Figure 4.8b and v_{11} in Figure 4.8c. This requires in Figure 4.8b that the labeling pattern of P_2 obtained from the net path S–X_1–X_2–X_5 is different from that obtained by the net path S–X_1–X_6–X_5. A simple Example 4.2 is shown in this chapter, but the subject is otherwise discussed in Chapter 5 where techniques and software involved in the analysis of realistic labeling experiments is treated in earnest.

It should be emphasized that it is of considerable practical interest to determine also the internal flux v_{10} in Figure 4.8b, and the two individual fluxes v_5 and v_{11} in Figure 4.8c. Using the alternative route $v_{10} - v_8$ rather than relying on $v_5 - v_8$ alone could lead to a large improvement in the yield of the product, especially if P_3 is undesirable. Knocking out the gene for the enzyme in internal rate v_9 and producing P_2 by combination of v_5 and v_{10} could give a much higher yield, $Y_{S,P2}$. For this reason, the internal rate v_{10} must be determined and thereafter increased by genetic engineering. Likewise *the net rate* of the reaction $X_2 \rightarrow X_5$ in Figure 4.8c may be low, although both v_5 and v_{11} are high. This is the sign of a thermodynamically limited reaction, but unless both rates are known one may keep increasing the enzyme level for $X_2 \rightarrow X_5$ without any appreciable improvement in the net rate of the reaction.

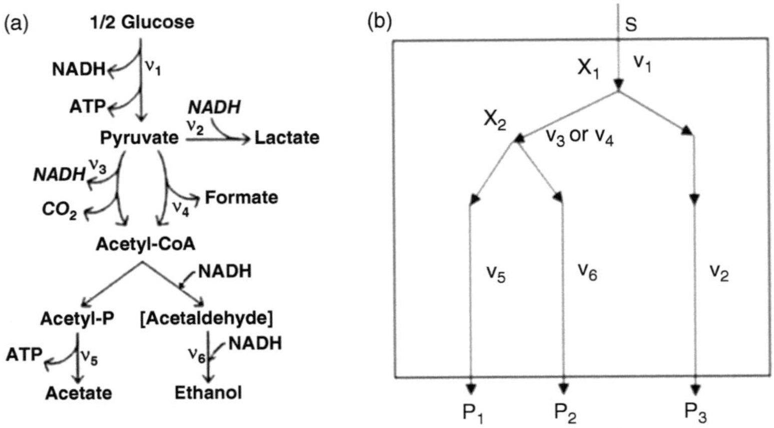

Figure 4.9 Pathways from glucose to fermentation products: (a) chart and (b) schematic representation.

In the remainder of Section 4.7, a number of examples will illustrate how the SP flux method can be used to analyze both trivial metabolic networks and also some networks of a moderate complexity. In the simplest examples, the stoichiometry for each V_i is defined on a C-mol basis, while a basis of one glucose molecule is used in more complex examples. The stoichiometry of the SP path V_i is as mentioned above written such that the yield coefficient $Y_{s,Pi} = 1$, in which case the SP flux V_i through the pathway equals r_{Pi}. This simplifies the solution of the redox and ATP balances to calculate the dependent rates from the measured rates.

Example 4.1 Distribution of carbon in heterofermentative metabolism of lactic bacteria

The fermentative metabolism of lactic bacteria is shown in Figure 4.4b, and the total anaerobic distribution of glucose to the independent products lactic acid, ethanol, and acetic acid is shown in Figure 4.9.

The two (diverging) nodes are $[X_1, X_2]$ at, respectively, PYR and AcCoA. The pathway from PYR to AcCoA is shown in two variants, *either* as v_3, catalyzed by pyruvate dehydrogenase, *or* as v_4, catalyzed by pyruvate formate lyase (see Figure 4.4b). $V_{total} = 2 \cdot (2 - 1) + 1 = P_{total} = 3$. $[P_1, P_2, P_3]$ are [HAc, ethanol = e, lactic acid = HLac]. Dependent reaction species are CO_2 in pathway v_3 and HCOOH in pathway v_4, and glucose, which is used in all three net pathways. There are $2 \cdot 2 + 1 = 5$ internal fluxes v_i and 2 internal mass balances (at X_1 and X_2).

The three SP pathways are $V_1 = S - X_1 - X_2 - P_1$, $V_2 = S - X_1 - X_2 - P_2$, and $V_3 = S - X_1 - P_3$. The reaction $X_1 \rightarrow X_2$ is catalyzed by either pyruvate dehydrogenase (variant (1)) or by pyruvate formate lyase (variant (2)). The stoichiometry of the three SP pathways is written based on one C-mol P_i in each net pathway, and $r_{Pi} = V_i$. In both variants, ATP is produced in all

SP pathways, and an ATP balance is not included. In simple MFA analysis, NAD^+ (or $NADP^+$) and ADP are not included. Nor is a nitrogen source when no products except biomass contains N.

$$\begin{aligned}
&V_1 : \frac{3}{2}CH_2O(\text{glucose}) \rightarrow CH_2O(\text{HAc}) + \frac{1}{2}CO_2 + NADH + (ATP)\\
&V_2 : \frac{3}{2}CH_2O(\text{glucose}) \rightarrow CH_3O_{1/2}(\text{ethanol}) + \frac{1}{2}CO_2 + \left(\frac{1}{2}ATP\right)\\
&\quad (1) : \text{SP pathways when pyruvatedehydrogenase}(v_3)\text{is used.}\\
&V_3 : CH_2O(\text{glucose}) \rightarrow CH_2O(\text{HLac}) + \left(\frac{1}{3}ATP\right).
\end{aligned} \tag{1}$$

$$\begin{aligned}
&V_1 : \frac{3}{2}CH_2O(\text{glucose}) \rightarrow CH_2O(\text{HAc}) + \frac{1}{2}HCOOH + \frac{1}{2}NADH + (ATP)\\
&V_2 : \frac{3}{2}CH_2O(\text{glucose}) \rightarrow CH_3O_{1/2}(\text{ethanol}) + \frac{1}{2}HCOOH - \frac{1}{2}NADH + \left(\frac{1}{2}ATP\right)\\
&\quad (2) : \text{ SP pathways when pyruvate formate lyase}(v_4)\text{is used.}\\
&V_3 : CH_2O \rightarrow CH_2O(\text{HLac}) + \left(\frac{1}{3}ATP\right).
\end{aligned} \tag{2}$$

In Eqs. (1), the NADH balance does not close, and unless some other pathway reaction that consumes NADH is included v_3 cannot be the pathway between PYR and AcCoA. Addition of small amounts of O_2 will, as discussed in Eqs. (4.7) and (4.8), make the pathway accessible when the organism contains the enzyme NADH oxidase. Lactic bacteria do express this enzyme, and the higher ATP yield in (1) can be harvested.

In Eqs. (2), the NADH balance closes when $V_1 = V_2$, and since $r_{HCOOH} = (1/2)(V_1 + V_2)$, one obtains that $r_{HAc} = r_e = r_{HCOOH}$, all on a C-mol basis. Let the total glucose consumption rate be $(-r_s) = 1$ C-mol (g X h)$^{-1}$ and let the glucose consumption rate in V_3 be α C-mol (g X h)$^{-1}$. $Y_{HAc} = Y_e = Y_{HCOOH} = (1/3)\,(1-\alpha)$ and the total stoichiometry is as follows:

$$\begin{aligned}
CH_2O(\text{glucose}) \rightarrow\; &\frac{1}{3}(1-\alpha)\left[CH_2O(\text{HAc}) + CH_3O_{1/2} + HCOOH\right]\\
&+ \alpha\, CH_2O(\text{HLac})\left(+\frac{1}{6}(3-\alpha)ATP\right).
\end{aligned} \tag{3}$$

When Eq. (3) is reformulated in terms of whole moles of reaction components, one obtains Eq. (4.8). The algebraic work involved to obtain Eq. (3) is much smaller (and it is obtained in a systematic fashion) than that involved in obtaining Eq. (4.8) without using the information from the metabolic pathways.

Example 4.2 Determining the flux through the PP pathway using labeled glucose as substrate

The influence of a loop in the network, such as that provided by v_{10} in Figure 4.8b, is illustrated by a combination of the PP pathway, Figure 4.7, and two glycolytic pathways, the EMP pathway or the Entner–Doudoroff pathway. The use of ^{13}C labeled glucose and the following calculation of label in the key product pools makes it possible to obtain the "missing" internal rate v_{10}. Calculation of the fraction of differently labeled products in the pools P_i is so simple that an ad-hoc nomenclature can be used. In Chapter 5, a *systematic* procedure is described that enables any labeling experiment to be analyzed.

Let the substrate be glucose $^{13}C_1$, that is, the carbon at the aldehyde group is labeled with ^{13}C. When the aldehyde is oxidized to 6-phosphogluconate in reaction (11) of Figure 4.7 and further decarboxylated to ribose-5-phosphate (R5P) in reaction (12), the labeled carbon is lost as $^{13}CO_2$. One may measure the production of $^{13}CO_2$ or more easily measure ^{13}C that appears in pyruvate or in some other products from the metabolic pathway, for example, in proteins. The carbon that is not entering the PP pathway will appear in the two trioses, DHAP (unlabeled) and GAP (labeled at C_3) that result from reaction (4) of Figure 4.2. *If* all DHAP is converted to GAP in reaction (5), 50% of the labeling is lost, and the remaining label is at C_3 (the carbon furthest away from aldehyde) in GAP. After combination of two R5P, the remaining carbon from 3 G6P is sent back to the EMP pathway either as unlabeled F6P or as unlabeled GAP, provided that the carbon is not used in other PP reactions, for example, for genetic material or E4P. But the carbon fraction that returns to the EMP pathway will be unmarked since both the F6P and the GAP are unmarked. Thus, if one measures that the fraction of pyruvate labeled in C_3 is 0.3 one obtains the following equation:

$$V_{EMP} \cdot 0.5 + V_{PP} \cdot 0 = 0.3, \quad \text{and with } V_{EMP} + V_{PP} = 1$$
$$\rightarrow V_{PP}/V_{EMP} = 0.4/0.6 = 2/3. \tag{1}$$

If carbon from glucose is converted to pyruvate via the Entner–Doudoroff (ED) pathway, the G6P is oxidized to 6-phosphogluconate, which as explained in Section 4.1 is converted to one PYR and one GAP by

$$\text{G6P} \rightarrow 6 - \text{phosphogluconate} + \text{NADPH}$$
$$\rightarrow \text{PYR} + \text{GAP} + \text{NADPH}. \tag{2}$$

In Eq. (2), the labeling ($^{13}C_1$) of G6P in C_1 is retrieved in PYR ($^{13}C_1$), and the GAP is converted to unlabeled PYR.

Thus, one would again obtain $V_{PP}/V_{EMP} = 2/3$ if the labeled fraction of PYR is 0.3, *but now labeled in C_1*.

One can consequently find *both* the contribution of each of the two glycolytic pathways *and* the PP pathway by the labeling experiment. Let the fraction of PYR marked at C_1 be 0.05 and that at C_3 be 0.40:

$$V_{EMP} \cdot 0.5 + V_{ED} \cdot 0 + V_{PP} \cdot 0 = PYR(C_3) = 0.40 = PYR(C_3) = 0.40.$$
$$V_{EMP} \cdot 0 + V_{ED} \cdot 0.5 + V_{PP} \cdot 0 = PYR(C_1) = 0.05.$$
$$V_{EMP} + V_{ED} + V_{PP} = 1 \rightarrow V_{EMP} = 0.8, \quad V_{ED} = 0.1, \quad V_{PP} = 0.1. \tag{3}$$

The example shows that the individual fluxes through irreversible pathways that are connected by several diverging and converging pathways can be calculated when labeled substrate is used. In this simple example, the calculations are trivial, but in realistic pathway analysis by labeling experiments, sophisticated computer software is needed. A relatively simple example of the complications that arise is the analysis of the fluxes through the three pathways (the reductive branch of the TCA cycle, the path via AcCoA to isocitric acid, and the pathway from AcCoA + GOX), see Figure 4.6 and Example 4.3. The "scrambling" of label that occurs when OAA is obtained by carboxylation of PEP complicates the analysis. The final labeling pattern of succinic acid *if it is assumed that all reactions are irreversible is*: Of the 24 C-moles in 6 succinic acid molecules, 12 carbon have a single labeled carbon, 11 carbon are unmarked, and 1 carbon is double-marked when the reaction is fed with $^{13}C_1$ glucose.

Example 4.3 Determination of the maximum yield of succinic acid and of 1,4 butane diol on glucose

Figure 4.10, simplified from Figure 4.6, is a schematic representation of the distribution of carbon from glucose to different pathways that finally end up in succinic acid, SA. There are three diverging nodes, at PEP, at oxaloacetate (OAA), and at isocitrate (ICA). There is one node for which out = in at AcCoA, one converging node at GOX, and a double converging node at SA (internal). $P_{total} = 3 \cdot (2-1) + 1 \cdot (1-2) + 1 \cdot (1-3) + 1 = 1$ key product, SA. If the rate of glucose consumption $(-r_s)$ is known, then r_{SA} can be calculated – that is, the yield coefficient $Y_{s,SA}$ can be directly calculated.

It is not difficult to set up a black-box model that satisfies the carbon and redox balances and where no product except SA is formed. A redox balance (1) will determine how much CO_2 can be taken up:

$$(1-x)CO_2 + x\,CH_2O(\text{glucose}) \rightarrow CH_{1.5}O(SA) + Y_{SA,w}H_2O.$$
$$(1-x) \cdot 0 + 4x = 3.5 + Y_{SA,w} \cdot 0 \rightarrow x = \frac{7}{8} \text{ and } Y_{SA,w} = \frac{1}{8}. \tag{1}$$

The maximum yield, that is, the yield of SA on glucose with CO_2 as the only other reactant, is 8/7. The question remains: Is it possible in the pathways that lead to SA to find the right ratio between the pathway rates to construct Eq. (1)?

Let the flow of glucose (Glc) be $(1+3x)/2$ mol (g biomass X h)$^{-1}$ or $1+3x$ mol PEP (g X h)$^{-1}$. x can have any value $> -1/3$, and the value of x

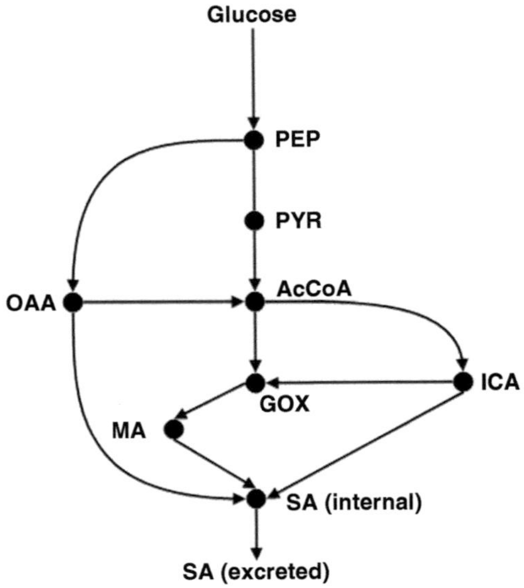

Figure 4.10 Production of succinic acid by a combination of the reductive pathway PEP → OAA → SA with the glyoxylate shunt PEP → AcCoA (+OAA) → ICA → SA + GOX, and GOX + AcCoA → MA → SA. The nodes are at PEP, OAA, AcCoA, ICA, GOX, and SA. As shown in the text, there is one key product, SA. When a redox balance is applied for the total network, the yield coefficient $Y_{\text{glucose, SA}} = Y_{\text{Glc, SA}}$ can be calculated.

will be determined by a total redox balance between the reductive path (PEP → OAA → SA) in which NADH is used and the oxidative pathway (PYR → AcCoA → SA either directly via ICA to SA or via the GOX shunt) in which NADH is produced. Since one molecule OAA and one molecule PEP is needed to form one molecule ICA, the stoichiometry for the reductive and the oxidative net pathways V_1 and V_2 are given by Eqs. (2) and (3), respectively, when V_1 is normalized to give one mol SA (see also Figure 4.6).

$$\begin{aligned} V_1 : & \frac{1+x}{2}\text{Glc} + (1+x)\text{CO}_2 \\ & \rightarrow \text{SA}(\text{C}_4\text{H}_6\text{O}_4) + x\,\text{OAA} + [(1+x)-2]\text{NADH}[+(1+x)\text{ATP}]. \end{aligned} \tag{2}$$

$$V_2 : 2x\frac{1}{2}\text{Glc} + x\,\text{OAA} \rightarrow 2x\,\text{CO}_2 + 2x\,\text{SA} + 3x\,\text{NADH}[+2x\,\text{ATP}]. \tag{3}$$

The value of x is determined by an NADH balance applied to Eqs. (2) and (3):

$$[(1+x)-2] + 3x = 0 \rightarrow x = \frac{1}{4}. \tag{4}$$

The two Eqs. (2) and (3) can now be combined to

$$\frac{7}{8}\text{Glc} + \frac{3}{4}\text{CO}_2 \rightarrow \frac{3}{2}\text{SA} + \frac{7}{4}\text{ATP} - \text{or on a C-mol basis}$$
$$: \text{CH}_2\text{O} + \frac{1}{7}\text{CO}_2 \rightarrow \frac{8}{7}\text{CH}_{1.5}\text{O} + \frac{1}{3}\text{ATP}\left(+\frac{1}{7}\text{H}_2\text{O}\right). \tag{5}$$

Note that for Eqs. (2) and (3), the carbon balance closes for any value of x, but the overall redox balance will close only for $x = ¼$. In Eq. (2), each of the $(1+x)$ PEP takes up $(1+x)$ CO_2, while each of the 2 x PEP will produce $xCO_2 + x$ NADH when converted to xAcCoA. In the further reactions:

$$x\,\text{AcCoA} + x\,\text{OAA} \rightarrow x\,\text{ICA}, \quad \text{and}\, x\,\text{ICA} + x\,\text{AcCoA} \rightarrow 2x\,\text{SA}. \tag{6}$$

An extra xNADH is produced as shown in Eq. (3).

The split of carbon at the PEP node is $\alpha = (1 + x/2\,x\,) = 5/2$ and the ratio of SA produced in the reductive pathway (2) to that produced in the oxidative pathway is $\beta = (1/2x) = 2$, that is, the reductive pathway must operate much faster than the oxidative pathway.

The net ATP production in the reactions (2) is $1 + x = 1.25$ ATP per molecule SA formed when PEP is carboxylated to OAA, since no ATP is produced or consumed when OAA is reduced to SA. The ATP generated in the reactions (3) is also one ATP per molecule SA formed. The ATP gain in the reactions from ½ Glc to PYR is one ATP per PYR molecule, and no ATP is produced or consumed in the further reactions to SA. It is somewhat doubtful whether in Eq. (2) the production of OAA proceeds directly from PEP or by carboxylation of PYR. In the latter case, the conversion of one PYR to OAA consumes one ATP, and there is no net gain of ATP. In the following Example 4.6 where ATP is used to sustain biomass production, it is assumed that $(1+x)$ ATP is gained by Eq. (2). Otherwise, the free energy available in an anaerobic fermentation process to SA is only $2x$ ATP per $(1+2x)$ SA produced, that is, 0.5 ATP per 1.5 molecules SA produced.

The result (5) proves that the maximum theoretical yield $Y_{s,SA} = 8/7$ can, indeed, be obtained solely by combination of glycolysis and TCA cycle reactions. The split of carbon at PEP is $\alpha = (1 + y/2y) = 5/2$.

Can the same yield of SA also be obtained if the glyoxylate shunt does not operate?

In this case, SA is produced by concerted action of the reductive and oxidative branch of the TCA cycle. Figure 4.11 is a schematic representation of the process.

There are $M = 4$ nodes: PEP, OAA, CIT, and SA_{int}.

The number of key products is $(2-1) + (2-1) + (1-2) + (1-2) + 1 = 1$.

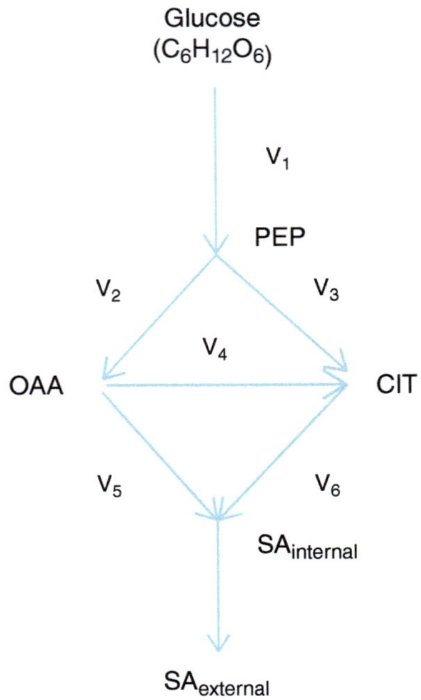

Figure 4.11 SA production via combination of the reductive and the oxidative branches of the TCA cycle.

For a given $(-r_{glc})$, the rate of SA formation $r_{SA} = Y_{glc,SA} \cdot r_{glc}$. $Y_{glc,SA}$ is calculated from the network structure and an overall NADH balance.

The productivity r_{SA} again depends on how a given flux of carbon, for example, $(-r_{glc}) = ½\ (3x+1)$ is distributed between the two SP pathways V_1: PEP-OAA-SA (the reductive branch) and PEP-AcCoA (+ OAA)-CIT-SA (the oxidative branch).

Due to the stoichiometry of forming CIT from one mol OAA + one mol AcCoA, the flux v_4 is equal to v_3.

v_5 is (arbitrarily) set to $1 + fx$, where f is a factor to be determined from an overall redox balance. In Eqs. (2)–(4), $(-r_{PEP})$ is $1 + 3x = 1.75$. When the extra parameter f is introduced, it will be shown below that the results hold for any value of x, that is, for an arbitrary feed $(-r_{PEP}) = (3x+1)$ mol PEP per unit time. Since v_6 (the production of SA in the oxidative branch) $= v_4 = v_3$ in Figure 4.11, one obtains:

$$v_1 = 3x + 1, v_3 = v_4 = v_6 = \left[3x + 1 - (1 + f\,x)\right]/2$$
$$= \frac{3-f}{2}x, \quad \text{and } v_2 = v_4 + v_5 = 1 + \frac{3+f}{2}x.$$

The NADH balance reads: consumed NADH in $v_5 = 2\ (1 + fx)$.

$$\text{Produced NADH in } v_1, v_3, \text{ and } v_6 = 3x + 1 + 3 \cdot \frac{3-f}{2}x. \tag{7}$$

Solution of the redox balance gives $f = (15/7) - (2/7x)$ and now the production of SA can be calculated:

$$r_{SA} = v_5 + v_6 = 1 + fx + \frac{3-f}{2}x = 1 + \frac{3+f}{2}x = 1 + \frac{3}{2}x + \frac{15}{14}x - \frac{1}{7}$$
$$= \frac{6}{7}(3x+1) = \frac{6}{7}r_{PEP}.$$

Hence, $Y_{PEP,SA} = (6/7)\ \text{mol mol}^{-1} = (8/7)\ \text{C -mol SA (C-mol Glc)}^{-1}$, the same yield found from the network (Figure 4.10).

The distribution of PEP at the PEP node is $a = \frac{v_2}{v_3} = \frac{(1+(3+f/2)x}{(3-f/2)x} = 6$, while $\beta = \frac{v_5}{v_6} = \frac{1+fx}{(3-f/2)x} = 5$.

Compared to the network shown in Figure 4.10, the distribution α of carbon at the PEP node is much more skewed, and most of the SA is produced in the reduced branch of the network.

In Figure 4.11, the internal fluxes v_i have been used rather than the net fluxes V_1 and V_2 of Eqs. (2) and (3). For the network in Figure 4.11, and with $(1 + x)$ PEP used in V_1 and x PEP used in V_2 would have been:

$$V_1 : \frac{1+x}{2}\text{Glc} + (1+x)CO_2$$
$$\rightarrow \text{SA}(C_4H_6O_4) + x\,\text{OAA} + [(1+x) - 2]\text{NADH} + (1+x)\text{ATP}.$$

$$V_2 : \frac{1}{2}x\,\text{Glc} + x\,\text{OAA} \rightarrow 3\,x\,CO_2 + x\,\text{SA} + 4x\,\text{NADH} + x\,\text{ATP}.$$

Solving the redox balance gives $x = 1/5$, $r_{SA} = 1 + x = 6/5$, $(-r_{PEP}) = 1 + 2x = 7/5$ and $Y_{Glc,SA} = 8/7\ \text{C-mol(C-mol)}^{-1}$. This is the same result as obtained by the internal flux method. The ratio between SA produced in the reductive branch and that produced in the oxidative branch is $\beta = (1/x) = 5$ as obtained above.

The calculation based on the internal fluxes v_i would have been simpler if v_1 had been defined as $v_1 = 3x$ and $v_5 = fx$ rather than $v_1 = 3x + 1$ and $v_5 = 1 + fx$ as used above. Now one immediately obtains $f = 15/7$ for any x value, $r_{SA} = (36/14)x = 6/7\ r_{PEP}$, and $Y_{Glc,SA} = 8/7\ \text{C-mol(C-mol)}^{-1}$. The more complicated calculation was chosen to demonstrate that the network gives the same $Y_{Glc,SA}$ for any choice of definition of the fluxes v_1 and v_5.

Biomass must also be produced, and as shown in Example 4.6, this leads to a smaller SA yield. One would wish that all fermentative by-products such as HAc or HCOOH could be avoided by knocking out the genes for the enzymes of reactions (4) and (6 or 7) in Figure 4.4a and b or (7) in Figure 4.4c. This is, however, not possible, since besides the loss of carbon to make biomass precursors, NADH is also used, for example, in the synthesis of glutamate. In a recent paper, van Heerden and Nicol [7] analyzed real fermentation data from a steady-state CSTR and they found an appreciable production of HAc ($Y_{HAc,SA} \approx 2$ mol/mol or 1 C-mol/C-mol). $Y_{s,SA} \approx 0.75$ g/g $= 0.75 \cdot (30/29.5) = 0.76$ C-mol (C-mol glucose)$^{-1}$. $Y_{sx} = 0.11$ g g^{-1} $= 0.13$ C-mol (C-mol glucose)$^{-1}$. The SA yield on glucose is far below the 8/7 C-mol/C-mol calculated above, and this is due to the formation of biomass and of HAc that is needed to supply enough NADH to reduce glucose to SA. One might hope that a better control of the fluxes through the different routes to SA may eventually lead to a higher SA yield.

When the final product is the very desirable 1,4 butane diol ($CH_{2.5}O_{0.5}$ = BDO) rather than SA, then by the stand-alone stoichiometry given by (8) with CO_2 as the only by-product. $Y_{s,BDO}{}^{max}$ is 8/11:

$$CH_2O \rightarrow \frac{8}{11} CH_{2.5}O_{0.5} + \frac{3}{11} CO_2 + \frac{1}{11} H_2O. \tag{8}$$

Since $\kappa_{BDO} = 5.5$ per C, versus 3.5 per C for SA ($CH_{1.5}O$), and consequently that four NADH must be used to reduce a molecule of SA to a molecule of BDO, the flow through the reductive SP path V_1 of the TCA cycle must be drastically reduced compared to that through the oxidative SP path V_2, if the goal of producing BDO with CO_2 as the only by-product shall be reached.

The metabolic network leading to BDO is the same as that shown above for SA production, but the result is now $BDO_{internal}$ and $BDO_{external}$. Again, the number of key products is 1, and with an arbitrary feed flux $v_1 = 3x$, one obtains $v_5 = fx$, $v_3 = v_4 = v_6 = (3 - f/2)x$, and $v_2 = (3 + f/2)x$. The redox balance is as follows:

Consumed $= 6fx + 4 \cdot \frac{3-f}{2} x =$ Produced $= 3x + 3 \cdot \frac{3-f}{2} x \rightarrow f = \frac{3}{11}$.

$$r_{BDO} = fx + \frac{3-f}{2} x = \frac{3+f}{2} x = \frac{6}{11} 3x = \frac{6}{11}(-r_{PEP}). \tag{9}$$

The carbon yield is 6/11 mol BDO (mol PEP)$^{-1}$ or 8/11 C-mol (C-mol Glc)$^{-1}$, the maximum yield predicted by (8).

$\alpha = v_2/v_3 = V_1/V_2 = v_2/v_3 = \alpha = v_2/v_3 = V_1/V_2 = v_2/v_3 = \frac{(3+f/2)x}{(3-f/2)x} = \frac{13}{9}$ and $\beta = v_5/v_6 = \frac{fx}{(3-f2)x} = \frac{4}{9}$. Clearly, there has been a major redistribution of carbon at the PEP node to favor the oxidative branch of the TCA cycle at the expense of the reductive branch. To make the TCA cycle operate alone for BDO production, one must upgrade the enzyme activities in the oxidative branch relative to the levels needed for SA production.

If any of the TCA cycle reactions to SA or BDO are to run at a yield approaching the theoretical, one must be sure that carbon is not drained off to other products, either in glycolysis or in the TCA cycle. In Table 4.2, the major losses of carbon to necessary cell precursors are listed (pyruvate, acetyl CoA, and 2-oxo glutarate). Some of these precursors (e.g., glutamate instead of 2-oxo glutarate and perhaps some other amino acids) could be supplied with the glucose feed, but some loss of glucose is unavoidable.

In practice, it is not possible to have only CO_2 as by-product, and some of the end products from the glycolysis (obtained to restore the NAD^+) are almost always found, typically HAc and HCOOH. In SA production and especially in BDO production much NADH is consumed, and if a second by-product is to be accepted, it should be HAc, since anaerobic oxidation of glucose to HAc and CO_2 releases 2/3 NADH per carbon by $CH_2O \rightarrow 2/3\ (CH_2O = HAc) + 1/3\ CO_2 + 2/3$ NADH – or 2 NADH per PEP.

If HAc is the *only* by-product besides CO_2, a combination of the ideal BDO stoichiometry and production of HAc in the fermentation reaction (Figure 4.4c if yeast is used as production organism) yields:

$$1.5\, C_6H_{12}O_6 \rightarrow 2\, C_4H_{10}O_2(\text{BDO}) + CO_2 + 3\, H_2O - 4\,\text{NADH}. \tag{9}$$

$$C_6H_{12}O_6 + 2\, H_2O \rightarrow 2\, C_2H_4O_2(\text{HAc}) + 2\, CO_2 + 4\,\text{NADH}, \tag{10}$$

when the yield coefficient $Y_{\text{BDO, HAC}} = 0.5$ C-mol HAc (C-mol BDO)$^{-1}$.

Combination of the two stoichiometries yields:

$$CH_2O \rightarrow \frac{8}{15} CH_{2.5}O_{0.5} + \frac{4}{15} CH_2O(\text{HAc}) + \frac{3}{15} CO_2 + \frac{1}{15} H_2O. \tag{11}$$

If in the TCA cycle, an extra 20% NADH is consumed for other purposes than making BDO, this amount of NADH must be supplied by reaction (10). The extra 20% amounts to

$0.2 \cdot \frac{1}{3}\ (6f + 4\ (3 - f/2)) = \frac{6}{11}$ NADH (C-mole PEP) since $f = \frac{6}{11}$ for reaction (8).

To compensate for this loss of NADH one combines Eq. (8) with (12):

$$\frac{3}{2} \cdot \frac{6}{11} \left[CH_2O(\text{glucose}) + \frac{1}{3} H_2O \rightarrow \frac{2}{3} CH_2O(\text{HAc}) + \frac{1}{3} CO_2 + \frac{2}{3}\text{NADH} \right]. \tag{12}$$

The yield coefficient of BDO decreases from $\frac{8}{11}$ to $\frac{(8/11)}{1+(9/11)} = \frac{2}{5}$.

In this way, the loss of BDO yield can be calculated for any loss of redox power (or carbon drained off the pathways).

Note: BDO is produced at 1 million tons/year at a sales price of about US$ 2000 a ton. In an important paper, Yim *et al.* [15] discuss BDO production from glucose in *E. coli*, either from succinate or from 2-oxoglutarate (Figure 4.6) via 4-hydroxybutyrate. Several patents for slightly different production routes are published. The trend is to produce SA in yeast rather than in bacteria. Hereby, one can work at considerably lower pH and thereby (as discussed in Chapter 17) reduce costs to neutralize the broth. Initially, yeast was not regarded as a good production organism since quite low titers were obtained, but this disadvantage is now overcome. An exciting SA industrial process based on yeast is described in EP 2207 297 (published May 23, 2011 by the Dutch company DSM.

In the following examples, biomass production is included. It will be represented as a single net pathway where substrate (glucose) is converted to biomass at a great expense in ATP (C-mol X)$^{-1}$:

$$\begin{aligned} V_x &: (1+\alpha)CH_2O \\ &\rightarrow X(\text{with degree of reduction}\,\kappa_x) + \alpha\,CO_2 + \frac{1}{2}[4(1+\alpha) - \kappa_x]\text{NADH} \\ &- \gamma\,\text{ATP}. \end{aligned} \tag{4.26}$$

It is, of course inconceivable that all the reactions needed to synthesize the macromolecules that together make up the biomass can be collected into one net flux reaction. In Section 3.4 a rough division of the biomass into macromolecules (Table 3.1) was discussed, and in genome-scale MFA (Chapter 6), these macromolecules are included. As indicated in Section 4.6, precursors for the macromolecules are drained off the primary pathways at many points. Still, a lot of experimental evidence has been collected to show that Eq. (4.26) with its completely unstructured picture of biomass X is a fairly good instrument to predict how much substrate is used to produce biomass.

Oura [8] was the first to suggest values for the two parameters [α, γ] in Eq. (4.26) based on a great number of fermentation studies with different microorganisms, and more recent experimental results with much better accuracy than what was possible 30–40 years ago can now be used to give more general statements concerning the value of especially γ for different cultivation conditions.

In Eq. (4.26), a CO_2 production of α mol (C-mol X)$^{-1}$ is included. Many of the reactions, both for building blocks and for polymerization, contain a decarboxylation step. In the analysis of a large number of fermentation reactions, Oura [8] found that $0.08 < \alpha < 0.14$ for cells with a "normal" metabolism (i.e., naturally growing cells with no abnormal accumulation of storage compounds). The stoichiometric coefficient for "NADH" is obtained by applying a redox balance to Eq. (4.26). The "NADH" is actually the difference between a term for production of NADH in the reactions that lead to biomass precursors and NADPH used in the anabolic reactions. Since both NADH and NADPH (and other redox carriers) are produced from the carbon source with the same stoichiometry ($CH_2O \rightarrow CO_2 + 4$ redox units), one may, in simplified calculations, collect all the redox carriers into "NADH = 2 redox units." In more detailed analysis, one needs to follow the fate of each redox carrier, but this is not necessary in the present, approximate but fairly accurate analysis of metabolic networks. The result obtained for the flux distribution, that is, the yield coefficients of a black-box model constructed on the basis of the analysis does not vary much with α. The value of γ has a much greater influence on the resulting black-box model. In Examples 4.5 and 4.8, it will be shown how an analysis of fermentation data obtained in a carefully conducted study can be used to find experimental values of γ that can be applied in similar cultivations.

Example 4.4 Heterofermentative lactic bacteria cultivation with biomass production

In Example 4.1, the distribution of metabolic products in lactic bacteria cultivation was analyzed. In Figure 4.12a, an extra net flux to $P_4 = X$ has been added. Since biomass is made of precursors drained off from many points in the metabolic network, it is best to display V_4 as a net flux path, parallel to but independent of the metabolic network diagram. In Figure 4.12b, V_4 is the separate SP flux to X. V_4 = internal flux v_3.

In Eq. (1), the net fluxes V_2, V_3, and V_4 are shown together with V_4 with stoichiometry given by Eq. (4.26)

$$
\begin{aligned}
&V_1 : \frac{3}{2}\ \mathrm{CH_2O(glucose)} \rightarrow \mathrm{CH_2O(HAc)} + \frac{1}{2}\mathrm{HCOOH} + \frac{1}{2}\mathrm{NADH} + \mathrm{ATP} \\
&V_2 : \frac{3}{2}\mathrm{CH_2O(glucose)} \rightarrow \mathrm{CH_3O_{1/2}(ethanol)} + \frac{1}{2}\mathrm{HCOOH} - \frac{1}{2}\mathrm{NADH} + \frac{1}{2}\mathrm{ATP} \\
&V_3 : \mathrm{CH_2O} \rightarrow \mathrm{CH_2O(HLac)} + \frac{1}{3}\mathrm{ATP} \\
&V_4 : 1.1\ \mathrm{CH_2O} \rightarrow \mathrm{X} = \mathrm{CH_{1.8}O_{0.5}N_{0.2}} + 0.1\ \mathrm{CO_2} + 0.1\ \mathrm{NADH} - 2\ \mathrm{ATP}
\end{aligned} \tag{1}
$$

The first three SP fluxes are from Example 4.1 using pyruvate formate lyase enzyme to convert PYR to AcCoA – the pyruvate dehydrogenase enzyme cannot be used unless HAc is *fed to and consumed* by the

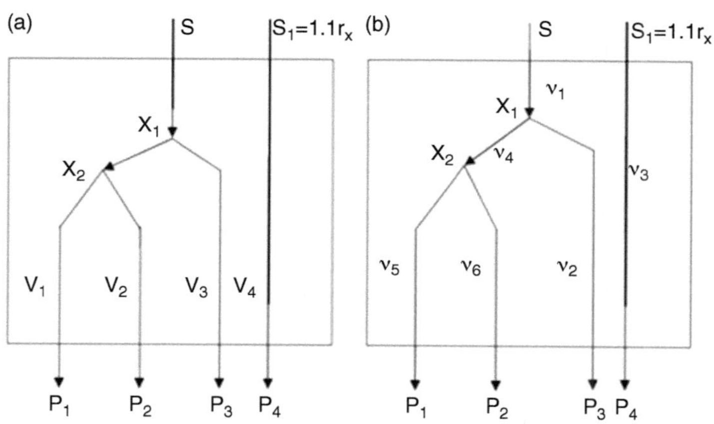

Figure 4.12 (a) Heterofermentative cultivation of lactic bacteria. Figure 4.9a has been expanded with a fourth SP flux to biomass X by a separate pathway V_4 with no nodes. The metabolic network that leads to $[P_1, P_2, P_3]$ = [HAc, ethanol (e), HLac] is the same as in Figure 4.9. (b) Representation of the network in terms of six internal fluxes v_i ($v_1 = (-r_s)$ (see Eq. (4.24)).

organism. $\gamma = 2$ ATP (C-mol biomass)$^{-1}$ in V_4 is chosen based on the insight obtained in Example 4.5. Balancing the four equations in (1) with respect to NADH and ATP gives two constraints on the four fluxes:

$$\begin{aligned} &\text{NADH balance}: \frac{1}{2}V_1 - \frac{1}{2}V_2 + 0.1\,V_4 = 0. \\ &\text{ATP balance}: V_1 + \frac{1}{2}V_2 + \frac{1}{3}V_3 - 2\,V_4 = 0. \end{aligned} \tag{2}$$

The SP fluxes have been normalized such that $[V_1, V_2, V_3, V_4] = [r_{HAc}, r_e, r_{HLac}, r_x]$.

Choose $[r_x, r_{HAc}]$ as the two "measured" rates. The two other rates $[r_e, r_{HLac}]$ are now found by solution of the two linear equations in (2):

$$r_e = r_{HAc} + 0.2\,r_x \text{ and } r_{HLac} = -4.5\,r_{HAc} + 5.7\,r_x. \tag{3}$$

The missing (the dependent or nonkey) rates are found as linear combinations of $\mathbf{r_m} = [r_x, r_{HAc}]$:

$$\begin{aligned} r_c &= 0.1\,r_x, r_{HCOOH} = \frac{1}{2}(r_{HAc} + r_e) = r_{HAc} + 0.1\,r_x \text{ and} (-r_s)_{total} \\ &= (-r_s) + (-r_s)_x = -1.5\,r_{HAc} + 7.1\,r_x. \end{aligned} \tag{4}$$

One can easily check that the expression for $(-r_s)_{total}$ equals that for $\sum P_i + r_c + r_{HCOOH}$.

The black-box model for anaerobic conversion of glucose to the six products is as follows:

$$\begin{aligned} CH_2O \rightarrow\ & Y_{sx}\,X + Y_{sHAc}\,CH_2O(HAc) + Y_{se}\,CH_3O_{0.5} \\ & + Y_{sHLac}\,CH_2O(HLac) + Y_{sHCOOH}\,HCOOH + Y_{sc}CO_2. \end{aligned} \tag{5}$$

The yield coefficients are determined from Eqs. (3) and (4) as $r_i/(-r_s)$. Neither the N source nor water has been included in Eq. (5). The corresponding yield coefficients can easily be derived (e.g., $Y_{sNH3} = 0.2Y_{sx}$).

It is seen that four of the six yield coefficients in Eq. (5) can be calculated based on only two measurements. Here $[r_x, r_{HAc}]$ was chosen, but other combinations of rates can also be used. Naturally, $[r_x, r_c]$ cannot be used since the two rates are linearly dependent (actually proportional). Since the ATP balance is added as an extra constraint together with the carbon and NADH balance and the normalization $(-r_s)_{total} = 1$, one measurement less is needed compared to the procedure used in Chapter 2. The uncertainty of this very easily applied method is that the two parameters $[\alpha, \gamma]$ in Eq. (4.26) are of empirical nature and may have different values under different environmental conditions. As indicated above, the sensitivity of the calculated yield coefficients on α is very small, and based on many experiments with different organisms, one can now fairly well state that $\gamma \sim 2$ for "normal" anaerobic cultivations. Still, better information to estimate γ under different environmental conditions is slowly gathered, and one can envisage that the details of Eq. (4.26) will eventually become fully known.

Example 4.5 Anaerobic fermentation of *S. cerevisiae*

In anaerobic cultivation of yeast, the redox produced in glycolysis and in biomass synthesis cannot be removed by ethanol formation alone since the conversion of glucose to ethanol and CO_2 is redox neutral. Hence, a drain of NADH in another pathway is needed, and the branch in Figure 4.2 from DHAP to glycerol and possibly to further reduced products must be operative. The occurrence of glycerol as a by-product in, for example, wine fermentation is, of course, well known, and here one NADH is oxidized per mole glycerol formed. In Figure 4.13, the key product P_4 is glycerol. Also a possibility of acetaldehyde production (key product P_2) is included, and since all NADH produced in glycolysis is not oxidized in the SP pathway V_2 to acetaldehyde, even more glycerol will have to be produced. The four SP pathways are given in Eq. (1).

$$\begin{aligned}
&V_1 = r_x : 1.12\,CH_2O \rightarrow X(CH_{1.74}O_{0.6}N_{0.12}) + 0.12\,CO_2 + 0.15\,NADH - 2.42\,ATP.\\
&V_2 = r_a : 1.5\,CH_2O \rightarrow CH_2O_{1/2}(a) + \tfrac{1}{2}\,CO_2 + \tfrac{1}{2}\,NADH + \tfrac{1}{2}\,ATP.\\
&V_3 = r_e : 1.5\,CH_2O \rightarrow CH_3O_{1/2}(e) + \tfrac{1}{2}\,CO_2 + \tfrac{1}{2}\,ATP.\\
&V_4 = r_g : CH_2O \rightarrow CH_{8/3}O(g) - \tfrac{1}{3}\,NADH - \tfrac{1}{3}\,ATP.
\end{aligned} \tag{1}$$

In the biomass formation pathway V_1, the biomass composition is taken to be $X = CH_{1.74}O_{0.60}N_{0.12}$, the same as that used in Sections 3.1 and 3.2 (the very accurate measurements of Duboc, von Stockar, and Villadsen [14]). The value of α in Eq. (4.26) is chosen as 0.12, but any α between 0.10 and 0.13 gives virtually the same result. The parameter γ is initially

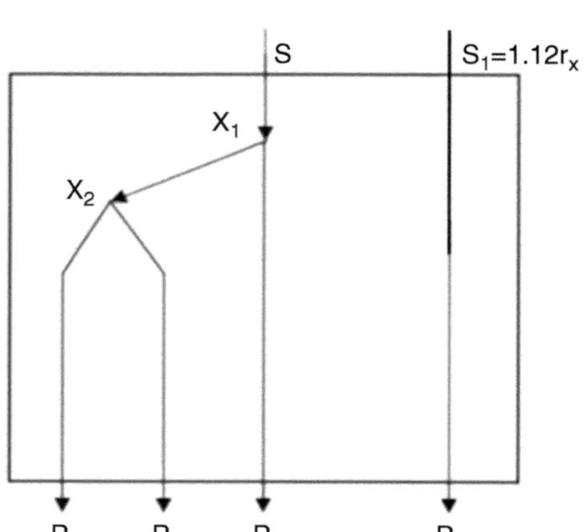

Figure 4.13 Conversion of glucose to biomass (x) P_1, acetaldehyde$_{excreted}$ (a) P_2, ethanol (e) P_3, and to glycerol (g) P_4, by anaerobic yeast fermentation. There are two nodes in the metabolic network: at $X_1 = F\ 1,6P$ and at $X_2 =$ acetaldehyde$_{interior}$. This corresponds to the $2 \cdot 1 + 1 = 3$ key products P_2 to P_4. P_1 is produced by the empirical stoichiometry (4.26).

set to 2.42 mol ATP (C-mol X)$^{-1}$. When results are compared with the experiments of Duboc, von Stockar, and Villadsen [14], γ is, however, found to be $\approx$1.8 mol ATP (C-mol biomass)$^{-1}$ as will be shown at the end of the example.

With four net-pathways and two constraints, a redox and an ATP balance, two rates $\mathbf{r_m}$ must be measured. Here, $\mathbf{r_m} = [r_x, r_a]$ is chosen, and $\mathbf{r_c} = [r_e, r_g]$, the remainder of the key product rates, can be calculated.

$$\text{NADH balance}: 0.15\, r_x + \tfrac{1}{2}\, r_a - \tfrac{1}{3}\, r_g = 0.$$

$$\begin{aligned}\text{ATP balance}: &-2.42\, r_x + \tfrac{1}{2}\, r_a + \tfrac{1}{2}\, r_e - 1/3\, r_g = 0\\ &\rightarrow r_g = 0.45\, r_x + 1.5\, r_a, \quad \text{and } r_e = 5.14\, r_x.\end{aligned}$$

The nonkey rates $(-r_s)_{total}$ and r_c can be calculated from Eq. (1): $(-r_s)_{total} = 9.28\, r_x + 3\, r_a$, and $r_c = 2.69\, r_x + ½\, r_a$.

Let us first assume that alcohol dehydrogenase has been knocked out. This means that $r_e = 0$, and since r_e is proportional to r_x no biomass is produced and the culture will not grow. The yield coefficients are as follows:

$$Y_{sa} = r_a/(3\, r_a) = \tfrac{1}{3}, Y_{sg} = 1.5\, r_a/(3\, r_a) = \tfrac{1}{2}, \text{and} = Y_{sc} = \tfrac{1}{2}\, r_a/(3\, r_a) = \tfrac{1}{6}:$$

$$CH_2O \rightarrow \frac{1}{2} CH_{8/3}O + \frac{1}{3} CH_2O_{1/2} + \frac{1}{6} CO_2. \tag{2}$$

The stand-alone stoichiometry (2) shows that glucose is converted to glycerol with a carbon yield of ½.

This process was used industrially in Germany during WW1 to produce glycerol for explosives. Sulfite was added to the culture, and ADH was poisoned. Today, one would use genetic engineering to produce an adh$^-$ (adh-negative) strain, but it is questionable if the strain would be viable.

Next, let r_a be zero, that is, no acetaldehyde escapes to the medium. The stoichiometry will become:

$$\begin{aligned}CH_2O &\rightarrow \frac{1}{9.28}(X + 5.14\, CH_3O_{1/2} + 0.45\, CH_{8/3}O + 2.69\, CO_2)\\ &= 0.108\, X + 0.554\, CH_3O_{1/2} + 0.049\, CH_{8/3}O + 0.289\, CO_2.\end{aligned} \tag{3}$$

When this is compared with the experimentally obtained stoichiometry, Eq. (3.1), it is seen that the yield coefficients are almost, but not quite, the same. Too small a production of biomass and glycerol is predicted, while the ethanol production is slightly too high. This can be accounted for if the ATP cost for biomass production is smaller than 2.42 ATP (C-mol X)$^{-1}$: less ethanol, but more biomass and glycerol will then be produced.

The best choice of γ is found by minimizing the sum of squared differences between the model and the experiments. The result is $\gamma = 1.8$, and now the stoichiometry (at $\alpha = 1.12$) will become:

$$CH_2O \rightarrow 0.136\,X + 0.524\,CH_3O_{1/2} + 0.061\,CH_{8/3}O + 0.278\,CO_2. \quad (4)$$

The result (4) is virtually indistinguishable from the experimentally obtained stoichiometry (3.1).

The analysis in Example 4.5 shows that one can also use MFA to find physiological parameters such as the ATP requirement for biomass production. One result is that $Y_{x,ATP}$ is more like 1.8 mol (C-mol X)$^{-1}$ for strictly anaerobic growth than the usually quoted value of 2.42 (= 0.1 mole (g biomass)$^{-1}$) that is assumed to hold for both anaerobic and aerobic growth. The value of $\gamma = 2$ chosen in Example 4.4 was based on this insight.

Example 4.6 Biomass growth with succinic acid as the only metabolic product

An SP pathway to biomass will now be added to the two catabolic pathways (2) and (3) in Example 4.3, Figure 4.10. Since the catabolic pathways lead to the same product, SA, they can be combined into one net pathway from glucose to SA. This single catabolic pathway is now coupled to the biomass production pathway V_1:

$$V_1 = r_x : 1.1\,CH_2O \rightarrow X(CH_{1.8}O_{0.5}N_{0.2}) + 0.1\,CO_2 + 0.1\,NADH - \gamma\,ATP.$$
$$V_2 = r_{SA} : (1 + 3y)Glc/2 + (1 - y)CO_2 \rightarrow (1 + 2y)SA + (4y - 1)NADH + (1 + 3y)ATP.$$

Redox balance : $0.1\,V_1 + (4y - 1)V_2 = 0$; ATP balance : $-\gamma V_1 + (1 + 3y)V_2 = 0$.

The solution of the two linear equations is $y = \dfrac{\gamma - 0.1}{4\gamma + 0.3}$ and $V_2 = \dfrac{4\gamma + 0.3}{7} V_1$.

$$(1)$$

Since r_{SA} in units of C-mol SA is 4 (1 + 2 y) V_2 one obtains:

$$r_{SA} = \frac{4(6\gamma + 0.1)}{(4\gamma + 0.3)} V_2 = \frac{4}{7}(6\gamma + 0.1)r_x. \quad (2)$$

$$(-r_s) = 1.1\,V_1 + 3(1 + 3\,y)V_2(\text{in C-mol glucose consumed})$$
$$= 1.1\,V_1 + 21\,\gamma(4\gamma + 0.3)^{-1}V_2 = (1.1 + 3\gamma)r_x. \quad (3)$$

By combination of (3) and (2), the yield coefficients of SA and X are obtained as functions of γ:

$$Y_{s,SA} = \frac{4(6\gamma + 0.1)}{7(3\gamma + 1.1)} \quad \text{and} \quad Y_{sx} = \frac{1}{3\gamma + 1.1}. \quad (4)$$

The result (4) may be checked by letting $\gamma \rightarrow \infty$. Then, $Y_{sx} \rightarrow 0$ and $Y_{s,SA} \rightarrow 8/7$ while from Eq. (1) $y \rightarrow 1/4$.

Since, according to the result of Example 4.5, it is unlikely that in anaerobic fermentation γ will exceed 2, then one would expect $Y_{s,SA}$ to be at least 0.97 C-mol SA (C-mol glucose)$^{-1}$.

If the ATP-yield in V_2 is not $(1+3y)$ but only $2y$, then less biomass will be produced per C-mole SA. The yield coefficient $Y_{s,SA}$ would be higher, but productivity probably lower due to a smaller value of x.

Example 4.7 Production of butanol and acetone by *Clostridium acetobutylicum*

The possibility to produce valuable solvents, for example, 1-butanol by fermentation has revived an interest in the Weizmann process that was invented just before WW I. It provided Great Britain with at that time expensive acetone that was used as solvent in production of smoke-free gun powder. The process is described in many papers (e.g., [9] and in many patents.

The metabolic pathways to acetone (ac) and *n*-butanol (BuOH) are shown in Figure 4.14.

The right-hand figure is a schematic representation of the metabolic network + a separate SP pathway to biomass. The network has four nodes

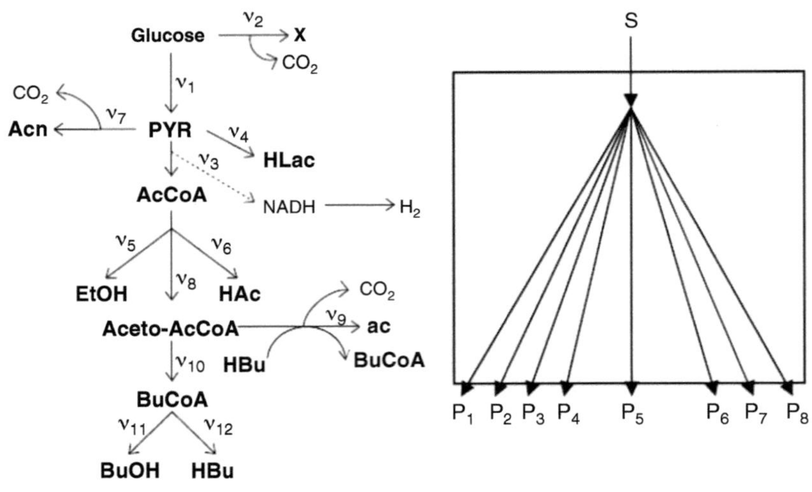

Figure 4.14 Pathways from glucose to solvents: ac = acetone, BuOH = 1- butanol. In v_7 a small quantity of acetoin (3-hydroxy butanone:CH_3-CHOH-CO-CH_3) is formed by condensation and decarboxylation of two PYR. Otherwise, the metabolic network before the condensation of two AcCoA in v_8 to the four-carbon aceto-AcCoA is the same as that of Figure 4.9. In v_{10} aceto-AcCoA, CH_3CO-CH_2CO-S-CoA, is reduced to butyryl-CoA using two NADH mol^{-1}. In v_9 aceto-AcCoA is split into CO_2 and acetone, CH_3COCH_3, while the CoA is transferred to butyric acid, HBu. Reactions v_{11} and v_{12} are similar to the reactions from AcCoA.

at PYR, AcCoA, aceto-AcCoA, and BuCoA. There are (2 + 2 + 1 + 1) + 1 key products from the network + biomass X, a total of eight key products: Biomass X, acetoin (acn), lactic acid (HLac), ethanol (e), acetic acid (HAc), acetone (ac), butyric acid (HBu), and 1-butanol (BuOH). HCOOH (or H_2) and CO_2 are produced in more than one net pathway. Their production rates are found as linear combinations of the production rates of the eight key products. There are (2 + 4 + 3 + 2 + 2) internal fluxes in the network, that is, a total of 13 internal fluxes when the flux v_2 to biomass is included. The 13 internal fluxes are connected through 5 mass balances.

The stoichiometries of the eight SP pathways is given in Eq. (1)

$$
\begin{aligned}
&V_1 = r_x : -1.1\,CH_2O + X(CH_2O_{0.5}N_{0.25}) + 0.1\,CO_2 + 0.075\,NADH - 2.48\,ATP = 0.\\
&V_2 = r_{acn} : -1.5\,CH_2O + CH_2O_{1/2} + \tfrac{1}{2}\,CO_2 + \tfrac{1}{2}\,NADH + 1/2\,ATP = 0.\\
&V_3 = r_{HLac} : -CH_2O + CH_2O(HLac) + 1/3\,ATP = 0.\\
&V_4 = r_e : -1.5\,CH_2O + \tfrac{1}{2}\,HCOOH + CH_3O_{1/2} - \tfrac{1}{2}\,NADH + \tfrac{1}{2}\,ATP = 0.\\
&V_5 = r_{HAc} : -1.5\,CH_2O + \tfrac{1}{2}\,HCOOH + CH_2O(HAc) + \tfrac{1}{2}\,NADH + ATP = 0.\\
&V_6 = r_{ac} : -2\,CH_2O + CH_2O_{1/3} + 2/3\,HCOOH + 1/3\,CO_2 + 2/3\,NADH + 2/3\,ATP = 0.\\
&V_7 = v_{HBu} : -1.5\,CH_2O + CH_2O_{1/2} + \tfrac{1}{2}\,HCOOH + \tfrac{3}{4}\,ATP = 0.\\
&V_8 = r_{BuOH:} - 1.5\,CH_2O + CH_{2.5}O_{0.25} + \tfrac{1}{2}\,HCOOH - \tfrac{1}{2}\,NADH + \tfrac{1}{2}\,ATP = 0.
\end{aligned}
\tag{1}
$$

The SP pathways V_4 and V_5 are written with HCOOH as by-product, that is, as if the oxidation of PYR to AcCoA is catalyzed by pyruvate formate lyase. No formate is, however, measured in the medium, but quite a large amount of H_2 is produced. This could result by decarboxylation of formic acid, but the final products, H_2 and CO_2, are the same as those that would result if the enzyme was pyruvate dehydrogenase, and the NADH subsequently converted to $NAD^+ + H_2$ catalyzed by the flavoprotein NADH dehydrogenase. The end result is the same, and "HCOOH" is only used as a representative of the oxidation of PYR.

The stoichiometry for the SP pathways is quite difficult to set up manually, so is keeping track of the redox and ATP yield of the long sequences of reactions to the end products. *But all the information is available in the data banks for metabolic networks, and the computer can construct the SP pathway stoichiometries V_1–V_8 just as easily as it can set up the stoichiometry for each of the 12 internal pathway reactions v_1–v_{12}.*

As already discussed when the SP flux method and the internal flux method were compared in the introduction to Section 4.7, the number of independent rates is the same in the two methods: When the five mass balances at the node points are used, one will also arrive at eight independent rates by the internal flux approach.

In Eq. (2) below, a redox and an ATP balance will also be applied. This reduces the number of independent key products and internal fluxes to six. There is, however, quite some difference in the computational work to find two dependent key products by (2) and to solve a 6×6 linear algebraic problem (4 mass balances at node points + redox and ATP balances) to find the six dependent internal fluxes. With experimental errors in the six measured rates, there is likely to be a large error magnification in the calculation of missing rates in the internal flux approach. This is discussed further in Section 4.9.

The NADH and the ATP balances provide two constraints on the eight SP fluxes:

$$\text{NADH balance}: 0.075\,V_1 + \tfrac{1}{2}\,V_2 - \tfrac{1}{2}\,V_4 + \tfrac{1}{2}\,V_5 + \tfrac{2}{3}\,V_6 - \tfrac{1}{2}\,V_8 = 0.$$
$$\text{ATP balance}: -2.48\,V_1 + \tfrac{1}{2}\,V_2 + \tfrac{1}{3}\,V_3 + \tfrac{1}{2}\,V_4 + V_5 + \tfrac{2}{3}\,V_6 + \tfrac{3}{4}\,V_7 + \tfrac{1}{2}\,V_8 = 0. \quad (2)$$

Two of the SP fluxes (equal to the production rates of the key products) can be calculated as linear combinations of the remaining six rates. Choose $[V_3, V_5] = [r_{HAc}, r_{HLac}]$ as the vector of calculated products. Solution of the two linear Eq. (2) gives:

$$V_5 = r_{HAc} = -0.150\,V_1 - V_2 + V_4 - 4/3\,V_6 + V_8.$$
$$V_3 = r_{HLac} = 7.89\,V_1 + 3/2\,V_2 - 9/2\,V_4 + 2\,V_6 - 9/4\,V_7 - 9/2\,V_8. \quad (3)$$

Inspection of (2) shows that a number of combinations, for example, $[V_2, V_6] = [r_{acn}, r_{ac}]$ and $[V_4, V_8] = [r_e, r_{BuOH}]$ cannot be selected for calculations, the reason being that the 2×2 coefficient matrix is singular. But these pairs can easily be detected by inspection, a situation very different from that in the internal flux method.

The dependent rates are next determined from the six key rates that remain after elimination of $[r_{HAc}, r_{HLac}]$:

$$-r_s = 1.1\,r_x + 3/2\,r_{acn} + r_{HLac} + 3/2\,r_e + 3/2\,r_{HAc} + 2\,r_{ac} + 3/2\,r_{HBu} + 3/2\,r_{BuOH} = 8.765\,r_x + 3/2\,r_{acn} - 3/2\,r_e + 2\,r_{ac} - 3/4\,r_{HBu} - 3/2\,r_{BuOH}.$$
$$r_{HCOOH} = \tfrac{1}{2}\,r_e + \tfrac{1}{2}\,r_{HAc} + 2/3\,r_{ac} + \tfrac{1}{2}\,r_{HBu} + \tfrac{1}{2}\,r_{BuOH} = -0.075\,r_x - \tfrac{1}{2}\,r_{acn} + r_e + \tfrac{1}{2}\,r_{HBu} + r_{BuOH} = \mathbf{r}_{H2}.$$
$$r_c = r_{HCOOH} + 0.1\,r_x + \tfrac{1}{2}\,r_{acn} + 1/3\,r_{ac} = 0.025\,r_x + r_e + 1/3\,r_{ac} + \tfrac{1}{2}\,r_{HBu} + r_{BuOH}.$$

The example shows that even rather complicated networks can be analyzed using the same concepts of the SP pathways that were introduced at the start of Section 4.7. It also demonstrates that the SP flux method and the internal flux method give the same results *for error free data*, but certainly not when the analysis is based on experimental data as discussed in Section 4.9.

Example 4.8 SP-pathways with more than one substrate, for example, glucose and oxygen

In Figure 4.15, the carbon drained off to make biomass is not written as a separate SP pathway, but is assumed to stem from PYR at node point X_1.

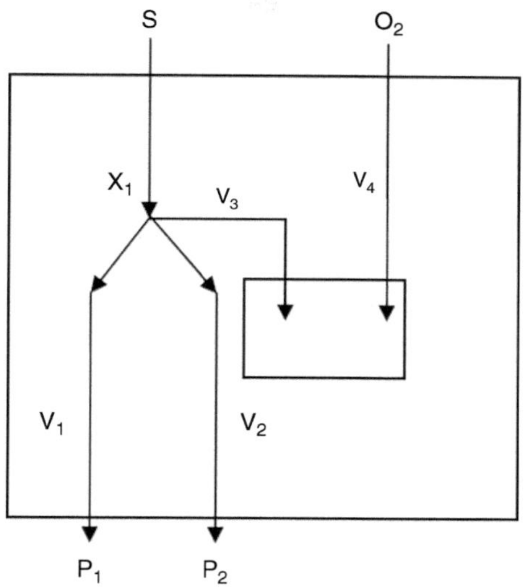

Figure 4.15 Schematic representation of aerobic yeast fermentation with two key products, biomass (P_1) and ethanol (P_2). The inner box represents diversion of pyruvate to the mitochondria where it is completely converted to CO_2 with an NADH yield of five NADH per mole PYR. A second substrate O_2 is fed to the inner box and the five NADH are converted to five P/O ATP. The only carbon product from the inner box is three CO_2 per mole PYR, and since CO_2 is not a key product (it is also produced in V_1 and V_2), there is no key carbon product from the inner box.

This is done to simplify the graph. The internal carbon flux v_3 also emanates from X_1 and ends up as CO_2. Clearly, the two SP fluxes to $P_1 = r_x$ and to $P_2 = \text{ethanol} = r_e$ are different, and V_1 and V_2 cannot be combined to one SP pathway V_3.

$$V_1 = r_x : 1.10\,CH_2O \rightarrow X(k_x = 4.20) + 0.10\,CO_2 + 0.10\,NADH - \gamma\,ATP. \tag{1}$$

$$V_2 = r_e : 1.5\,CH_2O \rightarrow CH_3O_{1/2} + 0.5\,CO_2 + 0.5\,ATP. \tag{2}$$

$$\text{The glucose spent to support } V_1 + V_2 \text{ is } (-r_s)_1 = 1.1\,r_x + 1.5\,r_e, \quad \text{and } (r_c)_1 = 0.1\,r_x + 0.5\,r_e. \tag{3}$$

The inner pathway reaction is given as follows:

$$\begin{aligned} &v_3 : CH_2O \rightarrow CO_2 + 2\,NADH + \epsilon\,ATP \text{ with rate } (-r_s)_2 \text{ and} \\ &v_4 : O_2 + 2\,NADH \rightarrow 2\,P/O\,ATP \text{ with rate } (-r_o). \end{aligned} \tag{4}$$

$$\begin{aligned} &\text{A redox balance for the total system gives } 0.1\,r_x + 2(-r_s)_2 - 2(-r_o) = 0 (\text{for } k_x = 4.20). \\ &\text{An ATP balance for the total system gives } -\gamma\,r_x + 0.5\,r_e + \epsilon(-r_s)_2 + 2P/O(-r_o) = 0. \end{aligned} \tag{5}$$

The NADH balance combines the redox produced in Eq. (1) with that produced from $(-r_s)_2$ in glycolysis and that produced in conversion of the glucose in v_3 all the way to CO_2 (2 NADH C-mol^{-1}). From this is subtracted that lost in $v_4 = (-r_o)$. ε is the ATP produced in the EMP pathway and further in the TCA cycle to convert glucose in $(-r_s)_2$ to CO_2. While it is certain that each carbon in $(-r_s)_2$ produces ⅓ ATP in the EMP pathway, it is uncertain how much ATP is produced in the rest of the pathway to CO_2. The GTP produced when succinyl CoA is converted to succinic acid on Figure 4.6 is, as explained in Section 4.3, equivalent to ATP (but used for special purposes). Thus, one would believe that $\epsilon = 2/3$, but the carbon that passes through the pyruvate shunt on Figure 4.4c must be reenergized in reaction (10) before it passes back into the mitochondria to be oxidized to CO_2. Since, even at low sugar consumption rate, a sizeable portion of the carbon passes the pyruvate shunt before entering the mitochondria, it is questionable if the ATP from the TCA cycle is effectively obtained. One can use an ε value anywhere in the interval [1/3, 2/3] – a compromise could be $\epsilon = 1/2$.

Whichever value of ε is used, it does not have any noticeable influence on the final flux distribution since the ATP production in respiration is far greater than the rest of the ATP produced. It is further assumed that all the redox produced when PYR is degraded to CO_2 is obtained as NADH, but one of the five "H_2" is, in fact, $FADH_2$, and since $FADH_2$ enters the electron transfer chain in oxidative phosphorylation (see Chapter 9) at complex II and not at complex I as does NADH, the energy yield from $FADH_2$ can be less than for NADH.

For the yeast *S. cerevisiae*, the NADH does, in contrast to the situation for many other yeasts, also enter at complex II, and the question of different energy yield from $FADH_2$ and NADH does not arise (the theoretical energy yield in oxidative phosphorylation is often taken to be P/O = 3 for NADH and P/O = 2 for $FADH_2$). More important, the *real* energy yield P/O in an actual fermentation process is far lower than either 2 or 3. For *S. cerevisiae*, van Gulik and Heijnen [4] found that $1.2 < P/O < 1.5$ in fermentations with different carbon substrates.

Solution of the two linear equations in (5) yields:

$$(-r_s)_2 = (r_c)_2 = \frac{\left(\gamma - 0.1\frac{P}{O}\right)r_x - 0.5\,r_e}{\epsilon + 2\,P/O} \quad \text{and} \quad (-r_o) = 0.05\,r_x + (-r_s)_2. \tag{6}$$

$$\text{The total glucose consumption is } (-r_s)_{total} = (-r_s)_1 + (-r_s)_2 = 1.1\,r_x + 1.5\,r_e + (-r_s)_2. \tag{7}$$

$$\text{The total } CO_2 \text{ production is } r_c = 0.1\,r_x + 0.5\,r_e + (-r_s)_2. \tag{8}$$

From Eqs. (6)–(8), the yield coefficients Y_{si} can be determined as functions of the key production rates $[r_x, r_e]$. With a set of experimental data obtained for different values of $(-r_s)_{total}$ – that is, for different dilution rates D in a steady-state CSTR – one can calculate the value of γ in Eq. (6) that best fits the experimental data. The data of von Meyenburg shown in Figure 2.3 will be used as an example. In Section 3.3 the stoichiometry for $D < 0.25\,h^{-1}$ was calculated in Eq. (3.11). Since $r_e = 0$ for $D < 0.25\,h^{-1}$, one obtains from Eq. (6) that

$$(-r_o) = 0.05\, r_x + \frac{\left(\gamma - 0.1\frac{P}{O}\right) r_x}{\epsilon + 2\,P/O}, \quad \text{i.e.,} (-r_o/r_x) = Y_{xo} = 0.05 + \frac{\left(\gamma - 0.1\frac{P}{O}\right)}{\epsilon + 2\,P/O}. \tag{9}$$

Using the value P/O = 1.25 suggested by van Gulik and Heijnen [4] and $\varepsilon = 0.5$, one obtains the following equation for γ:

$$\frac{\gamma - 0.125}{3} = Y_{xo} - 0.05. \tag{10}$$

When the value $Y_{xo} = 0.374/0.596 = 0.628$ is inserted from Eq. (3.11), one obtains $\gamma = 1.85$.

This value is far below the value for $Y_{xATP} \sim 2.40$ suggested in literature for fully respiratory cultivation.

It is possible that the value 1.25 used for P/O is too low. With P/O = 1.5, the result is $\gamma = 2.17$.

There is also another explanation for the low value of γ. The data analyzed in Eq. (3.11) and further in Villadsen and Jørgensen [10] used the biomass concentration $14\,g\,l^{-1}$ for $D < 0.25\,h^{-1}$ without correcting for ash in the reported biomass – von Meyenburg does not refer to an ash content. With 8 wt% ash in the biomass (this ash content is also used in Villadsen, Nielsen, and Lidén [11], p. 89) in their analysis of the same example) all the yield coefficients in Eq. (3.11) have to be recalculated. The result is $Y_{sx} = 0.92 \cdot 0.596 = 0.548$, $Y_{sc} = 0.452$, and $Y_{so} = (1/4)(4–4.2 \cdot 0.548) = 0.425$.

Based on the revised experimental data, one obtains $\gamma = 2.30$ ATP (C-mol biomass)$^{-1}$ for P/O = 1.25.

The analysis of the energy requirement for growth is another example where stoichiometry-based MFA is used to obtain values for physiological parameters – but as shown above the physiological parameters P/O and γ are closely correlated – when P/O is set to a higher value, one will also obtain a higher value of γ. Naturally, one must also trust the experimental data on which the analysis is based, and this can be tricky for literature data.

When Eqs. (7)–(9) were used by Villadsen, Nielsen, and Lidén [11] to analyze the data of von Meyenburg for $D = [0.3, 0.4]\,h^{-1}$, that is, for D values where ethanol is formed besides biomass, it was found that γ decreased with increasing D. The analysis of Example 4.5 showed that in fully anaerobic cultivation the γ is less than 2 mol ATP (C-mol biomass)$^{-1}$, and one may suspect that much of the ATP generated in aerobic cultivation is wasted in futile cycles.

4.8 Metabolic Flux Analysis Discussed in Terms of Individual Pathway Rates in the Network

The internal pathway fluxes v_i on Figure 4.12b can easily be calculated from the SP pathway fluxes $[V_1, V_2, V_3, V_4] = [r_{HAc}, r_e, r_{HLac}, r_x]$. Using the results from Example 4.4 one obtains:

$$\begin{aligned}\mathbf{v_5} &= \mathrm{V}_1 = \mathbf{r_{HAc}}, \mathbf{v_6} = \mathrm{V}_2 = \mathbf{r_e} = \mathbf{r_{HAc}} + \mathbf{0.2\,r_x}, \mathbf{v_2} = \mathrm{V}_3 = \mathbf{r_{HLac}} \\ &= -\mathbf{4.5\,r_{HAc}} + \mathbf{5.7\,r_x}, \mathbf{v_3} = \mathrm{V}_4 = \mathbf{r_x}.\end{aligned} \tag{4.27}$$

$$\begin{aligned}\text{From a mass balance at } X_2: \mathbf{v_4} &= \mathrm{v}_5 + \mathrm{v}_6 + r_{HCOOH} \\ &= \mathbf{r_{HAc}} + \mathbf{r_e} + \mathbf{r_{HAc}} + \mathbf{0.1\,r_x} = \mathbf{3r_{HAc}} + 0.3\,\mathbf{r_x}.\end{aligned} \tag{4.28}$$

$$\begin{aligned}\text{Finally, a mass balance at } X_1: \mathbf{v_1} &= \mathrm{v}_4 + \mathrm{v}_2 \\ &= 3\,r_{HAc} + 0.3\,r_x - 4.5\,r_{HAc} + 5.7\,r_x \\ &= -\mathbf{1.5\,r_{HAc}} + \mathbf{6r_x}.\end{aligned} \tag{4.29}$$

The results can be checked: $(-r_s)_{total} = v_1 + 1.1\ r_x = -1.5\ r_{HAc} + 7.1\ r_x$, the result (4) of Example 4.4.

The calculations in Eqs. (4.27)–(4.29) confirm that all six internal fluxes v_1–v_6 can be calculated from the specific rates of production $\mathbf{r} = [r_{HAc}, r_x, r_e, r_{HLac}, r_{HCOOH}, r_c]$ of carbon-containing products, and by the SP method with a redox and an ATP balance included only two of the rates $[r_{HAc}, r_x]$ need to be specified. The specific rates of production of the remaining key products $[r_e, r_{HLac}]$ were obtained from the two constraints provided by the redox and ATP balances. Finally, the specific rates of production of the dependent products $[r_{HCOOH}, r_c]$ were calculated based on r_i of the four key products. This demonstrates the one-to-one equivalence of MFA based on the SP flux method and the internal flux method that was stated in the introduction to Section 4.7.

In the following, the *same* problem will be solved based on the six internal fluxes v_1–v_6 instead of the specific rates of production r_1–r_4 of the four key products P_i. The problem formulation Eq. (4.30) and the resulting matrix algebra applied to find the internal flux vector **v** may look "complicated," but the computational procedure is backed up by easily available computer software. It is hoped that the perhaps intuitively "simpler" SP flux method may help the reader to understand what the computer programs actually do, and they *will* give the right answer if the matrix problem Eq. (4.30) is stated correctly (this requires accuracy, but guidance is given by the software).

$$\begin{bmatrix} -r_S \\ r_X \\ r_{HLac} \\ r_e \\ r_{HAC} \\ r_{HCOOH} \\ r_C \\ r_{PYR} \\ r_{AcCoA} \\ r_{NADH} \\ r_{ATP} \end{bmatrix} = \begin{bmatrix} 1 & 0 & 1.1 & 0 & 0 & 0 \\ 0 & 0 & 1 & 0 & 0 & 0 \\ 0 & 1 & 0 & 0 & 0 & 0 \\ 0 & 0 & 0 & 0 & 0 & 1 \\ 0 & 0 & 0 & 0 & 1 & 0 \\ 0 & 0 & 0 & 1/3 & 0 & 0 \\ 0 & 0 & 0.1 & 0 & 0 & 0 \\ 1 & -1 & 0 & -1 & 0 & 0 \\ 0 & 0 & 0 & 2/3 & -1 & -1 \\ 1/3 & -1/3 & 0.1 & 0 & 0 & -1 \\ 1/3 & 0 & -2 & 0 & 1/2 & 0 \end{bmatrix} \cdot \begin{bmatrix} v_1 \\ v_2 \\ v_3 \\ v_4 \\ v_5 \\ v_6 \end{bmatrix}$$

Figure 4.16 Matrix equation $\mathbf{r} = \mathbf{T}\ \mathbf{v}$. The transpose $\mathbf{T}^T$ of matrix $\mathbf{T}$ is usually called the *stoichiometric matrix*. Thus, the first column of $\mathbf{T}$ shows that one C-mol glucose is converted to one C-mol PYR in internal pathway v_1, with concomitant production of ⅓ NADH and ⅓ ATP. Independent of the metabolic network, X is produced from glucose by the empirical stoichiometry Eq. (4.26) and with stoichiometric coefficients stated in Example 4.4 and shown in column 3. Vector $\mathbf{r}$ contains the six production rates, one substrate consumption rate, and four more rates $\mathbf{r} = \mathbf{0}$ that express mass balances at the node points PYR and AcCoA on Figure 4.12, and redox + ATP balances over the whole network. These rates must be zero *at steady state*. Thus, PYR is obtained in v_1 and consumed in v_2 and v_4 (PYR → HLac and PYR → 2/3 AcCoA + 1/3 HCOOH). ATP is produced in $[v_1, v_5]$, namely, by the production of PYR from CH_2O and of HAc (CH_2O) from AcCoA, but ATP is consumed in the production of X. With four constraints ($r_i = 0$), only two of the v_i need to be measured; the other four can be calculated. Here we shall choose $\mathbf{v_m} = [v_3, v_5] = [r_x, r_{HAc}]$.

The arrangement of the calculation is shown in Figure 4.16 based on the system of Example 4.4:

$$\begin{bmatrix} -r_s \\ r_x \\ r_{HLac} \\ r_e \\ r_{HAc} \\ r_{HCOOH} \\ r_c \\ r_{PYR} \\ r_{AcCoA} \\ r_{NADH} \\ r_{ATP} \end{bmatrix} = \begin{bmatrix} 1 & 0 & 1.1 & 0 & 0 & 0 \\ 0 & 0 & 1 & 0 & 0 & 0 \\ 0 & 1 & 0 & 0 & 0 & 0 \\ 0 & 0 & 0 & 0 & 0 & 1 \\ 0 & 0 & 0 & 0 & 1 & 0 \\ 0 & 0 & 0 & 1/3 & 0 & 0 \\ 0 & 0 & 0.1 & 0 & 0 & 0 \\ 1 & -1 & 0 & -1 & 0 & 0 \\ 0 & 0 & 0 & 2/3 & -1 & -1 \\ 1/3 & -1/3 & 0.1 & 0 & 0 & -1 \\ 1/3 & 0 & -2 & 0 & 1/2 & 0 \end{bmatrix} \cdot \begin{bmatrix} v_1 \\ v_2 \\ v_3 \\ v_4 \\ v_5 \\ v_6 \end{bmatrix} \tag{4.30}$$

Multiplication of **v** onto the four last rows of **T** yields:

$$\mathbf{T}_0\mathbf{v} = \begin{pmatrix} 1 & -1 & 0 & -1 & 0 & 0 \\ 0 & 0 & 0 & 2/3 & -1 & -1 \\ 1/3 & -1/3 & 0.1 & 0 & 0 & -1 \\ 1/3 & 0 & -2 & 0 & 0.5 & 0 \end{pmatrix} \cdot \mathbf{v} \tag{4.31}$$

$$\mathbf{T}_1 \cdot \mathbf{v}_c + \mathbf{T}_2 \cdot \mathbf{v}_m = \mathbf{0} \text{ or : } \begin{pmatrix} 1 & -1 & -1 & 0 \\ 0 & 0 & 2/3 & -1 \\ 1/3 & -1/3 & 0 & -1 \\ 1/3 & 0 & 0 & 0 \end{pmatrix} \cdot \begin{pmatrix} v_1 \\ v_2 \\ v_4 \\ v_6 \end{pmatrix}$$

$$= -\begin{pmatrix} 0 & 0 \\ 0 & -1 \\ 0.1 & 0 \\ -2 & 0.5 \end{pmatrix} \cdot \begin{pmatrix} v_3 \\ v_5 \end{pmatrix} \tag{4.32}$$

Finally one obtains $\mathbf{v}_c$ as $\mathbf{v}_c = -(\mathbf{T}_1)^{-1} \cdot \mathbf{T}_2 \cdot \begin{pmatrix} v_3 \\ v_5 \end{pmatrix} \rightarrow \begin{pmatrix} v_1 \\ v_2 \\ v_4 \\ v_6 \end{pmatrix}$

$$= \begin{pmatrix} 6 & -1.5 \\ 5.7 & -4.5 \\ 0.3 & 3 \\ 0.2 & 1 \end{pmatrix} \cdot \begin{pmatrix} r_x \\ r_{HAc} \end{pmatrix} \tag{4.33}$$

The missing rates are obtained by multiplying **v** and the corresponding rows of **T:**

$$\begin{pmatrix} -r_s \\ r_{HLac} \\ r_e \\ r_{HCOOH} \\ r_c \end{pmatrix} = \begin{pmatrix} 7.1\ r_x - 1.5\, r_{HAc} \\ 5.7\, r_x - 4.5\, r_{HAc} \\ 0.2\, r_x + r_{HAc} \\ 0.1\, r_x + r_{HAc} \\ 0.1\, r_x \end{pmatrix} \tag{4.34}$$

The result Eq. (4.33) is seen to be identical to that obtained in Eqs. (4.27)–(4.29), and the calculated rates Eq. (4.34) are the same as obtained in Example 4.4. Two extra constraints are introduced to express that the C-mass balances must be satisfied at the nodes X_1 and X_2 in Figure 4.12b. With the NADH and the ATP balance, there are a total of four linear ties between the six internal fluxes v_i. In the SP-flux method, there are $P_{total} = \sum_1^M (\text{out} - \text{in})_i + 2$ key products from the metabolic network including X, that is, four key products in Figure 4.12a. With the NADH + ATP balances, this reduces to 2 SP-fluxes to be measured. In both cases, the network has two independent rates P_i or v_i, while the rest of the rates can be calculated. For manual calculations, the internal flux method is challenging, but simple software will solve the 4×4 algebraic problem just as easily as the 2×2 problem in the SP method. The difference in dimensionality of the two methods is more evident in Example 4.7, where a 7×7 matrix problem has to be solved in the internal flux method, while the SP method still only requires solution of a 2×2 problem. Still, with the use of perhaps even a pocket computer, Example 4.7 can also be solved by the internal flux method. One difficulty of the internal flux method is that linear dependency between rates becomes more difficult to detect (and if matrix $\mathbf{T_1}$ in Eq. (4.31) cannot be inverted the method fails, and another set of two v_i has to be used in $\mathbf{v_m}$). Linear dependency between rates is easy to detect in Eq. (3) of Example 4.7, but these linear dependencies can also be found (automatically) when a standard software is used to solve Eq. (4.30) by the internal flux method.

In Eqs. (4.31)–(4.34), a specific example, Example 4.4, was used to illustrate how the internal flux method works. A general procedure is given below:

The matrix $\mathbf{r}$ of Eq. (7.30) consists of M nonzero rates, rates of consumption of substrates and rates of production of products. The last N_c rates are zero. Hence, the dimension of the column vector $\mathbf{r}$ is $(M+N_c)\times1$. The flux vector $\mathbf{v}$ is of dimension $(J\times1)$. The value of J is calculated by Eq. (4.25) and based on a graph of the network as $v_{\text{total}} = \sum_1^M \text{out}_i + 1$, or as $v_{\text{total}} = \sum_1^M \text{out}_i + 2$ when biomass is included. In both cases, $J=N_c+N$, where N is the number of independent fluxes.

The N-independent (or "measured") fluxes are collected in vector $\mathbf{v_m}$ (dimension $(N\times1)$), while the remaining fluxes are collected in vector $\mathbf{v_c}$ with N_c elements and dimension $(N_c\times1)$.

In Example 4.4, $M=7$, $N_c=4$, $J=6$, and $N=2$. The matrix $\mathbf{T}$ has dimension $(M+N_c)\times J$, matrix $\mathbf{T_0}$: $(N_c\times J)$, matrix $\mathbf{T_1}$: $(N_c\times N_c)$, and matrix $\mathbf{T_2}$: $(N_c\times N)$.

When the stoichiometry of the internal reactions is normalized such that $\mathbf{v_m} = \mathbf{r_m}$ (as done in the SP flux method and illustrated in Example 4.4), the calculation of the nonmeasured rates from $\mathbf{v}$ is slightly simpler as illustrated in Eqs. (4.33) and (4.34).

As stated earlier, the stoichiometry-based MFA, whether done by the SP flux method or by the internal flux method, requires that the cells are in metabolic steady state. In Chapter 5, it is discussed to what extent the analysis of data from ^{13}C-labeled experiments are influenced by the use of pulse experiments rather than using steady-state experiments.

Example 4.9 The flux v_i through the internal pathways in succinic acid production

The application of the internal flux method to obtain v_i and thereafter r_{SA} is not straightforward due to the linear dependencies between the v_i values. In a simplified version of Figure 4.11, the (unnecessary) node at PEP has been deleted. With two nodes, at OAA and at AcCoA, and five internal fluxes v_i in the network Glc → SA.

$(-r_{glc}) = v_0 = v_1 + v_2$:
v_1 : glc($C_6H_{12}O_6$) + 2 CO_2 → 2 OAA + 2 NADH; v_2 : glc($C_3H_6O_3$)
→ AcCoA + CO_2 + 2 NADH.
v_3 : OAA → SA − 2 NADH; v_4 : OAA + AcCoA → SA + 2 CO_2 + 2 NADH.

$$\begin{pmatrix} -r_{glc} \\ r_c \\ r_{SA} \\ r_{OAA} \\ r_{AcCoA} \\ r_{NADH} \end{pmatrix} = \begin{pmatrix} -1 & -0.5 & 0 & 0 \\ -2 & 1 & 0 & 2 \\ 0 & 0 & 1 & 1 \\ 2 & 0 & -1 & -1 \\ 0 & 1 & 0 & -1 \\ 2 & 2 & -2 & 2 \end{pmatrix} \cdot \begin{pmatrix} v_1 \\ v_2 \\ v_3 \\ v_4 \end{pmatrix} \rightarrow \begin{pmatrix} v_2 \\ v_3 \\ v_4 \end{pmatrix}$$

$$= \begin{pmatrix} 0 & -1 & -1 \\ 1 & 0 & -1 \\ 2 & -2 & 2 \end{pmatrix}^{-1} - 1 \cdot \begin{pmatrix} -2 \\ 0 \\ -2 \end{pmatrix} v_1 = \begin{pmatrix} \frac{1}{3} \\ \frac{5}{3} \\ \frac{1}{3} \end{pmatrix} v_1$$

$$\rightarrow \begin{pmatrix} -r_{glc} \\ r_c \\ r_{SA} \\ r_{OAA} \\ r_{AcCoA} \\ r_{NADH} \end{pmatrix} = \begin{pmatrix} 1 + \frac{1}{6} \\ -2 + \frac{1+0+2}{3} \\ \frac{5+1}{3} \\ 0 \\ 0 \\ 0 \end{pmatrix} \quad \text{or}$$

$$: \frac{7}{6} \text{glc} + CO_2 \rightarrow 2\,\text{SA}.$$

This is the same result as obtained by the SP method, but there is rather more (manual) work involved.

The same result is obtained when also the node at AcCoA is deleted.

In the previous Sections 4.6 and 4.7, the metabolic network was analyzed with *exactly* the number of experimentally determined rates to permit calculation of the remaining rates. Although this is fine when the experimentally determined rates are error free, it will not work optimally when the experimental data are corrupted by experimental errors. Sometimes, considerable errors will occur in the calculated rates, especially when the measured rates are chosen in a nonoptimal way.

When more than N rates are included in $\mathbf{v_m}$, say $N+Q$ rates, and $(N_c - Q)$ rates are included in $\mathbf{v_c}$, *the system is overdetermined*. Now matrix $\mathbf{T_1}$ has $N_c - Q$ columns, but still N_c rows and the matrix cannot be inverted. This problem is easily resolved since the matrix $\mathbf{T_1}^T \mathbf{T_1}$ has dimension $[(N_c - Q) \times N_c] \cdot [N_c \times (N_c - Q)] = (N_c - Q) \times (N_c - Q)$ and can be inverted. Matrix $\mathbf{T_2}$ has dimension $N_c \times (N+Q)$.

Define matrix $\mathbf{T_1^\#}$, called the "pseudo-inverse" of $\mathbf{T_1}$ by Eq. (4.35):

$$\mathbf{T_1^\#} = (\mathbf{T_1}^T \cdot \mathbf{T_1})^{-1} \cdot \mathbf{T_1}^T \tag{4.35}$$

The dimension of $\mathbf{T_1^\#}$ is $[(N_c - Q) \times N_c \cdot N_c \times (N_c - Q)] \cdot [(N_c - Q) \times N_c] = [(N_c - Q) \times N_c]$

$$\text{Vector } \mathbf{v_c} \text{ is calculated by :} \quad \mathbf{v_c} = \mathbf{T_1}^\# \cdot \mathbf{T_2} \cdot \mathbf{v_m} \tag{4.36}$$

The dimension of $\mathbf{v_c}$ is $[(N_c - Q) \times N_c] \cdot [N_c \cdot (N + Q)] \cdot [(N + Q) \cdot 1]$ $[=[(N_c - Q) \times 1]$ as required.

The vector of internal fluxes $\mathbf{v}$ with J components is, of course, the combination of $\mathbf{v_m}$ and $\mathbf{v_c}$. Finally, the nonzero elements of vector $\mathbf{r}$ are found as the product of the first M rows of $\mathbf{T}$ and vector $\mathbf{v}$.

The calculation of the missing elements of $\mathbf{v}$ by *least squares* as described above will give a better accuracy of the now fewer elements of $\mathbf{v_c}$ when the experimental values of the components of $\mathbf{v_m}$ are subject to experimental inaccuracies. The only condition for success is that matrix $\mathbf{T_1}$ has full rank, and this is the case when the $N + Q$ columns of $\mathbf{T_2}$ that belong to $\mathbf{v_m}$ are not linearly dependent due to linear dependencies between the measured fluxes represented in $\mathbf{v_m}$.

Example 4.10 Including v_2 and v_6, that is, r_{HLac} and r_e in v_m

When in Example 4.4$\mathbf{v_m} = (v_2, v_3, v_5, v_6) = (r_{HLac}, r_x, r_{HAc}, r_e)$ are used to calculate $\mathbf{v_c} = (v_1, v_4)$:

$$\mathbf{T_1}^T \cdot \mathbf{T_1} = \begin{pmatrix} 1 & 0 & 1/3 & 1/3 \\ -1 & 2/3 & 0 & 0 \end{pmatrix} \cdot \begin{pmatrix} 1 & -1 \\ 0 & 2/3 \\ 1/3 & 0 \\ 1/3 & 0 \end{pmatrix}$$

$$= \begin{pmatrix} \frac{11}{9} & -1 \\ -1 & \frac{13}{9} \end{pmatrix} = \mathbf{A}$$

$$\mathbf{A}^{-1} = \frac{1}{62} \begin{pmatrix} 117 & 81 \\ 81 & 99 \end{pmatrix} \quad \text{and} \quad \mathbf{T_1^\#} = \mathbf{A}^{-1} \cdot \mathbf{T_1}^T$$

$$= \frac{1}{62} \begin{pmatrix} 36 & 54 & 39 & 39 \\ -18 & 66 & 27 & 27 \end{pmatrix}$$

$$\mathbf{v_c} = \begin{pmatrix} v_1 \\ v_4 \end{pmatrix} = \mathbf{T_1^\#} \cdot \mathbf{T_2} \cdot \mathbf{v_m} = \frac{1}{62} \begin{pmatrix} 49 & 74.1 & 34.5 & 93 \\ -9 & 51.3 & 52.5 & 93 \end{pmatrix} \cdot \begin{pmatrix} r_{HLac} \\ r_x \\ r_{HAc} \\ r_e \end{pmatrix}$$

It is easily seen that the result for, for example, $v_1{}^\# = 6\ r_x - 1.5\ r_{HAc}$ as was also obtained in Eq. (4.33):

$62\ v_1 = (49\ r_{HLac} + 74.1\ r_x + 34.5\ r_{HAc} + 93\ r_e) = (49 \cdot (5.7\ r_x - 4.5\ r_{HAc}) + 74.1\ r_x + 34.5\ r_{HAc} + 93 \cdot (0.2\ r_x + r_{HAc}) \rightarrow \mathbf{v_1} = \mathbf{6}\ \mathbf{r_x} - 1.5\ \mathbf{r_{HAc}}$. But the result is correct only when error-free measurements are used.

4.9 Propagation of Experimental Errors in MFA

Whether the calculated fluxes are determined from the minimum required measured fluxes or by least squares, the accuracy of the elements of $\mathbf{r_c}$ depends on the structure of the coefficient matrix $\mathbf{A}$ in the system of linear equations to be solved.

If $\mathbf{A}$ is *ill-conditioned*, its determinant is close to zero, and large errors in $\mathbf{r_c}$ are to be expected *unless the data in* $\boldsymbol{r_m}$ *are error free*. When exactly the required number (= N) of measurements are included in $\mathbf{A}$, then $\mathbf{A}$ is matrix $\mathbf{T_1}$. When $N+Q$ measurements are included in $\mathbf{r_m}$, matrix $\mathbf{A}$ is $\mathbf{T_1}^{\mathrm{T}} \cdot \mathbf{T_1}$.

One way of estimating how experimental errors in the elements of $\mathbf{r_m}$ are magnified in the elements of $\mathbf{r_c}$ is to calculate the *condition number* c_p of $\mathbf{A}$. c_p depends on the *norm* used to "measure" $\mathbf{A}$. The most common norm used is the "2-norm," the Euclidian or "least squares" norm. The condition number $c_2 = c_\mathrm{p}(\mathbf{A})_{2\text{-norm}}$ is calculated from the ratio of the largest and the smallest of the eigenvalues σ of $\mathbf{A} \cdot \mathbf{A}^{\mathrm{T}}$. Matrix $\mathbf{A} \cdot \mathbf{A}^{\mathrm{T}}$ is symmetric, the eigenvalues are real, and the error magnification factor is given by Eq. (4.37):

$$c_2(\mathbf{A}) = (|\sigma_{\max}|/|\sigma_{min}|)^{1/2}. \tag{4.37}$$

With $\mathbf{r_m} = \mathbf{v_m} = (\mathbf{r}_\mathrm{x}, r_\mathrm{HAc})$, where $\mathbf{T_1} = \mathbf{A}$ is given in Eq. (4.32).

$\mathbf{T_1}$ is not symmetric, and the eigenvalues are not guaranteed to be real. Hence, $\mathbf{T_1} \cdot \mathbf{T_1}^{\mathrm{T}}$ is calculated:

$$\begin{aligned}\mathbf{T}_1 \cdot \mathbf{T}_1^{\mathrm{T}} &= \begin{pmatrix} 1 & -1 & -1 & 0 \\ 0 & 0 & 2/3 & -1 \\ 1/3 & -1/3 & 0 & -1 \\ 1/3 & 0 & 0 & 0 \end{pmatrix} \cdot \begin{pmatrix} 1 & 0 & 1/3 & 1/3 \\ -1 & 0 & -1/3 & 0 \\ -1 & 2/3 & 0 & 0 \\ 0 & -1 & -1 & 0 \end{pmatrix} \\ &= \begin{pmatrix} 3 & -2/3 & 2/3 & 1/3 \\ -2/3 & 13/9 & 0 & -1/2 \\ 2/3 & 1 & 2/9 & 1/9 \\ 1/3 & -1/2 & 1/9 & 13/36 \end{pmatrix}.\end{aligned} \tag{4.38}$$

The eigenvalues of $\mathbf{T_1} \cdot \mathbf{T_1}^{\mathrm{T}}$ are $\sigma_i = [3.336, 2.341, 0.08128, 0.01945]$, and

$$c_2(\mathbf{A})_2 = (3.336/0.01945)^{1/2} = 13. \tag{4.39}$$

For $\mathbf{v_m}$: $(r_\mathrm{x}, r_\mathrm{e})$, $(r_\mathrm{x}, r_\mathrm{HLac})$, $(r_\mathrm{HAc}, r_\mathrm{HLac})$, $(r_\mathrm{HAc}, r_\mathrm{e})$, one obtains the following values for c_2:

$$c_2(\mathbf{A}) = 17, 9, 15, 206. \tag{4.40}$$

The smallest condition number is obtained when $\mathbf{v_m} = (r_\mathrm{x}, r_\mathrm{HLac})$, but $(\mathbf{r}_\mathrm{x}, r_\mathrm{HAc})$, $(r_\mathrm{x}, r_\mathrm{e})$ and $(r_\mathrm{HAc}, r_\mathrm{HLac})$ give approximately the same value of $c_2(\mathbf{A})$ as $(r_\mathrm{x}, r_\mathrm{HLac})$. The choice of $\mathbf{v_m} = (r_\mathrm{HAc}, r_\mathrm{e})$ is, however, very poor. Apparently, the two rates are strongly correlated. This is difficult to appreciate in advance, and the calculation shown above can decide which measurements should be included – at least to avoid that the condition number is inordinately large.

For $c_2(\mathbf{A}) = 10^Q$ one should work with data of at least Q digits in order to get meaningful results for the calculated rates $\mathbf{v_c}$. Consequently, two–three-digit accuracy in the measured rates is required when $\mathbf{v_m} = (r_{\text{HAc}}, r_e)$, but only one–two-digit accuracy is needed when the other pairs of $\mathbf{v_m}$ are used.

When extra measurements are included in $\mathbf{v_m}$, the accuracy of the calculated rates will be higher.

As an example, choose $\mathbf{v_m} = (v_2, v_3, v_5, v_6) = (r_{\text{HLac}}, r_x, r_{\text{HAc}}, r_e)$ to calculate $\mathbf{v_c} = (v_1, v_4)$

The matrix $\mathbf{T_1}^{\mathrm{T}} \cdot \mathbf{T_1}$ is symmetric and equal to $\begin{pmatrix} \frac{11}{9} & -1 \\ -1 & \frac{13}{9} \end{pmatrix} = \mathbf{A}$ from Example 4.10. Hence $c_2(\mathbf{A})$ is simply the ratio between the eigenvalues of $\mathbf{A}$, namely, $(12 + \sqrt{82}/9)$ and $(12 - \sqrt{82}/9)$.

$$c_2(\mathbf{A}) = \left(\frac{(12 + \sqrt{82})/9}{(12 - \sqrt{82})/9}\right)^{1/2} = 2.67$$

Consequently, c_2 has been much reduced, from its lowest value 9 for two measurements (r_x, r_{HLac}) to 2.67 with four measurements, and one-digit accuracy of the measurements of $(r_{\text{HLac}}, r_x, r_{\text{HAc}}, r_e)$ is enough to ensure that the last three nonzero rates $(-r_s, r_c, r_{\text{HCOOH}})$ are also calculated with one-digit accuracy.

When the SP flux method is compared with the method based on internal fluxes, the advantage of using the SP flux method becomes apparent. Let $\mathbf{V_m} = (r_x, r_{\text{HAc}})$ as in Example 4.4. Now

$$\begin{pmatrix} r_e \\ r_{\text{HLac}} \end{pmatrix} = \begin{pmatrix} \frac{1}{2} & 0 \\ \frac{1}{2} & -\frac{1}{3} \end{pmatrix}^{-1} = \begin{pmatrix} r_x \\ r_{\text{HLac}} \end{pmatrix} = \mathbf{A}^{-1} \cdot \mathbf{r}_c$$

$$= \begin{pmatrix} 0.2\, r_x + r_{\text{HAc}} \\ 5.7\, r_x - 4.5\, r_{\text{HAc}} \end{pmatrix} \cdot \begin{pmatrix} \frac{1}{2} & 0 \\ \frac{1}{2} & -\frac{1}{3} \end{pmatrix}^{-1}.$$

The eigenvalues of $\mathbf{T_1}\ \mathbf{T}^{\mathrm{T}}$ are 0.5617 and 0.04946 respectively, and $c(\mathbf{A})_2 = 3.37$, a remarkable improvement from $c(\mathbf{A})_2 = 13$ obtained for the same $\mathbf{r_m}$, but by the internal flux method. Also with the SP flux method $\mathbf{r_c} = (r_e, r_{\text{HLac}})$ will be a poor choice since $c_2(\mathbf{A}) = 123$.

Note: The method described above for estimation of the propagation of experimental errors is only one out of many methods – all based on linear algebra. The covariance matrix of Example 7.1 is another frequently used metric.

4.10 Conclusions

The review of biochemical pathways in Sections 4.0–4.5 does not even pretend to be sufficient for in-depth studies of metabolic flux analysis. The reader will need to consult standard textbooks on biochemistry in order to make genome-scale MFA. Even a number of pathways to important industrial metabolic products such as antibiotics (penicillins, polyketides, etc.), lipids, amino acids, and other valuable end products have not been mentioned. The material in Sections 4.0–4.5 is, however, sufficient to appreciate how stoichiometry-based MFA works in practice.

References

1 Nissen, T.L., Anderlund, M., Nielsen, J., Villadsen, J., and Kielland-Brandt, M.C. (2001) Expression of a cytoplasmic transhydrogenase in *Saccharomyces cerevisiae* results in formation of 2-oxogluterate due to depletion of the NADPH pool. *Yeast*, **18**, 19–32.

2 Marx, A., Eikmanns, B.J., Sahm, H., de Graaf, A.A., and Eggeling, L. (1999) Response of the central metabolism in *Corynebacterium glutamicum* to the use of an NADH dependent glutamate dehydrogenase. *Metab. Eng.*, **1**, 35–48.

3 Takeno, S., Murata, R., Kobayashi, R., Mitsuhashi, S., and Ikeda, M. (2010) Engineering of *Coryenebacterium glutamicum* with an NADPH-generating glycolytic pathway for L-lysine production. *Appl. Environ. Microbiol.*, **76**, 7154–7160.

4 van Gulik, W.M. and Heijnen, J.J. (1995) A metabolic network stoichiometry analysis of microbial growth and product formation. *Biotechnol. Bioeng.*, **48**, 681–698.

5 Lengeler, J.W., Drews, G., and Schlegel, H.C. (1999) *Biology of the Prokaryotes*, Thieme, Stuttgart.

6 Förster, J., Famili, I., Fu, P., Palsson, B., and Nielsen, J. (2003) Genome scale reconstruction of the *Saccharomyces cerevisiae* metabolic network. *Genome Res.*, **13**, 244–253.

7 van Heerden, C.D. and Nicol, W. (2013) Continuous and batch cultures of *Escherichia coli* KJ134 for succinic acid fermentation: metabolic flux distribution and production characteristics. *Microb. Cell Fact.*, **12** (80), 1–10.

8 Oura, E. (1983) *Biomass from Carbohydrates. Biotechnology*, vol. **3** (ed. H. Dellweg), Wiley-VCH Verlag GmbH, Weinheim, pp. 3–42.

9 Lee, S.Y., Park, J.H., Jang, S.H., Nielsen, L.K., Kim, J., and Jung, K.S. (2008) Fermentative butanol production by *Clostridia*. *Biotechnol. Bioeng.*, **101**, 209–224.

10 Villadsen, J. and Jørgensen, S.B. (2014) Reflections on the aerobic fermentation stoichiometry of Crabtree positive yeasts. *Biotechnol. Bioeng.*, **111**, 632–637.

11 Villadsen, J., Nielsen, J., and Lidén, G. (2011) *Bioreaction Engineering Principles*, 3rd edn, Springer.

12 Voet, D. and Voet, J.G. (1995) *Biochemistry*, 2nd edn, John Wiley & Sons, Inc., New York.

13 Berg, J.M., Tymoczko, J.L., and Stryer, L. (2007) *Biochemistry*, WH Freman & Company, New York.

14 Duboc, P., von Stockar, U., and Villadsen, J. (1998) Simple generic model for dynamic experiments with *Saccharomyces cerevisiae* in continuous culture. *Biotechnol. Bioeng.*, **60**, 180–189.

15 Yim, H. and Haselbeck, R. *et al.* (2011) Metabolic engineering of *E. coli* for direct production of 1,4-butane diol. *Nat. Chem. Biol.*, **7**, 445–452.

5
A Primer to ^{13}C Metabolic Flux Analysis

Wolfgang Wiechert, Sebastian Niedenführ, and Katharina Nöh

Summary

The experimental determination of metabolic reaction rates (fluxes) in a physiological steady state of a cell is called metabolic flux analysis (MFA) or, sometimes, "quantitative fluxomics". In the concert of the "omics" techniques, fluxomics plays a distinguished role because it characterizes the quantitative phenotype that emerges from regulation on all levels of cellular organization. Whereas the genome, proteome, or metabolome give the cells' inventory and reveal what a cell population can potentially achieve, a metabolic flux map describes the actual cellular function.

Instead of direct observation, MFA is based on a computational procedure in which intracellular fluxes in a given network model are calculated from measured uptake and production rates of extracellular substrates and products. This chapter provides an introduction to the principles of the currently most advanced MFA technique based on isotopic tracers (^{13}C MFA). It concentrates on modeling and data analysis by using a simple but highly instructive example network. With this example at hand, all steps of a typical ^{13}C-MFA evaluation workflow are discussed. A nontechnical overview of good practices as a guideline for choosing an appropriate software system finalizes the chapter.

5.1
Introduction

The experimental determination of metabolic fluxes in a physiological steady state of a cell is called metabolic flux analysis (MFA) or, sometimes, "(quantitative) fluxomics". In the context of other omics techniques, fluxomics plays a distinguished role because it characterizes the quantitative phenotype that emerges from regulation on all levels of cellular organization. Whereas the genome, proteome, or metabolome of a cell describe what a cell can do (potential), a

Fundamental Bioengineering, First Edition. Edited by John Villadsen.

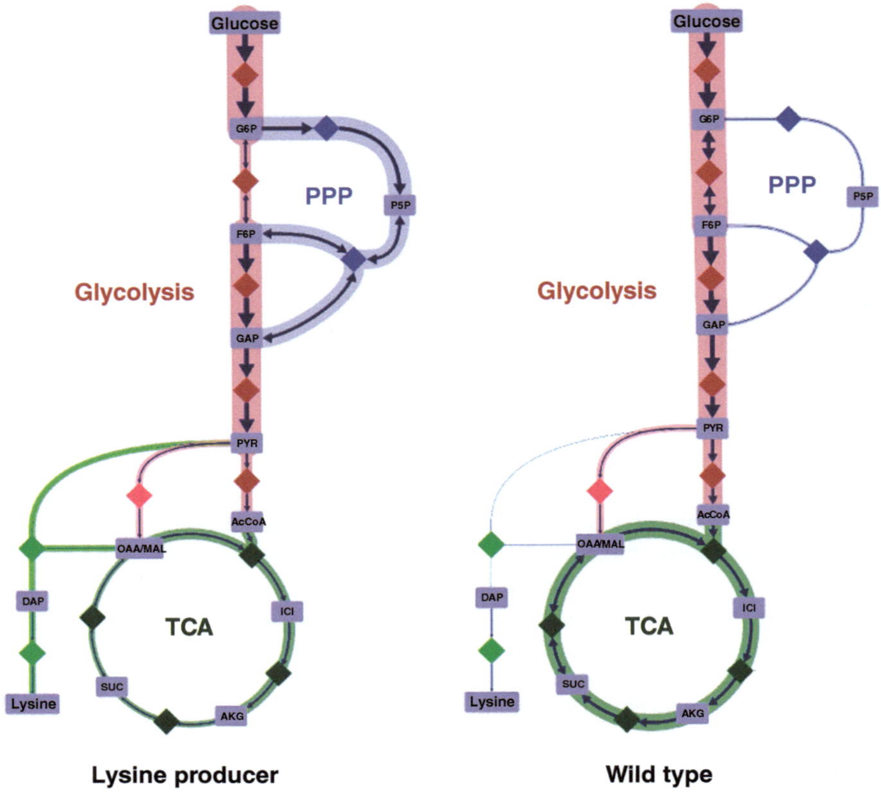

Figure 5.1 Schematic metabolic flux maps of *Corynebacterium glutamicum*: lysine producer compared with a wild type. Blue rounded rectangles and diamonds denote intracellular metabolites and enzymatic reactions, respectively. The widths of color-coded pathways correspond to the flux values (TCA, tricarboxylic acid cycle; PPP, pentose phosphate pathway). For details see Ref. [1].

metabolic flux map (cf. Figure 5.1) reveals what it really does (function). Here, the term "flux" (as used in the MFA community) can denote either reaction rates of enzymatic transformations or rates of transport steps over the cell wall or compartment boundary. The physical unit of a metabolic flux can be given on a molar (mol h^{-1}) or, most common, on a biomass specific (mol h^{-1} g_{DW}^{-1}) basis. This chapter gives an introduction to the principles of the currently most advanced MFA technique based on isotopic tracers (^{13}C MFA). It focuses on modeling and data analysis by using a simple but highly instructive example network.

Another important difference between fluxomics and other omics methods is already demonstrated by the examples in Chapter 4: MFA is not based on a direct measurement of intracellular quantities but is an indirect computational procedure in which intracellular fluxes in a given network model are calculated from measured uptake and production rates of extracellular substrates and products. Historically, the stoichiometric MFA method described in Chapter 4 was developed first [2]. However, it quickly turned out that the pure stoichiometric approach cannot resolve parallel internal pathways or certain metabolic cycles [3]. Most importantly, the balancing of energy metabolites such as ATP and NADH must be seen very critical because their intracellular sources and sinks, possible interconversions, or even futile cycles are far from being well understood [4]. Unfortunately, omitting the energy metabolite balances immediately leads to underdetermined equation systems. Thus, the stoichiometry alone becomes worthless.

5.2 Input and Output Data of ^{13}C MFA

This chapter focuses on the ^{13}C MFA method that extends classical stoichiometric MFA by using intracellular isotope labeling data as an additional source of measured information besides extracellular fluxes (cf. Figure 5.2). With ^{13}C MFA the shortcomings of classical flux-balancing are removed and the stoichiometric concepts described in Chapter 4 are greatly expanded. The large amount of extra isotopic data that is obtained makes it possible to find fluxes that could not be obtained based on extracellular fluxes only.

Experimental protocols are a very important issue in ^{13}C MFA [5]. Instead of being exhaustive, only the very essentials will be given here, because all aspects of cell cultivation, sampling, metabolite extraction, mass spectrometry, and so on

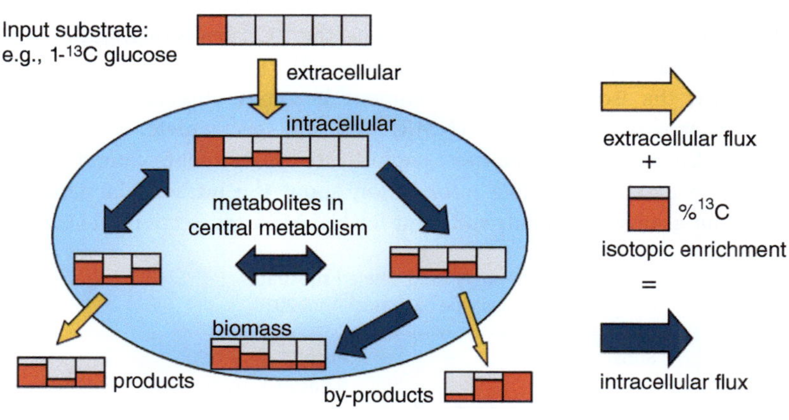

Figure 5.2 Input and output data of ^{13}C MFA: the isotopic composition of the input substrate is known. Further input data is given by measurements of some of the extracellular fluxes and intracellular isotopic enrichments. From this information, the intracellular fluxes are computed (output).

are comprehensively covered by comprehensive reviews [6–11]. Three types of input data play a role in ^{13}C MFA [3]:

1) *Extracellular fluxes* (including biomass accumulation) connect the cell with its environment. Substrate uptake, growth rate, product formation, CO_2 formation, or oxygen uptake all belong to this category. They are calculated from compound balances around the cultivation device as discussed in Chapter 2. This is simple for classical bioreactor setups [12] but can be a nontrivial task for special cultivation regimes, for more sophisticated cultivation devices, and for miniaturized equipment [13,14]. The fluxes to the biomass components cover cellular growth requirements. Because the intracellular rates depend on these fluxes, knowledge of the biomass composition (proteinogenic amino acids, lipids, nucleotides, cell wall, storage pools, and other carbohydrates) of the organism is a prerequisite for flux determination.
2) *Isotope labeling data* is generated by performing an isotope labeling experiment (ILE). Herein, the cells are fed with a specifically isotope-labeled substrate or substrate mixture. Most common are the stable ^{13}C isotopes. Specifically labeled substrates are available, for example, glucose being ^{13}C-labeled at the first carbon atom position ([1-^{13}C] glucose). Differently labeled molecules can be used in mixture or in multiple parallel labeling experiments in their pure form [15,16]. Table 5.1 gives some examples. The prices of isotope-labeled substrates depend strongly on the tracer molecule as well as its labeling pattern and its purity.

   A necessary requirement is that the ILE is performed by keeping the cells in a metabolic steady state over the whole duration of the experiment. This is best realized in steady-state continuous culture or in the exponential growth phase of a batch culture. Here, it is assumed that enzymatic reactions do not distinguish between unlabeled and labeled molecules. Summarizing, the cellular physiology must not be disturbed by the ILE. By the joint activity of all metabolic reactions, the ^{13}C label is distributed over all intracellular metabolite pools (cf. Figure 5.2).

   Finally, the isotopic label enrichment becomes observable by using nuclear magnetic resonance (NMR) or mass spectrometer (MS) techniques. Currently, even the most sophisticated instruments cannot determine the maximal possible information about the intracellular labeling state, but a subset is already sufficient for ^{13}C MFA (cf. Section 5.11).

Table 5.1 Commonly used labeled substrates taken from shop.isotope.com as of April 2014.

Substrate	Price [$ g^{-1}] 99 atom% purity
[1-^{13}C]-Glucose	212
[U-^{13}C]-Glucose	190
[1,2-^{13}C]-Glucose	750

3) *Intracellular pool sizes* – meaning the concentration of metabolites in cells – are not needed for the standard ^{13}C MFA method. However, they play an important role in the advanced isotopically nonstationary INST MFA approach that will be briefly explained in Section 5.21.

The output data of ^{13}C MFA is given by a metabolic flux map visualizing the estimated flux rates in the context of the metabolic network (see Figure 5.1). Mathematically, the vector composed of all single metabolic fluxes in a flux map is called "flux distribution". The complete procedure of ^{13}C MFA is technically and mathematically complex. It involves cell cultivation technology, biomass sampling and quenching, validated quantitative bioanalytics, mathematical modeling, and multivariate statistics. The objective of this chapter is to explain the very essentials of ^{13}C MFA modeling and data evaluation as a primer for reading more specialized papers.

5.3
A Brief History of ^{13}C MFA

Today, ^{13}C MFA is established as an indispensable analytical tool for systems biology, metabolic engineering, and synthetic biology. Lots of applications to different organisms and cultivation regimes are published and the reader is referred to some recent review papers for a broader survey [17–20]. However, it was a long way from the first applications of isotopic labeling in biochemistry to state-of-the-art ^{13}C MFA. The most important milestones are given as a reference to some important papers:

- The isotope labeling technique has its ancestors in the pioneering work of early biochemists who unraveled the main metabolic pathways by tracer experiments [21–24].
- Quantitative ^{13}C MFA has some early predecessors dealing with simplified systems that do not need extensive mathematics [25,26].
- Already in the 1980s Blum and coworker developed a fully model-based ^{14}C-labeling method being extremely laborious at that time but including all ingredients of a modern strategy [27].
- Sherry, Malloy, and coworkers were the first to extensively apply *in vivo* NMR to perfused organs doing isotope labeling studies. Particularly, they were the first who introduced isotopic balancing in order to describe complex NMR measurements [28].
- Stoichiometric MFA as a predecessor of ^{13}C MFA was rigorously applied for the first time by Vallino and Stephanopoulos [2]. However, their approach relied on some critical assumptions on energy metabolite balancing, an ingredient that is not needed for ^{13}C MFA.
- For bacteria the idea to extract sufficient amounts of labeled material from proteinogenic amino acids made quantitative measurement of isotopic enrichment in low-sensitivity NMR instruments possible [1,29,30].

- The first full central metabolism application of ^{13}C MFA based on NMR data and proteinogenic amino acids complemented by comprehensive mathematical modeling and statistical analysis was accomplished by Marx *et al.* [1].
- In the late 1990s, the information gain was substantially increased by exploiting isotopomer data and advanced nuclear magnetic resonance and mass spectrometry techniques [31–34]. Additionally, experimental design methods for planning informative experiments are available since then [35].
- At the advent of the new millennium, MS-based techniques outperformed NMR and MS fragmentation techniques by a significantly increased selectivity and sensitivity. This made labeling data from intracellular metabolites available and raised considerably the information gain of ^{13}C MFA [32,36–40].
- The development of a mathematical theory for ILEs led to the implementation of highly efficient computational algorithms that, nowadays, can simulate very comprehensive and complex networks within short time [41–46].
- In 2007, the first isotopically nonstationary MFA experiment (INST MFA) was designed, performed, and evaluated followed by several others [47,48].
- To date, several software tools for simulating and evaluating carbon labeling experiments are available (cf. Section 5.21) [49–58].

5.4 An Illustrative Toy Example

From now on, a toy example (cf. Figure 5.3) taken from [59] is used to explain how ILEs are modeled and evaluated. Although being strongly reduced, the

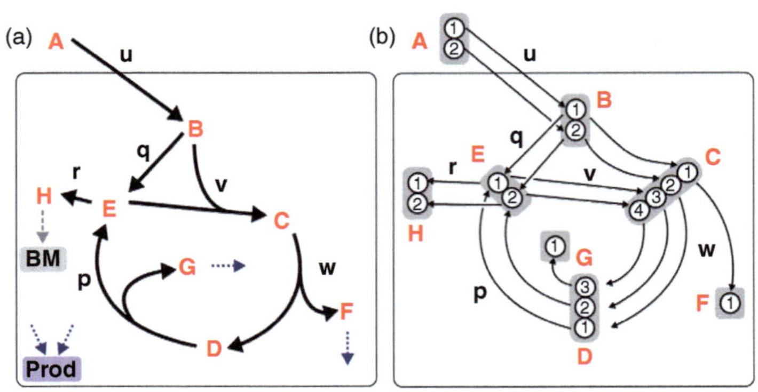

Figure 5.3 A simple network example to illustrate modeling concepts of ^{13}C MFA. (a) The metabolic network with reaction steps (black) and metabolites (red) and (b) carbon atom transition network. The numbers indicate carbon positions of the metabolites. The outer rounded boxes indicate the cell boundary. The extracellular input pool A presents an unlimited substrate reservoir, whereas the compounds F and G are products (e.g., CO_2, blue dotted arrows). The metabolite H is a biomass building block contributing to cellular growth (dotted gray arrow).

example mimics glycolysis, citric acid cycle, and anaplerosis (e.g., replenishing TCA cycle intermediates – see Figure 4.1). However, this analogy should not be taken too seriously. The network shown in Figure 5.3a consists of eight metabolite pools (A, B, C, D, E, F, G, and H) linked by six reactions (p, q, r, u, v, and w). Whereas A denotes the extracellular substrate pool, the pools F, G, and H represent products and cellular biomass components.

As a notational convention, for a reaction step u, *u* (italic) denotes its associated flux value. Likewise, the intracellular concentration of the metabolite pool B is denoted by *B*. In contrast to classical stoichiometric MFA (cf. Chapter 4), which deals with net fluxes only, it is highly important to realize that ^{13}C MFA always assumes unidirectional fluxes proceeding exclusively in one direction. If a reaction takes place in both directions at the same time – as it is the case for many reactions operating close to thermodynamic equilibrium [60] – then two different flux symbols for the forward and backward flux are needed (as in Figure 4.8c). For the moment, all six reactions in the example network are assumed unidirectional, whereas the more general case of bidirectionality will be treated in Section 5.15. For better readability fluxes are scaled relative to the substrate uptake and physical units are omitted in the following.

As a necessary precondition of all MFA methods, the system is considered to reside in a metabolic steady state. Thus, the following stoichiometric balance equations (see similar examples in the discussion of Section 4.8 of the Net Flux Method versus the Internal Flux Method) must hold:

$$\begin{array}{llll} \mathrm{B}: & u - v - q = 0 & \mathrm{D}: & w - p \qquad\qquad = 0. \\ \mathrm{C}: & v - w \qquad = 0 & \mathrm{E}: & q + p - (r + v) = 0. \end{array} \tag{5.1}$$

These are four linear equations balancing the inner metabolites B, C, D, and E for six flux contributions u, v, w, p, q, and r. Because none of the equations in 5.1 can be expressed by the others, the equation system has full row rank (cf. Chapter 4). Therefore, the stoichiometric system has two (=6-4) degrees of freedom. In other words, the complete flux distribution $(p, q, r, u, v, w)^{\mathrm{T}}$ is determined if two "free fluxes" are fixed to a value. To this end, we choose the substrate uptake flux u as a directly measurable extracellular flux and the intracellular "anaplerosis" flux q. Having defined the metabolic network by Eq. (5.1), the dependent fluxes p, r, v, w are determined by

$$p = v = w = u - q\,, \qquad r = q\,. \tag{5.2}$$

Our next task is to determine the free fluxes u, q from measured data. While typically the substrate uptake u is directly measurable, the intracellular fluxes q, v and the biomass effluxes w, q, r are not observable. This means that not enough flux measurements are available to determine both free fluxes. Clearly, the central problem lies in the flux q. For the moment, positive net fluxes (nominal directions in Figure 5.3) at least constrain the flux value of q:

$$0 \le u, 0 \le w, 0 \le p, 0 \le r \Rightarrow 0 \le q \le u. \tag{5.3}$$

5.5
The Atom Transition Network

To determine q, an alternative source of information is needed that is provided by isotopic data. As a consequence for modeling, another set of variables and equations must be introduced to describe the system's labeling state. These are the labeling fractions of metabolite pools and the corresponding balances that are explained next.

As a prerequisite, more detailed information about the metabolic network is needed which is given by the atom transition network (cf. Figure 5.3b). Assume that in the example the carbon backbones of the metabolites have one (G and F), two (A, B, E, and H), three (D), and four (C) carbon atoms. The atom transition network then defines the fate of each carbon atom in every reaction step. For example, in reaction step w the third atom of C is mapped to the second atom of metabolite D. In practice, this knowledge can be obtained from biochemistry textbooks, some tools, and databases (e.g., ReMatch [61] and the BioCyc database collection (biocyc.org)) or published network models like [62] (see Section 5.22 for more details).

A formal notation for carbon atom transitions has been introduced in [42]. It is frequently used to specify network stoichiometry together with the carbon atom transitions. Here, atom letters per reaction like i, j, k, l are introduced that indicate the unique location and position change of every carbon atom between compounds. The reader may verify the correctness of the following formal network specification from Figure 5.3b:

$$\begin{array}{llllllll} \mathrm{u}: & \mathrm{A\#ij} & \rightarrow & \mathrm{B\#ij} & \mathrm{w}: & \mathrm{C\#ijkl} & \rightarrow & \mathrm{F\#i + D\#jkl}. \\ \mathrm{q}: & \mathrm{B\#ij} & \rightarrow & \mathrm{E\#ij} & \mathrm{p}: & \mathrm{D\#ijk} & \rightarrow & \mathrm{E\#ij + G\#k}. \\ \mathrm{v}: & \mathrm{B\#ij + E\#kl} & \rightarrow & \mathrm{C\#ijkl} & \mathrm{r}: & \mathrm{E\#ij} & \rightarrow & \mathrm{H\#ij}. \end{array} \tag{5.4}$$

5.6
Isotopomers and Isotopomer Fractions

Every metabolite in the system can occur in different labeling states that are determined by those carbon atom positions labeled with a ^{13}C. For example, metabolite B occurs in four different forms (cf. Figure 5.4) that are conveniently enumerated in a binary notation ("0" and "1" means ^{12}C and ^{13}C, respectively):

$$\mathrm{B\#00}, \quad \mathrm{B\#01}, \quad \mathrm{B\#10}, \quad \mathrm{B\#11} \tag{5.5}$$

Generally, a metabolite with n carbon atoms occurs in 2^n different labeling states called isotopomers. The percentage of molecules in a certain labeling state relative to the total pool size is called isotopomer fraction (cf. Figure 5.4). The isotopomer fractions are written by using lower case letters for the metabolites and binary indices for the labeling patterns:

$$b_{00}, b_{10}, b_{01}, b_{11} \quad \text{with} \quad 0 \leq b_{ij} \leq 1, \quad \sum b_{ij} = 1. \tag{5.6}$$

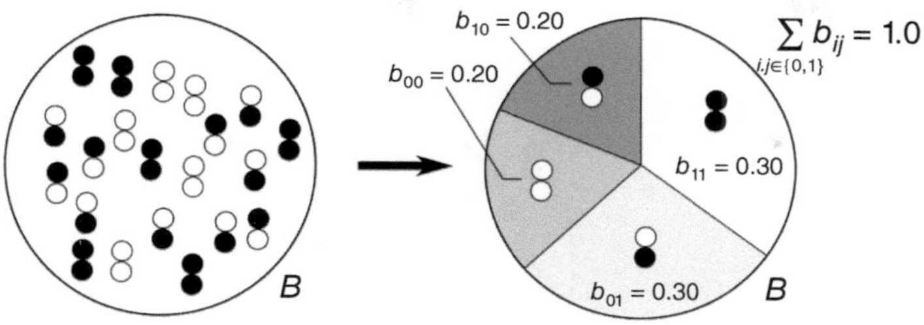

Figure 5.4 The metabolite pool B with two carbon atoms consists of four different isotopomers. The pie chart on the right represents the fractions of the isotopomer species depicted on the left.

Practically, in an ILE the input substrate is a mixture of molecules with specified labeling that is predetermined by the experimental setting. Thus, the labeling fractions of the input substrate are known and constant. For example, if we solely use a substrate A labeled at the second carbon atom position, this is described by the mixture data:

$$a_{00} = 0\%, \quad a_{01} = 100\%, \quad a_{10} = 0\%, \quad a_{11} = 0\%. \tag{5.7}$$

If additionally A labeled at the first position is available and both substrates are present in equal amounts, then

$$a_{00} = 0\%, \quad a_{01} = 50\%, \quad a_{10} = 50\%, \quad a_{11} = 0\%. \tag{5.8}$$

To describe the complete labeling state of the system as a function of the input labeling and the prevailing fluxes, the isotopomer fractions of the input substrate and all intermediates B, C, D, E need to be known resulting in a total of

$$2^2 + 2^4 + 2^3 + 2^2 = 32 \tag{5.9}$$

different intracellular fractions. How these labeled forms are computed from the free fluxes and the input labeling fractions is explained in the next sections.

5.7
The Isotopomer Transition Network

Based on the atom transitions, the isotopomer transition network is constructed that explains the fate of the isotopomers in each reaction step. For instance, in the toy example the reaction step w induces not less than 16 different isotopomer transitions:

$$\begin{array}{lll} \mathrm{w}: & \mathrm{C}\#0000 & \rightarrow & \mathrm{F}\#0 + \mathrm{D}\#000 \\ \mathrm{w}: & \mathrm{C}\#0001 & \rightarrow & \mathrm{F}\#0 + \mathrm{D}\#001 \\ \mathrm{w}: & \mathrm{C}\#0010 & \rightarrow & \mathrm{F}\#0 + \mathrm{D}\#010 \\ & & \vdots & \\ \mathrm{w}: & \mathrm{C}\#1111 & \rightarrow & \mathrm{F}\#1 + \mathrm{D}\#111 \end{array} \tag{5.10}$$

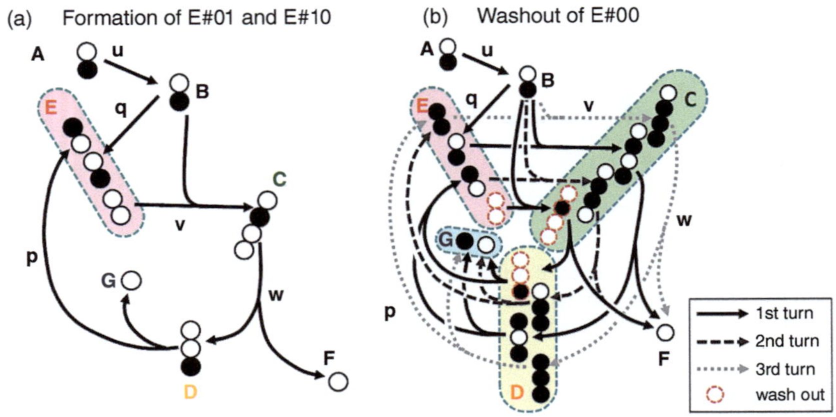

Figure 5.5 Stepwise formation of different intracellular labeling patterns in the toy example. (a) Initially, only the unlabeled intracellular E#00 can react with the labeling introduced by A#01 in reaction step v. Subsequently, C#0100, D#100, and E#10 are formed. At the same time, E#01 occurs via reaction step q. (b) Emergence of further isotopomers after some cycle turns (indicated by different line styles). Isotopomers marked in red are not replenished and, hence, are washed out over time. In the isotopic stationary state, only the isotopomers shown in black reside in the system.

The resulting transition network is already too complex for a complete graphical representation. For this reason, Figure 5.5 shows only a subnetwork that qualitatively explains the fate of the C-2-labeled input isotopomer A#01 when, initially, only unlabeled species B#00, C#0000, D#000, and E#00 are present (time $t = 0$).

The isotopomer transition network enables the qualitative simulation of an ILE: based on the input isotopomer species A#01, obviously the new isotopomer species B#01 shows up first (by reaction u). Then, fueled by E#00 in reaction v, C#0100 occurs. Later on D#100 and E#10 are formed via w and p, respectively. At the same time, E#01 is formed from B#01 via q. Clearly, this isotopomer tracing approach cannot explain the quantitative dynamics of the system because the isotopomer fractions depend on the concrete values of the fluxes and metabolite pool sizes (cf. Section 5.8). Nevertheless, by manually executing all reaction steps in a stepwise manner, we deduce that by the time all the species E#01, E#10, and E#11 emerge. On the other hand, E#00 – and likewise the unlabeled species of B, C, and D – are not formed anymore. Therefore, these only initially present species are washed out after some time. Finally, in the isotopic steady state, the isotopomers B#01, C#0111, C#0101, C#0110, D#100, D#101, D#111, E#11, E#01, and E#10 remain in the network. Keep in mind that by such manual atom tracing only the (non-)presence of labeled species can be predicted. The quantitative proportions in which the single isotopomers occur cannot be determined in this way.

5.8 Isotopomer Labeling Balances

The qualitative viewpoint is now changed to a quantitative one by introducing isotopomer labeling balances describing the dynamics of all isotopomer fractions in the system. This dynamics is described by one differential equation per intermediate isotopomer pool in the network. The general form of the mass balance equation for the time-dependent isotopomer fraction x is as follows:

$$\textit{pool size} \cdot \dot{x} = \textit{label influx} - \textit{label efflux}. \tag{5.11}$$

Here, "pool size" denotes the intracellular concentration of a metabolite irrespective of the labeling state of its single molecules (cf. Figure 5.4). Giving fluxes with the physical dimension mol h^{-1} g_{DW}^{-1}, the pool sizes must have the unit mol g_{DW}^{-1} because the fractional labeling enrichments x are dimensionless. In the following, units are left out to increase the readability of the text. In Eq. (5.11), the in- and out-fluxes on the right-hand side describe the introduction and removal of labeled material to and out of the balanced pool. Therewith, the equations strongly depend on the network topology. When formulating the balance equations, three types of reaction steps must be distinguished (cf. Figure 5.6): uni–uni – u, q, r, bi–uni – v, and uni–bi – w, p, where "uni" and "bi" denote the number of metabolites on the educt and product side of the reaction. Reactions with two or more substrates and products can be formally reduced to these three cases [44].

The different terms in the balance equations are now explained with the toy example at hand. We start with the four balance equations for pool B that is connected with the metabolic network by three reaction steps u, v, and q:

$$\mathrm{B\#ij}: B \cdot \dot{b}_{ij} = u \cdot a_{ij} - (v + q) \cdot b_{ij}, \quad i,j = 0,1. \tag{5.12}$$

The right-hand side is constructed as follows:

- The term "$u \cdot a_{ij}$" represents the influx into pool B from pool A. Here, the reaction rate u gives the (molar) amount of B formed from A per time irrespective of the molecular labeling state. Because a_{ij} is the fraction of A-isotopomers carried over to B#ij, the product $u \cdot a_{ij}$ gives the total amount of the species B#ij formed per time unit.
- Reactions v and q represent the effluxes of pool B. As before they correspond to two effluxes $-v \cdot b_{ij}$ and $-q \cdot b_{ij}$ removing certain amounts of B#ij from B. Generally, in all three cases effluxes lead to the same sort of negative product term, that is, $-\textit{efflux} \cdot \textit{fraction}$.

Next, in total 16 balances for C are considered:

$$\mathrm{C\#ijkl}: C \cdot \dot{c}_{ijkl} = v \cdot b_{ij} \cdot e_{kl} - w \cdot c_{ijkl}, \quad i,j,k,l = 0,1. \tag{5.13}$$

- A novel element in this balance is the influx term "$+v \cdot b_{ij} \cdot e_{kl}$." The product $b_{ij} \cdot e_{kl}$ of two labeling fractions occurs, because C is produced from B and E via the bimolecular reaction step v. The product $b_{ij} \cdot e_{kl}$ gives the probability to randomly draw the isotopomer pair (B#ij, E#kl) in the v reaction step.

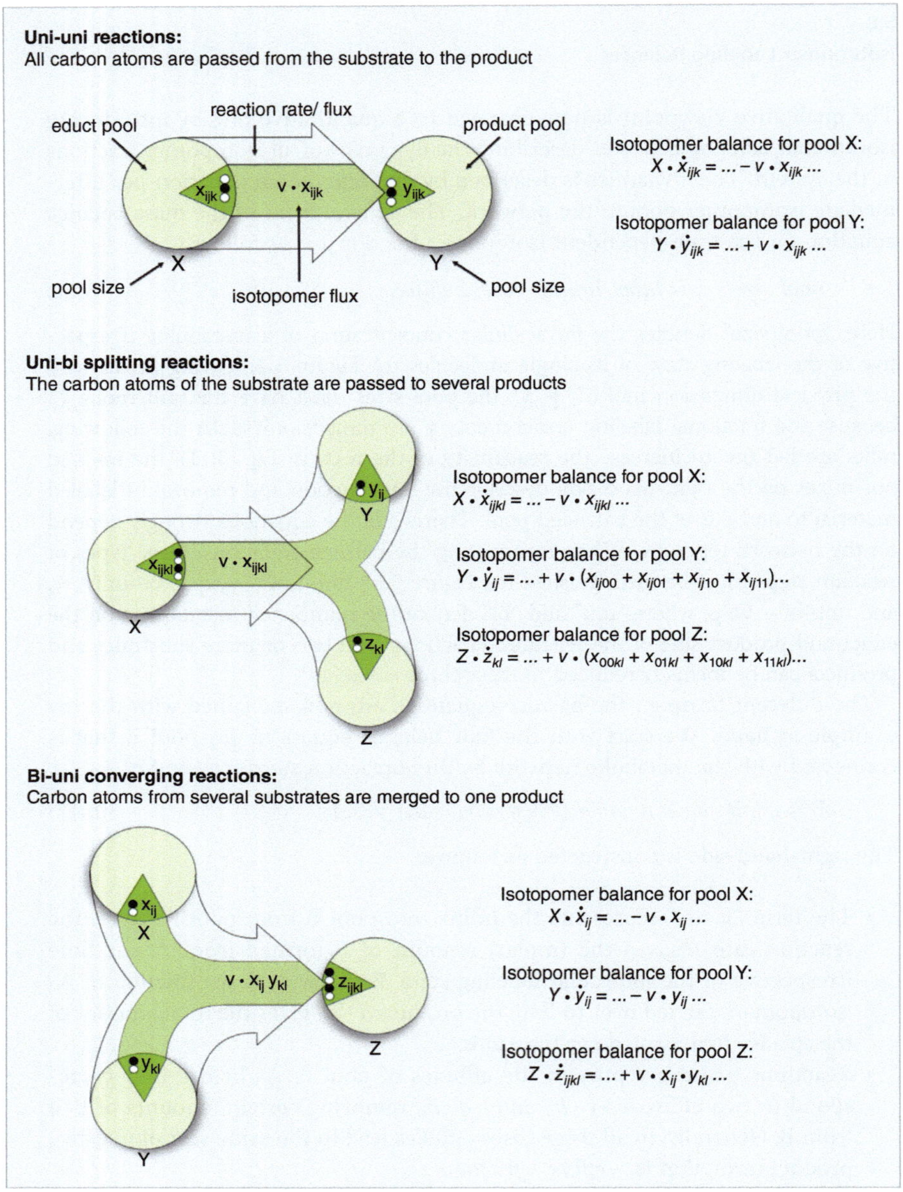

Figure 5.6 The three cases of isotopomer balance equations: uni–uni, uni–bi, and bi–uni. In each case, the shuffling of the intramolecular carbon atoms leads to a permutation in the indices $i,j,k,l = 0,1$.

Another situation occurs in the eight isotopomer balances of pool D fed by w and drained by p:

$$\mathrm{D}\#\mathrm{ijk}: \; D \cdot \dot{d}_{ijk} = w \cdot (c_{0ijk} + c_{1ijk}) - p \cdot d_{ijk}, \quad i,j,k = 0,1. \tag{5.14}$$

- The new term "$+w \cdot (c_{0ijk} + c_{1ijk})$" expresses the fact that in a uni–bi reaction the target isotopomers have more than one source. Here, the isotopomers C#0ijk and C#1ijk feed the same pool D#ijk.

The reader should now be able to verify the remaining four balances of metabolite pool E:

$$\mathrm{E}\#\mathrm{ij}: \; E \cdot \dot{e}_{ij} = p \cdot (d_{ij0} + d_{ij1}) + q \cdot b_{ij} - (r + v) \cdot e_{ij}, \quad i,j = 0,1. \tag{5.15}$$

Collecting all isotopomer balance equations for the toy system, we end up with 32 differential equations corresponding to the same number of fractions (cf. Eq. (5.9)). No balance equations have to be formulated for the external input labeling fractions a_{ij}. Notice that Eqs. (5.12)–(5.15) define a nonlinear differential equation system because product terms of state variables occur in Eq. (5.13). The general rules for formulating isotopomer balance equations are summarized in Figure 5.6 on a reaction step basis.

5.9 Simulating an Isotope Labeling Experiment

Once the isotopomer balance equations are derived and values for the free fluxes u and q as well as the pool sizes of all balanced metabolites are given, the dynamics of an ILE can be simulated (cf. Figure 5.7). Keep in mind that over the time span of the ILE all intracellular metabolite pool sizes are constant because

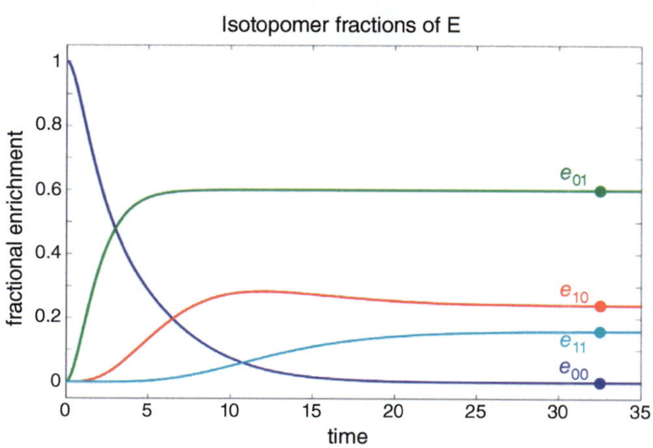

Figure 5.7 Simulation results for the isotopomer fractional enrichments of E over time with the toy example (see Eqs. (5.12)–(5.20) for parameter and initial values). The circles indicate the steady-state values.

the system is in a metabolic steady state. This means that pool sizes, input labeling fractions and fluxes are constant, whereas the labeling fractions are time dependent. Consequently, the following data is needed to completely specify a system:

1) The free fluxes must be fixed, for example, to the constant values:

$$u = 1, \quad q = 0.6 \tag{5.16}$$

From Eq. (5.2) all other fluxes are then calculated:

$$p = v = w = 0.4, \qquad r = 0.6 \tag{5.17}$$

2) The input labeling must be specified. Here we chose

$$a_{00} = 0\%, \quad a_{01} = 100\%, \quad a_{10} = 0\%, \quad a_{11} = 0\%. \tag{5.18}$$

3) Clearly, a reasonable initial condition for simulating the ILE is to assume an unlabeled initial state:

$$\begin{array}{llll} b_{00}(0) = 100\%, & c_{0000}(0) = 100\%, & d_{000}(0) = 100\%, & e_{00}(0) = 100\%, \\ b_{ij}(0) = 0\%, & c_{ijkl}(0) = 0\%, & d_{ijk}(0) = 0\%, & e_{ij}(0) = 0\% \text{ for other } i,j = 0,1. \end{array} \tag{5.19}$$

4) Finally, we have to specify values for the total metabolite pool sizes *B*, *C*, *D*, *E*. Pool sizes describe storage capacities for labeled material. In other words, with the same influx, large pools will need longer time to become labeled than small pools. Thus, together with the fluxes the pool sizes determine the overall time constants of the system. For simplicity (and good reasons given below) we assume:

$$B = C = D = E = 1.0. \tag{5.20}$$

Based on this information a differential equation solver (e.g., the MATLAB ode45 solver) can compute the labeling dynamics from the 32 differential equations Eqs. (5.12)–(5.15).

The simulation results precisely coincide with the qualitative findings from Section 5.5. Only pool E should be discussed here. Three labeling states of metabolite E are formed. Comparing the velocity of formation of the isotopomers E#01 emerges first, followed by E#10, and, finally, E#11 is produced. The fourth possible isotopomer E#00 is washed out during the experiment (cf. Figure 5.6). What cannot be deduced with the qualitative analysis is that E#10 shows a slight overshoot, because intermittently it is replenished via two routes simultaneously: E#00 → C#0100 → G#100 and E#01 → C#0101 → G#101.

5.10
Isotopic Steady State

From the simulation results shown in Figure 5.7 it becomes obvious that the labeling fractions approach a steady state after some time. As generally proven in [44], any isotopic labeling system converges to a unique equilibrated state that depends on the network topology, the metabolic flux values, and the input labeling but not on the initial labeling state, the choice of free fluxes, or the metabolite pool sizes. This metabolic and isotopic steady state is the basis for all classical ^{13}C MFA scenarios. The isotopic steady state is described by a set of algebraic equations that are immediately obtained from the differential equation system Eqs. (5.12)–(5.15) by replacing the time derivatives with zeros:

$$\begin{aligned}
\mathrm{B}: &\quad 0 = u \cdot a_{ij} & &-(v+q)\cdot b_{ij} \\
\mathrm{C}: &\quad 0 = v \cdot b_{ij} \cdot e_{kl} & &-w \cdot c_{ijkl} \\
\mathrm{D}: &\quad 0 = w \cdot (c_{0ijk} + c_{1ijk}) & &-p \cdot d_{ijk} \\
\mathrm{E}: &\quad 0 = p \cdot (d_{ij0} + d_{ij1}) + q \cdot b_{ij} & &-(r+v)\cdot e_{ij}, \quad i,j,k,l = 0,1.
\end{aligned} \tag{5.21}$$

This means that the metabolic pool sizes are eliminated from the isotopic steady-state equation system. Therefore, metabolite concentration data is not needed for isotopically stationary ^{13}C MFA. Note that the steady-state equation system has exactly as many equations as there are unknown intracellular labeling fractions. From now on, the isotopomer fraction symbols $b_{ij}, c_{ijkl}, d_{ijk}, e_{ij}$ mean the steady-state constant values, that is, these quantities (state variables) are no more time dependent.

The computed final labeling state is only a function of the free metabolic flux values and the input substrate labeling fractions, whereas the influence of the initial labeling state vanishes:

$$\begin{aligned}
&b_{ij} = b_{ij}(u,q;a_{00},a_{01},a_{10},a_{11}), && d_{ijk} = d_{ijk}(u,q;a_{00},a_{01},a_{10},a_{11}), \\
&c_{ijkl} = c_{ijkl}(u,q;a_{00},a_{01},a_{10},a_{11}), && e_{ij} = e_{ij}(u,q;a_{00},a_{01},a_{10},a_{11}), \quad i,j,k,l = 0,1.
\end{aligned} \tag{5.22}$$

These functions can, at least in principle, be computed by performing a dynamic simulation based on the original differential equation system Eqs. (5.12)–(5.15) and waiting until the system becomes stationary. Such an approach is, however, much too slow for today's applications. See Section 5.21 for details on state-of-the-art solution algorithms.

It is now investigated whether the isotopic steady state contains the desired information on the metabolic flux q. Using again 100% A#01 and $u = 1$, the second free flux value q is varied over the whole possible value range $0 \leq q \leq 1$ (cf. Eq. (5.2)). The resulting steady-state isotopomer fractions of pool E are shown in Figure 5.8. These fractions, indeed, strongly depend on the free flux value q. The reader might verify by qualitative network analysis that for $q = 0$ and $q = 1$ implying $v = 1$ and $v = 0$, respectively, only a single isotopomer can occur (i.e., either E#11 or E#01) and that E#00 does not reside in the system independent of q. Also the steady state for $q = 0.6$ (cf. Figure 5.7) can be verified from Figure 5.8.

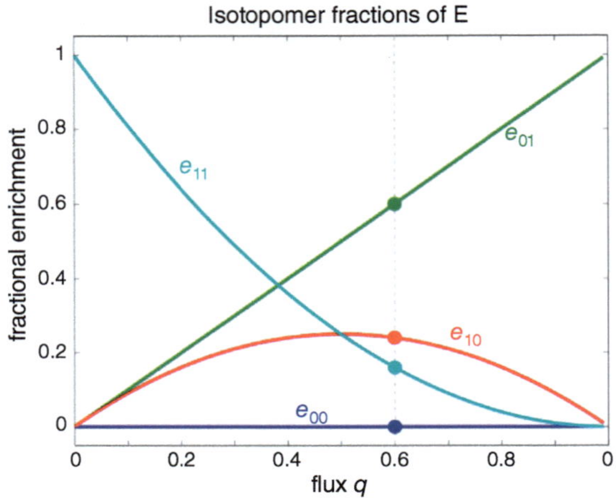

Figure 5.8 Steady-state labeling enrichments for the toy example with 100% A#01 input labeling, $u = 1.0$ and q ranging from 0 to 1. The value of $q = 0.6$ used for previous simulations shown in Figure 5.7 is marked by a circle.

5.11 Flux Identifiability

So far, we are able to predict the stationary labeling state of the system from a given intracellular flux distribution. This comparatively easy flux-to-labeling *forward* calculation exercise does not provide us with the flux map that is the final end point of a ^{13}C MFA study. Instead of the flux-to-labeling direction, the reverse labeling-to-flux route needs to be taken. Mathematically, this poses the so-called *inverse problem*. Unfortunately, compared to the forward simulation for the inverse direction, no similarly efficient solution is available.

Concerning the example system, Figure 5.8 answers the question whether the steady-state labeling fractions really contain the information required to identify all metabolic fluxes. In fact, the labeling fractions of E#01 and E#11 show a one-to-one correspondence between flux and fraction. Thus, if one of these two can be measured (cf. Section 5.10), then the reverse ^{13}C MFA problem is easily solved.

Notably, this does not hold for the other two fractions of E, that is, E#00 and E#10. On the one hand, E#00 is not influenced at all by q, a phenomenon we call *flux nonidentifiability*. E#10, on the other hand, is a very interesting case because the labeling-to-flux mapping is not one-to-one. In fact, each fraction in the range $0 \leq e_{10} < 0.25$ can be produced by two different fluxes q that is difficult to see by qualitative analysis (try it!). In the special case of $e_{10} = 0.25$, only the flux $q = 0.5$ explains this labeling pattern. Thus, for E#10 the flux identifiability depends on the flux value and is therefore called *local flux identifiability*. Only the additional information whether q is "low" ($q < 0.5$) or "high" ($q > 0.5$) will allow deriving a unique correspondence, that is, global flux identifiability.

A final remark is needed concerning the substrate uptake u that has been fixed to $u = 1$ in this investigation. Clearly, if a measured value is available, all fluxes can be scaled such that $u = 100\%$. In other words, the mappings shown in Figure 5.8 are the same if $u \neq 1$, but the x-axis shows q/u instead of q.

5.12
Measurement Models

It has been shown that measuring the substrate uptake rate u and the complete (or at least a certain part of the) labeling distribution of the metabolite E allows identifying all fluxes of the toy example. Nevertheless, there are some shortcomings: (i) measuring all isotopomer fractions of a metabolite is cumbersome, if not impossible for large molecules, and (ii) measurements are never exact. Thus, to deal with more realistic situations, measurement models are needed.

As an example, we will show how a typical MS measurement of metabolite E is described based on its isotopomer distribution. Mass spectrometry can only distinguish isotopomers by the number of contained labeled atoms (mass isotopomers [63]). Assume that either the metabolite E can be directly measured in the cellular cytosol or its successor metabolite H can be extracted from the biomass (cf. Figure 5.3). Relying on the retro-biosynthetic approach, H has the identical label enrichment as E [30]. However, using MS, only the fractions of completely unlabeled (E#00), one-time labeled (E#10 + E#01), and twofold labeled (E#11) isotopomers can be observed. In other words, a single MS measurement can provide only partial isotopomer information. The MS measurement is modeled by a set of measurement equations defining the mass isotopomer fractions:

$$\begin{aligned} e_{m+0} &= e_{00} \\ e_{m+1} &= e_{10} + e_{01} \\ e_{m+2} &= e_{11}. \end{aligned} \tag{5.23}$$

Clearly, for the complete set of mass isotopomer fractions always the conservation equation $e_{m+0} + e_{m+1} + e_{m+2} = 1$ must hold. This equation implies that only two of the mass fractions contain nonredundant information (see also Figure 5.4). Figure 5.9a shows the measurable mass isotopomers as a function of q. Because $e_{m+0} = 0$ implies $e_{m+1} = 1 - e_{m+2}$, it immediately becomes clear that in the present example only one single piece of nonredundant information remains. Fortunately, this does not destroy identifiability, meaning that also in the case of a practically relevant measurement model the flux determination problem can be solved.

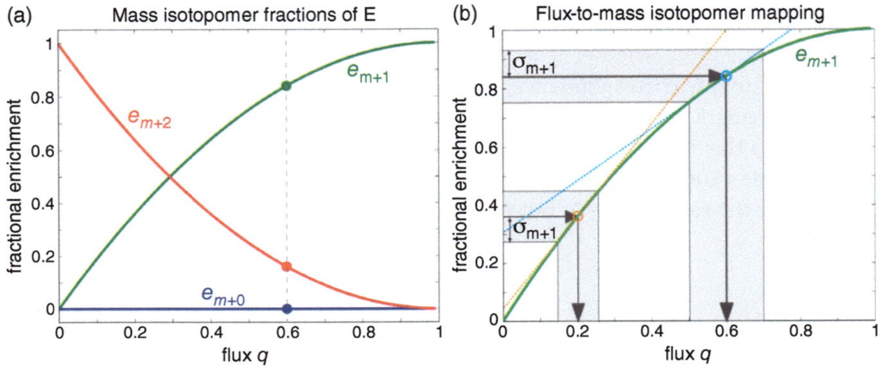

Figure 5.9 (a) MS measurements for metabolite E as functions of the free flux value q. Parameters are chosen as in Figure 5.8. (b) The green line gives the nonlinear mapping between the flux q and the measurement e_{m+1} . From the flux-to-labeling viewpoint, the flux values $q = 0.2$ and 0.6 are mapped to the measurements $e_{m+1} = 0.36$ and 0.84 as indicated by the gray arrows. Linear tangent lines (dotted blue and orange) at the two flux measurement pairs are used to map the standard deviations of the measurements to those of the fluxes. For demonstration purposes, both measurement standard deviations are $\sigma_{m+1} = \pm 0.1$, which translate to flux standard deviations as indicated by the gray shaded regions. As a result, low fluxes can be determined with better precision than high fluxes.

5.13 Statistical Considerations

So far, measurements are considered to be exact values although, in practice, they can be quite noisy. Using the MS measurement e_{m+1} for illustration, Figure 5.9b exemplifies the statistical influence of experimental errors. Here, the inverse problem (labeling-to-flux) is represented by a nonlinear relation between the MS fraction space (y-axis) and the flux space (x-axis). Exemplarily, for the two cases $e_{m+1} = 0.36$ and $e_{m+1} = 0.84$, the exact flux values $q = 0.2$ and $q = 0.6$, respectively, are indicated. The fact that the measurement value is in practice not exact is treated by adding some noise:

$$\tilde{e}_{m+1} = e_{m+1} + \epsilon_{m+1} \quad \text{with} \quad \mathrm{Var}[\epsilon_{m+1}] = \sigma^2_{m+1}. \tag{5.24}$$

Here, e_{m+1} is the true value, $\tilde{e}_{m+1}$ its measured (i.e., noise affected) value, and $\epsilon_{m+1}$the measurement noise with standard deviation σ_{m+1}. In Figure 5.9b, upper and lower boundaries (shaded regions) indicate measurement uncertainties of $\pm\sigma_{m+1}$. In practice, measurement standard deviations are known and specific for the equipment in use. Figure 5.9b illustrates how measurement noise translates to flux estimation uncertainty.

Error propagation is a statistical tool that calculates the estimation error from the measurement error [64,65]. A basic ingredient for error analysis is the ability to determine the sensitivity of the measurements with respect to the fluxes. In our case, the sensitivity describes how differing values of q impact the measurement e_{m+1}. This influence is mathematically expressed by the partial derivative $\partial e_{m+1}/\partial q$. Figure 5.9b indicates that the sensitivity of the measured fraction with respect to the investigated flux value can depend on the respective flux. In fact, the relation between the measurement standard deviation σ_{m+1} and the standard deviation σ_q of the corresponding flux estimate is approximately given by

$$\sigma_q \approx \left(\frac{\partial}{\partial q} e_{m+1}\right)^{-1} \cdot \sigma_{m+1}. \tag{5.25}$$

In other words, by the reciprocal partial derivative, a low measurement sensitivity (i.e., dependency of the labeling pattern on the flux value) leads to a large flux error and vice versa. For example, from Figure 5.9b it can be derived that for small fluxes, for example, $q = 0.2$, the statistical quality of the flux estimate will be better than for large fluxes such as $q = 0.6$. In Section 5.19, we show that the inverse relation between sensitivities and statistical quality given by Eq. (5.25) holds also in more general cases.

5.14 Experimental Design

Awareness of the correspondence between measurement error and flux uncertainty leads to a very important question for ^{13}C MFA: can we influence the statistical quality of the estimated fluxes by a proper design of the experiment? Here, the following measures can be taken:

1) Improving the measurement precision (improved equipment, repeated measurements, and exploiting data redundancy).
2) Improving the measurement configuration (other equipment, additional measurements).
3) Changing the input label composition (other labeled substrates or mixtures of them).

The first two options directly feedback to experimental equipment and can be influenced only in rare cases. Therefore, we focus on the third measure, the composition of labeled species. The important point is that a targeted input label design can prevent noninformative ILEs. The choice of the input label can strongly influence the costs and/or duration of a ^{13}C MFA. Hence, the cost of the input substrate might become a practically limiting factor (cf. Table 5.1). Therefore, the modeler has to control the trade-off between maximizing the information gain and limiting the costs.

The influence of the input substrate composition is now exemplarily investigated with the toy example. As a preliminary insight, the reader should verify by qualitative path tracing (cf. Section 5.7) that for 100% pure A#00, A#11, or A#10

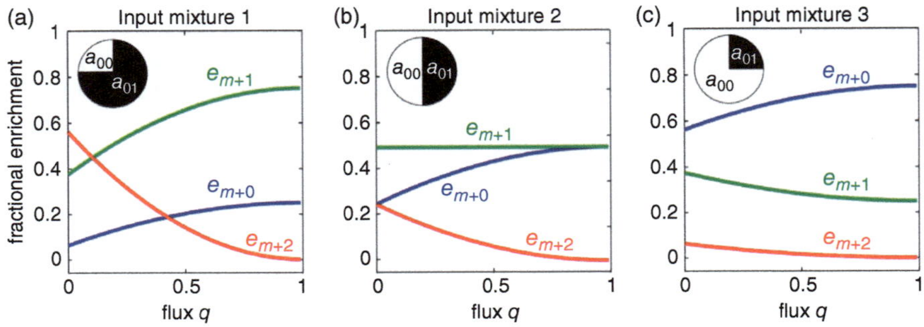

Figure 5.10 Influence of input labeling mixture composition on the information content of measurements. (a–c) Three different mixtures of the input isotopomers A#00 and A#01 are compared in the scenario of Figure 5.9. The slope of the curves indicates the measurement sensitivity that is high for small fluxes and decreases with larger flux values of q. For $q \approx 1$ all measurements become insensitive.

substrates the experiment contains no flux information at all because in each of these cases only a single labeled form arises in each metabolite pool of the network.

For varying flux values q, Figure 5.10 shows the stationary fractional enrichments emerging from three different mixtures of A#01 and A#00. From the statistical "lesson" given before, it is clear that e_{m+1} and e_{m+2} are most informative in the case of pure A#01, because their sensitivities are the highest, whereas e_{m+0} is completely noninformative. This situation changes in the presence of A#00. Now unlabeled species E#00 occur to differing amounts (e_{m+0}). While e_{m+0} becomes informative, the sensitivity of e_{m+2} decreases with the percentage of A#00 that is fed. Interestingly, for the 50%:50% A#00:A#01 mixture composition, e_{m+1} is noninformative. Summarizing, the best choice of the input mixture is to take pure A#01.

Figure 5.10 shows that for high values of q the labeling measurement becomes insensitive toward the flux, whereas for low values of q the highest sensitivities are found. Although the result is apparent after the previous exercises, the important message is that the optimal experimental design depends on the flux value q itself because the sensitivities depend on the fluxes (see also Eq. (5.22)). As a consequence of this dependency, experimental design should test the whole range of reasonable flux values to obtain a reliably informative input mixture. Clearly, having an educated guess on the range of expected free flux values q is of great benefit (cf. Section 5.20) because it narrows down the suggestions for informative input mixtures.

5.15
Exchange Fluxes

So far all metabolic fluxes in the toy system are regarded as unidirectional, that is, there are only one-way routes from substrates to products. In reality, reactions can proceed in forward and backward direction [42]. In particular, reactions operating close to thermodynamic equilibrium may have notably large fluxes in both directions [60,66]. This holds, for example, for the transaldolase and transketolase reactions in the pentose phosphate pathway. When setting up ^{13}C MFA models, we must account for this bilateral material exchange between substrates and products. As an illustrative (counter) example it was shown in [67] that the high labeling enrichment of up to 25% in the pentose-5-phosphate pools observed in an ILE with pure [1-^{13}C] glucose cannot be predicted by a unidirectional carbon transition model.

The toy example can also be used to illustrate this effect. We assume from now on that the reaction q operates in both directions which is denoted by q: B ↔ E. To distinguish between the forward and the backward fluxes of q, the notations $q^{\rightarrow}$ and $q^{\leftarrow}$ are introduced. Correspondingly, in the toy reaction network, the reaction step q must be substituted by two separate reactions (cf. Eq. (5.4)):

$$\begin{aligned} q^{\rightarrow} &: \quad \mathrm{B\#ij} \rightarrow \mathrm{E\#ij} \\ q^{\leftarrow} &: \quad \mathrm{E\#ij} \rightarrow \mathrm{B\#ij} \end{aligned} \tag{5.26}$$

Consequently, the isotope labeling balance equations must be reformulated, too (left as an exercise to the reader). Although formally straightforward, forward and backward fluxes cannot be separately accessed by any measurement instrument and are therefore hard to interpret in the context of a biological network. To facilitate human interpretation of these confounding variables, they are converted to a different flux system, the system of net and exchange fluxes. The transformation is as follows:

$$\begin{aligned} q^{\mathrm{net}} &= q^{\rightarrow} - q^{\leftarrow} \\ q^{\mathrm{xch}} &= \min(q^{\rightarrow}, q^{\leftarrow}). \end{aligned} \tag{5.27}$$

The exchange flux provides a measure of the reversibility of the reaction. An exchange flux of zero corresponds to a unidirectional reaction, whereas an exchange of infinity indicates that the reaction is in a fast equilibrium. Clearly, the stoichiometric equations hold for net fluxes in the same way as for unidirectional fluxes. Thus, it follows directly from Eq. (5.2):

$$p = v = w = u - q^{\mathrm{net}}, \qquad r = q^{\mathrm{net}}. \tag{5.28}$$

In contrast to the net fluxes, stoichiometry does not constrain any of the exchange fluxes [42]. In the toy example, q^{net} must be positive as a consequence of Eq. (5.3). Thus, in this particular case q^{xch} simplifies to $q^{\mathrm{xch}} = q^{\leftarrow}$. In general, this relation need not hold because net fluxes can also take negative values in realistically complex networks. In any case, as a trade-off and because q^{xch} cannot be determined from the stoichiometry alone, the system now has two degrees of freedom. Precisely, the two fluxes q^{net} and q^{xch} need to be determined from the labeling measurements instead of one net flux q.

To motivate why there is a realistic chance to determine both fluxes q^{net} and q^{xch}, a qualitative analysis is done first. Clearly, because the assumption of a

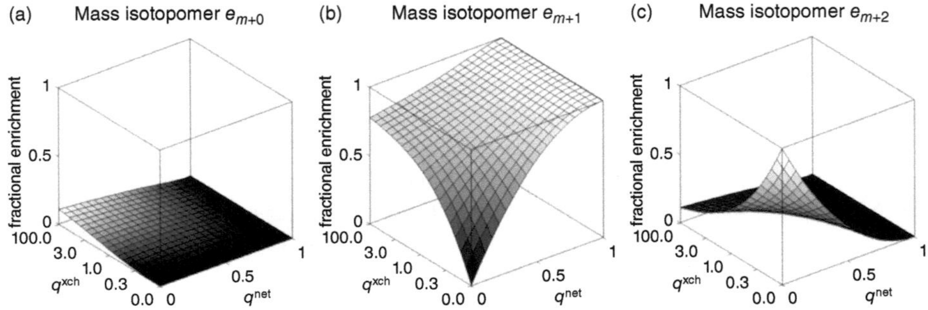

Figure 5.11 MS measurements $e_{m+0}, e_{m+1}, e_{m+2}$ for the toy example as functions of the fluxes q^{net} and q^{xch}. All remaining model parameters are the same as in the scenario of Figure 5.8.

reversible reaction step q just extends the formerly discussed carbon transition network by the new step q$^{\leftarrow}$, the isotopomers E#01, E#10, and E#11 found in Section 5.10 with a pure A#01 input labeling will still be formed in the augmented network. Additionally, if q$^{\leftarrow}$ is active, the following reaction sequence becomes possible:

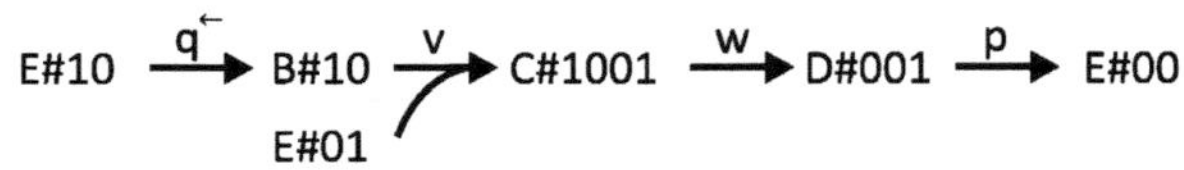

Surprisingly, by this route the unlabeled E#00 species is now continually regenerated, circumventing the washout of E#00. As a direct consequence, the presence of the unlabeled E#00 after label equilibration proves the activity of q$^{\leftarrow}$. This means that a network with exchange fluxes produces a broader variety of information in terms of different mass isotopomers compared to a network without the exchange flux. The influence of q^{xch} on measured quantities is also quantitatively obvious by repeating the simulation underlying Figure 5.9 for different exchange flux levels (cf. Figure 5.11). It turns out that the $m+0$ mass isotopomer e_{00} shows a low but measurable response in the presence of the q^{xch} flux.

As a quintessence, by falsely neglecting q^{xch} in the network model, the determination of q^{net} from Figure 5.11 leads to gross errors or even noninterpretable flux values. This is exactly the problem when pentose phosphate pathway labeling enrichments are interpreted under the false assumption of unidirectional fluxes [42].

5.16
Multidimensional Flux Identifiability

We now deal with the multidimensional analog of Section 5.13. Given $u = 1$ and two mass isotopomer measurements of pool E (keep in mind that the third measurement is always redundant), are we able to uniquely determine both fluxes q^{net} and q^{xch}? To answer this question we can analyze e_{m+0}, e_{m+1}, and e_{m+2} as two-dimensional functions of the investigated fluxes q^{net} and q^{xch} (cf. Figure 5.11).

From these plots, it becomes clear that both fluxes independently influence the measurements in a nonlinear manner. In the following, we restrict to the non-redundant measurements e_{m+0}, e_{m+1}.

With Figure 5.11 at hand, the flux identifiability problem can be graphically solved as follows:

1) Select two hypothetical measurement values (e_{m+0}, e_{m+1}). These define the height of horizontal cutting planes in Figure 5.11a and b that intersect with the measurement surfaces of the mass isotopomer fractions.
2) The intersections determine the contour lines of the measurement planes corresponding to the chosen fractions e_{m+0} and e_{m+1}, respectively. They define all value pairs (q^{net}, q^{xch}) that produce the measured values. The result for the scenario $e_{m+0} = 0.01$, $e_{m+1} = 0.87$ is given in Figure 5.12a and b.

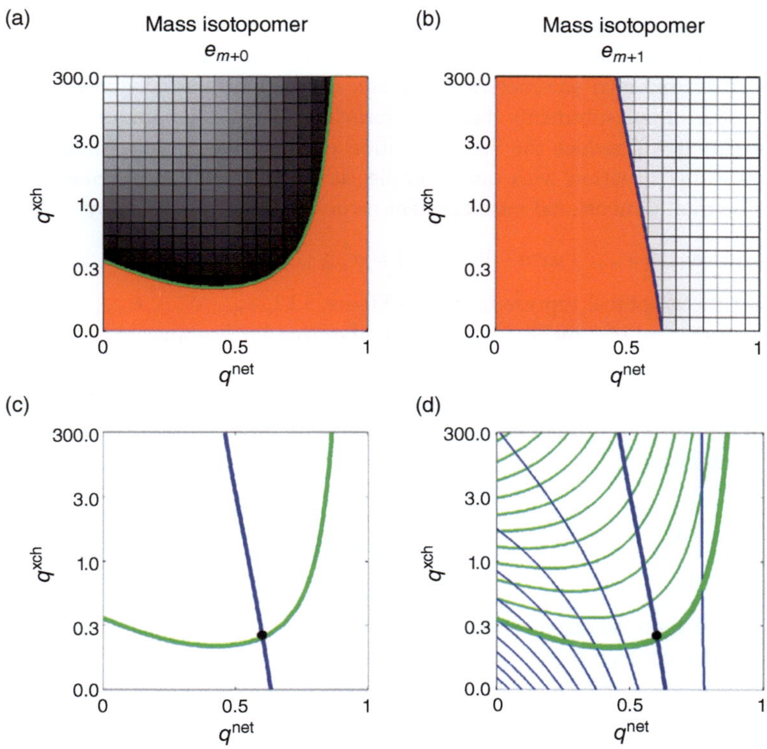

Figure 5.12 Two-dimensional flux identification analysis for (q^{net}, q^{xch}) from the measured values (e_{m+0}, e_{m+1}). Parameters are chosen as in Figure 5.11. (a)–(b) Contour lines correspond to the assumed measured data pair $e_{m+0} = 0.01$ (green, a) and $e_{m+1} = 0.87$ (blue, b); (c) intersection of the contour lines shown in (a)–(b) is indicated by the black circle; (d) global identifiability analysis by displaying multiple contour lines (e_{m+0} – green, e_{m+1} – blue).

3) If these contour lines intersect at one single point (cf. Figure 5.12c), a unique solution for $(q^{\text{net}}, q^{\text{xch}})$ exists that exactly reproduces both measured values.

For deciding about a priori global identifiability, the procedure must be repeated for all possible pairs of fractions (e_{m+0}, e_{m+1}). This can be most easily achieved by displaying all contour lines into a single plot as it is done in Figure 5.12d. As a result, the checkerboard-like pattern reveals a global unique identification of $(q^{\text{net}}, q^{\text{xch}})$ from the (e_{m+0}, e_{m+1}) data.

5.17 Multidimensional Flux Estimation

Up to now, identifiability as discussed in Section 5.11 assumed error-free measurements. In practice, we have to address the problem of multidimensional flux estimation in the presence of measurement noise. Moreover, in contrast to multidimensional identifiability problems where the number of measured data equals the number of unknown parameters, parameter estimation is usually based on a higher number of measurements than of parameters. Using a maximum of available information will reduce the flux estimation error due to data redundancies. To illustrate this situation with our example, at least three measurements are needed. The multidimensional measurement error model is now given by

$$\tilde{e}_{m+i} = e_{m+i} + \epsilon_{m+i} \quad \text{with} \quad \text{Var}[\epsilon_{m+i}] = \sigma^2_{m+i} \quad \text{for} \quad i = 0, 1, 2. \tag{5.29}$$

Underlying the graphical representation in Figure 5.11, $e_{m+0}, e_{m+1}, e_{m+2}$ are also functions of q^{net} and q^{xch} by Eq. (5.22). Remember that for keeping things simple we still assume the uptake rate to be fixed ($u = 1.0$). For quantifying the discrepancy between the label enrichments $e_{m+0}, e_{m+1}, e_{m+2}$ predicted by the model and their actually measured analogs $\tilde{e}_{m+0}, \tilde{e}_{m+1}, \tilde{e}_{m+2}$, the sum of squares residual (SSR) is formulated:

$$\text{SSR}(q^{\text{net}}, q^{\text{xch}}) = \left(\frac{e_{m+0} - \tilde{e}_{m+0}}{\sigma_{m+0}}\right)^2 + \left(\frac{e_{m+1} - \tilde{e}_{m+1}}{\sigma_{m+1}}\right)^2 + \left(\frac{e_{m+2} - \tilde{e}_{m+2}}{\sigma_{m+2}}\right)^2. \tag{5.30}$$

Obviously, the three squared terms of the SSR penalize the deviation between each single predicted value and the corresponding measurement. Each deviation is weighted by the corresponding measurement standard deviation σ. Hence, precise measurements have a higher impact on the SSR compared to noisy data.

Before drawing the SSR landscape for an *in silico* ILE (cf. Figure 5.13), the toy example needs some modification. Because of the equation $e_{m+0} + e_{m+1} + e_{m+2} = 1$, the three measurements (observed and simulated) are always redundant. As a consequence, when two of them fit perfectly to the measured data (which is always possible due to identifiability) the third will also fit. For this reason, the resulting minimal SSR value is zero. In order to

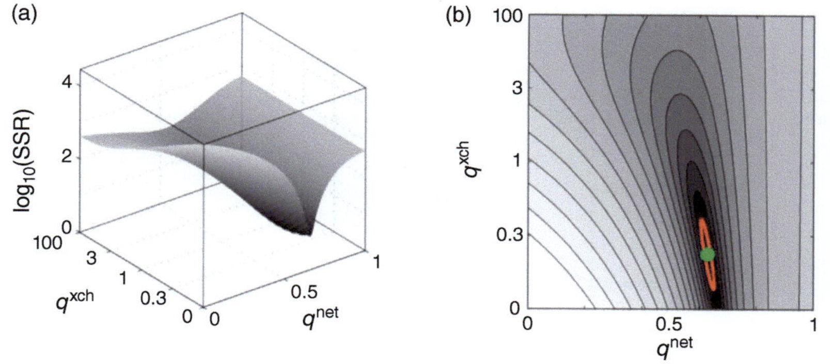

Figure 5.13 (a) 3D representation of the SSR landscape (logarithmic scale) and (b) contour plots. The residual sum of squares function $\text{SSR}(q^{net}, q^{xch})$ provides a measure of how well the model fits the measured data. For the input label A#01 and an influx of $u = 1$, two independent sets of measured E mass isotopomers are assumed to be available: $(\tilde{e}_{m+0,1}, \tilde{e}_{m+1,1}, \tilde{e}_{m+2,1}) = (0.005, 0.895, 0.100)$ and $(\tilde{e}_{m+0,2}, \tilde{e}_{m+1,2}, \tilde{e}_{m+2,2}) = (0.0125, 0.8675, 0.1200)$. Each single measurement in both sets is supposed to have an absolute standard deviation of 0.01. For this setup, the global optimum is located at $\hat{q}^{net} = 0.63$ and $\hat{q}^{xch} = 0.23$ with $\text{SSR}(\hat{q}^{net}, \hat{q}^{xch}) = 6.1$ (green dot, b).

change this oversimplified situation, we assume from now on that two data sets $\tilde{e}_{m+0,i}, \tilde{e}_{m+1,i}, \tilde{e}_{m+2,i}, i = 1, 2$ with slightly different values are available from two repeated measurements for the same experiment (mimicking biological replicates). For this configuration, the SSR (Eq. (5.30)) has to be extended to six terms (not shown here). Figure 5.13 shows the resulting SSR landscape. From this optimization landscape, the flux estimates that minimize the SSR can be directly obtained. For the investigated scenario, the minimal SSR is greater than zero ($\text{SSR}(\hat{q}^{net}, \hat{q}^{xch}) = 6.1$).

Notice that by adding either three or six independent squared terms in SSR we implicitly assume statistical independency of every three measurement sets $e_{m+0}, e_{m+1}, e_{m+2}$ that is obviously not fulfilled due to $e_{m+0} + e_{m+1} + e_{m+2} = 1$. We refer the interested reader to the literature providing details on how the SSR function can be formulated for correlated measurements [35,67].

5.18 The General Parameter Fitting Procedure

Up to now, a toy example and a graphical method have been used to demonstrate how the fluxes are obtained from the measured isotopomer fractions (cf. Figure 5.13). In reality, ^{13}C MFA deals with high-dimensional systems with more than 20 free fluxes and often hundreds of measurements. Solving such high-dimensional flux determination problems requires another, more systematic approach. In practice, the minimum of the SSR function must be computed

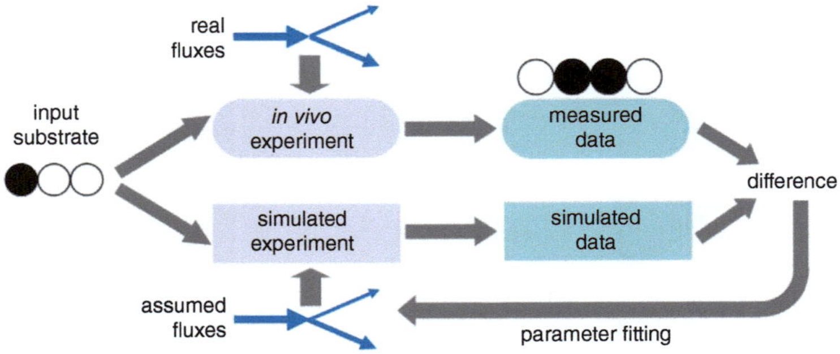

Figure 5.14 The workflow of flux estimation. The real ILE (upper branch) is paralleled *in silico* (lower branch). Starting with an initial guess for the fluxes, labeling data are simulated and compared with the observed ones. By adapting the flux values, this difference is iteratively minimized during the parameter fitting process.

using numerical optimization algorithms, for example, the MATLAB built-in function `fmincon`.

The general workflow for estimating the (real) fluxes from experimental measurements is shown in Figure 5.14. First, an ILE is performed in the wet-lab with a well-chosen substrate mixture (cf. Section 5.14) and the stationary labeling patterns in the cell are measured (top branch). Then, in the dry-lab, the estimation of the unknown *in vivo* metabolic fluxes from the labeling patterns is started by choosing initial values for the set of free fluxes. With the same input substrate as in the ILE, the labeling measurements are simulated (bottom branch) and compared to the real measurements (cf. Section 5.17). By minimizing the SSR, that is, the discrepancy between simulated data and measurements (Eq. (5.30)) through systematically changing the flux values in the direction of their highest sensitivity (cf. Section 5.13), the simulated flux distribution is approaching the *in vivo* fluxes (identifiability supposed). Once the difference is minimal, an estimate for the real *in vivo* flux distribution is obtained.

This procedure is well known as nonlinear parameter fitting [68]. Although it is conceptually simple, the solution of an optimization problem needs care: nonlinear parameter estimation problems with dozens of parameters (i.e., fluxes) are not easy to solve and no absolutely reliably algorithms for its solution exist. The SSR can possess not only a single global optimum (like it is the case in the scenario of Figure 5.13), but several local optima from which the lowest one has to be chosen as the solution. The standard optimization algorithms will in general only find a single local minimum, typically depending on the chosen initial free flux values.

Moreover, depending on the shape and "roughness" of the optimization landscape, search algorithms tend to prematurely terminate in poorly determined valleys or get stuck in suboptimal local minima. To avoid such behavior, advanced optimization techniques like multiple restarting of the fitting procedure from randomly chosen initial flux values or even more sophisticated globalized optimization strategies have been developed [69]. Still, a ^{13}C MFA may have multiple solutions, in particular, if free flux parameters are nonlinearly correlated. Here, the interested reader is referred to specialized literature [70].

5.19 Multidimensional Statistics

In Section 5.13, the error propagation from a single measured value to a single estimated flux has been discussed. In a general multidimensional setting things are less simple and the error propagation from a set of measured values to a (usually smaller) set of flux estimates must be analyzed. Without going too deep into nonlinear regression theory [68], only the general idea is sketched. Figure 5.13a shows the optimization landscape resulting from SSR of the toy example. Close to the optimum it turns out that the SSR contour lines are approximately concentric ellipses. In fact, multivariate nonlinear regression theory tells that these ellipses (or, more generally, ellipsoids) define approximate confidence regions for the estimated fluxes $(\hat{q}^{\text{net}}, \hat{q}^{\text{xch}})$. The smaller the ellipse, the higher is the confidence that the true parameter is actually located close to the computed minimum. The interested reader is referred to classical textbooks or papers for more details on multidimensional regression theory [67,68]. We focus on the problem to calculate the confidence ellipsoids.

To this end, the multidimensional analog of the derivative in Eq. (5.25) needs to be formulated. This generalization is the so-called Jacobian matrix that collects the sensitivities of each measured value with respect to each flux parameter. The Jacobian must be computed with respect to the previously computed estimates $(\hat{q}^{\text{net}}, \hat{q}^{\text{xch}})$ (shown for three measurements here):

$$\mathbf{J} = \mathbf{J}(\hat{q}^{\text{net}}, \hat{q}^{\text{xch}}) = \left.\begin{pmatrix} \partial e_{m+0}/\partial q^{\text{net}} & \partial e_{m+1}/\partial q^{\text{net}} & \partial e_{m+2}/\partial q^{\text{net}} \\ \partial e_{m+0}/\partial q^{\text{xch}} & \partial e_{m+1}/\partial q^{\text{xch}} & \partial e_{m+2}/\partial q^{\text{xch}} \end{pmatrix}\right|_{(\hat{q}^{\text{net}}, \hat{q}^{\text{xch}}) = (q^{\text{net}}, q^{\text{xch}})}. \tag{5.31}$$

A second matrix generalizes the concept of the one-dimensional standard deviations. If all measurements are independent of each other, the multidimensional uncertainty in the data is described by the measurement covariance matrix (again for three measurements):

$$\boldsymbol{\Sigma} = \begin{pmatrix} \sigma_{m+0}^2 & 0 & 0 \\ 0 & \sigma_{m+1}^2 & 0 \\ 0 & 0 & \sigma_{m+2}^2 \end{pmatrix}. \tag{5.32}$$

Measurement correlations can be considered by replacing the off-diagonal zero values of the matrix Σ by nonzero values (covariances). This is not done here to keep things simple (cf. comment in Section 5.18). Putting things together, the general multidimensional version of the error transduction Eq. (5.25) is given by the parameter covariance matrix:

$$\text{Cov}\,(\hat{q}^{\text{net}}, \hat{q}^{\text{xch}}) = \left(\mathbf{J} \cdot \mathbf{\Sigma}^{-1} \cdot \mathbf{J}^{T}\right)^{-1}. \tag{5.33}$$

From the diagonal entries of the covariance matrix, the variances (i.e., squared standard deviations) of the two flux estimates $\hat{q}^{\text{net}}, \hat{q}^{\text{xch}}$ are obtained. Moreover, the off-diagonal elements contain information on their correlation. Importantly, the shape of the confidence ellipsoids can be directly derived from the covariance matrix by calculating its principal axes. The details can be taken from [35].

In the example with 100% A#01 input labeling, the resulting covariance matrix based on the free flux estimates $\hat{q}^{\text{net}} = 0.63$, $\hat{q}^{\text{xch}} = 0.23$ and measurement standard deviations of 0.01 is given by the following equation:

$$\text{Cov}\,(\hat{q}^{\text{net}}, \hat{q}^{\text{xch}}) = \begin{pmatrix} \sigma_{\text{net}}^2 & \tau_{\text{net,xch}} \\ \tau_{\text{net,xch}} & \sigma_{\text{xch}}^2 \end{pmatrix} = \begin{pmatrix} 0.0004 & -0.0018 \\ -0.0018 & 0.0102 \end{pmatrix}. \tag{5.34}$$

With a standard deviation of $\sigma_{\text{net}} = \sqrt{0.0004} = 0.02$, the net flux is better determined than the exchange flux ($\sigma_{\text{xch}} = \sqrt{0.0102} = 0.10$). The correlation coefficient is given by $\rho = \tau_{\text{net,xch}}/(\sigma_{\text{net}} \cdot \sigma_{\text{xch}}) = -0.89$. These values correspond to the confidence ellipse found by graphical analysis of the optimization landscape in Figure 5.13.

5.20
Multidimensional Experimental Design

As it becomes clear from Eqs. (5.31)–(5.34), the statistical quality of the flux estimates depends on the real fluxes ($q^{\text{net}}, q^{\text{xch}}$), the true measurement standard deviations ($\sigma_{m+0}, \sigma_{m+1}, \sigma_{m+2}$), and the chosen input substrate mixture ($a_{00}, a_{01}, a_{10}, a_{11}$):

$$\text{Cov}(q^{\text{net}}, q^{\text{xch}}) = f\left(q^{\text{net}}, q^{\text{xch}}; \sigma_{m+0}, \sigma_{m+1}, \sigma_{m+2}; a_{00}, a_{01}, a_{10}, a_{11}\right). \tag{5.35}$$

This is the clue to optimal experimental design: $\sigma_{m+0}, \sigma_{m+1}, \sigma_{m+2}$ and $a_{00}, a_{01}, a_{10}, a_{11}$ can be influenced in the experimental setup. Because the influence of $\sigma_{m+0}, \sigma_{m+1}, \sigma_{m+2}$ is rather obvious, we focus on the input mixture design. Like in the one-dimensional case, the major problem when aiming at minimizing flux confidence intervals is that the parameter covariance matrix cannot be calculated *a priori* because it depends on the unknown, real free fluxes ($q^{\text{net}}, q^{\text{xch}}$). Let us assume for the moment that we have a good initial guess ($\hat{q}^{\text{net}}, \hat{q}^{\text{xch}}$) for these fluxes. Then, the following strategy can be applied:

1) Choose an input mixture ($a_{00}, a_{01}, a_{10}, a_{11}$) and calculate the resulting covariance matrix from Eq. (5.33).
2) Define a criterion that summarizes the overall quality of the experiment in one single number. Because the covariance matrix (Eq. (5.33)) determines the shape of the confidence ellipsoid for ($\hat{q}^{\text{net}}, \hat{q}^{\text{xch}}$) (cf. Figure 5.13), properties of this matrix can be used to define the criterion. A frequently used design criterion is the determinant of the

covariance matrix which (up to some constant) is the volume of the confidence ellipse (cf. Figure 5.13b) [71]:

$$I_D = \det \text{Cov}(\hat{q}^{\text{net}}, \hat{q}^{\text{xch}}; a_{00}, a_{01}, a_{10}, a_{11}). \tag{5.36}$$

3) Determine the criterion for all available mixtures $(a_{00}, a_{01}, a_{10}, a_{11})$. Then, select the mixture composition with the smallest D-criterion value I_D.

In practice, it is hardly sensible to choose any theoretically possible mixture composition out of $(a_{00}, a_{01}, a_{10}, a_{11})$. In most cases, it is sufficient to restrict to three different labeled species of the input substrate, of which one is often the unlabeled isotopomer (A#00) and one is the fully labeled isotopomer (A#11). As a third isotopomer in the toy example, we chose A#01 as the promising candidate for obvious reasons. Because A#10 is absent, we obtain from $a_{00} + a_{01} + a_{11} = 1$ that only two degrees of freedom remain in the mixture design that can be best represented in a mixture triangle (cf. Figure 5.15). In the triangle, the

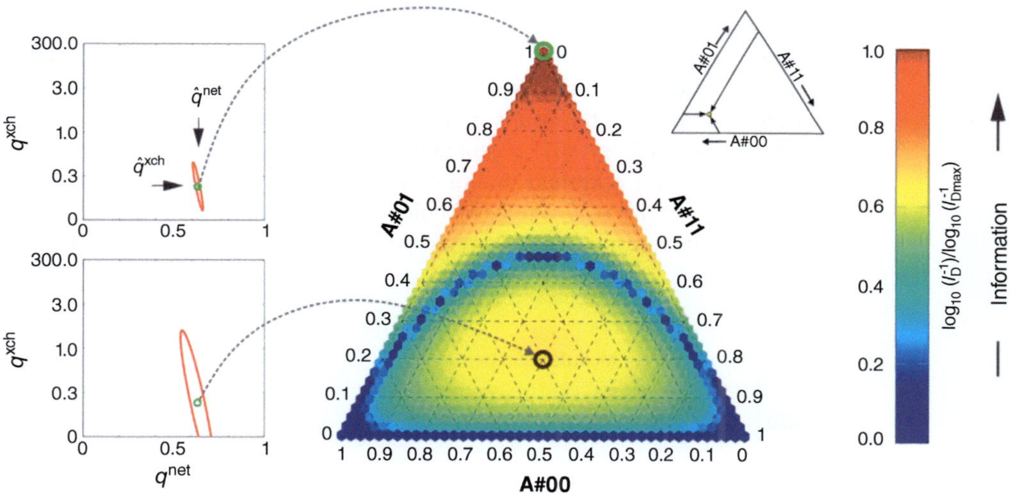

Figure 5.15 Experimental design study for the toy example with reference fluxes $\hat{q}^{\text{net}} = 0.63$, $\hat{q}^{\text{xch}} = 0.23$ based on the D-criterion. The full MS measurement information $(m+0, m+1, m+2)$ of metabolite E was used with measurement standard deviations of 0.01 (as in Figure 5.14). $1/I_D$ is plotted on a logarithmic scale and values are scaled to [0,1]. Therefore, large values represent a high information content of the respective experiment, that is, a small confidence ellipse. Two confidence ellipses of the free fluxes are shown for separate points in the mixing triangle. A value of 1 (red) corresponds to maximal possible statistical identifiability, whereas noninformative input mixtures are characterized by values approximatively 0 (blue). The inset diagram on the upper right shows how to read the mixture triangles. Although there is a second local optimum in the middle (orange area, black cycle: 40% A#00, 20% A#01, 40% A#11) the corresponding information is much lower than that of the top corner (green cycle: 100% A#01). The blue zero information half-circle line corresponds to cases where the measurements produce redundant information (not discussed in detail here).

color codes for the statistical information. At each point a covariance matrix calculation is performed (Eq. (5.35)), using the respective mixture and the guessed flux values. This is exemplified in Figure 5.15 by showing two of the calculated confidence ellipsoids. For the flux estimates $(\hat{q}^{\text{net}}, \hat{q}^{\text{xch}}) = (0.63, 0.23)$, a distinct information optimum can be found for the pure A#01 "mixture." All other mixtures show a minor potential to achieve a similar statistical confidence as indicated by lower information criterion values. This, on a higher level of abstraction, confirms all the results presented before.

It has to be mentioned that many variants of experimental design criteria exist. The second frequently used criterion is the A-criterion which is the trace of the covariance matrix, that is, the sum of all flux variances. Hence, in contrast to the D-criterion, the A-criterion neglects correlations:

$$I_{\text{A}} = \text{trace Cov}(\hat{q}^{\text{net}}, \hat{q}^{\text{xch}}). \tag{5.37}$$

Both, the D- and the A-criterion (5.36) and (5.37), respectively, can also target a subset of fluxes by putting weighted emphasis on them. The reader is referred to [71,72] for an in-depth treatment of the topic.

It remains to answer the question what to do if no good initial guess of the fluxes $(q^{\text{net}}, q^{\text{xch}})$ is available. In this case, the mixture triangle can be calculated for a representatively chosen set of different values for $(q^{\text{net}}, q^{\text{xch}})$ that cover all expected outcomes. The results of such a screening are shown in Figure 5.16. The top corner (100% A#01) is generally a good choice because it represents the global information optimum in almost all cases. Nevertheless, it must be said that for $(q^{\text{net}}, q^{\text{xch}}) \approx (0, 0)$ the best choice is in the middle, whereas for $q^{\text{net}} \approx 1$ there is a poor information for any possible input mixture. However, these are

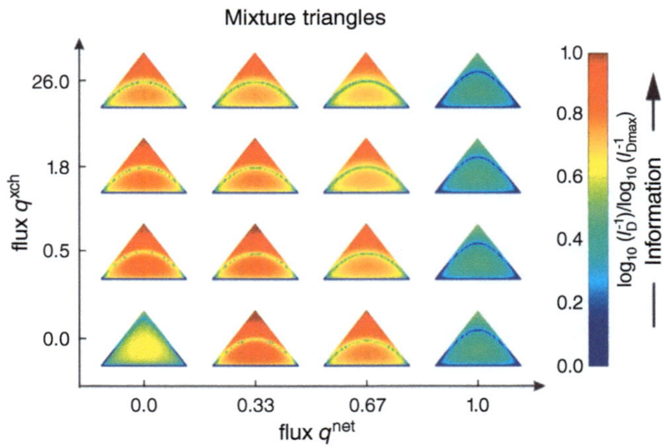

Figure 5.16 Dependency of experimental design on flux values. The analysis in Figure 5.15 was repeated for different flux values. $1/I_D$ is plotted as in Figure 5.15 and all mixture triangles were equally scaled.

pathological flux arrangements (cf. Figures 5.8 and 5.11). Summarizing, a 100% A#01 "mixture" is recommended.

5.21 The Isotopically Nonstationary Case

We are now going all the way back to the end of Section 5.9 where the focus has been on the isotopic steady state. Although the isotopic steady-state condition is today the most frequently used ^{13}C MFA configuration, relying on the isotopic steady state entails a principal limitation: it needs the time until the system becomes isotopically stationary. Essentially, in the retrobiosynthetic approach we have to wait with sampling until every molecule in the biomass becomes labeled and the labeling states are not changing any more. To fulfill the isotopic equilibrium condition at least approximately, cells must be fed with labeled compounds for a duration that is in the order of several cell-doubling times, an endeavor that easily becomes expensive [73]. Advances in mass spectrometry have enabled the measurement of labeling patterns and pool sizes for central carbon intermediates (e.g., [74]). This obviates the need for the retro-biosynthetic approach and shortens the experimental timescales by an order of magnitude. Switching to the isotopic transient state leads to the so-called isotopically nonstationary (INST) MFA technique, which is at least another order of magnitude faster. Because there is no free lunch, INST ^{13}C MFA, however, involves a series of additional steps [75,76]:

- Because the transient dynamics of labeling enrichment in the cellular metabolites has to be monitored, a repeated sampling schedule must be established (cf. Figure 5.17). Particularly, in the first few seconds after switching to the labeled substrate fractions change most rapidly and, thus, the most informative samples can (and must) be taken very quickly in the isotopic transient phase. For some bacteria, where the patterns change in subsecond scale, this requires an automated sampling device [9,77].
- Because the transient dynamics of the system – in contrast to the isotopic steady state – does indeed depend on the metabolic pool sizes (cf. Figure 5.18), information about intracellular metabolite concentrations is needed. This means that quantitative metabolomics has to be performed in addition to labeling analytics [78,79]. Typically, a pool size sample is drawn shortly before the label is switched [47].
- As a consequence, compared to the isotopic steady state, multiple samples must be analyzed. This calls for automation of routine tasks, quality measures for peak integration, and embedding evaluation steps in scientific workflows [80].
- Finally, the complete computational machinery must be reestablished and enhanced to the INST case. First of all, this means to solve differential equations (Eqs. (5.11)–(5.15)) instead of "simple" steady-state equations

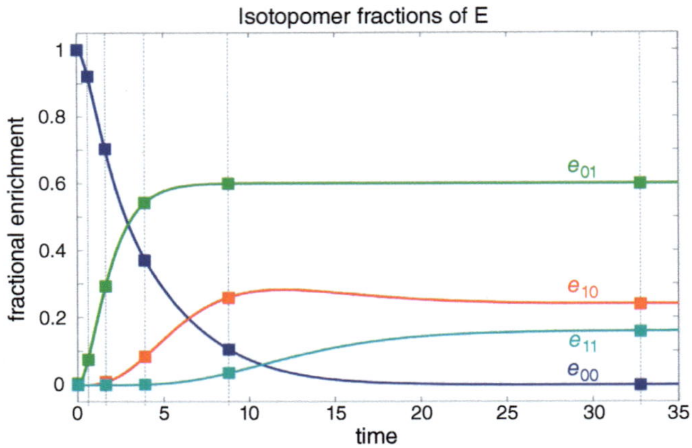

Figure 5.17 Sampling time schedule of an INST MFA experiment. At each time point (vertical lines), measurements are taken. Typically, the sampling time points are nonuniformly distributed to cover both the isotopic transient and the stationary phases. Concerning the number of samples, a compromise has to be made weighting the effort for experimentation, quick sampling, and raw data evaluation.

making the calculations much more expensive and requiring for advanced numerical solvers. The parameter fitting scheme has to be extended by terms representing the single sampling time points as well as the observed pool sizes. Moreover, with the sampling times, new design parameters enter the stage.

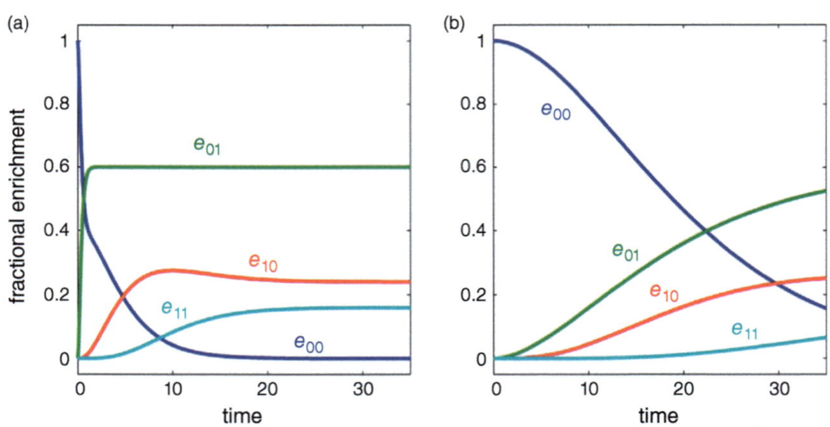

Figure 5.18 INST ILE simulation of E-isotopomer fractions with the toy example for two pool size scenarios (a) $E = B = 0.2$ and (b) $E = B = 10.0$. All other pool sizes are set to 1.0. The remaining parameters are taken from Eqs. (5.16)–(5.19).

Although the investments to be made are high, INST MFA is worth the effort [75,76] for the following reasons:

1) INST MFA experiments are orders of magnitude faster than classical ^{13}C MFA experiments. For microbes a flux map can be generated from ILEs that take no longer than a few minutes compared to hours or days because the early transient data is already sufficient to determine the fluxes.
2) Particularly, the measurement of fluxes in transient growth phases can be approached only by INST ILEs. This also holds true for genetically nonstable systems and cells growing under less controllable environmental conditions, for example, diauxic growth on multiple substrates. Briefly, in all cases where the cellular metabolism is (approximately) stationary only for a short time, INST is the only option.
3) INST MFA sheds some light on the validity of common hypotheses on isotopic steady states, protein turnover, inactive pathways, reaction irreversibility, and so on. Thus, INST helps to critically assess the achieved results.
4) INST MFA produces a wealth of informative data and, thus, can statistically better resolve the metabolic fluxes than classical ^{13}C MFA. Particularly, certain fluxes (e.g., cycling bidirectional anaplerosis fluxes) still remain poorly determinable with classical methods. A well-designed INST MFA can help to tackle structural and statistical nonidentifiability problems.
5) In case of single-carbon substrates such as carbon dioxide or methanol, it is generally impossible to resolve fluxes with the classical ^{13}C MFA method because the steady-state labeling enrichment is not influenced by metabolic fluxes [44]. In this situation, INST MFA provides a unique opportunity.
6) Last but not least, labeling experiments are expensive. In case of exotic substrates, an INST MFA experiment may be the only affordable MFA variant.

These benefits are the reason why in recent years the complete computational machinery of ^{13}C MFA has indeed been extended to the isotopic nonstationary case [81–83]:

- The modeling frameworks have been enhanced.
- High-performance differential equation solvers have been implemented.
- All required Jacobian matrices can be computed.
- Experimental design was generalized to the transient case.
- Parameter fitting works.

These aspects can be developed along the same lines of the present introduction but would exceed the scope of this primer. The reader may consult the publications referred to above that should be understandable with the knowledge built up so far.

5.22
Some Final Remarks on Network Specification

From the running example, the reader might obtain the impression that in ^{13}C MFA the specification of the metabolic network and the carbon atom transitions is a rather straightforward routine task. In practice, this impression is not correct. Network modeling is a highly demanding and essential step because a couple of assumptions for network simplification have to be made depending on the biological system under investigation. Particularly, some pitfalls have to be avoided.

Essentially, the metabolic network data can be collected from information provided by biological databases such as the Kyoto Encyclopedia of Genes and Genomes (KEGG, http://www.genome.jp/kegg), the BioCyc Database Collection (http://biocyc.org), MetRxn [84], and literature resources ([85] and references therein), or model repositories such as the BioModels database [86] in a more or less automated way. After resolving incompatibilities in data representation such as metabolites with multiple names, stoichiometric errors (i.e., elemental imbalances), or missing cofactor specificity, the network model must be polished and simplified for flux analysis:

- A reasonable decision on the extent of the network must be made. Clearly, the pathways of interest must be represented in the model but it makes no sense to include detailed representation of any isoenzyme or biosynthetic pathway. Particularly this holds if corresponding measurement information for resolving these spots in the metabolic network is not available. Typical networks are restricted to the central metabolic pathways and the biosynthesis routes for major biomass precursors (e.g., amino acids). Furthermore, the catalog of reactions should also include all extracellular transport steps (substrate uptake of all potential carbon sources from and product secretion into the medium).
- In classical ^{13}C MFA, linear reaction sequences containing no intermediate branching points can be condensed to one single virtual reaction step. For example, this is typically done for parts of glycolysis or the oxidative pentose phosphate pathway (cf., Figure 5.1). However, lumping of reaction sequences is correct only in the isotopically stationary state. In the isotopically nonstationary situation, such lumping operations are an approximation to a physical reality because the dynamics of a reaction sequence is different from a one-step reaction.
- If there is good evidence from other data (i.e., transcriptomic or proteomic analyses), some reaction steps can be assumed to be inactive, which strongly aids to identify the remaining active fluxes. A prominent example is the glyoxylate shunt in *C. glutamicum* for growth on glucose [87], which contributes to the statistical determinacy of the anaplerotic reactions.
- Finally, an assumption on uni- or bidirectionality must be made for each single reaction step. Obviously, the more bidirectional reaction steps are present, the more information must be obtained from measured data.

Generally, reaction directions are governed by thermodynamic driving forces and those reaction steps that operate far from thermodynamic equilibrium such as phosphofructokinase and pyruvate kinase in the glycolysis or the glucose-6-phosphate dehydrogenase in the oxidative pentose phosphate pathway are assumed to operate in only one direction. However, the decision to fix a reaction direction is not static meaning that it must be reconsidered each time the environmental conditions change. Prominent examples are the "acetate switch" in bacteria and the switch from respiration to fermentation in yeasts [88,89].

The next nontrivial step after network specification is the assignment of carbon atom transitions to each reaction step. Clearly, this data must be 100% correct because otherwise the carbon flow model predicts physical nonsense. The following things must be considered:

- Although some databases are already available that contain information about atom transitions (e.g., KEGG RPAIR and MetaCyc databases), the modeler should never rely on database information as sole information source because this data is automatically generated, still not complete, fully curated, and validated. A very good tool to get confidence is the graphical visualization of carbon atom transitions on the molecular structures (cf., Figure 5.19).

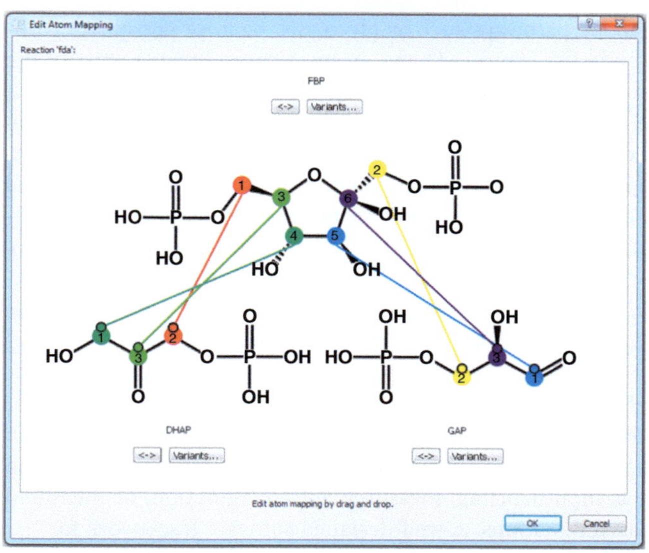

Figure 5.19 Atom transition for the fructose-bisphosphate aldolase (fda) catalyzed reaction in the pentose phosphate pathway. Carbon atoms of metabolites are colored. The fate of the carbon atoms is indicated by connection between the educts' (fructose-1,6-bisphosphate – FBP) and products' (dihydroxy-acetone phosphate – DHAP; glyceraldehyde 3-phosphate – GAP) carbon atoms (see also Figure 4.2).

- Although an IUPAC standard for the unique enumeration of carbon atoms in any molecule exists, not all published carbon transition data maintain this standard. This means that mixing data from different sources will introduce enumeration inconsistencies. Here, powerful tools such as the InChI format are available that can be used to enforce a unique atom numbering [90,91].
- All this does not help to cure one common problem in ^{13}C MFA. There are some symmetric molecules in metabolic pathways that lead to the so-called label scrambling. For instance, for succinate and fumarate, C-1 and C-2 positions are indistinguishable from C-4 and C-3 positions. Up to now, databases do not account for this phenomenon. A "scrambling reaction" is usually modeled by a virtual mirroring reaction of the following type:

$$\begin{array}{ccc} \text{succ} & \leftrightarrow & \text{succ} \\ \#\text{abcdef} & & \#\text{fedcba} \end{array}$$

Concerning the flux values, it is still an open debate if the (standard) assumption of 50% scrambling is appropriate because enzymatic preferences might lead to different values.

5.23 Algorithms and Software Frameworks for ^{13}C MFA

In practice, the metabolic networks investigated in the context of ^{13}C MFA contain the central metabolic pathways such as glycolysis, pentose phosphate pathway, and citric acid cycle (cf. Figure 5.1) and, in most cases, include parts of the biosynthesis of, for example, amino acids or organic acids. Thus, realistic networks contain more than 80 metabolites and, as a consequence, the number of isotopomers easily exceeds 10 000. This explains why extensive software systems are needed that support ^{13}C MFA in all phases of modeling, experimental design, and data evaluation. Over the recent years, several such software frameworks have been published such as ^{13}CFLUX, ^{13}CFLUX2, influx_s, Metran, NMR2FLUX, or OpenFlux that serve the standard tasks of the ^{13}C MFA workflow [49–56,92]. No recommendation for a specific software system is made here. Instead, the following checklist can be used to make a choice:

1) *Network modeling*: Unlike the simple toy network discussed in this introductory text, the size of real-life networks prohibits a manual formulation of all the equations needed to describe network stoichiometry, carbon atom transitions, input labeling, measurement configuration, or measured data with standard deviations. A well-designed software framework for ^{13}C MFA should guide the user through all these steps and, at the same time, give full flexibility to specify organism-specific networks and even exotic measurement configurations. Optimally, the whole configuration process is guided by graphical tools allowing for a visual verification. As an example for a modern graphical modeling environment, Figure 5.19 shows the

carbon atom transition specification window of the Omix visualization tool. The reader can refer to [93] for further examples of visual ^{13}C MFA modeling.

2) *Model generation*: A software tool should automatically generate all the required equations and computational primitives from the model specification. Moreover, it should analyze and simplify the equations on-the-fly to allow for high-speed computations. To avoid errors, the user should not be involved in this procedure. Moreover, the framework should do extensive tests for possible formal errors in the network model and experimental specification.
3) *Simulation*: As explained, the simulation algorithm predicting the outcome of an ILE based on the generated model, specified input labeling, and assumed intracellular fluxes is the core computational step of any ^{13}C MFA workflow. It should calculate the predictions as fast as possible because this procedure has to be repeated many times in all subsequent computational tasks. Several algorithms for simulation have been published, partly improving existing methods, partly introducing new ideas [41,52,59,94,95].

   To answer the question about the fastest simulator, it must be clearly said that not only the underlying algorithm itself but also the used benchmark example, the programming language of the software (e.g., MATLAB, Python, C++, or even assembler code), the implemented data structures to represent the model, the computing hardware (desktop PCs versus supercomputers), and, last but not least, the professionalism of the software implementation play a crucial role. Notably, to find the fastest algorithm in practice requires also having a deep look at the mathematical formulation itself. Comparing, for instance, the very similar cumomer and EMU methods (cf. Glossary) in a single computing environment shows, in fact, that the winner depends on the experimental setup [55]. After all, as a rough number even highly dimensional systems on the order of 10^5 isotopomer fractions can nowadays be solved on a consumer PC in a fraction of a second. This puts the ongoing discussion on the fastest algorithm into perspective because computation time is not the limiting factor for classical ^{13}C MFA. It has to be said that the situation is quite different for INST MFA where differential equation systems instead of algebraic equations have to be solved.
4) *Sensitivity analysis*: A simulation algorithm should not only calculate model predictions but also communicate their sensitivities with respect to the input labeling and the intracellular fluxes. Unfortunately, most implementations still approximate these sensitivities by simple finite difference formulae, although this procedure is known to be unreliable and costly for complex system simulation. It is still not widely known that sensitivities can be exactly computed by the so-called automatic differentiation in about the same time as needed for the simulation itself [96]. This is not only much faster than finite differences but also highly reliable even in

poorly conditioned cases. Based on the calculated sensitivity matrices, the flux covariance matrix can be quickly and easily calculated which is the basis for experimental design and statistical data evaluation.

5) *Experimental design*: Standard optimal experimental design algorithms maximize some information criterion that condenses the expected statistical quality of the flux estimates in a single average number (cf. Section 5.20). Clearly, this is a drastic reduction of the multidimensional complexity and, thus, the resulting condensed results must be treated with care. For this reason, a software tool should allow to use different information criteria besides the D-criterion, for example, A-, E-, G-criteria, to weight the importance of the wanted fluxes or even to put the focus exclusively on a subset of all fluxes. The interested reader can consult excellent contributions covering this topic in depth [71,72,97].
6) *Flux estimation*: The majority of published tools perform flux estimation on the basis of an iterative parameter fitting algorithm. Mathematically speaking, this is a multidimensional nonlinear constrained minimization problem. Depending on the available measurements, the solution of this estimation problem is not *per se* well posed. This means that neither the uniqueness of the solution nor the fact that slight variations in the measurements shift the estimated fluxes only slightly can be guaranteed. Moreover, it is known that no deterministic strategy exists to solve such problems reliably in finite time. As a result, it is rather a common experience that various algorithms or even one single algorithm with different initial points converge to quite diverse solutions with almost identical SSR values. Technically, only the simultaneous usage of different optimization algorithms, the frequent multistarting with initial values distributed over the whole theoretically possible flux space and a careful evaluation of thousands of simulation runs help us to prevent misinterpretations. Nevertheless, even after exclusion of numerical artifacts, the uniqueness of the nonlinear parameter fitting problem is not assured. The case of distinct solutions, that is, multiple optima (like for e_{10} in Figure 5.8), are indeed encountered in practice (see [98] for an example). Unfortunately, most tool users are not aware of all these numerical issues that are ubiquitous and, in consequence, just use one single algorithm once. The good news is that nowadays this fitting procedure takes only several minutes even for large networks. Thus, exhaustive multistarts can easily be run over night time on a computer cluster.
7) *Statistical data evaluation*: After parameter fitting, the statistical quality of the obtained flux estimates needs to be assessed. In practice, there are more measured data than degrees of freedom; occationally the measurement information is redundant. This fact helps to correct for gross measurement errors (e.g., outliers) or to find inconsistencies between the network model and the measured data. Using the parameter covariance matrix is the common way to judge flux confidence regions and therefore should always be the starting point of the analysis. From the matrix, all flux

standard deviations, correlations, and approximate confidence regions can be rapidly calculated.

8) *Advanced statistical methods*: It is well known that in case of poorly determined fluxes, the covariance matrix calculated from the sensitivity matrix can produce misleading results. In this situation, Monte Carlo methods like parametric bootstrapping help to better describe the set of all those parameters that explain the experimental data well [99]. The interpretation of raw covariance matrices or even large-scale Monte Carlo results is laborious. Therefore, a user friendly graphical display of all statistical results in form of histograms and scatter plots is mandatory. At best, these plots are linked to the flux map.
9) *Supplementary service tools*: When talking about ^{13}C MFA, it should be noticed that the described toolbox is still not complete. In fact, a considerable amount of time is spent on the evaluation of raw data for producing the input data needed by flux analysis tools. For example, measured labeling fractions have to be derived from peak integrations of NMR or MS spectra. This, however, is a problem in itself and requires highly sensitive devices and access to the raw chromatograms [100]. Then in laborious painstaking manual work, use-data for ^{13}C MFA is extracted no matter whether commercial or noncommercial tools are used. Likewise, the problem of natural isotope correction is well known in the community [101–105]. Finally, the extracellular fluxes that are determined from mass balances around the cultivation device can be backed up only by proper confidence intervals if the respective statistics based on bioprocess models is done well (see [106,107] for examples). For this reason, a complete ^{13}C MFA tool suite also needs several auxiliary software tools accelerating, facilitating, and managing the complete data analysis pipeline that is, in the optimal case, embedded in a service-oriented scientific workflow framework [80].

Glossary

Metabolic fluxes	Synonymously used for reaction rates in a metabolic network.
^{13}C metabolic flux analysis or ^{13}C MFA	An experimental technique used to determine intracellular metabolic fluxes. Based on extracellular consumption and production rates augmented by isotopic labeling data from intracellular metabolites, the technique uses a mathematical model consisting of stoichiometric and isotopic balance equations. Knowledge of the metabolic network, its atom transitions, and reaction's reversibility is needed. The final result of a ^{13}C MFA, the flux distribution, is visualized in a flux map.

Isotope labeling experiment (ILE)	The experimental procedure underlying ^{13}C MFA. In an ILE, cells are cultivated under steady-state conditions where they are fed with isotopically labeled substrates. By metabolic activity, the isotopes are distributed all over the intracellular metabolite pools. The corresponding isotopic fractional enrichments are measured and provide the basis of flux estimation.
Metabolic and isotopic (pseudo-) steady state	A metabolic network operates in steady state if the intracellular concentrations (i.e., pool sizes) do not change over time. In practical terms, in (pseudo-) steady states any change of metabolite pool sizes is orders of magnitude below the fluxes through the pool and, hence, approximately zero. The second steady state type concerns the isotopic label enrichment; here, isotopic enrichments in a metabolite pool are constant. ^{13}C MFA, as discussed in this chapter, always relies on the metabolic (pseudo-) steady-state assumption. In contrast, a distinction is made between the isotopic steady-state experiments (^{13}C MFA) and the more advanced isotopic nonstationary experiments (INST MFA).
Proteinogenic amino acids	These amino acids are integrated into proteins, that is, the cell mass. By extracting and hydrolyzing proteins, a "retro-biosynthetic" access to amino acids in central metabolic pathways is obtained. Compared to free intracellular amino acids, proteinogenic amino acids have significantly longer turnover times and are also present in higher concentrations in the cell.
Experimental design	A statistical term denoting the targeted planning of experimental conditions with the goal to obtain the maximal information yield. Precisely, a designed ^{13}C MFA experiment strives for the smallest possible confidence intervals of the fluxes that are to be estimated in the study.
Isotopomer, cumomer, elementary metabolite unit	The term isotopomer is a synonym for "isotopic isomer." Isotopomers have a chemically identical structure but carry different carbon isotopes in their backbone. A metabolite with n carbon atoms can occur in 2^n differently labeled forms (i.e., isotopomers). Cumomers, short form of "cumulative isotopomers," and EMUs ("elementary metabolite units") are alternative concepts to describe measured labeling patterns. Both cumomers and EMUs are virtual molecules or fragments of metabolites not really needed to understand ^{13}C MFA. However, the formulation of mass balances in terms of cumomers or EMUs is computationally highly advantageous.

References

1 Marx, A., de Graaf, A.A., Wiechert, W., Eggeling, L. *et al.* (1996) Determination of the fluxes in central metabolism of *Corynebacterium glutamicum* by NMR spectroscopy combined with metabolite balancing. *Biotechnol. Bioeng.*, **49**, 111–129.

2 Vallino, J.J. and Stephanopoulos, G. (1993) Metabolic flux distributions in *Corynebacterium glutamicum* during growth and lysine overproduction. *Biotechnol. Bioeng.*, **41**, 633–646.

3 Wiechert, W. (2001) Minireview:^{13}C metabolic flux analysis. *Metab. Eng.*, **3**, 195–206.

4 Petersen, S., de Graaf, A.A., Eggeling, L., Möllney, M. *et al.* (2000) *In vivo* quantification of parallel and bidirectional fluxes in the anaplerosis of *Corynebacterium glutamicum*. *J. Biol. Chem.*, **275**, 35932–35941.

5 Wiechert, W., Möllney, M., Petersen, S., and de Graaf, A.A. (2001) A universal framework for ^{13}C metabolic flux analysis. *Metab. Eng.*, **3**, 265–283.

6 Zamboni, N., Fendt, S.M., Rühl, M., and Sauer, U. (2009) ^{13}C-based metabolic flux analysis. *Nat. Protoc.*, **4**, 878–892.

7 Winder, C.L., Dunn, W.B., and Goodacre, R. (2011) TARDIS-based microbial metabolomics: time and relative differences in systems. *Trends Microbiol.*, **19**, 315–322.

8 Millard, P., Massou, S., Wittmann, C., Portais, J.-C. *et al.* (2014) Sampling of intracellular metabolites for stationary and non-stationary ^{13}C metabolic flux analysis in *Escherichia coli*. *Anal. Biochem.*, **465C**, 38–49.

9 van Gulik, W.M. (2010) Fast sampling for quantitative microbial metabolomics. *Curr. Opin. Biotechnol.*, **21**, 27–34.

10 Oldiges, M., Lütz, S., Pflug, S., Schroer, K. *et al.* (2007) Metabolomics: current state and evolving methodologies and tools. *Appl. Microbiol. Biotechnol.*, **76**, 495–511.

11 You, L., Page, L., Feng, X., Berla, B. *et al.* (2012) Metabolic pathway confirmation and discovery through ^{13}C-labeling of proteinogenic amino acids. *J. Vis. Exp.*, **59**, e3583.

12 Nielsen, J.H. and Villadsen, J. (1994) *Bioreaction Engineering Principles*, Kluwer Academic Pub., New York.

13 Martzolff, A., Cahoreau, E., Cogne, G., Peyriga, L. *et al.* (2012) Photobioreactor design for isotopic non-stationary ^{13}C-metabolic flux analysis (INST ^{13}C-MFA) under photoautotrophic conditions. *Biotechnol. Bioeng.*, **109**, 1–11.

14 Nanchen, A., Schicker, A., and Sauer, U. (2006) Nonlinear dependency of intracellular fluxes on growth rate in miniaturized continuous cultures of *Escherichia coli*. *Appl. Environ. Microbiol.*, **72**, 1164–1172.

15 Crown, S.B. and Antoniewicz, M.R. (2013) Parallel labeling experiments and metabolic flux analysis: past, present and future methodologies. *Metab. Eng.*, **16**, 21–32.

16 Kiefer, P., Heinzle, E., Zelder, O., and Wittmann, C. (2004) Comparative metabolic flux analysis of lysine-producing *Corynebacterium glutamicum* cultured on glucose or fructose. *Appl. Environ. Microbiol.*, **70**, 229–239.

17 Tang, Y.J., Martin, H.G., Myers, S., Rodriguez, S. *et al.* (2009) Advances in analysis of microbial metabolic fluxes via ^{13}C-isotopic labeling. *Mass Spectrom. Rev.*, **28**, 362–375.

18 Zamboni, N. (2011) ^{13}C-metabolic flux analysis in complex systems. *Curr. Opin. Biotechnol.*, **22**, 103–108.

19 Toya, Y. and Shimizu, H. (2013) Flux analysis and metabolomics for systematic metabolic engineering of microorganisms. *Biotechnol. Adv.*, **31**, 818–826.

20 Klein, S. and Heinzle, E. (2012) Isotope labeling experiments in metabolomics and fluxomics. *Wiley Interdiscip. Rev. Syst. Biol. Med.*, **4**, 261–272.

21 Britton, G., Goodwin, T.W., Lockley, W.J.S., Mundy, A.P. *et al.* (1979) Stereochemistry of cyclization in carotenoid biosynthesis; use of ^{13}C-labelling to elucidate the stereochemical behaviour of the C-1

methyl substituents during zeaxanthin biosynthesis in a Flavobacterium. *J. Chem. Soc. Chem. Commun.*, 27–28.

22 Skipper, H.E., Bryan, C.E., White, L., and Hutchison, O.S. (1948) Techniques for *in vivo* tracer studies with radioactive carbon. *J. Biol. Chem.*, **173**, 371–381.

23 Schepartz, B. and Gurin, S. (1949) The intermediary metabolism of phenylalanine labeled with radioactive carbon. *J. Biol. Chem.*, **180**, 663–673.

24 Schoenheimer, R. and Rittenberg, D. (1935) Deuterium as an indicator in the study of intermediary metabolism. *Science.*, **82**, 156–157.

25 Walsh, K. and Koshland, D.E. (1984) Determination of flux through the branch point of two metabolic cycles. The tricarboxylic acid cycle and the glyoxylate shunt. *J. Biol. Chem.*, **259**, 9646–9654.

26 Sonntag, K., Eggeling, L., de Graaf, A.A., and Sahm, H. (1993) Flux partitioning in the split pathway of lysine synthesis in *Corynebacterium glutamicum*. Quantification by ^{13}C- and ^{1}H-NMR spectroscopy. *Eur. J. Biochem.*, **213**, 1325–1331.

27 Blum, J.J. and Stein, R.B. (1982) On the analysis of metabolic networks, in *Biological Regulation and Development*, Plenum Press, pp. 99–124.

28 Malloy, C.R., Sherry, A.D., and Jeffrey, F.M. (1988) Evaluation of carbon flux and substrate selection through alternate pathways involving the citric acid cycle of the heart by ^{13}C NMR spectroscopy. *J. Biol. Chem.*, **263**, 6964–6971.

29 Szyperski, T. (1995) Biosynthetically directed fractional ^{13}C-labeling of proteinogenic amino acids. An efficient analytical tool to investigate intermediary metabolism. *Eur. J. Biochem.*, **232**, 433–448.

30 Eisenreich, W., Strauss, G., Werz, U., Fuchs, G. *et al.* (1993) Retrobiosynthetic analysis of carbon fixation in the phototrophic eubacterium *Chloroflexus aurantiacus*. *Eur. J. Biochem.*, **215**, 619–632.

31 Christensen, B. and Nielsen, J. (1999) Isotopomer analysis using GC–MS. *Metab. Eng.*, **1**, 282–290.

32 van Winden, W., Schipper, D., Verheijen, P., and Heijnen, J. (2001) Innovations in generation and analysis of 2D [^{13}C, ^{1}H] COSY NMR spectra for metabolic flux analysis purposes. *Metab. Eng.*, **3**, 322–343.

33 Szyperski, T. (1998) ^{13}C-NMR, MS and metabolic flux balancing in biotechnology research. *Q. Rev. Biophys.*, **31**, 41–106.

34 Wittmann, C. and Heinzle, E. (1999) Mass spectrometry for metabolic flux analysis. *Biotechnol. Bioeng.*, **62**, 739–750.

35 Möllney, M., Wiechert, W., Kownatzki, D., and de Graaf, A.A. (1999) Bidirectional reaction steps in metabolic networks: IV. Optimal design of isotopomer labeling experiments. *Biotechnol. Bioeng.*, **66**, 86–103.

36 van Ooyen, J., Noack, S., Bott, M., Reth, A. *et al.* (2012) Improved L-lysine production with *Corynebacterium glutamicum* and systemic insight into citrate synthase flux and activity. *Biotechnol. Bioeng.*, **109**, 2070–2081.

37 Luo, B., Grönke, K., Takors, R., Wandrey, C. *et al.* (2007) Simultaneous determination of multiple intracellular metabolites in glycolysis, pentose phosphate pathway and tricarboxylic acid cycle by liquid chromatography–mass spectrometry. *J. Chromatogr. A*, **1147**, 153–164.

38 Rühl, M., Rupp, B., Nöh, K., Wiechert, W. *et al.* (2012) Collisional fragmentation of central carbon metabolites in LC–MS/MS increases precision of ^{13}C metabolic flux analysis. *Biotechnol. Bioeng.*, **109**, 763–771.

39 Dauner, M. and Sauer, U. (2000) GC–MS analysis of amino acids rapidly provides rich information for isotopomer balancing. *Biotechnol. Prog.*, **16**, 642–649.

40 Choi, J. and Antoniewicz, M.R. (2011) Tandem mass spectrometry: a novel approach for metabolic flux analysis. *Metab. Eng.*, **13**, 225–233.

41 Antoniewicz, M.R., Kelleher, J.K., and Stephanopoulos, G. (2007) Elementary metabolite units (EMU): a novel framework for modeling isotopic distributions. *Metab. Eng.*, **9**, 68–86.

42 Wiechert, W. and de Graaf, A.A. (1997) Bidirectional reaction steps in metabolic

networks: I. Modeling and simulation of carbon isotope labeling experiments. *Biotechnol. Bioeng.*, **55**, 101–117.

43 van Winden, W.A., van Gulik, W.M., Schipper, D., Verheijen, P.J.T. *et al.* (2003) Metabolic flux and metabolic network analysis of *Penicillium chrysogenum* using 2D [^{13}C,^{1}H] COSY NMR measurements and cumulative bondomer simulation. *Biotechnol. Bioeng.*, **83**, 75–92.

44 Wiechert, W. and Wurzel, M. (2001) Metabolic isotopomer labeling systems. Part I: global dynamic behavior. *Math. Biosci.*, **169**, 173–205.

45 Schmidt, K., Marx, A., de Graaf, A.A., Wiechert, W. *et al.* (1998) ^{13}C tracer experiments and metabolite balancing for metabolic flux analysis: comparing two approaches. *Biotechnol. Bioeng.*, **58**, 6–9.

46 Weitzel, M., Wiechert, W., and Nöh, K. (2007) The topology of metabolic isotope labeling networks. *BMC Bioinformatics*, **8**, 315.

47 Nöh, K., Grönke, K., Luo, B., Takors, R. *et al.* (2007) Metabolic flux analysis at ultra short time scale: isotopically non-stationary ^{13}C labeling experiments. *J. Biotechnol.*, **129**, 249–267.

48 Schaub, J., Mauch, K., and Reuss, M. (2008) Metabolic flux analysis in *Escherichia coli* by integrating isotopic dynamic and isotopic stationary ^{13}C labeling data. *Biotechnol. Bioeng.*, **99**, 1170–1185.

49 Sokol, S., Millard, P., and Portais, J.-C. (2012) influx_s: increasing numerical stability and precision for metabolic flux analysis in isotope labelling experiments. *Bioinformatics*, **28**, 687–693.

50 Quek, L.-E., Wittmann, C., Nielsen, L.K., and Krömer, J.O. (2009) OpenFLUX: efficient modelling software for ^{13}C-based metabolic flux analysis. *Microb. Cell Fact.*, **8**, 25.

51 Zamboni, N., Fischer, E., and Sauer, U. (2005) FiatFlux: a software for metabolic flux analysis from ^{13}C-glucose experiments. *BMC Bioinformatics*, **6**, 209–216.

52 Srour, O., Young, J.D., and Eldar, Y.C. (2011) Fluxomers: a new approach for ^{13}C metabolic flux analysis. *BMC Syst. Biol.*, **5**, 129.

53 Cvijovic, M., Olivares-Hernández, R., Agren, R., Dahr, N. *et al.* (2010) BioMet toolbox: genome-wide analysis of metabolism. *Nucleic Acids Res.*, **38**, W144–W149.

54 Yoo, H., Antoniewicz, M.R., Stephanopoulos, G., and Kelleher, J.K. (2008) Quantifying reductive carboxylation flux of glutamine to lipid in a brown adipocyte cell line. *J. Biol. Chem.*, **283**, 20621–20627.

55 Weitzel, M., Nöh, K., Dalman, T., Niedenführ, S. *et al.* (2013) 13CFLUX2: high-performance software suite for ^{13}C-metabolic flux analysis. *Bioinformatics*, **29**, 143–145.

56 Zamboni, N., Fendt, S.-M., Rühl, M., and Sauer, U. (2009) ^{13}C-based metabolic flux analysis. *Nat. Protoc.*, **4**, 878–892.

57 Young, J.D. (2014) INCA: a computational platform for isotopically non-stationary metabolic flux analysis. *Bioinformatics*, **30**, 1333–1335.

58 Kajihata, S., Furusawa, C., Matsuda, F., and Shimizu, H. (2014) OpenMebius: an open source software for isotopically nonstationary ^{13}C-based metabolic flux analysis. *Biomed Res. Int.*, **2014**, 627014.

59 Wiechert, W., Möllney, M., Isermann, N., Wurzel, M. *et al.* (1999) Bidirectional reaction steps in metabolic networks: III. Explicit solution and analysis of isotopomer labeling systems. *Biotechnol. Bioeng.*, **66**, 69–85.

60 Wiechert, W. (2007) The thermodynamic meaning of metabolic exchange fluxes. *Biophys. J.*, **93**, 2255–2264.

61 Pitkänen, E., Akerlund, A., Rantanen, A., Jouhten, P. *et al.* (2008) ReMatch: a web-based tool to construct, store and share stoichiometric metabolic models with carbon maps for metabolic flux analysis. *J. Integr. Bioinform.*, **5**.

62 Ravikirthi, P., Suthers, P.F., and Maranas, C.D. (2011) Construction of an *E. coli* genome-scale atom mapping model for MFA calculations. *Biotechnol. Bioeng.*, **108**, 1372–1382.

63 Katz, J., Wals, P., and Lee, W.N. (1993) Isotopomer studies of gluconeogenesis and the Krebs cycle with ^{13}C-labeled lactate. *J. Biol. Chem.*, **268**, 25509–25521.

64 Ratkowsky, D.A. (1983) *Nonlinear Regression Modeling: A Unified Practical Approach*, Marcel Dekker, Inc.

65 Taylor, J.R. (1997) *Introduction to Error Analysis: The Study of Uncertainties in Physical Measurements*, Univ. Science Books, Revised edition.

66 Kümmel, A., Panke, S., and Heinemann, M. (2006) Systematic assignment of thermodynamic constraints in metabolic network models. *BMC Bioinformatics*, **7**, 512.

67 Wiechert, W., Siefke, C., de Graaf, A.A., and Marx, A. (1997) Bidirectional reaction steps in metabolic networks: II. Flux estimation and statistical analysis. *Biotechnol. Bioeng.*, **55**, 118–135.

68 Bates, D.M. and Watts, D.G. (1988) *Nonlinear Regression Analysis and Its Applications*, John Wiley & Sons.

69 Moles, C.G., Mendes, P., and Banga, J.R. (2003) Parameter estimation in biochemical pathways: a comparison of global optimization methods. *Genome Res.*, **13**, 2467–2474.

70 Gill, P.E., Murray, W., and Wright, M.H. (1982) *Practical Optimization*, Emerald Group Publishing Limited.

71 Pukelsheim, F. (1993) *Optimal Design of Experiments*, John Wiley & Sons.

72 Franceschini, G. and Macchietto, S. (2008) Model-based design of experiments for parameter precision: state of the art. *Chem. Eng. Sci.*, **63**, 4846–4872.

73 Wiechert, W. and Nöh, K. (2005) From stationary to instationary metabolic flux analysis. *Adv. Biochem. Eng. Biotechnol.*, **92**, 145–172.

74 Bennett, B.D., Kimball, E.H., Gao, M., Osterhout, R. *et al.* (2009) Absolute metabolite concentrations and implied enzyme active site occupancy in *Escherichia coli*. *Nat. Chem. Biol.*, **5**, 593–599.

75 Nöh, K. and Wiechert, W. (2011) The benefits of being transient: isotope-based metabolic flux analysis at the short time scale. *Appl. Microbiol. Biotechnol.*, **91**, 1247–1265.

76 Wiechert, W. and Nöh, K. (2013) Isotopically non-stationary metabolic flux analysis: complex yet highly informative. *Curr. Opin. Biotechnol.*, **24**, 979–986.

77 Schädel, F. and Franco-Lara, E. (2009) Rapid sampling devices for metabolic engineering applications. *Appl. Microbiol. Biotechnol.*, **83**, 199–208.

78 Tillack, J., Paczia, N., Nöh, K., Wiechert, W. *et al.* (2012) Error propagation analysis for quantitative intracellular metabolomics. *Metabolites*, **2**, 1012–1030.

79 Noack, S. and Wiechert, W. (2014) Quantitative metabolomics: a phantom? *Trends Biotechnol.*, **32**, 238–244.

80 Dalman, T., Droste, P., Weitzel, M., Wiechert, W. *et al.* (2010) Workflows for metabolic flux analysis: data integration and human interaction. *Lect. Notes Comput. Sci.*, **6415**, 261–275.

81 Nöh, K., Wahl, A., and Wiechert, W. (2006) Computational tools for isotopically instationary ^{13}C labeling experiments under metabolic steady state conditions. *Metab. Eng.*, **8**, 554–577.

82 Nöh, K. and Wiechert, W. (2006) Experimental design principles for isotopically instationary ^{13}C labeling experiments. *Biotechnol. Bioeng.*, **94**, 234–251.

83 Young, J.D., Walther, J.L., Antoniewicz, M.R., and Yoo, H. (2008) An elementary metabolite unit (EMU) based method of isotopically nonstationary flux analysis. *Biotechnol Bioeng*, **99** (3), 686–699.

84 Kumar, A., Suthers, P.F., and Maranas, C.D. (2012) MetRxn: a knowledgebase of metabolites and reactions spanning metabolic models and databases. *BMC Bioinformatics*, **13**, 6.

85 Kim, T.Y., Sohn, S.B., Kim, Y. Bin, Kim, W.J. *et al.* (2012) Recent advances in reconstruction and applications of genome-scale metabolic models. *Curr. Opin. Biotechnol.*, **23**, 617–623.

86 Chelliah, V., Laibe, C., and Le Novère, N. (2013) BioModels database: a repository of mathematical models of biological processes. *Methods Mol. Biol.*, **1021**, 189–199.

87 Eggeling, L. and Bott, M. (2005) *Handbook of Corynebacterium Glutamicum*, CRC Press.

88 Wolfe, A.J. (2005) The acetate switch. *Microbiol. Mol. Biol. Rev.*, **69**, 12–50.

89 Otterstedt, K., Larsson, C., Bill, R.M., Ståhlberg, A. *et al.* (2004) Switching the mode of metabolism in the yeast *Saccharomyces cerevisiae*. *EMBO Rep.*, **5**, 532–537.

90 Mu, F., Williams, R.F., Unkefer, C.J., Unkefer, P.J. *et al.* (2007) Carbon-fate maps for metabolic reactions. *Bioinformatics*, **23**, 3193–3199.

91 Heller, S., McNaught, A., Stein, S., Tchekhovskoi, D. *et al.* (2013) InChI: the worldwide chemical structure identifier standard. *J. Cheminform.*, **5**, 7.

92 Sriram, G., Fulton, D.B., Iyer V, V., Peterson, J.M. *et al.* (2004) Quantification of compartmented metabolic fluxes in developing soybean embryos by employing biosynthetically directed fractional ^{13}C labeling, two-dimensional [^{13}C, ^{1}H] nuclear magnetic resonance, and comprehensive isotopomer balancing. *Plant Physiol.*, **136**, 3043–3057.

93 Nöh, K., Droste, P., and Wiechert, W. (2015) Visual workflows for ^{13}C metabolic flux analysis. *Bioinformatics*, **31**, 346–354.

94 Sriram, G. and Shanks J, V. (2004) Improvements in metabolic flux analysis using carbon bond labeling experiments: bondomer balancing and boolean function mapping. *Metab. Eng.*, **6**, 116–132.

95 van Winden, W.A., Heijnen, J.J., and Verheijen, P.J.T. (2002) Cumulative bondomers: a new concept in flux analysis from 2D [^{13}C, ^{1}H] COSY NMR data. *Biotechnol. Bioeng.*, **80**, 731–745.

96 Naumann, U. (2012) *The Art of Differentiating Computer Programs: An Introduction to Algorithmic Differentiation*, SIAM.

97 Pázman, A. (1986) *Foundations of Optimum Experimental Design*, Springer, Boston, MA, USA.

98 Beste, D.J.V., Bonde, B., Hawkins, N., Ward, J.L. *et al.* (2011) ^{13}C metabolic flux analysis identifies an unusual route for pyruvate dissimilation in Mycobacteria which requires isocitrate lyase and carbon dioxide fixation. *PLoS Pathog.*, **7**, e1002091.

99 Joshi, M., Seidel-Morgenstern, A., and Kremling, A. (2006) Exploiting the bootstrap method for quantifying parameter confidence intervals in dynamical systems. *Metab. Eng.*, **8**, 447–455.

100 Bueschl, C., Krska, R., Kluger, B., and Schuhmacher, R. (2013) Isotopic labeling-assisted metabolomics using LC–MS. *Anal. Bioanal. Chem.*, **405**, 27–33.

101 Fernandez, C.a., Des Rosiers, C., Previs, S.F., David, F. *et al.* (1996) Correction of ^{13}C mass isotopomer distributions for natural stable isotope abundance. *J. Mass Spectrom.*, **31**, 255–262.

102 Millard, P., Letisse, F., Sokol, S., and Portais, J.-C. (2012) IsoCor: correcting MS data in isotope labeling experiments. *Bioinformatics*, **28**, 1294–1296.

103 Wahl, S.A., Dauner, M., and Wiechert, W. (2004) New tools for mass isotopomer data evaluation in ^{13}C flux analysis: mass isotope correction, data consistency checking, and precursor relationships. *Biotechnol. Bioeng.*, **85**, 259–268.

104 van Winden, W.A., Wittmann, C., Heinzle, E., and Heijnen, J.J. (2002) Correcting mass isotopomer distributions for naturally occurring isotopes. *Biotechnol. Bioeng.*, **80**, 477–479.

105 Wittmann, C. (2007) Fluxome analysis using GC–MS. *Microb. Cell Fact.*, **6**, 6.

106 Noack, S., Nöh, K., Moch, M., Oldiges, M. *et al.* (2011) Stationary versus non-stationary ^{13}C-MFA: a comparison using a consistent dataset. *J. Biotechnol.*, **154**, 179–190.

107 Hanke, T., Nöh, K., Noack, S., Polen, T. *et al.* (2013) Combined fluxomics and transcriptomics analysis of glucose catabolism via a partially cyclic pentose phosphate pathway in *Gluconobacter oxydans* 62^{1}H. *Appl. Environ. Microbiol.*, **79**, 2336–2348.

6
Genome-Scale Models

Basti Bergdahl, Nikolaus Sonnenschein, Daniel Machado, Markus Herrgård, and Jochen Förster

Summary

An introduction to genome-scale models, how to build and use them, will be given in this chapter. Genome-scale models have become an important part of systems biology and metabolic engineering, and are increasingly used in research, both in academica and in industry, both for modeling chemical production and for fundamental studies aiming at describing and explaining phenotypic behavior of microbial and mammalian cells.

6.1
Introduction

Chapter 4 introduced the concept of metabolic flux analysis, a powerful tool for modeling metabolic behavior and estimating intracellular and extracellular fluxes. In metabolic flux analysis, reduced or lumped models of the central carbon metabolism, or parts of metabolism, such as glycolysis, are applied, the metabolic behavior is described, and intracellular fluxes are simulated. The models are supplemented with experimental data such as production rates and sugar consumption rates. Chapter 6 is a primer on genome-scale models. It gives an introduction to building metabolic reconstructions and shows examples of the application of genome-scale models, that is, for chemical production or drug compound identification. Genome-scale models became available shortly after the complete genomes were sequenced, and today with the availability of next-generation sequencing methods that allow overnight sequencing of complete microbial genomes, they are in the repertoire of many research groups that focus on systems biology and metabolic engineering for chemical and fuel production using microorganisms or higher eukaryotic cells. Genome-scale models contain "all" metabolic reactions or as many reactions as the user finds meaningful, and they are initially based on an annotated genome sequence combined with experimental evidence on stoichiometry and reversibility of metabolic reactions. Here,

Fundamentals of Bioengineering, First Edition. Edited by John Villadsen.

genes are directly related to reactions and vice versa. A genome-scale model relies on mass balances around metabolites and for simplicity one does – at least initially – neglect any regulation or information from transcript-level kinetics, metabolite concentration, and so on and assumes pseudo-steady state in the simulation. Typically, the outcome is linear models or equation systems that are underdetermined, but can be solved with standard tools of linear algebra, that is, using linear programming, mixed integer linear programming, or convex analysis. After introduction of the concepts of reconstruction of metabolic networks and genome-scale models in Section 6.2, tools for solving genome-scale models are presented in Section 6.3. Various examples of bacterial as well as eukaryotic models and model simulations will be given in Sections 6.4 and 6.5. Today, there is much focus on integration of data from other levels of metabolism, such as the proteome, transcriptome, and metabolome, and Section 6.6 gives an overview of methods that combine proteome, transcriptome, and metabolome data with genome-scale models. In the future, such methods will be very useful both for genome-scale data analysis and for building models with more predictive power.

6.2
Reconstruction Process of Genome-Scale Models

The reconstruction of a metabolic network or the design of a genome-scale models starts with the annotated genome sequence. When a genome sequence is not available, it is today possible to rapidly sequence and prelimarily annotate a genome, and the model construction starts from there. All genes that are identified to encode enzymes are collected into a database and the corresponding enzymatic reactions are identified. A relation between genes and reactions is established (sometimes referred to as a genotype–phenotype relation), and one has to take into account that enzyme complexes may be encoded by two or several genes and that genes often encode enzymes that are promiscuous or utilize different cofactors and hence can catalyze more than one reaction (Figure 6.1). Also, as mentioned in the introduction, genome-scale models focus solely on establishing a link between all fluxes (the fluxome) and the corresponding genes and information from intermediate levels of metabolism such as proteome, transcriptome, and the metabolome are neglected.

Establishing a clear link between genes and reaction is very important in order to be able to simulate the (potential) impact of gene deletion(s) or gene overexpression(s) on the phenotypic behavior (i.e., growth or chemical production). The reconstruction process is further detailed in Figure 6.2.

Next to the information from the annotated genome sequence, the metabolic reconstruction and the collection of reactions is supported by consulting standard biochemistry textbooks, pathway databases such as the KEGG database at www.genome.jp or the MetaCyc database at www.metacyc.com, and so on, and scientific articles that give clear-cut evidence of the properties of the reactions

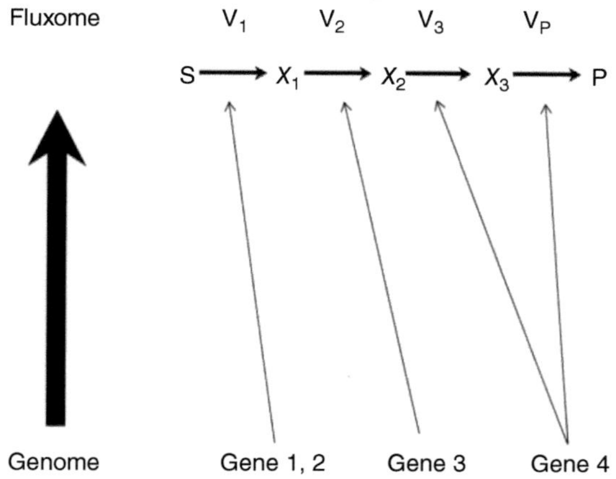

Figure 6.1 The genotype–phenotype relationship in genome-scale models. Genome-scale models exclusively include information about the stoichiometry of metabolic reactions, and they neglect any information at other levels of the metabolism, that is, transcriptome, proteome, regulation, and so on. Here some of the typical scenarios for the genotype–phenotype relationship are presented: gene 1 and gene 2 encode proteins, which build an enzyme complex that catalyzes the reaction from substrate S to intermediate metabolite X_1 with the corresponding flux v_1; gene 3 encodes an enzymes that catalyzes the reaction from metabolite X_1 to X_2 with flux v_2; and gene 4 encodes one enzyme that can carry out multiple reactions with fluxes v_3 and v_p catalyzing the conversion of metabolite X_2 to X_3 and of metabolite X_3 to product P, respectively.

that should be included in the model. Information from biochemistry textbooks should be considered with caution as textbooks are often not specific to the organism that one wishes to model. The stoichiometry of metabolic reactions is usually known except for transport reactions (where one has to estimate the number of protons that are transported together with an intermediate metabolite). Reactions that can use two cofactors such as NADH or NADPH (ATP or GTP) should be included as two reactions, one utilizing NADH (ATP) and one utilizing NADPH (GTP). Information on reversibility can be difficult to obtain, and while one can argue that all reactions are reversible, it is beneficial to identify reactions that are overwhelmingly irreversible, for example, the phosphorylation of glucose to glucose-6-phosphate in *Saccharomyces cerevisiae.* Localization has to be taken into account. The mass balance of metabolites has to be considered for each compartment and need to close. Hence, it is important to differentiate metabolites that are chemically identical but are contained in different compartments, that is, malate can be found both in the cytosol and in the mitochondria in *Saccharomyces cerevisiae.*

To model growth, a drain of precursor metabolites for macromolecules RNA, DNA, and proteins and for other cell constituents such as lipids and carbohydrates needs to be estimated. This drain can be calculated from the experimental determination of the biomass composition. As the determination of the biomass composition is cumbersome and tedious, most models make rough assumptions,

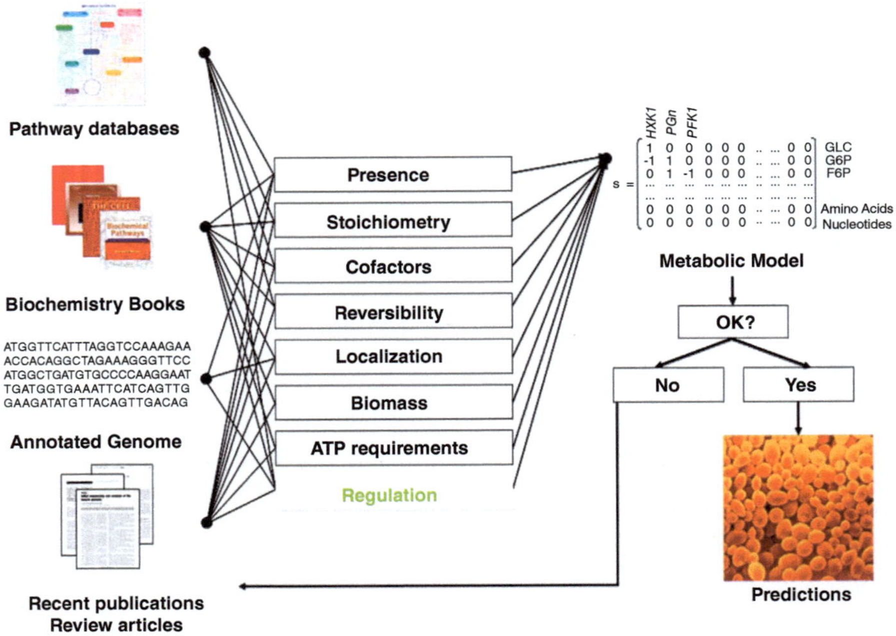

Figure 6.2 Metabolic reconstruction of genome-scale models. A genome-scale reconstruction starts by consulting the annotated genome sequence, and subsequently the metabolic reactions will be collected that are (hypothetically) known to be available in the metabolic network of the organism to be reconstructed. When determining whether a reaction is present in a cell, typical questions are as follows: what is the stoichiometry? What are the cofactors? Is the reaction reversible, in which compartment is the reaction localized? Furthermore, information on the biomass composition must be available to define a drain of precursor metabolite into biomass, so should be good estimates of growth- and nongrowth-dependent ATP requirements. Generally, the more experimental evidence is available for each reaction, the better is the model. Hence, it is important to search beyond the annotated genome sequence with regard to the presence and properties of a reaction. (adopted from Ref. [1].)

that is, comparing compositions that are known from other organisms, and use these for defining a drain into biomass. Last but not least, in order to be able to calculate fluxes and flux distributions that are physiologically meaningful, one must include well-estimated values for growth- and nongrowth-dependent ATP requirements (see Chapter 4 for estimation of ATP requirements). Once all the information has been collected, the set of reactions can be translated into a mathematical formulation that allows simulation of metabolic behavior (see Figure 6.3). Initially, it will be important to verify the reaction set by simulating known results, compare the model simulation with the experimental results, and adjust the reaction set until it appears biologically meaningful.

While the reconstruction of genome-scale models is often done manually, there is now much focus on automizing the reconstruction process [2–5]. While

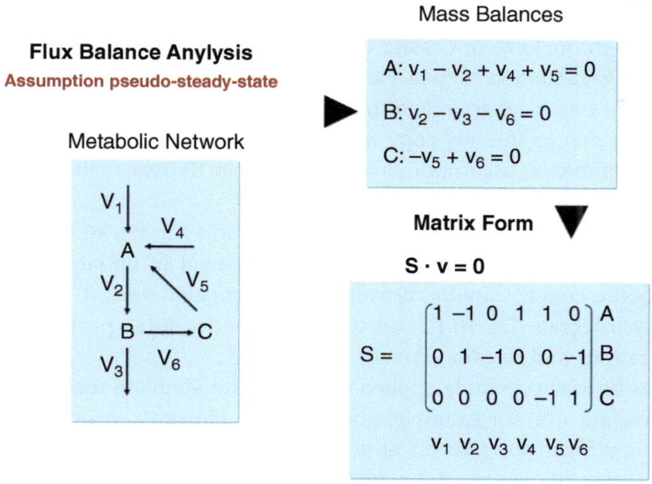

Figure 6.3 Metabolic reconstruction and flux balance analysis A metabolic network with three metabolites A, B, and C and 6 fluxes v_1–v_6. Assuming pseudo-steady state, mass balances around each metabolite lead to an underdetermined linear equation system with six variables, the fluxes, and three equations, the mass balances around each of the metabolites. The fluxes v_i typically have the unit [mmol/gDW/h]. The linear equation system can be written in matrix form $S \cdot v = 0$, where S is the stoichiometric matrix containing the stoichiometric coefficients of the metabolites.

many of the strategies are very promsing, a manual curation of such automatic or semiautomatic reconstructions is still necessary.

6.3 Genome-Scale Model Prediction

6.3.1 Mathematical Description of Biochemical Reaction Systems

A commonly utilized mathematical formalism for describing biochemical reaction systems is in terms of coupled ordinary differential equations (ODEs):

$$\frac{d\boldsymbol{x}(t)}{dt} = \boldsymbol{S}\,\boldsymbol{v}(\boldsymbol{x}(t), \boldsymbol{p}, \ldots). \tag{6.1}$$

Vector $\boldsymbol{x}$ represents time-dependent concentrations of the chemical species. Vector $\boldsymbol{v}$ contains the rate expressions that are functions of $\boldsymbol{x}$ and also time-invariant parameters $\boldsymbol{p}$ that represent net reaction fluxes. Matrix $\boldsymbol{S}$ captures the stoichiometry of the system and interlinks fluxes $\boldsymbol{v}$ to form a set of equations that describe the concentration changes.

In order for Eq. (6.1) to be an appropriate description of the system, the following assumptions have to hold: (i) molecules diffuse quickly and are homogeneously distributed, thus making subcellular localization unimportant; and (ii) copy numbers are reasonably high, making stochastic effects irrelevant (see Mahmutovic *et al.* [6] for an extensive discussion of spatial and stochastic aspects).

Although certainly not an ideal approach to the analysis of transcriptional and translational processes due to the low copy number and spatial organization of the participating components, assumptions (i) and (ii) can be reasonably well applied to cellular metabolism.

Generally, it is not possible to find an analytical solution for Eq. (6.1) (see Halasz *et al.* [7] for a notable exception). However, given a set of initial conditions, that is, metabolite concentrations measured at a particular state, it is possible to numerically integrate Eq. (6.1) and to obtain time–course profiles of metabolite concentrations and reaction fluxes.

This approach has been successfully applied to model and simulate metabolic pathways of intermediate size, for example, glycolysis [8]. However, parameterization of Eq. (6.1) is difficult for systems at genome scale. Consequently, other approaches that use simplifications of Eq. (6.1) have been utilized to scale up modeling and will be discussed in the following section.

6.3.2
Constraint-Based Modeling

In this section, we will focus on structural approaches that consider the steady-state behavior of biochemical reaction systems and incorporate constraints, for example, thermodynamic relations or flux capacities of transport reactions, to narrow down the space of feasible solutions. This mathematical framework for the analysis of genome-scale metabolic systems is commonly referred to as *constraint-based modeling* [9].

Under steady-state conditions (i.e., $d\mathbf{x}(t)/dt = 0$) Eq. (6.1) reduces to

$$\boldsymbol{S}\,\boldsymbol{v} = 0 \tag{6.2}$$

This is a system of homogeneous linear equations (see also Figure 6.3). Usually, the number of reactions exceeds the number of species in a biochemical system, rendering Eq. (6.2) underdetermined. So, no unique solution exists, but the correct $\boldsymbol{v}$ lies somewhere in the null space of $\boldsymbol{S}$ (i.e., $Null(\boldsymbol{S})$). The introduction of constraints on $\boldsymbol{v}$, for example, irreversibility and capacity constraints,

$$\boldsymbol{v}_{\mathrm{lb}} \leq \boldsymbol{v} \leq \boldsymbol{v}_{\mathrm{ub}} \tag{6.3}$$

will narrow down the space of all possible solutions into a convex shape (flux cone) that is a subspace of $Null(\boldsymbol{S})$ (see Figure 6.4).

6.3.3
Pathway Analysis

The mathematical groundwork for the analysis of Eqs. (6.2) and (6.3) has been laid out by Bruce Clarke in his seminal book chapter in *Advances in Chemical*

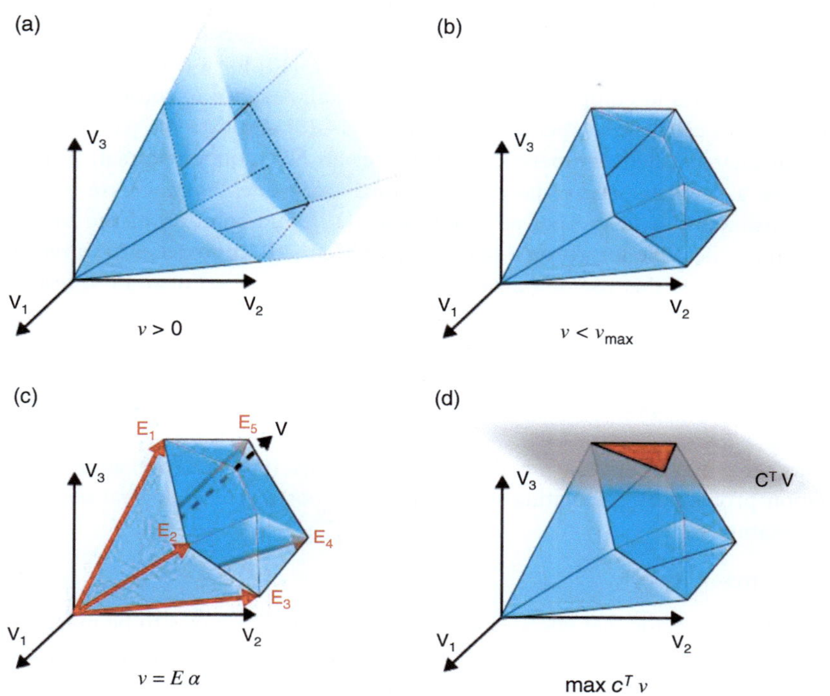

Figure 6.4 (a) The imposition of irreversibility constraints renders the solution space a convex hull, which is a subspace of the null space of S. (b) Capacity constraints (e.g., known uptake and secretion rates) eventually make the solution space bounded. (c) Extreme currents (in dark gray) constitute the basis solutions of $S\,v = 0$. (d) Biologically motivated objectives (gray surface) can be utilized to predict *in vivo* flux distributions. Computed solutions are not necessarily unique as exemplified in this schema (the surface spanned by the objective touches the cone at multiple positions covering the whole facet).

Physics [10] and is usually referred to as *stoichiometric network analysis* (SNA). Developed by Clarke to investigate stability properties of reaction systems, SNA contains one particular concept of special interest to the readers of this chapter, namely, *extreme currents*.

Extreme currents constitute the basic solutions of Eqs. (6.2) and (6.3) and represent the vertices of a convex flux cone (Figure 6.4a). Furthermore, every steady-state flux distribution can be decomposed in the following way:

$$\boldsymbol{v} = \boldsymbol{E\alpha} \tag{6.4}$$

In Eq. (6.4), $\boldsymbol{E}$ is a matrix whose columns capture all extreme currents of the system.

Two groups realized later that these basic solutions provide biologically relevant pathway descriptions and developed slightly different definitions of Eq. (6.4), that is, elementary flux modes (EFM) [11] and extreme pathways

(ExPa) [12], where EFM allows for reversible reactions, while in the calculation of ExPa each reversible reaction is split into two irreversible reactions for the forward and backward directions (see Papin *et al.* [13] for a comprehensive historical review and comparison of EFM and ExPa). A series of software tools are available to calculate elementary flux modes [14,15] and extreme pathways [16]. Unfortunately, their enumeration constitutes an NP (nonpolynomial) problem, that is, a problem where the computational work does not increase with the dimensionality d to the power d but faster [17], and solution of the problem becomes infeasible for genome-scale models. Heuristic approaches have recently been developed that use optimization to sample from the set of all EFM (e.g., the k-shortest EFM [18]).

6.3.4 Flux Balance Analysis

Another powerful approach is the treatment of Eqs. (6.2) and (6.3) as an optimization problem by calculating $\boldsymbol{v}$ that optimizes postulated biological objectives, for example,

$$\max \boldsymbol{c}^T \boldsymbol{v}, \tag{6.5}$$

where $\boldsymbol{c}$ is a vector that encodes a linear combination of fluxes to be maximized (see Figure 6.4d). Examples of often utilized objectives include the maximization of ATP production or yield, growth rate, production of desired compounds, and minimization of fluxes (minimization of protein cost). For a detailed assessment of a selection of objectives and a detailed review of the popular biomass objective function, see Schuetz, Kuepfer, and Sauer [19] and Feist and Palsson [20].

The formulation and solution of problems similar to Eq. (6.4) are usually referred to as *flux balance analysis* (FBA; at steady state all internal fluxes have to balance each other to fulfill mass conservation – the concept was introduced in Chapter 4 and illustrated by a toy example). The global optimum of Eq. (6.4) can be efficiently calculated using linear programming since the objective function is linear and equations in (6.2) and (6.3) constitute linear equalities and inequalities involving only linear terms. Solutions to Eq. (6.4) are often degenerate for genome-scale models (see Figure 6.4d). Flux variability analysis (FVA) [21] (computing all minimal and maximal fluxes for $\boldsymbol{c}^T\boldsymbol{v} = \max \boldsymbol{c}^T\boldsymbol{v}$) and Monte Carlo sampling can be used to estimate the amount of slack [22].

Flux balance analysis has proven to be a powerful tool for predicting gene lethality [23], flux distributions that (as illustrated in Chapter 5) match experimental ^{13}C flux measurements [24], and the outcome of adaptive laboratory evolution experiments [25]. Furthermore, the simplicity and success of constraint-based modeling has led to applications in a wide range of fields and to a large number of methods that tailor FBA to specific problems [26]. A detailed description of methods that incorporate experimental data is given in Section 6.6. In the following, we will discuss applications of constraint-based modeling to the redesign of biological systems.

6.3.5 Engineering Applications of Constraint-Based Modeling

The application of constraint-based modeling in metabolic engineering projects is straightforward. Theoretical yields can readily be assessed using FBA to calculate the minimum and maximum production rates of a specific product while varying growth rate (see production envelope in Figure 6.5a). Furthermore, methods based on pathway analysis (e.g., EFM) have been particularly popular with biologists [27], as they allow the enumeration and prioritization (by yield) of all possible flux routes from substrates to products, allowing the evaluation of metabolic engineering strategies on a pathway basis.

Furthermore, rational strain design methods have been developed that compute interventions (gene knockouts, modulation of gene expression, etc.) that will drive a metabolic system from a reference state to a desired production state. They fall into three categories:

i) Prediction of gene knockout combinations that either couple the production of a desired product to growth (see Figure 6.5a) or are likely to lead to a mutant flux distribution with a certain property.
ii) Comparison of flux ranges of a desired flux state and a reference flux distribution using flux variability analysis (see Figure 6.5b) to determine knockout, overexpression, and downregulation targets.
iii) Enumeration and scoring of optimal heterologous pathways for non-native compounds from a database of known reactions.

Multiple methods have been developed that enumerate (i) growth-coupled strain designs, that is, sets of reaction/gene deletions that force a growing

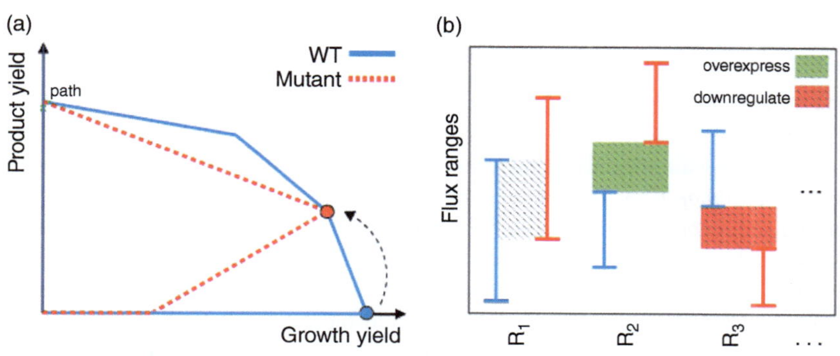

Figure 6.5 (a) Production envelopes for a wild-type and a mutant strain. The production rate r_p of P is coupled to the growth rate of the mutant. (b) Flux variability analysis-based prediction of overexpression and downregulation targets. The flux ranges of the wild type and mutant overlap for reaction R_1 so no clear recommendation can be made. Gaps are visible for the flux ranges of reaction R2 and R3 suggesting overexpressing and downregulating their corresponding genes in the mutant.

organism to produce a desired by-product (see Figure 6.4a). For example, the method OptKnock [28] uses mixed integer linear programming (MILP) to formulate a nested optimization problem, that is, an inner problem maximizing a biological objective and an outer problem maximizing the production of the desired product by reaction or gene deletions (using integer decision variables). Since the publication of OptKnock, a series of modifications and slightly different approaches have been published; for example, OptGene [29] uses genetic algorithms to solve the outer problem and FBA for the inner problem to determine growth-coupled designs. Other methods, like MOMA (minimization of metabolic adjustment: minimizing the Euclidean distance of the mutant flux distribution to the wild type) [30] or ROOM (regulatory on/off minimization: minimizing the number of flux changes in the mutant) can be used instead of FBA to predict likely mutant flux distributions that can be scored in terms of the production goal. Furthermore, an EFM-based method exists that uses minimal cut sets to exhaustively enumerate growth-coupled designs [31].

Methods like OptKnock predict knockout strategies but do not provide any input concerning overexpression or downregulation of target reactions. OptForce is an (ii) FVA-based method that compares flux ranges of a wild type and a desired production state (see Figure 6.5b) and uses a MILP formulation to compute a minimal set of interventions to push the system to the desired state (assuming that the rest of the reactions will follow to satisfy the steady-state condition).

Methods for the prediction of (iii) heterologous pathways include OptStrain [32], formulating a MILP that will compute a minimal set of reactions to include from a universal reaction matrix (compiled from pathway databases like KEGG and MetaCyc).

Available tools for (i–iii) range from the open distributions of source files (GAMS, Matlab scripts, etc.) to full-fledged tools with graphical user interfaces (e.g., OptFlux from [33]).

6.4 Genome-Scale Models of Prokaryotes

First genome-scale models of metabolism were built for prokaryotic organisms such as *Haemophilus influenzae* and *Escherichia coli* for which annotated genome sequences were available before eukaryotic genome sequencing projects were completed [34]. Prokaryotic metabolism has been extensively studied for over half a century. This made it possible to use both genome annotation information and detailed genetic, physiological, and biochemical studies in the metabolic network reconstruction process. Furthermore, much of this information for key model prokaryotes such as *E. coli* has been collected in handbooks and extensive review articles facilitating rapid development of draft reconstructions. In this chapter, we will briefly review the history of prokaryotic genome-scale metabolic models, current status, and new uses of these models, as well as the more recently established field of metabolic modeling of bacterial communities.

6.4.1
Escherichia coli

E. coli not only has served as the major model organism for prokaryotic biology but also has been used as a test bench for the development of most new genome-scale modeling methods. Extensive physiological and genetic studies of *E. coli* have been performed, and its genome was among the first to be sequenced [35]. For this reason, the current genome-scale metabolic model of *E. coli* iJO1366 [36] is the most well-curated and comprehensive metabolic model for any species. The over 20 year history of genome-scale models of *E. coli* has been extensively reviewed elsewhere [37] and we will provide only a brief summary here.

Initially, large-scale models of *E. coli* metabolism appeared in the pregenomic era of the early 1990s and were built solely based on published genetic, biochemical, and physiological characterizations. The largest of these models incorporated central carbon, amino acid, nucleotide, and cofactor metabolism as well as realistic biomass functions [38]. The models were already predictive of metabolic phenotypes, but their scope was limited by the availability of data on specific genes, proteins, and enzymes. The complete genome sequence of *E. coli* became available in 1997 [35] and since then four major versions of ever more complete GSMs of *E. coli* have been published [23,36,39,40]. These models have increased both in their coverage of metabolic pathways and in terms of the accuracy of representation of cellular metabolism [41]. For example, the early models did not include compartmentalization or charge and elemental balancing that are now standard features of new genome-scale models.

Genome-scale models of *E. coli* metabolism have been extensively used in a wide variety of application areas [37]. The most high-profile application of *E. coli* models has been in the study of basic system-wide biological mechanisms such as epistatic interactions, systemic structure of metabolism, and the role of different cellular objectives in bacterial physiology. An example of this type of studies is the recent exploration of what biological objectives are consistent with experimentally observed metabolic flux distributions in *E. coli* [42]. This study found that metabolic flux data obtained in a wide variety of combinations is consistent with three possible objective functions: maximization of biomass formation, maximization of ATP production, and minimization of total flux through the metabolic network. All experimentally measured flux distributions were located very close to the Pareto (multiobjective optimal) surface defined by these three objectives. This finding indicates that at least the space of central metabolic fluxes appears to be much smaller than what is allowed by the basic constraints included in genome-scale metabolic models.

The second major application of *E. coli* genome-scale models has been the prediction of genetic targets or media manipulation to enhance production of a biotechnologically relevant chemical or protein [42]. These approaches will be reviewed in more detail later in this chapter, but here we briefly mention some of the relevant work. Modeling methods have been established for both prediction of gene deletion targets [28] (to redirect flux) and for identification of overexpression targets [44] (to open up bottlenecks in pathways). These approaches have been used to successfully predict targets that enhance overproduction of desirable products such as amino acids and carotenoids in *E. coli* [45,46]. The model-based strategies usually need to be combined with other genetic manipulations, for example, to relieve allosteric inhibition of enzymes or optimizing expression of heterologous enzymes.

The third major application of *E. coli* genome-scale models highlighted here is the use of the models to identify gaps in our biological knowledge [47]. This typically requires measuring either metabolic fluxes or gene deletion strain growth phenotypes and then comparing model predictions of fluxes or growth phenotypes with experimental data. Major discrepancies in these comparisons can be used to suggest new reactions that should be added to the model and can also possibly indicate what genes could code for the corresponding enzymatic activities. Despite extensive characterization, there are still hundreds of poorly understood metabolic enzymes in *E. coli*, and model-driven gap filling approaches together with targeted experimentation have allowed identifying specific roles for some of these enzymes [48]. These studies have also paved the way for similar studies in other organisms including higher eukaryotes [49].

In addition to having served as a test bench for new applications, *E. coli* genome-scale models have also been used to model approaches that take genome-scale models beyond metabolism. These models have progressed furthest in incorporating processes related to protein biosynthesis (transcription and translation) into genome-scale models. The most recent generation of these types of models includes comprehensive descriptions of transcription, translation, protein turnover, and protein complex formation processes [50]. These metabolic/expression (ME) models expand the predictive capability of "metabolic-only" models (M) to, for example, more accurately predict the effects of overexpressing heterologous proteins by accounting for the cost of protein expression. ME models also allow more readily the integration of gene or protein expression data into genome-scale models since the corresponding processes are explicitly represented in the models. The major drawback of ME models is that they have very large numbers of variables and making even simple predictions requires much more powerful computing resources than M models. In addition to protein expression, there have also been efforts to incorporate transcriptional regulation into genome-scale metabolic models [51–54]. However, the representation of transcriptional regulatory networks has usually been done at the qualitative level through Boolean or Bayesian network formalisms, as the biochemical details about individual transcription factor–DNA binding events are not known except for specific cases.

The other area where *E. coli* has been used to aid development of new approaches for genome-scale metabolic modeling is in the reconstruction of strain-specific models. Advances in next-generation sequencing have made it possible to rapidly and accurately sequence the genomes of thousands of isolates of bacteria of the same species from, for example, different patients or environments [55,56]). In the case of *E. coli*, these isolates are usually found to be quite different from each other in their gene content, indicating that strain-specific differences in metabolic phenotypes should also exist [57]. Initial efforts for strain-specific model reconstruction were focused on industrial isolates of *E. coli* for which detailed physiological data exists, and thus the individual models could be validated thoroughly [58]. However, recently more automated strain-specific reconstruction methods have been developed and the resulting strain-specific models have been successfully used to predict more coarse-grained metabolic phenotypes such as the utilization of specific carbon sources in specific strains [59]. Development of strain-specific reconstructions for thousands of strains of *E. coli* is now in principle possible, and comparative analysis of the metabolic models will allow a deeper understanding of the role of metabolic physiology in pathogenesis as well as open up the possibility for developing more targeted treatments for pathogenic *E. coli*.

The history of *E. coli* genome-scale models and their applications highlights the overall progression in metabolic modeling that has taken place in the past 15 years (Figure 6.6). Genome-scale models have improved significantly in (1) the completeness of coverage of metabolic reaction space, (2) the accuracy of representation of metabolism (e.g., compartmentalization and charge balancing), and (3) the representation of strain-specific metabolic functions. At the same time, the application areas of *E. coli* genome-scale models have increased due to

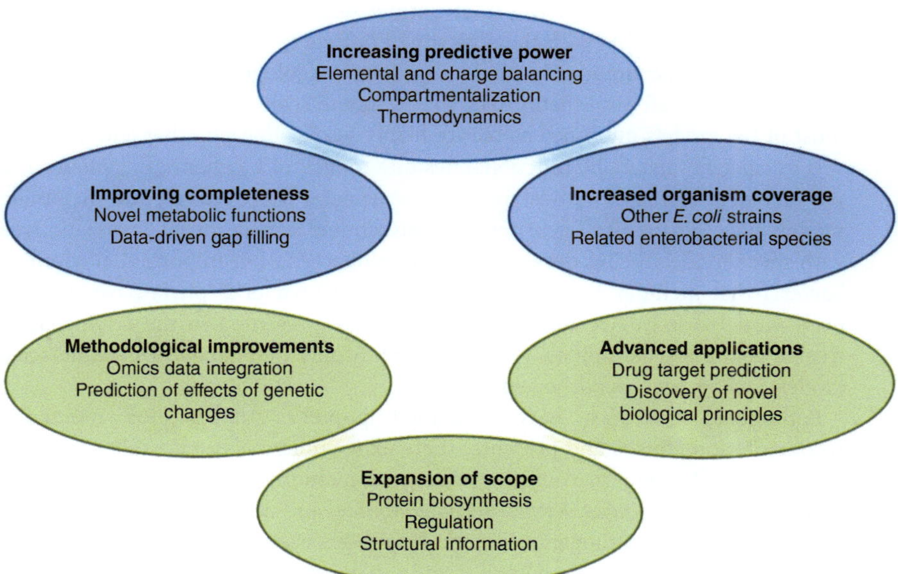

Figure 6.6 Six areas of development that have improved the utility of *E. coli* genome-scale models as predictive models for practical applications. The three developments highlighted in blue are primarily related to improving the accuracy and completeness of GSMs. The three developments shown in green are related to expanding the scope of GSMs to allow covering new application areas.

(1) expanding the scope of models by incorporation of, for example, protein biosynthesis, (2) improvements in computational methods for interrogating models, and (3) more innovative applications of genome-scale models to novel medically and biotechnologically relevant areas. Such recently proposed innovative applications of *E. coli* genome-scale models include accurate prediction of effects of nutrient perturbations on thermotolerance of *E. coli* enabled by integrating structural information with genome-scale models [60] and prediction of novel antimicrobial strategies by incorporating production of reactive oxygen species in genome-scale models [61]. These applications highlight novel uses of genome-scale models that can be exploited also in biotechnology through the rational development of stress-resistant strains.

6.4.2
Other Prokaryotes

The first genome-scale metabolic model reconstructed was actually not for *E. coli* but rather for the pathogenic bacterium *H. influenza* whose genome was sequenced before the *E. coli* genomic sequence became available. The number of prokaryotic species with sequenced genomes continues to grow at an ever increasing pace, and at the same time the number of curated prokaryotic genome-scale models has increased significantly (Figure 6.7). Prokaryotic models include more than 70% of the total unique reaction content in all metabolic models, which is in line with the diversity of different metabolic capabilities present in prokaryotic organisms inhabiting diverse ecological niches. However, a recent review of all published prokaryotic genome-scale analyses has shown that most of the models published so far are highly similar to each other and may lack species-specific metabolic functionalities due to lack of biochemical, genetic, and physiological studies for most prokaryotic species [62]. Prokaryotic metabolic models have also so far been unevenly distributed across the phylogenetic tree with most models being developed for proteobacteria (an important phylum of bacteria that includes *E. coli*), and more than 10 bacterial phyla have still not been modeled [62]. Archaea are also sparsely represented in the phylogenetic tree of genome-scale models (Figure 6.7) with only four archaeal genome-scale models having been published so far.

Pathogenic microbes have remained popular subjects for metabolic reconstruction efforts, and presently there exist curated genome-scale metabolic models for all major prokaryotic pathogens including *Staphylococcus aureus*, *Pseudomonas aeruginosa*, *Klebsiella pneumoniae*, and *Mycobacterium tuberculosis* [63]. Models of pathogenic bacteria have been used to identify possible drug targets and to interpret physiological and other data. One of the challenges with building high-quality genome-scale models for pathogens is the lack of experimental studies focusing on metabolic physiology and the lack of good genetic manipulation methods for many pathogens. However, recent developments in high-throughput genotype-to-phenotype mapping, using libraries of loss-of-function mutants [64] and phenotyping arrays [65], enable rapid generation of

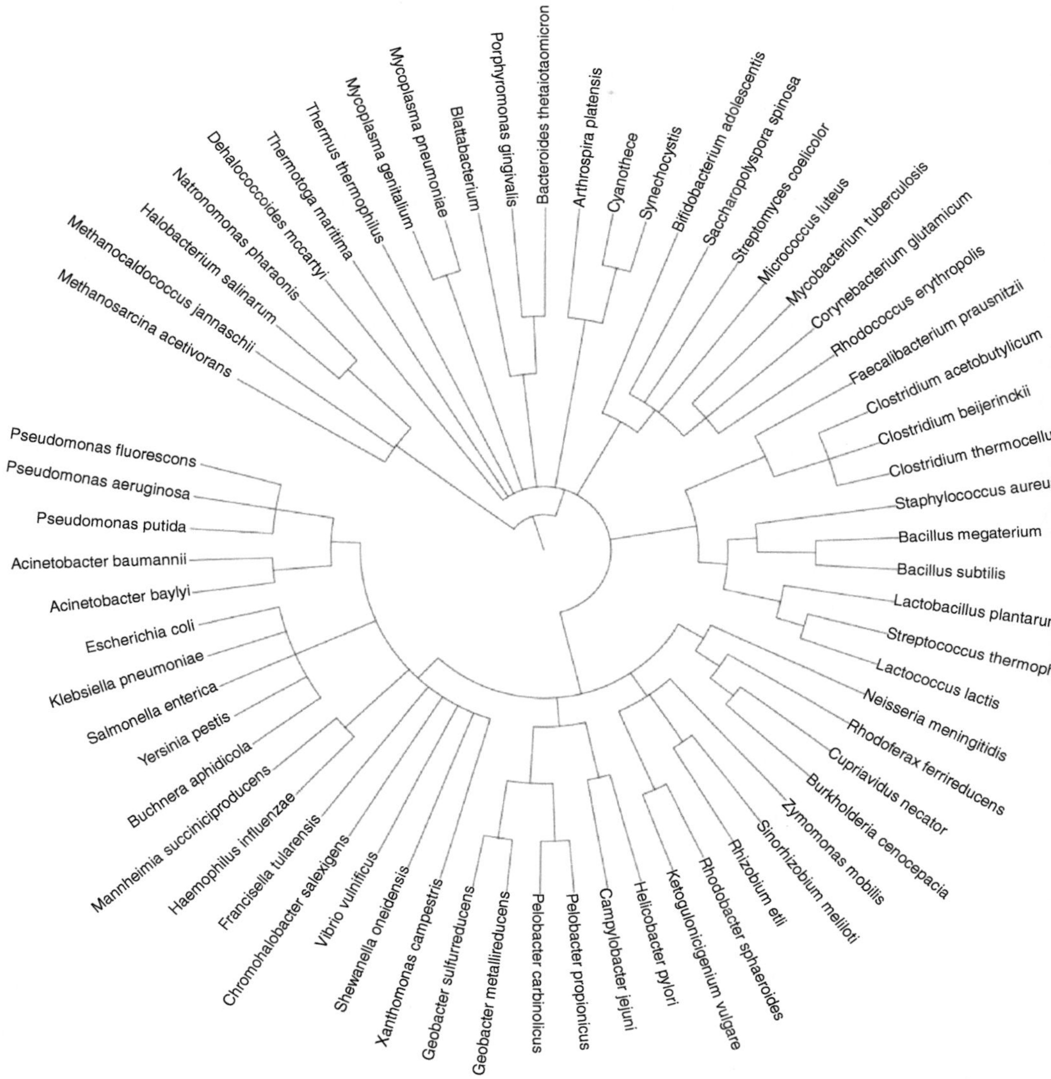

Figure 6.7 Phylogenetic tree of 64 published prokaryotic genome-scale models. Four of the major clades have been labeled. The tree was generated using phyloT (phylot.biobyte.de/) and iTOL (itol.embl.de/) using official species names for the organisms for which the genome-scale model has been built. In a handful of cases, the species name was replaced by a genus name as the species name was not found in the NCBI taxonomy.

the type of physiological and genetic data that can be applied to improve draft metabolic reconstructions.

Next to pathogenic microorganisms, industrially relevant microorganisms that are used for chemical or protein production as well as for bio-remediation receive strong attention and there exist, among others, genome-scale metabolic models of biotechnologically important prokaryotes such as *Bacillus subtilis, Corynebacterium glutamicum,* and *Lactococcus lactis* [66,67]. Recent years have also seen an increased interest in building models of photosynthetic bacteria to produce biofuels or biochemicals directly from CO_2 and sunlight by cyanobacteria [68]. Models of industrially relevant prokaryotes have been used to assess the potential for production of specific chemicals or proteins, design genetic manipulations to increase production, and model fermentation conditions to optimize product formation [66,67]. For industrial production of organisms, ample physiological data are usually available, and at least in principle this allows building of high-quality genome-scale models as is the case for *E. coli*.

As described in Section 6.2, the reconstruction and model development still requires significant efforts for each target organism. For this reason, multiple automated and semiautomated methods and tools have been developed to speed up the reconstruction process [69]. These tools are available both as part of standard genome-scale metabolic modeling toolboxes and as Web-based platforms coupled to genome annotation tools. The models produced by the automated methods without the incorporation of any experimental data are not of very high quality, but many tools also allow data to be included in the reconstruction process, such as results from growth assays on multiple carbon and nitrogen sources. This will improve the quality of the reconstruction [70]. Even with data-driven automated reconstruction, there is still a need to curate resulting draft models, carefully based on published literature for the target organism. Despite these challenges, automated reconstruction methods can still be useful for generating consistent draft models. Overall, there is still much work to be done in order to build high-quality genome-scale models of species representing all major prokaryotic phyla and unique metabolic functions.

6.4.3
Prokaryotic Communities

The importance of metabolic and other interactions between prokaryotes in many ecological habitats has been known for decades [71]. However, the study of prokaryotic communities has truly flourished only recently for two different reasons. First, the development of next-generation sequencing technologies has allowed sequencing of metagenomes of entire communities without the need for isolation of specific organisms [72,73]. Second, it has become clear that microbial communties living in association with eukaryotic hosts such as humans or plants play a key role in both normal eukaryotic physiology and in disease pathology [74]. Metagenomic sequencing has provided the basis for building models of prokaryotic communties either by integrating existing genome-scale models of individual prokaryotic species or by direct reconstruction of community-wide models from annotated metagenomes.

Genome-scale metabolic models have been applied to understanding prokaryotic communities both at the level of small, well-defined communities and as a tool to interpret large-scale metagenomic data on more complex communities [75]. Some of the early modeling efforts for prokaryotic communities were focused on small communities consisting of a handful of species. Examples of such studies include static modeling of mutualistic communities of sulfate reducers and methanogens where each species is essentially represented as a separate compartment in the community model [76]. Studies that use metabolic models to interpret metagenomic data for complex communities typically rely on models created by automated pipelines like those mentioned above. This type of approach has been utilized to predict the types of metabolic interactions that could occur between species in complex communities such as the human gut microbiome [77,78].

More realistic modeling of microbial community metabolism can be achieved by incorporating dynamic flux balance analysis methods allowing the prediction of temporal dynamics of species composition in a community [79]. Dynamic models of communities have been used to study both natural communities relevant for bio-remediation and synthetic, designed communities tailored for specific biotechnological applications such as fermentation of complex feedstocks. Dynamic modeling approaches have been extended to account for community-level (as opposed to species level) objective functions [80] and by incorporating spatial heterogeneity of prokaryotic communities [81]. These models that incorporate both dynamic growth of individual species and diffusion of metabolites between them can be used to predict *de novo* the structure of prokaryotic communities.

The increased interest in the microbiomes of humans, plants, and other environments has also led to a resurgence of research activity in basic and applied prokaryotic microbiology. These activities contribute to a better understanding of prokaryotic metabolism in two distinct ways. First, the range of bacterial and archaeal species that are studied is expanding, which inevitably leads to the discovery of novel enzymatic activities and modes of prokaryotic metabolism. Second, there will be an increasing emphasis on in-depth experimental studies of metabolism in both well-characterized and less well-known prokaryotic species. These developments, in turn, will fuel the development of genome-scale metabolic models for additional prokaryotic species and the improvement of quality and predictive ability of current models.

6.5 Genome-Scale Models of Eukaryotes

The first eukaryotic genome to be sequenced was that of *Saccharomyces cerevisiae* S288C in 1996 [82]. This was followed by the genome of *Caenorhabditis elegans* in 1998 [83] that consequently became the first multicellular organism with an annotated genome. *C. elegans* is a nematode and one of the simplest organisms with a nervous system [84]. As such, *C. elegans* is an important model organism for studies of neural development in animals and work related to this organism has been awarded with three Nobel prizes. *Arabidopsis thaliana* was

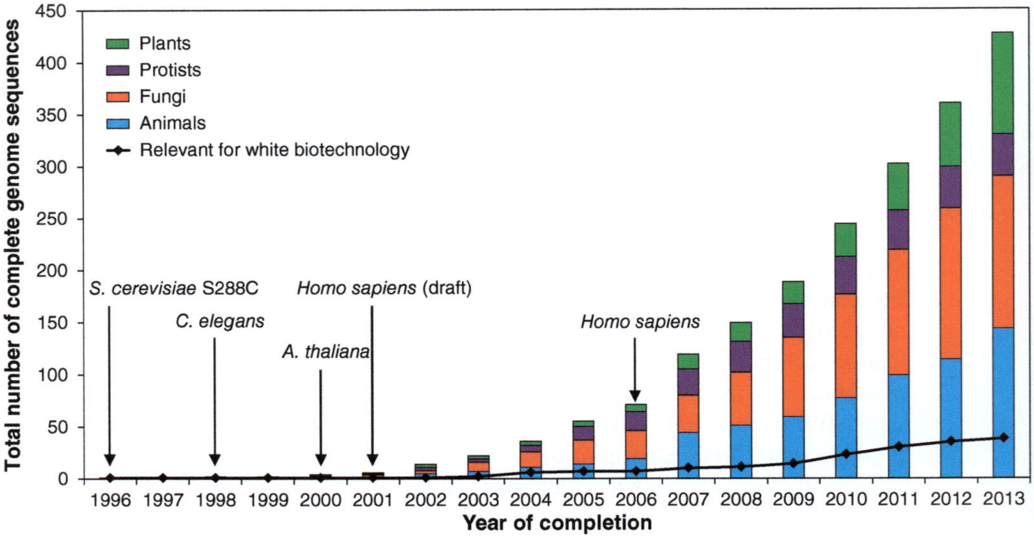

Figure 6.8 Annual increase in the total number of publicly available complete nuclear and organelle genome sequences that have been assembled, annotated, and published. The draft genome of *Homo sapiens* is included to give a historical perspective. The total number of genomes relevant for industrial applications is also shown.

the first plant genome to be sequenced in 2000 [85] and only a year later came the first draft of the human genome [86,87]. Since its completion, the number of annotated genomes of eukaryotic organisms, ranging from fungi and protists to plants and animals, has increased steadily on an annual basis and exceeded 400 genomes (Figure 6.8). With a large number of ongoing sequencing projects, this number is expected to increase even further in coming years (www.ncbi.nlm.nih.gov/genome).

The genomic data of eukaryotic organisms generated so far forms a basis for the construction of a multitude of specific eukaryotic genome-scale models. Most of these organisms are of interest mainly in biomedical and agricultural research areas. Still, approximately one-tenth of the available sequences are related to industrial biotechnological applications (Figure 6.8) and the models developed so far are described in the following paragraphs.

6.5.1 Saccharomyces cerevisiae

The yeast *S. cerevisiae* has become one of the most important model organisms for studies of genetics, molecular biology, metabolism, and physiology in eukaryotes. As approximately 66% of the verified yeast ORFs have an associated ortholog (i.e. they share the same ancestral gene sequence) in the human genome (YeastMine data; yeastmine.yeastgenome.org/yeastmine), it is well suited for studies of human diseases [88,89] and screening of novel drugs [90,91]. The long tradition of using *S. cerevisiae* in large-scale production processes (i.e., baking, brewing, and wine making) has generated a wealth of information regarding its physiology. The physiological characteristics, such as exceptionally fast hexose utilization, anaerobic growth capability, and high tolerance toward environmental stress factors (e.g., low pH, high temperature, and inhibitors), combined with straightforward techniques for genetic engineering, has made *S. cerevisiae* the organism of choice for bio-based production of nutraceuticals, pharmaceuticals, chemicals, and fuels [92]. Access to genome-scale metabolic models aid in the development of new and efficient production strains through advanced metabolic engineering and contributes to an increased understanding of cellular metabolism.

The first formulation of a stoichiometric reaction network in a eukaryote was published in 1995 ([93] – an important paper that is also mentioned in Chapter 4). The authors constructed five core models with 70–81 reactions and 85–88 metabolites that described central carbon metabolism in *S. cerevisiae.* The models were used to estimate the growth-associated ATP requirement during growth on different substrates as well as the effective P/O ratio in aerobic cultivations. Using these values, predictions of the biomass yields correlated well with yields experimentally determined in continuous carbon-limited cultivations using glucose, ethanol, and acetate. Due to the demonstrated usefulness and predictive capability, such metabolic core models have since been applied in several studies focusing on different aspects of yeast metabolism and physiology: aerobic growth in mixed substrate cultivations [94], growth on glucose [95,96], comparison between wild-type and glucose repression mutants [97], and analysis of recombinant strains producing heterologous proteins [98], PHB [99], or fermenting xylose [100,101]. The work presented by Gombert *et al.* [97] also demonstrated that the application of a core model in metabolic network analysis [102] using ^{13}C-labeled substrate was very robust and suitable for quantitating the metabolic fluxes in central carbon metabolism. In this respect, smaller core models are still relevant in the field of systems biology since all the interactions in the metabolic network ultimately generate a distinct flux profile that is directly connected with the cellular phenotype [103,104].

The first genome-scale model of a eukaryote was based on the annotated genome of *S. cerevisiae* S288C and was presented in 2003 [1]. This model, named iFF708, described a network of 1379 reactions, of which 944 were associated with one or more genes, and 555 unique metabolites distributed over two cellular compartments (i.e., the cytosol and the mitochondrion). The authors showed that the model was capable of predicting the P/O ratio, the growth-associated ATP requirement and specific substrate uptake rates, metabolic product yields, and growth rate under a wide range of conditions [105]. It was also shown that the qualitative impact of single-gene deletions on growth corresponds with experimental results in 88% of 599 considered cases [106]. Being the first genome-scale model of *S. cerevisiae*, iFF708 became a scaffold for the construction of more detailed models (Table 6.1). Two models that were directly derived

Table 6.1 Genome-scale metabolic models of *S. cerevisiae*.

Model	Year	Genes	Reactions with genes	Gap-filling reactions	Unique metabolites	Cellular compartments	Reference
iFF708[a),b)]	2003	619	944	435	555	2	[1]
iND750[b),c)]	2004	750	805	341	646	7	[107]
iLL672[d)]	2005	660	872	223	636	2	[108]
iMH805[b)]	2006	805	805	341	646	7	[52]
iIN800[a),c)]	2008	661	834	221	682	2	[109]
iMM904[b),c)]	2009	904	1043	370	713	7	[110]
Yeast 1.0[e)]	2008	888	1407	450	664	14	[111]
iAZ900	2010	900	1049	548	717	7	[49]
Yeast 4[e)]	2010	924	1243	622	763	15	[112]
Yeast 5[e)]	2012	918	1217	893	791	14	[113]
Yeast 6[e)]	2013	900	1180	708	768	14	[114]
Yeast 7[e)]	2013	910	2310	1188	770	14	[115]
iTO977[a)]	2013	961	1046	516	817	3	[116]

a) Available from BioMet Toolbox (www.sysbio.se).
b) Available from UCSD Systems Biology Research Group (http://gcrg.ucsd.edu/InSilicoOrganisms/Yeast).
c) Available from The Model SEED (http://www.theseed.org/models).
d) Available from The Yeast Systems Biology Network YSBN (http://www.gmm.gu.se/globalysbn/models.htm).
e) Available from YEASTNET: A consensus reconstruction of yeast metabolism (http://sourceforge.net/projects/yeast/files).

from iFF708 were iLL672 [108] and iIN800 [109]. The iLL672 model was constructed by extensive curation and was used to improve the functional classification of duplicate genes. The iIN800 model was expanded with a more detailed description of lipid metabolism and introduced updated biomass equations to improve the predictive power in FBA simulations.

As illustrated in Table 6.1, the available models can be divided into three classes with regard to the number of cellular compartments: (i) only the primary organelles (i.e., cytoplasm, mitochondrion, and peroxisome), (ii) all organelles (i.e., cytoplasm, mitochondrion, peroxisome, endoplasmic reticulum, Golgi apparatus, vacuole, and nucleus), and (iii) all organelles and their membranes regarded as separate compartments. The iND750 was the first model to include all organelles as well as completely balanced reactions with regard to both elements and charge and hence introducing a cell-wide proton balance [107]. Despite the higher degree of compartmentalization and expansion of the functional assignment of gene products and the relationship between ORFs, proteins, and reactions, the ability of iND750 to predict the effect of gene deletions was not improved compared to iFF708. The iND750 model was later expanded with

Table 6.2 Genome-scale metabolic models of unicellular eukaryotes.

Organism	Group	Model	Genes	Reactions/metabolites/compartments	Reference
Aspergillus nidulans	Filamentous fungi	iHD666[a)]	666	794/732/3	[117]
A. niger	Filamentous fungi	iMA871[a)]	871	1190/1045/2	[118]
A. oryzae	Filamentous fungi	iWV1314[a)]	1314	1053/1073/2	[119]
A. terreus	Filamentous fungi	iJL1454	1454	1451/1155/2	[120]
C. glabrata	Yeast	iNX804	804	1287/1025/5	[66,67]
Chlamydomonas reinhardtii	Algae	iRC1080	1080	2190/1068/9	[121]
Leishmania major	Protozoa	iAC560	560	1112/1101/7	[122]
P. chrysogenum	Fungi	iAL1006[a)]	1006	1471/1235/3	[2]
Pichia pastoris	Yeast	PpaMBEL1254	540	1254/1147/7	[123]
P. pastoris	Yeast	iPP668	668	1361/1177/7	[124]
P. pastoris	Yeast	iLC915[a)]	915	1423/899/6	[125]
S. stipitis (formely *P. stipitis*)	Yeast	iSS884[a)]	884	1332/992/3	[125]
S. stipitis (formely *P. stipitis*)	Yeast	iBB814	814	1371/971/2	[126]
S. stipitis (formely *P. stipitis*)	Yeast	iTL885	885	1240/870/2	[127]
Schizosaccharomyces pombe	Yeast	SpoMBEL1693	605	1693/1744/7	[128]

a) Available from BioMet Toolbox (www.sysbio.se).

a regulatory network consisting of 837 regulatory interactions [52]. Apart from the 750 metabolic genes in the iND750 model, the new model, iMH805/837, also contained genes encoding 55 nutrient-regulated transcription factors. The model, therefore, enabled distinct intra- and extracellular metabolites to act as signals for the regulation of all the metabolic subsystems included in iND750. This was the first description of a large-scale integration of metabolism and its transcriptional regulation in a eukaryotic organism. The predictive power of the iMH805/837 model was evaluated using phenotypic data from 10 transcription factor deletion strains cultivated on 12 different carbon sources. Apart from a few cases where the model overestimated the growth rate, the predictions were in good agreement with the experimental data.

The second fully compartmentalized reconstruction was the iMM904 model [110]. The iMM904 model represents a comprehensive review of the iND750 model with regard to reactions (substrate and cofactor specificity, reversibility, and compartmentalization) and gene–protein–reaction associations. The reconstruction was created using dedicated modeling software (SimPheny, Genomatica Inc., San Diego, CA, USA) and introduced a new nomenclature for reactions and metabolites. The iMM904 model was later revised based on gene essentiality analysis, which resulted in model iAZ900 [49]. The iMM904 and iLL672 models were also used to create a consensus model that aimed at resolving discrepancies between the two models as well as introducing annotations of metabolites and reactions consistent with external databases and defining standards for curation of models [111]. The resulting reconstruction was named Yeast 1.0, which has gone through several updates and is currently at version 7 (Table 6.1). The consensus network was recently merged with the iIN800 model and current knowledge of yeast metabolism that generated the model iTO977 [116]. This latest model has a higher number of genes and unique metabolites than previous models and was designed to allow a connection with a separate genome-scale model of the protein secretion process in *S. cerevisisae* [129].

6.5.2
Other Unicellular Eukaryotes

Although *S. cerevisiae* possesses several traits that make it exceptionally well suited for industrial applications, there are other organisms with native abilities of equal interest. The Aspergillus species *nidulans, niger, oryzae,* and *terreus* are natural producers of hydrolytic enzymes (needed to convert sugar polymers from cellulose into fermentable monomers), organic acids (e.g., citrate, malate, and succinate), and natural products (e.g., Lovastatin) [130]. These are the most well-known species, and *A. oryzae* has been used for centuries in Japan for the production of Sake, soy sauce, and miso. Even so, the first genome-scale model constructed for *in silico* analysis of an Aspergillus species was made for *A. nidulans* [117] (Table 6.2). However, a model for *A. oryzae* was published only a month later and contained nearly twice the number of genes [119] (Table 6.2). The *A. nidulans* model iHD666 was created using available reaction networks of *S. cerevisiae* and *A. niger*, while the *A. oryzae* model iWV1314 was created using a annotation strategy based on assembling of EST sequences. Both models were validated by comparing predicted biomass yields on different carbon sources with experimental data. The iHD666 model was later used as a scaffold for creating a model for *A. niger* [118] (Table 6.2). The latest model of a Aspergillus species describes the metabolic network of *A. terreus* [120] (Table 6.2). *A. terreus* is capable of producing a wide range of chemicals, from organic acids to enzymes and high-value molecules such as itaconate (a starting monomer in chemical synthesis of biomedical compounds), lovastatin, geodin, and cyclosporine A [120].

The methylotrophic yeast *Pichia pastoris* has several native characteristics that make it one of the most common hosts for production of recombinant proteins [131]. *P. pastoris* has a natural tendency to secrete proteins and can grow to very high cell densities under aerobic conditions without producing byproducts such as ethanol and acetate. The production of specific proteins can be easily induced using methanol by coupling of the ORF with the *AOX1* promoter. This is one of the strongest and most tightly regulated promoters known in eukaryotes. The natural lack of α-1,3-linked mannosyl transferase reduces the immunogenic response toward proteins and through further engineering more humanized glycosylation has been achieved [132]. This opens the possibility for industrial production of therapeutic proteins. In 2010, two genome-scale models of *P. pastoris* metabolism were published: the PpaMBEL1254 [123] and iPP668 [124] models. Both models were constructed based on the genome sequence of *P. pastoris* DSMZ [133] and were of roughly the same size except for a slightly higher number of genes in the latter (Table 6.2). The PpaMBEL1254 model was validated and expanded with unannotated reactions using physiological data on 25 different carbon sources [123]. Model simulations showed that the maximum amino acid production capacity was the same in *P. pastoris* and *S. cerevisiae*, which suggests that other characteristics (e.g., post-translational modifications and secretion efficiency) have a more important role in host selection. The model was also able to capture the relationship between optimal protein production rate and oxygen uptake rate and can thus be used as a tool to optimize fermentation conditions [123]. The iPP668 model was validated using data from chemostat cultivations with glucose or glycerol/methanol mixture as carbon sources and identified a need for better experimental analysis of the biomass composition during methanol induction [124]. The model was further used to evaluate different feeding strategies beneficial for protein production. Based on criteria such as high growth rate and high capacity to produce amino acids, sorbitol was found as a suitable cosubstrate with methanol. An expanded model of *P. pastoris* was recently constructed using a semiautomatic approach with the RAVEN Toolbox [2]. The new model, iLC915 [125], contained more genes and reactions than the two previous models and showed better predictive power of growth phenotypes with methanol or glycerol/methanol mixtures.

Scheffersomyces stipitis (formerly belonging to genus *Pichia* [134]) is a natural pentose-utilizing yeast [135] and therefore of interest for the production of bio-based chemicals from lignocellulosic raw materials. Compared to *S. cerevisiae*, however, *S. stipitis* requires microaeration and higher pH for optimal fermentation performance, and also a lower cultivation temperature [136]. Three genome-scale models of *S. stipitis* metabolism were independently published in 2012 (Table 6.2). All models were of equal size and validated by comparing *in silico* predictions of substrate utilization with experimental data, except for iTL885. The iCL915 model was validated using phenotypic data from xylose-limited chemostat cultivations [125]. The iBB914 model was validated by qualitative comparison with high-throughput substrate utilization data generated using Biolog Phenotype MicroArrays (i.e., growth on 160 carbon sources, 93 nitrogen sources, 58 phosphorous sources, and 28 sulfur sources). This enabled the identification of network gaps, and after model expansion the overall prediction accuracy increased from 56 to 74% [126]. The iBB914 model was further validated by the successful identification of dihydroorotate dehydrogenase as a growth-enabling enzyme under anaerobic conditions [126]. Although all three models were used to optimize aeration and substrate feeding strategies for high ethanol productivity, information regarding regulatory interactions needs to be incorporated into these models to enable rational design of a strain capable of growing anaerobically on xylose.

Penicillium chrysogenum and *Candida glabrata* are two fungi with high industrial relevance. *P. chrysogenum* has been used for more than 50 years in large-scale production of the β-lactam antibiotic penicillin [137]. The worldwide annual sales of this type of antibiotics are estimated at ~US$ 15 billion, which represents 65% of the total market of antibiotics [137]. Although the genome sequence of *P. chrysogenum* became available in 2008 [138], a genome-scale metabolic model of this important industrial producer was constructed only recently [2]. The model (iAL1006) was reconstructed by a semiautomatic approach with the RAVEN Toolbox and contains 1471 reactions and 1235 metabolites distributed over 3 cellular compartments (Table 6.2). The model was used to investigate the relationship between high penicillin yield and provision of ATP and NADPH. The analysis showed that increasing the supply of NADPH, or reducing its consumption, may be a valid strategy for achieving a higher product yield. *C. glabrata* is an important cell factory for the industrial production of pyruvate [139], but it is, unfortunately, also an opportunistic pathogen in individuals with compromised immune system (e.g., from HIV) [140]. Pyruvate is currently utilized in the industry for biosynthesis of chemicals, food additives, and pharmaceuticals and the demand is increasing. A genome-scale model of *C. glabrata* metabolism was recently constructed [66,67] (Table 6.2) and may be a valuable tool in the engineering of this relatively unexplored organism.

6.5.3 Other Multicellular Eukaryotes

The first genome-scale model of a mammalian cell based on genomic data was constructed in 2005 and described the carbon, nitrogen, and energy metabolism of a generic mouse cell [141]. The mouse is a fundamental experimental animal for conducting research on human diseases, drug screening, and other biomedical applications. The availability of strains with specific gene knockouts enables rigorous validation of mammalian reconstructions. Several models of mouse metabolism have been published over the years (Table 6.3), and the latest model iMM1415 contains 3726 reactions and 2774 metabolites distributed over 7 cellular compartments [142]. The model was created by homology mapping to the reconstructed network of human metabolism ("Recon 1" [143]) and refined to satisfy 260 validation tests. Mouse myeloma cells, together with Chinese hamster ovary (CHO) cells, are the most commonly used cell lines for the production of monoclonal antibodies [144]. This class of biological drugs represented 20% of

Table 6.3 Genome-scale metabolic models of multicellular eukaryotes.

Organism/tissue	Group	Model	Genes	Reactions/metabolites/compartments	Reference
Mus musculus (mouse)	Mammals	N/A	473	1220/872/2	[141]
M. musculus (mouse)	Mammals	N/A	1399	2037/2179/2	[145]
M. musculus (mouse)	Mammals	N/A	724	1494/1287/2	[146]
M. musculus (mouse)	Mammals	iMM1415	1415	3726/2774/7	[142]
A. thaliana	Plants	AraGEM	1419	1567/1748/5	[147]
Mesophyll + bundle sheath cells	Plants	C4GEM	3557–11 623	1588/1755/5	[148]
Homo sapiens (humans)	Mammals	Recon 1	1496	3311/2766/7	[143]
H. sapiens (humans)	Mammals	EHMN	2322	4804/2671/8	[149,150]
H. sapiens (humans)	Mammals	iHuman1512	1512	5535/4137/7	[151]
Generic model of a cancer cell	Mammals	N/A	683	772/–/–	[152]
69 different human cell types + 16 associated cancer types	Mammals	INIT models	*var.*	*var./var./var.*	[151]
5 different cancer types associated with human cells	Mammals	INIT models	*var.*	*var./var./var.*	[153]
Human adipocytes	Mammals	N/A	831	649/554/5	[154]
Human adipocytes	Mammals	iAdipocytes 1809	1809	6160/4550/7	[155]
Human hepatocytes	Mammals	HepatoNet1	N/A	2539/777/6	[156]
Human hepatocytes	Mammals	iLJ1335	1335	1827/1360/	[157]
Human hepatocytes	Mammals	N/A	953	776/659/5	[154]
Human hepatocytes	Mammals	iHepatocytes 2322	2322	7930/5686/7	[158]
Human myocytes	Mammals	N/A	979	734/615/5	[154]
Multitissue model of human hepatocyte, myocyte, and adipocyte	Mammals	N/A	N/A	2208/1771/4	[154]
Human alveolar macrophage + *M. tuberculosis*	Mammals + bacteria	iAB-AMØ-1410-Mt-661	2071	3951/3400/3	[159]
Human brain astrocytes + various neuron types	Mammals	iNL403	403	Glutamatergic: 1066/983/5 GABAergic: 1067/983/5 Cholinergic: 1070/987/5	[160]

the biologics sales in 2011 – a market with an annual turnover of US$ 120 billion [144]. A large part of these sales comes from products produced by the CHO cell line. It is the workhorse of complex human protein production due to several advantageous traits such as the ability to grow to high cell density, the adaptability to serum-free conditions, and genomic variability [144]. The increasing availability of genomic data for different CHO cell lines [161–163] combined with omics data sets will enable improvements of strain capability and bioprocess design [164].

The human metabolic network was reconstructed and published by two independent groups in 2007 [143,150]. The Recon 1 model contained 1496 genes, 3311 reactions, and 2766 metabolites distributed over 7 cellular compartments [143]. The EHMN model included more genes and reactions than Recon 1, but contained fewer metabolites (Table 6.3), and compartmentalization was introduced in a later version [149]. Both reconstructions were generated from available information in the databases KEGG, EntrezGene, H-Invitational Database, Uniprot, and HGNC. Both models have been used as scaffolds for the creation of several tissue-specific and multiple interaction models (e.g., iLJ1335, HepatoNet1, and iAB-AMØ-1410-Mt-661; Table 6.3). In 2012, the Human Metabolic Reaction (HMR) database was created that combined the information in Recon 1 and EHMN with current information on human metabolism available in HumanCyc and KEGG [151]. The structure of the HMR was designed to be hierarchical where genes are linked to proteins via splicing variants and further to reactions and metabolites. The database also includes tissue-specific expression profiles of genes via BioGPS, protein abundances via HPA, and metabolite information collected from HMDB. The HMR is part of the Human Metabolic Atlas, which is regularly updated with new information and can be accessed via www.metabolicatlas.com. The information in HMR was used to construct a new network model (iHuman1512) that contained more reactions and metabolites than previous models (Table 6.3) [151]. The iHuman1512 was used as a template to automatically generate 69 tissue-specific models with normal physiology and 16 models describing an associated cancer type [151]. The automatically generated iHepatocytes2322 was compared with the manually created iHepatoNet1 model and discrepancies between the two models were supported by experimental evidence in favor of the automatic reconstruction algorithm [151].

The large number of human tissue-specific reconstructions opens up new possibilities in clinical research. The alterations in cell physiology that ultimately lead to malignant growth are traditionally explained by progressive accumulation of mutations in regulatory genes that govern normal cell proliferation and homeostasis [165]. More than a decade ago, these alterations were used as a basis in the declaration of the so-called six "hallmarks of cancer": sustaining proliferative signaling, evading growth suppressors, resisting cell death, enabling replicative immortality, inducing angiogenesis, and activating invasion and metastasis [166]. In 2011, these cancer hallmarks were reevaluated and, among others, the major reprogramming of cellular energy metabolism was added as a new hallmark [167]. Concurrent with this revision, the first generic genome-scale model of a cancer cell was published [152]. The model contained 683 genes and 772 reactions (Table 6.3) and was capable of correctly predicting the effect of 11 out of 14 FDA-approved metabolic anticancer drugs. The study also demonstrated the possibility of using the model for identifying new candidates as potential drug targets [152]. Recently, several tissue-specific models describing the associated cancerous metabolism have been developed [151,153] that can provide a higher predictive power of specific cancer types and also help in the task of finding new therapeutic treatments.

6.6 Integration of Polyomic Data into Genome-Scale Models

Genome-scale models are usually applied to describe the space of solutions for the metabolic flux distribution at steady state. Unlike kinetic models, which require extensive parameter characterization, flux calculation is based on network topology and reaction stoichiometry, without any explicit references to enzyme or metabolite concentrations. For that reason, this modeling approach is quite scalable and facilitates the rapid development of metabolic models at the genome scale.

The trade-off between model size and detail does not come without disadvantages. In order to understand the cellular response to environmental and genetic perturbations, it is fundamental to analyze the interplay between genes, proteins, and metabolites. Although their levels can be experimentally measured, they cannot be directly integrated into genome-scale models. Therefore, novel approaches to integrate multiple sources of experimental data into these models are currently being developed [168]. A few examples have already been presented above. An overview of further approaches is presented next (Figure 6.9).

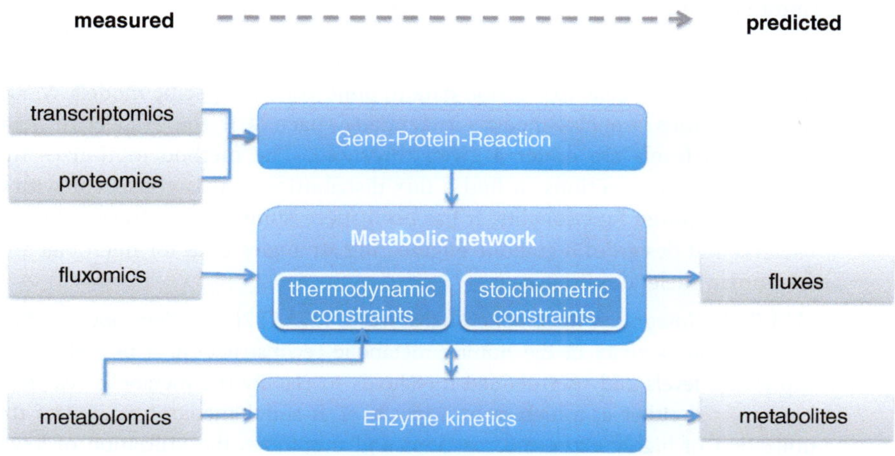

Figure 6.9 Multiple approaches for integration of different sources of *omics* data into large-scale metabolic models.

6.6.1 Integration of Transcriptomics and Proteomics Data

With the recent advancements in high-throughput sequencing methods, the cost of DNA and RNA sequencing has rapidly decreased [169]. For that reason, transcriptomics data is now one of the most abundant types of *omics* data available and also the focus of many research projects to integrate experimental data into genome-scale models.

In the context of genome-scale modeling, transcript levels are used as indicators of enzyme activity. Although transcript and enzyme levels are not perfectly correlated, the quantity of expressed mRNA is assumed to reflect the abundance of the respective enzyme. This provides an indirect measure of the flux-carrying capacity of the respective metabolic pathways under given experimental conditions. Transcript levels are mapped into the metabolic network using gene–protein–reaction (GPR) associations present in genome-scale metabolic reconstructions.

Naturally, methods for integration of transcriptomics data into metabolic models can also use proteomics data. This results in a more accurate estimate of the respective enzyme activity levels. However, proteomics data are less abundant and generally provide a smaller coverage of metabolism compared to the transcriptomics data, which are usually available at the genome scale.

The integration of transcriptomics (or proteomics) data into genome-scale models can follow two main directions. First, transcript levels can be used to improve the prediction of metabolic fluxes by searching for a flux distribution that matches the gene expression pattern. Second, they can be used to create context-specific models from a generic model for a given organism. For instance, one can use expression data from different human tissues to create tissue-specific models from the global human metabolic reconstruction [143].

The following are some of the most relevant methods for integration of transcriptomics data into genome-scale models:

GIMME: Gene inactivity moderated by metabolism and expression [170] is a method that uses gene expression data to build context-specific models. Metabolic reactions are classified as potentially inactive if the respective gene expression levels are under a given threshold. The method minimizes the activity of these reactions to find a flux distribution that is consistent with a minimal required functionality. It has been successfully applied to evolution of bacteria and design of metabolic engineering strategies both for microbial and for human skeletal muscle cells.

iMAT: the integrative metabolic analysis tool [171,172] was developed to find tissue-specific activity in the human metabolic reconstruction. It uses the gene expression levels and user-defined thresholds to classify the metabolic reactions as expressed either at a high or at a low level. It simultaneously maximizes the utilization of highly expressed reactions and minimizes the utilization of lowly expressed reactions such that a feasible flux distribution can be obtained.

MADE: Metabolic adjustment by differential expression [173] is a method that uses differential levels of gene expression under multiple conditions to determine the state of each gene without using arbitrary thresholds. Genes are classified as active (resp. inactive) for a given condition when their level significantly increases (resp. decreases) relative to the previous condition. The method searches for a gene expression pattern across all conditions that are consistent with the metabolic model and results in a feasible flux distribution for each condition.

INIT: Integrative network inference for tissues [151] is a method developed to use proteomic data from the Human Protein Atlas, in order to create tissue-specific models from the global human model reconstruction. It maximizes or minimizes the utilization of metabolic reactions based on a calculated evidence score for their respective proteins. It can also use evidence for the presence of given metabolites. If a metabolite must be present, it relaxes the steady-state condition to allow a small net accumulation for those metabolites, in order to guarantee the presence of all required reactions.

mCADRE: Metabolic context-specificity assessed by deterministic reaction evaluation [174] is a method for building tissue-specific models, which combine gene expression data with network topology to calculate evidence scores for all reactions in the model. In this case, an expression-based evidence level is used to calculate a connectivity-based evidence score among neighboring reactions. This approach allows for a fast generation of functional tissue-specific models.

E-flux: A method for combining gene expression and flux using a pipeline capacity analogy [175]. In this case, the transcript levels are normalized to dimensionless quantities and directly mapped into the metabolic model as upper bounds for the reaction fluxes. One advantage of E-flux is that the gene expression levels are used as continuous variables and not simply reduced to binary levels.

Lee-12: This method, developed by Lee and coworkers, replaces the cellular objective function of a given model with one that minimizes the absolute difference between the gene expression vector and the flux vector [176]. Also in this case, transcript levels are used in absolute values rather than discretized to on/off states. Another advantage of this method is that it is used to simulate the metabolic phenotype under conditions where the original cellular objective assumption does not apply.

6.6.2 Metabolomics Data

Metabolomics data is a critical source of information regarding the cellular metabolism, which can be best characterized by a snapshot of the metabolome. Since genome-scale models are usually applied at steady state, it is not possible to use metabolite concentrations in the models. Nonetheless, metabolomics data is often used to calculate thermodynamic reversibility constraints and further reduce the solution space of constraint-based models.

The directionality of a biochemical reaction is determined by the change in Gibbs free energy (ΔG), which can be calculated from standard change in Gibbs free energy (ΔG^0), and the concentrations of its substrates and products (see Chapter 9). Hoppe and coworkers implemented a method for integration of metabolite concentrations into flux balance analysis using thermodynamic constraints [177]. Two related approaches have also been implemented, namely, the thermodynamics-based metabolic flux analysis (TMFA) [178], and the network-embedded thermodynamic analysis (NET) [179].

6.6.3
Integration of Multiple *Omics*

Comprehensive studies can provide multiple sources of omics data for an organism under the same experimental conditions [180,181]. However, the methods mentioned so far only integrate a single kind of omics data into genome-scale models. The quantitative relation between enzyme levels, metabolite concentrations, and reaction fluxes is given by mechanistic rate laws, which currently have not been successfully applied at the genome scale.

The integrative omics-metabolic analysis (IOMA) method uses a generalized form of Michaelis–Menten rate equations (see Chapter 7) to integrate proteomics and metabolomics data into constraint-based models [182]. For enzymes with available protein and metabolite measurements, the enzyme saturation is calculated and inserted into the constraint-based formulation. This calculation requires dissociation constants for the respective intermediates (such as ES or ESS_1), which can be obtained from the BRENDA database.

In place of extending constraint-based models with new types of constraints, an interesting alternative is to use these models as a scaffold to generate genome-scale kinetic models. This can be performed using approximate kinetic rate laws with canonical forms. Jamshidi and Palsson established a protocol for building genome-scale kinetic models derived from mass–action kinetics [183]. Using a different method, based on lin–log kinetics (Chapter 8), Smallbone *et al.* generated the first genome-scale kinetic model of yeast [184]. With this kind of approach, experimental omics data can be directly integrated into genome-scale models for quantification of concentrations and calibration of kinetic parameters.

Acknowledgment

The authors wish to thank John Villadsen for comments on the manuscript as well as for the invitation to write this chapter.

References

1 Förster, J., Famili, I., Fu, P., Palsson, B.Ø., and Nielsen, J. (2003) Genome-scale reconstruction of the *Saccharomyces cerevisiae* metabolic network. *Genome Res.*, **13**, 244–253.

2 Agren, R., Liu, L., Shoaie, S., Vongsangnak, W., Nookaew, I., and Nielsen, J. (2013) The RAVEN toolbox and its use for generating a genome-scale metabolic model for *Penicillium chrysogenum*. *PLoS Comput. Biol.*, **9**, e1002980.

3 Devoid, S., Overbeek, R., DeJongh, M., Vonstein, V., Best, A.A., and Henry, C. (2013) Automated genome annotation and metabolic model reconstruction in the SEED and Model SEED. *Methods Mol. Biol.*, **985**, 17–45.

4 Pitkänen, E., Jouhten, P., Hou, J., Syed, M.F., Blomberg, P., Kludas, J., Oja, M., Holm, L., Penttilä, M., Rousu, J., and Arvas, M. (2014) Comparative genome-scale reconstruction of gapless metabolic networks for present and ancestral species. *PLoS Comput Biol.*, **6**, 10.

5 Vitkin, E. and Shlomi, T. (2012) MIRAGE: a functional genomics-based approach for metabolic network model reconstruction and its application to cyanobacteria networks. *Genome Biol.*, **13**, R111.

6 Mahmutovic, A., Fange, D., Berg, O.G., and Elf, J. (2012) Lost in presumption: stochastic reactions in spatial models. *Nat. Methods*, **9**, 1163–1166.

7 Halasz, A.M., Lai, H.-J., McCabe, M.M., Radhakrishnan, K., and Edwards, J.S. (2013) Analytical solution of steady-state equations for chemical reaction networks with bilinear rate laws. *IEEE/ACM Trans. Comput. Biol. Bioinform.*, **10**, 957–969.

8 Smallbone, K. *et al.* (2013) A model of yeast glycolysis based on a consistent kinetic characterisation of all its enzymes. *FEBS Lett.*, **587**, 2832–2841.

9 Bordbar, A., Monk, J.M., King, Z.A., and Palsson, B.O. (2014) Constraint-based models predict metabolic and associated cellular functions. *Nat. Rev. Genet.*, **15**, 107–120.

10 Clarke, B.L. (1980) Stability of complex reaction networks, in *Advances in Chemical Physics*, vol. 43 (eds I. Prigogine and S.A. Rice), pp. 1–215.

11 Schuster, S. and Hilgetag, C. (1994) On elementary flux modes in biochemical reaction systems at steady state. *J. Biol. Syst.*, **2**, 165–182.

12 Schilling, C.H., Letscher, D., and Palsson, B.O. (2000) Theory for the systemic definition of metabolic pathways and their use in interpreting metabolic function from a pathway-oriented perspective. *J. Theor. Biol.*, **203**, 229–248.

13 Papin, J.A., Price, N.D., Wiback, S., Fell, D., and Palsson, B.O. (2003) Metabolic pathways in the post-genome era. *Trends Biochem. Sci.*, **28**, 250–258.

14 Terzer, M. and Stelling, J. (2008) Large-scale computation of elementary flux modes with bit pattern trees. *Bioinformatics*, **24**, 2229–2235.

15 Von Kamp, A. and Schuster, S. (2006) Metatool 5.0: fast and flexible elementary modes analysis. *Bioinformatics*, **22**, 1930–1931.

16 Bell, S.L. and Palsson, B.O. (2005) Expa: a program for calculating extreme pathways in biochemical reaction networks. *Bioinformatics*, **21**, 1739–1740.

17 Papin, J.A., Stelling, J., Price, N.D., Klamt, S., Schuster, S., and Palsson, B.O. (2004) Comparison of network-based pathway analysis methods. *Trends Biotechnol.*, **22**, 400–405.

18 de Figueiredo, L.F., Podhorski, A., Rubio, A., Kaleta, C., Beasley, J.E., Schuster, S., and Planes, F.J. (2009) Computing the shortest elementary flux modes in genome-scale metabolic networks. *Bioinform.*, **25**, 3158–3165.

19 Schuetz, R., Kuepfer, L., and Sauer, U. (2007) Systematic evaluation of objective functions for predicting intracellular fluxes in *Escherichia coli*. *Mol. Syst. Biol.*, **3**, 119.

20 Feist, A.M. and Palsson, B.O. (2010) The biomass objective function. *Curr. Opin. Microbiol.*, **13**, 344–349.

21 Mahadevan, R. and Schilling, C.H. (2003) The effects of alternate optimal solutions in constraint-based genome-scale metabolic models. *Met. Eng.*, **5**, 264–276.

22 Price, N.D., Schellenberger, J., and Palsson, B.O. (2004) Uniform sampling of steady-state flux spaces: means to design experiments and to interpret enzymopathies. *Biophys. J.*, **87**, 2172–2186.

23 Edwards, J.S. and Palsson, B.O. (2000) The *Escherichia coli* MG1655 *in silico* metabolic genotype: its definition, characteristics, and capabilities. *Proc. Natl. Acad. Sci. USA*, **97**, 5528–5533.

24 Almaas, E., Kovács, B., Vicsek, T., Oltvai, Z.N., and Barabási, A.-L. (2004) Global organization of metabolic fluxes in the bacterium *Escherichia coli*. *Nature*, **427**, 839–843.

25 Ibarra, R.U., Edwards, J.S., and Palsson, B.O. (2002) *Escherichia coli* K-12 undergoes adaptive evolution to achieve *in silico* predicted optimal growth. *Nature*, **420**, 186–189.

26 Lewis, N.E., Nagarajan, H., and Palsson, B.O. (2012) Constraining the metabolic genotype–phenotype relationship using a phylogeny of *in silico* methods. *Nat. Rev. Microbiol.*, **10**, 291–305.

27 Becker, J., Zelder, O., Häfner, O., Schröder, H., and Wittmann, C. (2011) From zero to hero: design-based systems metabolic engineering of *Corynebacterium glutamicum* for L-lysine production. *Metab. Eng.*, **13**, 159–168.

28 Burgard, A.P., Pharkya, P., and Maranas, C.D. (2003) Optknock: a bilevel programming framework for identifying gene knockout strategies for microbial strain optimization. *Biotechnol. Bioeng.*, **84**, 647–657.

29 Patil, K.R., Rocha, I., Forster, J., and Nielsen, J. (2005) Evolutionary programming as a platform for *in silico* metabolic engineering. *BMC Bioinformatics*, **6**, 308.

30 Segrè, D., Vitkup, D., and Church, G. (2002) Analysis of optimality in natural and perturbed metabolic networks. *Proc. Natl. Acad. Sci. USA*, **99**, 15112–15117.

31 Von Kamp, A. and Klamt, S. (2014) Enumeration of smallest intervention strategies in genome-scale metabolic networks. *PLoS Comput. Biol.*, **10**, e1003378.

32 Pharkya, P., Burgard, A.P., and Maranas, C. (2004) OptStrain: a computational framework for redesign of microbial production systems. *Genome Res.*, **14**, 2367–2376.

33 Rocha, I., Maia, P., Evangelista, P., Vilaça, P., Soares, S., Pinto, J.P., Nielsen, J., Patil, K.R., Ferreira, E.C., and Rocha, M. (2010) OptFlux: an open-source software platform for *in silico* metabolic engineering. *BMC Syst. Biol.*, **4**, 45.

34 Fleischmann, R.D. *et al.* (1995) Whole-genome random sequencing and assembly of *Haemophilus influenzae* Rd. *Science*, **269**, 496–512.

35 Blattner, F.R. *et al.* (1997) The complete genome sequence of *Escherichia coli* K-12. *Science*, **277**, 1453–1462.

36 Orth, J.D. *et al.* (2011) A comprehensive genome-scale reconstruction of *Escherichia coli* metabolism. *Mol. Syst. Biol.*, **7**, 535.

37 McCloskey, D., Palsson, B.O., and Feist, A.M. (2013) Basic and applied uses of genome-scale metabolic network reconstructions of *Escherichia coli*. *Mol. Sys. Bio.*, **9**, 661.

38 Pramanik, J. and Keasling, J.D. (1997) Stoichiometric model of *Escherichia coli* metabolism: incorporation of growth-rate dependent biomass composition and mechanistic energy requirements. *Biotechnol. Bioeng.*, **56**, 398–421.

39 Reed, J.L., Vo, T.D., Schilling, C.H., and Palsson, B.O. (2003) An expanded genome-scale model of *Escherichia coli* K-12 (iJR904 GSM/GPR). *Genome Biol.*, **4**, R54.

40 Feist, A.M. *et al.* (2007) A genome-scale metabolic reconstruction for *Escherichia coli* K-12 MG1655 that accounts for 1260 ORFs and thermodynamic information. *Mol. Syst. Biol.*, **3**, 121.

41 Edwards, J.S., Ibarra, R.U., and Palsson, B.O. (2001) *In silico* predictions of *Escherichia coli* metabolic capabilities are consistent with experimental data. *Nat. Biotechnol.*, **19**, 125–130.

42 Schuetz, R., Zamboni, N., Zampieri, M., Heinemann, M., and Sauer, U. (2012) Multidimensional optimality of microbial metabolism. *Science*, **336**, 601–604.

43 Lee, J.W. *et al.* (2012a) Systems metabolic engineering of microorganisms for natural and non-natural chemicals. *Nat. Chem. Biol.*, **8**, 536–546.

44 Park, J.M. *et al.* (2012) Flux variability scanning based on enforced objective flux for identifying gene amplification targets. *BMC Syst. Biol.*, **6**, 106.

45 Choi, H.S., Lee, S.Y., Kim, T.Y., and Woo, H.M. (2010) *In silico* identification of gene amplification targets for improvement of lycopene production. *Appl. Environ. Microbiol.*, **76**, 3097–3105.

46 Park, J.H., Lee, K.H., Kim, T.Y., and Lee, S.Y. (2007) Metabolic engineering of *Escherichia coli* for the production of L-valine based on transcriptome analysis and *in silico* gene knockout simulation. *Proc. Natl. Acad. Sci. USA*, **104**, 7797–7802.

47 Orth, J.D. and Palsson, B.O. (2011) Systematizing the generation of missing metabolic knowledge. *Biotechnol. Bioeng.*, **107**, 403–412.

48 Reed, J.L. *et al.* (2006) Systems approach to refining genome annotation. *Proc. Natl. Acad. Sci. USA*, **103**, 17480–17484.

49 Zomorrodi, A.R. and Maranas, C.D. (2010) Improving the iMM904 *S. cerevisiae* metabolic model using essentiality and synthetic lethality data. *BMC Syst. Biol.*, **4**, 178.

50 O'Brien, E.J., Lerman, J.A., Chang, R.L., Hyduke, D.R., and Palsson, B.O. (2013) Genome-scale models of metabolism and gene expression extend and refine growth phenotype prediction. *Mol. Syst. Biol.*, **9**, 693.

51 Covert, M.W., Knight, E.M., Reed, J.L., Herrgard, M.J., and Palsson, B.O. (2004) Integrating high-throughput and computational data elucidates bacterial networks. *Nature.* May 6; 429(6987): 92–6.

52 Herrgård, M.J., Lee, B., Portnoy, V., and Palsson, B.O. (2006) Integrated analysis of regulatory and metabolic networks reveals novel regulatory mechanisms in *Saccharomyces cerevisiae*. *Genome Res.*, **16**, 627–635.

53 Chandrasekaran, S. and Price, N.D. (2010) Probabilistic integrative modeling of genome-scale metabolic and regulatory networks in *Escherichia coli* and *Mycobacterium tuberculosis*. *Proc. Natl. Acad. Sci. USA*, **107**, 17845–17850.

54 Akesson, M., Forster, J., and Nielsen, J. (2004) Integration of gene expression data into genome-scale metabolic models. *Met. Eng.*, **6**, 285–293.

55 Nasser, W. *et al.* (2014) Evolutionary pathway to increased virulence and epidemic group A Streptococcus disease derived from 3, 615 genome sequences. *Proc. Natl. Acad. Sci. USA*, **111**, E1768–E.1776.

56 Chewapreecha, C. *et al.* (2014) Dense genomic sampling identifies highways of pneumococcal recombination. *Nat. Genet.*, **46**, 305–309.

57 Kaas, R.S., Friis, C., Ussery, D.W., and Aarestrup, F.M. (2012) Estimating variation within the genes and inferring the phylogeny of 186 sequenced diverse *Escherichia coli* genomes. *BMC Genomics*, **13**, 577.

58 Archer, C.T. *et al.* (2011) The genome sequence of *E. coli* W (ATCC 9637): comparative genome analysis and an improved genome-scale reconstruction of *E. coli*. *BMC Genomics*, **12**, 9.

59 Monk, J.M. *et al.* (2013) Genome-scale metabolic reconstructions of multiple *Escherichia coli* strains highlight strain-specific adaptations to nutritional environments. *Proc. Natl. Acad. Sci. USA*, **110**, 20338–20343.

60 Chang, R.L., Andrews, K., Kim, D., Li, Z., Godzik, A., and Palsson, B.O. (2013) Structural systems biology evaluation of metabolic thermotolerance in *Escherichia coli*. *Science*, **340**, 1220–1223.

61 Brynildsen, M.P., Winkler, J.A., Spina, C.S., MacDonald, I.C., and Collins, J.J. (2013) Potentiating antibacterial activity by predictably enhancing endogenous microbial ROS production. *Nat. Biotechnol.*, **31**, 160–165.

62 Monk, J., Nogales, J., and Palsson, B.O. (2014) Optimizing genome-scale network reconstructions. *Nat. Biotechnol.* **32**(5), 447–452.

63 Chavali, A.K., D'Auria, K.M., Hewlett, E.L., Pearson, R.D., and Papin, J.A. (2012) A metabolic network approach for the identification and prioritization of antimicrobial drug targets. *Trends Microbiol.*, **20**, 113–123.

64 Van Opijnen, T. and Camilli, A. (2013) Transposon insertion sequencing: a new tool for systems-level analysis of microorganisms. *Nat. Rev. Microbiol.*, **11**, 435–442.

65 Bochner, B.R. (2009) Global phenotypic characterization of bacteria. *FEMS Microbiol. Rev.*, **33**, 191–205.

66 Xu, C. *et al.* (2013) Genome-scale metabolic model in guiding metabolic engineering of microbial improvement. *Appl. Microbiol. Biotechnol.*, **97**, 519–539.

67 Xu, N., Liu, L., Zou, W., Liu, J., Hua, Q., and Chen, J. (2013) Reconstruction and analysis of the genome-scale metabolic network of *Candida glabrata*. *Mol. Biosyst.*, **9**, 205–216.

68 Kim, T.Y., Sohn, S.B., Kim, Y.B., Kim, W.J., and Lee, S.Y. (2011) Recent advances in reconstruction and applications of genome-scale metabolic models. *Curr. Opin. Biotechnol.*, **23**, 617–623.

69 Hamilton, J.J. and Reed, J.L. (2014) Software platforms to facilitate reconstructing genome-scale metabolic networks. *Environ. Microbiol.*, **16**, 49–59.

70 Henry, C.S. *et al.* (2010) High-throughput generation, optimization and analysis of genome-scale metabolic models. *Nat. Biotechnol.*, **28**, 977–982.

71 Klitgord, N. and Segrè, D. (2011) Ecosystems biology of microbial metabolism. *Curr. Opin. Biotechnol.*, **22**, 541–546.

72 Tyson, G.W. *et al.* (2004) Community structure and metabolism through reconstruction of microbial genomes from the environment. *Nature*, **428**, 37–43.

73 Venter, J.C. *et al.* (2004) Environmental genome shotgun sequencing of the Sargasso Sea. *Science*, **304**, 66–74.

74 Turnbaugh, P.J. *et al.* (2007) The Human Microbiome Project. *Nature*, **449**, 804–810.

75 Mahadevan, R. and Henson, M.A. (2012) Genome-based modeling and design of metabolic interactions in microbial communities. *Comput. Struct. Biotechnol. J.*, **3**, e201210008.

76 Stolyar, S. *et al.* (2007) Metabolic modeling of a mutualistic microbial community. *Mol. Syst. Biol.*, **3**, 92.

77 Freilich, S. *et al.* (2011) Competitive and cooperative metabolic interactions in bacterial communities. *Nat. Commun.*, **2**, 589.

78 Levy, R. and Borenstein, E. (2013) Metabolic modeling of species interaction in the human microbiome elucidates community-level assembly rules. *Proc. Natl. Acad. Sci. USA*, **110**, 12804–12809.

79 Zhuang, K. *et al.* (2011) Genome-scale dynamic modeling of the competition between *Rhodoferax* and *Geobacter* in anoxic subsurface environments. *ISME J.*, **5**, 305–316.

80 Zomorrodi, A.R., Islam, M.M., and Maranas, C.D. (2014) d-OptCom: dynamic multi-level and multi-objective metabolic modeling of microbial communities. *ACS Synth. Biol.*, **3**, 247–257.

81 Harcombe, W.R. *et al.* (2014) Metabolic resource allocation in individual microbes determines ecosystem interactions and spatial dynamics. *Cell Rep.*, **7**, 1104–1115.

82 Goffeau, A., Barrell, B., Bussey, H., Davis, R., Dujon, B., Feldmann, H., Galibert, F., Hoheisel, J., Jacq, C., Johnston, M., Louis, E., Mewes, H., Murakami, Y., Philippsen, P., Tettelin, H., and Oliver, S. (1996) Life with 6000 genes. *Science*, **274**, 563–567.

83 The *C. elegans* Sequencing Consortium (1998) Genome sequence of the nematode *C. elegans*: a platform for investigating biology. *Science*, **282**, 2012–2018.

84 White, J.G., Southgate, E., Thomson, J.N., and Brenner, S. (1986) The structure of the nervous system of the nematode *Caenorhabditis elegans*. *Philos. Trans. R. Soc. Lond. B*, **314**, 1–340.

85 The Arabidopsis Genome Initiative (2000) Analysis of the genome sequence of the flowering plant *Arabidopsis thaliana*. *Nature*, **408**, 796–815.

86 Venter, J.C., Adams, M.D., Myers, E.W., Li, P.W., Mural, R.J., Sutton, G.G., Smith, H.O., Yandell, M., Evans, C.A., Holt, R.A., Gocayne, J.D., Amanatides, P., Ballew, R. M., Huson, D.H., Wortman, J.R., Zhang, Q., Kodira, C.D., Zheng, X.H., Chen, L., Skupski, M., Subramanian, G., Thomas, P.D., Zhang, J., Gabor Miklos, G.L., Nelson, C., Broder, S., Clark, A.G., Nadeau, J., McKusick, V.A., Zinder, N., Levine, A.J., Roberts, R.J., Simon, M., Slayman, C., Hunkapiller, M., Bolanos, R., Delcher, A., Dew, I., Fasulo, D., Flanigan, M., Florea, L., Halpern, A., Hannenhalli, S., Kravitz, S., Levy, S., Mobarry, C., Reinert, K., Remington, K., Abu-Threideh, J., Beasley, E., Biddick, K., Bonazzi, V., Brandon, R., Cargill, M., Chandramouliswaran, I., Charlab, R., Chaturvedi, K., Deng, Z., Di Francesco, V., Dunn, P., Eilbeck, K., Evangelista, C., Gabrielian, A.E., Gan, W., Ge, W., Gong, F., Gu, Z., Guan, P., Heiman, T.J., Higgins, M.E., Ji, R.R., Ke, Z., Ketchum, K.A., Lai, Z., Lei, Y., Li, Z., Li, J., Liang, Y., Lin, X., Lu, F., Merkulov, G. V, Milshina, N., Moore, H.M., Naik, A.K., Narayan, V. A., Neelam, B., Nusskern, D., Rusch, D.B., Salzberg, S., Shao, W., Shue, B., Sun, J., Wang, Z., Wang, A., Wang, X., Wang, J., Wei, M., Wides, R., Xiao, C., Yan, C., Yao, A., Ye, J., Zhan, M., Zhang, W., Zhang, H., Zhao, Q., Zheng, L., Zhong, F., Zhong, W., Zhu, S., Zhao, S., Gilbert, D., Baumhueter, S., Spier, G., Carter, C., Cravchik, A., Woodage, T., Ali, F., An, H., Awe, A., Baldwin, D., Baden, H., Barnstead, M., Barrow, I., Beeson, K., Busam, D., Carver, A., Center, A., Cheng, M.L., Curry, L., Danaher, S., Davenport, L., Desilets, R., Dietz, S., Dodson, K., Doup, L., Ferriera, S., Garg, N., Gluecksmann, A., Hart, B., Haynes, J., Haynes, C., Heiner, C., Hladun, S., Hostin, D., Houck, J., Howland, T., Ibegwam, C., Johnson, J., Kalush, F., Kline, L., Koduru, S., Love, A., Mann, F., May, D., McCawley, S., McIntosh, T., McMullen, I., Moy, M., Moy, L., Murphy, B., Nelson, K., Pfannkoch, C., Pratts, E., Puri, V., Qureshi, H., Reardon, M., Rodriguez, R., Rogers, Y.H., Romblad, D., Ruhfel, B., Scott, R., Sitter, C., Smallwood, M., Stewart, E., Strong, R., Suh, E., Thomas, R., Tint, N.N., Tse, S., Vech, C., Wang, G., Wetter, J., Williams, S., Williams, M., Windsor, S., Winn-Deen, E., Wolfe, K., Zaveri, J., Zaveri, K., Abril, J.F., Guigó, R., Campbell, M.J., Sjolander, K. V, Karlak, B., Kejariwal, A., Mi, H., Lazareva, B., Hatton, T., Narechania, A., Diemer, K., Muruganujan, A., Guo, N., Sato, S., Bafna, V., Istrail, S., Lippert, R., Schwartz, R., Walenz, B., Yooseph, S., Allen, D., Basu, A., Baxendale, J., Blick, L., Caminha, M., Carnes-Stine, J., Caulk, P., Chiang, Y.H., Coyne, M., Dahlke, C., Mays, A., Dombroski, M., Donnelly, M., Ely, D., Esparham, S., Fosler, C., Gire, H., Glanowski, S., Glasser, K., Glodek, A., Gorokhov, M., Graham, K., Gropman, B., Harris, M., Heil, J., Henderson, S., Hoover, J., Jennings, D., Jordan, C., Jordan, J., Kasha, J., Kagan, L., Kraft, C., Levitsky, A., Lewis, M., Liu, X., Lopez, J., Ma, D., Majoros, W., McDaniel, J., Murphy, S., Newman, M., Nguyen, T., Nguyen, N., Nodell, M., Pan, S., Peck, J., Peterson, M., Rowe, W., Sanders, R., Scott, J., Simpson, M., Smith, T., Sprague, A., Stockwell, T., Turner, R., Venter, E., Wang, M., Wen, M., Wu, D., Wu, M., Xia, A., Zandieh, A., Zhu, X., (2001). The sequence of the human genome. *Science* **291**, 1304–1351.

87 International Human Genome Sequencing Consortium (2001) Initial sequencing and analysis of the human genome. *Nature,* **409** (6822), 860–921.

88 Barrientos, A. (2003) Yeast models of human mitochondrial diseases. *IUBMB Life,* **55**, 89–95.

89 Petranovic, D. and Nielsen, J. (2008) Can yeast systems biology contribute to the understanding of human disease? *Trends Biotechnol.*, **26**, 584–590.

90 Simon, J.A. and Yen, T.J. (2003) Novel approaches to screen for anticancer drugs using *Saccharomyces cerevisiae. Methods Mol. Biol.*, **223**, 555–576.

91 Sturgeon, C.M., Kemmer, D., Anderson, H.J., and Roberge, M. (2006) Yeast as a tool to uncover the cellular targets of drugs. *Biotechnol. J.*, **1**, 289–298.

92 Borodina, I. and Nielsen, J. (2014) Advances in metabolic engineering of

yeast *Saccharomyces cerevisiae* for production of chemicals. *Biotechnol. J.*, **9**, 609–620.

93 Van Gulik, W.M. and Heijnen, J.J. (1995) A metabolic network stoichiometry analysis of microbial growth and product formation. *Biotechnol. Bioeng.*, **48**, 681–698.

94 Vanrolleghem, P.A., de Jong-Gubbels, P., van Gulik, W.M., Pronk, J.T., van Dijken, J.P., and Heijnen, S. (1996) Validation of a metabolic network for *Saccharomyces cerevisiae* using mixed substrate studies. *Biotechnol. Prog.*, **12**, 434–448.

95 Förster, J., Gombert, A.K., and Nielsen, J. (2002) A functional genomics approach using metabolomics and *in silico pathway* analysis. *Biotechnol. Bioeng.*, **79**, 703–712.

96 Nissen, T.L., Schulze, U., Nielsen, J., and Villadsen, J. (1997) Flux distributions in anaerobic, glucose-limited continuous cultures of *Saccharomyces cerevisiae. Microbiology*, **143**, 203–218.

97 Gombert, A., dos Santos, M., Christensen, B., and Nielsen, J. (2001) Network identification and flux quantification in the central metabolism of *Saccharomyces cerevisiae* under different conditions of glucose repression. *J. Bacteriol.*, **183**, 1441–1451.

98 Jin, S., Ye, K., and Shimizu, K. (1997) Metabolic flux distributions in recombinant *Saccharomyces cerevisiae* during foreign protein production. *J. Biotechnol.*, **54**, 161–174.

99 Carlson, R., Fell, D., and Srienc, F. (2002) Metabolic pathway analysis of a recombinant yeast for rational strain development. *Biotechnol. Bioeng.*, **79**, 121–134.

100 Jin, Y.-S. and Jeffries, T.W. (2004) Stoichiometric network constraints on xylose metabolism by recombinant *Saccharomyces cerevisiae. Metab. Eng.*, **6**, 229–238.

101 Wahlbom, C.F., Eliasson, A., and Hahn-Hägerdal, B. (2001) Intracellular fluxes in a recombinant xylose-utilizing *Saccharomyces cerevisiae* cultivated anaerobically at different dilution rates and feed concentrations. *Biotechnol. Bioeng.*, **72**, 289–296.

102 Christensen, B. and Nielsen, J. (2000) Metabolic network analysis: a powerful tool in metabolic engineering. *Adv. Biochem. Eng. Biotechnol.*, **66**, 209–231.

103 Sauer, U. (2006) Metabolic networks in motion: 13C-based flux analysis. *Mol. Syst. Biol.*, **2**, 62.

104 Tang, Y.J., Martin, H.G., Myers, S., Rodriguez, S., Baidoo, E.E.K., and Keasling, J.D. (2009) Advances in analysis of microbial metabolic fluxes via (13)C isotopic labeling. *Mass Spectrom. Rev.*, **28**, 362–375.

105 Famili, I., Förster, J., Nielsen, J., and Palsson, B.O. (2003) *Saccharomyces cerevisiae* phenotypes can be predicted by using constraint-based analysis of a genome-scale reconstructed metabolic network. *Proc. Natl. Acad. Sci. USA*, **100**, 13134–13139.

106 Förster, J., Famili, I., Fu, P., Palsson, B.Ø., and Nielsen, J. (2003b) Large-scale evaluation of *in silico* gene deletions in *Saccharomyces cerevisiae. OMICS*, **7**, 193–202.

107 Duarte, N.C., Herrgård, M.J., and Palsson, B.Ø. (2004) Reconstruction and validation of *Saccharomyces cerevisiae* iND750, a fully compartmentalized genome-scale metabolic model. *Genome Res.*, **14**, 1298–1309.

108 Kuepfer, L., Sauer, U., and Blank, L.M. (2005) Metabolic functions of duplicate genes in *Saccharomyces cerevisiae. Genome Res.*, **15**, 1421–1430.

109 Nookaew, I., Jewett, M.C., Meechai, A., Thammarongtham, C., Laoteng, K., Cheevadhanarak, S., Nielsen, J., and Bhumiratana, S. (2008) The genome-scale metabolic model iIN800 of *Saccharomyces cerevisiae* and its validation: a scaffold to query lipid metabolism. *BMC Syst. Biol.*, **2**, 71.

110 Mo, M.L., Palsson, B.O., and Herrgård, M.J. (2009) Connecting extracellular metabolomic measurements to intracellular flux states in yeast. *BMC Syst. Biol.*, **3**, 37.

111 Herrgård, M.J., Swainston, N., Dobson, P., Dunn, W.B., Arga, K.Y., Arvas, M., Blüthgen, N., Borger, S., Costenoble, R., Heinemann, M., Hucka, M., Le Novère, N., Li, P., Liebermeister, W., Mo, M.L.,

Oliveira, A.P., Petranovic, D., Pettifer, S., Simeonidis, E., Smallbone, K., Spasić, I., Weichart, D., Brent, R., Broomhead, D.S., Westerhoff, H.V., Kirdar, B., Penttilä, M., Klipp, E., Palsson, B.Ø., Sauer, U., Oliver, S.G., Mendes, P., Nielsen, J., and Kell, D.B. (2008) A consensus yeast metabolic network reconstruction obtained from a community approach to systems biology. *Nat. Biotechnol.*, **26**, 1155–1160.

112 Dobson, P.D. *et al.* (2010) Further developments towards a genome-scale metabolic model of yeast. *BMC Syst. Biol.*, **4**, 145.

113 Heavner, B.D., Smallbone, K., Barker, B., Mendes, P., and Walker, L.P. (2012) Yeast 5: an expanded reconstruction of the *Saccharomyces cerevisiae* metabolic network. *BMC Syst. Biol.*, **6**, 55.

114 Heavner, B.D., Smallbone, K., Price, N.D., and Walker, L.P. (2013) Version 6 of the consensus yeast metabolic network refines biochemical coverage and improves model performance. Database (Oxford). 2013, bat059.

115 Aung, H.W., Henry, S.A., Walker, L.P., (2013). Revising the Representation of Fatty Acid, Glycerolipid, and Glycerophospholipid Metabolism in the Consensus Model of Yeast Metabolism. *Ind. Biotechnol.* **9**, 215–228

116 Österlund, T., Nookaew, I., Bordel, S., and Nielsen, J. (2013) Mapping condition-dependent regulation of metabolism in yeast through genome-scale modeling. *BMC Syst. Biol.*, **7**, 36.

117 David, H., Ozçelik, I.S., Hofmann, G., and Nielsen, J. (2008) Analysis of *Aspergillus nidulans* metabolism at the genome-scale. *BMC Genomics*, **9**, 163.

118 Andersen, M.R., Lehmann, L., and Nielsen, J. (2009) Systemic analysis of the response of *Aspergillus niger* to ambient pH. *Genome Biol.*, **10**, R47.

119 Vongsangnak, W., Olsen, P., Hansen, K., Krogsgaard, S., and Nielsen, J. (2008) Improved annotation through genome-scale metabolic modeling of *Aspergillus oryzae*. *BMC Genomics*, **9**, 245.

120 Liu, J., Gao, Q., Xu, N., and Liu, L. (2013) Genome-scale reconstruction and *in silico* analysis of *Aspergillus terreus* metabolism. *Mol. Biosyst.*, **9**, 1939–1948.

121 Chang, R.L., Ghamsari, L., Manichaikul, A., Hom, E.F.Y., Balaji, S., Fu, W., Shen, Y., Hao, T., Palsson, B.Ø., Salehi-Ashtiani, K., and Papin, J.A. (2011) Metabolic network reconstruction of *Chlamydomonas* offers insight into light-driven algal metabolism. *Mol. Syst. Biol.*, **7**, 518.

122 Chavali, A.K., Whittemore, J.D., Eddy, J.A., Williams, K.T., and Papin, J.A. (2008) Systems analysis of metabolism in the pathogenic trypanosomatid Leishmania major. *Mol. Syst. Biol.*, **4**, 177.

123 Sohn, S.B., Graf, A.B., Kim, T.Y., Gasser, B., Maurer, M., Ferrer, P., Mattanovich, D., and Lee, S.Y. (2010) Genome-scale metabolic model of methylotrophic yeast *Pichia pastoris* and its use for *in silico* analysis of heterologous protein production. *Biotechnol. J.*, **5**, 705–715.

124 Chung, B.K., Selvarasu, S., Andrea, C., Ryu, J., Lee, H., Ahn, J., Lee, H., and Lee, D.-Y. (2010) Genome-scale metabolic reconstruction and *in silico* analysis of methylotrophic yeast *Pichia pastoris* for strain improvement. *Microb. Cell Fact.*, **9**, 50.

125 Caspeta, L., Shoaie, S., Agren, R., Nookaew, I., and Nielsen, J. (2012) Genome-scale metabolic reconstructions of *Pichia stipitis* and *Pichia pastoris* and *in silico* evaluation of their potentials. *BMC Syst. Biol.*, **6**, 24.

126 Balagurunathan, B., Jonnalagadda, S., Tan, L., and Srinivasan, R. (2012) Reconstruction and analysis of a genome-scale metabolic model for *Scheffersomyces stipitis*. *Microb. Cell Fact.*, **11**, 27.

127 Liu, T., Zou, W., Liu, L., and Chen, J. (2012) A constraint-based model of *Scheffersomyces stipitis* for improved ethanol production. *Biotechnol. Biofuels*, **5**, 72.

128 Sohn, S.B., Kim, T.Y., Lee, J.H., and Lee, S.Y. (2012) Genome-scale metabolic model of the fission yeast *Schizosaccharomyces pombe* and the reconciliation of *in silico*/*in vivo* mutant growth. *BMC Syst. Biol.*, **6**, 49.

129 Feizi, A., Österlund, T., Petranovic, D., Bordel, S., and Nielsen, J. (2013) Genome-scale modeling of the protein

secretory machinery in yeast. *PLoS One*, **8**, e63284.

130 Knuf, C. and Nielsen, J. (2012) Aspergilli: systems biology and industrial applications. *Biotechnol. J.*, **7**, 1147–1155.

131 Celik, E. and Calık, P. (2012) Production of recombinant proteins by yeast cells. *Biotechnol. Adv.*, **30**, 1108–1118.

132 Hamilton, S.R., Davidson, R.C., Sethuraman, N., Nett, J.H., Jiang, Y., Rios, S., Bobrowicz, P., Stadheim, T.A., Li, H., Choi, B.-K., Hopkins, D., Wischnewski, H., Roser, J., Mitchell, T., Strawbridge, R.R., Hoopes, J., Wildt, S., and Gerngross, T.U. (2006) Humanization of yeast to produce complex terminally sialylated glycoproteins. *Science*, **313**, 1441–1443.

133 De Schutter, K., Lin, Y.-C., Tiels, P., Van Hecke, A., Glinka, S., Weber-Lehmann, J., Rouzé, P., Van de Peer, Y., and Callewaert, N. (2009) Genome sequence of the recombinant protein production host *Pichia pastoris*. *Nat. Biotechnol.*, **27**, 561–566.

134 Kurtzman, C.P. and Suzuki, M. (2010) Phylogenetic analysis of ascomycete yeasts that form coenzyme Q-9 and the proposal of the new genera *Babjeviella*, *Meyerozyma*, *Millerozyma*, *Priceomyces*, and *Scheffersomyces*. *Mycoscience*, **51**, 2–14.

135 Jeffries, T.W. (1983) Utilization of xylose by bacteria, yeasts, and fungi. *Adv. Biochem. Eng. Biotechnol.*, **27**, 1–32.

136 Rudolf, A., Baudel, H., Zacchi, G., Hahn-Hägerdal, B., and Lidén, G. (2008) Simultaneous saccharification and fermentation of steam-pretreated bagasse using *Saccharomyces cerevisiae* TMB3400 and *Pichia stipitis* CBS6054. *Biotechnol. Bioeng.*, **99**, 783–790.

137 Elander, R.P. (2003) Industrial production of beta-lactam antibiotics. *Appl. Microbiol. Biotechnol.*, **61**, 385–392.

138 Van den Berg, M.A., Albang, R., Albermann, K., Badger, J.H., Daran, J.-M., Driessen, A.J.M., Garcia-Estrada, C., Fedorova, N.D., Harris, D.M., Heijne, W.H.M., Joardar, V., Kiel, J.A.K.W., Kovalchuk, A., Martín, J.F., Nierman, W.C., Nijland, J.G., Pronk, J.T., Roubos, J.A., van der Klei, I.J., van Peij, N.N.M.E., Veenhuis, M., von Döhren, H., Wagner, C., Wortman, J., and Bovenberg, R.A.L. (2008) Genome sequencing and analysis of the filamentous fungus *Penicillium chrysogenum*. *Nat. Biotechnol.*, **26**, 1161–1168.

139 Li, S., Chen, X., Liu, L., and Chen, J. (2013) Pyruvate production in *Candida glabrata*: manipulation and optimization of physiological function. *Crit. Rev. Biotechnol.*, Early Online: 1–10.

140 Brunke, S. and Hube, B. (2013) Two unlike cousins: *Candida albicans* and *C. glabrata* infection strategies. *Cell. Microbiol.*, **15**, 701–708.

141 Sheikh, K., Förster, J., and Nielsen, L.K. (2005) Modeling hybridoma cell metabolism using a generic genome-scale metabolic model of *Mus musculus*. *Biotechnol. Prog.*, **21**, 112–121.

142 Sigurdsson, M.I., Jamshidi, N., Steingrimsson, E., Thiele, I., and Palsson, B.Ø. (2010) A detailed genome-wide reconstruction of mouse metabolism based on Human Recon 1. *BMC Syst. Biol.*, **4**, 140.

143 Duarte, N.C., Becker, S.A., Jamshidi, N., Thiele, I., Mo, M.L., Vo, T.D., Srivas, R., and Palsson, B.Ø. (2007) Global reconstruction of the human metabolic network based on genomic and bibliomic data. *Proc. Natl. Acad. Sci. USA*, **104**, 1777–1782.

144 Bandaranayake, A.D. and Almo, S.C. (2014) Recent advances in mammalian protein production. *FEBS Lett.*, **588**, 253–260.

145 Quek, L.E. and Nielsen, L.K. (2008) On the reconstruction of the *Mus musculus* genome-scale metabolic network model. *Genome Inform.*, **21**, 89–100.

146 Selvarasu, S., Karimi, I.A., Ghim, G.-H., and Lee, D.-Y. (2010) Genome-scale modeling and *in silico* analysis of mouse cell metabolic network. *Mol. Biosyst.*, **6**, 152–161.

147 De Oliveira Dal'Molin, C.G., Quek, L.-E., Palfreyman, R.W., Brumbley, S.M., and Nielsen, L.K. (2010) AraGEM, a genome-scale reconstruction of the primary metabolic network in *Arabidopsis*. *Plant Physiol.*, **152**, 579–589.

148 Dal'Molin, C.G., de, O., Quek, L.-E., Palfreyman, R.W., Brumbley, S.M., and Nielsen, L.K. (2010) C4GEM, a genome-scale metabolic model to study C4 plant metabolism. *Plant Physiol.*, **154**, 1871–1885.

149 Hao, T., Ma, H.-W., Zhao, X.-M., and Goryanin, I. (2010) Compartmentalization of the Edinburgh Human Metabolic Network. *BMC Bioinformatics*, **11**, 393.

150 Ma, H., Sorokin, A., Mazein, A., Selkov, A., Selkov, E., Demin, O., and Goryanin, I. (2007) The Edinburgh Human Metabolic Network reconstruction and its functional analysis. *Mol. Sys. Biol.*, **3**, 135.

151 Agren, R., Bordel, S., Mardinoglu, A., Pornputtapong, N., Nookaew, I., and Nielsen, J. (2012) Reconstruction of genome-scale active metabolic networks for 69 human cell types and 16 cancer types using INIT. *PLoS Comput. Biol.*, **8**, e1002518.

152 Folger, O., Jerby, L., Frezza, C., Gottlieb, E., Ruppin, E., and Shlomi, T. (2011) Predicting selective drug targets in cancer through metabolic networks. *Mol. Syst. Biol.*, **7**, 501.

153 Gatto, F., Nookaew, I., and Nielsen, J. (2014) Chromosome 3p loss of heterozygosity is associated with a unique metabolic network in clear cell renal carcinoma. *Proc. Natl. Acad. Sci. USA*, **111**, E866–E875.

154 Bordbar, A., Feist, A.M., Usaite-Black, R., Woodcock, J., Palsson, B.O., and Famili, I. (2011) A multi-tissue type genome-scale metabolic network for analysis of whole-body systems physiology. *BMC Syst. Biol.*, **5**, 180.

155 Mardinoglu, A., Agren, R., Kampf, C., Asplund, A., Nookaew, I., Jacobson, P., Walley, A.J., Froguel, P., Carlsson, L.M., Uhlen, M., and Nielsen, J. (2013) Integration of clinical data with a genome-scale metabolic model of the human adipocyte. *Mol. Sys. Bio.*, **9**, 649.

156 Gille, C., Bölling, C., Hoppe, A., Bulik, S., Hoffmann, S., Hübner, K., Karlstädt, A., Ganeshan, R., König, M., Rother, K., Weidlich, M., Behre, J., and Holzhütter, H.-G. (2010) HepatoNet1: a comprehensive metabolic reconstruction of the human hepatocyte for the analysis of liver physiology. *Mol. Syst. Biol.*, **6**, 411.

157 Jerby, L., Shlomi, T., and Ruppin, E. (2010) Computational reconstruction of tissue-specific metabolic models: application to human liver metabolism. *Mol. Syst. Biol.*, **6**, 401.

158 Mardinoglu, A., Agren, R., Kampf, C., Asplund, A., Uhlen, M., and Nielsen, J. (2014) Genome-scale metabolic modelling of hepatocytes reveals serine deficiency in patients with non-alcoholic fatty liver disease. *Nat. Commun.*, **5**, 3083.

159 Bordbar, A., Lewis, N.E., Schellenberger, J., Palsson, B.Ø., and Jamshidi, N. (2010) Insight into human alveolar macrophage and *M. tuberculosis* interactions via metabolic reconstructions. *Mol. Syst. Biol.*, **6**, 422.

160 Lewis, N.E., Schramm, G., Bordbar, A., Schellenberger, J., Andersen, M.P., Cheng, J.K., Patel, N., Yee, A., Lewis, R.A., Eils, R., König, R., and Palsson, B.Ø. (2010) Large-scale *in silico* modeling of metabolic interactions between cell types in the human brain. *Nat. Biotechnol.*, **28**, 1279–1285.

161 Brinkrolf, K., Rupp, O., Laux, H., Kollin, F., Ernst, W., Linke, B., Kofler, R., Romand, S., Hesse, F., Budach, W.E., Galosy, S., Müller, D., Noll, T., Wienberg, J., Jostock, T., Leonard, M., Grillari, J., Tauch, A., Goesmann, A., Helk, B., Mott, J.E., Pühler, A., and Borth, N. (2013) Chinese hamster genome sequenced from sorted chromosomes. *Nat. Biotechnol.*, **31** 694–695.

162 Lewis, N.E., Liu, X., Li, Y., Nagarajan, H., Yerganian, G., O'Brien, E., Bordbar, A., Roth, A.M., Rosenbloom, J., Bian, C., Xie, M., Chen, W., Li, N., Baycin-Hizal, D., Latif, H., Förster, J., Betenbaugh, M.J., Famili, I., Xu, X., Wang, J., and Palsson, B.O. (2013) Genomic landscapes of Chinese hamster ovary cell lines as revealed by the *Cricetulus griseus* draft genome. *Nat. Biotechnol.*, **31**, 759–765.

163 Xu, X., Nagarajan, H., Lewis, N.E., Pan, S., Cai, Z., Liu, X., Chen, W., Xie, M., Wang, W., Hammond, S., Andersen, M.R., Neff, N., Passarelli, B., Koh, W., Fan, H.C., Wang, J., Gui, Y., Lee, K.H.,

Betenbaugh, M.J., Quake, S.R., Famili, I., Palsson, B.O., and Wang, J. (2011) The genomic sequence of the Chinese hamster ovary (CHO)-K1 cell line. *Nat. Biotechnol.*, **29**, 735–741.

164 Kildegaard, H.F., Baycin-Hizal, D., Lewis, N.E., and Betenbaugh, M.J. (2013) The emerging CHO systems biology era: harnessing the omics revolution for biotechnology. *Curr. Opin. Biotechnol.*, **24**, 1102–1107.

165 Cordon-Cardo, C. (1995) Mutations of cell cycle regulators. Biological and clinical implications for human neoplasia. *Am. J. Pathol.*, **147**, 545–560.

166 Hanahan, D. and Weinberg, R.A. (2000) The hallmarks of cancer. *Cell*, **100**, 57–70.

167 Hanahan, D. and Weinberg, R.A. (2011) Hallmarks of cancer: the next generation. *Cell*, **144**, 646–674.

168 Reed, J.L. (2012) Shrinking the metabolic solution space using experimental datasets. *PLoS Comput. Biol.*, **8**, e1002662.

169 Shendure, J. and Aiden, E.L. (2012) The expanding scope of DNA sequencing. *Nat. Biotechnol.*, **30**, 1084–1094.

170 Becker, S.A. and Palsson, B.O. (2008) Context-specific metabolic networks are consistent with experiments. *PLoS Comput. Biol*, **4**, e1000082.

171 Shlomi, T., Cabili, M.N., Herrgård, M.J. *et al.* (2008) Network-based prediction of human tissue-specific metabolism. *Nat. Biotechnol.*, **26**, 1003–1010.

172 Zur, H., Ruppin, E., and Shlomi, T. (2010) iMAT: an integrative metabolic analysis tool. *Bioinformatics*, **26**, 3140–3142.

173 Jensen, P.A. and Papin, J.A. (2011) Functional integration of a metabolic network model and expression data without arbitrary thresholding. *Bioinformatics*, **27**, 541–547.

174 Wang, Y., Eddy, J.A., and Price, N.N.D. (2012) Reconstruction of genome-scale metabolic models for 126 human tissues using mCADRE. *BMC Syst. Biol.*, **6**, 153.

175 Colijn, C., Brandes, A., Zucker, J. *et al.* (2009) Interpreting expression data with metabolic flux models: predicting *Mycobacterium tuberculosis* mycolic acid production. *PLoS Comput Biol*, **5**, e1000489.

176 Lee, D., Smallbone, K., Dunn, W.B. *et al.* (2012b) Improving metabolic flux predictions using absolute gene expression data. *BMC Sys. Biol.*, **6**, 73.

177 Hoppe, A., Hoffmann, S., and Holzhütter, H.G. (2007) Including metabolite concentrations into flux balance analysis: thermodynamic realizability as a constraint on flux distributions in metabolic networks. *BMC Syst. Biol.*, **1**, 23.

178 Henry, C.S., Broadbelt, L.J., and Hatzimanikatis, V. (2007) Thermodynamics-based metabolic flux analysis. *Biophys. J.*, **92**, 1792–1805.

179 Kümmel, A., Panke, S., and Heinemann, M. (2006) Putative regulatory sites unraveled by network-embedded thermodynamic analysis of metabolome data. *Mol. Syst. Biol.*, **2**, 0034.

180 Ishii, N., Nakahigashi, K., Baba, T. *et al.* (2007) Multiple high-throughput analyses monitor the response of *E. coli* to perturbations. *Science*, **316**, 593–597.

181 Buescher, J.M., Liebermeister, W., Jules, M. *et al.* (2012) Global network reorganization during dynamic adaptations of *Bacillus subtilis* metabolism. *Science*, **335**, 1099–1103.

182 Yizhak, K., Benyamini, T., Liebermeister, W. *et al.* (2010) Integrating quantitative proteomics and metabolomics with a genome-scale metabolic network model. *Bioinformatics*, **26**, i255–i260.

183 Jamshidi, N. and Palsson, B.O. (2008) Formulating genome-scale kinetic models in the post-genome era. *Mol. Syst. Biol.*, **4**, 171.

184 Smallbone, K., Simeonidis, E., Swainston, N., and Mendes, P. (2010) Towards a genome-scale kinetic model of cellular metabolism. *BMC Syst. Biol.*, **4**, 6.

7
Kinetics of Bio-Reactions

John Villadsen

Summary

Mechanistically founded rate expressions are derived for enzyme reactions, and in Linewaever–Burk plots of $1/r$ versus. $1/s$, it is shown how the kinetic and thermodynamic parameters are calculated from experiments. Kinetic expressions for cell reactions are set up by analogy with the enzyme kinetic expressions. Conditions for which the cell kinetic models are valid are stated. In an industrially relevant example, the kinetics for succinic acid production is derived from experiments. Examples show that simple rate expressions for cell reactions are invalid. Kinetic models and models for transport of substrate into pellets are combined. Finally, an introduction is given to metabolic control analysis.

In Chapter 2, the rates of bioreactions were defined both for enzymatically catalyzed reactions and for reactions with living cells. The rates of the reactions were measured in a steady-state CSTR through measurement of the outlet concentrations of reactants. These rates were empirical quantities, valid only for the environmental conditions of the experiment, for example, the dilution rate D.

In this chapter, the *specific* rates of reaction will be *predicted* by means of a mathematical expression, *the kinetics of the reaction.* This expression can be derived through a *mechanistic* interpretation of an enzymatically catalyzed reaction, but it is essentially of *empirical* nature for cell reactions. The models can be used in mass balances for *design* of processes under process conditions not yet studied experimentally. The value of the predictive kinetic *model* depends on the quality of the experimental data on which the model is based, and well-founded kinetic models for enzyme reactions have a considerable predictive power. This is also true for cell reaction models, *when the model is used in its proper context.*

Fundamental Bioengineering, First Edition. Edited by John Villadsen.

7.1 Simple Models for Enzymatic Reactions and for Cell Reactions with Unstructured Biomass

The key feature of the kinetic model is that it *predicts* the reaction rate at given values of the concentrations of substrate(s) and product(s).

$$r = f(s_i, p_j) \tag{7.1}$$

s_i and p_j are the concentrations of substrates and products *in the reactor*, respectively, and for enzymatically catalyzed reactions, they are equal to the effluent concentrations. $[s_i, p_1]$ are constant in time when the reactor is a CSTR at steady state – as defined in Chapter 2 – but they can also be functions of time *when the reactor is operated in a transient state*, for example, when the substrate concentration is abruptly changed from s_0 to $s_0{}^1$. In design of reactors with living cells $[s_i, p_j]$ are also the concentrations measured in the well-stirred medium, but when kinetic analysis is used to study reaction rates of metabolic pathway reactions $[s_i, p_j]$ denotes the *internal* concentrations of, respectively, the substrate and product of a given pathway reaction. The concentration of any metabolite X_k in the whole pathway can influence the rate of reaction, that is, the flux of carbon through the pathway, and even concentrations of metabolites from *other* pathways (the so-called *effectors*) can influence $f(s_i, p_j)$.

We shall begin with kinetics for enzymatically catalyzed reactions ("enzyme reactions"). In this and the following sections, the kinetics can be derived from a mechanistic model. Later, in (7.21)–(7.23) empirical expressions will be derived for cell reactions by analogy with enzyme reactions. Whereas the rate expressions for enzyme reactions hold true both for steady state and for transient operation, this is rarely the case for cell reactions. In cell reactions, the effluent concentrations can also be different from the concentrations in the reactor. This is the case for cell recirculation, a subject discussed in Section 10.3, and in many other respects cell reactions are different from enzyme reactions mainly because the cell reaction is autocatalytic – the volumetric rates q_i are proportional to the cell concentration x – and biomass X is a product of cell reactions. Still, as seen in Section 2.1, the basic mass balances for enzyme reactions and for cell reactions are the same, and when enzyme reactions and cell reactions are treated in this chapter – just as the *measured* rates were treated together in Chapter 2 – both the similarities between all bioreactions and their differences will become much clearer.

The simplest expression $f(s_i, p_j)$ to be used in Eq. (7.1) is as follows:

$$r = (-r_s) = f(s_i, p_j) = \frac{ks}{s + K_m}. \tag{7.2}$$

Here, r, the specific consumption rate of substrate S, is assumed to depend on the concentration of S in the medium, but it is independent of concentrations of other substrates that could also be consumed in the reaction, and r is independent of all product concentrations.

The rate constant k has units of, for example, (g substrate S) converted per h and per unit of catalyst. For enzymatically catalyzed processes, this will give (g substrate converted) (unit enzyme h)$^{-1}$, and for cell reactions where the catalyst is the biomass X, k has units of (g substrate consumed) (g X h)$^{-1}$. Whereas the unit for X is straightforward, the "unit enzyme" has many definitions, for example, "the amount of a given catalase preparation that will convert 1 g of H_2O_2 to ½ $O_2 + H_2O$ in one hour."

For enzymatic reactions, Eq. (7.2) has a mechanistic interpretation, first proposed by Michaelis and Menten [1]:

Let the amount of enzyme added to the reactor be 1 unit. The conversion of S to product P proceeds in two steps.

$$\mathrm{S + E \leftrightarrow SE \rightarrow P + E} \tag{7.3}$$

In the reversible reaction, substrate S is docked on the active site of the free enzyme E and forms an enzyme–substrate complex SE. In the second step, the substrate complex reacts to form P with regeneration of the free enzyme E. Let the fraction of free enzyme in the added 1 unit of enzyme preparation be e and the bound fraction be se. The free and bound enzyme fractions add up to 1:

$$e + \mathrm{se} = 1. \tag{7.4}$$

The decomposition of SE to P + E is assumed to be governed by first-order kinetics $r = k \cdot \mathrm{se}$, and let the reaction of S and E to form SE be proportional to the substrate concentrations in both the forward direction (rate constant $k_{\rightarrow}$) and the reverse direction (rate constant $k_{\leftarrow}$). *If* both $k_{\rightarrow}$ and $k_{\leftarrow}$ are very large compared to the rate constant k for decomposition of SE, then

$$\frac{s \cdot e}{se} = K_{\mathrm{eq}} \rightarrow \mathrm{se} = \frac{s \cdot e}{K_{\mathrm{eq}}} = \frac{(1 - \mathrm{se}) \cdot s}{K_{\mathrm{eq}}} \rightarrow \mathrm{se} = \frac{s}{s + K_{\mathrm{eq}}} \text{ and finally } r = \frac{k \cdot s}{s + K_{\mathrm{eq}}}. \tag{7.5}$$

Exactly, the same arguments were proposed a few years later by Langmuir [2] to explain that a rate expression identical to Eq. (7.5) holds for catalytic reactions on a surface: The substrate (reactant S) adsorbs by an equilibrium reaction to the free catalyst sites (E) on the surface, and the occupied sites ES decompose by a first-order reaction to product with regeneration of the free sites. When S is fully adsorbed, the number of occupied sites on the surface is high and the rate of dissociation of P from the sites by ES → P + free sites is high. The total rate of conversion of reactant to product per surface unit is proportional to the total concentration of sites e_0 per unit surface. In complete analogy, the *volumetric* rate of conversion of substrate to product is given by

$$q(\text{g substrate converted to product})(\text{l reactor h})^{-1} = r \cdot e_0. \tag{7.6}$$

In Eq. (7.6), e_0 is the number of units of the enzyme preparation per l reactor volume.

Briggs and Haldane [3] relaxed the assumption that the first reaction in Eq. (7.3) should be at equilibrium. Instead they proposed that the concentration of ES would be in a pseudo-steady state, that is, the rate of production of ES should equal its rate of consumption.

$$\frac{d(\mathrm{es})}{dt} = 0 = k_{\rightarrow}(e\cdot\ s) - k_{\leftarrow}(\mathrm{es}) - k(\mathrm{es}) = k_{\rightarrow}s\cdot(1-\mathrm{es}) - k_{\leftarrow}(\mathrm{es}) - k(\mathrm{es}) \rightarrow$$

$$\mathrm{es} = \frac{k_{\rightarrow}s}{k + k_{\rightarrow}s + k_{\leftarrow}} = \frac{s}{s + (k + k_{\leftarrow}/k_{\rightarrow})}$$

$$r = k\cdot\mathrm{es} = \frac{ks}{s+(k+k_{\leftarrow}/k_{\rightarrow})} \approx \frac{ks}{s+\ (k_{\leftarrow}/k_{\rightarrow})} = \frac{ks}{s+K_{\mathrm{m}}} \quad \text{for } k \ll k_{\leftarrow}\,. \tag{7.7}$$

When the substrate is tightly bound to the enzyme, then $k_{\leftarrow} \ll k_{\rightarrow}$ and the Michaelis constant is small, that is, $r \approx \mathrm{k}$. In this case, the *affinity* of the enzyme for the substrate is high, a favorable situation since the overall rate of conversion of S to P approaches its maximum value k.

In a *reversible* enzyme-catalyzed process, the intermediate is decomposed to product + E by a reversible reaction. In this case,

$$\mathrm{E} + \mathrm{S} \leftrightarrow \mathrm{ES} \leftrightarrow \mathrm{P} + \mathrm{E}. \tag{7.8}$$

For the first reaction, the rate constants are $[k_{1\rightarrow}, k_{1\leftarrow}]$, and for the second reaction, they are $[k_{2\rightarrow}, k_{2\leftarrow}]$. By repeating the derivation in Eq. (7.7), one obtains for $e + \mathrm{es} = 1$ that:

$$\mathrm{es} = \frac{k_{1\rightarrow}s + k_{2\leftarrow}p}{k_{1\leftarrow} + k_{2\rightarrow} + k_{1\rightarrow}s + k_{2\leftarrow}p} \rightarrow r = k_{2\rightarrow}\,\mathrm{es} - k_{2\leftarrow}p(1-\mathrm{es})$$

$$= \frac{k_{1\rightarrow}k_{2\rightarrow}s - k_{1\leftarrow}k_{2\leftarrow}p}{k_{1\leftarrow} + k_{2\rightarrow} + k_{1\rightarrow}s + k_{2\leftarrow}p}. \tag{7.9}$$

The equilibrium reaction for the overall reaction S ↔ P is $K_{\mathrm{eq}} = \dfrac{k_{1\rightarrow}k_{2\rightarrow}}{k_{1\leftarrow}k_{2\leftarrow}}$. (7.10)

Let the feed have the concentrations $[s, p] = [s_0, 0]$ in C-mol l^{-1}.

$$s_0 = s + p = s_{\mathrm{eq}} + p_{\mathrm{eq}} = s_{\mathrm{eq}}\left(1 + \frac{p_{\mathrm{eq}}}{s_{\mathrm{eq}}}\right) = s_{\mathrm{eq}}(1 + K_{\mathrm{eq}}). \tag{7.11}$$

When Eq. (7.10) and (7.11) are inserted in Eq. (7.9), one obtains

$$r = (-r_{\mathrm{s}}) = \frac{k_{\max}(s - s_{\mathrm{eq}})}{K_{\mathrm{m}} +\ (s - s_{\mathrm{eq}})}. \tag{7.12}$$

The structure of Eq. (7.12) is the same as that of Eq. (7.7), but the parameters $k_{\max}$ and K_{m} are complicated combinations of the rate constants in Eq. (7.9). When the enzyme has a low affinity for the substrate, $K_{\mathrm{m}} \gg (s - s_{\mathrm{eq}})$ and $(-r_{\mathrm{s}}) = (k_{\max}/K_{\mathrm{m}})(s - s_{\mathrm{eq}})$, a relation that also holds for (7.7): $(-r_{\mathrm{s}}) = (k/K_{\mathrm{m}})s$.

When Eq. (7.2) is used to predict the specific rate of substrate consumption $(-r_{\mathrm{s}})$ for a cell reaction, a number of assumptions listed in (1) to (3) below must be satisfied.

First a convention that dates back to the early days of quantitative analysis of cell reaction kinetics must be introduced. In cell kinetics, the specific rate *of biomass formation* is given by Eq. (7.2):

$$r_x = \mu = \frac{\mu_{max}\, s}{s + K_m} \text{ and now} (-r_s) = Y_{xs}\, \mu = \frac{\mu_{max}\, s}{s + K_m} Y_{xs}. \tag{7.13}$$

In many references, the term μ is used for r_x, but in the present text r_x will be used unless the meaning of μ is obvious from the context. $\mu_{max} = (r_x)_{max}$, the maximum specific rate of biomass production in units of, for example, g biomass produced h^{-1} (g biomass present)$^{-1}$, and this is obtained when $s \gg K_m$.

The relation (7.13) has absolutely no mechanistic background. It expresses that the rate of any chemical reaction, including conversion of substrate to products by living organisms, is first order in the substrate concentration for $s \to 0$. The rate is $(\mu_{max}/K_m)Y_{xs}s$ (unit, for example, (g S produced)(g X h)$^{-1}$). Also Eq. (7.13) expresses that the specific rate of substrate conversion cannot increase indefinitely. The biomass X contains a certain maximum fraction of machinery that is able to run the conversion of substrate to more biomass and to other products. This limitation of the specific rate of cell reactions is the same as that discussed for enzymatic reactions: Enzymes have a maximum turn over frequency – the turnover rate of a single enzyme molecule is extraordinarily high, but not infinitely high. Catalytic surfaces have a limited number of active sites per m^2. If these are all occupied by allowing the pressure of a gaseous reactant to increase unlimited, then the overall conversion rate per m^2 reaches a maximum.

The very persuasive empirical model (7.13) – it is not possible to devise a two-parameter kinetic model that would better fit obvious laws of Nature – was first derived in the seminal work of Jacques Monod [4], reviewed in Villadsen, Nielsen, and Lidén [5]. The constant K_m is named the *Monod constant* and it has obviously no physical relation to the Michaelis constant. The reason is that it is inconceivable that all the reactions in the metabolic pathways that generate X from S can be described with only the concentration of S as variable. Despite its empirical foundation, Eq. (7.13) gives excellent predictions of r_x (and of $(-r_s)$) also for s values considerably outside the range of values used in experimental studies to determine the kinetics.

The assumptions that must be satisfied in order to get good results from Eq. (7.13) are as follows:

1) Eq. (7.13) is only an acceptable kinetic model for *steady-state experiments*. Steady state is here interpreted as *metabolic steady state*, that is, the composition of the cell does not change with time. This assumption holds for steady-state experiments in a stirred tank where the steady-state assumption includes that the cell machinery for converting cells to biomass is constant in time. The assumption may also be correct for batch experiments, since if s_0, the initial substrate concentration, is high enough, then the uptake rate of substrate is constant, independent of s. In the beginning of a batch fermentation, there is a lag phase where cell machinery is built up,

and at the very end of a batch fermentation, there may be a change in stoichiometry and Eq. (7.13) does not hold. Changes of metabolism occur in aerobic yeast fermentation: For $D > D_{crit}$ ethanol is a product, and the kinetic expression derived for $D < D_{crit}$ *is invalid* for $D > D_{crit}$. Likewise, in aerobic batch fermentation with yeast, the cells take up the ethanol produced in the beginning of the fermentation when $s_{glc} \rightarrow 0$. The long lag phase in *diauxic* growth and the ensuing growth on ethanol must be modeled by a new kinetic expression. The so-called *fed-batch* operation of a stirred tank reactor (Section 10.5) can often be modeled by Eq. (7.13), despite the unsteady state nature of fed-batch fermentation. The reason is that in this mode of operation, the specific growth rate $\mu = r_x$ is often kept constant, at least during the initial phase of the fermentation.

Truly transient cultivations, for example, when the dilution rate D is changed in an abrupt step, cannot at all be modeled by Eq. (7.13) as will be shown in several examples.

The failure of Eq. (7.13) to represent transient experiments is due to the very long time (several hours) used to change the cell machinery. In this respect, enzymatically catalyzed reactions are very different. The enzyme is not much different from traditional catalysts. The enzyme almost immediately responds to a change in the environment, for example, when a pulse of substrate is added to the reactor.

2) When Eq. (7.13) is used as a kinetic model for cell reactions, one must be certain that S is *the limiting substrate*. It is of no avail to apply Eq. (7.13) with S = glucose if NH_3, HPO_4^{-2}, or one of the many growth enhancing substrates (cations such as Mg^{+2} or Cu^{+2}, hormones, an essential amino acid that cannot be produced by the organism, or even compounds that need to be present only in ppm concentrations) is the limiting substrate. The best indication that one has chosen the wrong limiting substrate is that the K_m value calculated from the experiments is above 250–400 mg l^{-1}. Table 7.3 lists typical values for K_m which is often in the 1–10 mg l^{-1} range when the correct limiting substrate has been used in the fitting of the parameters to experiments.
3) Whereas substrate utilization for maintenance is not an issue when Eq. (7.13) is applied as a model for enzyme reactions, it can have a considerable influence on cell reactions. Monod kinetics is often applied with an implicit assumption that the specific growth rates are proportional, for example, in Eq. (7.13) where $(-r_s)$ is calculated as $Y_{xs}\, r_x$. The true model for $(-r_s)$ and (r_p) to be used together with Eq. (7.13) for r_x is (3.13):

$$(-r_s) = Y_{xs}^{true}\, r_x + m_s, \quad \text{and} \quad r_p = Y_{xp}^{true}\, r_x + m_p. \tag{7.14}$$

Example 3.6 shows why "simple Monod kinetics" (7.13) cannot be used when substrate is used for maintenance: The stoichiometry changes with μ (= D in steady-state continuous cultivations), and Eq. (7.13) has to be modified as shown in Figure 7.5 and in Example 7.2.

In conclusion, Eq. (7.13) is a good model both for predicting the specific rates of enzyme reactions and for cell reactions. It is based on a classic mechanistic model in enzyme reactions, and although it is of empirical nature it also represents many cell reactions when the assumptions (1)–(3) are satisfied. Variants of Eq. (7.13) have been constructed for product and substrate-inhibited reactions. These are the subject of Section 7.2, and for enzyme reactions they are rigorously derived, while the corresponding expressions (7.21)–(7.23) for cell reactions are empirical, but quite accurate when (1) to (3) are satisfied.

The *volumetric* reaction rates q_i are shown in Eq. (7.15):

$$(-q_s) = \frac{\mu_{max}\, s}{s + K_m} e_0 \text{ for enzyme reactions and}$$
$$q_x = \frac{\mu_{max}\, s}{s + K_m} x \text{ for cell reactions,} \tag{7.15}$$

with $e_0 = \text{g}$ enzyme preparation (l medium)$^{-1}$ and $x = \text{g}$ biomass (l medium)$^{-1}$ $(-q_s)$ and q_x have units of, respectively, g substrate consumed (l medium · h)$^{-1}$ and g X produced (l medium · h)$^{-1}$.

7.2 Variants of Michaelis–Menten and Monod kinetics

In this section, the Michaelis–Menten kinetics (7.13) and the analogous Monod kinetics will be extended to hold for reactions where some form of substrate or product inhibition adds more detail to the kinetic model. As is the case for (7.13), the extended models are of mechanistic nature for enzyme reactions while being of strictly empirical nature for cell reactions.

In an enzyme reaction, there may be more than one substrate, and the substrates may compete for the active site on the enzyme. If the specificity of the enzyme for substrate is low, several substrate–enzyme complexes are produced, and in some cases different products can result.

Here, it will be assumed that a competing substrate S_1 cannot be converted to the by-product P_1, but the presence of S_1 will reduce the rate of formation of the main product P.

First, assume that S_1 competes with S for free enzyme E:

$$E + S_1 \leftrightarrow ES_1 \text{ while } E + S \leftrightarrow ES \rightarrow P + S. \tag{7.16}$$

The "oral" interpretation of Eq. (7.16) is that some of the free enzyme is trapped in ES_1 – a "dead-end." Since e_0 (= 1) is distributed between E, ES, and ES_1, the rate of formation of P will be severely reduced when the binding of E to S_1 is strong, and the fraction of e_0 tied up in ES_1 is large. The resulting kinetic expression (7.17) is derived by implying the steady-state concept for ES, ES_1 that was used in the Briggs–Haldane mechanistic approach.

$$r = (-r_s) = \frac{ks}{s + K_m\left(1 + (s_1/K_{eq1})\right)}. \tag{7.17}$$

Here, s_1 is the concentration of the inhibitor S_1 in the medium. K_m is the Michaels constant for the formation of ES, the intermediate in the main reaction, and K_{eq1} is the corresponding equilibrium constant for dissociation of ES_1. Obviously, if S_1 is closely bound to E, the equilibrium constant K_{eq1} is small, and the inhibition is severe. When the concentration s of S is large, the inhibition becomes less severe. The last term in the denominator of Eq. (7.17) of the so-called *competitive inhibition* is a modified and larger Michaelis constant, and the rate of conversion of S is reduced. A typical example of competitive inhibition is the interaction between many compounds and the enzyme β-galactosidase. The enzyme takes up lactose and cleaves it to glucose and galactose. Glucose is converted in glycolysis and galactose is usually excreted to the medium. Isopropyl-β-D-thiogalactoside (IPTG) is an inducer for lactose uptake, but it is not a substrate for the enzyme. At too high levels of IPTG, the rate of hydrolysis of lactose is reduced.

When ES_1 in Eq. (7.16) can react with S to form ES_1S that can lose S_1 by $ES_1S \leftrightarrow ES + S_1$, the "active" intermediate ES is regenerated. Again, the rate expression (7.18) can be derived using the Briggs–Haldane approach (now with ES_1S as an additional form of the enzyme):

$$r = (-r_s) = \frac{ks}{s + K_m}\left(1 + \frac{s_1}{K_{eq1}}\right)^{-1}. \tag{7.18}$$

In the *noncompetitive* inhibition mechanism (7.18). the rate constant of the reaction is reduced by a factor $\left(1 + (s_1/K_{eq1})\right)^{-1}$, and it does not help to increase s, since K_m remains the same.

Finally, S_1 can attack ES rather than E. Here, the resulting rate expression is Eq. (7.19):

$$r = (-r_s) = \frac{ks}{K_m + \left(1 + (s_1/K_{eq1})\right)s}. \tag{7.19}$$

For lack of a better word, the kinetics (7.19) is named *uncompetitive.* If it is S that attacks ES to form ESS, then Eq. (7.19) illustrates *substrate inhibition* of the reaction. For a small K_{eq1}, both the rate constant and the apparent value of K_m decrease rapidly with increasing s_1.

The product P can play the role of S_1 in all the kinetic expressions (7.17)–(7.19). Product inhibition is, in fact, quite common since the structure of P is not too different from that of S (otherwise S and P could not be docked simultaneously on the enzyme).

Each of the kinetic expressions (7.13) and (7.17)–(7.19) can be transformed into linear expressions, and thereafter the parameters can be obtained by linear plots:

$$\frac{1}{r} = \frac{1}{k} + \frac{K_m}{k}\frac{1 + (s_1/K_{eq1})}{s}, \tag{7.17}$$

$$\frac{1}{r} = \frac{1 + (s_1/K_{\text{eq1}})}{k} + \frac{K_{\text{m}}}{k} \frac{1 + (s_1/K_{\text{eq1}})}{s}, \tag{7.18}$$

$$\frac{1}{r} = \frac{1 + (s_1/K_{\text{eq1}})}{k} + \frac{K_{\text{m}}}{ks}, \tag{7.19}$$

$$\frac{1}{r} = \frac{1}{k} + \frac{K_{\text{m}}}{ks}. \tag{7.13}$$

The linear plots are shown in Figure 7.1. With reasonably accurate measurements, one is able to distinguish between different types of inhibition, but once estimates of the parameters have been obtained, better estimates should be found by least squares fitting, using the structure of the kinetics that has been obtained by the linear plots. Since the rates $(-r_s)$ (or the volumetric rates $(-q_s) = (-r_s)\, e_0$) have been determined experimentally, probably with a constant analysis error on s and on $q = D(s_0 - s)$, one must count on very large errors in the points where s is small.

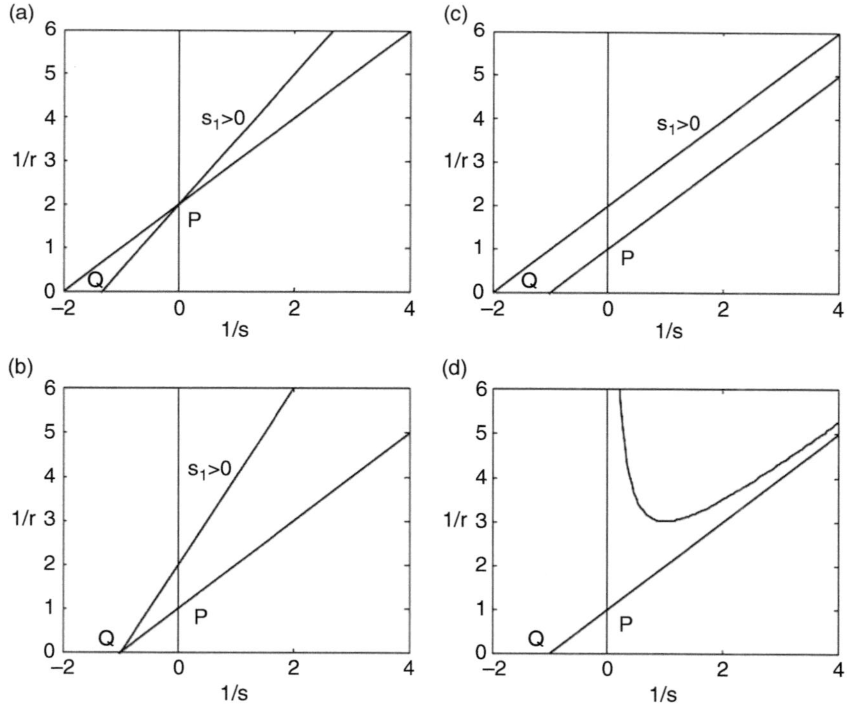

Figure 7.1 Double reciprocal plots of $1/r$ versus $1/s$ for Michaelis–Menten kinetics with different types of inhibition of the specific rate r. s_1 is the concentration of the inhibitor. The parameters to be estimated are $[k, K_m]$, the Michaelis–Menten parameters, and K_{eq1}, an inhibition parameter. (a) Competitive inhibition (7.17). (b) Noncompetitive inhibition (7.18). (c) and (d) $(s = s_1)$: Uncompetitive inhibition (7.19). All plots have two lines, one for $s_1 = 0$ and one for $s_1 > 0$. $Q = -K_{\text{apparent}}$, and the slope of the lines is $K_{\text{apparent}}/k_{\text{apparent}}$.

a) Competitive inhibition (7.17). The lines intersect at $(1/r) = (1/k)$ and $(1/r) = 0$ for $\frac{1}{s} = -K^{-1}_{\text{apparent}} = -\left[K_m\left(1 + (s_1/K_{eq1})\right)\right]^{-1}$.
b) Noncompetitive inhibition (7.18). The slope of the lines $= \frac{K_m}{k}\left(1 + \frac{s_1}{K_{eq1}}\right)$; $\frac{1}{r} = 0$ for $\frac{1}{s} = -K_m^{-1}$.
c) Uncompetitive inhibition (7.19). The lines are parallel with slope $= (K_m/k)$; $(1/r) = 0$ for $\frac{1}{s} = -\frac{1 + (s_1/K_{eq1})}{K_m}$.
d) Uncompetitive inhibition (7.19) with $s_1 = s$ (substrate inhibition). $(1/r) \rightarrow \infty$ for $(1/s) = 0$ and $\approx (K_m/ks)$ for $(1/s) \rightarrow \infty$. The value of K_{eq1} is obtained from the value of $1/s$ at the minimum of $(1/r)$: $((1/s))_{\min} = (K_m K_{eq})^{1/2}$.

Parameters used: **a** $[K_m, k] = \left[\frac{1}{2}, \frac{1}{2}\right]$, $K_{eq1} = 2$. In **b** – **d** $[K_m, k] = [1, 1]$, $K_{eq1} = 1$. $s_1 = 1$ in **a** – **c**, and $K_{eq1} = 1$ in **d**.

Kinetic expressions with a structure that is different from that of Michaelis–Menten kinetics must often be used to explain important phenomena in metabolic pathways, especially the observation that $(-r_s)$ does not increase toward $(-r_s)_{\max}$ right from $s = 0$, but rather in an S-interval $[s_{\text{lower}}, s_{\text{upper}}]$. Figure 7.2 shows how $(-r_s)$ increases with s for the so-called *Hill kinetics*:

$$(-r_s) = \frac{ks^n}{s^n + K^n} = \frac{r_{\max} s^n}{s^n + K^n}. \tag{7.20}$$

The type of kinetics exhibited in Figure 7.2 is rarely seen in industrial enzyme reactions, but many enzymes in the living cell work only in a narrow substrate concentration range, and within this range the enzyme exhibits a rapid change of rate with changing substrate concentration. Note that $r/r_{\max} = ½$ at $s = K$ for any value of n. $r/r_{\max}$ increases from 0.5 to 0.9 in the range $K < s < \sqrt[n]{9}\,$K. For $n = 1$ (Michaelis–Menten kinetics) $r/r_{\max}$ increases from ½ to 0.9 in the s interval $K_m < s < 9\,K_m$, and for $n = 4$ in the s interval $K < s < \sqrt[4]{9}\,$K. Clearly, the enzyme with $n = 4$ is much more sensitive to changes in s than an enzyme that obeys Michaelis–Menten kinetics (see Figure 7.2).

Phosphofructokinase, an enzyme shown in Figure 4.2, is a typical example of a *positive cooperative* enzyme. It consists of four subunits and all four subunits

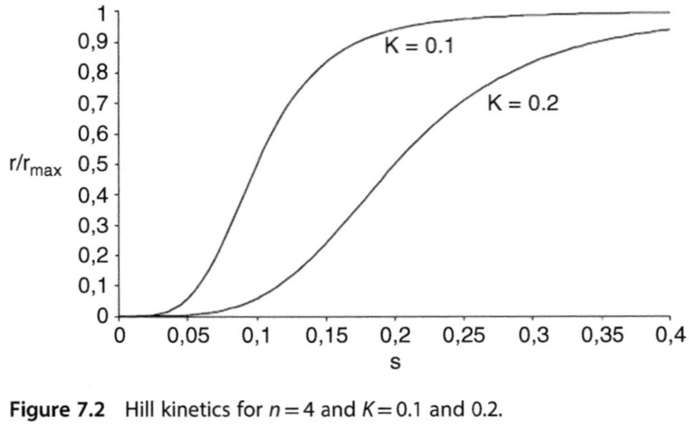

Figure 7.2 Hill kinetics for $n = 4$ and $K = 0.1$ and 0.2.

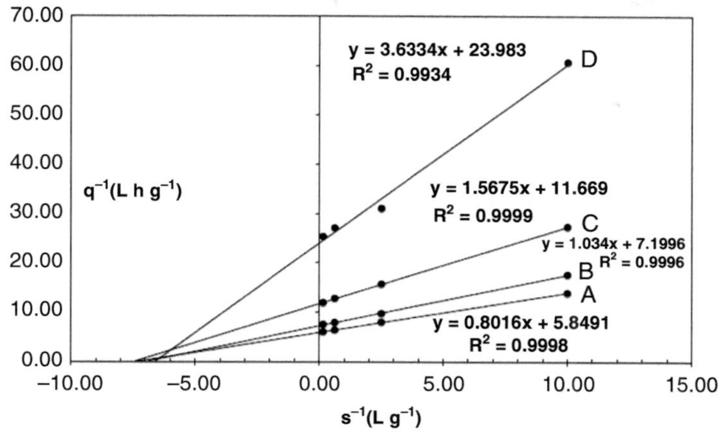

Figure 7.3 Reciprocal plots of r as a function of s for different product levels (A: $p = 3\ \text{mg l}^{-1}$).

need to have F6P attached before the conversion of F6P to F1,6 diphosphate occurs. The literature on cooperativity in enzymology is vast, and specialist text must be consulted to appreciate the complexity of the corresponding reactions. Cornish-Bowden [6] and Fell [7] are good references to *allosteric enzymes*, a collective name for enzymes that exhibit positive (or negative) cooperativity.

Example 7.1 Kinetic parameters in Michaelis–Menten-type rate expressions

Table 7.1 and Figure 7.3 show steady-state reaction rates (g (l h)$^{-1}$) for an enzymatic reaction in which S is irreversibly converted to P in a steady-state CSTR. The feed has been "spiked" with P such that the concentrations of P in the effluent (and in the reactor) have the values 3, 9, 27, or 81 mg l^{-1}. The effluent concentration of S is given in the first column and the corresponding values of r at the different product levels are shown in the next four columns.

Table 7.1 Volumetric rates q (g (l h)$^{-1}$) at different substrate and product concentrations.

s (g l^{-1})	Product concentration p (mg l^{-1})			
	3	9	27	81
0.1	0.073	0.058	0.036	0.017
0.4	0.128	0.102	0.064	0.032
1.6	0.158	0.126	0.079	0.037
6.4	0.168	0.134	0.084	0.039

A comparison with Figure 7.1b strongly indicates that the product P (= S_1) acts as a noncompetitive inhibitor of the conversion of S to P. Straight lines are fitted to the data for each level of P, and the linear expressions for $1/r$ as function of $1/s$ are shown in the figure for each of the p values in Table 7.1.

First, K_m is determined from the average of the intersection with the $1/s$ axis. From Figure 7.3, one obtains $K_m^{-1} = 7.04 \rightarrow K_m = 0.142(\text{g l}^{-1})$. Next, the slopes of $1/r$ versus $1/s$ are plotted as a function of $s_1 = p$:

$$\text{Slope} = \frac{K_m}{k}\left(1 + \frac{p}{K_{eq1}}\right) \text{ for the kinetics of Eq. (7.18).} \tag{1}$$

From Figure 7.4, one determines the intersection with the p-axis: $\frac{K_m}{k} = 0.670 \rightarrow k = \frac{0.142}{0.670} = 0.212\ \text{g S(l h)}^{-1}$.

Next from the slope: $\frac{K_m}{k \cdot K_{eq1}} = 0.0363 \rightarrow K_{eq1} = \frac{0.670}{0.0363} = 18.45\ \text{mg l}^{-1}$.

The parameter values obtained by the graphical method are next used as starting values in a least squares fitting procedure using Eq. (7.18) as the model. The result is:

$$\begin{aligned}[K_m, k, K_{eq1}] = [0.139\ \text{g l}^{-1},\quad k = q_{max} = 0.197\ \text{g(l h)}^{-1},\\ K_{eq1} = 20.97\ \text{mgl}^{-1}].\end{aligned} \tag{2}$$

The correlation matrix between the parameters is as follows:

$$M = \begin{pmatrix} 1 & 0.44 & -0.75 \\ 0.44 & 1 & 0.0012 \\ -0.75 & 0.0012 & 1 \end{pmatrix}. \tag{3}$$

Although the result (2) is quite satisfactory, the parameters (especially K_m and k) are strongly correlated: When k and K_m are positivly correlated, k and K_{eq1} are negatively correlated.

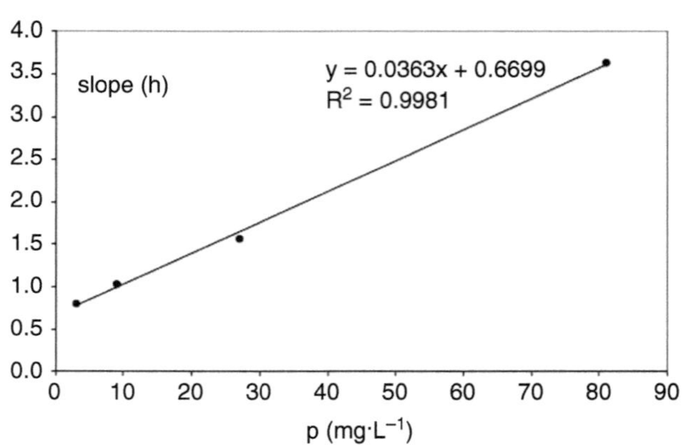

Figure 7.4 Determination of k and K_{eq1} from Eq. (1).

For cell reactions, the simple Monod model $r_x = (ks/s + K_m)$ can also be expanded to include both substrate inhibition (7.21) and product inhibition (7.22) or (7.23):

$$r_x = \mu = \frac{\mu_{max}\, s}{(s^2/K_i) + s + K_m}. \tag{7.21}$$

$$r_x = \mu = \frac{\mu_{max}\, s}{s + K_m}\left(1 - \frac{p}{p_{max}}\right). \tag{7.22}$$

Or.

$$r_x = \mu = \frac{\mu_{max}\, s}{s + K_m}\frac{K_i}{K_i + p}. \tag{7.23}$$

Eq. (7.21) is a close analog of Eq. (7.19), and (7.23) is an analog of Eq. (7.18), both with $p = s_1$. All three equations are empirical in nature, and their form just mimics oral models for, respectively, substrate and product inhibition of cell reactions. Any other function that for small s gives increasing r_x, but for large s decreasing r_x (e.g., $r_x = ks \exp(-k_1 s)$) could be used to fit experimental data for either substrate or product inhibition. In Eq. (7.23) the influence of $p = Y_{sp}$ $(s_0 - s)$ leads to a more gradual decrease of r_x with increasing $(s_0 - s)$ than does Eq. (7.22) where r_x drops to zero at $p = p_{max}$. In wine fermentation, Eq. (7.22) gives a good representation of fermentation data when $p \to p_{max} \approx 120$–$130$ g ethanol (l^{-1}) for most wine yeasts.

The expressions (7.21)–(7.23) can be used as reliable predictive models for cell kinetics when the assumptions (1) to (3) for using the basic Monod kinetics (7.13) are satisfied: The cell needs to be in metabolic steady state and the limiting substrate has to be used for S.

Many other empirical models have been proposed to find r_x for an unstructured biomass X:

$\mu = \mu_{max}(1 - \exp(-s/K_m))$ proposed by Monod's teacher Tessier. (7.24a)

$\mu = \mu_{max}\dfrac{s}{s + K_s x}$ proposed by Contois. (7.24b)

$\mu = \mu_{max}\left(1 - \dfrac{x}{K_s}\right)$ "the logistic law" proposed in the 19th-century for population growth. (7.24c)

Eq. (7.24a) simplifies to the Monod expression for small $s(r_x = (\mu_{max}/K_m)s)$ and for large s $(r_x \to \mu_{max})$.

It also gives a steeper increase of r_x with decreasing K_m, and in all respects it is as reasonable as Monod's model. The two expressions (7.24b) and (7.24c) should not be used. There is no reason why the usually small biomass concentration should influence r_x, and Eq. (7.24b) is more likely to be an expression that indicates product inhibition since p and x will increase proportionally when the feed to a steady-state CSTR does not contain biomass or product. Eq. (7.24c) is based on a false analogy with population statistics.

It is true that a batch reactor experiment can give a time profile for r_x that looks like Eq. (7.24c). The specific rate is high when x is small, and $r_x \to 0$ when the substrate is used up. The expression does, however, not take into account that the maximum value of x depends on the initial substrate concentration, and this would make K_s depend on s.

Maintenance substrate consumption by Eq. (7.14) can be included as a variant of the Monod model with unstructured biomass:

For a feed concentration of biomass $x_f = 0$ and $s = s_f$, one obtains the following relation between x and D:

$$\left(Y_{xs}^{\text{true}} D + m_s\right)x = (-q_s) = D(s_f - s) = D\left(s_f - \frac{D K_m}{\mu_{\max} - D}\right) \to$$
$$x = \frac{D\left(s_f - (D K_m/\mu_{\max} - D)\right)}{Y_{xs}^{\text{true}} D + m_s}. \tag{7.25}$$

$$\text{If } m_s = 0 \text{ (7.25) simplifies to (2.2)} : x = Y_{sx}^{\text{true}}\left(s_f - \frac{D K_m}{\mu_{\max} - D}\right). \tag{7.26}$$

Whereas Eq. (7.26) gives $x = Y_{sx}^{\text{true}} s_f$ for $D = 0$, Eq. (7.25) gives $x = 0$ for $D = 0$. For large D where $Y_{xs}^{\text{true}} D \gg m_s$ Eq. (7.25) does, of course, simplify to Eq. (7.26), but for small D the shape of $x(D)$ is quite different for $m_s > 0$. This follows from Figure 7.5 for $\mu_{\max} = 1\,\text{h}^{-1}$, $K_m = 0.01\,\text{g l}^{-1}$, $Y_{xs} = 1.82\,\text{g g}^{-1}$, and $m_s = 0.08\,\text{g S (g X h)}^{-1}$. For $D > 0.5\,\text{h}^{-1}$ $x(D)$ is approximately $= (1/1.82)(s_f - s) \approx (s_f/1.82) = 5.5\,\text{gl}^{-1}$ when $s_f = 10\,\text{g l}^{-1}$, but for $D < 0.3\,\text{h}^{-1}$ the biomass concentration is much smaller than $5.5\,\text{g l}^{-1}$.

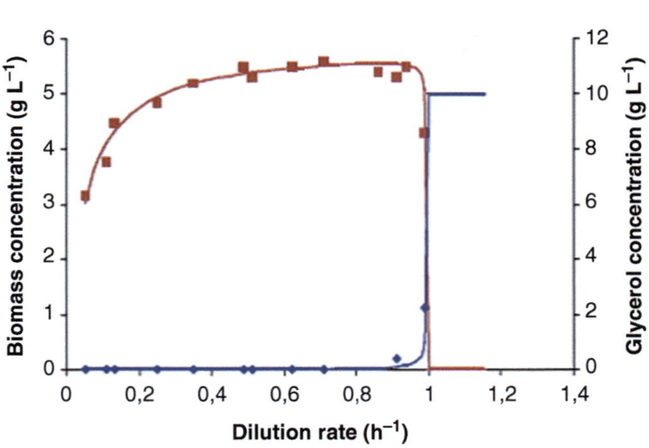

Figure 7.5 Biomass concentration as a function of dilution rate for growth of *Aerobacter aerogenes* on glycerol. The data are from Herbert [8]. $\mu_{\max} = 1\,\text{h}^{-1}$, $K_m = 0.01\,\text{g l}^{-1}$, $Y_{xs} = 1.82$, $m_s = 0.08\,\text{g glycerol (g X h)}^{-1}$.

Example 7.2 Kinetics of succinic acid production from glucose

In Example 4.6, the yield coefficients were calculated for a process where glucose is converted to succinic acid (SA) + biomass (X) with no metabolic by-products such as acetic acid. The stoichiometry is as follows:

$$\begin{aligned} &CH_2O + (3\gamma - 0.3)N^{-1}\, CO_2 + 1.4\, N^{-1}\, NH_3 \rightarrow \\ &4 \cdot (6\gamma + 0.1)N^{-1}\, CH_{1.5}O + 7\, N^{-1}\, CH_{1.8}O_{0.5}N_{0.2} + (3\gamma + 3.2)\, N^{-1}\, H_2O; \\ &N \equiv 7 \cdot (3\gamma + 1.1). \end{aligned} \tag{1}$$

The rate of conversion of glucose to SA $= (CH_{1.5}O)_4$ will be derived based on the unstructured kinetic models for cell reactions discussed above and using the data of Table 7.2 to fit the parameters.

First, the data are qualitatively examined:

1) The "washout dilution rate" must be close to $D = 0.75\,h^{-1}$ since $s \rightarrow s_0 = 20\,g\,l^{-1}$.
2) The biomass concentration rapidly decreases for small D. This indicates maintenance substrate utilization, as seen in Figure 7.5.
3) The SA concentration p decreases for all D. This is an indication of product inhibition, and Eq. (7.23) (or (7.22)) should be chosen as the best model structure.

To quantify the influence of maintenance, the specific rates $[(-r_s), r_p]$ are plotted against D on Figure 7.6.

Table 7.2 Effluent concentrations[1] from a steady-state CSTR for succinic acid (SA) production operated at different dilution rates. Feed concentration of biomass and of SA is zero, and of glucose $s_0 = 20\,g\,l^{-1}$.

D (h^{-1})	x (g biomass l^{-1})	SA (g l^{-1})	s (g glucose l^{-1})
0.05	0.38	21.91	0.01
0.15	0.81	21.22	0.07
0.25	1.02	20.11	0.80
0.35	0.74	12.71	7.74
0.45	0.49	7.71	12.51
0.55	0.31	4.50	15.60
0.65	0.16	2.27	17-77
0.75	0.05	0.64	19.37

1) The data were kindly provided by Prof. Willie Nicol, University of Pretoria, South Africa. They are not obtained from fermentation experiments, but the kinetics that results from analysis of the data is representative of what would actually be found for a real succinic acid production process.

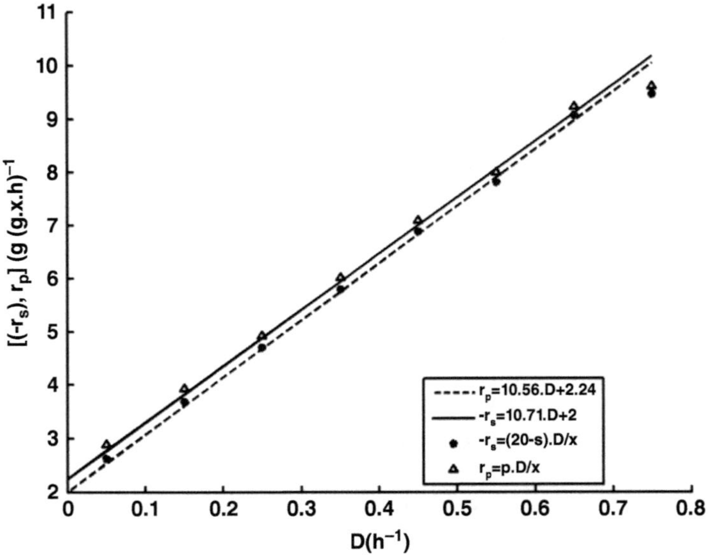

Figure 7.6 Specific rates $[(-r_s), r_p]$ (g (g X h)$^{-1}$) as functions of dilution rate D (h^{-1}).

Except for the very last points at $D = 0.75\,\mathrm{h}^{-1}$ where x and p are close to zero the data are well represented by straight lines:

$$(-r_s) = \frac{(20 - s)D}{x} = 10.71\,D + 2.00, \quad \text{and} \quad r_p = \frac{pD}{x}$$
$$= 10.56\,D\ + 2.24. \tag{1}$$

Comparison with Eq. (7.14) shows that $Y_{xs}^{true} = 10.71$ g glucose (g X h)$^{-1}$, $Y_{px}^{true} = 10.56$ g SA (g X h^{-1}).

The maintenance coefficients are $m_s = 2.00$ glucose (g X h)$^{-1}$ and $m_p = 2.24$ g SA (g X h)$^{-1}$.

$$\frac{\mathrm{m_p}}{\mathrm{m_s}} = 1.12\ \mathrm{g\ SA(g\ glucose)^{-1}}$$
$$= \frac{1.12 \cdot 30}{29.5}\mathrm{C\text{-}mol\ SA(C\text{-}mol\ glucose)^{-1}} \rightarrow \mathrm{Y_{s,SA}}$$
$$= 1.14\ \mathrm{C\text{-}mol(C\text{-}mol)^{-1}}$$

Since no metabolic product is produced besides succinic acid, the maintenance reaction is the same as that obtained in Example 4.3 for succinic acid production without biomass formation:

$$\mathrm{CH_2O} + \frac{1}{7}\mathrm{CO_2} \rightarrow \frac{8}{7}\mathrm{CH_{1.5}O} + \frac{1}{7}\mathrm{H_2O} \tag{2}$$

The value 1.14 obtained from the experiments is very close to 8/7, and this is a nice confirmation of the hypothesis that the maintenance reaction is, indeed, (2).

From $Y_{\rm px}^{\rm true}$ and $Y_{\rm xs}^{\rm true}$ one obtains that $Y_{\rm sp}^{\rm true} = \frac{10.56}{10.71} = 0.986\,{\rm g/g} = 1.002\,\text{C-mol(C-mol)}^{-1}$.

Again, one may refer to Example 4.6 where $Y_{\rm s,SA}^{\rm true}$ was found to be $4(6\gamma + 0.1)/7(3\gamma + 1.1)$.

Inserting the experimentally determined value 1.00 for $Y_{\rm xs}^{\rm true}$, one obtains $\gamma = 2.43$ mol ATP C-mol biomass. This value is reasonable also for an anaerobic cultivation, since the cell is stressed by the production of a considerable quantity of succinic acid.

One finally obtains $Y_{\rm sx} = \frac{1}{3\gamma+1.1} = 0.119$ C-mol X (C-mol glucose).

Consequently, the complete set of stoichiometric equations for the process is as follows:

$$\text{Main reaction}: \mathrm{CH_2O} + 0.119\,\mathrm{CO_2} + 0.0238\,\mathrm{NH_3} \rightarrow 0.119\,\mathrm{CH_{1.8}O_{0.5}N_{0.2}} + 1.00\,\mathrm{CH_{1.5}O} + 0.179\,\mathrm{H_2O}. \tag{3}$$

$$\text{Maintenance reaction}: \mathrm{CH_2O} + \frac{1}{7}\mathrm{CO_2} \rightarrow \frac{8}{7}\mathrm{CH_{1.5}O} + \frac{1}{7}\mathrm{H_2O}. \tag{4}$$

The observed stoichiometry is derived as shown in Example 3.6:

$$\mathrm{CH_2O} + \frac{D + \frac{1}{7}m_s}{DY_{xs}^{true} + m_s}\mathrm{CO_2} + \frac{0.2\,D}{DY_{xs}^{true} + m_s}\mathrm{NH_3} \rightarrow \frac{D}{DY_{xs}^{true} + m_s}\mathrm{X} + \frac{8.403\,D + \frac{8}{7}m_s}{DY_{xs}^{true} + m_s}\mathrm{SA} + \frac{1.504\,D + \frac{1}{7}m_s}{DY_{xs}^{true} + m_s}\mathrm{H_2O} \tag{5}$$

$Y_{\rm xs}^{\rm true} = 8.403$ C-mol glucose (C-mol X)$^{-1}$ and $m_{\rm s} = 2.00$ g glucose (g X h)$^{-1} = 1.64$ C-mol glucose (C-mol X h)$^{-1}$.

$$\mathrm{CH_2O} + \frac{D + 0.2342}{8.403\,D + 1.64}\mathrm{CO_2} + \frac{0.2\,D}{8.403\,D + 1.64}\mathrm{NH_3} \rightarrow \frac{D}{8.403\,D + 1.64}\mathrm{X} + \frac{8.403\,D + 1.874}{8.403\,D + 1.64}\mathrm{SA} + \frac{1.504\,D + 0.2342}{8.403\,D + 1.64}\mathrm{H_2O}. \tag{6}$$

The stoichiometry (6) permits a calculation of x and p for any value of D, for example:

$$\text{At}\,D = 0.25\,\mathrm{h^{-1}},\quad s = 0.8\,\mathrm{gl^{-1}}.\ Y_{\rm sx}^{\rm obs} = \frac{0.25}{3.741}\cdot\frac{24.6}{30} = 0.0548\,\mathrm{gg^{-1}} \rightarrow x = 0.0548(20 - 0.8) = 1.05\,\mathrm{gl^{-1}}.$$

$$Y_{\rm sp} = 1.063\cdot\frac{29.5}{30} = 1.045\,\mathrm{gg^{-1}} \rightarrow p = \mathrm{SA} = 20.06\,\mathrm{gl^{-1}}.$$

At $D = 0.35\,\mathrm{h^{-1}}$, $s = 7.74\,\mathrm{g\,l^{-1}}$, the results are: $x = 0.77\,\mathrm{g\,l^{-1}}$ and $p = \mathrm{SA} = 12.67\,\mathrm{g\,l^{-1}}$.

All the concentration data of Table 7.2 can be calculated using Eq. (6). This is a truly remarkable result: Only one reactant concentration, for example, the glucose concentration, needs to be measured and the effluent concentrations of X, SA (and the CO_2 uptake) can be calculated. This means that Eq. (6) can be used in Chapter 10 to design a reactor for the process, with S representing all the reactants. The level of understanding the system increases dramatically above the level of Example 3.6, and this new insight is based on the application of MFA to analyze the metabolic pathway reactions.

Eq. (7.23) can be reformulated to

$$K_p = p\left(\frac{s\,\mu_{max}}{\mu(s+K_m)} - 1\right)^{-1} = p\left(\frac{s\,\mu_{max}}{D(s+K_m)} - 1\right)^{-1}. \tag{7}$$

Here, μ_{max} is determined by extrapolating $D(s)$ to $s = s_0 = 20\,\mathrm{g\,l^{-1}}$. The value obtained is $\mu_{max} = 0.80 \pm 0.01\,\mathrm{h^{-1}}$.

K_m is very small compared to K_p, usually in the range of a few mg $\mathrm{l^{-1}}$. To avoid correlation between the two parameters, K_m is arbitrarily set to $25\,\mathrm{mg\,l^{-1}}$. Now K_p can be calculated from Eq. (7). Thus, for $K_m = 0.025\,\mathrm{g\,l^{-1}}$:

D ($\mathrm{h^{-1}}$)	0.05	0.15	0.25	0.35	0.45	0.55	0.65	0.75
K_p ($\mathrm{g\,l^{-1}}$)	6.13	7.24	9.56	9.94	9.95	9.96	9.93	9.84

The values of s for small D are very uncertain, and at large D values, p is uncertain. Thus K_p is determined to $10\,\mathrm{mg\,l^{-1}}$ for D values in [0.45, 0.65].

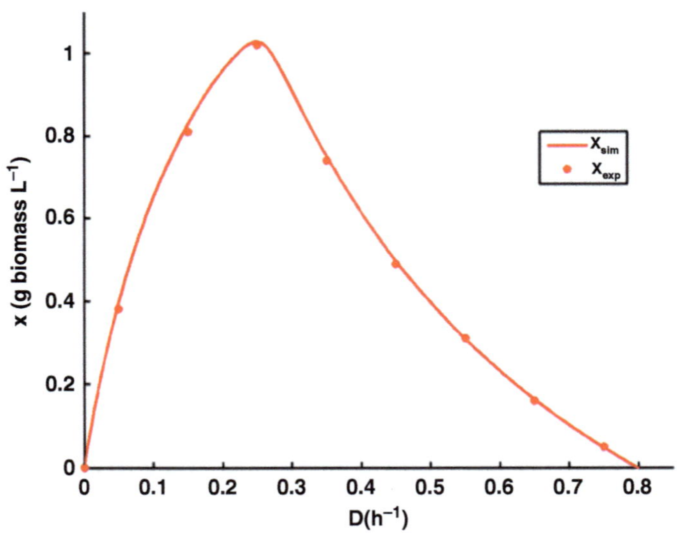

Figure 7.7 Biomass concentration x ($\mathrm{g\,L^{-1}}$) as a function of D ($\mathrm{h^{-1}}$).

For $K_m = 50\,\text{mg}\,l^{-1}$, one obtains:

D (h^{-1})	0.05	0.15	0.25	0.35	0.45	0.55	0.65	0.75
K_p ($g\,l^{-1}$)	13.1	10.05	9.99	10.00	10.00	10.00	9.99	10.01

One concludes that the best set of parameters is [μ_{max}, K_p, K_m] is [$0.8\,h^{-1}$, $10.0\,g\,l^{-1}$, $50\,\text{mg}\,l^{-1}$].

When these data are used in a plot of $x(D)$, one obtains Figure 7.7. The figure displays the main features of the product-inhibited kinetics + maintenance substrate utilization. $x \to 0$ for $D \to 0$, and due to strong product inhibition, x decreases long before washout occurs at $D = 0.80\,h^{-1}$. The contrast to Figure 7.5 is striking.

7.3 Summary of Enzyme Kinetics and the Kinetics for Cell Reactions

1) Enzyme reactions can be modeled by variants of Michaelis–Menten kinetics (7.13) and (7.17)–(7.19). The models can be used for inhibition by another substrate S_1. Substrate and product inhibition can be modeled when S_1 is identified as, respectively, S and P.

   Hill kinetics (7.20) and similar (much more complicated) expressions are used to model the function of allosteric enzymes where more than one site on the enzyme complex interacts with the substrate.

   The parameters are typically found as explained in Example 7.1. First, the double reciprocal plots (often called Linewaever–Burk plots after Linewaever and Burk [9], who used the method for graphical interpretation of the Michaelis–Menten kinetics (7.13)), are used to find the structure of the kinetic model. After first estimates of the parameters have been found, the data are submitted to computer analysis, and the best parameter values are obtained by linear least squares.

   To find the influence of an inhibitory substrate S_1, the experiment is performed with a set of predetermined values of s_1. If product or substrate inhibition is suspected, the experiment is performed at different feed (or initial) values of S.

2) Kinetic models for cell reactions are selected by analogy with enzyme reactions. Any resemblance with the mechanistically derived models for enzyme kinetics is incidental, but the cell kinetics should represent phenomena that can be explained in an oral model. Thus, the Monod model gives a logical explanation for the kinetic behavior of cell kinetics for both small and large substrate concentration.

   Whereas the stoichiometry of an enzyme reaction is not likely to change with the operating conditions – the enzyme is very specific in its catalytic function and is rarely able to convert the substrate S by different reactions –

a change of stoichiometry with changing environment is more a rule than an exception for cell reactions. The influence of maintenance substrate consumption is not seen in enzyme reactions but often in cell reactions, and the stoichiometry varies with D as seen first in Example 3.6 and then in Example 7.2. Analysis of the metabolic pathways by metabolic flux analysis (MFA) in Chapter 4 is *also a valuable tool when the kinetics of the cell reaction is studied*. The occurrence of a by-product for certain environmental conditions (low pH, high temperature, or in different ranges of the dilution rate) can be envisaged by reference to metabolic charts or computer software. As seen in Example 7.2, the number of independent reactants can be substantially reduced after a careful analysis of the stoichiometry by MFA.

It should *again* be stressed that *none* of the kinetic models presented here for cell reactions will give meaningful results unless the cell is in metabolic equilibrium, that is, when the composition of macromolecules in the cell remains constant during the reaction. In a truly transient experiment where, for example, the substrate concentration in a stirred tank reactor is changed abruptly (by adding a pulse of substrate or a pulse of strong acid) or very rapidly (after a change of dilution rate), the Monod-type kinetics (7.13) or (7.21)–(7.23) will not be applicable. The only way to study transient situations is to examine one little bit of the metabolism after the other as shown in Section 7.4. The whole picture of the transient metabolism of cells will not for many years be completely understood.

It is useful to have an indication of the parameter values to be expected in cell reaction kinetics.

Table 7.3 gives approximate values of the two parameters μ_{max} and K_m in the Monod model. If an analysis of fermentation data by the Monod model (7.13) gives a K_m value substantially different from the values shown in the table, there is reason to believe that the wrong structure of the kinetic expression has been chosen. Thus, if K_m is found to be $1\,g\,l^{-1}$, this is a sign

Table 7.3 Values of the Monod constant K_m. When $s = K_m$, $r/r_{max} = ½$.

Organism	Substrate	K_m mg l^{-1}
S. cerevisiae	Glucose	150
Escherichia coli	Glucose	4
Klebsiella oxytoca	Glucose	10
	Fructose	10
L. cremoris	Arabinose	50
P. chrysogenum	Glucose	2
Aspergillus oryzae	Lactose	10
	Fructose	3
	Glucose	4
	Glucose	1

Table 7.4 Typical values for the maximum specific growth rate of cells.

Organism	*E. coli*	*L. lactis*	*S. cerevisiae*	*A. oryzae*	*P, crysogenum*	CHO cells	Pear cells
μ_{max} (h^{-1})	0.8–1.0	0.7–0.85	0.38–0.44	0.20–0.25[a]	0.20–0.25[a]	0.033[b]	0.016[c]

a) Values from industrial strains (Novozymes).
b) [10].
c) (suspended cultures of the plant cells) [11].

that the wrong limiting substrate has been chosen or the reaction suffers from product inhibition.

Table 7.4 gives values for the maximum growth rate μ_{max}. Only approximate values can be obtained from the literature since the growth rate depends on: (1) The environment. At low pH, μ_{max} may be much smaller than at pH = 7, and with an insufficient substrate recipe the growth rate can also be very small. (2) When the organism is loaded with gene copies for a heterologous protein, the growth rate is much smaller than for the wild type strain. This also happens in strains that are overproducing a metabolite, for example, industrial strains of *Penicillium chrysogenum* grow much slower than wild-type strains, and when even a few cells of the wild-type fungus appear (e.g., by mutation) the whole culture will rapidly revert to the almost "bald" (nonproducing) strain.

7.4 Cell Reactions with Unsteady State Kinetics

The kinetic models for cell reactions discussed in Sections 7.1–7.3 are not valid for transient experiments. In the following, some models that have no mechanistic foundation but express accepted features of cell growth in an empirical way will be discussed. They are capable of modeling the lag phase of batch cultivations, and they give an acceptable quantitative description of the metabolic changes after abrupt changes in, for example, substrate availability.

The main feature of these models is that the cell mass X is divided into *compartments*, X_1, X_2, and the rate of formation of each compartment has a different stoichiometry and kinetic expression.

It is reasonable to formulate an oral model that expresses that "before any biomass growth can take place there must be available machinery in X_1 to produce the macromolecules that make up the cell." RNA and certain proteins are essential for building cell machinery. DNA is, of course, also essential, but experiments show that the DNA content of fast growing cells is not much different from that of resting cells, whereas the cell content (wt %) of RNA (especially m-RNA) and of proteins is high for very active cells (large μ or large D values for continuous cultures). Since anabolic reactions in compartment X_2 cannot proceed unless building blocks and sufficient ATP for synthesis of macromolecules is available, it is likely that just after an upward change of D in a continuous culture the catabolic machinery X_1 will be the first to be built from substrate.

The models described below are based on these general principles.

A very simple compartment model was formulated independently by Ramkrishna, Fredrickson, and Tsuchiya [12] and by Williams [13] – corrections of the Williams model were given by Fredrickson [14] – and by Roels and Kossen [15]. The concepts will be illustrated by a model of Nielsen, Nikolajsen, and Villadsen [16,17] for lactic bacteria cultivation. The Nielsen, Nikolajsen, and Villadsen model is analogous to the Roels and Kossens model. In later compartment models, and such studies appear right up to the present, the original idea of having two (or at most three compartments) has been expanded to five or even more compartments. With increasing number of compartments, the model becomes a better data fitter, but the simplicity of the original approach gradually vanishes.

The model of Nielsen, Nikolajsen, and Villadsen [16,17] has two compartments: (1) an active compartment X_A that contains the protein synthesis system (PSS) and small building blocks and (2) a passive compartment X_G that makes up the rest of the cell material.

X_A is synthesized by the substrate S, while X_G is synthesized at the expense of X_A:

$$\mathrm{S} \rightarrow \gamma_{11}\,\mathrm{X_A}, \text{and thereafter}: \mathrm{X_A} \rightarrow \gamma_{22}\,\mathrm{X_G} \tag{7.27}$$

The pool of X_A is likely to be much smaller than that of X_G, and γ_{11} is in the range 0.2–0.5 (g X_A/g S). γ_{22} (g X_G/g X_A) is much smaller.

The kinetics of both reactions is assumed to be proportional to X_A: the rate of substrate uptake depends on the amount of receiving machinery X_A and so does the rate of formation of X_G.

$$(-r_s) = \frac{k_1\,s}{s+K_1}, \quad r_{XA} = \left(\frac{\gamma_{11}k_1\,s}{s+K_1} - \frac{k_2 s}{s+K_2}\right)X_A \quad \text{and} \quad r_{XG} = \frac{\gamma_{22}k_2 s}{s+K_2}X_A. \tag{7.28}$$

$$\text{The specific growth rate } r_x = \mu = r_{XA} + r_{XG} = \left(\frac{\gamma_{11}k_1 s}{s+K_1} + (1-\gamma_{22})\frac{k_2 s}{s+K_2}\right)X_A. \tag{7.29}$$

If $K_1 > K_2$, then the formation of X_A is favored at high substrate concentration, that is, the active compartment increases with increasing growth rate μ. This is well known from experiments.

Figure 7.8 illustrates that the machinery for treating substrate is large for large D in a steady-state CSTR and small for small D, but if the initial D is small then X_A will grow and catch up with X_G as time increases.

In Figure 7.8, the transient mass balance is as follows:

$$\frac{dx}{dt} = q_x - Dx = (\mu - D)x. \tag{7.30}$$

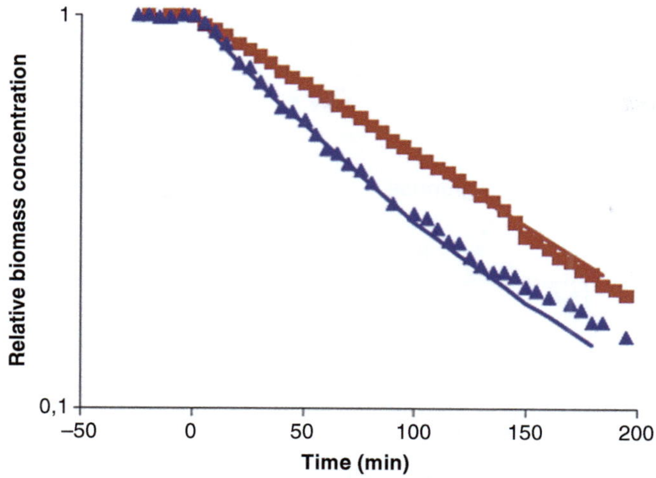

Figure 7.8 Decrease of biomass concentration in a CSTR when D is increased above the washout value $D_{washout}$, where $D > (r_x)_{max}$. At time = 0, D is increased to $0.99\,h^{-1}$ which is above $D_{washout}$. For the upper curve, $D = 0.50\,h^{-1}$ at the steady state before $t = 0$. For the lower curve, $D = 0.10\,h^{-1}$ before $t = 0$. The data are from cultivations of *Lactococcus lactis* by Nielsen, Nikolajsen, and Villadsen [17].

Since D for $t > 0$ is very high one can assume that $\mu = \mu_{max}$, and the solution of the differential equation (7.30) is

$$\ln(x/x(t = 0)) = (\mu_{max} - D)t. \tag{7.31}$$

The upper curve in Figure 7.8 can be approximated by the straight line $\ln(x/x_0) = -kt = -265\,min^{-1}\ t = -0.226\,h^{-1}\ t$. Consequently, $\mu_{max} = 0.99 - 0.226 = 0.76\ h^{-1}$. Apparently, the cell machinery is almost fully expressed, also at $D = 0.50\,h^{-1}$, the steady-state value of D for the upper curve. This is not the case for the lower curve that starts at $D = 0.1\,h^{-1}$. It takes about 100 min = 1.7 h to build up a cell machinery that is capable of utilizing the sugar offered to the culture at $D = 0.99\,h^{-1}$. After about 100 min, the two curves run parallel in the semilog plot.

The slow build-up of cell machinery during a transient where D is increased in a step from $D = 0.075\,h^{-1}$ to a much higher value of $D = 0.409\,h^{-1}$ is also recognized in Figure 7.9. Before $t = 0$, the specific growth rate μ as measured from biomass concentration in the effluent is equal to the dilution rate $0.075\,h^{-1}$, typical for the steady-state continuous cultivation with no biomass in the feed. For several hours after the stepwise increase of D to $0.409\,h^{-1}$ at $t = 4.3\,h$, the specific growth rate is lower than D. At $t = 6.2\,h$, the specific growth rate becomes higher than D, and after $t = 9\,h$, the specific growth rate approaches $D = 0.409\,h^{-1}$ asymptotically.

This time profile for μ is understood when the curves for glucose concentration and RNA content of the cells are considered. The sudden increase of D at

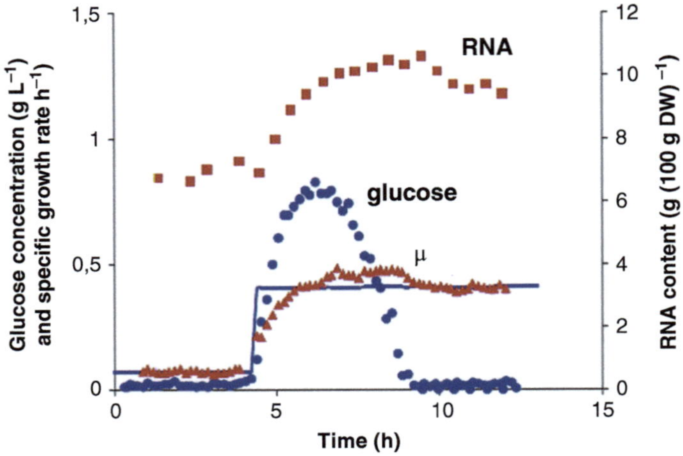

Figure 7.9 Transient cultivation of *L. lactis* between 4.3 and 11 h. *D* is increased from 0.075 to 0.409 h^{-1} at t = 4.3 h. Inlet substrate (glucose) concentration *s*0 = 20 g l^{-1}. Data from Benthin, Nielsen, and Villadsen [18].

4.3 h "saturates" the medium with glucose from an initial value of 5.5 mg l^{-1}, and s reaches a maximum of 750 mg l^{-1} at $t = 5.4$ h. At first, there is no machinery available to metabolize the extra glucose, but slowly – with a time constant of about 1.3 h^{-1} – this machinery is built up and the accumulated glucose is metabolized. The RNA concentration is a typical measure of the "size" of the cell machinery, and it peaks at 8–9 h, where after it decreases to a value ≈40% higher than the initial value before the transient. At the same time, μ also decreases to its new steady-state value of 0.409 h^{-1}, and here the glucose concentration has decreased to 11 mg l^{-1}.

The details of the kinetic expression during the transient can, of course, not be obtained when only a few variables (RNA, μ, s) are measured as functions of time, but the essential features of the transient are clearly understood in terms of the "active" machinery represented by RNA that increases or decreases as a function of the limiting substrate concentration. It is worth noting that the substrate concentration varies only from 5.5 to 11 mg l^{-1} when the RNA content of the cells increases from a low value to almost the value at washout. Comparison with Figure 7.8 shows that at $D = 0.50$ h^{-1}, the machinery is at its maximum value. The constant slope of the $\ln(x/x_0)$ versus t line shows that $\mu = \mu_{max}$ even at $t = 0$. At $D = 0.409$ h^{-1}, the machinery appears to be almost fully built up, and during the transient the activity r_x of the cell reaches even higher values.

Whereas in Figure 7.9 RNA is identified as the driver in changing the composition of the biomass, the study of anaerobic yeast fermentation [19] indicates that as a result of a sudden change of dilution rate, the whole catabolic machinery changes relative to the anabolic machinery. The two rates, $r_{catabolic}$ and

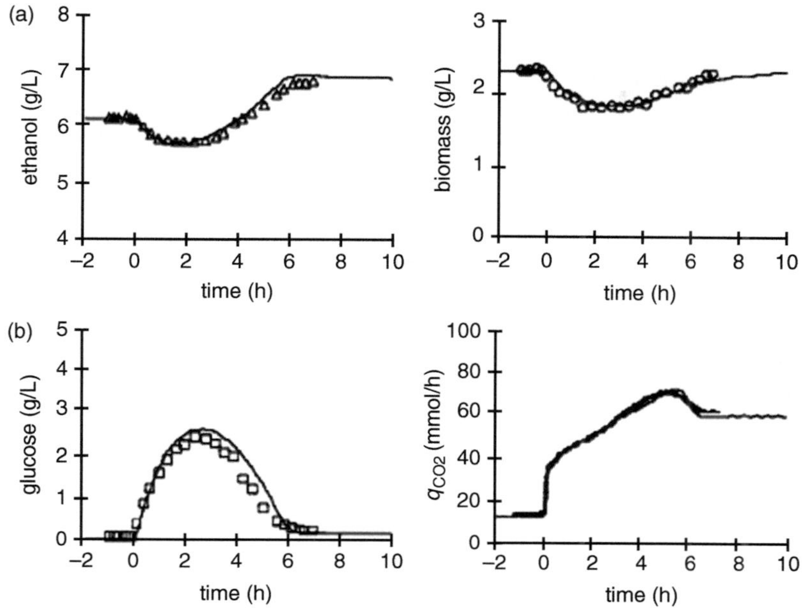

Figure 7.10 Anaerobic cultivation of *S. cerevisiae* with glucose (feed concentration 20 g l^{-1}) as limiting substrate. At $t = 0$, the dilution rate is shifted from 0.053 to 0.25 h^{-1}, and after 6.7 h the new steady state is reached. (a) Effluent concentrations of ethanol and biomass. (b) Glucose concentration and volumetric rate of CO_2 production $q_{CO2}\,V$ (mmol h)$^{-1}$ with $V = 1.65$ l. (Data from Duboc, von Stockar, and Villadsen [19].)

$r_{anabolic}$, are shown in Figure 7.11 as functions of time. They will not again be equal before the end of the transient toward the new D value.

The key feature of Figure 7.10 is the sudden jump in $q_{CO2} \cdot V = r_{CO2} \cdot x \cdot 1.65$ (l) from 12 to 40 mmol CO_2 h^{-1} right at the start of the transient when D is changed

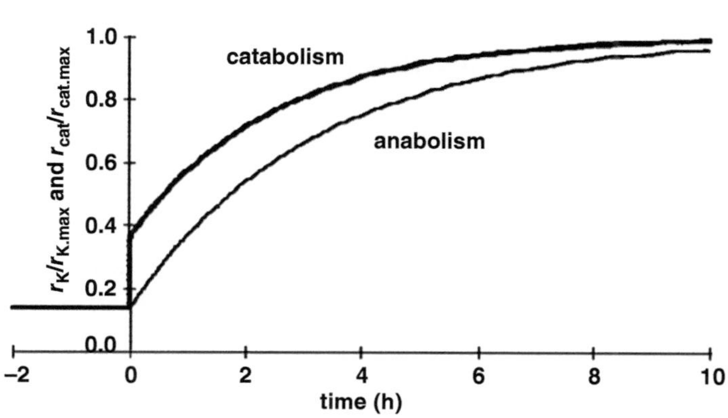

Figure 7.11 Specific rates of catabolic and of anabolic reactions during a transient from $D = 0.053\,h^{-1}$ to $D = 0.25\,h^{-1}$. The curves are simulated by (7.32) using $[\tau_{catabolic}, \tau_{anabolic}] = [2.5, 3.2\text{ (h)}]$.

from $0.053\,h^{-1}$ to $0.25\,h^{-1}$. This is clearly associated with the catabolic machinery since the specific rate of ethanol production $q_e/x = D \cdot e/x$ increases almost throughout the transient, whereas x decreases rapidly at first (indicating an incipient washout) and only approaches its final, somewhat lower value (the steady-state x value decreases slightly with increasing D for $D \ll D_{washout} = 0.39\,h^{-1}$) at the end of the transient.

From $t = 0$ to 7 h, the glucose concentration is very high, reaching a peak of $2.7\,g\,l^{-1}$ at about 2 h. The rate of CO_2 production increases significantly from its initial value ≈40 $mmol\,h^{-1}$ to almost 80 $mmol\,h^{-1}$ at 5.5 h. When the production of anabolic machinery becomes significant after about 2.8 h, there is a further increase of CO_2 production above that due to the increase of the catabolic machinery. All the diagrams in Figure 7.10 point to an offset between the catabolic and the anabolic machinery. Eq. (7.32) is a simple (empirical) model that explains the observed data:

For $t < 0$: $D = D_0 = 0.053\,h^{-1}$ and $r_x = r_{anabolic} = r_{catabolic}$. At $t = 0$: $r_x = D_0$ and $r_{catabolic} = r_{catabolic}\,(D_0) + \Delta$.

$$\text{For } t > 0: \frac{dr_x}{dt} = \frac{r_{x,max} - r_x}{\tau_{anabolic}} \quad \text{and} \quad \frac{dr_{catabolic}}{dt} = \frac{r_{catabolic,max} - r_{catabolic}}{\tau_{catabolic}}. \tag{7.32}$$

The time constant for anabolic anabolism (associated with biomass production) is determined to $\tau_{anabolic} = 3.2$ h, while the time constant for catabolic metabolism (associated with the CO_2 production rate alone for $t < 2.5$ h) is determined to 2.5 h.

Simulations of the glucose, the ethanol, and the biomass concentration and of the CO_2 production rate is shown on Figure 7.10 together with the experimental data. The quality of the kinetic model appears to be quite good. In Figure 7.11, the time profiles of $r_{catabolic}$ and of $r_{anabolic}$ are shown as simulated by (7.32).

Duboc, von Stockar, and Villadsen [19] also studied aerobic cultivation of *Saccharomyces cerevisiae*. While they found that a jump from dilution rate D_0 to D_1 could be described by Eq. (7.32) when both D_1 and D_0 are below the limit D_{crit} where fermentative growth occurs, very large and apparently random variations of the jump Δ in q_{CO2} at $t = 0$ were obtained when D_0 was below D_{crit} and D_1 was above D_{crit}. Perhaps, these erratic results are related to the instability of the steady-state around D_{crit} reported for *S. cerevisiae* by [20] and discussed by [21].

As the last example, a study by Melchiorsen *et al.* [22] of the transition between homolactic and heterolactic fermentation (Section 4.2, Figure 4.4b) is discussed in Figures 7.12 and 7.13. A specific set of reactions is discussed, namely, (1) lactic acid production catalyzed by lactic acid dehydrogenase and (2) production of formic acid, HAc, and ethanol, using lactate dehydrogenase (LDH), and pyruvate formate lyase (PFL), respectively, as enzymes. At very low glucose concentrations, pyruvate formate lyase is active, but the enzyme is inhibited by the glycolysis metabolite GAP (Figure 4.1), and the concentration of this metabolite is high at high glucose concentration.

Figure 7.12 shows results from a continuous stirred tank experiment where D was changed from 0.10 to $0.50\,h^{-1}$. As usual, the biomass concentration

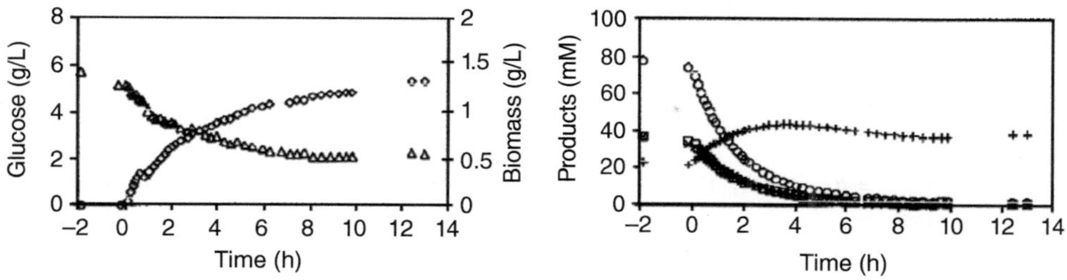

Figure 7.12 Transient cultivation of *L. lactis* on a feed with glucose (10 g l^{-1}) + MS10 medium (containing degraded protein). Increase of dilution rate from 0.095 to 0.50 h^{-1}. (a) Glucose (◇) and biomass (△), both in g l^{-1}. (b) HCOOH (○), HAc (□), ethanol (□) and lactic acid (+). Note that the ratios of effluent concentrations of HCOOH, HAc, and ethanol (equal to r_i/D when the feed does not contain any of the products) are [1 : 1 : 1] on a C-mol basis as predicted in Example 4.4 Eq. (4). (Data from Melchiorsen *et al.* [22].)

decreases (from 1.5 to 0.5 g l^{-1}) with increasing D, and the glucose concentration increases (from virtually zero to 4.3 g l^{-1}). The most interesting feature is that production of the mixed acid fermentation products *immediately* stops, and the effluent concentrations (in mM) of (HCOOH, HAc, and ethanol) decrease exponentially with rate constant $k = D = 0.50\,h^1$, that is, by washout of the products initially present in the tank. The immediate and complete inhibition of PFL is caused by an increase of the trioses GAP and DHAP. The change of the concentrations of [GAP, DHAP] was from [0.1, 2 mM] to [0.2, 6 mM], which is probably the cause of the rapid deactivation of PFL. Also at $D = 0.095\,h^{-1}$, a considerable production of HLac occurs, but after the shift of D to the higher value of 0.50 h^{-1}

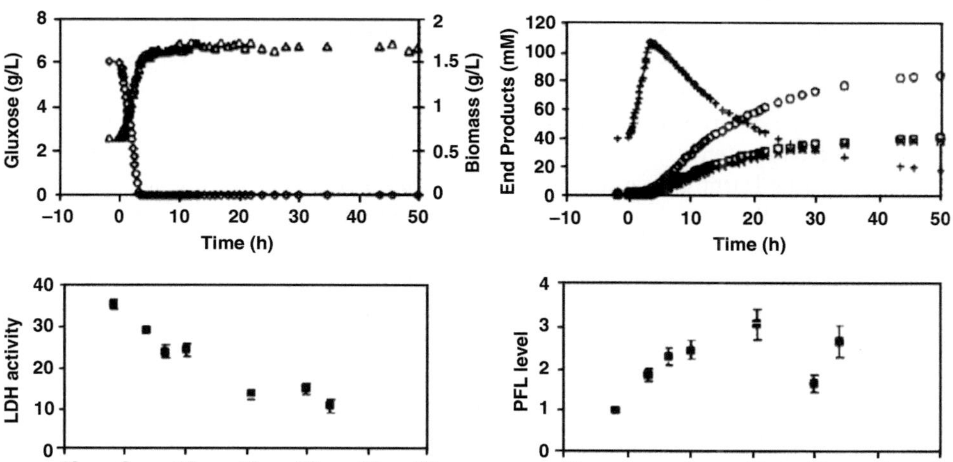

Figure 7.13 Reversal of the experiment in Figure 7.12. (a) Substrates and products (using the symbols of Figure 7.12). (b) Activity of LDH and PFL during the transient.

[HLac] increases through a transient maximum to a new steady state. The behavior of the time profile of HLac can easily be simulated by the model (7.33):

$$\frac{d[\mathrm{HLac}]}{dt} = r_{\mathrm{HLac}}x - 0.50[\mathrm{HLac}], \quad \text{where} \quad [\mathrm{HLac}] = 22\,\mathrm{mM} \text{ at } t = 0. \tag{7.33}$$

$r_{\mathrm{HLac}} = f(s, [\mathrm{LDH}], t)$ where [LDH] is the pyruvate dehydrogenase content of the cell.

Figure 7.13 shows the time course of glucose, biomass, and the four metabolites, HLac, HAc, ethanol, and HAc, after a downshift of D from 0.49 to 0.095 h^{-1} in a reversal of the experiment in Figure 7.12. Also shown are time profiles of the decrease of [LDH] and the increase of [PFL], the key enzymes involved.

The biomass and glucose concentrations return to the values at the start of the transient in Figure 7.12, but with a time constant smaller than in Figure 7.12 – the dilution terms $D \cdot x$ and $D \cdot s$ are much smaller than in the previous experiment. Net production of HLac is first positive, again due to the smaller dilution term D · [HLac], but eventually [HLac] decreases due to the decreasing value of [LDH]. Production of the other metabolites starts as soon as the accumulated glucose in the medium has been metabolized (after 2–3 h), and again [HCOOH, HAc, ethanol] are produced in the C-mol ratio of [1, 1, 1].

The lower panel indicates that the time constant for change of [PFL] is smaller than the time constant for [LDH], but it is remarkable that even after a long transient some of the enzyme is left. This is very advantageous for the cell as it can start working on the different metabolism right after a change of environment, in this case from low to high sugar concentration and vice versa. Hence the enzymes are *always expressed constitutively at a certain level*, but under the right environmental conditions extra enzyme can be synthesized. Pathway enzymes are consequently *inhibited* or *activated* as soon as the environment changes, and their synthesis – at a much smaller rate – is determined by up- or downregulation of the corresponding gene expression.

The different experimental studies discussed in Section 7.3 reveal that to a certain extent transient kinetics can be modeled, also for cell reactions. Much can be learnt concerning the influence of a particular macromolecule (RNA), a key to the syntesis of proteins, on the up- or downregulating of genes (for example *pfl*). This causes a long time effect on metabolism along with a fast inhibition/activation of the enzymes. Finally, the different dynamics of the two main parts of metabolism, catbolism and anabolism, can be explored.

These insights into transient metabolism are, however, only of qualitative nature. As yet, one is far from understanding the *regulatory network* of the living cell, a network that works at a level below the *mass flow* network discussed in Chapters 2–6 – always with the tacital understanding that the cell is *in metabolic steady state*. The crucial importance of understanding the structure and function of the regulatory network of living cells is presently addressed in an "-omics" discipline developing in parallel with the "classical" genomics, proteomics, and metabolomics, see for example Gompert and Nielsen [23].

7.5 Cybernetic Modeling of Cellular Kinetics

Cybernetics, a word coined by Norbert Wiener in the mid-twentieth century for the "self-regulating," "self-governing" nature of systems, whether of sociological, economic, organizational, or biological origin, has revolutionized our concept of all these areas. A simple device was invented more than 2000 years ago to keep the water level constant in a tank that served a water wheel, and today countless self-regulating control systems are used both in industry and in our daily life. The living cell regulates its function on many different levels, from DNA synthesis to the flow of carbon in the metabolic network. It is conceivable that we will come to understand the function of the regulatory network of cells by applying the knowledge accumulated during the past 50 years on control strategies for mechanical systems – a discipline that has advanced hugely, both with respect to understanding the mathematical foundation of complicated control systems and with respect to constructing "optimal" control systems.

Ramkrishna [24] and Kompala, Ramkrishna, and Tsao [25] were the first to use the concepts of cybernetics to biotechnology.

In a series of papers, starting with Kompala *et al.* [26], Ramkrishna and his students modeled *diauxic growth* of cells: The leavening of bread is sharply divided into two phases. The dough contains both glucose from hydrolyzed starch and maltose. Glucose is the *preferred substrate* and CO_2 is produced while glucose is consumed. After a *lag phase* that may last for several hours, maltose is metabolized by the yeast and more CO_2 is produced in the second leavening phase. Most microorganisms have glucose as the preferred substrate, and when the substrate contains many sugars (e.g., from hydrolysis of lignocellulosic biomass), pentoses such as xylose and arabinose are (slowly) metabolized in batch operation *after* the glucose is used up. In continuous culture, both substrates can be metabolized simultaneously at low D, but at high D only the preferred substrate is utilized. In fundamental biology, the studies of Jacques Monod on diauxic growth was the key to his famous explanation for the functioning of the *lac operon.*

The models for diauxic growth developed by the Ramkrishna group contain empirical models for the production of biomass X from substrate S_i and the *key enzyme* X_{Ei} that mediates conversion of S_i to biomass and metabolic products:

$$S_i \rightarrow \gamma_i \, X \text{ with rate } r_i = r_{i,\max} \, w_i = \frac{k_i s_i}{s_i + K_i} x_{Ei} w_i.$$

$$X \rightarrow X_{Ei} \text{ with rate } r_{Ei} = r_{Ei,\max} u_i = \frac{k_{Ei} \, s_i}{s_i + K_{Ei}} u_i. \tag{7.34}$$

$X_{Ei} \rightarrow X$ (degradation of the enzyme to X) with rate $r_{Ei,\text{degrade}} = k_{\text{degrade},i} \, x_{Ei}$.

Eq. (7.34) is a compartment model where substrate S_i, one of N substrates, is used to make biomass X, with yield coefficient γ_i. A fraction x_{Ei} of the biomass machinery produces the enzyme E_i. Finally, the enzyme degrades to bulk biomass X by a first-order reaction in x_{Ei}. The new concept of the model is that of the two *cybernetic variables* w_i and u_i. The kinetics of uptake of substrate S_i to make X is assumed to be proportional to w_i, and the part of X that is used to make X_{Ei}, that is, *the allocation of cell machinery to make the enzyme that controls the uptake of* S_i is the other cybernetic variable u_i.

Clearly w_i controls the degree to which S_i contributes to X (through γ_i), while u_i determines how fast this contribution is delivered (by the rate of synthesis of E_i). The quality of the cybernetic model depends on how well the user replicates the function of natural processes in the cell.

In early cybernetic models, u_i was based on the so-called "matching law" model. This expresses that the resources (the cell machinery) allocated to produce the different enzymes should be distributed such that the total machinery gives the highest possible specific growth rate of X. This looks like an eminent oral model for the way we believe (from an antropomorphistic analogy) that cells will work. The definition of w_i is not so clear, but based on a large amount of experimental evidence Kompala *et al.* [26] proposed w_i to be defined by the second expression in Eq. (7.35):

$$\text{Allocation of machinery}: u_i = \frac{r_{\text{i,max}}}{\sum_1^{\text{N}} r_{\text{i,max}}}; \quad \text{allocation of } \text{S}_\text{i}: w_i = \frac{r_{\text{i,max}}}{\max_j(r_{\text{i,max}})}. \tag{7.35}$$

In the version of cybernetic modeling outlined above, a steady-state CSTR fed with only substrate S_i would stop synthesis of all enzymes E_j different from E_i and the enzymes E_j would be degraded by the last reaction of Eq. (7.34). That this is not the case is illustrated by Figure 7.13: neither LDH nor PFL is washed out even after many hours of cultivation with total inhibition of one of the enzymes – this would make it impossible for the cell to react immediately to a change in substrate level. The "hidden capacity" of the catabolic machinery in Figure 7.11 is another example of the adaptability of the cell to react fast to changes in the environment.

These observations have led to more complicated definitions of the cybernetic variables, and it seemed that the usefulness of cybernetic models had reached its limit.

Recently published cybernetic models have, however, again demonstrated how well adapted cybernetic modeling is to get at least a semiquantitative insight into the functioning of the regulatory system on the scale of the whole cell. Young *et al.* [27] is a particularly brilliant application of cybernetic modeling on the overall level of the cell.

Young *et al.* [27] applied metabolic flux analysis, metabolic control analysis (MCA) (Section 7.6) with "log–log kinetics" for individual network reactions (see Chapter 8), and cybernetic modeling to study the steady-state flux through a metabolic network. A simplified *E. coli* network is used as example.

A simple two-step linear pathway is used to illustrate how different choices of the cybernetic variables lead to very different simulation results.

7.6 Bioreactions with Diffusion Resistance

In the previous sections of this chapter, it has been assumed that the concentration of all reactants is the same everywhere in the medium. This is not necessarily true: the cells may agglomerate to form flocs, and in the cultivation of filamentous fungi the hyphae may be entangled to form what looks like a pellet of appreciable dimensions. The cells may aggregate in large clumps on any solid interface in the reactor, for example, the stirrer, and in water-cleaning systems, biofilms often form in the pipes. Consequently, the notion of a volumetrically homogeneous suspended cell culture may sometimes be wrong, and mass transport of substrates and products to and from clumps of cells or biofilms is a rate process that needs to be considered in parallel with the chemical kinetics of the bioreaction. Besides liquid-phase diffusion of reactants discussed here, transport of reactants from a gas phase to a liquid phase is of considerable importance. Transport of O_2 from the gas phase by the mass balance, Eq (2.6), may well become the rate-limiting step rather than the aqueous phase reactions between dissolved gases and other substrates. Mass transport of this kind will be discussed in Chapter 11, and it will be shown that in the macroscopic rate of mass transport, the parameter k_la in (2.6) depends on the mixing intensity of the medium. The liquid-phase mass transfer processes discussed in this chapter depend on molecular diffusivity, a transfer process with a different scale of operation than the larger scale mixing by input of mechanical energy.

Enzymes and cells may deliberately be embedded in a porous polymer pellet or in a film. This makes it possible to retain the "catalyst" in the reactor, also at high volumetric throughput of the medium, for example, as used in a liquid–solid fluid bed reactors.

This section will provide the necessary mathematical background for calculation of the combined processes of chemical reaction and mass transport at the molecular scale. This subject is well known from the literature on heterogeneous catalysts that has been part of the core curriculum for students of chemical engineering for the past 50 years. Consequently, the subject needs not to be discussed at length here, and a short description with a few illustrative examples will suffice. The new parameter is the molecular mass diffusion coefficient $\mathcal{D}$ for a reacting species in the medium that is dispersed in the porous interior of a biofilm or in an enzyme-containing particle. Experimental values for liquid-phase diffusion coefficients are found in most chemical engineering tables – or they can be calculated from well-established theoretical models.

The model for diffusion and reaction in a particle or in a biofilm is formulated as a mass balance, in which substrate flows into the particle by molecular diffusion and is consumed by chemical reaction. In Eq. (7.36a) and (7.36b), the mass balance is written for plane-parallel and for spherical symmetry, the two geometries relevant for bioreactions. Only steady-state

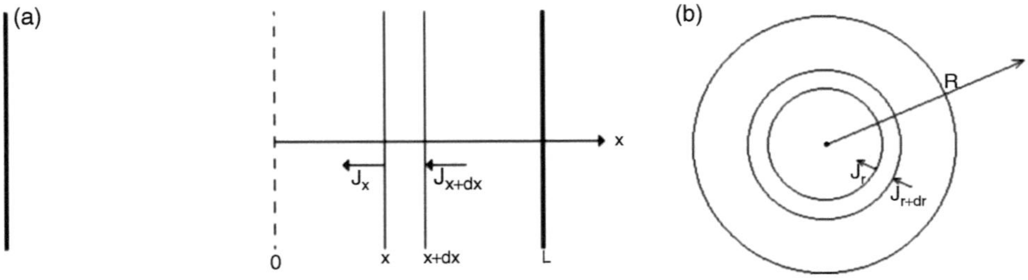

Figure 7.14 Mass balances for diffusion and chemical reaction: (a) slab symmetry and (b) spherical symmetry.

operation is considered:

$$\mathcal{D}A\left(\frac{ds}{dx}\right)_{x+dx} - \mathcal{D}A\left(\frac{ds}{dx}\right)_x = J_{x+dx} - J_x = (-q_s)\cdot A\,dx \rightarrow \frac{d^2s}{dx^2} = \mathcal{D}^{-1}(-q_s). \tag{7.36a}$$

$$\mathcal{D}\left(-4\pi r^2\frac{ds}{dr}\right)_{r+dr} - \mathcal{D}\left(4\pi r^2\frac{ds}{dr}\right)_r = (-q_s)\cdot 4\pi r^2 dr \rightarrow \frac{1}{r^2}\left(\frac{d}{dr}\left(r^2\frac{ds}{dr}\right)\right) = \mathcal{D}^{-1}(-q_s). \tag{7.36b}$$

The terms on the left-hand side represent the flux of substrate S at $x+dx$ (or $r+dr$) by molecular diffusion down the concentration gradient from the pellet surface (see Figure 7.14). The difference between the two terms is equal to the loss of S by chemical reaction in the segment. Define $\xi \equiv x/L$ and $\xi \equiv r/R$ in Eq. (7.36a) and (7.36b), respectively, and define $y = s/s_0$, where s_0 is the *surface concentration* of S.

$$\frac{d^2y}{d\xi^2} = \mathcal{D}^{-1}(-q_s)\frac{L^2}{s_0} \quad \text{and} \quad \left(\frac{1}{\xi^2}\frac{d}{d\xi}\left(\xi^2\frac{ds}{d\xi}\right)\right) = \mathcal{D}^{-1}(-q_s)\frac{R^2}{s_0}. \tag{7.37}$$

An nth-order rate expression $(-q_s) = k_0\, e_0\, s^n = k\, s^n$ will be chosen for further study:

$$\frac{d^2y}{d\xi^2} = \mathcal{D}^{-1}\, k y^n\, L^2 s_0^{n-1} = \Phi_n^2\, y^n \quad \text{and} \quad \frac{1}{\xi^2}\left(\frac{d}{d\xi}\left(\xi^2\frac{dy}{d\xi}\right)\right) = \Phi_n^2\, y^n. \tag{7.38}$$

$\mathcal{D}$ is diffusivity of the substrate in the porous enzyme pellet or in the biofilm. For diffusion of substrate (e.g., glucose or HAc) in a thin biofilm, $\mathcal{D}$ can be taken as the tabular value for molecular diffusion of, for example, glucose in water ($0.66\cdot 10^{-9}\,\text{m}^2\,\text{h}^{-1}$ at 25 °C), perhaps corrected a little for the semisolid gelatinous film. An enzyme can be entrapped in insoluble beads of, for example, calcium alginate or even covalently bonded in an inactive matrix by reactions that preserve the active site of enzyme. In both cases, the *effective* diffusivity of substrate in the medium is much smaller than the molecular diffusion coeficient:

$$\mathcal{D} = \mathcal{D}_{\text{molecular}}\frac{\varepsilon}{\tau}. \tag{7.39}$$

In Eq. (7.39), ε is the porosity of the pellet and τ is the tortuosity of the diffusion path through the porous system of the pellet. Typical values for $[\varepsilon, \tau]$ are [0.2–0.5, 2–3].

The parameters $\mathcal{D}^{-1}\,kL^2 s_0^{n-1}$ of Eq. (7.38) are collected in the dimensionless *Thiele modulus* Φ_n:

$$\Phi_n = \lambda(k/\mathcal{D}\,s_0^{n-1})^{1/2},\ \text{where}\ \lambda = L\ \text{or R for, respectively, slab and spherical symmetry.} \tag{7.40}$$

The two values $n=0$ and 1 in Eq. (7.38) will be treated in more detail, since a typical rate expression for enzyme reactions $q_s = (k_0\,se_0/s + K_m) = (ks/s + K_m)$ is bounded between first-order kinetics for small s (or large K_m) and zero-order kinetics for large s (or small K_m).

Example 7.3 The influence of diffusion limitation on enzyme reactions: first and zero-order kinetics

First-order kinetics

For slab symmetry $\Phi_1 = L(k/\mathcal{D})^{1/2}$ and the linear boundary value problem (7.38) with $y=1$ at $\xi=1$ (the pellet or film surface) and with $\frac{dy}{d\xi}=0$ at $\xi=0$ (the center of the slab) is as follows:

$$\frac{d^2y}{d\xi^2} = \Phi_1^2 y \rightarrow y = \frac{\cosh\Phi\,\xi}{\cosh\Phi} = \frac{\exp(\Phi\xi)+\exp(-\Phi\xi)}{\exp(\Phi)+\exp(-\Phi)}. \tag{1}$$

For spherical symmetry, one substitutes $y = z/\xi$, and the left-hand side of Eq. (7.38) simplifies to the following equation:

$$\frac{1}{\xi^2}\left(\frac{d}{d\xi}\left(\xi^2\frac{dy}{d\xi}\right)\right) = \frac{1}{\xi^2}\left(\frac{d}{d\xi}\xi^2\left(\frac{1}{\xi}\frac{dz}{d\xi} - \frac{z}{\xi^2}\right)\right) = \frac{1}{\xi}\frac{d^2z}{d\,\xi^2} = \Phi_1^2\frac{z}{\xi},\quad \text{i.e.,}\ \frac{d^2z}{d\xi^2} = \Phi_1^2 z. \tag{2}$$

$$z = \text{A}\cosh(\Phi\xi) + \text{B}\sin\text{h}(\Phi_1\,\xi)\ \text{and}\ y = \frac{z}{\xi} = \frac{\text{A}\cosh(\Phi_1\xi)}{\xi} + \frac{\text{B}\sin\text{h}(\Phi_1\xi)}{\xi}. \tag{3}$$

The boundary conditions of (2) are $y=1$ for $\xi=1$ and y *finite* at $\xi=0$. Since $\cos\text{h}\ \xi\ \Phi_1 = 1$ for $\xi=0$ and $\sin\text{h}\ \xi\ \Phi_1 = 0$ for $\xi=0$, one obtains: $A=0$ and $B=(\sin\text{h}\ \Phi_1)^{-1}.2$

$$\text{Hence for spherical symmetry:}\quad y = \frac{\sin\text{h}(\Phi_1\xi)}{\xi\sin\text{h}(\Phi_1)} = \frac{\exp(\Phi_1\xi)-\exp(-\Phi_1\xi)}{\xi(\exp(\Phi_1)-\exp(-\Phi_1))}. \tag{4}$$

The volumetric rate of consumption of S in the pellet is equal to the rate of transport of S into the pellet:

$$
\begin{aligned}
(-q_s) &= A_{\text{pellet}}\, \mathcal{D}\left(\frac{\mathrm{d}s}{\mathrm{d}r}\right)_{r=R} \\
&= 4\,\pi R^2\, \mathcal{D}\frac{s_0}{R}\left(\frac{\xi\Phi_1 \cos h(\Phi_1\xi) - \sin h(\Phi_1\xi)}{\xi^2 \sin h(\Phi_1)}\right)_{\xi=1} \\
&= (\Phi_1 \coth(\Phi_1) - 1)4\,\pi R\,\mathcal{D}\, s_0.
\end{aligned}
$$

Without diffusion limitation $(-q_s)_0 = \frac{4}{3}\pi R^3 \cdot k s_0$, and one obtains the *effectiveness factor* η:

$$
\begin{aligned}
\eta = (-q_s)/(-q_s)_0 &= (\Phi_1 \coth(\Phi_1) - 1) \cdot \frac{3}{R^2 k}\mathcal{D} \\
&= \frac{3(\Phi_1 \coth(\Phi_1) - 1)}{\Phi_1^2}.
\end{aligned}
\tag{5}
$$

For large Φ_1 the function cot $(\Phi_1) = \frac{\cosh(\Phi_1)}{\sinh(\Phi_1)} \to 1$ and $\eta \to \frac{3}{\Phi_1}$.

The McLaurin series for coth (Φ_1) is $\coth(\Phi_1) = \Phi_1^{-1} + \frac{1}{3}\Phi_1 - \frac{1}{45}\Phi_1^3 \ldots$, and consequently for $\Phi_1 \to 0$:

$$
\begin{aligned}
\eta &= \frac{3(\Phi_1 \coth(\Phi_1) - 1)}{\Phi_1^2} \to \frac{3\left(\left(\Phi_1^{-1} + \frac{1}{3}\Phi_1^1 - \frac{1}{45}\Phi_1^3\right)\Phi_1 - 1\right)}{\Phi_1^2} \\
&= 1 - \frac{1}{15}\Phi_1^2.
\end{aligned}
\tag{6}
$$

For slab symmetry and following the derivation in Eqs. (5) and (6), one obtains:

$$
(-q_s) = \frac{\Phi_1 \sinh \Phi_1}{\cosh \Phi_1}\frac{s_0}{L}A\mathcal{D} = \Phi_1 \tanh \Phi_1 \text{ and } \eta = \frac{\tanh \Phi_1}{\Phi_1}. \tag{7}
$$

$$
\begin{aligned}
&\text{For } \Phi_1 \to \infty \tanh \Phi_1 \to 1 \text{ and } \eta \approx \frac{1}{\Phi_1} \text{ and for } \Phi_1 \to 0, \\
&\tanh \Phi_1 = \Phi_1 - \frac{1}{3}\Phi_1^3 \ldots, \quad \text{and } \eta = 1 - \frac{1}{3}\Phi_1^2.
\end{aligned}
\tag{8}
$$

Zero-order kinetics

Integration of $(d^2y/\mathrm{d}\xi^2) = \Phi_0^2$ with boundary conditions $y = 1$ at $\xi = 1$ and $(\mathrm{d}y/\mathrm{d}\xi) = 0$ at $\xi = 0$ yields

$$
\begin{aligned}
y &= \frac{1}{2}\Phi_0^2\,\xi^2 + B, \text{ where } B = 1 - \frac{1}{2}\Phi_0^2 \text{ since } y = 1 \text{ at } \xi = 1 \to y \\
&= 1 - \frac{1}{2}\Phi_0^2(1 - \xi^2).
\end{aligned}
$$

For $\Phi_0^2 < 2$, the dimensionless substrate concentration $y > 0$ for $0 \leq \xi \geq 1$, and the effectiveness factor $\eta = 1$ since as long as $y \geq 0$ the rate is independent of y. For $\Phi_0^2 > 2$, an inner part of the slab $0 < \xi \leq \xi_0$ is, however, empty of substrate, and the arbitrary constants must be found from

$(\mathrm{d}y/\mathrm{d}\xi) = y = 0$ at $\xi = \xi_0$:

$$y = \frac{1}{2}\Phi_0^2\,\xi^2 + A\,\xi + B,\ \text{where}\ A = -\Phi_0^2\,\xi_0,\quad B = \frac{1}{2}\Phi_0^2\,\xi_0^2 \rightarrow y$$
$$= \frac{1}{2}\Phi_0^2\,\xi^2 - \Phi_0^2\,\xi_0\,\xi + \frac{1}{2}\Phi_0^2\,\xi_0^2.$$

When the boundary condition $y = 1$ at $\xi = 1$ is applied, one obtains the following equation for ξ_0:

$$1 = \frac{1}{2}\Phi_0^2 - \Phi_0^2\,\xi_0 + \frac{1}{2}\Phi_0^2\,\xi_0^2 \rightarrow \xi_0 = 1 - \Phi_0^{-1}\sqrt{2}.$$

The part of the slab for $\xi = 1$ to $\xi = 1 - \Phi_0^{-1}\sqrt{2}$, that is, $\Phi_0^{-1}\sqrt{2}$ is fully active, the rest is inactive.

Hence, for $\Phi_0^2 > 2$ the effectiveness factor is $\eta = \Phi_0^{-1}\sqrt{2}$, and for $\Phi_0^2 \leq 2$, $\eta = 1$.

The same calculation is made for spherical symmetry where again the substitution $z = \xi\,y$ is made:

$$\frac{d^2 z}{\mathrm{d}\xi^2} = \xi\Phi_0^2 \rightarrow z = \frac{1}{6}\xi^3\Phi_0^2 + A\,\xi + B \rightarrow y = \frac{1}{6}\xi^2\,\Phi_0^2 + A + \frac{B}{\xi}. \tag{9}$$

$$\text{When}\ y \geq 0\ \text{at}\ \xi = 0,\ \text{the arbitrary constant}\ B = 0,$$
$$\text{and}\ A = 1 - \frac{1}{6}\Phi_0^2 \ \rightarrow y = 1 - \frac{1}{6}\Phi_0^2(1 - \xi^2). \tag{10}$$

For $\Phi_0 > \sqrt{6}$ calculation of y by (10) would give $y < 0$ at $\xi = 0$. Now, the arbitrary constants A and B, and a ξ value ξ_0 between 0 and 1 are found from the following:

$$\frac{\mathrm{d}y}{\mathrm{d}\xi} = y = 0\ \text{at}\ \xi = \xi_0,\quad \text{and}\quad y = 1\ \text{at}\ \xi = 1.$$

Solving for A and B gives: $y = \frac{1}{6}\Phi_0^2\,\xi^2 - \frac{1}{2}\Phi_0^2\,\xi_0^2 + \frac{1}{3}\Phi_0^2\,\xi_0^3\,\xi^{-1}$. Using $y = 1$ at $\xi = 1$ allows ξ_0 to be found:

$$\xi_0\ \text{is the solution of the equation}\ 6\,\Phi_0^{-2} = 2\,\xi_0^3 - 3\,\xi_0^2 + 1 \tag{11}$$

For $\Phi^2 \leq 6$, the whole sperical particle is active, and for a zero-order reaction $\eta = 1$.

For $\Phi^2 > 6$, the effectiveness factor gradually decreases from 1. For large Φ, the active part of the sphere is small, and ξ_0 can be approximated by $1 - a$. Inserting this approximation in Eq. (11) gives the following:

$$6\,\Phi^{-2} \approx 2(1 - a)^3 - 3(1 - a)^2 + 1 \rightarrow 6\,\Phi^{-2} = a^2(3 - 2a)\ \text{and for}$$
$$a \rightarrow 0\ \text{as}\ \sqrt{2}\Phi^{-1}.$$

The active part of the sphere is $1 - (1 - a)^3 = a^3 - 3a^2 + 3a \approx 3a$ for $a \rightarrow 0$. Hence, $\eta \approx 3\sqrt{2}\,\Phi^{-1}$.

Just as was the case for first-order reaction one obtains, the same asymptotic behavior for slab and for spherical symmetry when the length parameter for the sphere is chosen as $R/3 = L$. This has given rise to a neat approximation for the behavior of $\eta(\Phi)$ for any n in Eq. (7.38). In slab symmetry (7.38) can be converted into a first-order differential equation in $p = (dy/d\xi)$, and the differential equation is integrated from $y_0 = y$ $(\xi = 0)$, where $(dy/d\xi) = 0$ to $y = y(\xi)$ as shown in (12):

$$\frac{d^2y}{d\xi^2} = \frac{d}{d\xi}\left(\frac{dy}{d\xi}\right) = \frac{d}{dy}\left(p\frac{dp}{dy}\right) = \frac{1}{2}\frac{d(p^2)}{dy} = \Phi_n^2 y^n \rightarrow p = \left(\frac{dy}{d\xi}\right)_{\xi} = \sqrt{\frac{2}{n+1}}\Phi_n\sqrt{y^{n+1} - y_0^{n+1}}. \tag{12}$$

Eq. (12) cannot be used to find an analytical solution for $y(\xi)$, but for y_0^{n+1} small (i.e., for large Φ_n):

$$\left(\frac{dy}{d\xi}\right)_{\xi} \approx \sqrt{\frac{2}{n+1}}\Phi_n\, y^{((n+1)/2)}, \quad \text{and for} \quad \xi = 1: \quad \left(\frac{dy}{d\xi}\right)_{\xi=1} = \sqrt{\frac{2}{n+1}}\Phi_n. \tag{13}$$

In analogy with the derivation in (7)–(8) for a first-order reaction in slab symmetry, one obtains:

Flux of S into the pellet = reacted S per volume pellet = $(-q_s) = \mathcal{D}/\lambda s_0$ $= \left(\frac{dy}{d\xi}\right)_{\xi=1} = \sqrt{\frac{2}{n+1}}\Phi_n \rightarrow$

$$(-q_s) = \sqrt{\frac{2\,k\mathcal{D}}{(n+1)\lambda^2}s_0^{(n+1)/2}} \tag{14}$$

The rate expression (14) has a distinctly different behavior from that of the reaction with no diffusion limitation, namely, the reaction order is changed for large values of k or for a large pellet dimension λ. Only the first-order reaction ($n = 1$) remains a first-order reaction (with unchanged k value). The zero-order reaction becomes of order $n = ½$ in the diffusion-limited region, and the reaction of order 2 becomes of order 3/2 (with a rate constant that is different from the one that applies in the kinetic controlled region).

It is worth noting that a reaction that was first studied with a low loading of enzyme can perform very differently for an "improved" enzyme with higher activity. For example, it may happen that the observed rate $(-q_s)$ in the diffusion-limited region increases with increasing medium concentration of S.

In Figure 7.15b, an example of $\Phi_0 > \sqrt{6}$ is shown. $y = (dy/d\xi) = 0$ for $\xi = \xi_0 = 0.387$.

For reaction rate expressions more complicated than Eq. (7.38), for example, $(-q_s) = \frac{ks}{K_m + (1 + (s/K_{eq1}))s}$, Eq. (7.19) with $s_1 = s$, one has to solve the

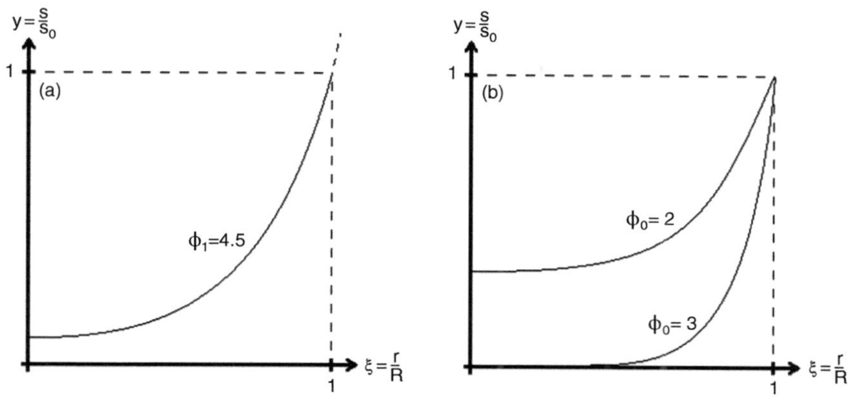

Figure 7.15 Spherical symmetry. Profiles for first-order kinetics (a) and zero order kinetics (b).

boundary value problem by numerical integration. Note that $(-q_s)$ in Eq. (7.19) has a maximum for a certain value of s. For $s < s_{max}$, $(-q_s)$ will increase with *decreasing* s, and the effectiveness factor increases with Φ for small λ when the reaction has an apparent *negative* reaction order.

The *orthogonal collocation method* is a handy numerical method that today is available as easy-to-use computer software. The method was first proposed by Villadsen and Stewart [28] and further developed by Villadsen and Michelsen [29]. Many subsequent researchers have applied the method for all sorts of catalytic reactions.

Example 7.4 Suspended enzymes or immobilized enzymes?

The user of enzymes will be interested in retaining the enzyme in the reactor, while the enzyme producer will sell more enzyme if a suspended enzyme is washed out with the product and lost in the purification process. Diffusion limitation determines which choice is the best for the customer.

Nordkvist *et al.* [30] studied conversion of lactose to lactobionic acid. Lactobionic acid is able to chelate large amounts of Ca^{+2}, whereby precipitation of Ca soaps is avoided even when washing in water with a high concentration of Ca^{+2} ("hard water"). Detergent enzymes may, therefore, contain lactobionic acid instead of polyphosphates, also a chelating compound, but one which gives rise to environmental problems.

A Novozymes carbohydrate oxidase has a very low K_m for lactose ($23\ mg\ l^{-1}$) and virtually all lactose can be converted at maximum specific rate μ_{max}. The limiting reactant is oxygen, where K_m is close to

$s^*_{O2} = 1.0$ mM, corresponding to 1 bar O_2 pressure at the reaction temperature 38 °C. In a measurement at 38 °C, it was found that for a given concentration of freely suspended enzyme in the medium, the reaction was first order in O_2 and that 38 mmole O_2 $(l\ h^{-1})$ is consumed. The oxygen concentration in the liquid is $s_0 = 0.1\, s^*_{O2}$. This gives a first-order rate $k = 38/0.1 = 380\ h^{-1} = 0.105\ s^{-1}$.

The molecular diffusitivity of O_2 in water is 2.42 $10^{-9}\ m^2\ s^{-1}$ at 38 °C.

Assume that $[\varepsilon, \tau] = [0.3, 4]$ for the porous pellets (see Eq. (7.39)). This gives $\mathcal{D} = 1.9\ 10^{-10}\ m^2\ s^{-1}$.

$\Phi_1^2 = \frac{R^2 \cdot 0.105}{1.9\ 10^{-10}} = 552$ when spherical pellets of radius $R = 1$ mm are used.

Clearly, the effectiveness factor is in the asymptotic region $\eta \approx 3/\sqrt{552} = 0.127$.

It is questionable if one would accept a reduction of $(-q_s)$ by a factor 8 to "save" the enzyme from being washed out of the reactor with the product.

When cells that are immobilized in a pellet or in a biofilm, it becomes more difficult to simulate the substrate profile. Substrate is used to create more cells, and since $(-q_s) = (-r_s)\ x$, the volumetric rate increases with time if the cell concentration in the film increases. Frequently, the cells will also produce metabolites in the form of polysaccharides and x may even decrease in time while the biofim thickness increases.

The transport of substrate into the biofilm is, however, much faster than the increase of cell concentration in a film of constant thickness or than the increase of film thickness. When the substrate is glucose, the molecular diffusivity is $\mathcal{D} = 0.6\ 10^{-9}\ m^2\ s^{-1}$. The rate at which S is transported through a surface of area A is $A\ \mathcal{D} s_0\ l^{-1}$ moles $(m^2\ s)^{-1}$. It is consumed by a chemical reaction in the volume $A \cdot L\ m^3$. Let the reaction be first order in S and the consumption rate is $A \cdot L \cdot ks \cdot x$. The ratio between the two rates is:

$$\frac{\text{Rate of transport}}{\text{Rate of reaction}} = \mathcal{D}\frac{1}{kL^2}\frac{s_0}{sx} = \frac{s_0}{sx\,\Phi_1^2}.$$

For a film of thickness $L = 1$ mm and a volumetric rate of biomass formation $q_x = 0.2\ x\ h^{-1}$ one obtains $\Phi_1^2 = 10.8$, and unless x is very high, the film is practically saturated with substrate at $s = s_0$. Very slowly x increases and diffusion limitation sets in. This also happens when x remains constant while L increases due to conversion of substrate by maintenance to more biofilm.

Consequently, the model for a biofilm involves a pseudo-steady-state consumption of substrate overlaid with a slow transient change in the film thickness or in the reaction rate. Models of this kind have been treated in chemical reaction engineering as "increasing/shrinking core models." They usually require numerical solution of the differential equation model, but guidelines of the structure of the solution were given by Levenspiel (1972 and later editions, for example, [31]).

7.7 Sequences of Enzymatic Reactions: Optimal Allocation of Enzyme Levels

Metabolic pathways are characterized as often long sequences of consecutive reactions, sometimes with branch points to other pathways. A key objective of metabolic engineering is to create an organism where the flux J of carbon source S toward the desired end product P is maximized. This is seldom possible by increasing the enzyme levels of all the enzymes in the pathway. The reason is twofold: (1) a high level of the enzyme that catalyzes the reaction from internal metabolite M_i to internal metabolite M_{i+1} increases the concentration in the pool of M_{i+1}, and this could give rise to product inhibition of the reaction $M_i \rightarrow M_{i+1}$, or even worse a feed-forward inhibition of the reaction $M_{i+k} \rightarrow M_{i+k+1}$. (2) Increasing the rate of $M_i \rightarrow M_{i+1}$ may lead to loss of S due to enhancement of a side reaction at a branch point in the reaction sequence.

This is the reason why the *structure* of the metabolic network has become a subject for research. The name given to this discipline is metabolic control analysis, and it requires both a considerable level of mathematical insight and clever computer coding to analyze complicated networks. A large number of papers from the past 15 years are devoted to the topic. A good introductory text is Stephanopoulos *et al.* [32], and the journals *Metabolic Engineering* and *Biotechnology & Bioengineering* are key sources for the numerical tools and computer programs needed to solve the mathematical models.

Chapter 8 of this book will further discuss how the *sensitivity* of the flux through a network is influenced by the *structure* of the network.

As an introduction to Chapter 8, the concepts of metabolic control analysis are presented below. Simple, but instructive examples dealing only with the sequence $S \rightarrow S_1 \rightarrow P$ will be used – the so-called "toy examples," an unnecessesarily derogative term. The rate expressions of the individual pathway reactions are chosen from those discussed in Sections 7.1 and 7.2. With only one intermediate metabolite S_1 between the substrate S and the product P, the mathematics involved in calculating the sensitivities is manageable, and the key objectives and methods of MFA are highlighted.

The primary goal of the analysis is to calculate the allocation of ressources (here the level of enzyme) to each of the two reactions between S and P. *Only the steady-state flow of carbon between S and P is considered.* Also, the *concentration levels* [s, p] *of S and of P are fixed.* The concentration s_1 of the internal metabolite is a dependent variable, while the thermodynamic constants such as the Michaelis–Menten constant K_m, and especially the rate constants k_1 and k_2 of the two reactions, are independent (adjustable) variables. The concentrations $[s, p]$ of S and of P are as mentioned above *fixed*, but the values of s and p can influence the flux J through the pathway. The influence of the *effectors* s and p, on the optimal structure of the pathway is studied separately. Finally, note that although we consider only the steady-state flow between S and P (possibly to $P_1 \ldots P_N$ in pathways with branches), the sensitivities are often found by *transient* experiments in a stirred tank reactor, as will be shown in Chapter 8.

Example 7.5 Analysis of the structure of a linear pathway $S \rightarrow S_1 \rightarrow P$ for given values of [s, p]

$$\textbf{A1}.\ \text{The rate of } S \rightarrow S_1 \text{ is} (-r_1) = \frac{k_1 s}{K_1 + s} \text{ and that of } S_1 \rightarrow P \text{ is} (-r_2) = \frac{k_2 s_1}{K_2 + s_1}. \tag{1}$$

Typical units of $[-r_1, -r_2]$ are, for example, C-mol S or S_1 converted (g enzyme preparation h)$^{-1}$; $[k_1, k_2]$: h^{-1} (or more precisely, C-mol S converted (g enzyme preparation h)$^{-1}$; $[K_1, K_2]$: C-mol l^{-1}. Volumetric rates are obtained when the expressions (1) are multiplied by e_i in units of (g enzyme preparation)(l medium)$^{-1}$.

The *steady-state* dimensionless flux $J' = (J/k_2)$ through the pathway is $\frac{(k_1/k_2)\,s}{K_1+s} = \frac{xs}{K_1+s} = \frac{s_1}{K_2+s_1}$. Since s is a constant, the parameters of r_2 have no influence on J' – the steady-state flux is determined by how fast carbon passes through $S \rightarrow S_1$.

Next, the sensitivity of J on changes of the two structural parameters in r_1 and on s is determined. Partial derivatives are used, since the influence of each of the three variables is calculated at constant values of the others. Since the units of J and of the parameters are different, *dimensionless* sensitivities are calculated.

$$\frac{\partial J}{\partial k_1}\frac{k_1}{J} = 1; \quad \frac{\partial J}{\partial K_1}\frac{K_1}{J} = -\frac{K_1}{K_1 + s}; \quad \frac{\partial J}{\partial s}\frac{s}{J} = \frac{K_1}{K_1 + s}. \tag{2}$$

When k_1 is increased by a factor f, the flux J through the pathway increases by the same factor.

The important result of the calculation (2) is 1. The *control* of the flux lies entirely in r_1, and an increase of the rate constant k_1 of the first reaction leads to a proportional increase of J. When Eq. (1) is multiplied by e_1 [g enzyme (or units of enzyme) per l medium] one obtains $(-q_s)_1 =$ the volumetric rate of conversion of S. An increase of k_1 in Eq. (1) by a factor f is supposed to correspond to an increase of the enzyme loading by the same factor f. Hence, the first result of Eq. (2) can also be interpreted as follows:

$$\frac{\partial J}{\partial e_1}\frac{e_1}{J}, \text{ the sensitivity of J to a small change of the concentration of } E_1 \text{ from a set point } e_1^0 \tag{3}$$

The quantity $\frac{\partial J}{\partial e_1}\frac{e_1}{J}$ is defined as the *flux control coefficient* C_1^J *of* r_1. $C_1^J = 1$ and $C_2^J = 0$ for the consecutive Michaelis–Menten reactions (1) at *any* level of the enzyme concentrations e_i.

Increasing K_1 from K_1^0 to a value slightly larger than K_1^0 *decreases* J by the factor $\frac{K_1^0}{K_1^0+s}$, that is, not by very much when K_1^0 is large or when s is large. Increasing s from s^0 to s^1 *increases* J by the factor $\frac{K_1}{K_1+s^0}$, again not by much if K_1 is large. All the results are intuitively reasonable.

A2. A complete study of the pathway structure also includes calculation of the concentration s_1 of the intermediate S_1 to find, for example, how the level of S influences s_1 and how changes in the structural parameters of $[r_1, r_2]$ will change s_1. One obtains:

$$J = \frac{k_1 s}{K_1 + s} = \frac{k_2 s_1}{K_2 + s_1} \rightarrow s_1 = \frac{s}{\frac{1}{K_2}\left(\frac{k_2}{k_1} - 1\right)s + \frac{k_2}{k_1} \cdot \frac{K_1}{K_2}}. \tag{4}$$

For $(k_2/k_1) \rightarrow 0$ the intermediate metabolite has the concentration $s_1 = \frac{K_2 s}{(k_2/k_1) \cdot K_1 - s}$. s_1 can be much higher than s, and no acceptable solution is obtained when $s \geq (k_2/k_1)K_1$. A large value of K_1, that is, a poor affinity of E_1 for S, and a small K_2, that is, a high affinity of E_2 for S_1, will decrease s_1.

These considerations are, as stated above, of secondary importance for MFA since the flux control coefficients C_i^J are the all important quantities. They do, however, contribute to an understanding of consecutive enzymatic reactions, whether they take place in a steady state CSTR or in a metabolic pathway.

B. The rate of $S \rightarrow S_1$ is $(-r_1)$

$$= \frac{k_1 s}{K_1(1 + (s_1/K_{\text{eq}})) + s}, \quad \text{and that of} \tag{5}$$

$$S_1 \rightarrow P \text{ is} (-r_2) = \frac{k_2 s_1}{K_2 + s_1}.$$

Clearly, the more complex form of $(-r_1)$ – the expression is that for competitive inhibition, Eq. (7.17) – will make an analytic calculation of the flux control coefficients C_i^J more difficult since both k_1 and s_1 appear in $(-r_1)$. For fixed values of $[K_1, K_2, K_{\text{eq}}]$, the value of s_1 depends on $[k_1, k_2]$ through the relation $J = (-r_1) = (-r_2)$:

$$k_1 = \frac{k_2(s_1/s)}{K_2 + s_1}\left(K_1\left(1 + \frac{s_1}{K_{eq}}\right) + s\right). \tag{6}$$

An analytical calculation (supported by numerical simulation) of $C_1^J = \frac{\partial J}{\partial k_1}\frac{k_1}{J} = \frac{\partial J}{\partial e_1}\frac{e_1}{J}$ will be shown below.

For small values of k_1 (or e_1), the concentration of S_1 will not be high unless k_2 is very small. Consequently, C_i^J is 1 as in (A) when $k_1 \rightarrow 0$. s_1 will, however, increase with k_1 for a fixed k_2, and "the return on investment," that is, the increase of $(-r_1)$ with increasing k_1 decreases. Consequently, J becomes more and more dependent on the ability of $S_1 \rightarrow P$ to remove S_1. One concludes that eventually the flux control will switch from $(-r_1)$ to $(-r_2)$, that is, k_2 should be increased. When the feed concentration s is set at a higher value, s_1 becomes larger and although the flux J through the pathway increases, a shift of control from $S \rightarrow S_1$ to $S_1 \rightarrow P$ occurs at a smaller value of k_1. The sensitivity of J on the effector S is positive: $\frac{\partial J}{\partial s}\frac{s}{J} > 0$.

The derivation of expressions for C_i^J will now be shown, but in order to reduce the number of parameters dimensionless quantities are introduced:

$$y = \frac{s_1}{s}; \quad a = \frac{K_{\text{eq}}}{s}; \quad b = \frac{K_1}{s}; \quad c = \frac{K_2}{s} \rightarrow \quad (-r_1) = J = \frac{k_1}{1 + b(1 + (y/a))} \quad \text{and} \quad (-r_2) = J = \frac{k_2 y}{y + c}. \tag{7}$$

The value of y as a function of k_1 for a fixed value of k_2 can be found by solution of the algebraic equation:

$$F = F(k_1, k_2) = 0 = k_1 - \frac{k_2 y}{y + c}\left(1 + b\left(1 + \frac{y}{a}\right)\right) \text{or } x = \frac{k_1}{k_2} = \frac{y}{y + c} \cdot \left(1 + b\left(1 + \frac{y}{a}\right)\right). \tag{8}$$

In the following $y(k_1)$ for fixed, k_2 will, however, be found as the solution of the differential equation (13).

The value of C_i^J is found by using the "chain rule for differentiation:" $C_i^J = \frac{\partial J}{\partial k_i}\frac{k_1}{J} = \frac{\partial J}{\partial y}\frac{\partial y}{\partial k_i}\frac{k_i}{J}\frac{\partial y}{\partial k_i}$ is calculated from (8), that is, from: $F = F(k_1, k_2) = 0 = k_1 - \frac{k_2 y}{y+c} \cdot (1 + b(1 + \frac{y}{a}))$:

$$0 = \frac{\partial F}{\partial k_1} = 1 - k_2\left\{\frac{c}{(y + c)^2}\left(1 + b\left(1 + \frac{y}{a}\right)\right) + \frac{b}{a}\frac{y}{y + c}\right\}\frac{\partial y}{\text{d}k_1} \rightarrow$$

$$\frac{\partial y}{\text{d}k_1} = \frac{\frac{1}{k_2}(y + c)}{\frac{c}{y+c}\left(1 + b\left(1 + \frac{y}{a}\right)\right) + \frac{b}{a}y} = \frac{\frac{1}{k_2}(y + c)}{c\frac{k_1}{k_2 y} + \frac{b}{a}y} \tag{9}$$

and similarly:

$$\frac{\partial y}{\text{d}k_2} = -\frac{\frac{1}{k_2}(y + c)}{c\frac{k_1}{k_2 y} + \frac{b}{a}y}. \tag{10}$$

$$C_1^J = \frac{k_2 c}{(y + c)^2}\frac{k_1(c + y)}{k_2 y}\frac{\frac{1}{k_2}(y + c)}{c\frac{k_1}{k_2 y} + \frac{b}{a}y} = \frac{c k_1}{k_2 y\left(c\frac{k_1}{k_{2y}} + \frac{b}{a}y\right)} = \frac{1}{1 + \frac{b}{a}\frac{k_2}{c}\frac{k_2}{k_1}y^2}. \tag{11}$$

$$C_2^J = \left(\frac{y}{y + c} - \frac{c}{(y + c)^2}\frac{\frac{1}{k_2}(y + c)}{c\frac{k_1}{k_{2y}} + \frac{b}{a}y}\right)\frac{y + c}{y} = \frac{\frac{b}{a}\frac{k_2}{c}\frac{k_2}{k_1}y^2}{1 + \frac{b}{a}\frac{k_2}{c}\frac{k_2}{k_1}y^2} = 1 - C_1^J. \tag{12}$$

From Eqs. (9) and (10) one obtains: $\text{d}y = \frac{\partial y}{\text{d}k_1}\text{d}k_1 + \frac{\partial y}{\text{d}k_2}\text{d}k_2$, and for constant value of k_2 a differential equation is obtained from which $y(k_i)$ is found as a function of k_i:

$$\frac{\text{d}y}{\text{d}k_1} = \frac{\frac{1}{k_2}(y + c)}{c\frac{k_1}{k_2 y} + \frac{b}{a}y} \quad \text{with} \quad y(k_1 = 0) = 0. \tag{13}$$

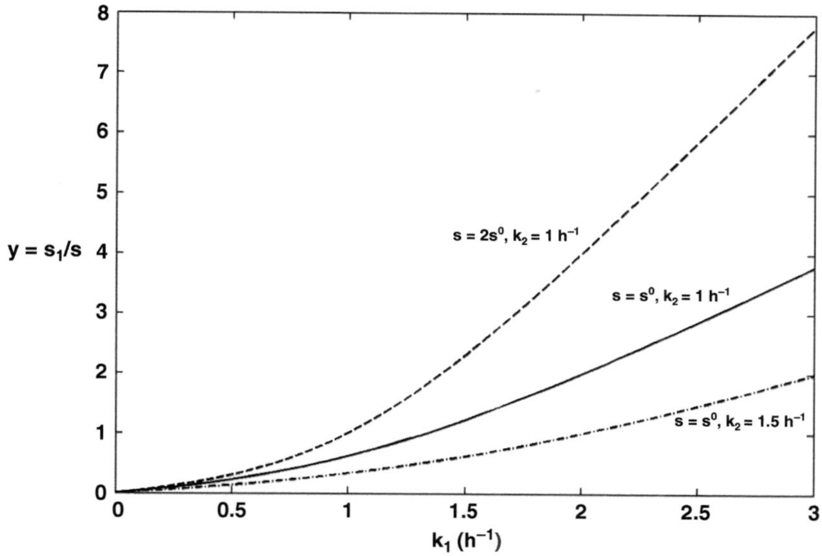

Figure 7.16 $y = s_1/s$ for feed concentration $s = s^0$ and $s^1 = 2\,s^0$ as a function of k_1 for $k_2 = 1\,h^{-1}$ and $[a, b, c] = [1, ½, ½]$. Also shown is y as a function of k_1 for $s = s^0$ and $k_2 = 1.5\,h^{-1}$. The concentration of S_1 increases more slowly with k_1 for $k_2 = 1.5\,h^{-1}$ since the second enzyme removes S_1 more rapidly.

Since $y = 0$ at $k_1 = 0$, one cannot start exactly at $k_1 = 0$. For $y \to 0$, the expression (8) can, however, be approximated by $k_1 = k_2 y \frac{1+b}{c}$, that is, $y = \frac{c}{k_2(1+b)} k_1$. Thus, with the choice of parameters $a = 1, \quad b = c = 1/2$ in (7), $y = (2/3\,k_2)k_1$, an excellent approximation of y for $0 < k_1 < 0.0001$ is obtained. The parameters $[a, b, c]$ have the same values in all three Figures 7.16–7.18.

Once $y(k_1)$ is determined for a fixed k_2, the values of $J = (k_2\,y/y + c)$ (or of the dimensionless $J'(k_1) = (y/y + c)$) and of C_1^J given by (11) can easily be found. $J'(k_1)$ is the most practical to depict since it varies between 0 and 1.

In Figure 7.16, the solution of $(dy/dk_1) = f(k_1)$ is shown for two values of $k_2 = [1, 1.5]\ (h^{-1})$ and $[a, b, c] = [1, ½, ½]$. $y(k_1)$ is also shown for $k_2 = 1\,h^{-1}$ and a feed concentration $s = s^1 = 2\,s^0$. Figure 7.17 shows $J\prime(k_1) = (y/y + c)$ for two values of $k_1 = [1, 1.5]$ and for $s = s^1 = 2\,s^0$.

In Figure 7.18, $C_1^J(k_1)$ is shown for the two values of k_2 and for $s = s^1 = 2\,s^0$. Note that due to equation (8) y is only a function of $x = (k_1/k_2)$. Thus, the same value of y $(= 0.618)$ is obtained for $[k_1, k_2] = [1, 1]$ and $[1.5, 1.5]$, while $y = 2$ for $[k_1, k_2] = [2, 1]$ and $[3, 1.5]$.

The sensitivity of J to a change in the inlet concentration s from s^0 to s^1 is calculated as follows:

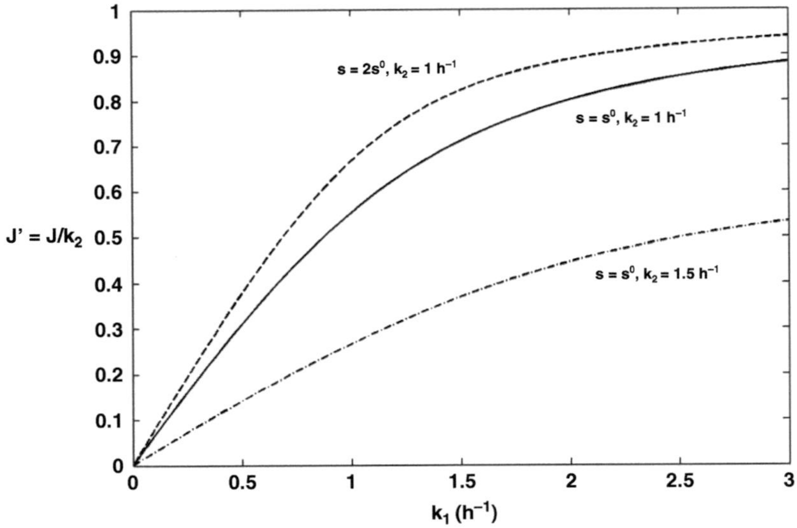

Figure 7.17 Dimensionless flux $J'(k_1) = (J/k_2) = (y/y + c)$ for the three cases shown in Figure 7.16.

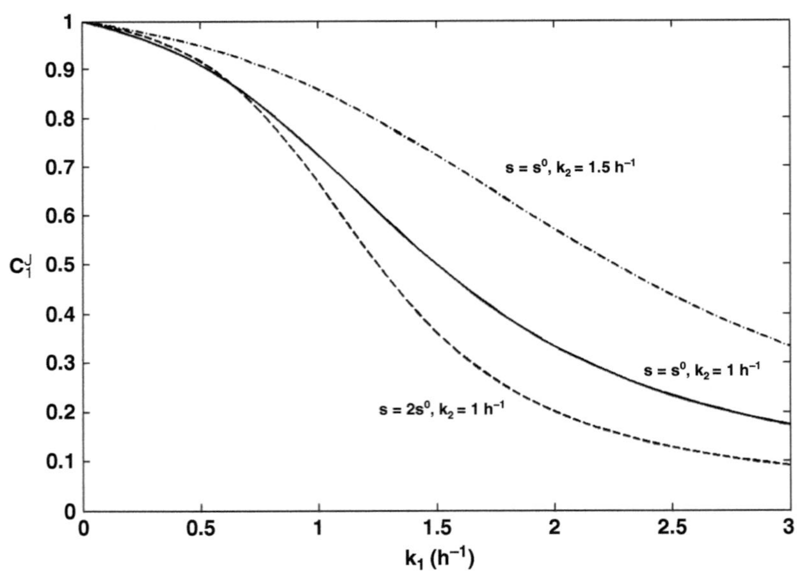

Figure 7.18 The flux control coefficient C_1^J as a function of k_1. $k_2 = 1\ h^{-1}$ and $1.5\ h^{-1}$. $[a, b, c] = [1, ½, ½]$. When s^0 increases to 2 times the set-point value s^0 S_1 accumulates to a higher concentration and the control switches to the second reaction at a smaller k_1 value. Also, at the higher value of $k_2 = 1.5\ h^{-1}$, C_1^J decreases more slowly with k_1. According to Eq. (11) C_1^J has the same value = 0.723 for $[k_1, k_2] = [1,1]$ and $[1.5, 1.5]$, and 1/3 for $[k_1, k_2] = [2, 1]$ and $[3, 1.5]$.

Define y, a, b, and c as before: $y = \frac{s_1}{s^0}$; $a = \frac{K_{eq}}{s^0}$; $b = \frac{K_1}{s^0}$; $c = \frac{K_2}{s^0}$, and add the definition $s' = \frac{s^1}{s^0}$.

Equation (8) now reads: $(-r_1) = \dfrac{k_1 s'}{s' + b\left(1 + \frac{y}{a}\right)} = \dfrac{k_2 y}{y + c} \rightarrow k_1 = \dfrac{k_2 y\left(s' + b\left(1 + \frac{y}{a}\right)\right)}{s'(y + c)} = \dfrac{k_2 y\left(1 + \frac{b}{s'}\left(1 + \frac{y}{a}\right)\right)}{(y + c)}$

When the calculations (8)–(12) are repeated with $b' = b\,(s^0/s^1)$, one obtains the results for a feed concentration of s^1. For $s^1 = 2\ s^0$ and $[k_2, a, b, c] = [1\ (\mathrm{h}^{-1}), 1, ½, ½]$. $b\prime = \frac{1}{4}$ and consequently, integration of $\frac{dy}{dk_1} = \dfrac{(1/k_2)(y + c)}{c(k_1/k_2 y) + (b'/a)\, y}$ with $k_2 = 1\,\mathrm{h}^{-1}$ starts with $y \approx y = \frac{c}{k_2(1+b')} k_1 = \frac{2}{5} k_1$.

The analysis in Example 7.5 has demonstrated the following main features of MCA:

1) The steady-state flux J through a linear pathway is equal to any of the rates r_i of the individual reactions that convert S to P in a steady-state process. The concentrations of S and P are fixed quantities.
2) The steady-state concentration s_i of any of the internal metabolites S_i can be found as a function of the concentration of S by solution of $N+1$ (often nonlinear) algebraic equations that express equality of the $N+1$ internal reaction rates r_i. The solution s_i, $i = 1, 2 \ldots N$ will be functions of the feed concentration s and of the values of the thermodynamic and kinetic parameters.
3) The sensitivity of the flux to a change in any of the enzyme loadings in the $N+1$ reactions between S and P is determined in dimensionless form as $C_i^J \equiv \frac{\partial J}{\partial e_i}\frac{e_i}{J}$. Assuming that the rate constant k_i is proportional to e_i, the definition of C_i^J can be rewritten as $C_i^J \equiv \frac{\partial J}{\partial k_i}\frac{k_i}{J}$, a formula that is useful when the rate expressions of r_i are known.

   The example has shown that $C_1^J = 1 - C_2^J$ for the rate expressions r_i of the example. This can be proved to be true for any *linear* pathway $S \rightarrow S_1 \rightarrow S_2 \rightarrow \ldots S_N \rightarrow P$ and the result is as follows:

$$\text{The } \textit{flux control summation} \text{ theorem}: \sum_1^{N+1} C_i^J = 1 \text{ and all } C_i^J \text{ are } \geq 0. \tag{7.41}$$

The *optimal* allocation of enzyme dose to r_i is that for which all control coefficients C_i^J are equal, that is, $C_i^J = (1/N + 1)$. In Example 7.5, e_1 (or k_1) should be increased until C_i^J has decreased to 0.5. This happens for $k_1 \approx 1.55\,\mathrm{h}^{-1}$ for $s = s^0$, $k_2 = 1\,\mathrm{h}^{-1}$, but for a lower value $k_1 \approx 1.30\,\mathrm{h}^{-1}$ at the higher feed concentration $s = 2\ s^0$, where the concentration of S_1 is higher for any k_1, and this would require a higher k_2 to remove S_1. For $k_2 = 1\ (\mathrm{h}^{-1})$, $s = s^0$, the best enzyme allocation policy would be to increase k_1 and k_2 in the ratio 1.55 for $k_1 > 1.55$. For $k_2 = 1.5\,\mathrm{h}^{-1}$, k_1 has to be increased to a value higher than $1.55\,\mathrm{h}^{-1}$ before the point of equal control of the two reactions is reached.

4) The flux control coefficients are *local* quantities. They depend both on the state $[e_j, j=1, 2, \ldots N]$ selected for all the other enzyme levels and on the *given* values of the *effectors* – in Example 7.5, the feed concentration s. The example shows that *when the rate expressions r_i are known*, it is possible to calculate C_i^J and consequently to calculate the set $[e_{i,}\ i=1, 2 \ldots N+1]$ from which a proportional allocation of e_i should be started. If the rate expressions are unknown, one has to conduct a large experimental investigation in which *each* enzyme level is perturbed at given levels for the other enzyme levels as well as a given level of the effectors. Each experiment involves genetic engineering of the organism, *and the result holds only in the vicinity of the set point values for all the other variables.* Hence, experimental determination of C_i^J is a frustrating experience. Large steps in the enzyme levels have to be made in order to get a reasonable accuracy when the control coefficients are calculated from experiments corrupted by experimental inaccuracies, but then the local character of the determination of C_i^J is invalidated.

Small and Kacser [33] devised an approximate solution to this dilemma, which is discussed next.

When $J'(x) = \frac{J}{k_2}$ can be approximated by $J'(x) = \frac{x}{x+K} = \frac{(k_1/k_2)}{(k_1/k_2)+K}$, that is, a downward concave function that starts at $J'(0)=0$ and approaches $J'(\infty)=1$ for large x, one may determine $(C_i^J)^0$, the control coefficient at a given set point for the variables, *also when a large step in x is used*:

$$(C_i^J)^0 = \left(\frac{\partial J'}{\partial x}\frac{x}{J'}\right)^0 = \frac{K}{(x^0+K)^2}\frac{x^0}{J'(x^0)} = KJ'(x^0)/x^0 = \frac{x}{J'}\frac{J'-J'(x^0)}{x-x^0}$$
$$= \frac{x}{J'}\frac{K(x-\ x^0)}{(x-\ x^0)(x+K)(x^0+K)} = KJ'(x^0)/x^0.$$

First $(C_i^J)^0$ is calculated exactly, and next by the slope of the chord between $[x^0, x]$ and $[J'(x^0), J'(x)]$. The two values are the same, and $(C_i^J)^0$ can be calculated by taking a large step in x $(= k_1/k_2)$ and measuring $J'(x)$.

In Example 7.5, the control coefficient C_1^J at $[k_1, k_2]^0 = [1, 1]$ can be found from $[J(1.5), 1.5]$, $[J(1), 1]$, which are read from Figure 7.17 to be $\approx[0.71, 1.5]$, $[0.55, 1]$. $C_1^J(k_1 = 1) \approx \frac{1.5}{0.71}\cdot\frac{0.71-0.55}{1.5-1} = 0.68$. From the simulations in Figure 7.15, one obtains $C_1^J(k_1 = 1) = 0.72$. The step Δk_1 from 1 to 1.5 is large enough to obtain J values that are sufficiently apart to give a sufficient accuracy in the calculation of $C_1^J(k_1 = 1)$.

Although the approximate method described above permits a reasonably accurate value of the control coefficients to be obtained experimentally, the number of experiments needed to cover the full range of set points $k_{1,}$ – to be repeated for other values of k_2 and for different values of s – is daunting. One will have to conclude that experimental determination of C_i^J by the use of a series of strains with different enzyme loading is impractical. The solution is to use the structure of r_i that is analyzed in terms of a series of *elasticity coefficients* ε_{ji}. These are defined as the (dimensionless) sensitivity of the ith reaction rate r_i to the concentration s_j of the jth internal metabolite S_j:

$$\varepsilon_{ji} = \frac{\partial r_i}{\partial s_j}\frac{s_j}{r_i}, \quad i = 1,2\ldots N+1 \quad \text{and} \quad j = 1,2\ldots N. \tag{7.42}$$

The flux control coefficients and the elasticities are connected by the flux connectivity theorem:

$$\sum_{1}^{N+1}(C_i^J \varepsilon_{ji}) = 0, j = 1,2,\ldots N. \tag{7.43}$$

Eqs. (7.41) and (7.43) can be connected in the following linear matrix equation:

$$\begin{pmatrix} 1 & 1 & 1 & 1 & 1 \\ \varepsilon_{11} & \varepsilon_{12} & \varepsilon_{13} & .. & \varepsilon_{1,N+1} \\ \varepsilon_{21} & \varepsilon_{22} & .. & .. & \varepsilon_{2,N+1} \\ .. & .. & .. & .. & .. \\ \varepsilon_{N,1} & \varepsilon_{N,2} & .. & .. & \varepsilon_{N,N+1} \end{pmatrix} \cdot \begin{pmatrix} C_1^J \\ C_2^J \\ \ldots \\ \ldots \\ C_{N+1}^J \end{pmatrix} = \begin{pmatrix} 1 \\ 0 \\ 0 \\ .. \\ 0 \end{pmatrix}. \tag{7.44}$$

Consequently, if the $(N \times N+1)$ matrix $\mathbf{E}$ of elasticities is known, the $N+1$ flux control coefficients can be calculated by solution of the $N+1$ linear algebraic equations symbolized in (7.44).

In Example 7.5, there are two elasticity coefficients:

$$(\varepsilon_{11} \quad \varepsilon_{12}) = \left(-\frac{y\,\frac{b}{a}}{b\left(1+\frac{y}{a}\right)+1}, \frac{c}{y+c}\right):$$

$$\begin{pmatrix} 1 & 1 \\ \varepsilon_{11} & \varepsilon_{12} \end{pmatrix} \cdot \begin{pmatrix} C_1^J \\ C_2^J \end{pmatrix} = \begin{pmatrix} 1 \\ 0 \end{pmatrix} \rightarrow \begin{pmatrix} C_1^J \\ C_2^J \end{pmatrix} = \frac{1}{\varepsilon_{12}-\varepsilon_{11}} \begin{pmatrix} \varepsilon_{12} \\ -\varepsilon_{11} \end{pmatrix} = \begin{pmatrix} 1/(1+\frac{by^2}{acx^{-1}}) \\ 1-C_1^J \end{pmatrix},$$

where $x = \frac{k_1}{k_2}$. The result is the same as that obtained in (11) of the example.

The kinetics of r_i is, of course, not known in advance, but in Chapter 8 a general method of approximating the rate expressions (the so-called lin-log approximation that for Example 7.5 would read: $r_1/k_2 \approx \alpha_1 + \beta_1 \ln y + \gamma_1 \ln s'$ and $r_2/k_2 \approx \alpha_2 + \beta_2 \ln y$, where the five parameters are fitted to measurements of the two rates) is introduced. With this structure of the approximation for r_i, the calculation of ε_{ji} and thereafter C_i^J by Eq. (7.44), as well as the sensitivity coefficients related to effectors such as s, can easily be done.

Before ending this introduction to metabolic control analysis, it must be emphasized that only a very sketchy picture has been drawn. The crucial theorems (7.41) and (7.43) are proved in the texts mentioned earlier, for example, Stephanopoulos *et al.* [32].

Real MCA problems are certainly not constrained to linear pathways, the only examples treated here. In branched pathways – which may be much more important to analyze since the yield of the desired metabolic product depends on the architecture around the branch point – the flux control summation theorem (7.41) still holds, but the C_i^J are not all positive. Heijnen *et al.* [34] illustrates how the flux control coefficients C_i^J around a branch point can be obtained from transient measurements and using the lin-log approximation method of the rates. It is especially important to utilize all the constraints that apply for the C_i^J around a branch point (besides their summation to 1), as illustrated by the simple examples in the paper that can be followed without resorting to computer calculation.

This chapter does, albeit in an introductory form, demonstrate how the architecture of the pathway reactions can be modified to obtain a high carbon flux, especially toward the desired metabolic product. Together with the derivation in Chapter 8 of flux control coefficients by approximation of the rates r_i obtained by transient experiments (e.g., after adding a pulse of substrate to a steady-state CSTR), one will obtain a tool-box to analyze metabolic networks with the ulitimate pupose of improving an industrial bioprocess.

References

1 Michaelis, L. and Menten, M. (1913) Die kinetik der invertinwirkung. *Biochem. Z.*, **49**, 333–369.
2 Langmuir, I. (1918) The constitution and fundamental properties of solid and liquid surfaces, 1. Solids. *J. Am. Chem. Soc.*, **38**, 142–153.
3 Briggs, G.E. and Haldane, J.B.S. (1925) A note on the kinetics of enzyme action. *Biochem. J.*, **19**, 339–349.
4 Monod, J. (1942) *Recherches Sur la Croissance de Cultures Bacteriennes. (Diss Dr Sci)*, Hermann et Cie, Paris.
5 Villadsen, J., Nielsen, J., and Lidén, G. (2011) *Bioreaction Engineering Principles*, 3rd edn, John-Wiley & Sons, Inc, New York.
6 Cornish-Bowden, A. (1995) *Fundamentals of Enzyme Kinetics*, Portland Press, London, UK.
7 Fell, D. (1997) *Understanding the Control of Metabolism*, Portland Press, London, UK.
8 Herbert, D. (1959) Some principles of continuous culture. *Recent Prog. Microbiol.*, **7**, 381–396.
9 Linewaever, H. and Burk, D. (1934) The determination of enzyme dissociation constants. *J. Am. Chem. Soc.*, **56**, 658–666.
10 Ahn, W.S. and Antoniewicz, M.R. (2011) Metabolic flux analysis of CHO cells at growth and non-growth phases using isotopic tracers. *Metabolic. Eng.*, **13**, 598–609.
11 Ruy, D.D.Y., Lee, S.O., and Romani, R. (1990) Determination of the growth rate for plant cells. *Biotechnol. Bioeng.*, **35**, 305–311.
12 Ramkrishna, D., Fredrickson, A.G., and Tsuchiya, H.M. (1967) Dynamics of microbial propagation. Models considering inhibitors and variable cell composition. *Biotechnol. Bioeng.*, **9**, 129–170.
13 Williams, F.M. (1967) A model of cell growth dynamics. *J. Theoretical Biology*, **15**, 190–207.
14 Fredrickson, A.G. (1976) Formulation of structured models in bioengineering. *Biotechnol. Bioeng.*, **18**, 1481–1486.
15 Roels, J.A. and Kossen, N.W. (1978) On the modeling of microbial metabolism. *Prog. Ind. Microbiol.*, **14**, 95–204.
16 Nielsen, J., Nikolajsen, K., and Villadsen, J. (1991) Structured modeling of a microbial system: a theoretical study of the lactic acid fermentation. *Biotechnol. Bioeng.*, **38**, 1–10.
17 Nielsen, J., Nikolajsen, K., and Villadsen, J. (1991b) Structured modeling of a microbial system: verification of a structured lactic acid fermentation model. *Biotechnol. Bioeng.*, **38**, 11–23.
18 Benthin, S., Nielsen, J., and Villadsen, J. (1991) A simple and reliable method for determination of cellular RNA content. *Biotechnol. Techniques*, **5**, 39–42.

19 Duboc, P., von Stockar, U., and Villadsen, J. (1998) Simple generic model for dynamic experiments with *Saccharomyces cerevisiae* in continuous culture. *Biotechnol. Bioeng.*, **90**, 180–189.

20 Lei, F., Olsson, L., and Jørgensen, S.B. (2003) Experimental investigations of multiple steady states in aerobic continuous cultivations of *Saccharomyces cerevisiae. Biotechnol. Bioeng.*, **82**, 766–777.

21 Villadsen, J. and Jørgensen, S.B. (2014) Reflections on the aerobic fermentation stoichiometry of Crabtree positive yeasts. *Biotechnol. Bioeng.*, **111**, 632–637.

22 Melchiorsen, C.R., Siemsen, N.B., Christensen, B., Jokumsen, K.V., and Villadsen, J. (2001) Dynamics of pyruvate metabolism in *Lactococcus lactis. Biotechnol. Bioeng.*, **74**, 271–279.

23 Gompert, A.K. and Nielsen, J. (2000) Mathematical modeling of metabolism. *Curr. Opin. Biotechnol.*, **11**, 180–186.

24 Ramkrishna, D. (1982) A cybernetic perspective of microbial growth, in *Foundations of Biochemical Engineering: Kinetics and Thermodynamics in Biological Systems* (eds. E. Papoutsakis, G.N. Stephanopoulos, and H.W. Blanch), American Chemical Society, Washington, pp. 161–178.

25 Kompala, D.S., Ramkrishna, D., and Tsao, G.T. (1984) Cybernetic modeling of microbial growth on multiple substrates. *Biotechnol. Bioeng.*, **26**, 1272–1281.

26 Kompala, D.S., Ramkrishna, D., Jansen, N.B., and Tsao, G.T. (1986) Investigation of bacterial growth on mixed substrates: experimental evaluation of cybernetic models. *Biotechnol. Bioeng.*, **28**, 1044–1055.

27 Young, J.D., Henne, K.L., Morgan, J.A., Konopka, A.E., and Ramkrishna, D. (2008) Integrating cybernetic modeling with pathway analysis provides a dynamic, systems-level description of metabolic control. *Biotechnol. Bioeng.*, **100**, 542–559.

28 Villadsen, J. and Stewart, W.E. (1967) Solution of boundary value problems by orthogonal collocation. *Chem. Eng. Sci.*, **22**, 1483–1501.

29 Villadsen, J. and Michelsen, M.L. (1978) *Solution of Differential Equation Models by Polynomial Approximation*, Prentice Hall, Englewood Cliffs, NJ.

30 Nordkvist, M., Nielsen P.M., and Villadsen, J. (2006). Oxidation of lactose to lactobionic acid by a Mitochondrium nivale carbohydrate oxidase, kinetics and operational stability. *Biotechnol. Bioeng.*, **97**, 694–707.

31 Levenspiel, O. (1999) *Chemical Reaction Engineering*, 3rd edn, John-Wiley & Sons, Inc, New York.

32 Stephanopoulos, G., Aristidou, AA., and Nielsen, J. (1998) Metabolic Engineering, Academic Press, San Diego, USA.

33 Small, J.R. and Kacser, H. (1993) Response of metabolic systems to large changes in enzyme activities and effectors. *Eur. J. Biochem.*, **213**, 613–640.

34 Heijnen, J.J., van Gulik, W.M., and Shimizu, H., and Stephanopoulos, G. (2004) Metabolic flux control analysis of branch points: an improved approach to obtain flux control coefficients from large perturbation data. *Metabol. Eng.*, **6**, 391–400.

8
Application of Dynamic Models for Optimal Redesign of Cell Factories

Matthias Reuss

Summary

This chapter is about the systems-oriented analysis of the dynamics of cellular systems. This is an important step toward model-predicted rational design for reshaping cellular systems. Whereas the potential and promise of biological systems modeling is substantial, several obstacles are still encountered when addressing the issue of predictive design based on dynamic models. This chapter describes the development of experimental methods that are essential for the identification of kinetic models. On the basis of experiments, following the strategy of the stimulus–response methodology, two different approaches for identification of the dynamic models are described. At the core of the modular concept is the decomposition of the network into manageable subunits. The emphasis of such an approach is the identification of kinetic expression comparable to elucidation of mechanisms in classical enzyme kinetics. In the second approach, canonical approximate models are used emphasizing the systemic context of the individual reactions. With the aid of three examples, strategies for optimal reshaping of networks based on different models are presented. The objective function of these optimizations is first of all the maximization of specific production rates. The optimizations are subject to important constraints, such as crowding, homeostasis, and stable solutions for the dynamic network of the redesigned strain. The last example – in Section 8.7 – aims at reducing the glycolytic flux, thereby addressing an important issue of drug target identification for cancer treatments.

8.1
Introduction

The use of high-throughput and efficient omics platforms has monopolized systems biology research in recent years. Given this high priority of the top-down approach, it is not surprising that network reconstruction and clustering of network components as well as multivariate analysis for assessing similarities

Fundamental Bioengineering, First Edition. Edited by John Villadsen.

between omics profiles from different samples are gaining more and more attention. This data-driven attempt to extract biological knowledge through integration of information from genome-level data sources is developing particularly fast in the emerging field of metabolomics. It is certainly true that the static view captures some important aspects of the structural properties of a system. The heart of systems behavior, however, which lies in the complex dynamics, cannot be covered by this approach. The omics platforms are also essential for this challenging goal, but focusing on elucidation of the dynamic behavior of the system is more than just collecting and correlating observables.

Dynamic models of metabolic networks can be written as

$$\frac{\mathrm{d}\vec{c}(t)}{\mathrm{d}t} = \mathbf{S} \cdot \mathbf{r}(\vec{c}(t), \vec{p}), \tag{8.1}$$

with $\mathbf{S}$ the stoichiometric matrix of dimension $m \times n$, and $\mathbf{r}(\vec{c}(t), \vec{p})$ the n-dimensional vector of reaction rates that depend on the concentration vector $\vec{c}(t)$. The kinetic or dynamic modeling can be clearly distinguished from the large number of the so-called genome-scale models, which are focused on the stoichiometry of the reactions in the network. These approaches consider the steady-state situation represented by

$$\mathbf{S}\,\mathbf{r} = 0. \tag{8.2}$$

However, in this chapter, we will see that the solution of Eq. (8.2) representing the steady-state flux distribution is an essential contribution to the dynamic modeling.

Quantitative analysis of the dynamic behavior of metabolic networks depends on the mathematical description of the multiple interactions between the individual components (kinetic modeling). Establishing kinetic models of cellular metabolism has always been in the realm of dispute whether (1) according to the traditional biochemistry point of view, models can be designed by aggregation of *in vitro* enzyme kinetics [1–4] or (2) *in vivo* measurements of metabolites should be applied for identification of kinetic properties. For *in vitro* kinetics, we have to take the systems out of their closed metabolic cages and the observed kinetics usually describes the behavior of test-tube isolates. There are two pivotal questions regarding the application of *in vitro* kinetics for studying the dynamics of metabolic network as real-life phenomena.

1) To what extent does the multitude of interacting processes inside the cell lead to a kinetic behavior that differs from *in vitro* conditions?
2) What is the influence of the functioning of the entire ensemble in the complex architecture of living cells? (Or, can we simply sum up every enzyme reaction to understand the system quantitatively?).

By comparison of *in vitro, in situ* (permeabilized cells), and *in vivo* results for enzyme kinetics of the phosphofructokinase I system in *Saccharomyces cerevisiae* (see Section 8.4.2), it has been demonstrated that remarkable differences in the structure of the kinetic expressions, as well as in parameter values, can be observed in the case of such a complex metabolically regulated enzyme system [5]. Further evidence for pronounced differences between *in vitro* and *in vivo* kinetics has been presented by Aon and Cortassa [6]. These authors discussed several examples of glycolytic enzymes, in which modulation of activities and kinetic parameters was caused by interactions of the enzymes with structural polymers, such as cytoskeleton components (tubulin and actin). For detailed analysis of the important issue of the impact of physiological enzyme concentrations on kinetic analysis, the reader is referred to the contribution of Aragon and Sanchez [7]. Recent investigations that address the issue of mathematical modeling of metabolic channeling in the crowded environment of eukaryotic cells are helpful for a fundamental understanding of the complex territory, but they cannot as yet provide predictive tools for solving the problem Klann *et al.* [8].

Focusing attention on metabolomics, the key to generate dynamic metabolic network models is, therefore, to extract the kinetics of the biochemical reactions from these data and, as such, consider the reactions in their "systemic" context. This chapter is intended to summarize tools and methods for identification of *in vivo* kinetics from the metabolite measurements. The tight link between the experimental observations on the one hand and modeling and simulation on the other hand is the critical issue in this model-driven strategy. From there, computational experimentation – particularly in regard to appropriate perturbations in the sense of optimal experimental design [9] – is of major importance. An important outcome of this discussion may be summarized in the key message: any success in the identification of systems dynamics critically depends on the continued interaction between simulations and wet-lab experiments.

Provided that progress in the dynamic modeling and simulation of metabolic pathways and networks critically depends on such a tight link between the dry- and wet-lab, the organization of this chapter is as follows: first a summary of the most important tools for quantitative measurement of intracellular metabolites along with typical results is given, and this is followed by an introduction of approaches for identification of the kinetics based upon different methods for decomposition of the network. Next, a critical assessment of attempts to identify systems dynamics for individual reactions and whole-cell network models will be presented. The final part of the chapter will be focused on the application of these models to technologically important tasks of model-based design by reshaping components of the network for the purpose of optimization of production strains and identification of drug targets.

8.2
Kinetics of Pathway Reactions: the Need to Measure in a Very Narrow Time Window

The major challenge pertaining to the task of measurement of metabolites originates from the fact that the half-life of metabolites is extremely short and, therefore, rapid sampling and fast and effective inactivation of the metabolic activities (quenching) is required. A large number of sampling devices for measurement of intra- and extracellular metabolites in narrow time windows have been described in the literature [5,10–25]. However, it is not within the scope of this chapter to provide the reader with a detailed and comprehensive treatment on the various concepts and tools. Instead, the main principles will be discussed briefly to help the reader gain a deeper insight into the related techniques, which will be introduced with a few examples.

For a quantitative analysis of the dynamic behavior of metabolic networks, it is essential to define the physiological state of the cells used for experimental observations. Of course, this imperative requires experimental conditions and related process operations, which are defined and reproducible. This is not a trivial task. As indicated by Wu *et al.* [26] and by unpublished results from the authors' laboratory, pools of metabolites can significantly change during long-time operation of continuous cultures at a given dilution rate because of various adaptation phenomena. Thus, the corresponding constant specific growth rate, equal to the dilution rate under steady-state conditions in continuous cultures, does not provide a guarantee of a defined physiological state. We also need to devise such methods for sampling, quenching, and extraction as ensure the results of the subsequent analysis accurately reflect the status of the living cell. The concrete design of the appropriate tools and operations is in the following sequence:

- Process operation (steady state, fed-batch, batch, and transient conditions)
- Sampling (steady or quasi-steady state, time span, and sampling frequency during transient stimulus–response experiments)
- Quenching
- Extraction
- Analysis

The methods depend on (a) the biotic variables to be measured and (b) the information to be derived from the observations. Attributes related to (a) include chemical, thermal, and biological stability of the compound, turnover times, analytical methods applied, and so on. The focus of the second point is the purpose of the measurements. The interest in such measurements may be related to dynamic responses of metabolite pools to extracellular disturbances (stimulus–response methodology) for identification of *in vivo* kinetics including modulation at the metabolite level. Another focal point of these measurements could be studies of the dynamics of regulation phenomena involved in transcription, translation, or post-translational processing. In the following, examples of this sequence of operations are provided from work conducted in the authors' laboratory. The main focus of these measurements is investigations under transient conditions and application of the data for the *in vivo* diagnosis of intracellular reactions [5,11,13,15,19,27–30].

The approach pictured in Figure 8.1a, for which the name "stimulus–response methodology" has been coined, is based on disturbance of a steady state during continuous operation by a pulse of glucose. The response of the biological

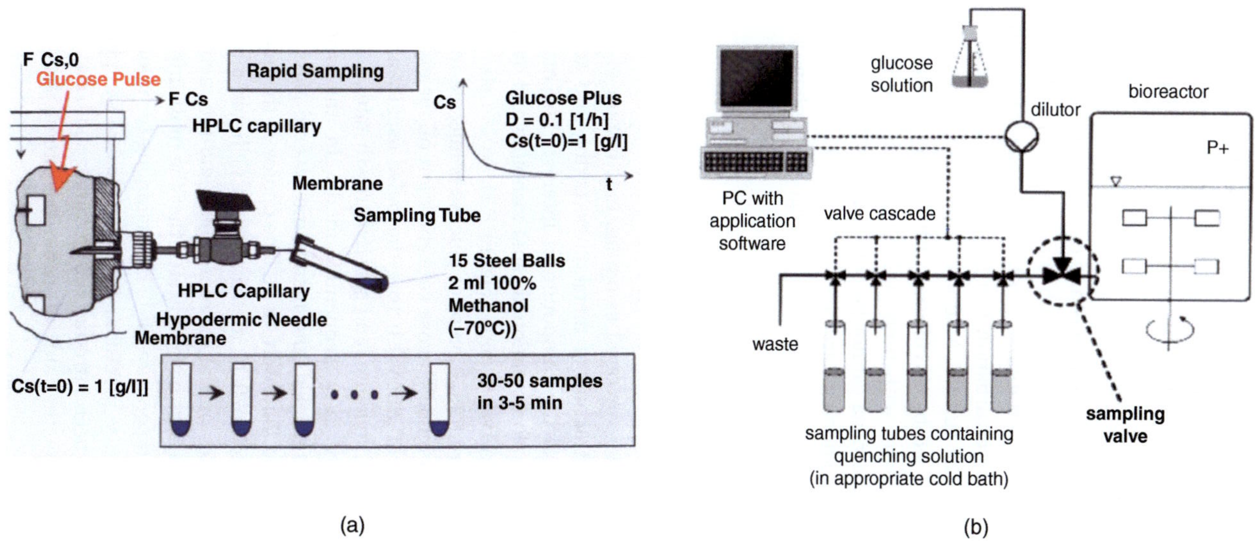

Figure 8.1 Two different sampling techniques for measurement of intracellular metabolites at transient conditions: (a) manual sampling after a pulse of glucose is injected into the bioreactor and (b) continuous pulse experiments with a stopped flow technique and shift of glucose concentrations within the sampling valve.

system is quantitatively characterized by measuring extra- and intracellular concentrations with high frequencies within milliseconds, seconds, or minutes after the disturbance. There are two reasons for choosing the particular timescale:

- Regulation at the metabolic level occurs within seconds or even faster.
- Within this timescale, changes are caused only by metabolic regulation including post-translational modification such as phosphorylation [31]. Biosynthetic reactions, for example, protein biosynthesis, can be regarded as staying in a "frozen" state.

Depending on the purpose of the measurements and derived kinetic models, it is also possible to apply other perturbations. Schaub and Reuss [32] and Canelas *et al.* [33] measured metabolite concentrations at different steady-state dilution rates in continuous cultures. The perturbation of the system is related to the change in specific growth rate of the culture. A similar approach can be applied to quasi-steady-state conditions of fed-batch cultures with constant feeding rate, where specific growth rates decreases with time [34]. Under these conditions, transcription and translation phenomena cannot be considered to be in a frozen steady state. In contrast, these phenomena are the focus of attention, and the measurements provide the basis for dynamic modeling of regulation phenomena.

8.2.1
Sampling

A precise measuring of intracellular concentrations in the time window of seconds requires appropriate techniques for rapid sampling, inactivation of metabolic enzymes (quenching), and extraction of metabolites, taking into account the high metabolic turnover rates of the compound of interest. The manifold sampling systems developed for the aforementioned rapid sampling and quenching problems may be classified into two groups (Figure 8.1). In the stimulus–response approach depicted in Figure 8.1a, a pulse of glucose is injected into the bioreactor with a syringe to give an initial glucose concentration of, for example, $1\,\mathrm{g\,l^{-1}}$ (steady-state concentrations of glucose before the pulse usually are less than $20\,\mathrm{mg\,l^{-1}}$, depending on strain, medium, and specific growth rate). Samples are then rapidly taken aseptically in a sequential mode. In the original approach [13,15], vacuum-sealed, precooled, and membrane-covered glass tubes containing an appropriate quenching fluid were used. The choice of the quenching fluid (e.g., perchloric acid solution: −20 °C; methanol: −70 °C; liquid nitrogen: −196 °C: hot water: +100 °C) depends on the microorganism and the metabolite to be measured. Systematic investigations have indicated that the most important quenching effect is caused by the temperature [35]. This sampling technique can be automated to increase the frequency, precision, and robustness of sampling. A semiautomated sampling device has been introduced by Schmalzriedt *et al.* [22].

Fully automated sampling with the aid of robotic systems has been suggested [18,36].

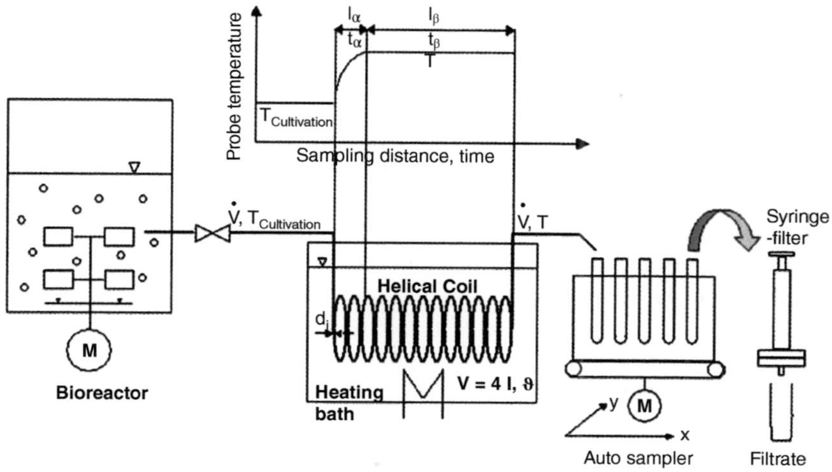

Figure 8.2 Integrated sampling system with heat exchanger and a robotic system for sample collection [25].

A disadvantage of most of the sampling/extraction techniques is the dilution of the sample during the treatment and the contamination of the sample by solvents or other acids/bases with subsequent pH adjustment and formation of salts. This kind of treatment with chemical reagents may generate specific problems in the analytical determination of the concentration. Schaub *et al.* [25] suggested a new integrated sampling technique, which overcomes most of these problems (Figure 8.2). Simultaneous quenching and quantitative extraction of intracellular metabolites is realized by short-time exposure of cells to temperatures of ~95 °C, where intracellular metabolites are released quantitatively. Based on these findings, a sampling procedure has been developed that is based on a coiled single tube heat exchanger. The samples are collected with an x–y robotic driven rack of sampling tubes and filtered before analysis of metabolites.

The second approach illustrated in Figure 8.1b is based on the stopped flow method used for fast measurements during enzymatic reactions. In its application to sampling from bioreactors ("BioScope:" [21,24] and the "stopped flow sampling technique:" [20]), a continuous stream of biosuspension leaving the bioreactor is mixed with a concentrated glucose solution in a mixing chamber. The position of the valves in the cascade illustrated in Figure 8.1b determines the residence time of the biosuspension before it is quenched in the corresponding sampling tube. The main features of this sampling device may be characterized as follows:

1) Very sharp stimuli, easy to be extended to, for example, temperature, pH, and other stress factors.
2) The culture remains at steady state because the microorganisms are stimulated by glucose in the mixing valve.

3) The sampling time and reaction time are decoupled. The volume of the individual samples can be chosen independently.
4) The time span between glucose stimulus and first sample can be less than 100 ms [20].

The only limitation of the technique is the problem of oxygen limitation at aerobic growth. However, this problem has been overcome by the development of the BioScope sampling system [21], which uses oxygen permeable silicon tubing. The design of this system has been further improved in the so-called "second-generation" BioScope [24].

8.2.2 Quenching and Extraction

Figure 8.3 summarizes some aspects of the various techniques used for the simultaneous or sequential operations of quenching and extraction for extra- and intracellular metabolites. In utilizing any of these methods, attention should be paid to ensure that

- metabolic activity is rapidly and irreversibly inactivated
- all the metabolites to be measured are completely extracted
- the quenching/extraction solutions used do not interfere with the analytical method used for the analysis.

Particular attention is required when the same metabolite is present in the intracellular and extracellular space. If this situation is the result of an export under physiological conditions, the problem can be solved by an additional measurement of the concentration in the extracellular environment with the aid of special filter units at low temperatures [15,36]. More critical are those cases where the export of metabolites is caused by leakage during quenching. Such leakage has been, *inter*

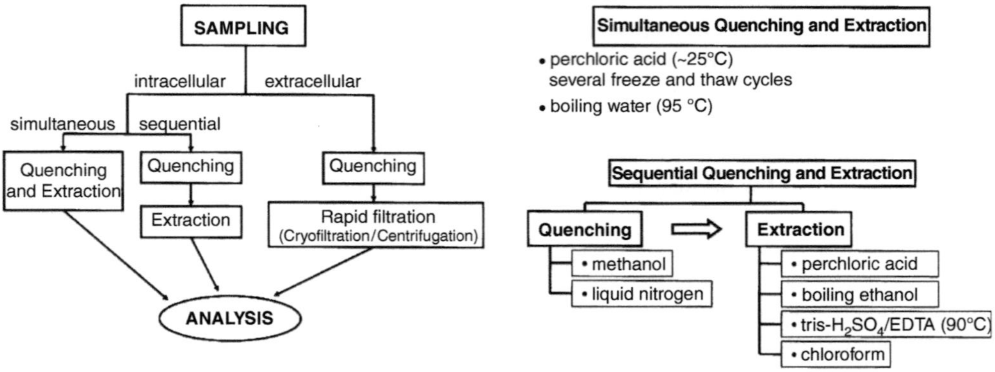

Figure 8.3 Simultaneous and sequential techniques for quenching and extraction.

alia, observed during the most popular quenching method using methanol. Systematic investigations [37,38] with *Saccharomyces cerevisiae* have demonstrated the key role of the methanol concentration. These investigations recommend the use of pure methanol. From systematic comparisons of methanol quenching and simultaneous quenching and extraction with the aid of the heat exchanger sampling device (Figure 8.2) [36], it was concluded that even when pure methanol is used the leakage of a large number of metabolites is observed. Also, the use of methanol may lead to precipitation of certain metabolites [39]. These investigations recommend a different technique: simultaneous quenching and extraction via heat, and additional measurement of extracellular metabolites after filtering. The unknown intracellular concentration is then calculated from

$$c_{\mathrm{intracellular}} = c_{\mathrm{total}} - c_{extracellular}.$$

Further quenching techniques are discussed by Van Gulik [38].

8.2.3
Analysis

The analytical tools utilized in the earlier targeted, model-driven metabolite measurements were either enzymatic assays or HPLC. The application of more powerful instrumental separation technology such as LC–MS, GC–MS, as well as capillary electrophoresis and NMR was an important extension of the approach. An overview of the extensive list of relevant literature can be found in the paper of Vielhauer *et al.* [36].

Liquid chromatography coupled to mass spectroscopy (LC–MS) is currently the most widely used analytical technology in absolute quantification. In contrast to LC–MS, gas chromatography–mass spectrometry (GC–MS) has rather established itself as a tool for qualitative and semiquantitative strategies in metabolomics (e.g., metabolic fingerprinting). The need to derivatize polar metabolites in the presence of varying biological matrices prior to analysis is probably one of the many reasons for observed uncertainties in the reliability of quantification of metabolites utilizing GC–MS. On the other hand, GC–MS analysis is attractive because of the superior separation power after proper derivatization of the metabolites that often coelute in LC–MS. Higher sensitivities with simple quadrupole technology, excellent ionization efficiency for a wide spectrum of derivatized metabolites, the information-rich EI mass spectra as well as comparatively easy operation and robust technology all contribute as further benefits. An example of successful application to quantitative measurements of metabolites in cultivations of *S. cerevisiae* has been presented by Vielhauer *et al.* [36].

An important issue in the quantitative measurement of metabolites concerns the use of appropriate standards. Wu *et al.* [40] suggested a method in which *S. cerevisiae* is cultivated in a small-scale fed-batch fermentation on 100% ^{13}C-labeled substrates, from which the labeled metabolites were extracted using proper rapid sampling and quenching procedures. A similar approach has been used by Benett *et al.* [41] for cell cultures. Vielhauer *et al.* [36] successfully used

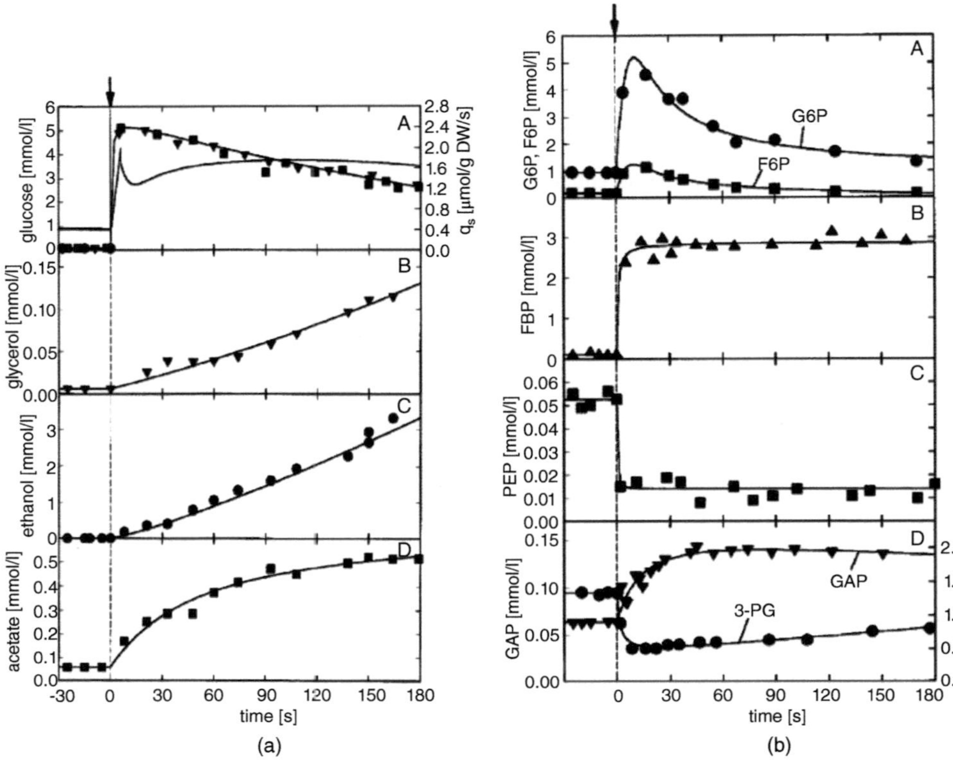

Figure 8.4 Metabolite concentrations during stimulus–response experiments with *S. cereviae*. (a) Extracellular and (b) intracellular concentrations [15]. The arrows indicate the pulse of glucose.

commercially available ^{13}C-labeled algal cells as a convenient source for the preparation of internal standards.

8.2.4
Examples for Quantitative Measurements of Metabolites in Stimulus–Response Experiments

Figures 8.4 and 8.5 illustrate examples of quantitative measurements of extra- and intracellular metabolites during cultivation of *S. cerevisiae* [15]. The figures summarize the metabolite concentration during steady state of a continuous operation and the dynamic pattern of the responses in a stimulus (glucose pulse)–response experiment. The analysis has been performed with the aid of enzymatic assays.

The extracellular measurements show the consumption of glucose after the pulse and excretion of glycerol, ethanol, acetic acid, and pyruvate. The intracellular concentrations indicate some interesting phenomena. Whereas the first two metabolites of the glycolysis show a maximum in their concentrations

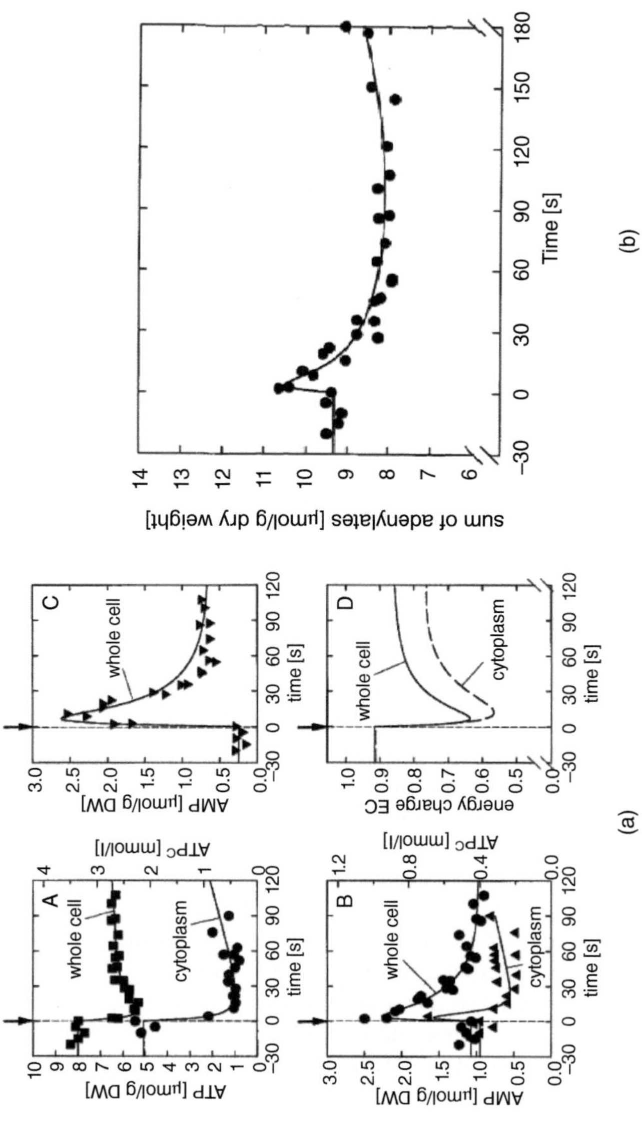

Figure 8.5 (a) Changes of levels of adenine nucleotides ATP, ADP, and AMP as well as energy charge after the glucose pulse [15]. Analytical methods for the measurement of cytoplasmic concentrations have been described by Theobald, Mailinger, and Rizzi [14]. (b) Total concentrations of adenine nucleotides. Time window: ~100 s.

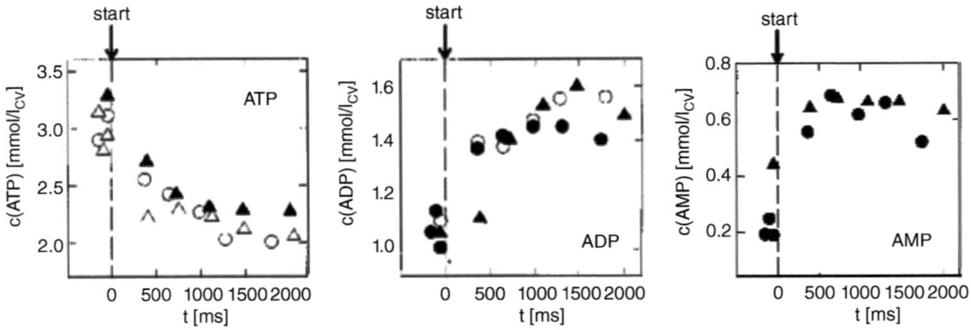

Figure 8.6 Dynamic response of adenine nucleotides due to increased glucose concentration in a stopped flow experiment with a continuous culture of *S. cerevisiae* (open and closed symbols represent the results from different experiments). Time window: ms.

shortly after the pulse of glucose, FBP and GAP remain at high levels for a long time. This behavior may result from the interaction between glycolysis and the pentose-phosphate (PP) cycle. The initial increase of G6P concentration is accompanied by a delayed increase of 6PG [29], the first metabolite in the PP cycle. This will cause an increasing flux through the PP cycle, which most likely will serve to keep FBP and GAP concentrations at higher levels. The most striking results, however, concern the opposite trend of the succeeding metabolites, 3-PG and PEP. This dynamic behavior may be explained by the well-known activation of the pyruvate kinase through fructose-1,6-bisphosphate (FBP) [42]. Thus, increasing FBP levels activate the conversion of PEP to Pyr resulting in a sharp decrease in the PEP concentration. If we assume that the reactions catalyzed by enolase and mutase are very close to equilibrium, then 3-PG must show the same trend. The qualitative discussion of the measured data and development of corresponding hypotheses are extremely important in the setup of the kinetic model, because the associated regulation phenomena must be conveniently pictured in the kinetic expression. In so doing we make use of the enormous fund of knowledge created by systematic research dedicated to enzyme kinetics.

The intracellular adenine nucleotides respond extremely rapidly (Figures 8.5 and 8.6). At first glance, the sharp decrease of ATP and corresponding increase in the ADP level may be associated with the two ATP-consuming reactions catalyzed by hexokinase and phosphofructokinase. However, ATP remains at a low level for quite a long time that cannot be explained by the characteristic time constants of glycolysis. A possible explanation of this strange phenomenon will be presented below.

Figure 8.6 shows the results of adenine nucleotide measurements performed with the aid of the stop-flow sampling technique (see Figure 8.1b). The figure shows in an impressive way how fast the yeast cells can respond to fast changes in the extracellular environment. This could be relevant for simulations of scale-up of bioreactors based on integrating Computational Fluid Dynamics (CFD) and intracellular kinetics as illustrated by Lapin *et al.* [43].

The drop of ATP concentration after the pulse of glucose as well as the strange behavior of the energy charge and total concentration of adenine nucleotides (Figures 8.5 and 8.6) have been debated ever since the original work of Theobald *et al.* [13]. Similar results have been reported by Visser *et al.* [44], Masegho *et al.* [24], and Kresnowati *et al.* [45]. It was the pioneering work of Loret, Pedersen, and Francois [46] and Walther *et al.* [47], which indicated a fast activation of the purine salvage pathway responsible for accumulation of the intermediate inosine and thus accounting for the apparent loss of adenine nucleotides. The analysis has been performed with several yeast mutants in shake flask cultivations with a shift from substrate trehalose to glucose. To integrate these observations into the stimulus–response experiments starting with the steady-state conditions of the chemostat, this analysis has been repeated in continuous cultivations of *S. cerivisiae* growing under anaerobic conditions. The results of the analysis of a stimulus–response experiment have been partially presented by Vermer *et al.* [48] and are summarized in Figure 8.7. The results are in agreement with the conclusions of Walther *et al.* [47] and can be interpreted as follows: immediately after the pulse, ATP drops and ADP increases due to the phosphorylation reactions. ADP is converted to AMP and ATP with the aid of the adenylate kinase. The increased AMP concentration is converted to inosine monophosphate (IMP). The high concentration of AMP inhibits the adenylatesuccinate synthase that converts IMP via adenylate succinate to AMP. The inhibition of this reaction leads to an accumulation of IMP and its conversion to the dephosphorylated inosine (I). The inosine formation serves to prevent the AMP accumulation and to store the adenylate nucleotides in a neutral form, thus preventing unwanted regulation phenomena in glycolysis.

The example of the important energy metabolism already demonstrates the problems of a partition of metabolic networks into relatively autonomous subunits. This issue will be further addressed in Section 8.4.

8.3 Tools for *In Vivo* Diagnosis of Pathway Reactions

To describe the dynamic systems behavior, deterministic kinetic rate equations of the form

$$r_i = r_i^{\max} f(\mathbf{c}, \mathbf{p})) \tag{8.3}$$

are formulated. The capacity of the reaction is characterized by its maximal rate, and the kinetic function represents the kinetic properties of the reaction. Substrates, products, and other metabolic effectors influencing the rate of reaction are represented by the state vector of metabolite concentrations **c**. The parameters of reaction i are summarized in vector **p**.

In what follows, different strategies are outlined for the identification of structure of the kinetic expression and of the values for the parameters by using metabolite measurements.

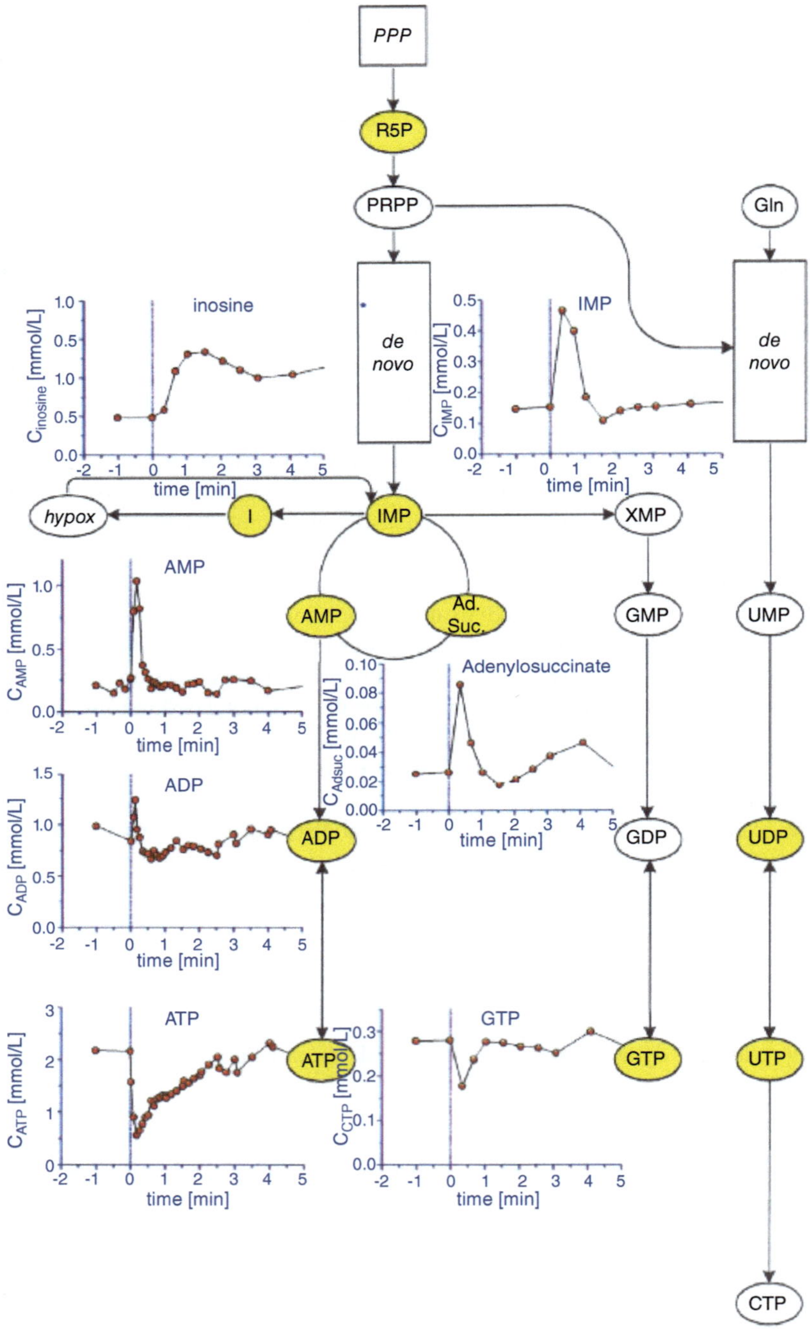

Figure 8.7 The purine salvage pathway responsible for the purine nucleotide cycling. Experiments are performed by the pulse response method with *S. cerevisiae* in continuous culture and under anaerobic conditions.

8.3.1
Modular Decomposition of the Network: the Bottom-Up Approach

The approach is similar to classical pathway modeling in that individual rate expressions are combined to describe the dynamic behavior of subsystems. A major difficulty when applying this strategy is, however, the definition of criteria for the demarcation of the combined modules to guarantee a certain level of autonomy [49]. Although a multitude of methods for decomposition of networks have been suggested, the specification and proof of existence of these modules are still a great challenge. For the time being, these modules are most often defined from an empirical, textbook-driven decomposition of the network into subsystems that perform particular physiological functions. For the difficult task of dynamic modeling, the only justification of the modular concept is that the approach allows for a decomposition of complex networks into manageable units. An important characteristic of the strategy of *in vivo* diagnosis, discussed in the following, is the partial integration of the subsystem into the network as a whole. This is an essential prerequisite for extension of the submodules and the reassembling of modules for the design of whole-cell models.

The first step to embed the behavior of the subsystem into the metabolic network as a whole is provided by the estimation of the maximum rates of the individual reactions. Applying the rate Eq. (8.3) to the steady state leads to

$$\tilde{r}_j = \tilde{r}_j^{\max} f_j\left(\tilde{c}_j, \mathbf{p}_j\right). \tag{8.4}$$

Let us assume that reaction rate $\tilde{r}_j$ at steady state has been estimated from metabolic flux analysis. Let us further assume that the first estimate of the structure of the kinetics f_j and the parameter vector $\mathbf{p}_j$ is available from *in vitro* measurements. If the components of the concentration vector $\mathbf{c}_j$ influencing the rate of the reaction have been measured at steady state, the unknown maximal rates are given as

$$\tilde{r}_j^{\max} = \frac{\tilde{r}_j}{f_j\left(\mathbf{c}_j, \mathbf{p}_j\right)}. \tag{8.5}$$

Further details of the strategy to identify the *in vivo* kinetics from the measured stimulus–response data are discussed in the context of concrete example presented in the following.

8.4
Examples: The Pentose-Phosphate Shunt and Kinetics of Phosphofructokinase

8.4.1
Kinetics of the Irreversible Reactions of the Pentose-Phosphate Shunt

Figure 8.8 summarizes the metabolites and enzymes of the pentose-phosphate shunt in *S. cerevisiae* along with the upper part of glycolysis [29].

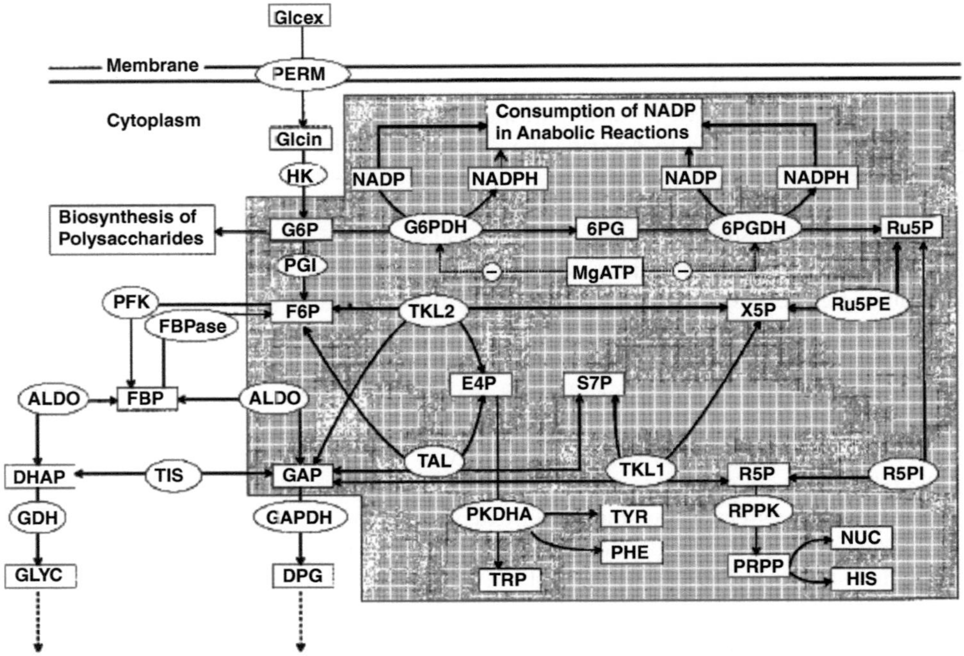

Figure 8.8 The pentose-phosphate pathway.

The first step toward quantitative analysis of the shunt involves the assumption that the first two reactions catalyzed by glucose-6-phosphate dehydrogenase (G6PDH) and 6-phospho-gluconate dehydrogenase (6PGDH) are irreversible, whereas the other reactions are assumed to be reversible and in near-equilibrium. The balance equation for 6-phospho-gluconate is given by

$$\frac{dc_{6PG}}{dt} = r_{\mathrm{G6PDH}} - r_{\mathrm{6PGDH}} - \mu c_{\mathrm{6PG}}. \tag{8.6}$$

The last term in Eq. (8.6) represents the dilution caused by the growth of yeast.

The measurements required for the identification of the kinetics of the two enzymes are summarized in Figure 8.9.

In addition to the substrates G6P and 6PG, we need to know the dynamic response of the concentrations of the cosubstrate $NADP^+$. Because NADPH is known as a product inhibitor for both reactions, this cometabolite must be measured. A careful inspection of the measured data illustrates that the level of G6P at steady state already results in a substrate saturation of G6P-dehydrogenase. Consequently, the increase in the concentration of NADPH after the pulse of glucose cannot be explained by an increase of the flux caused by the substrate G6P. This observation strongly supports the strategy to treat the pentose-phosphate shunt as a semiautonomous unit as far as the link to its substrate

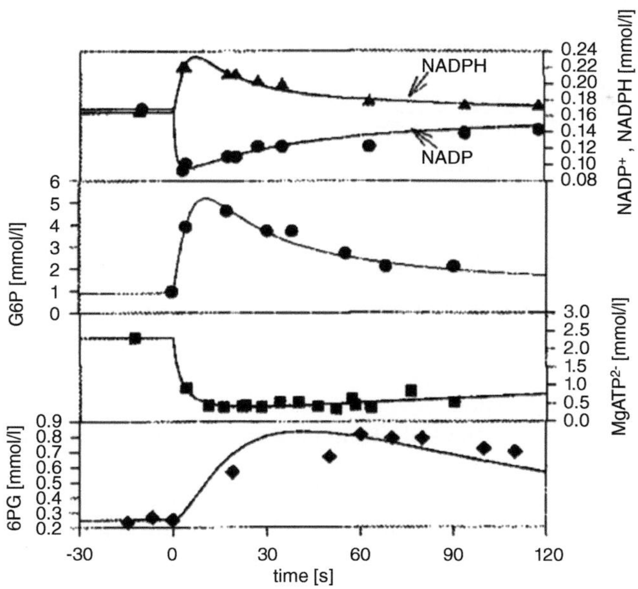

Figure 8.9 Time course of NADP, NADPH, ATP, and 6PG after a pulse of glucose is injected into a glucose-limited continuous culture of *S. cerevisiae*. The solid line in the 6PG-time profile results from the model simulations [29].

G6P is concerned. The increase of NADPH in response to the pulse of glucose still remains to be explained. A thorough consideration of the literature and additional *in vitro* measurements with the isolated G6P dehydrogenase provides an interesting solution to the problem. ATP turns out to be a strong inhibitor of not only G6P-DH but also 6-PGDH. The increased flux through the PP shunt can, therefore, be easily explained by the decrease of ATP after the pulse of glucose [13,15,50]. Summarizing the aforementioned characteristics of the two reactions, the following rate expressions are suggested in the balance equation for 6PG (Eq. (8.6)):

$$\frac{\mathrm{d}c_{6\mathrm{PG}}}{\mathrm{d}t} = r_{\mathrm{G6P-DH}}^{\max} \frac{c_{\mathrm{NADP}}(t)}{\left[c_{\mathrm{NADP}}(t) + K_{\mathrm{NADP},1}\left(1 + \frac{c_{\mathrm{NADPH}}(t)}{K_{i,\mathrm{NADPH},1}}\right)\right]\left[1 + \frac{c_{\mathrm{MgATP}}(t)}{K_{i,\mathrm{MgATP},1}}\right]} - r_{\mathrm{6PG-DH}}^{\max} \frac{c_{\mathrm{NADP}}(t)}{\left[c_{\mathrm{NADP}}(t) + K_{\mathrm{NADP},2}\left(1 + \frac{c_{\mathrm{NADPH}}(t)}{K_{i,\mathrm{NADPH},2}}\right)\right]\left[1 + \frac{c_{\mathrm{MgATP}}(t)}{K_{i,\mathrm{MgATP},2}}\right]} - \mu c_{6\mathrm{PG}}(t) \quad (8.7)$$

Notice that the concentrations on the right-hand side of the equation are time dependent. A comprehensive solution to the problem would require additional balance equations for NADP, NADPH, and ATP. To reduce the complexity of the problem, measured data for the cometabolites are approximated by approximate analytical functions. The following functions have been used to fit the observed time series of data after the glucose pulse:

$$\hat{c}_{\mathrm{NADP}}(t) = 0.17 - \frac{1.48t}{9.7 + 16.1t + 0.48t^2}. \tag{8.8}$$

$$\hat{c}_{\mathrm{NADPH}}(t) = 0.16 - \frac{0.516t}{25.4 + 0.37t + 0.5t^2}. \tag{8.9}$$

$$\hat{c}_{\mathrm{MgATP}}(t) = 2.3 - \frac{29.8t}{29.8 + 13.4t + 0.05t^2}. \tag{8.10}$$

These functions are only used to fit the data and do not have any mechanistic background. However, they serve to couple the dynamics of the autonomous balance equation with the dynamic response of the cometabolites, thus mirroring the behavior of the cell as a whole.

A serious drawback of this method of fitting data with empirical functions is the effort required to find an appropriate function, and then in some cases, there are problems to extrapolate the fitting function to a stable steady state. An alternative is based on penalized regression splines. Figure 8.10a shows such a function and Figure 8.10b is an example of fitting measured concentrations for glucose-6-phosphate in response to a glucose pulse in anaerobic cultivations with *S. cerevisiae.*

The next step toward the estimation of parameters is the calculation of maximum rates $r_j^{\max}$. Estimates from measured enzyme activities under *in vitro* conditions are questionable because of the possible removal of effectors during cell disruption, shear sensitivity, effects of ion strength, protein–membrane and protein–protein interactions, and so on. An alternative estimate is based on the rate equation and on measured intracellular concentrations under steady-state conditions as described above (Eq. (8.5)). This approach, again, guarantees that the isolated system is tightly linked to the behavior of the whole cell, because the steady-state metabolic flux distribution is a holistic attribute of the cell and does not depend on individual modules.

Eventually, the dynamic balance equation for 6-phospho-gluconate (Eq. (8.4)) can be numerically integrated and the sum of least squares minimized as follows:

$$\Phi_{\mathrm{rel,6PG}} = \sum_k \left(\frac{c_{\mathrm{6PG}}^{\mathrm{calc}}(t_k) - c_{\mathrm{6PG}}^{\mathrm{meas}}(t_k)}{c_{\mathrm{6PG}}^{\mathrm{calc}}(t_k)} \right)^2. \tag{8.11}$$

The unknown kinetic parameters can be estimated with the aid of appropriate parameter estimation algorithms. It is important to emphasize the iterative character of the estimation of maximum rates. After each step of iteration in the optimization procedure, the parameters need to be readjusted with the aid of Eq. (8.2).

This example illustrates the strategy for the *in vivo* identification of intracellular kinetics. For a more detailed description of this approach, the reader is referred to the original publications of Rizzi *et al.* [51] and Vaseghi *et al.* [29], which also indicate an interesting result regarding the issue of modular structures, which is important for more complex network analysis.

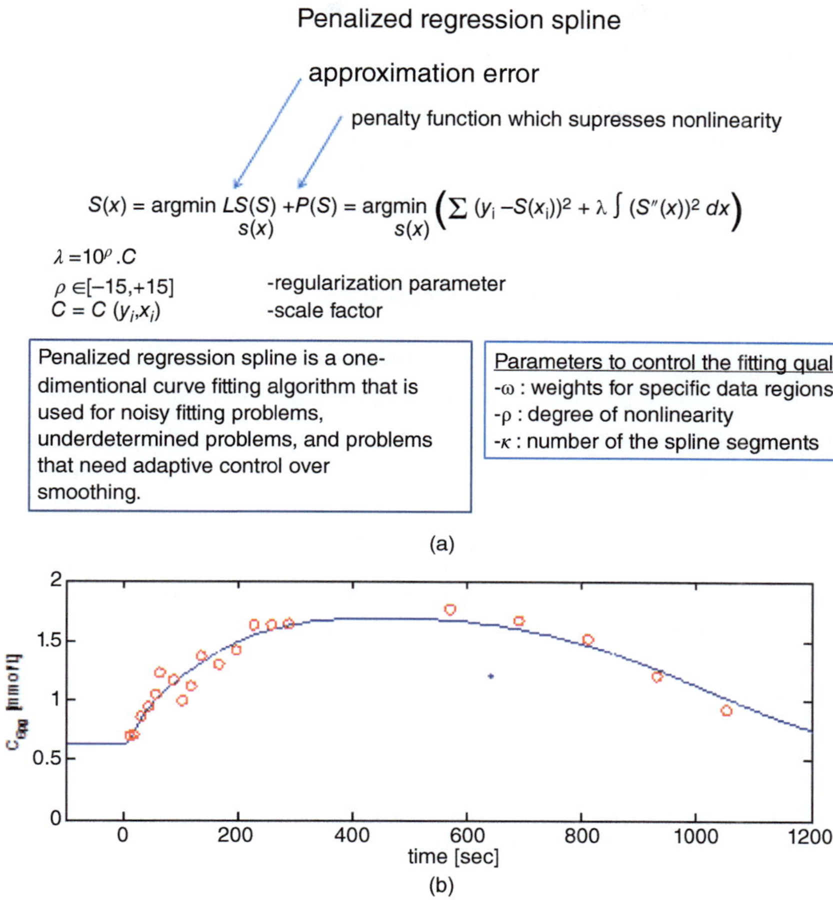

Figure 8.10 (a) Penalized regression splines applied to fit the dynamic response of (b) glucose-6-phosphate in *S. cerevisiae*.

According to the results of the identification procedure, the flux through the PP shunt is independent of the concentration of the substrate G6P and depends both on the ATP pool and on the NADP/NADPH ratio. It, therefore, seems that the flux is adjusted to the energy state and balanced to the demand for biosynthesis. Thus, the PP shunt apparently acts only as a semi-autonomous functional unit, which is carefully modulated by the energy state of the cell and the demand for biosynthesis. As such, the module is wired with the hubs of the cellular network, which participate in a very large number of links [52]. These hubs eventually integrate all substrates into a single integrated Web in which the existence of fully autonomous modules is prohibited.

8.4.2 Kinetics of the Phophofructokinase I (PFK1)

The kinetic behavior of this enzyme has attracted a lot of attention because of its important role in the regulation of glycolysis. The allosteric enzyme catalyzes the phosphorylation of fructose-6-phosphate (F6P) to fructose-1,6-bisphosphate (F1,6BP) and it is modulated by several effectors as schematically illustrated in Figure 8.11.

The identification of the kinetic rate expression and the parameters is based on the balance equation of F6P, which reads as follows:

$$\begin{aligned}\frac{dc_{\mathrm{F6P}}}{dt} &= -r_{\mathrm{PFK1}}^{\max} f_{\mathrm{PFK1}}\left(c_{\mathrm{F6P}}(t), c_{\mathrm{AMP}}(t), c_{\mathrm{ADP}}(t), c_{\mathrm{ATP}}(t), c_{\mathrm{F2,6,BP}}(t), \mathbf{p}_{\mathrm{PFK}}\right) \\ &\quad + r_{\mathrm{PGI}}^{\max} f_{\mathrm{PGI}}\left(c_{\mathrm{G6P}}(t), c_{\mathrm{F6P}}(t), \mathbf{p}_{\mathrm{PGI}}\right) + r_{\mathrm{TAL}}(t) + r_{\mathrm{TKL2}}(t) - \mu c_{\mathrm{F6P}}(t).\end{aligned}$$

The reactions involved in the balance are shown in Figure 8.11. The contribution from the reactions transaldolase (TAL) and transketolase (TKL2) are computed from the model of the pentose shunt [29]. The rate expression for the glucose isomerase (PGI) is described in the original paper of Rizzi *et al.* [51]. Again the time course of the concentrations of the various effectors in response to a pulse of glucose has been measured (Figure 8.12) and fitted with approximate functions as described above. The details of the identification of the kinetic expression for the PFK1 of the yeast *S.cerevisiae* have been comprehensively described by Mauch, Vaseghi, and Reuss [5] and Reuss, Aguilera-Vazquez, and Mauch [53]. An important aspect of this analysis concerns the comparison between the kinetic rate expression identified from *in vitro, in situ,* and *in vivo* measurements. The *in situ* measurements were performed via permeabilization of the yeast cells with a toluol-ethanol mixture according to a protocol suggested by Serrano, Gancedo, and Gancedo [54]. For a discussion on the pronounced differences between *in vivo, in situ,* and *in vitro* results, it is necessary to introduce some characteristics of the kinetic properties of the enzyme.

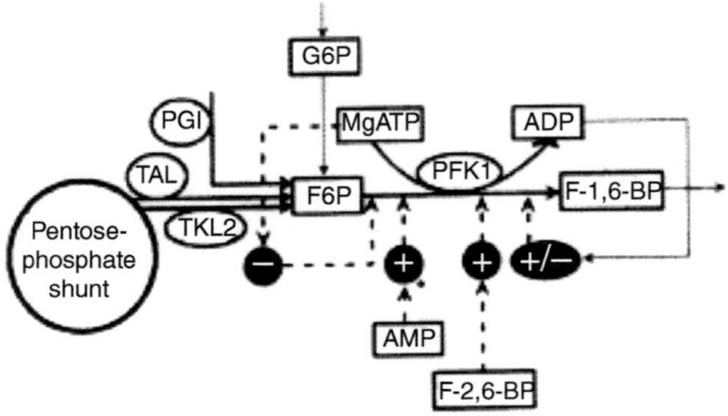

Figure 8.11 Reactions around PFK1.

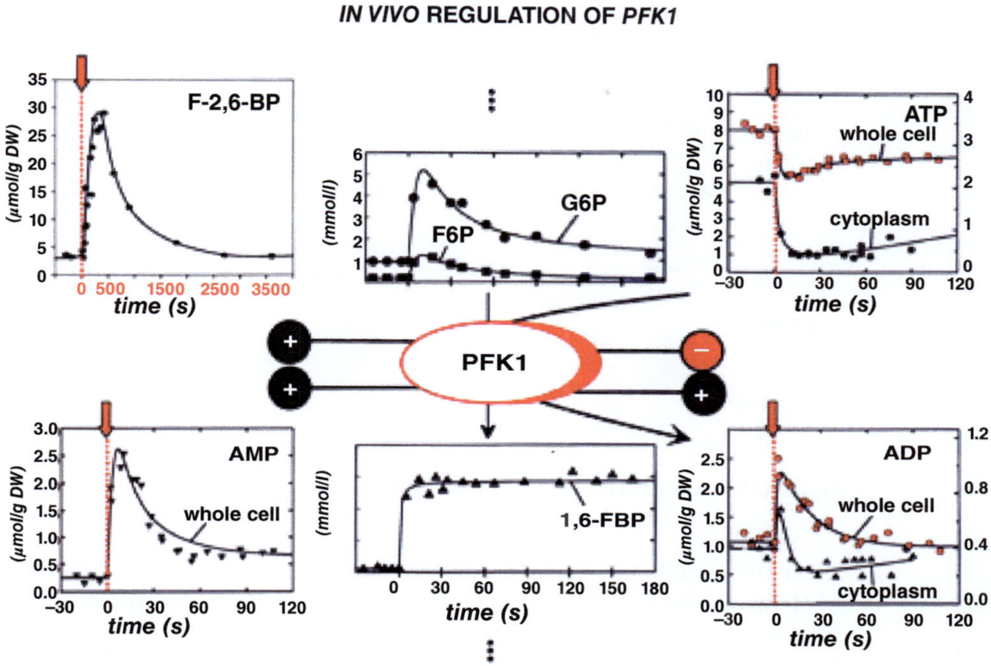

Figure 8.12 Time series of data from stimulus–response experiments for identification of the kinetics of PFK1.

The most important attempts to model the allosteric properties of PFK1 were made long ago by Monod, Wyman, and Changeux [55], who considered two conformations of the enzyme: conformation T with a low and conformation R with a high affinity to the substrate F6P. Each enzyme molecule consists of eight subunits. The basic model of Monod, Wyman, and Changeux [55] considers only the allosteric behavior with respect to the substrate F6P and a nonallosteric effect of ATP. The so-called Reuter model [56] additionally considers the modulation through AMP, ADP, and ATP and their influence upon the allosteric effects of F6P. In a similar way, the more sophisticated models by Vaseghi [5,29,35] illustrated in Figure 8.13 assume that

- each promoter of the octameric protein appears in two basic conformations R1 and R2 as well as two subconformations R2 and T2.
- substrate F6P is bonded to R1 and T1 with different affinities.
- ATP binds to all different conformations with the same affinity.
- ATP binds as inhibitor to R1 and T1.
- AMP and ADP bind as activators to R1/T1 and R2/T2, respectively.

As illustrated in Figure 8.13, the model of Vaseghi (V-model) also considers the strong activation through F-2,6 BP as well as a competitive inhibition between F-2,6 BP and F6P.

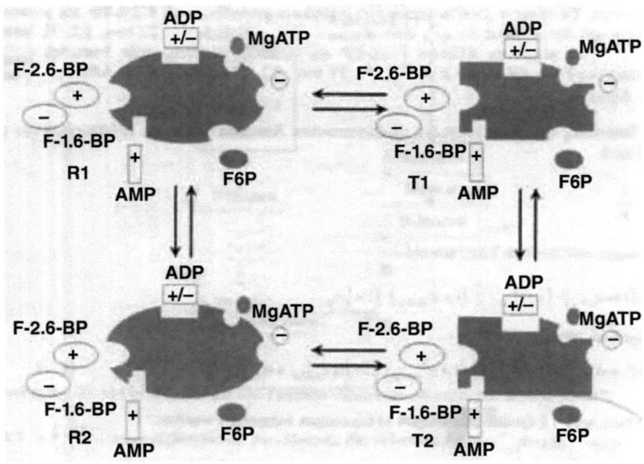

Figure 8.13 The mechanism of the V-model for PFK1.

For comparison, the kinetic parameters were estimated from the *in vitro* data of Hofmann and Kopperschlager (personal communication) [57] and compared with our own *in situ* and *in vivo* measurements. The differences in the values of the parameters are pronounced [5]. There are two striking phenomena that attract attention:

- Compared to the *in vitro* situation, the modulation strength of all the effectors are less pronounced under *in situ* and *in vivo* conditions. This is exemplarily shown in Figure 8.14, in which the modulation of the enzyme activity through ATP is depicted for the *in vitro* and *in situ* conditions.
- The parameter value L0 (ratio of the tensed to the relaxed state) tends to zero under *in situ* and *in vivo* conditions. This result leads to the interesting

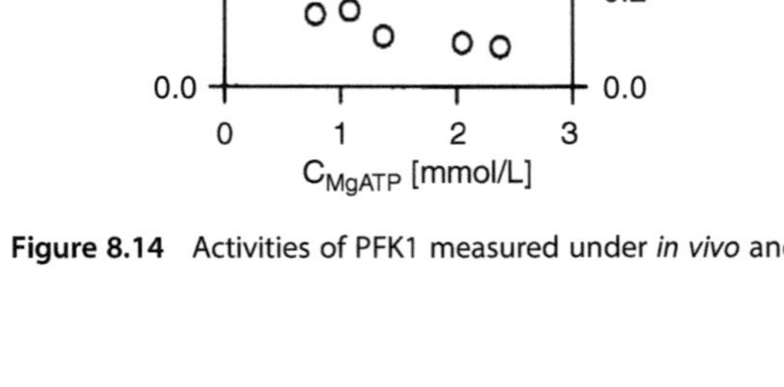

Figure 8.14 Activities of PFK1 measured under *in vivo* and *in situ* conditions.

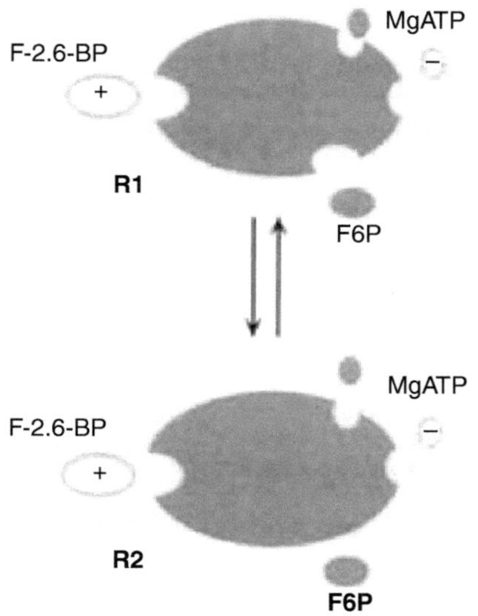

Figure 8.15 The reduced V-model for PFK1 under *in vivo* conditions.

conclusion that the enzyme acts only in the R conformation (high affinity). On the basis of these observations, the model can be reduced to a much more simple structure illustrated in Figure 8.15.

To extrapolate the results obtained from the kinetic analysis of PFKI to other enzymes, it is worthwile to speculate about possible clues for these pronounced differences between *in vivo* and *in vitro* conditions. A reasonable explanation needs to consider the well-known effects of homogeneous and heterologous protein interactions inside the cell. The first effect is related to the concentration of the enzyme. Srere [58] indicated that concentrations of glycolytic enzymes inside the cells are usually 100 times higher than those normally applied to *in vitro* assays. According to Aragon and Sanchez [7], the concentration of PFKI in *S. cerevisiae* is of the order of 190–550 $\mu g\,ml^{-1}$, whereas the concentration of *in vitro* assay is 1–10 $\mu g\,ml^{-1}$. The influence of high intracellular concentrations of the enzyme on the regulation of PFKI has been shown by Aragon and Sanchez [7]. The second effect – heterologous interactions – is concerned with associations between enzymes and structure proteins of the cytoskeleton [59]. Particularly for PFKI in *S. cerevisiae*, (G. Kopperschlager (1999), personal communication) has shown an organized association between the enzyme molecule and the microtubules under *in vivo* conditions. The impact of such associations on the kinetic behavior of the enzymes is not known yet. However, because of the strong relationship between structure and function, these are strong candidates for the explanation of the aforementioned differences between *in vitro* and *in vivo* conditions. Because of the closer agreement between *in situ* and *in vivo* observations, the *in situ* experiments could be viewed as interesting and alternative tools for the kinetic analysis of intracellular reactions.

The approach is characterized by a decomposition of the metabolic network in smaller modules for identification of structures and parameters of individual reactions. The modules can afterward be aggregated to create dynamic models for pathways and smaller networks. With the aid of appropriate identification methods, the aggregated system can be further improved by fine-tuning the parameter space.

In spite of its modularity, the approach guarantees, at least to a certain extent, cross-talk between the individual modules and the cellular network as a whole. These important dependencies are manifested in the approach for estimating the maximum rates from steady-state metabolic flux analysis and the incorporation of cofactors by fitting their dynamic response. The two examples pictured in Figures 8.16 and 8.17 illustrate applications of this approach for the dynamic modeling of the central metabolism of *S. cerevisiae* [51] and *Escherichia coli* [19].

8.5
Additional Approaches for Dynamic Modeling Large Metabolic Networks

The modular approach discussed so far is based on a partition of metabolic networks into relatively autonomous subunits. It is widely believed that the modular structure of the network plays a critical role in its functionality. Usually, the decomposition of metabolic networks is performed on an intuitive basis. In spite of the fact that several algorithms, purely based on network topology, have been proposed, the development of methods for exploiting the modularity, the accurate definition of the modules, and questions about their autonomy are still unresolved. There is a clear need to develop alternative algorithms for identifying modules accurately [60–67]. Because of a missing rigorous definition of the subunits, the question remains if the fundamental organization principle of metabolic networks is modular at all or distributed – or if it can best be described as having a little of both [68].

In the context of the discussion of modular approaches to dynamic modeling, the issue of ubiquitous substrates – or sometimes, by analogy to economy, termed currency metabolites – such as AMP, ADP, ATP, NAD, NADH, NADP, NADPH, and so on needs to be specifically considered because these metabolites are involved in a very large number of reactions. Even if we restrict the discussion to only structural properties of modularity, in most cases the data are preprocessed by removing the high degree node (hubs) compounds before a decomposition algorithm is applied. As nicely elaborated by Huss and Holme [68], an effective modularity will increase when a currency metabolite is deleted from the network. The question is to what extent the results of, for example, "cartographic representation" of complex networks for identification of modules can be applied to the problem of dynamic modeling. This remains a completely open question.

What is actually required for a systematic and rigorous decomposition of metabolic networks applied to dynamic modeling are considerations of

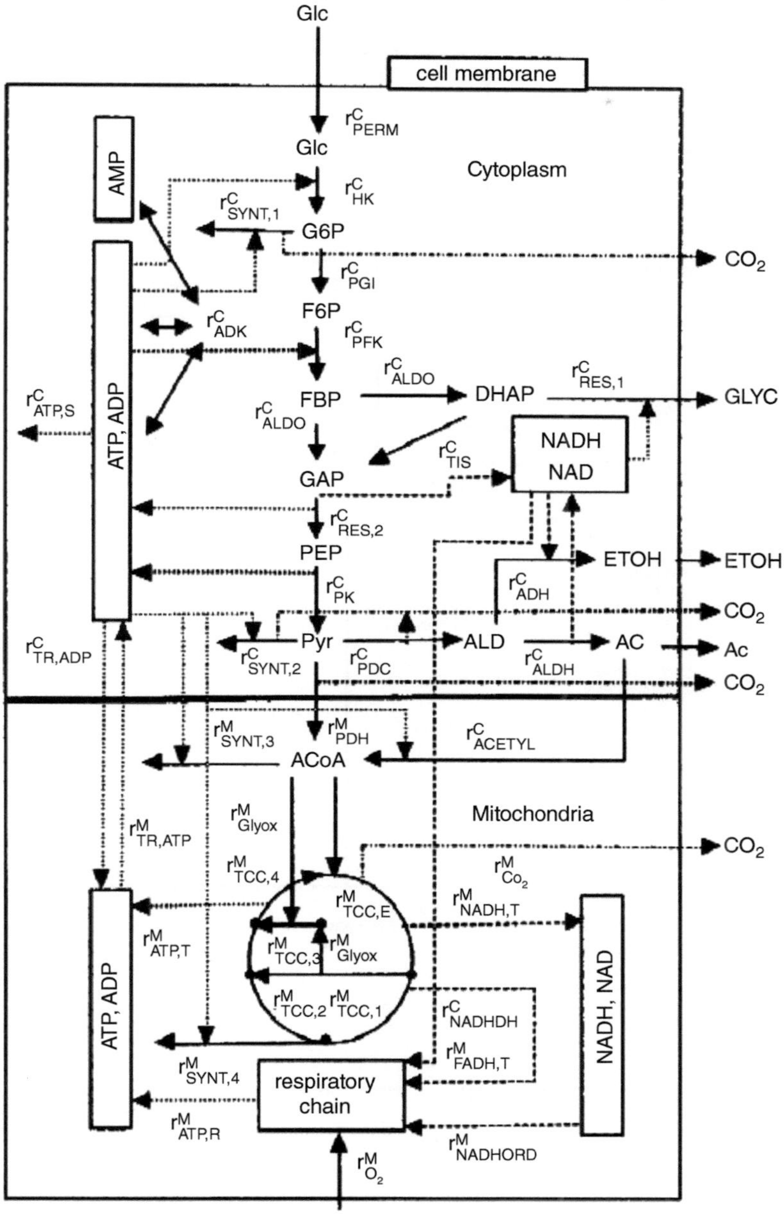

Figure 8.16 Dynamic model of the central metabolism of *S. cerevisiae* [51].

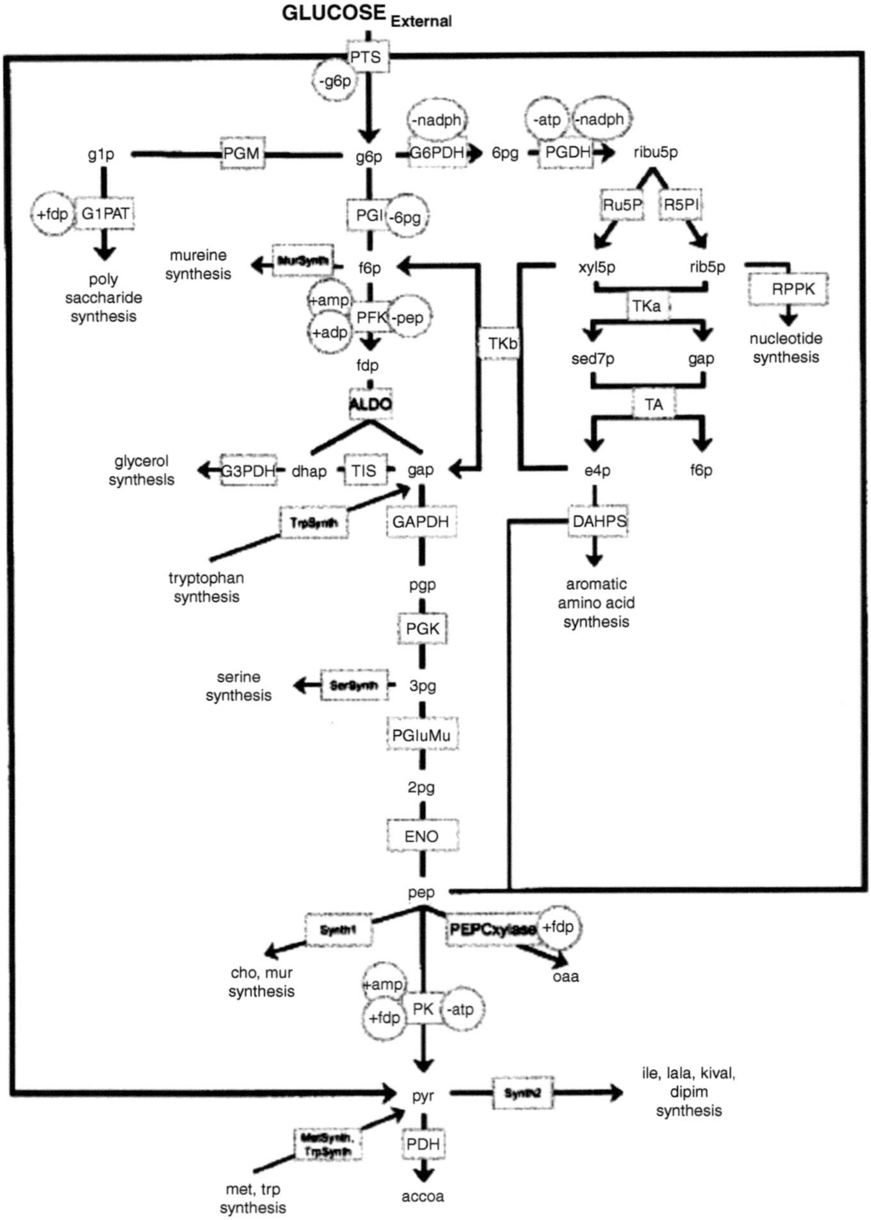

Figure 8.17 Dynamic model of the central metabolism of *E. coli* [19].

biochemical modules as dynamical entities, taking into account the impact of common cometabolites occurring in and between almost all modules. Even if we take care of a partial integration of modules in the network as a whole, as demonstrated in the aforementioned examples, there are several applications in which the predictive power of the dynamic models would be necessary (e.g., to optimal redesign of metabolic systems [69] or to estimation of optimal drug dosages). Here a rigorous balancing of the cometabolites and metabolites in overlapping modules is required. The setup of large-scale dynamic models based on a canonical representation of reaction kinetics is, therefore, an important issue.

The aforementioned methods are based on the decomposition of the metabolic network in different modules and may be characterized by the phrase: "bottom-up approach." An advantage of this approach is the manageability and better control of the well-known problems of parameter sensitivity, that is, the linear dependency between the parameters together with their identifiability. An alternative method is a kind of "top-down approach," in which canonical, approximate kinetic representations are used.

The tendency to derive approximate dynamic models reflects the increasing awareness of systems biology approaches to understand and describe the holistic behavior of biological systems. As far as the kinetic behavior of enzymes is concerned, what is required is a description of the kinetics of enzymes that takes into account the systemic context in which each enzyme is found. As such, the focus of enzyme kinetics has shifted. The exact enzyme catalytic mechanisms are unimportant as long as the response to reactants and effectors is adequately described. The overall dynamic behavior of the pathways and networks is thus described by mathematical models instead of by exact mechanistic enzymatic rate expressions.

Because of the canonical presentation of the kinetics, the models can be built automatically from the reaction scheme (used in genome-scale stoichiometric models). The most popular canonical or appropriate kinetic expressions are the generalized mass action kinetics derived for use in classical chemical catalysis by [70,71], the power law or S-system approach [72,73], the convenience rate law (Liebermeisterr and Klipp [74]), and the log–lin and lin–log kinetics [75,76].

8.5.1 Generalized Mass Action

The general mass action kinetics is a simple rate law with the rate constant (k), the thermodynamic equilibrium constant K_{eq}, and additional constants for activation and inhibition phenomena. The rate expression for the individual reactions in the network is given by

$$r = k\prod_{l=1}^{nA}\left(\frac{c_{A,l}}{c_{A,l}+K_{\text{A}}^{l}}\right)^{w_l^+}\prod_{i=1}^{nI}\left(\frac{1}{1+K_I^i c_{I,i}}\right)^{w_i^-}\left(\prod_{j=1}^{nS}c_{S,j}^{g_j}-\frac{1}{K_{\text{eq}}}\prod_{k=1}^{nP}c_{P,k}^{h_k}\right). \tag{8.12}$$

The indices A, I, S, and P stand for activators, inhibitors, substrates, and products of the reaction of interest, respectively.

8.5.2
S-Systems Approach

The S-system form focuses on the metabolic pools [77]. The balance equation for the pool i is given by (8.13):

$$\frac{\mathrm{d}c_i}{\mathrm{d}t} = r_i^+ - r_i^- = \alpha_i \prod_{j=1}^{n} c_j^{g_{ij}} - \beta_i \prod_{j=1}^{n} c_j^{h_{ij}} \quad i = 1, \ldots, n. \tag{8.13}$$

The multiplier α_i and β_i are rate constants, the exponents g_{ij} and h_{ij} are real-valued kinetic orders.

8.5.3
Convenience Kinetics

The semimechanistic rate law that is a particular case of the generalized mass action kinetics is formulated from a random-order enzyme mechanism and is given by

$$r = c_{\mathrm{E}} \prod_{l=1}^{nA} \left(\frac{c_{A,l}}{c_{A,l} + K_{A,l}} \right)^{w_l^+} \prod_{i=1}^{nI} \left(\frac{1}{1 + K_{I,i} c_{I,i}} \right)^{w_i^-} \times \frac{k_{\mathrm{cat}}^+ \prod_{j=1}^{nS} \left(\frac{c_{S,j}}{K_{M,S,i}} \right)^{\alpha_i} - k_{\mathrm{cat}}^- \prod_{k=1}^{nP} \left(\frac{c_{P,k}}{K_{M,P,k}} \right)^{\beta_j}}{\prod_{j=1}^{nS} \left(\sum_{m=0}^{\alpha_j} \left(\frac{c_{S,j}}{K_{M,S,j}} \right)^m \right) + \prod_{k=1}^{nP} \left(\sum_{m=0}^{\beta_k} \left(\frac{c_{P,k}}{K_{M,P,k}} \right)^m \right) - 1}. \tag{8.14}$$

For details on the derivation of the equation and discussion of the parameter see Liebermeister and Klipp [74]. Apparently, this equation requires comparatively many more parameters than the other approximate approaches.

8.5.4
Log–Lin and Lin–Log Approaches

The rates of the individual reactions in a log–lin model [75] are represented as

$$\frac{r}{J^0} = 1 + \ln\left(\frac{c_{\mathrm{E}}}{c_{\mathrm{E}}^0}\right) + \sum_i \epsilon_{S,i} \ln\frac{c_{S,i}}{c_{S,i}^0} + \sum_j \epsilon_{P,j} \ln\frac{c_{P,j}}{c_{P,j}^0} + \sum_k \epsilon_{A,k} \ln\frac{c_{A,k}}{c_{A,k}^0} + \sum_l \epsilon_{I,l} \ln\frac{c_{I,l}}{c_{I,l}^0}. \tag{8.15}$$

while the rates in a lin–log model [76] are given as

$$\frac{r}{J_0} = \frac{c_{\mathrm{E}}}{c_{\mathrm{E}}^0} \left(1 + \sum_i \epsilon_{S,i} \ln\frac{c_{S,i}}{c_{S,i}^0} + \sum_j \epsilon_{P,j} \ln\frac{c_{P,j}}{c_{P,j}^0} + \sum_k \epsilon_{A,k} \ln\frac{c_{A,k}}{c_{A,k}^0} + \sum_l \epsilon_{I,l} \ln\frac{c}{c_{I,l}^0} \right). \tag{8.16}$$

Superscripts (^{0}) in Eqs. (8.15) and (8.16) denote the reference state (e.g., steady state or wild type). J^0 is the reference flux through this reaction, c_E is the concentration of the enzyme of the reaction and the elasticity coefficients (see Section 7.6) are defined as follows:

$$\epsilon_M = \frac{c_{M_0}}{r_0}\left(\frac{\partial r}{\partial c_M}\right)_0 \tag{8.17}$$

The background of the generalized mass action, the S-systems, and the convenience kinetics are obvious. They are characterized by generalizations of well-known enzyme kinetics (see Chapter 7) and/or mass–action kinetics.

The settings of the log–lin and lin–log approaches are less obvious and they will, therefore, be further discussed below. The derivations are intended for better understanding and moreover, they provide valuable information on limits of applicability.

Unlike the approaches discussed so far, the roots of the lin–log approach lie in *nonequilibrium thermodynamics*. For systems not in thermodynamic equilibrium, Onsager [77] proposed an approach, according to which a reaction rate is linearly related to the driving thermodynamic forces X_j:

$$r_i = \sum_j L_{ij} X_j. \tag{8.18}$$

For a chemical reaction, the thermodynamic force is equal to the reaction affinity

$$\begin{aligned} X_j &= -\Delta G_j = -\Delta G_j^{0'} - RT\ln\left(\prod_k c_k^{g_{jk}}\right) \\ &= -\Delta G_j^{0'} - RT\ln\sum_k g_{jk}\ln(c_k). \end{aligned} \tag{8.19}$$

The dependence of the reaction rate on the concentration of reactants is, therefore, linear in the logarithms of the concentrations. Theoretically, the validity of these equations is limited to the vicinity of the thermodynamic equilibrium. However, despite this limitation, these relationships can be applied to numerous biological reactions, which usually operate far from equilibrium (see [78–81] for a more detailed discussion of this issue).

In an extension of this approach, Nielsen [81] described the additional influence of other effectors like activating and inhibiting compounds in analogy to the impact of reactants. Thus, the following generalized representation of the effect of different compounds was suggested:

$$r_i = b_i + a_i \sum_j k_{ij}\ln\left(c_j\right). \tag{8.20}$$

While k_{ij} is an empirical kinetic parameter in case of effectors, the meaning of these parameters in case of reactants remains a stoichiometric coefficient.

Modeling of enzymatic reactions is based on the assumption that the rate of the reaction is proportional to the concentration of the enzyme. Applying this assumption Visser and Heijnen [76] suggested the following relationship:

$$r_i = c_{\mathrm{E},i}\left(b_i + \sum_j a_{ij}\ln(c_j)\right). \tag{8.21}$$

At the same time, these authors proposed to generalize the relations by allowing all kinetic parameters a_{ij} to be freely chosen.

Equation (8.21) can be applied to the reference state characterized by the flux J_0 for the reaction rate. According to Eq. (8.17), the elasticity coefficient can be written as follows:

$$\epsilon_{ij}^0 = \frac{\partial J_0}{\partial c_j}\frac{c_j}{J_0}. \tag{8.22}$$

With J_0 given by (8.23)

$$J_0 = c_{\mathrm{E},i}^0\left(b_i + \sum_j a_{ij}\ln\left(c_j^0\right)\right), \tag{8.23}$$

one obtains

$$a_{ij} = \epsilon_{ij}^0\frac{J_i^0}{c_{\mathrm{E},i}^0} \tag{8.24}$$

and

$$b_j = \frac{J_i^0}{c_{\mathrm{E},i}^0}\left(1 - \sum_j \epsilon_{ij}^0\ln\left(c_j^0\right)\right). \tag{8.25}$$

Finally,

$$\frac{r_i}{J_i^0} = \frac{c_{\mathrm{E},i}}{c_{\mathrm{E},i}^0}\left(1 + \sum_j \epsilon_{ij}^0\ln\frac{c_j}{c_j^0}\right), \tag{8.26}$$

which is the same as Eq. (8.16) introduced by Visser and Heijnen [76].

The log–lin approach suggested by Hatzimanikatis and Bailey [75] starts from a completely different background – the mass balances on the metabolites of the system written as

$$\frac{\mathrm{d}\mathbf{c}}{\mathrm{d}t} = \mathbf{f}(\mathbf{r}(\mathbf{c};\mathbf{p}(t)), \mathbf{c};\mathbf{p}(t), t). \tag{8.27}$$

The system of equations is afterward linearized around the steady state with the aid of Taylor series. In what follows, the derivation of the log–lin model is performed directly from the rate equation of a single reaction of the system. This allows a more consistent comparison with the lin–log approach of (8.15). Thus, with a nonlinear rate equation given by

$$r_i = f_i\left(c_{ij}, c_{\mathrm{E}};\mathbf{p}\right).$$

Linearization of r_i from r_i^0 by a Taylor series yields

$$r(c_{ij}, c_{\mathrm{E},i}, \mathbf{p}) = r_i^0 + \left(c_{ij} - c_{ij}^0\right)\left(\frac{\partial r_i}{\partial c_{ij}}\right)_0 + \left(c_{\mathrm{E},i} - c_{\mathrm{E},i}^0\right)\left(\frac{\partial r_i}{\partial c_{\mathrm{E},i}}\right)_0 + \sum_{k=1}^{mP} (p_k - p_k^0)\left(\frac{\partial r_i}{\partial p_k}\right)_0. \tag{8.28}$$

Replacing r_i^0 with J_0, introducing the elasticity coefficients

$$\left(\frac{\partial r_i}{\partial c_{ij}}\right)_0 = \epsilon_{ij}\frac{J_0}{c_{ij}^0} \qquad \left(\frac{\partial r_i}{\partial c_{\mathrm{E},i}}\right)_0 = \epsilon_{\mathrm{E},i}\frac{c_{\mathrm{E},i}^0}{J_0},$$

where it is assumed that the parameters do not change $(p_k = p_0)$. Finally, the following expression for linearization of $\ln y$ from $y = y_0$ is used:

$$\ln(y) \cong \ln(y_0) + \frac{y - y_0}{y_0} \quad \text{or} \quad \ln\left(\frac{y}{y_0}\right) \cong \frac{y - y_0}{y_0},$$

and the representation of r_i suggested by Hatzimanikatis and Bailey [75] is obtained:

$$r_i = J_0\left(1 + \sum_j \epsilon_{ij} \ln\frac{c_{i,j}}{c_{i,j}^0} + \epsilon_{\mathrm{E},i} \ln\frac{c_{\mathrm{E},i}}{c_{\mathrm{E},i}^0}\right) = J_0\left(1 + \ln\frac{c_{\mathrm{E},i}}{c_{\mathrm{E},i}^0} + \sum_j \epsilon_{ij} \ln\frac{c_{i,j}}{c_{i,j}^0}\right). \tag{8.29}$$

In Eq. (8.29), it is assumed that the rate of reaction is proportional to the concentration of enzyme and hence that the elasticity coefficient $\epsilon_{\mathrm{E},i}$ is equal to 1.

Thus, in case of stimulus–response experiments, the two approaches (lin–log (Eq. (8.26)) and log–lin (Eq. (8.29)) give the same result when the concentration of the enzyme remains constant $\left(c_{\mathrm{E},i} = c_{\mathrm{E},i}^0\right)$. In all those applications in which the enzyme concentration is no longer constant, for example, continuous cultures with varying dilution rates, fed-batch cultures with constant feeding rate and application of the models for predictive design of metabolic networks, the two models provide different results.

When carefully studying the relevant literature, an ongoing debate with conflicting opinions regarding the quality of the various approximate approaches is discerned. In a profound overview based on theoretical arguments, Heijnen [82] evaluated six different approximate formats (linear, logarithmic–linear, power law GMA, power law S-systems, thermokinetic, and linear–logarithmic) and concluded that the lin–log approach is the most appropriate approximate kinetic format. The most important criterion is the proportionality between reaction rate and enzyme concentration, and the quality of the approximation of the hyperbolic kinetics (irreversible Michaelis–Menten). Also, the possibility to obtain analytical solutions at steady state is a criterion. Based on a concrete, experimentally observed system (recombinant *E. coli*), representatives of the S-systems approach have resulted in contradictory findings [83]. The dispute focuses on the problem of the lin–log approach to correctly describe the systems behavior for very small concentrations and there is no definition of rates at concentrations equal to zero. An added weakness of the S-systems approach is the fact that power law representations of individual reactions grow without bound if substrate concentration grow to infinity. The well-known homeostasis – the tendency of a regulated system to maintain itself close to some fixed point – is used to argue that these extreme situations may be unimportant in reality. On the other hand, approximate models are just "models" that turn out to be more easy to work with than theoretical (and sometimes speculative) enzymatic reaction kinetics. ("Remember that all models are wrong; the practical question is how wrong do they have to be, to not be useful." [84])

There are, however, other deficits of both approaches that deserve to be mentioned. The focus on the comparison is most often restricted to simple Michaelis–Menten kinetics and other important phenomena like substrate inhibition and sigmoidal Hill kinetics (see Section 7.2) have received far too little attention. Also, enzymes with much more complicated kinetic behavior, such as the phosphofructokinase (described in Section 8.2), introduce additional problems. ATP is both a substrate and an inhibitor, however, with different binding sites and therefore unequal elasticity coefficients. A possible solution of these problems could be the use of hybrid models, composed of mechanistic and approximate kinetics. Costa *et al.* [85] have successfully applied such an approach and compared their results with the full mechanistic reference model of Chassagnole *et al.* [19].

The real challenge for the approximate models is the modeling of large networks with the vision of dynamic simulations of the genome-scale models of Chapter 6. Initial attempts have been undertaken to address this issue through the development of dynamic metabolic flux analysis [86–88]. For more details, the reader is referred to the review of Antoniewicz [89]. The usefulness of this technique for cell factory improvements remains to be proven [88].

The following study [53] is an attempt to settle the important question if approximative formats such as the (lin–log) approach can be used for these important tasks.

Figure 8.18 shows a graphical representation (software: Insilico Discovery/ Insilico Biotechnology AG) of the reaction network of *E. coli* introduced by [19]. The network comprises both catabolic and anabolic routes with protein, DNA, RNA, polysaccharides, murein, and lipids building up the biomass. Sequential reaction steps and parallel routes are lumped. With 126 reactions, 130 balanced metabolites, and 7 conserved moieties, the degree of freedom of the network is fixed to $126 - 130 + 7 = 3$. The relatively small number of free fluxes in combination with a fixed direction of about 50% of all reactions ("irreversible reactions") leads to a total of 23 elementary flux modes. The complete list of reactions can be found in [19]. Additional information used for the dynamic modeling takes into account the network of the metabolic modulation. The inhibitions and

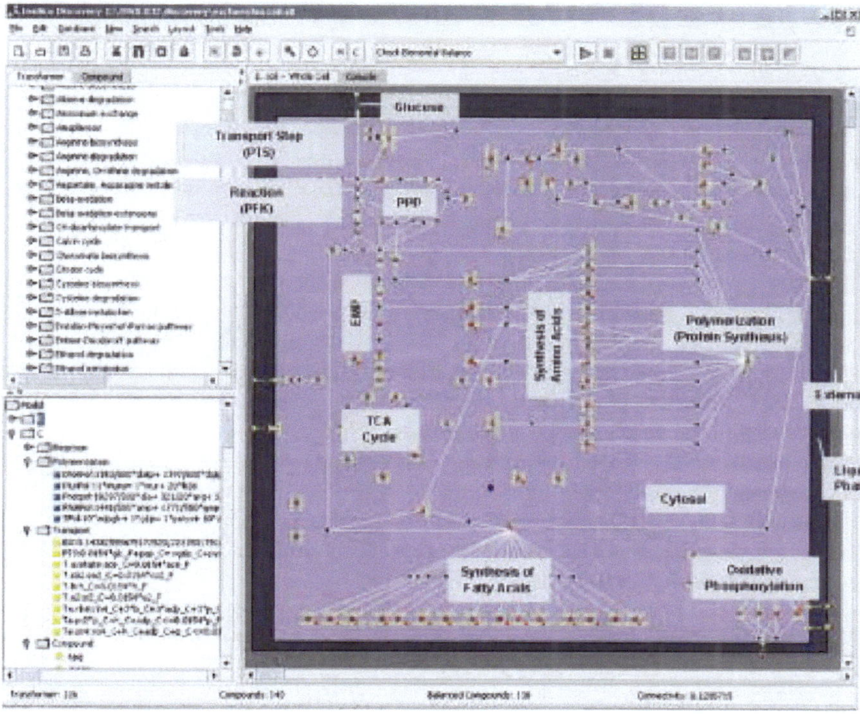

Figure 8.18 Metabolic reaction network of *E. coli* [19,53].

activations for the individual reactions have been gathered from the MetaCyc database (www.metacvc.org, [90]) (Figure 8.19).

Typical results of comparisons between model simulations and experimentally observed time series of metabolites are shown in Figure 8.20. The trajectories represent simulations with kinetic parameters (elasticities) resulting from 10 independent optimization runs. With exception of PEP, NAD, and NADP, the trend of the experimental data is described well by the various simulation runs. Highly connected metabolites, for example, ATP, and feedback mechanisms from monomer building blocks show how the interaction of anabolism and catabolism constitutes important structural determinants for metabolic network dynamics. Consequently, the predictive strength of whole-cell metabolic networks is assumed to be more pronounced compared to metabolic submodels, where in most of the cases highly linked metabolites are kept constant in applications of model-based metabolic redesign. Although a rather simple linear–logarithmic kinetic behavior has been assigned to the individual reaction steps, the whole-cell network describes the experimentally observed complex metabolic behavior quite well. This observation is in line with similar comparisons between lin–log kinetics, more sophisticated kinetic modeling, and experimental observations for the submodel [19] of the central metabolism of *E. coli* [44]. Due to

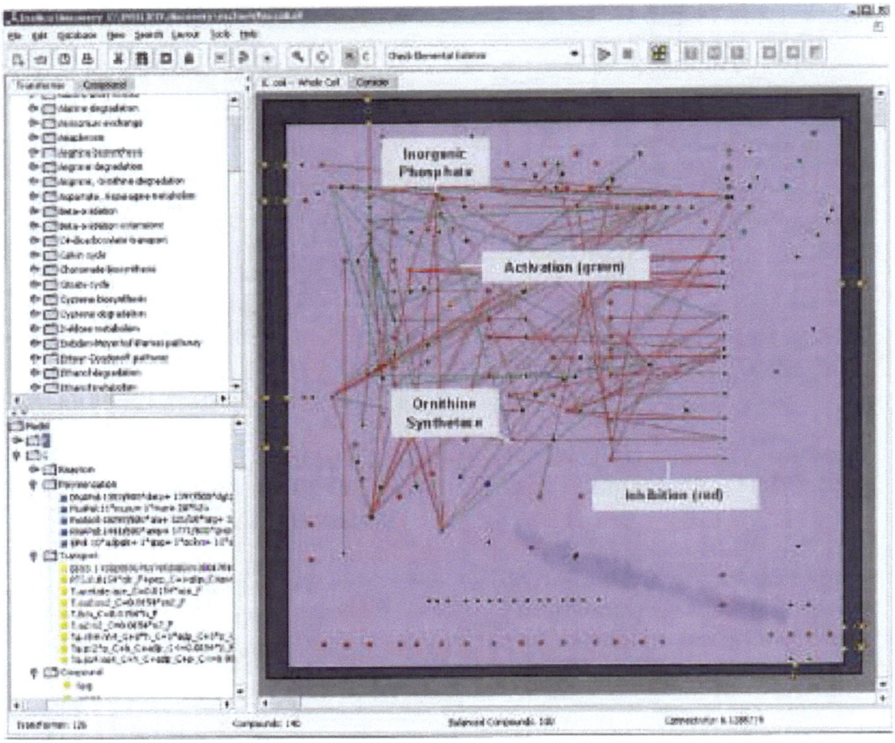

Figure 8.19 Modulation network of *E. coli*. Inhibition and activation in accordance with the MetaCyc database.

self-generating network kinetics and the application of evolutionary strategies on computer clusters, the time from data collection to validated network models can be remarkably reduced.

To obtain a true picture of the dynamics of metabolic networks and to predict targets for relevant genetic manipulations or drug discovery, we need quantitative descriptions of the behavior of the relevant reactions embedded in the network of interest. The identification of reliable and accurate structure and parameters of the kinetic expressions imposes severe challenges, and there is a great need for more rigorous approaches. It should be noted that in the past a great deal of research has been based on bottom-up methods in which kinetics from *in vitro* measurements have been aggregated to obtain models of pathways or modules. There are two major problems related to this approach. First of all, there are numerous indications that *in vitro* kinetics may considerably differ in structure and parameters from *in vivo* situations. Moreover, despite impressive progress made in more rigorous definitions of modularity in context with biochemical networks, open questions remain as "whether and to what extent large-scale dynamic models generated from aggregation of these modules reflect reality." It needs to

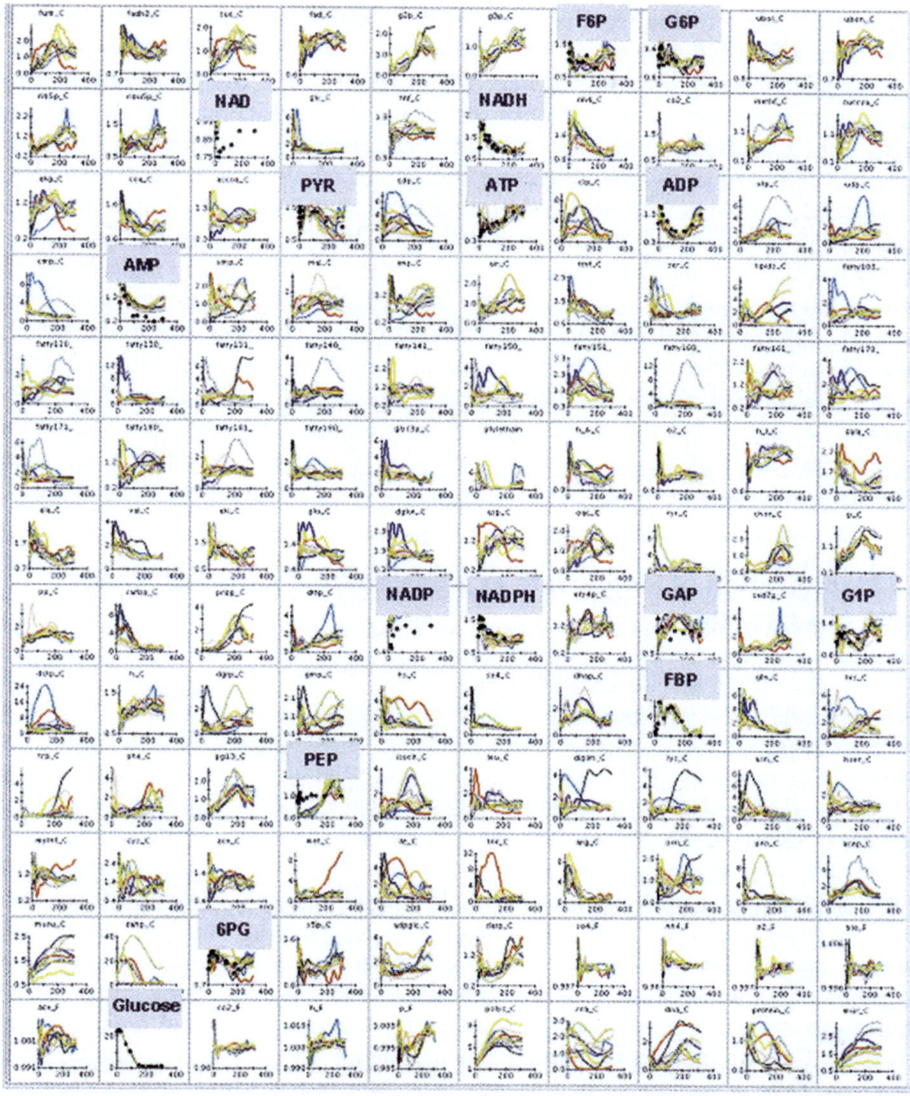

Figure 8.20 Whole-cell metabolic network dynamics. The trajectories represent simulations with kinetic parameters (elasticities) resulting from 10 different optimization runs [53]. Time series data represented by circles were obtained by stimulating *E. coli* at a growth rate of $D = 0.1\ \text{h}^{-1}$ with glucose [19].

be stressed that while the definition of functional modules greatly facilitates model development for cellular processes, one must not forget that the definition of these modules is usually not unique and mainly serves to delineate systems boundaries when developing models for small subsystems. Assigning highly connected compounds, such as adenine or pyridine nucleotides, to a specific module represents a further challenge, considering that these metabolites are produced and consumed in a vast number of intracellular reactions, in processes as diverse as respiration, biosynthetic reactions, solute transport, cytoskeletal dynamics, or cell cycle progression. Consequently, when formulating a model for one of these processes as an isolated functional module, it will frequently be difficult to obtain a satisfactory description of, for example, ATP dynamics solely on the basis of its role within the module compared to *in vivo* data.

The construction of large-scale dynamic models based on a canonical representation of reaction kinetics is, as illustrated in this chapter, a promising concept to overcome the aforementioned limitations. The approach has been illustrated by identifying whole-cell network dynamics for *E. coli* based on time series data from stimulus–response experiments. A model of such a large dimension necessarily extends beyond the module boundaries. However, much remains to be done with respect to the problems of observability and optimal experimental design. The fact that in this chapter only a passing reference has been given to these problems should not obscure their importance. In fact, these extraordinarily important aspects of model verification are not confined to the sensitivity of the parameters but, among other things, also concern issues of model reductions based on thermodynamic considerations (equilibrium and near-equilibrium reactions) and on possible correlations between different model parameters. It is recommended to carefully study the relevant literature, for example, Bertoumieux *et al.* [91], Nikerel *et al.* [92], Bucher *et al.* [93], Canelas *et al.* [33], Pozo *et al.* [73], Di Maggio *et al.* [94], and Almquist *et al.* [88].

In the long perspective, it may prove useful to combine the detailed representation of a functional module with a less detailed large-scale representation of the dynamics of highly connected module compounds as illustrated in the discussion by Costa *et al.* [85] of the still comparatively small model of Chassagnole *et al.* [19]. Key challenges of these and other rigorous large-scale dynamic network models will be further development of appropriate computational methods and the critical linkage between modeling and experiment.

8.6
Dynamic Models Used for Redesigning Cell Factories. Examples: Optimal Ethanol Production in Yeast and Optimal Production of Tryptophan in *E. Coli*

Experimental observations and mathematical modeling together form the best tool we have for understanding the dynamics of living systems. Utilizing a quotation from Francis Bacon's posthumously published (1627) "A natural History in Ten Centuries," there are experiments of two kinds: "solitary experiments and experiments in consort," that is, those that bring light on a single phenomenon and those that explain a wider truth ("models"), this chapter aims at discussing examples for application of the models for predictive design ("bringing fruits"). The problems of application of dynamic metabolic models to predictive design concern not only the reshaping of cellular networks but also the optimization of operation conditions in the context of scale-up of bioreactors [43]. In other words, we are addressing issues beyond the use of mathematical models for description of experimental observations (deriving knowledge). These tasks pose additional challenges for the quality of the models because they are to be used for predictions beyond the judgment based on the available experiments. As such, mathematical modeling is assumed to be about both explaining and predicting.

The driving force for selection of an optimal pathway is usually the maximization of the yield of the product. However, economic considerations (size of the plant) also require optimization of the product formation rate (productivity). This problem – to be discussed in more general terms in Chapter 16 – leads to the question of an optimal modulation of enzyme activities in metabolic networks.

At first glance, the example chosen for this discussion – ethanol production with the yeast *S. cervisiae* [95] – appears to be rather boring because it has been tackled so many times, and not only by yeast geneticists. The primary aim of these empirical attempts to modulate (mostly amplify) the key enzymes within glycolysis is the maximization of ethanol production rate that correlates with carbon dioxide evolution and, in turn, with the leavening power of yeast. While addressing this problem, it should be emphasized that the interest in answering the question of an optimal redistribution of enzyme activities is much broader. Generally, knowledge of the rate controlling steps in the central metabolism (glycolysis and PP shunt) is of central importance for cell cultures used both for the production of pharmaproteins and for the analysis of potential targets in cancer cells (see Section 8.7). Another field of interest is the identification of potential targets for antitrypanosomal drugs, important for treatment of the African sleeping sickness [96]. The last named authors concluded: "Despite the great interest, it is not yet known completely for any organism how the control of the glycolytic flux is distributed."

8.6.1
Dynamic Model

Similar to the strategy of Bakker *et al.* [96], the problem can be approached on the basis of experimentally determined kinetic properties of the key enzymes that are then aggregated to a dynamic model. Individual rate expressions including their kinetic parameters have been identified *in vivo* by the stimulus–response methodology [15,29,51] (see Section 8.3): a pulse of glucose or of an alternative stimulus is introduced into a continuous culture operating at steady state and the transient response of several intracellular metabolite and cometabolite pools is experimentally determined in time spans of seconds. Within these relatively short time spans, enzyme concentrations are considered to be in a "frozen" state. Figure 8.21 summarizes some of the experimental observations of metabolites and cometabolites from the yeast *S. cerevisiae* growing under anaerobic conditions.

The response of the metabolome due to dynamic system excitation has been used to identify the dynamic system behavior by a stepwise internalization of

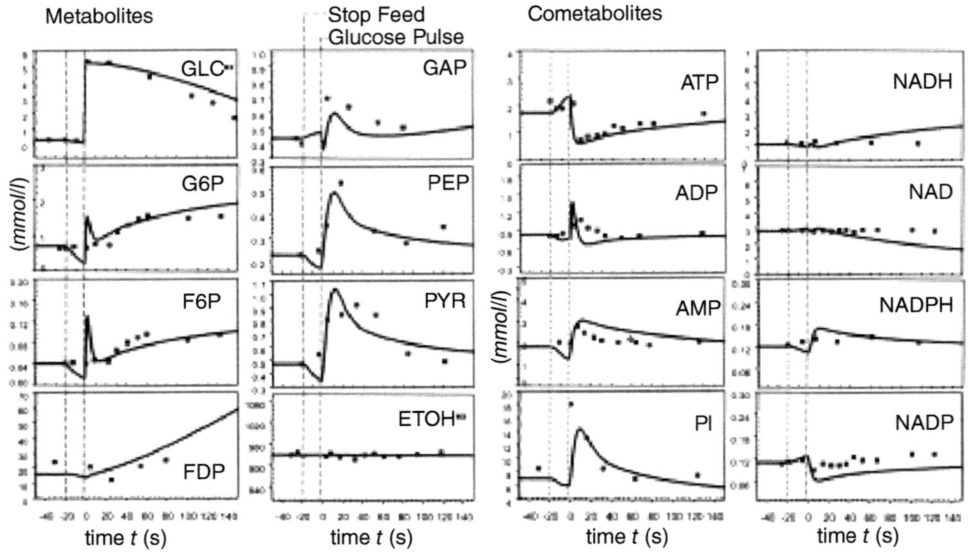

Figure 8.21 Comparison between model simulation and measured concentrations of glycolytic metabolites and cometabolites after a dynamic system excitation. Data shown to the left of the broken lines represents steady-state values at a growth rate of $\mu = 0.1\ h^{-1}$ [22].

metabolites similar to the method proposed by Rizzi *et al.* [51]. The details of this method have been described in Section 8.3 for *S. cerevisiae* growing under aerobic conditions.

Results of the comparison between the experimental observation during the pulse-response experiments and the model predictions under anaerobic conditions are shown in Figure 8.21.

8.6.2 Metabolic Control (Sensitivity) Analysis

The next step toward solving the envisaged design problem is to calculate the so-called flux control coefficients or – in the terminology of engineering – sensitivity coefficients. The flux control coefficient C_E^J has been defined as the fractional change of the network flux J caused by a fractional change in the level of enzyme activity E_i (see Section 7.6).

Thus,

$$\text{Flux control coefficient} = \frac{dJ(E_i)}{dE_i} \tag{8.30}$$

or normalized

$$C_{E_i}^J = \frac{dJ/J}{dE_i/E_i} = \frac{d\ln J(E_i)}{d\ln E_i}. \tag{8.31}$$

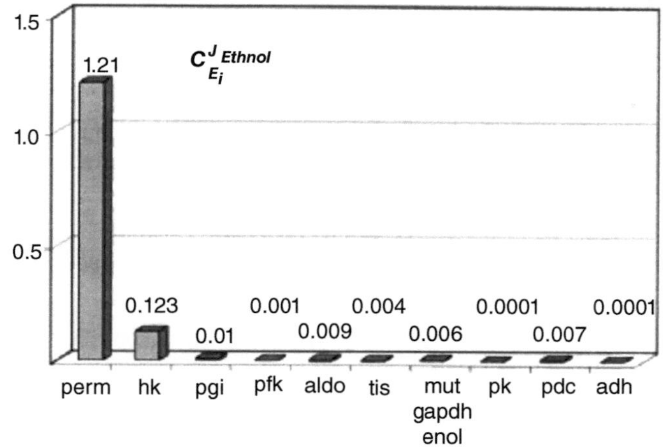

Figure 8.22 Distribution of flux control coefficients on ethanol formation route. Enzymes involved: hexose transporter (perm), hexokinase(hk), phosphoglucose isomerase (pgi), phosphofructokinase (pfk), aldolase (aldo), triose phosphate isomerase (tis), gap dehydrogenase (gapdh), phospoglyceromutase (mut), enolase (enol), pyruvate kinase (pk), pyruvatedecarboxylase (pdc), and alcohol dehydrogenase (adh) [95] (see also Figures 4.2 and 4.4c).

Figure 8.22 depicts the results of these calculations for those enzymes involved in the path from glucose to ethanol based on the dynamic model.

Since only a subset of the flux control coefficients with respect to ethanol formation is shown in Figure 8.22 (excluding, for example, the coefficients of enzymes involved in the PPP shunt), flux coefficients of this subset do not necessarily sum up to one. From the hierarchy of sensitivities, it can be concluded that the enzyme responsible for the transport of glucose via the cell membrane (permease) in reality controls the ethanol production rate. Consequently, amplification of this enzyme should lead to an increased flux from glucose to ethanol. The result is in agreement with what has been discussed by the yeast genetists when trying to amplify individual enzymes to increase the flux to ethanol. The concluding sentence "However, no increase in the glycolytic flux could be achieved with the overexpressing strain" is in line with the extremely small flux control coefficients shown in Figure 8.22 for all the enzymes except the transporter that was not tackled in the mentioned work [97].

8.6.3 Synthesis Amplification of Hexose Transporters

Experimental verification of the above-named design proposal has been performed in a collaboration project with Institute of Microbiology from the University of Frankfurt (Prof. (Dr.) Boles). Within this project, the kinetics of the 3 most important hexose transporters out of the 17 transporters identified from

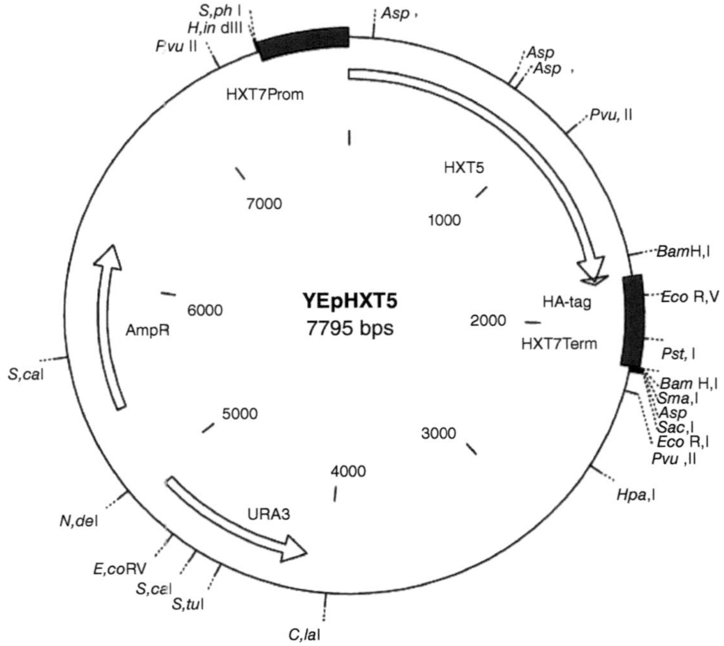

Figure 8.23 HXT5 multicopy plasmid.

the yeast genome project have been investigated [50,98]. One of these – HXT5 – has also been expressed with a multicopy plasmid, shown in Figure 8.23.

The transporter gene in HXT5 is flanked by an enhanced HXT7 promoter and HXT7 terminator, respectively. HXT7 is a high-affinity transporter expressed at low glucose concentrations. The construct shown in Figure 8.23 ensures expression of HXT5 under target growth conditions. The same transporter has been integrated as single copy in the chromosome of a yeast strain in which all the genes of the other transporters were knocked out [99]. As a result of these genetic constructions (summarized in Figure 8.24), it is possible to compare the flux of glucose through glycolysis between two strains that differ only in the amount of the hexose transporter. According to the hierarchy of flux control coefficients discussed in context with Figure 8.22, one would expect a noticeable increase of ethanol flux.

Experiments with three different strains were performed in continuous cultures at a dilution rate of $D = 0.07\,\text{h}^{-1}$. The results of these experiments are summarized in Figure 8.25.

Compared to the wild type, ethanol excretion rate and substrate uptake rate of the strain with higher transporter activity was found to be 10 and 25% higher, respectively. The discussion of the relevance of these results from an industrial point of view is beyond the scope of this chapter. However, the rather modest effect prompts the question: are there any other alternative design possibilities resulting in a substantial increase of the ethanol production rate?

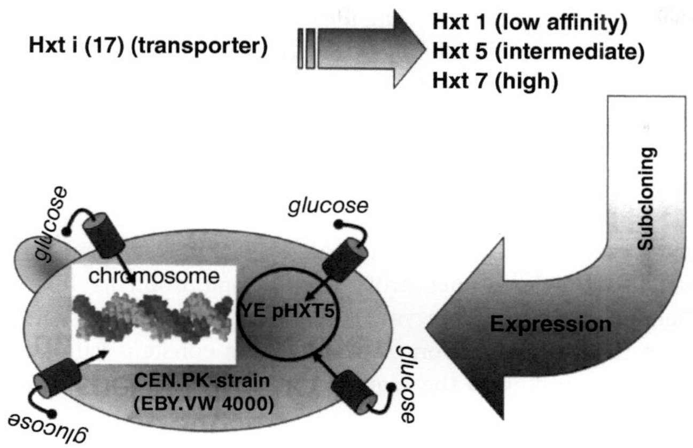

Figure 8.24 Strain construction with HXT5 integrated into the chromosome and on multicopy plasmids.

8.6.4
Objective Function

The apparent failure to produce a significant increase of the glycolytic flux points to the fundamental question: are the underlying assumptions that lead to the design suggestion really adequate? Keeping in mind that the pathway of

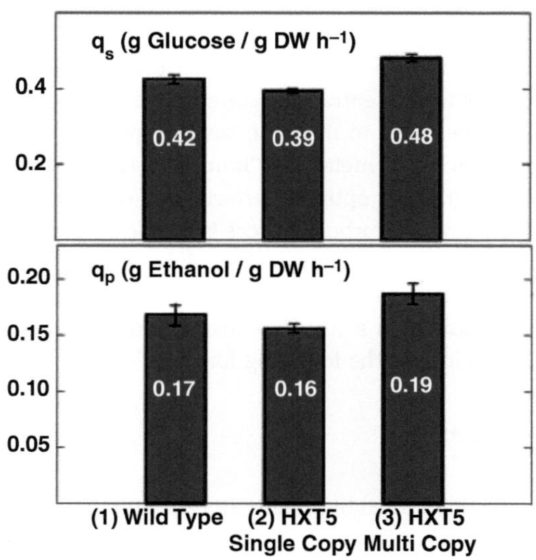

Figure 8.25 Specific substrate uptake rate q_S and specific product formation rate q_P during anaerobic chemostat cultivation. (1) Wild strain, (2) HXT5 with a single copy of the gene integrated into the chromosome and constitutively expressed, and (3) multicopy plasmid with HXT5.

interest is part of a whole – the living cell – the idea that one needs to amplify a single or several enzymes according to the hierarchy of flux control coefficients may not correspond to physiological reality. There are two aspects that should attract attention. First, an increased expression of enzymes is linked with energetically expensive protein synthesis. Glycolytic enzymes in yeast are known to contribute approximately 30% to the total amount of cellular proteins. Thus, it seems likely that overexpression of these enzymes will result in a stress situation with an unforeseeable impact on cell physiology. Also, the unconstrained overexpression may create unwanted and uncontrollable crowding problems. Instead of a single or simultaneous elevation of enzyme activities, a more robust strategy should try to keep the total concentration of proteins at a constant value and redistribute the activities according to the required objective. The optimization problem may then be stated as

$$\text{Maximize}\, J\left(\mathbf{r}_{max}^{Path}\right) \tag{8.32}$$

subject to

$$\frac{1}{w}\sum_{i=1}^{w}\frac{r_i^{\max}}{r_{i,\text{reference}}^{\max}} \leq \Omega. \tag{8.33}$$

In writing Eq. (8.32), we assume that the maximum rate $r_i^{\max}$ is proportional to the enzyme amount. Therefore, Eq. (8.32) specifies a fixed level for the total enzyme activity.

Another relevant issue concerns the pool concentration of the metabolites within the cell. Attempts to increase metabolic fluxes by changing individual enzyme concentrations may lead to substantial changes in metabolite concentrations. This may either prove to be cytotoxic or at least lead to an undesired flux diversion [100]. Therefore, the preservation of metabolite concentrations close to the steady-state values of the wild strain is at any rate desirable to meet the basic property of well-established metabolic systems for which the metaphor "homeostasis" has been coined [101]. It is also known that part of an optimal performance of cells with respect to the use of, for example, energy and carbon sources is a process of adaptation leading to a change of structure to control homeostasis. Interestingly, such structural changes may include a change of enzyme concentrations via regulation of enzyme synthesis. Hence, homeostasis puts a further constraint on the metabolic redesign. Mathematically, this could take the following form:

$$\frac{1}{m}\sum_{i=1}^{m}\frac{\left|c_{i,\text{optimum}}^{\text{steady } state} - c_{i,\text{reference}}^{\text{steady } state}\right|}{c_{i,\text{reference}}^{\text{steady } state}} \leq \Theta. \tag{8.34}$$

Finally, the optimum must be constrained to stable steady states, leading to condition (8.35)

$$\frac{d}{dt}\mathbf{c} = 0 \quad \text{and} \quad Re\left(\lambda_{i,\text{optimum}}\right) \leq 0, \tag{8.35}$$

with $Re(\lambda_i)$ denoting the real part of the system's eigenvalues.

8.6.5
Optimal Solutions

Retaining the total activity at the system's initial state at $\Omega = 1$, nonlinear optimization of enzyme activities yielded significantly higher ethanol formation rates (see Figure 8.26).

As shown in Figure 8.26, amplification of ethanol formation largely depends on the allowed deviations from the initial pool concentrations Θ. Since the ratios of enzyme levels on the elementary flux mode glucose $\rightarrow$ ethanol are subject to modifications, no amplification can be expected for $\Theta = 0\%$, while the maximal possible amplification of ethanol formation is found to be as much as 144% at $\Theta = 210\%$. No stable steady states have been detected above this value. Figure 8.27 shows the optimal distribution of enzyme activities for the selected example, with 40% increase of pool concentrations resulting in an amplification of ethanol formation of 63%.

The optimized modulation of the enzyme activities [95] results in an unexpected and interesting redistribution of enzyme activities (Figure 8.27). Due to its large share in the control of ethanol formation rate, the activity of the hexose transporter (perm) is found to be significantly higher in the optimized metabolic system. In contrast, even though the control coefficients of tis, gapdh, mut, enol, and adh are relatively small, activities of those enzymes are also found to be larger compared to the nonoptimized system.

This outcome can be explained by the ability of the enzymes to counteract an increase of pool concentrations provoked by an increased glucose influx. Obviously, this is the result of the superposition of the three objectives: maximization of flux under more or less homeostatic conditions and unchanged total amount of enzymes.

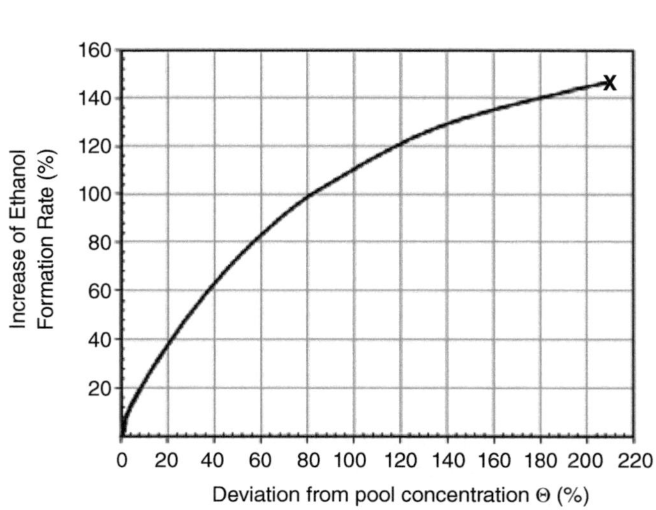

Figure 8.26 Increase of the ethanol formation rate as a function of the maximum deviation from the initial pool concentration.

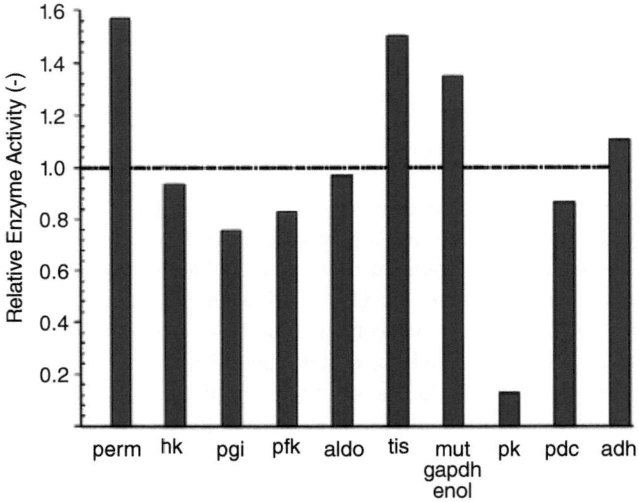

Figure 8.27 Optimal distribution of enzyme activities. All activities are related to their initial values.

The advantage of the aforementioned approach [95] is twofold: (1) in contrast to the application of metabolic control analysis effects of large changes in enzyme concentrations can be easily studied and taken into account for optimization and (2) constraints such as limits on the total amount of enzymes and deviations from steady-state metabolite pool concentrations can be easily taken into account. Also, the condition of stable steady states for the redesigned strains turned out to be very important.

To complete the above investigations regarding the maximization of the flux to ethanol, it should be mentioned that Hatzimanikatis and Bailey [75] have tackled the same problem with the aid of the approximative log–lin approach. The example shows the great advantages of the analytical solution of the log–lin approach.

8.6.6
Flux Optimization of Tryptophan Production with *E. Coli* [69]

Tryptophan is a very important amino acid that is widely used both in medicine and as feed additive. In addition, industrially relevant end products can be derived from tryptophan synthesis intermediates, such as indigo and the anti-influenza drug oseltamivir phosphate [102].

From an industrial point of view, high production rates are desirable, and by applying recombinant DNA technology, many attempts have been made to improve productivities and yields [103,104].

Figure 8.28 shows an overview of the metabolic processes contributing to tryptophan production in *E. coli*. The most important precursor is

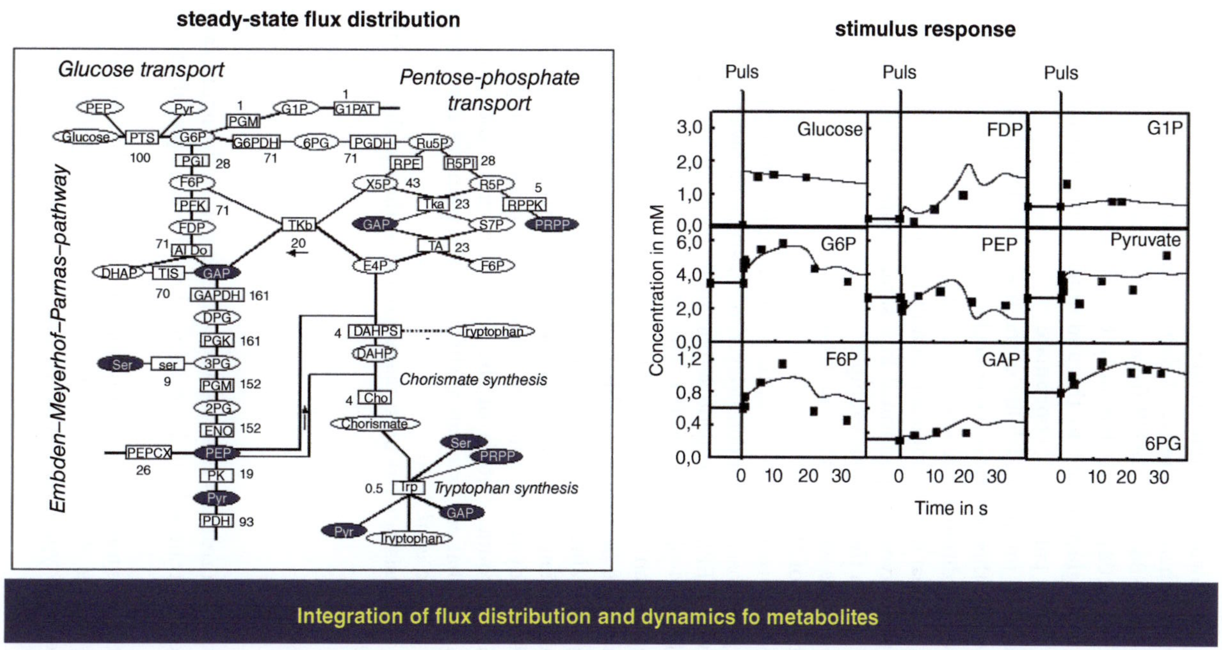

Figure 8.28 Synthesis of tryptophan from glucose in *E. coli*. Summary of the results from experimental investigations and comparison between model prediction and observations. (Data from Ref. [19].)

chorismate that in turn requires the two precursors, phosphoenolpyruvate (PEP) and erythrose-4-phosphate (E4P). After eliminating the feedback inhibitions and repressions, as well as amplification of the operons, emphasis has been given to improve the supply of the precursors PEP and E4P, as reviewed by Bongaerts *et al.* [105] and Nielsen [106]. Second, catabolic and anabolic parts of the network of precursor supply and product formation can be exemplified by the following example: while the maximum yield of the first metabolite of tryptophan synthesis, 3-deoxy-D-arabino-heptulosonate-7-phosphate (DAHP), is 0.86 mole DAHP from 1 mole glucose [107], Schuster, Dandekaar, and Fell [108] found that only 0.45 mole tryptophan can be obtained from 1 mole glucose due to the additional consumption of phosphoenol pyruvate in the pathway of chorismate.

Previous work focused on the availability of the precursor DAHP (see reviews of [105,106]). However, this one-sided view neglects the fact that product synthesis requires several additional precursors (see Figure 8.28).

In what follows, a global optimization for identification of the required amplification of individual enzymes is performed in a similar approach as illustrated for ethanol production in the previous example.

The investigation starts with the dynamic model of Chassagnole *et al.* [19] for the metabolic network of the central carbon metabolism of *E. coli* wild-type strain W3110 (see Figure 8.17). It comprises the phophotransferase system (PTS) transporting glucose, the Emden–Meyerhoff–Parnas pathway providing phosphoenol pyruvate (PEP), and the pentose-phosphate pathway (see Figure 4.7) that supplies erythrose-4-phosphate (E4P). The original model was developed based on the measurement of metabolites in a continuous culture perturbated with a glucose pulse (see Section 8.1). This model is expanded with the shikimate pathway, synthesizing chorismate, and the tryptophan pathway, each lumped in a single reaction step. Enzyme activity in the tryptophan production pathway is considered a dynamic variable in order to take possible amplification of the operon into account. Details of the differences from the model of Chassagnole *et al.* [19] that are associated with this expansion are described in the publication of Schmid *et al.* [69].

The tryptophan synthesis rate has been maximized according to the optimization criterion

$$\max_{r_i^{\max}} r_{TrpSynth} \tag{8.36}$$

by variation of maximum reaction rates for a set of reactions. In the optimization process, the same constraints have been used as in the example of ethanol production.

The results of the integrated optimization of the variation of maximal rates are presented in Figures 8.29 and 8.30. Figure 8.29 summarizes the results obtained with the wild-type strain W3110. The figure depicts two different

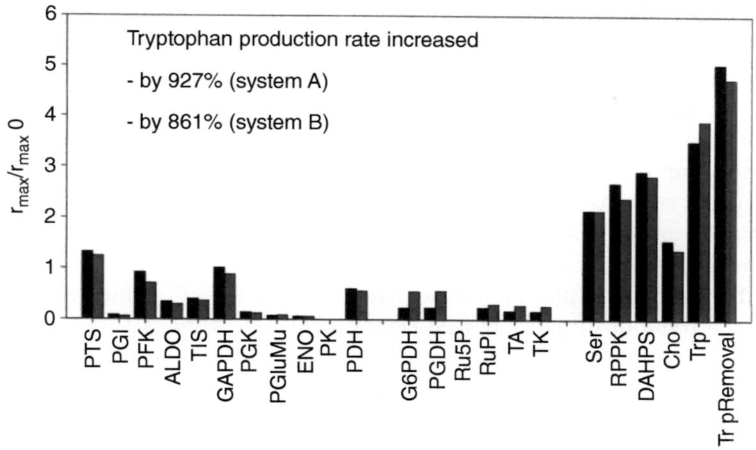

Figure 8.29 Improvement of tryptophan production rate achieved by optimization of enzyme activities for the wild strain W3110.

optimization studies. Systems A and B differ in the assumption of the coenzyme dependency of the isocitrate dehydrogenase. During simulation with system A, it is assumed that NADH is the coenzyme, whereas NADPH is the coenzyme in system B.

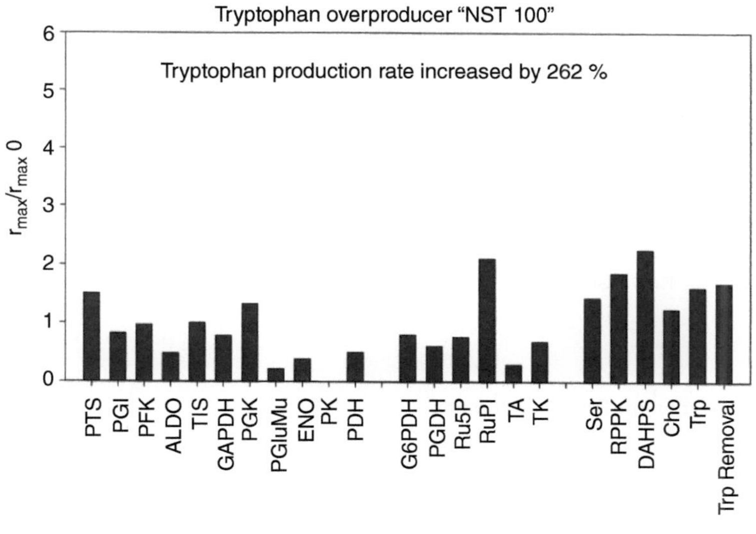

Figure 8.30 Improvement of tryptophan production achieved by optimization of enzyme activities in the production strain NST 100.

The results in Figure 8.29 can be interpreted as follows: there is a strong effect of "tryptophan removal" that reflects the strong impact of inhibition and repression of the enzymes of the pathway caused by the end product. Additional targets for the increase of the production rate are the enzymes of the tryptophan and chorismate pathways and the DAHP synthetase. These targets were already pinpointed in earlier successful applications of recombinant DNA technology and serve as a proof of concept. In other words, the model-based optimization recommends similar targets as were already selected by trial and error. A new target is the ribosephosphate-pyrophosphokinase (RPPK) responsible for the synthesis of phosphoribose pyrophosphate (PRPP), a further precursor in the tryptophan pathway (see Figure 8.28). The same is true for the serin synthesis pathway, with end product serine as another precursor in the tryptophan pathway. With the exception of the sugar transport system, the optimization study recommends a decrease of activities of several enzymes in the central metabolism.

The figure shows the results of an optimization study performed with the strain NST 100 from the contribution of Triebe and Pittard [109]. The metabolic flux analysis at steady state is carried out based on the fluxes of glucose uptake, biomass production, and tryptophan excretion during the exponential growth phase of the strain. The model also differs from the other models by the suppression of the synthesis of phenylalanine and tyrosine. Since there is an upregulated copy of the *trp* operon in strain NST 100, gene expression regulation is removed in the model. In NST 100, feedback inhibition of DAHP synthetase is removed as well. All these assumptions are based on the description of the strain in the original paper [109].

An interesting result from the optimization studies is that a further increase in the production rate is possible by appropriate selection of the target enzymes. The fact that these targets differ in comparison to the wild strain is not surprising.

Visser *et al.* [44] have used the dynamic model of Chassagnole *et al.* [19] to estimate the parameters of a lin–log approximation of the central metabolism *of E. coli.* The optimization studies concerned the maximization of the flux through the PTS system. These authors applied the same constraints for the optimization as illustrated in the yeast example.

8.7
Target Identification for Drug Development

In contrast to the examples shown so far, which have dealt with questions of the redesign of industrial production strains, the following study focuses on a medically relevant system. The task concerns the important drug development process in the context with new cancer therapies.

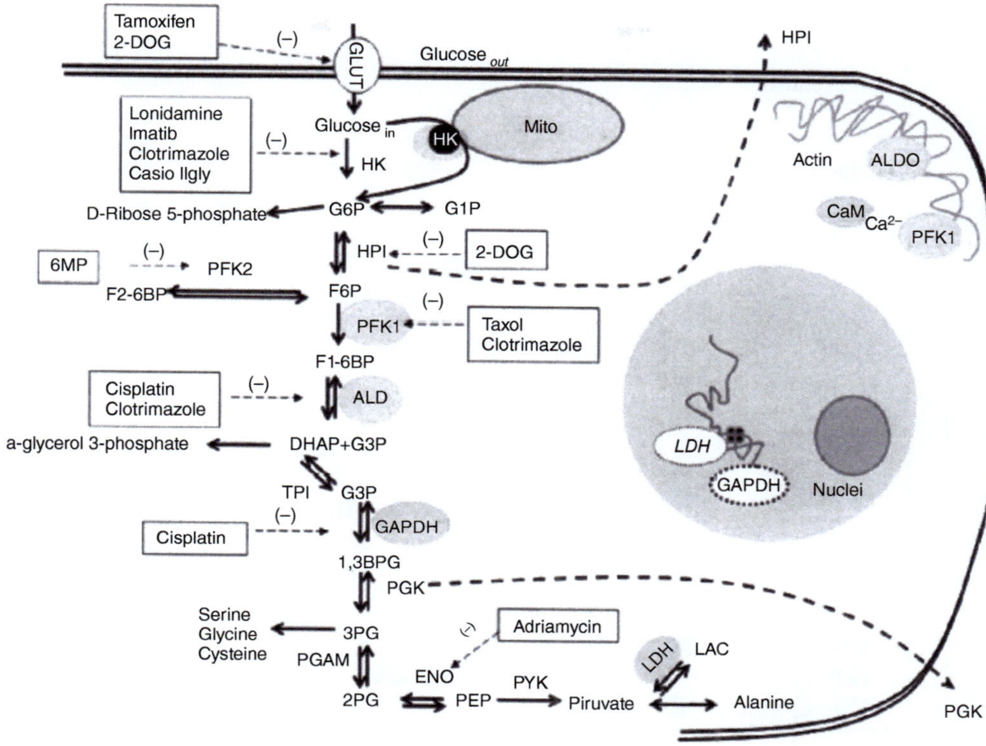

Figure 8.31 Glycolytic enzyme targeting by anticancer drugs [111].

Warburg showed in the 1920s that per unit time and under aerobic conditions, tumor tissue metabolizes approximately 10 times as much glucose to lactate as does a healthy cell, a phenomenon known as the Warburg effect. This very old discovery is the driving force for a recent renaissance of drug discovery processes focusing on targets in the energy metabolism of cancer cells (e.g., [110,111]). In contrast to the two previous examples of redesign of production strains, the task is not related to increase of fluxes but to decrease of fluxes, aimed at a lower ATP production in the cancer cells. At a first glance, the overview of glycolytic enzyme targeting summarized in Figure 8.31 [111] looks like a random walk through the energy metabolism. However, many of the mentioned drugs are already under study in phase I–III clinical trials.

In what follows, a more systematic approach for target identification is discussed that is based on the strategies of model-based design of metabolic networks discussed earlier. Figure 8.32 depicts the part of the metabolic network used for the modeling approach. The model includes glycolysis, the pentose-

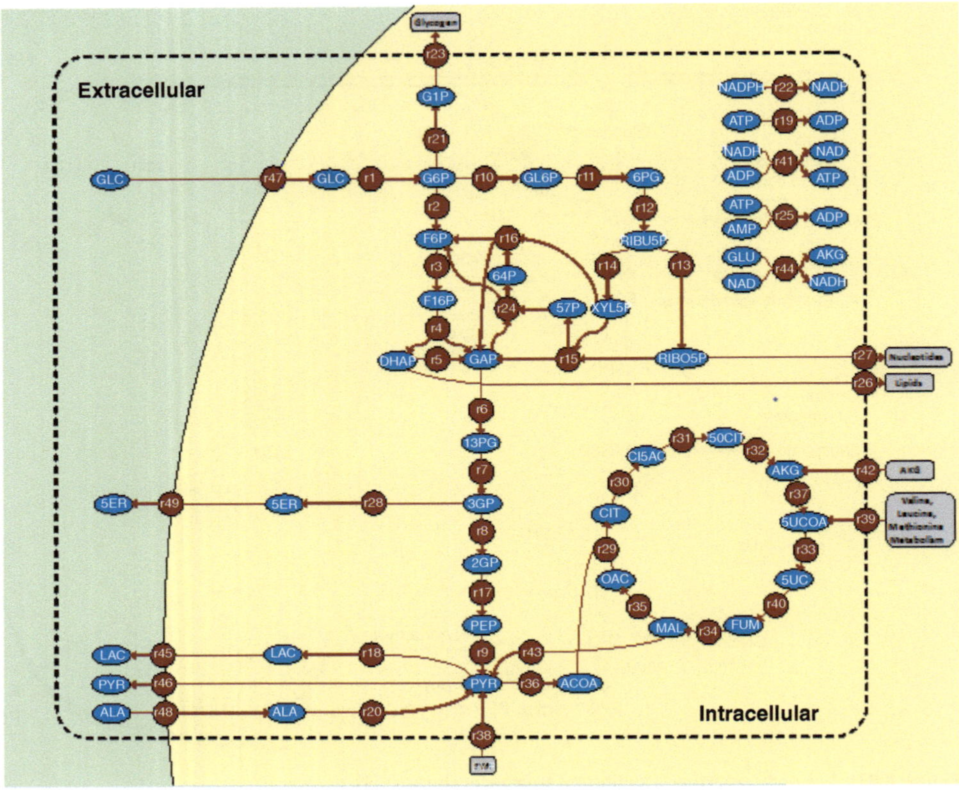

Figure 8.32 Metabolic network model (HEPG2: liver cancer cells) [112].

phosphate shunt, the citric acid cycle, and respiration. The experiments required for model verification have been performed with HEPG2 cells (human liver cancer cells) cultivated on six-well plates. Again, the kinetic modeling is performed on the basis of the universal canonical approach of the lin–log kinetics introduced in Eq. (8.16).

The quantitative information about the steady-state metabolic flux analysis was derived with the aid of an instationary C-13 metabolic flux analysis [113,114] (see Chapter 5). Results of the estimated flux distribution are summarized in Figure 8.33. As expected, the Warburg effect is manifested in the extremely high flux to lactate as well as the very low flux into the TCA.

Figure 8.34 summarizes the comparison of the measured data and model simulations for the dynamic model on the basis of the universal lin–log approach. The perturbation of the steady state was not performed with a

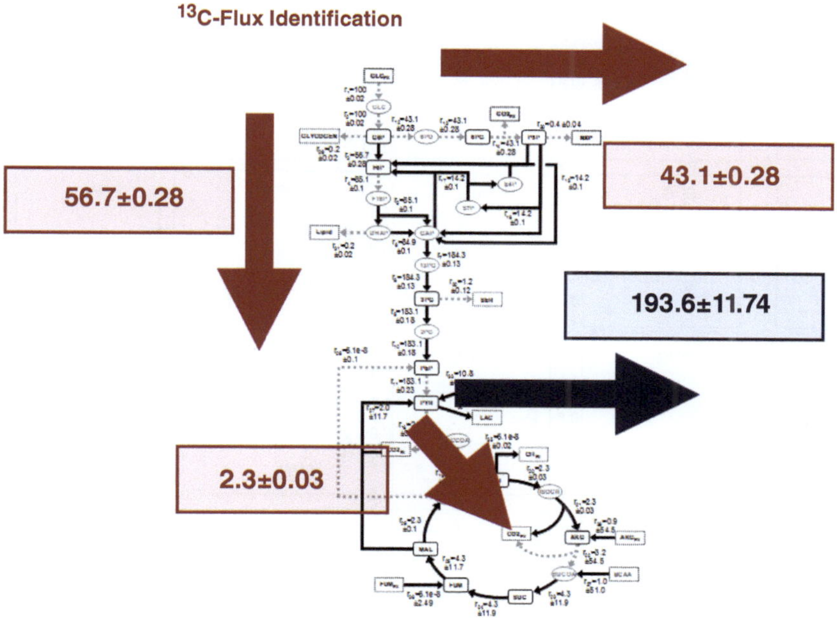

Figure 8.33 Metabolic flux distribution for HEPG cells obtained from transient ^{13}C flux analysis [113,114].

pulse of glucose. Due to the saturation of the glucose uptake rate in the HEPG2 cells, it was necessary to stimulate perturbation by changing the medium for the adherently growing cells from glucose-rich conditions to a medium without glucose.

The application of the dynamic model to the task of identifying drug targets to reduce the flux through glycolysis is at present restricted to the metabolic flux control analysis. However, particularly for solving problems of promising combinatorial drug treatment, it would be also possible to apply the global optimization strategy as illustrated in Section 8.6.

The results of the analysis are summarized in Figure 8.35, showing the impact of the concentration of an enzyme i upon the flux to reaction j. The indices i and j correspond to the reaction shown in Figure 8.32.

This last example of application of top-down dynamic modeling shows that it is possible to perform a rigorous analysis of the control pattern. The most interesting result concerns the identification of glucose-6-phosphate dehydrogenase as a possible target for negative control of the flux through glycolysis. This result is in accordance with the observation of Boren *et al.* [115] based on a completely different approach.

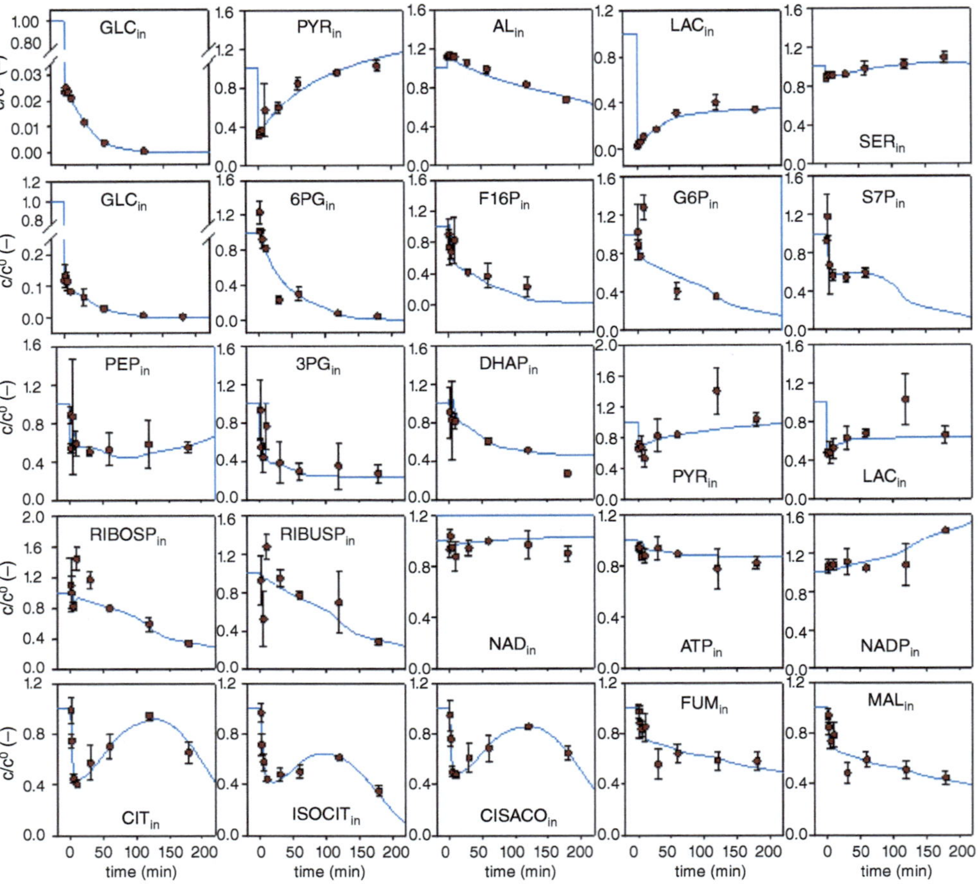

Figure 8.34 Comparison between measured intra- and extracellular metabolite and dynamic simulations with the aid of lin–log kinetics. The subscripts "in" and "ex" denote intracellular and extracellular metabolites, respectively. The concentration values were normalized with respect to their reference values, that is, the concentrations directly before the stimulus. The error bars indicate standard deviations of the experimental data. To perturb the central metabolism, the glucose-containing culture medium was exchanged with glucose-free medium. As a consequence, the extracellular glucose level dropped, and this stimulated significant intracellular metabolite dynamics [112].

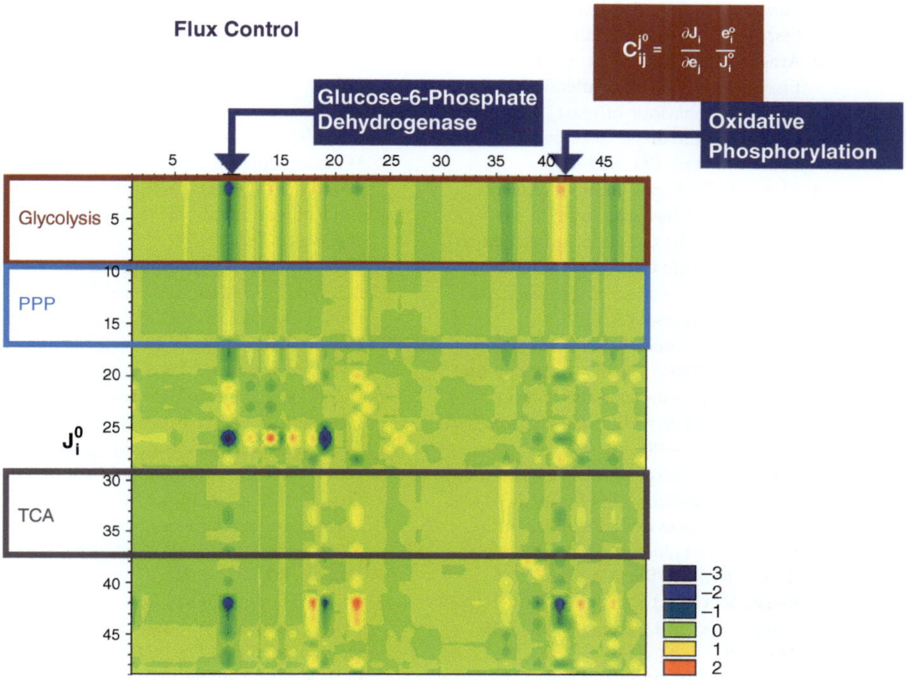

Figure 8.35 Flux control coefficients. The colors in row i and column j indicate the control that enzyme j exerts over the steady-state flux through reaction i (i and j correspond to the reaction number in Figure 8.32). The glucose-6-phosphate dehydrogenase (r10) was found to have significant negative control over all glycolytic fluxes, whereas oxidative phosphorylation (r41) exerted positive control (Warburg effect) [112].

References

1 Wright, B.E. and Kelly, P.J. (1981) Kinetic models of metabolism in intact cells, tissues and organisms. *Curr. Top. Cell. Regul.*, **19**, 103–158.

2 Teusink, B., Passarge, J., Reijenga, C.A., Esgalhado, E., van der Weijden, C.C., Schepper, M., Walsh, M.C., Bakker, B.M., van Dam, K., Westerhoff, H.V., and Snoep, J.L. (2000) Can yeast glycolysis be understood in terms of *in vitro* kinetics of the constituent enzymes? Testing biochemistry. *Eur. J. Biochem.*, **267**, 5313–5329.

3 Snoep, J.L. and Westerhoff, H.V. (2005) From isolation to integration, a systems biology approach for building the silicon cell. *Top. Curr. Genet.*, **13**, 13–30.

4 Snoep, J.L., Bruggeman, F., Olivier, B.G., and Westerhoff, H.V. (2006) Towards building the silicon cell: a modular approach. *Biosystems*, **83**, 207–216.

5 Mauch, K., Vaseghi, S., and Reuss, M. (2000) Quantitative analysis of metabolic and signalling pathways in Saccharomyces cerevisiae, in *Bioreaction Engineering* (eds. K. Schügerl and K.H. Bellgardt), Springer, Berlin, pp. 435–477.

6 Aon, M.A. and Cortassa, S. (1997) Dynamic biological organization, in

Fundamentals as Applied to Cellular Systems, Chapman & Hall, London.

7 Aragon, J.J. and Sanchez, V. (1985) Enzyme concentration affects the allosteric behaviour of yeast phosphofructokinase. *Biochem. Biophys. Res. Commun.*, **131**, 849–855.

8 Klann, M.T., Lapin, A., and Reuss, M. (2011) Agent-based simulation of reactions in the crowded and structured intracellular environment: influence of mobility and location of the reactants. *BMC Systems Biol.*, **5**, 71.

9 Baltes, U., Schneider, R., Sturm, C., and Reuss, M. (1994) Optimal experimental design for parameter estimation in unstructured growth models. *Biotechnol. Prog.*, **10**, 480–488.

10 Weibel, K.E., Mor, J.R., and Fiechter, A. (1974) Rapid sampling of yeast cells and automated assays of adenylate, citrate, pyruvate and glucose-6-phosphate pools. *Anal. Biochem.*, **58**, 208–216.

11 Reuss, M. (1991) Structured modeling of bioreactors. *Ann. NY Acad. Sci.*, **646**, 284–299.

12 De Koning, W. and van Dam, K. (1992) A method for determination of changes of glycolytic metabolites in yeast on a subsecond time scale using extraction at neutral pH. *Anal. Biochem.*, **204**, 118–123.

13 Theobald, U., Mailinger, W., Reuss, M., and Rizzi, M. (1993) *In vivo* analysis of glucose-induced fast changes in yeast adenine nucleotide pool applying a rapid sampling technique. *Anal. Biochem.*, **214**, 31–37.

14 Theobald, U., Mailinger, W., and Rizzi, M. (1994) Use of HgCl2 to investigate dynamic phenomena in yeast cytoplasm. *Biotechnol. Techn.*, **8**, 723–728.

15 Theobald, U., Mailinger, W., Baltes, M., Rizzi, M., and Reuss, M. (1997) *In vivo* analysis of metabolic dynamics in *Saccharomyces cerevisiae*: I. Experimental observations. *Biotechnol. Bioeng.*, **55**, 305–316.

16 Weuster-Botz, D. and de Graaf, A.A. (1996) Reaction engineering methods to study intracellular metabolite concentrations. *Adv. Biochem. Eng.*, **54**, 75–108.

17 Gonzales, B., Francois, J., and Renaud, M. (1997) A rapid and reliable method for metabolite extraction in yeast using boiling buffered ethanol. *Yeast*, **13**, 1347–1356.

18 Schaefer, U., Boos, W., Takors, R., and Weuster-Botz, D. (1999) Automated sampling device for monitoring intracellular metabolite dynamics. *Anal. Biochem.*, **270**, 88–96.

19 Chassagnole, C., Noisommit-Rizzi, N., Schmid, J.W., Mauch, K., and Reuss, M. (2002) Dynamic modelling of the central carbon metabolism of *Escherichia coli*. *Biotechnol. Bioeng.*, **79**, 53–73.

20 Buziol, S., Bashir, I., Baumeister, A., Claassen, W., Noisommit-Rizzi, N., Mailinger, W., and Reuss, M. (2002a) New bioreactor-coupled rapid stopped-flow sampling technique for measurements of metabolite dynamics on a subsecond time scale. *Biotechnol. Bioeng.*, **80**, 632–636.

21 Visser, D., van Zuylen, G.A., van Dam, J.C., Oudshoorn, A., Eman, M.R., Ras, C., van Gulik, W.M., Frank, J., van Dedem, G.W.K., and Heijnen, J.J. (2002) Rapid sampling for analysis of *in vivo* kinetics using the BioScope: a system for continuous-pulse experiments. *Biotechnol. Bioeng.*, **79**, 674–681.

22 Schmalzriedt, S., Jenne, M., Mauch, K., and Reuss, M. (2003) Integration of physiology and fluid dynamics. *Adv. Biochem. Eng. Biotechnol.*, **80**, 19–68.

23 Castrillo, J.I., Hayes, A., Mohammed, S., Gaskell, S.J., and Oliver, S.G. (2003) An optimized protocol for metabolome analysis in yeast using direct infusion electrospray mass spectrometry. *Phytochemistry*, **62**, 929–937.

24 Masegho, M.R., van Gulik, W.M., Vinke, J.L., Visser, D., and Heijnen, J.J. (2006) *In vivo* kinetics with rapid perturbation experiments in *Saccharomyces cerevisiae* using a second-generation BioScope. *Metabol. Eng.*, **8**, 370–383.

25 Schaub, J., Schiesling, C., Reuss, M., and Dauner, M. (2006) Integrated sampling procedure for metabolome analysis. *Biotechnol. Progr.*, **22**, 1434–1442.

26 Wu, L., Masegho, M.R., Proell, A.M., Vinke, J.L., Ras, C., van Dam, J.C., van

Winden, W.A., van Gulik, W.M., and Heijnen, J.J. (2006) *In vivo* kinetics of primary metabolism in *S. cerevisiae* studied through prolonged chemostat cultivation. *Metabol. Eng.*, **8**, 160–171.

27 Rizzi, M., Theobald, U., Querfurth, E., Rohrhirsch, T., Baltes, M., and Reuss, M. (1996) *In vivo* investigations of glucose transport in *Saccharomyces cerevisiae*. *Biotechnol. Bioeng.*, **52**, 316–327.

28 Mailinger, W., Baumeister, A., Reuss, M., and Rizzi, M. (1998) Rapid and highly automated determination of adenine and pyridine nucleotides in extracts of *Saccharomyces cerevisiae* using a micro robotic sample preparation-HPLC system. *J. Biotechnol.*, **63**, 155–166.

29 Vaseghi, S., Baumeister, A., Rizzi, M., and Reuss, M. (1999) *In vivo* dynamics of the pentose phosphate pathway in *Saccharomyces cerevisiae*. *Metabol. Eng.*, **1**, 128–140.

30 Vaseghi, S., Macherhammer, F., Zibek, S., and Reuss, M. (2001) Signal transduction dynamics of the protein kinase A/ phosphofructokinase-2-system in *Saccharomyces cerevisiae*. *Metabol. Eng.*, **3**, 163–217.

31 Stephanopoulos, G.N., Aristidou, A.A., and Nielsen, J. (1998) *Metabolic Engineering*, Academic Press.

32 Schaub, J. and Reuss, M. (2008) *In vivo* dynamics of glycolysis in *Escherichia coli* shows need for growth-rate dependent metabolome analysis. *Biotechnol. Progr.*, **24**, 1402–1407.

33 Canelas, A.B., Ras, C., ten Pierick, A., van Gulik, W.M., and Heijnen, J.J. (2011) An *in vivo* data driven framework for classification and quantification of enzyme kinetics and determination of apparent thermodynamic data. *Metabol. Eng.*, **13**, 294–306.

34 Hardiman, T., Meinhold, H., Hofmann, J., Ewald, J.C., Siemann-Herzberg, M., and Reuss, M. (2010) Prediction of kinetic parameters from DNA-binding site sequences for modeling global transcription dyxnamics in *Escherichia coli*. *Metabol. Eng.*, **12**, 196–211.

35 Vaseghi, S. (2000) Modellgestützte Analyse der Dynamik des Phosphofructokinase-Systems in *Saccharomyces cerevisiae*. PhD Thesis: University Stuttgart.

36 Vielhauer, O., Zakhartsev, M., Horn, T., Takors, R., and Reuss, M. (2011) Simplified absolute metabolite quantification by gas-chromatography-isotope dilution mass spectrometry on the basis of commercially available source material. *J. Chromatography B*, **879**, 3859–3870.

37 Canelas, A.B., Ras, C., ten Pierick, A., van Dam, J.C., Heijnen, J.J., and Van Gulik, WM. (2008) Leakage free rapid quenching technique for yeast metabolomics. *Metabolomics*, **4**, 226–239.

38 Van Gulik, W.M. (2010) Fast sampling for quantitative microbial metabolomics. *Cur. Opinion Biotechnol.*, **21**, 27–34.

39 Zakhartsev, M., Vielhauer, O., Horn, T., Yang, X., and Reuss, M. (2014) Fast sampling for quantitative microbial metabolomics: new aspects on cold methanol quenching: metabolite co-precipitation. *Metabolomics*. doi: 10.1007/s11306-014-0700-8

40 Wu, L., Mashego, M.R., van Dam, J.C., Proell, A.M., Vinke, J.L., Ras, C., van Winden, W.A., van Gulik, W.M., and Heijnen, J.J. (2005) Quantitative analysis of the microbial metabolome by isotope dilution mass spectrometry using uniformly ^{13}C-labeled cell extracts as internal standards. *Anal. Biochem.*, **336**, 164–171.

41 Benett, B.D., Yuan, J., Kimball, H., and Rabinowitz, J.D. (2008) Absolute quantification of intracellular metabolite concentrations by an isotope ratio-based approach. *Nat. Protoc.*, **3**, 1299–1311.

42 Johannes, K.J. and Hess, B. (1973) Allosteric kinetics of pyruvate kinase of *Saccharomyces cerevisiae*. *J. Mol. Biol.*, **76**, 181–205.

43 Lapin, A., Schmid, J., and Reuss, M. (2006) Modeling the dynamics of *E. coli* populations in the three-dimensional turbulent field of a stirred-tank bioreactor: a structured-segregated approach. *Chem. Eng. Sci.*, **61**, 4783–4797.

44 Visser, D., Schmid, J., Mauch, K., Reuss, M., and Heijnen, J.J. (2004) Optimal redesign of primary metabolism in *Escherichia coli* using linlog kinetics. *Metabol. Eng.*, **6**, 378–390.

45 Kresnowati, M.T.A.P., van Winden, W.A., Aering, M.J.H., ten Pierick, A., Ras, C., Knijnenburg, T.A., Daran-Lapujade, P., Pronk, J.T., Heijnen, J.J., and Daran, J.M. (2006) When transcriptome meets metabolome: fast cellular responses of yeast sudden relief of glucose limitation. *Mol. Systems Biol.* doi: 10.1038/msb4100083

46 Loret, M.O., Pedersen, L., and Francois, J. (2007) Revised procedures for yeast metabolites extraction: application to a glucose pulse to carbon-limited yeast cultures, which reveals a transient activation of the purine salvage pathway. *Yeast*, **24**, 47–60.

47 Walther, T., Novo, M., Rössger, K., Letisse, F., Loret, M.-O., Portais, J.-C., and Francois, J.-M. (2010) Control of ATP homeostasis during the respire-fermentative transition in yeast. *Mol. Syst. Biol.*, **6**, 344.

48 Vermer, M., Zakhartsev, M., Reuss, M., and Westerhoff, HV. (2013) 'Domino' systems biology and the 'A' of ATP. *Biochim. Biophys. Acta.*, **1827**, 19–29.

49 Kremling, A., Stelling, J., Bettenbrock, S., Fischer, S., and Gilles, ED. (2005) Metabolic networks: biology meets engineering sciences, in *Topics in Current Genetics: Systems Biology – Definitions and Perspectives*, vol. **13** (eds. L. Alberghina and H.V. Westerhoff), Springer, Berlin, pp. 215–234.

50 Buziol, S., Becker, J., Baumeister, A., Jung, S., Mauch, K., Reuss, M., and Boles, E. (2002) Determination of *in vivo* kinetics of the starvation-induced Hxt5 glucose transporter of *Saccharomyces cerevisiae*. *FEMS Yeast Res.*, **2**, 283–291.

51 Rizzi, M., Baltes, M., Theobald, U., and Reuss, M. (1997) *In vivo* analysis of metabolic dynamics in *Saccharomyces cerevisiae*: II. Mathematical model. *Biotechnol. Bioeng.*, **55**, 592–608.

52 Müller, D., Aguilera-Vazquez, L., Reuss, M., and Mauch, K. (2005) Integration of metabolic and signaling networks, in *Systems Biology: Topics in Current Genetics*, vol. **13** (eds. L. Alberghina and H.V. Westerhoff), Springer, Berlin, pp. 235–256.

53 Reuss, M., Aguilera-Vazquez, L., and Mauch, K. (2007) Reconstruction of dynamic models from metabolite measurements, in *Topics in Current Genetics – Metabolomics – A Powerful Tool in Systems Biology*, vol. **18** (eds. J. Nielsen and M.C. Jewett), Springer, Berlin, pp. 97–127.

54 Serrano, R., Gancedo, J.M., and Gancedo, C. (1973) Assay of yeast enzymes *in situ*. *Eur. J. Biochem.*, **34**, 479–482.

55 Monod, J., Wyman, J., and Changeux, J.-P. (1965) On the nature of allosteric transitions: a plausible model. *J. Mol. Biol.*, **12**, 88–118.

56 Reuter, R., Eschrich, K., Schellenberger, W., and Hofman, E. (1979) Kinetic modelling of yeast phosphofructokinase. *Acta. Biol. Med. Germ.*, **38**, 1067–1107.

57 Hofmann, E. and Kopperschlager, G. (1982) Phosphofructokinase from yeast. *Meth. Enzymol.*, **90**, 49–60.

58 Srere, P.A. (1967) Enzyme concentrations in tissues. *Science*, **158**, 936–937.

59 Ovadi, J. and Srere, PA. (1999) Macromolecular compartmentation and chanelling. *Int. Rev. Cytol.*, **192**, 255–280.

60 Murray, W. (1999) From molecular to modular cell biology. *Nature*, **402**, C47–C52.

61 Girvan, M. and Newman, M.E.J. (2002) Community structure in social and biological networks. *Proc. Natl. Acad. Sci. USA*, **99**, 7821–7826.

62 Ravasz, E., Somera, A.L., Mongru, D.A., Oltvai, Z.N., and Barabasi, A.L. (2002) Hierarchical organisation of modularity in metabolic networks. *Science*, **297**, 1551–1555.

63 Alon, U. (2003) Biological networks: the tinkerer as an engineer. *Science*, **26**, 1866–1867.

64 Holme, P., Huss, M., and Jeong, H. (2003) Subnetwork hierarchies of biochemical pathways. *Bioinformatics*, **19**, 532–538.

65 Newman, M.E.J. and Girvan, M. (2004) Finding and evaluating community structure in networks. *Phys. Rev.*, **E69**, 026113.

66 Radicchi, F., Castellano, C., Cecconi, F., Loreto, V., and Parisi, D. (2004) Defining and identifying communities in networks. *Proc. Natl. Acad. Sci. USA*, **101**, 2658–2663.

67 Guimera, R. and Amaral, L.A.N. (2005) Functional cartography of complex metabolic networks. *Nature*, **433**, 895–900.

68 Huss, M. and Holme, P. (2006) Currency and commodity metabolites: their identification and relation to the modularity of metabolic networks. *Quant. Bio.*, MN/0603038, arXiv.org.

69 Schmid, J.W., Mauch, K., Reuss, M., Gilles, E.D., and Kremling, A. (2004) Metabolic design based on a coupled gene expression-metabolic network model of tryptophan production in *Escherichia coli*. *Metabol. Eng.*, **6**, 364–377.

70 Horn, F. and Jackson, R. (1972) General mass action kinetics. *Arch. Rational Mech. Anal.*, **47**, 81–116.

71 Schauer, M. and Heinrich, R. (1979) Analysis of the quasi-steady-state approximation for an enzymatic one-substrate reaction. *J. Theor. Biol.*, **79**, 425–442.

72 Savageau, M.A. and Voit, E.O. (1982) Power-law approach to model biochemical systems. *J. Ferment. Technol.*, **60**, 221–228.

73 Pozo, C., Marin-Sanguino, A., Alves, R., Guillen-Gsalbez, G., Jimenez, L., and Sorribas, A. (2011) Steady-state global optimization of metabolic non-linear dynamic models through recasting into power-law canonical models. *BMC Syst. Biol.*, **5**, 137.

74 Liebermeister, W. and Klipp, E. (2006) Bringing metabolic networks to life: convenience rate law and thermodynamic constraints. *Theor. Biol. Med. Model.*, **3**, 41.

75 Hatzimanikatis, V. and Bailey, J.E. (1997) Effects of spatiotemporal variations on metabolic control: approximate analysis using (log)linear kinetics'models. *Biotechnol. Bioeng.*, **54**, 91–104.

76 Visser, D. and Heijnen, J.J. (2003) Dynamic simulation and metabolic redesign of a branched pathway using linlog kinetics. *Metabol. Eng.*, **5**, 164–176.

77 Onsager, L. (1931) Reciprocal relations in irreversible processes. *Phys. Rev.*, **37**, 405–426.

78 Rottenberg, H. (1973) The thermodynamic description of enzyme catalyzed reactions. The linear relationship between the reaction rate and the affinity. *Biophys. J.*, **13**, 503–511.

79 Van der Meer, R., Westerhoff, H.V., and van Dam, K. (1980) Linear relation between rate and thermodynamic force in enzyme catalyzed reactions. *Biochim. Biophys. Acta.*, **591**, 488–493.

80 Westerhoff, H.V. and Van Dam, K. (1987) *Thermodynamics and Control of Biological Free Energy Transduction*, Elsevier, Amsterdam.

81 Nielsen, J. (1997) Metabolic control analysis of biochemical pathways based on a thermokinetic description of reaction rates. *Biochem. J.*, **321**, 133–138.

82 Heijnen, J.J. (2005) Approximative kinetic formats used in metabolic network modeling. *Biotechnol. Bioeng.*, **91**, 534–545.

83 Wang, F.S., Ko, C.L., and Voit, E.O. (2007) Kinetic modeling using S-systems and lin–log approaches. *Biochem. Eng. J.*, **33**, 238–247.

84 Box, G.E.P. and Draper, N.R. (1987) *Empirical Model-Building and Response Surfaces*, Wiley-Blackwell, p. 424.

85 Costa, R.S., Machado, D., Rocha, I., and Ferreira, C. (2010) Hybrid dynamic modeling of *Escherichia coli* central metabolic network combining Michaelis–Menten and approximate kinetic equations. *BioSystems*, **100**, 150–157.

86 Leighty, R.W. and Antoniewicz, MR. (2011) Dynamic metabolic flux analysis (DMFA): a framework for determining fluxes at metabolic non-steady state. *Metabol. Eng.*, **13**, 745–755.

87 Jamshidi, N. and Palsson, B.O. (2010) Mass action stoichiometric simulation models: incorporating kinetics and regulation into stoichiometric models. *Biophys. J.*, **98**, 175–185.

88 Almquist, J., Cvijovic, M., Hatsimanikatis, V., Nielsen, J., and Jirstrand, M. (2014) Kinetic models in industrial

biotechnology: improving cell factory performance. *Metabol. Eng.*, **24**, 38–60.

89 Antoniewicz, M.R. (2013) Dynamic metabolic flux analysis: tools for probing transient states of metabolic networks. *Curr. Opinion in Biotechnol.*, **24**, 973–978.

90 Karp, P.D., Arnaud, M., Collado-Vides, J., Ingraham, J., Paulsen, I., and Saier, M. (2004) The *E. coli* ecocyc data base: no longer just a metabolic pathway database. *ASM News*, **70**, 25–30.

91 Bertoumieux, S., Brilli, M., Kahn, D., de Jong, H., and Cinquemani, E. (2013) On the identifiability of metabolic network models. *J. Math Biol.*, **77**, 1795–1832.

92 Nikerel, I.E., van Winden, W.A., Verheijen, P.T.J., and Heijnen, J.J. (2011) Model reduction and *a priori* parameter identifiability analysis using metabolome time series for metabolic reaction networks with linlog kinetics. *Metabol. Eng.*, **11**, 20–30.

93 Bucher, J., Riedmaier, S., Schnabel, A., Marcus, K., Vacun, G., Weis, T.S., Thasler, W.E., Nüssler, A.K., Zanger, U.M., and Reuss, M. (2011) A systems biology approach to dynamic modeling and inter-subject variability of statin pharmocokinetics in human hepatocytes. *BMC Syst. Biol.*, **5**, 66.

94 Di Maggio, J., Paulo, C., Estrada, V., Perotti, N., Ricci, D.J.C., and Diaz, M.S. (2014) Parameter estimation in kinetic models for large scale biotechnological systems with advanced mathematical programming techniques. *Biochem. Eng. J.*, **83**, 104–115.

95 Mauch, K., Buziol, S., Schmid, J., and Reuss, M. (2002) Computer-aided design of metabolic networks. *AIChE Symp. Ser.*, **98**, 82–91.

96 Bakker, B.M., Michels, P.A.M., Opperdoes, F.R., and Westerhoff, H.V. (1999) What controls glycolysis in bloodstream from *Trypanosoma brucei*? *J. Biol. Chem.*, **274**, 14551–14559.

97 Hauf, J., Zimmermann, F.K., and Müller, S. (2000) Simultaneous genomic overexpression of seven glycolytic enzymes in the yeast *Saccharomyces cerevisiae. Enzyme Microbial. Technol.*, **26**, 688–698.

98 Boles, E. and Hollenberg, C.P. (1997) The molecular genetics of hexose transport in yeast. *FEMS Microbial. Rev.*, **166**, 283–292.

99 Wieczorke, R., Krampe, S., Weierstall, T., Freidel, K., Hollenberg, C.P., and Boles, E. (1999) Concurrent knock-out of at least 20 transporter genes is required to block uptake of hexose in *Saccharomyces cerevisiae. FEBS Lett.*, **464**, 123–128.

100 Kell, D.B. and Mendes, P. (2000) Snapshots of systems: metabolic control analysis and biotechnology in the post-genomic era, in *Technology and Medical Implications of Metabolic Control Analysis*, Kluwer, Dordrecht, pp. 3–25.

101 Reich, J.G. and Selkov, E.E. (1981) Energy metabolism of the cell, in *A Theoretical Treatise*, Academic Press, New York.

102 Rasor, J.P. and Voss, E. (2001) Enzyme-catalyzed processes in pharmaceutical industry. *Appl. Catalysis A*, **221**, 145–158.

103 Ikeda, M. (2006) Towards bacterial strains oveproducing L-tryptophan and other aromatics by metabolic engineering. *Appl. Microbiol. Biotechnol.*, **69**, 615–626.

104 Shen, T., Liu, Q., Xie, C.X., Xu, Q., and Chen, N. (2012) Improved production of tryptophan in genetically engineered *Escherichia coli* with TktA and PpsA over expression. *J. Biomed. Biotechnol.* Article ID 605219, doi: 10.1155/2012/605219

105 Bongaerts, J., Krämer, M., Müller, U., Raeven, L., and Wubbolds, M. (2001) Metabolic engineering for microbial production of aromatic amino acids and derived compounds. *Metabol. Eng.*, **3**, 289–300.

106 Nielsen, J. (2001) Metabolic engineering. *Appl. Microbiol. Biotechnol.*, **55**, 263–283.

107 Liao, J.C., Hou, S.Y., and Chao, Y.P. (1996) Pathway analysis, engineering, and physiological considerations for redirecting central metabolism. *Biotechnol. Bioeng.*, **100**, 129–140.

108 Schuster, S., Dandekaar, T., and Fell, D.A. (1999) Detection of elementary flux modes in biochemical networks: a promising tool for pathway analysis and metabolic engineering. *TIBTECH*, **17**, 53–60.

109 Triebe, D.E. and Pittard, J. (1979) Hyperproduction of tryptophan by *Escherichia coli*: genetic manipulation of the pathway leading to tryptophan formation. *Appl. Environ. Microbiol.*, **38**, 181–190.

110 Tennant, D.A., Duran, R.V., and Gottlieb, E. (2010) Targeting metabolic transformation for cancer therapy. *Nat. Rev. Cancer*, **10**, 267–277. doi: 10.1038/nrc2817

111 Rodriguet-Enriquez, S., Marin-Hernandez, A., Gallardo-Perez, J.C., Carreno-Fuentes, L., and Moreno-Sanchez, R. (2009) Targeting of cancer energy metabolism. *Mol. Nutr. Food Res.*, **53**, 29–48.

112 Maier, K., Hofman, U., Reuss, M., and Mauch, K. (2010) Dynamics and control of the central carbon metabolism in hepatoma cells. *BMC Systems Biol.*, **4**, 54.

113 Hofman, U., Maier, K., Niebel, A., Vacun, G., Reuss, M., and Mauch, K. (2008) Identification of metabolic fluxes in hepatic cells from transient 13C-labeling experiments: Part I. Experimental observations. *Biotechnol. Bioeng.*, **83**, 400–415.

114 Maier, K., Hofman, U., Reuss, M., and Mauch, K. (2008) Identification of metabolic fluxes in hepatic cells from ^{13}C-labeling experiments: Part II. Flux estimation. *Biotechnol. Bioeng.*, **100**, 344–354.

115 Boren, J., Montoya, A.R., de Atauri, P., Comin- Anduix, B., Cortes, A., Centelles, J.J., Fredericks, W.M., Van Noorden, C.J., and Cascante, M. (2002) Metabolic control analysis aimed at the ribose synthesis pathways of tumor cells: a new strategy for antitumor drug development. *Mol. Biol. Rep.*, **29**, 7–12.

9
Chemical Thermodynamics Applied in Bioengineering

John Villadsen

Summary

The analysis of bioreactions in Chapters 4, 7, and 8 is incomplete without including the thermodynamics of the reactions. The chemical equilibrium determines how far a reaction will convert substrates to products, and the heat of reaction must be determined in order to design heat exchangers. Here, the most important elements of chemical thermodynamics are reviewed. After definition of the state functions of thermodynamics and showing how thermodynamic properties can be calculated from tables, the feasibility of the EMP pathway from glucose to pyruvate is discussed and the importance of having large values of $(-\Delta G_R)$ for the first and last reactions in a pathway is noted. Examples of calculations of $(-\Delta G_R)$ for a number of biochemical reactions are given, first using thermodynamic tables and next using the methods of electrochemistry. The exceedingly complicated energy transformation from reduced substrates to ATP in mitochondria is reviewed. Finally, simple methods for calculation of the heat of reaction of bioreactions are discussed. These methods are based on the same principles as was used in the redox balances of Chapter 3, and their application is illustrated by several examples.

9.0
Introduction

Equilibrium thermodynamics is concerned with systems that are in *equilibrium*. The saturation concentration s_o^* of oxygen in an aqueous phase sparged with air of a given partial pressure π_{O2} of O_2 is primarily a function of temperature and to a smaller extent a function of the liquid composition, for example, the ion strength or the glucose concentration:

$$s_o^* = f(p_{O2}, T, \text{liquid phase properties}). \tag{9.1}$$

$$s_o^* = 0.0270\, p_{O2} \exp\left(\frac{1142}{T(\text{K})}\right) \text{mmol } O_2\, l^{-1} (298 < T < 324\,\text{K}). \tag{9.2}$$

Fundamental Bioengineering, First Edition. Edited by John Villadsen.

The general formula (9.1) shows that s_o^* is a function of π_{O2}, of temperature T, and of the composition of both the gas phase and the liquid phase. Equation (9.2) [1] is an explicit expression for the relation between s_o^* and the *independent* or *state* variables, T, π_{O2}, and the composition of gas and liquid phases. For moderate values of π_{O2}, it is accurate to within 1% relative in the indicated temperature interval and with a liquid phase of pure water.

The equilibrium vapor pressure of H_2O over pure water is a function of temperature T alone:

$$p_{sat}(H_2O)\text{mm Hg} = \exp\left(20.386 - \frac{5132}{T(\text{K})}\right). \tag{9.3}$$

From Eq. (9.3) $p_{sat}(H_2O) = 49.1$ mm Hg at 38 °C. Thus, when a liquid phase of pure H_2O is sparged with pure (dry) O_2 at 38 °C and total pressure $P = 1$ atm, the partial pressure of O_2 above the liquid phase is $760 - 49.1 = 710.9$ mm Hg = 0.935 atm. By *Henry's law* s_o^* is *proportional* to the gas-phase partial pressure of O_2 for low to moderate values of the total pressure P.

$$\text{Hence by(9.2)}: \ s_o^* = 0.935 \cdot 0.0270 \exp\left(\frac{1142}{311.16}\right)$$
$$= 0.991 \text{ mM at } 38\,^\circ\text{C and P} = 1 \text{ atm}$$

From the tables of Geng and Duan [2], one obtains the influence of different salt solutions on s_o^* calculated by Eq. (9.2). Thus, for $0 < [\text{NaCl}] < 2$ M the preexponential factor of Eq. (9.2) should be multiplied by exp(−0.375 [NaCl] M). Glucose and other substrates have a similar effect on the oxygen solubility – the solubility s_o^* decreases approximately linearly with the molarity of the sugar concentration from $0 < s_{glucose} < 1$ M [3], p. 483).

Common to all processes in which components in several distinct phases can be distributed between the phases is that the equilibrium distribution of the components is calculated by empirical relations such as (9.1) or (9.2). This leads to a set of algebraic equations from which dependent variables can be calculated for given values of the other state variables. During a transient from one state to another, the relations derived from equilibrium thermodynamics cannot be applied.

The number F of intensive variables (T, P, concentrations, etc.) that needs to be specified to characterize the equilibrium state is given by the phase law of Gibbs:

$$F(= \text{degrees of freedom}) = C - \Phi + 2. \tag{9.4}$$

In (9.4) C is the number of components and Φ the number of phases in the system.

Thus, if two components A and B are distributed between two phases (e.g., an organic phase and an aqueous phase), the *equilibrium* distribution of A and B in the two phases is determined when T and P are given. The molar fraction of A in both phases could also be used to determine the value of P and T, but in any case two equations stipulate equality of the *chemical potential* of each component A or B in the two phases.

For Eq. (9.2), the phase rule gives $F = 3$ with three components (H_2O, O_2, and NaCl) and two phases. The system is described using the three variables

P, T, and [NaCl]. For given [P,T, [NaCl]] the vapor pressure of H_2O in the gas phase $p_{sat,H2O}$ is known, and hence $\pi_o = P - p_{sat,H2O}$. This allows s_o^* to be calculated by Eq. (9.2).

During a transient from one state to another, and also in continuous processes conducted at *steady state* (rather than at or very close to equilibrium), the relations derived from equilibrium thermodynamics must be modified. This is the subject of *nonequilibrium thermodynamics.*

The focus of this chapter will be on chemical reactions. Chemical equilibrium will be assumed in the calculation of relations between reacting species in a given *aqueous* medium by the laws of chemical *equilibrium* thermodynamics. This leads to discussion of the feasibility of a sequence of metabolic reactions in metabolic pathways, the topic of Chapter 4. The question of feasibility is of great importance when cells or combinations of enzymes are used to produce a desired product P from substrate S. If the *affinity* of the process is not positive, that is, if for thermodynamic reasons the *equilibrium* conversion of S to P is small, then no amount of genetic engineering will help to achieve a higher conversion at the given T. Enzyme levels can be increased, and enzymes that exhibit product inhibition can be replaced by modified enzymes, but the thermodynamic limitation illustrated in Figure 4.8c cannot be removed. In cell reactions, there is a *possibility* that the thermodynamic limitation of certain reactions in a pathway can be circumvented by choosing a different pathway from S to P – although the overall conversion of S to P is still determined by the (free) energy content of S and P. It will, however, be shown that energy can be donated from one pathway to another, and sometimes new steps can be introduced by means of which the free energy loss per step is smaller.

Another topic to be discussed in the following is the calculation of metabolic heat production in (mainly) aerobic cell reactions. Again the fundamental relations of chemical equilibrium thermodynamics are invoked, and simple formulas for calculating the heat of reaction based on the *redox level* of substrate(s) and product(s) will be derived in Section 9.4.

Finally, determination of the free energy change ΔG for a redox reaction will be discussed in terms of an *electrical potential change*, the intensive variable of electrical charge transfer. Other intensive variables are temperature T for caloric transfer to and from the system, concentrations for mass transfer, and pressure P for mechanical work on a volume V.

A large number of important processes in bioengineering are *not* amenable to the laws of classical equilibrium thermodynamics. The chemical processes discussed above can be modeled by equilibrium thermodynamics *only* when they move from one state to another by infinitely small and *reversible* steps. Hence, reaction kinetics, molecular diffusion in a porous support, membrane filtration, protein folding, agglomeration of cells, and numerous other important processes must be described by *nonequilibrium thermodynamics*, a subject outside the scope of this chapter. A large number of (reasonably) accessible textbooks are available on nonequilibrium thermodynamics – usually with a precise explanation of the difference between equilibrium and nonequilibrium thermodynamics. One such book is by Lebon *et al.* [4]. A recent text from one of the stalwarts of thermodynamics and with special emphasis on bioreactions is by Alberty [5].

9.1 Chemical Equilibrium and Thermodynamic State Functions

The state of the systems discussed in the following is determined by three *state functions*, the free energy G, the enthalpy H, and the entropy S. At equilibrium, the three state functions *for a given compound* are connected by the *second law of thermodynamics*:

$$G = H - TS. \tag{9.5}$$

For a chemical reaction that proceeds at constant T and pressure P from state 1 to state 2:

$$\Delta G_R = \Delta H_R - T\,\Delta S_R. \tag{9.6}$$

In Eq. (9.6), Δ stands for (property at state 2) minus (property at statel) as defined in Eqs. (9.8) and (9.9).

Substrates are spontaneously converted to products as long as ΔG_R *is negative.* This means that the free energy G of the system continuously decreases until a minimum is reached at the equilibrium conversion of substrates. Spontaneous reactions have a *positive affinity.*

$(-\Delta H_R)$ is the *heat released to the surroundings by the chemical reaction.* For very exothermic reactions $(-\Delta H_R) >> 0$, and consequently, $(-\Delta G_R)$ is likely to be large and positive. ΔS_R is a measure of the *disorder of the system.* In metabolic reactions, "well-ordered" substrates such as glucose are converted to a "less well-ordered" mixture of products, ethanol, CO_2, and H_2O, and ΔS_R is positive.

This is not always so. The formation of biomass and methane from H_2 and CO_2 by anaerobic growth of the bacterium *Methanobacterium thermoautotrophicum* studied by Schill *et al.* [6] – see also Example 9.3 – is given by the stoichiometry (9.7):

$$H_2 + 0.260\,CO_2 + 0.0048\,NH_3 \rightarrow 0.02\,CH_{1.68}O_{0.39}N_{0.24} + 0.240\,CH_4 + 0.511\,H_2O. \tag{9.7}$$

The biomass yield is very small. The heat of reaction $Q = -\Delta H_R$ is very large, 63.4 kJ (mol H_2)$^{-1}$ at 333 K, but $-\Delta G_R$ is only 12.9 kJ (mol H_2)$^{-1}$. This is due to the large and *negative* value of ΔS_R in a process where very highly ordered biomass is to be synthesized as one of the products. Also, the small concentration of H_2 in the aqueous phase decreases $-\Delta G_R$ significantly.

Since both ΔH_R and ΔS_R are weak functions of T, it may happen that an endothermic $(-\Delta H_R < 0)$ process with positive ΔS_R has negative affinity at low T but positive affinity at higher T. Folded protein with a high degree of "order" is stable at low T, while the process of denaturation (unfolding) of the protein proceeds at a higher T, even though $-\Delta H_R < 0$.

The following rules apply to the calculation of changes of state functions by chemical reaction:

1) All *elements* in their stable form at $T = 298.16$ K and $P = 1$ atm (e.g., gaseous oxygen at $\pi_i = 1$ atm) are defined to have both G and $H = 0$. The entropy of a perfect crystal of any element is, however, equal to 0 at $T = 0$ K. At higher temperature, all elements have $S > 0$.
2) Based on the definitions of (1), the *free energy of formation* G_f *as well as* H_f *and* S_f *of any compound* can be calculated under the *standard conditions* from tables of thermodynamic properties or from the extensive data banks that are now available. Thus, at $[T, P] = [298.16\text{ K}, 1\text{ atm}]$ the free energy and the enthalpy of formation of gaseous H_2O is $[G_f, H_f] = [-228.72, -241.95\text{ kJ}]\text{ mol}^{-1}$. Also from tables one obtains $S^0 = [205, 130.6, 188.8]$ J (mol K)$^{-1}$ for $[O_2, H_2, H_2O_g]$, and $\Delta S_R = 188.8 - 130.6 - ½\ 205 = -44.3$ J (mol K)$^{-1}$ by the stoichiometry ½ $O_2 + H_2 \rightarrow H_2O$.

The consistency of these data can be checked by insertion in Eq. (9.6):

$$\begin{aligned}-228.72 &= \Delta G_R = G_{f,H_2O\text{ gas}} \approx -241.95 = \Delta H_R - T\,\Delta S_R \\ &= H_{f,H_2O\text{ gas}} - 298.16(-44.3)10^{-3} = -228.74\text{ kJ(mol H}_2\text{O)}^{-1}.\end{aligned}$$

If the final product is liquid water, H_2O_{liquid}, which is more ordered than H_2O_{gas}, one obtains $S^0 = 70.0$ J (mol K)$^{-1}$ rather than 188.8 J (mol K)$^{-1}$. $\Delta S_R = -163.1$ J (mol K)$^{-1}$. H_f (H_2O_{liquid}) is also more negative than H_f (H_2O_{gas}) due to the heat release $2.436 \cdot 18 = 43.84$ kJ (mol H_2O)$^{-1}$ by condensation of the vapor. Consequently,

$$\begin{aligned}G_f(H_2O_{liquid}) &= -241.95 - 43.84 - 298.16(-163.1)10^{-3} \\ &= -237.3\text{ kJ(mol H}_2\text{O}_{liquid})^{-1}.\end{aligned}$$

This value for G_f (H_2O_{liquid}) is also found in the tables of thermodynamic properties of compounds.

Since $[G_f, H_f] = [0, 0]$ for $[O_2, H_2]$ under the standard conditions [298.16 K, 1 atm], it follows that the reaction of H_2 and O_2, starting with no H_2O, has a very high affinity. At termination of the reaction when the equilibrium concentrations of $[O_2, H_2, H_2O]$ have been reached, this free energy is dissipated to a large heat release $-\Delta H_R$, less the contribution to make a more ordered H_2O molecule. The large negative values of $[G_f, H_f]$ for H_2O signify that a large input of free energy (or heat) is needed to convert H_2O back to its constituent elements.

3) All the previous calculations of free energy and enthalpy of formation of compounds *at standard conditions* are done using relations that formally look like the redox balances set up in Chapter 3:

$$(\Delta G)_{\mathrm{R}}^{0} = \left(\sum_{i} Y_{\mathrm{p}_i} G_{\mathrm{f},i}^{0}\right)_{\text{products}} - \left(\sum_{j} Y_{\mathrm{s}_j} G_{\mathrm{f},j}^{0}\right)_{\text{substrates}}. \tag{9.8}$$

$$(\Delta H)_{\mathrm{R}}^{0} = \left(\sum_{i} Y_{\mathrm{p}_i} H_{\mathrm{f},i}^{0}\right)_{\text{products}} - \left(\sum_{j} Y_{\mathrm{s}_j} H_{\mathrm{f},j}^{0}\right)_{\text{substrates}}. \tag{9.9}$$

To calculate the redox content of reactants, a set of redox neutral compounds are defined. The standard state [298.16 K, 1 atm] plays an analogous role in chemical thermodynamics. In Section 9.4, Eqs. (9.8) and (9.9) will be used in a similar way to calculate especially the heat of reaction by analogy to redox balances.

In a real chemical reaction that starts with $n_{i,0}$ moles of, respectively, substrates and products, the process stops when the reactants have reached their equilibrium values at the maximum extent $\xi_{\max}$ of the reaction. $n_i = n_{i,0} + Y_i\ \xi$, and $\xi = \xi_{\max}$, when $n_i = n_{i,\text{equilibrium}}$. When a chemical reaction is conducted at constant T and P, the value of G_{system} reaches a minimum for $\xi = \xi_{\max}$. The value of $\xi_{\max}$ depends on the values of T and P. ΔG_{R} is calculated by

$$\Delta G_{\mathrm{R}} = \Delta G_{\mathrm{R}}^{0} + RT \ln \frac{\prod a_{\mathrm{p}i}^{Y\mathrm{p}i}}{\prod a_{\mathrm{s}i}^{Y\mathrm{s}i}} = \Delta G_{\mathrm{R}}^{0} + RT \ln K_{\mathrm{a}}. \tag{9.10}$$

In Eq. (9.10), $a_{\mathrm{p}i}$ and $a_{\mathrm{s}i}$ are, respectively, the *activity* of products and substrates at a given value of ξ. Each factor in the two products is raised to the power $Y_{\mathrm{p}i}$ or $Y_{\mathrm{s}i}$ where $Y_{\mathrm{p}i}$ and $Y_{\mathrm{s}i}$ are the stoichiometric coefficients in mol/mol. For ideal gases $a_i = \pi_i$.

K_{a} is defined in terms of activities, while $K_{\mathrm{c}} \equiv \left(\prod c_{\mathrm{p}i}^{Y\mathrm{p}i} / \prod c_{\mathrm{s}i}^{Y\mathrm{s}i}\right)$ is defined in terms of the directly measurable reactant concentrations. With $a_i = \gamma_i\ c_i$, one obtains $K_{\mathrm{a}} = K_{\mathrm{c}} + \prod \gamma_i^{\nu_i}$, where $\nu_i = +Y_i$ for products and $-\ Y_i$ for substrates. In tables of thermodynamic properties for bioreactions, ln $\left(\prod \gamma_i^{\nu_i}\right)$ is often incorporated into $\Delta \mathrm{G}_{\mathrm{R}}^{0*} = \Delta \mathrm{G}_{\mathrm{R}}^{0} + \ln\left(\prod \gamma_i^{\nu_i}\right)$ for different ionic strengths. Other mass action functions are used, especially for gas-phase reactions, namely, K_{p} and K_{y}, which are based on, respectively, partial pressure π_i and mole fractions y_i (= volume fractions) of the reactants.

Since by the ideal gas law $\pi_i = \frac{RT\,n_i}{V} = c_i RT$, one obtains $K_{\mathrm{c}} = K_{\pi}\left(RT^{-\sum \nu_i}\right)$.

$$\text{Furthermore, } y_i = \frac{\pi_i}{P}, \text{ and } K_y = \mathrm{K}_{\pi} P^{-\sum \nu_i}, \text{ i.e., } K_{\mathrm{c}} = K_y \left(\frac{P}{RT}\right)^{\sum \nu_i}. \tag{9.11}$$

Hence, for the activity coefficients γ_i close to 1, Eq. (9.11) allows ΔG_{R} to be determined, also when partial pressures π_i or mole fractions y_i are used in the calculation of K:

$$\begin{aligned}\Delta G &= \Delta G_{\mathrm{R}}^{0} + RT \ln \left(\frac{\prod a_{\mathrm{p}i}^{Y\mathrm{p}i}}{\prod a_{\mathrm{s}i}^{Y\mathrm{s}i}}\right)_{\mathrm{eq}} \approx \Delta G_{\mathrm{R}}^{0} + RT \ln K_{\mathrm{c,eq}} \\ &= \Delta G_{\mathrm{R}}^{0} + RT \ln \left[K_{\pi}\left(\frac{1}{RT}\right)^{\sum \nu_i}\right] = \Delta G_{\mathrm{R}}^{0} + RT \ln \left[K_y \left(\frac{P}{RT}\right)^{\sum \nu_i}\right].\end{aligned} \tag{9.12}$$

Since $\Delta G_R^0 \equiv$ the free energy of reaction under standard conditions is known, Eq. (9.12) is used to calculate ΔG_R under arbitrary conditions. To obtain dimensionless variables in K, [a_i or π_i] are conventionally divided by [a_i, π_i] under standard conditions. Now ln, K can be calculated without worrying about units.

Example 9.1 Determination of the equilibrium conversion of $H_2 + (1/2)$ $O_2 \rightarrow H_2O$

For the reaction $H_2 + (½)\ O_2 \rightarrow H_2O_g$ with an initial gas mixture of composition $[y_{O2}, y_{H2}] = [(1/3), (2/3)]$, that is, with O_2 and H_2 in the correct stoichiometric ratio, one obtains the following:

$$\frac{-\Delta G_R^0}{RT} = \frac{228.72}{298.16 \cdot 8.314\ 10^{-3}} = 92.27 = RT \ln K_{c,eq}$$

$$= RT \ln \left[K_{y,eq} \left(\frac{P}{RT} \right)^{\sum \nu_i} \right]. \quad \text{where } K_{y,eq} = \left(\frac{y_{H_2O}}{y_{O2}^{0.5}\, y_{H2}} \right)_{eq}$$

Clearly, y_{H2O} is very close to 1 and y_{O2}, $y_{H2} \approx 0$. This result is true, unless the reaction temperature T is extremely high or $P \approx 0$. If y_{O2}^0 was less than 1/3 (and y_{H2} correspondingly > 2/3), the equilibrium composition would contain both H_2 and H_2O, but virtually no O_2.

If the end product is H_2O_{liquid}, the total reaction is a combination of the gas-phase reaction and a condensation of the gaseous H_2O by an equilibrium reaction for which $\Delta G = 0$.

Eq. (9.10) now reads : $(\Delta G_R)^0_{\text{liquid H20}}$

$$= (\Delta G_R)^0_{\text{gas phase}} + RT \ln \pi_{H_2O} \qquad (298.16, 1\ \text{atm})$$

$$= -228.72 - RT \ln \frac{23.7}{760}$$

$$= -228.72 - 298.16 \cdot 8.314\ 10^{-3} \cdot 3.467$$

$$= -237.3\ \text{kJ(mol)}^{-1}.$$

This is the same result as was obtained previously for $G_f(H_2O)_{liquid}$.

4) In biochemical reactions, the definitions of standard states are somewhat different from those used in classical chemical thermodynamics. Most tables use 25 °C, 1 atm as the standard temperature and total pressure, respectively. The concentration (or activities) of reacting species is set to 1 M, *except for* $[H^+]$ that is set to 10^{-7}. The very considerable difficulties in determination of ΔG_R for biochemical reactions in aqueous media arise because of the many reactions with dissociating species that occur simultaneously. In "ordinary" chemical reactions – for example, those involved in the refinery – the pressure can be high and the ideal gas law is replaced by $pV = \zeta\ RT$ where ζ is a compressibility factor. [π_i, T] are replaced by "reduced" variables $\pi_{r,i} = \pi_i/p_c$ and $T_r = T/T_c$, where [p_c, T_c] are, respectively, the pressure and temperature at the critical point. The compressibility factor ζ is read from charts that exhibit ζ (p_c, T_c) for gases at high pressure.

This is, however, a minor complication compared to the difficulty of determining the activity coefficient γ_i in the relation $a_i = \gamma_i\, c_i$ between the measured concentration c_i of reactant and its activity a_i that appears in the mass action function $K = \left(\prod a_{pi}^{Y_{pi}} / \prod a_{si}^{Y_{si}}\right)$.

In the calculation of ethanol stripping from an aqueous fermentation medium in Example 2.2, the mole fraction of ethanol y_p in the gas phase is only proportional to the liquid mole fraction x_p when x_p is close to 1. For small x_p, realistic in biofuel production, $y_p = \gamma_p\ (x_p \cdot \pi_p)$, and $\gamma_p = 6.6$ for $x_p < 0.01$. Added to this complication, one must consider the influence of ions on the medium.

The mechanistically derived Debye–Hückel formula for ionic strength $z \to 0$, or its empirical extensions to higher ionic strength, determines the influence of total ionic strength I (M) in the medium on the activity coefficient γ_i of a species in the liquid phase:

$$\log_{10}\gamma_i = -0.511\, z_i^2\, I^{1/2} \text{ for small } I, \quad \text{or}$$

$$\log_{10}\gamma_i = -\frac{0.511\, z_i^2\, I^{1/2}}{1 + 1.6\, I^{1/2}} \text{ for I} > 0.05\,\text{M}. \tag{9.13}$$

The activity coefficient of an undissociated compound is not influenced significantly by the ionic strength of the medium $I = \frac{1}{2}\sum_{j=1}^{M} z_j^2 c_j$, where $j = 1$, 2. M ions in the medium have concentrations c_j and charge z_j. This is not the case for ionic reactants with concentration c_i and activity coefficient γ_i. The fermentation medium contains a plethora of ions from dissociated organic acids, polyphosphates, and many cations such as Ca^{+2}, Mg^{+2}, Na^{+}, and K^{+} that partake in the formation of reacting complexes.

As an example, the difference between $C = -\log_{10}[H^+]_a = -\log(\gamma(H^+)) + (pH)_c$ to be used in equilibrium calculations and $(pH)_c = -\log_{10}[H^+]$ measured by a pH meter is given by Eq. (9.13). $pH_a = 0.107 + pH_c$ for $I = 0.1$ M. A solution with [NaCl] = 0.1 M (= 5.8 g l^{-1}) has $I = 0.5\ (1 \cdot 0.1 + 1 \cdot 0.1) = 0.1$ M, and this shows that even modest salt concentrations influence the values to be used in calculating K_{eq}.

Example 9.2 Free energy changes in the reactions of the EMP pathway

The data used in Table 9.1 are from [7], except for the redox cofactor concentrations [8] and the quite uncertain concentration interval (60–250 μM) for DPG obtained experimentally by [9]. The most correct charge at pH = 7 for the cofactors is indicated in parenthesis, that is, all four phosphate groups in "ATP" are deprotonated and $[ATP^{-4}] =$ 1850 μM (see text below). Concentrations of metabolites are stated for the total molarity of the undissociated compound and all its dissociated derivatives. Thus, phosphates are mostly present as HPO_4^{-2}. "Free NADH and NAD" is the amount of unbound cofactors that is available in the cytosol as redox carrier. Bekers, van Gulik, and Heijnen [8] give further details.

Table 9.2 lists values of ΔG^0 at $(pH)_c = 7$ and T = 298.16 K = 25 °C, while all other concentrations are at 1 M. The values of ΔG are estimated

Table 9.1 Metabolites in the EMP pathway.

Metabolite	Concentration (μM)	Metabolite (cofactor)	Concentration (μM)
Glucose (GLC)	5000	2 Phosphoglycerate (2PG)	29.5
Glucose 6 P (G6P)	83	Phosphoenolpyruvate (PEP)	23
Fructose 6 P (F6P)	14	Pyruvate (PYR)	51
Fructose 1,6 bis-P (FDP)	31	ATP (ATP^{-4})	1850
Dihydroxyacetone (DHAP)	138	ADP (ADP^{-3})	138
Glyceraldehyde 3 P (GAP)	18.5	P_i (HPO_4^{-2})	1000
1,3 Diphosphoglycerate (1,3 DPG)	60-250	NADH/NAD (NAD^+) (total) anaerobic, *S. cerevisiae*	2.5 (dimensionless)
3 Phosphoglycerate (3PG)	118	NADH/NAD (NAD^+) (free) anaerobic, *S. cerevisiae*	$0.6 \rightarrow 6$ for $D = 0.1 \rightarrow 0.3\ h^{-1}$

Table 9.2 Values of ΔG^0.

Reaction/enzyme	ΔG^0 (kJ mol^{-1})	ΔG (expression)	ΔG (kJ mol^{-1})
Hexokinase	−16.74	$\Delta G^0 + RT \ln \frac{[G6P][ADP]}{[GLC][ATP]}$	−33.3
Glucose 6 P isomerase	1.67	$\Delta G^0 + RT \ln \frac{[F6P]}{[G6P]}$	−2.7
Phosphofructokinase	−14.22	$\Delta G^0 + RT \ln \frac{[FDP][ADP]}{[F6P][ATP]}$	−18.7
Aldolase	23.97	$\Delta G^0 + RT \ln \frac{[DHAP][GAP]}{[FDP]}$	0.7
Triose P isomerase	7.66	$\Delta G^0 + RT \ln \frac{[GAP]}{[DHAP]}$	2.7
GAP dehydrogenase	6.28	$\Delta G^0 + RT \ln \frac{[NADH][H^+][1,3\,DPG]}{[NAD^+]\left[10^{-7}\right][GAP]}$	1.1 (4.6)
3 PG kinase	−18.83	$\Delta G^0 + RT \ln \frac{[ATP][3PG]}{[ADP][1,3\,DPG]}$	−10.7 (−14.3)
Phosphoglycerate mutase	4.44	$\Delta G^0 + RT \ln \frac{[2PG]}{[3PG]}$	1.0
Enolase	1.84	$\Delta G^0 + RT \ln \frac{[PEP]}{[2PG]}$	1.2
Pyruvate kinase	−31.38	$\Delta G^0 + RT \ln \frac{[PYR][ATP]}{[PEP][ADP]}$	−23.0

based on the metabolite concentrations of Table 9.1, and no correction is done for the difference between activities a_i and concentrations c_i.

Table 9.2 provides approximate $\Delta G°$ and ΔG values for EMP pathway reactions in the human erythrocyte.

The results in Table 9.2 indicate that although ΔG^0 is large and positive for some of the reaction steps, negative values of ΔG are obtained for almost all steps. It is noted that $\Delta G°$ is large and negative both for the first and the last step – a common feature of all primary metabolic pathways that have to operate at very different inlet substrate concentrations and final product concentrations. Thus, the last reaction must also produce PYR when large amounts of PEP is drained off for carboxylation to OAA (Figure 4.6).

The numbers given in the table are, as mentioned, no more than indicative. The EMP pathway does operate, and this means that *under the conditions in the cytosol of cells* ΔG for all steps is negative or at least only slightly positive. The calculation of $\Delta G°$ is reasonably accurate, although some of the reactions actually are sums of several reactions, each with a specific standard free energy change. The main difficulty is to assess the second term in ΔG since some of the internal metabolite concentrations are determined only with a substantial uncertainty, and the influence of other reactants such as Mg^{+2} that is known to play a part in the reaction "ATP → ADP" is not easily quantified – see the very detailed analysis of [10]. Actually for the reaction by which ATP is hydrolyzed to ADP,

$$ATP^{-4} + H_2O \rightarrow ADP^{-3} + HPO_4^{-2} + H^+[pH = 7, T = 298.2\,K]$$

$$RT \ln K = RT \ln \frac{a(ADP^{-3})\, a(H^+) a(HPO_4^{-2})}{a(ATP^{-4})}$$
$$= 2.48 \ln \left(\frac{1000\,10^{-6} \cdot 1 \cdot 50\,10^{-6}}{138\,10^{-6}} \right) = -19.6,$$

when the activities of the other reactants are inserted from Table 9.1. Consequently, under physiological conditions, the free energy released by hydrolysis of ATP is much larger than the value $-\Delta G^0 = 31\text{–}35\,kJ\,mol^{-1}$ predicted under standard conditions. Considering the uncertainty in the measured concentrations, the uncertainty of relating activities to concentrations, and of the role of Mg^{+2} in ATP formation, the best one can conclude is that the free energy release by hydrolysis of ATP under physiological conditions is in the range 51–55 kJ mol^{-1}. The very large free energy that theoretically could be harvested by hydrolysis of ATP under physiological conditions is certainly not retrieved at a thermodynamic efficiency that approaches 1 – some of the energy is released as metabolic heat. In Section 9.3, it is, however, argued that the energy *input* to synthesize ATP from ADP is of the order of 54 kJ (mol ATP)$^{-1}$ that is, of course, better related to the theoretical release of free energy by hydrolysis of the molecule.

For a reaction taking place in the aqueous phase with reactants transported from the gas phase, the calculation of ΔG must incorporate the solubility of the gas-phase reactant.

Thus, G_f^0 $(O_2)_g$ (298.2 K, $\pi_{O2} = 1$ atm) = 0 by definition. Under standard conditions, the molar fraction of O_2 in water is $x_{O2} = 2.293 \cdot 10^{-5}$, that is, the solution contains $55.5 \cdot 2.293 \cdot 10^{-5} = 1.27$ mmol O_2 l^{-1}. G_f^0 $(O_2)_{aq}$ is based on 1 M concentrations of all reactants, and consequently, for $\pi_{O2} = 1$ atm:

$$\Delta G((O_2)_g \rightarrow (O_2)_{aq}) = 0 + RT \ln(1/0.00127)$$
$$= 16.53 \text{ kJ/mol dissolved } O_2. \qquad (9.14)$$

To transport 1 mol O_2 from the gas phase, one needs a mechanical energy input $\ln(P/1) = RT \ln 784 = 16.53$ kJ that is equivalent to the increase of free energy of the reacting system by transporting O_2 from the gas phase to the liquid phase.

The same calculation can be done for dissolution of gaseous H_2S in water:

$$\Delta G_R^0 \text{ for the reaction } S_{solid} + (H_2)_g \rightarrow G_f(H_2S)_g$$
$$= -33.6 \text{ kJ mol}(H_2S)_g(1 \text{ atm}, 298.2 \text{ K}).$$

The molar concentration of H_2S in water for $\pi_{H2S} = 1$ atm is 0.102 mol l^{-1} and $G_f(H_2S)_{aq} = -33.6 + RT \ln 9.802 = -27.9$ kJ. In the aqueous phase, H_2S is slightly dissociated by the reaction $H_2S \rightarrow HS^- + H^+$ with acidity constant $K_a = [H^+] \cdot [HS^-]/[H_2S]_{aq}$, and for standard conditions $[H_2S]_{aq} = 1$ M. Hence $K_a = [H^+] \cdot [HS^-] = 10^{-6.89}$, and ΔG^0 for the reaction $(H_2S)_{aq} \rightarrow HS^- + H^+$ is $G_f(H_2S)_{aq} + RT \ln 10 \cdot pK_a = -27.9 + 39.34 = 11.4 \text{ kJ} = G_f\,(HS^-)$.

In the same way, the standard free energy of formation of any ion by

$$HA \rightarrow A^- + H^+ \text{ or } B + H^+ \rightarrow HB^+ \qquad (9.15)$$

is found from the G_f for the undissociated compound under standard conditions: [HA] or [B] = 1 M:

$$G_f(A^-) = G_f(HA) + RT \ln 10 \cdot pK_a \text{ and } G_f(HB^+) = G_f(B) - RT \ln 10 \cdot pK_a. \qquad (9.16)$$

Some examples derived from Eq. (9.17) are as follows:

$$G_f^0(OH^-) = G_f(H_2O)_l + RT \ln 10 \cdot 14 = -237.3 + 80.0 = -157.3 \text{ kJ mol}.$$

$$G_f^0(NH_4^+) = G_f(NH_3)_{aq} - RT \ln 10 \cdot 9.25 = -26.50 - 52.81$$
$$= -80.3 \text{ kJ(mol } NH_4^+)^{-1}. \qquad (9.17)$$

$$G_f^0(CH_3COO^-) = G_f(CH_3COOH)_{aq} + RT \ln 10 \cdot 4.75 = -396.46 + 27.1$$
$$= -369.3 \text{ kJ(mol } CH_3COO^-)^{-1}.$$

$$G_f^0(HCO_3^-) = G_f(H_2CO_3)_{aq} + RT \ln 10 \cdot 6.37 = -623.08 + 36.4$$
$$= -586.7 \text{ kJ(mol } HCO_3^-)^{-1}.$$

Values for the standard free energy of formation are of considerable value to find ΔG_R^0 for the many reactions both in the cell and in the design of bio-remediation units. Sometimes, the ion can further dissociate, for example, $HCO_3^- \rightarrow H^+ + CO_3^{-2}$, but for a given pH of the solution one can probably always find the dominant ion present.

Of special interest are the ions involved in converting the nucleotides ATP to ADP (ATP^{-4} to ADP^{-3}) and NADH to NAD^+ according to Figure 4.3. In both cases, the least energetic species is *defined* to have $G_f^0 = 0$ – a concept similar to that used in Chapter 4 where ADP and NAD^+ were not included in the stoichiometry for a net pathway.

Thus, ATP^{-4} has a free energy content of 31–35 kJ mol^{-1} as discussed in Example 9.2, while the free energy of NADH is obtained from ΔG^0 for the reaction:

$$\frac{1}{2}(O_2)_g + (H_2)_g \rightarrow H_2O_l \text{ subtracted from that of } NAD^+ + (H_2)_{aq} \rightarrow NADH + H^+$$

$$NAD^+ + (H_2)_{aq} - \frac{1}{2}(O_2)_g - (H_2)_g \rightarrow NADH + H^+ - H_2O_l. \tag{9.18}$$

In Eq. (9.18), all species except $(H_2)_{aq}$, H_2O_l, and NADH have $G_f^0 = 0$ and one obtains G_f^0 (NADH) $- G_f^0(H_2O)_l = G_f^0((H_2)_{aq})$. The solubility of H_2 in water at $T = 298.2$ K and $\pi(H_2)_g = 1$ atm is $7.83 \cdot 10^{-4}$ mol l^{-1}, and consequently

$$G_f^0(\text{NADH}) = G_f^0(H_2O)_l + RT \ln\left(\frac{1}{7.83 \cdot 10^{-4}}\right) = -237.3 + 17.73 = -219 \text{ kJ mol}^{-1}. \tag{9.19}$$

Example 9.3 Synergism of several microbial consortia in biogas production

In Section 3.6, a very simplified description was given of the anaerobic digestion of waste (exemplified by cellulose) to methane and carbon dioxide. The process is in reality the sum of several subprocesses in which bacteria with different but mutually supporting metabolism cooperate.

First, the cellulose is hydrolyzed to small carbohydrates, and these are degraded by fermentative metabolism, for example, to lactate, ethanol, and acetate following Figures 4.2 and 4.4 (a or b, with 2 H_2 produced instead of 2 NADH or 2 HCOOH). Let acetate and ethanol be the only metabolic products. In the pH interval 5.5 to 8 needed to obtain a high yield of CH_4, the redox balanced stoichiometry is as follows:

$$\frac{1}{n}(C_6H_{12}O_6)_n(+H_2O) \rightarrow C_6H_{12}O_6 + 3\,H_2O \rightarrow CH_3COO^{-1} + CH_3CH_2OH + 2\,HCO_3^{-1} + 2(H_2)_{aq} + 3H^+. \tag{1}$$

Alcohols and carboxylic acids with three or four carbon atoms are produced by other bacteria (e.g., propionibacteria).These are present in the digester, but ethanol + acetic acid are representative metabolites.

In the next step, the three and four carbon carboxylic acids and the alcohols are converted to acetic acid that is the only substrate for CH_4 production by decarboxylation. This is *acidogenesis* and finally *acetogenesis*:

$$CH_3CH_2OH + H_2O \rightarrow CH_3COO^- + H^+ + 2(H_2)_{aq}. \tag{2}$$

CH_4 is produced either by reaction between $HCO_3{}^{-1}$ and H_2 or by decarboxylation of acetate:

$$4(H_2)_{aq} + HCO_3^{-1} + H^+ \rightarrow CH_4 + 3\,H_2O \quad \text{or} \quad CH_3COO^{-1} + H^+ \rightarrow CH_4 + CO_2. \tag{3}$$

The first of the two reactions in (3) is part of the metabolism of *methanogens.*

The acetate produced in (1) can always be converted to CH_4 by decarboxylation, but to obtain a high yield of CH_4 the large amount of H_2 produced in (1) and (2) must be taken up by the methanogens – otherwise, H_2 is "lost" (assuming that CH_4 is the desired product). Hence, the operation conditions must be favorable to production of methanogens. In the *syntrophic* process, H_2 is transported over a short distance 10–100 nm between the H_2 producing and the H_2 consuming microbial consortia that are intimately mixed on microflocs.

ΔG^0 for $(1) = (-369.3 - 2 \cdot 586.7 - 174.8$ (ethanol) $+ 2 \cdot 17.73$ $(H_2)_{aq}$, $\pi_{H2} = 1$ atm)) $- (-917.2 - 3 \cdot 237.3) = -52.9$ kJ mol^{-1} glucose^{-1}. ΔG^0 for $(2) = -369.3 + 2 \cdot 17.73 - (-174.8 - 237.3) = 78.4$ kJ mol^{-1}.

The two reactions in (3), and especially the one in which methanogens reduce HCO_3^{-1} to CH_4, have positive affinities under standard conditions. ΔG_R^0 is, respectively, $-50.7(CH_4) - 3 \cdot 237.3 - (-586.7 + 4 \cdot 17.73) = -246.7$ kJ mol^{-1}, and $-50.7 - 394.3$ $(CO_2) - (-369.3) = -75.7$ kJ mol^{-1}.

Unless the methanogens are present and active, methane is produced only by decarboxylation of acetate, while the large potential for methane production from HCO_3^{-1} and ethanol produced in (1) is lost, and 2 $(H_2)_{eq}$ is transported to the gas phase with $\Delta G^0 = -17.73$ kJ (mol $H_2)^{-1}$, while the final medium will contain ethanol (and perhaps remaining digestible carbon from the feed if the rate of (1) is too low).

Note that the above calculations could also have been done with NADH as the redox carrier instead of $(H_2)_{aq}$, but the transport of dissolved H_2 between the acidogenetic and methanogenetic bacteria has been proven by experiments.

9.2 Thermodynamic Properties Obtained from Galvanic Cells

The galvanic cell can be visualized as two half cells, separated by a porous membrane (or a salt bridge). Consider the reversible reaction $H_2 + ½\ O_2 \leftrightarrow H_2O$. Read from left to right, the reaction specifies generation of electrical energy by oxidation of H_2, while read from right to left the reaction specifies water splitting by *electrolysis*. In the first scenario, positively charged protons, *cations*, flow from half cell 1 to half cell 2. Negatively charged ions, *anions*, flow in the opposite direction from half cell 2 to half cell 1 at a rate that conserves elctroneutrality. An outside, conducting wire permits the electrons liberated at the *anode* in half cell 1 to flow back toward the *cathode* in half cell 2. The flow of electrons in the outside wire may be used to charge a battery. To increase the conductivity in the cell, SO_4^{-2} (pH low) or Na^+ (pH high) is added. Neither Na^+ nor SO_4^{-2} is, respectively, reduced or oxidized at the cathode/anode in preference to H^+/H_2O.

Read from right to left, water splitting is achieved by an input of electrical energy from a battery. In the water splitting mode, the reactions are as follows:

$$\text{Half cell 1}: 2\,H^+ + 2\,e^- \rightarrow H_2(\text{pH low})\text{ or }2\,H_2O + 2\,e^- \rightarrow H_2 + 2\,OH^-(\text{high pH})\text{at the cathode.}$$

$$\text{Half cell 2}: H_2O \rightarrow {}^1\!/_2\,O_2 + 2\,H^+ + 2\,e^{-1}(\text{pH low})\text{ or }2\,OH^- \rightarrow {}^1\!/_2\,O_2 + H_2O + 2\,e^-(\text{high pH})\text{at the anode.}$$

When the battery is discharged, the cathode is in half cell 1 where H_2 is produced, while the anode is in half cell 2 where O_2 is produced. The electrons now flow from half cell 2 back to the positive electrode of the battery and return to the cathode in half cell 1.

The electromotoric force, emf (with unit V), of the galvanic cell can be measured on a voltmeter placed in the outside connection between the half cells and the meter shows the same potential, 1.229 V whether the process produces $O_2 + H_2$ or consumes the gases. The flow of electrons in the outside wire from the anode to the cathode of the cell is an electric current. Each electron transferred liberates an energy quantum $1.602 \cdot 10^{-19}$ J per V potential difference. Hence for 1 mol H_2 + ½ mol O_2 converted to H_2O_l, the system loses 237.3 kJ. For thermodynamically ideal energy conversion from chemical to electrical energy, this energy is used to recharge the battery. The energy production rate is $237.3 \cdot 3.600^{-1} = 66\,W = 2 \cdot 1.602 \cdot 10^{-19} \cdot 6.022 \cdot 10^{23} \cdot \text{emf} = 2 \cdot 96472 \cdot 1.229 \cdot 3600^{-1}\,W$ when 1 mol H_2 is converted h^{-1}. Thus in summary, for the half-cell reactions

$$H_2 \rightarrow 2\,H^+ + 2\,e^- \quad \text{and} \quad {}^1\!/_2\,O_2 + 2\,e^- + 2\,H^+ \rightarrow H_2O_l.$$

$$-\Delta G^0 = 237.3\text{ kJ(mol }H_2O\text{ formed)}^{-1}\text{ and emf} = \frac{237.3}{2 \cdot 96,472} = 1.229\text{ V}. \qquad (9.20)$$

The emf is a combination of the electrode potentials E^0 in the two half cells. As long as the stoichiometries of the two half-cell reactions are written such that production of electrons in one half cell balances the consumption in the other half cell, the emf of the galvanic cell is the same, that is, one would get the same emf for the total stoichiometry $O_2 + 2\,H_2 \rightarrow 2\,H_2O$ as for the overall reaction used above.

In *chemical thermodynamics*, the reaction $H_2 \rightarrow 2\,H^+ + 2\,e^-$ is *defined* to have an electrode potential $E^0 = 0$ at pH = 0, where $[H^+]$ as well as the other ions have activity 1. Hence, the electrode potential of the other half cell where O_2 is reduced is $E^0 = 1.229$ V under standard conditions. In *Bio-thermodynamics*, the standard condition specifies pH = 7. Now, according to Eq. (9.16), the electrode potential for the cathode reaction is $RT\ln(10^{-7}) = -(8.314 \cdot 298 \cdot 7 \cdot 2.303/96490) = -0.414$ V. Correspondingly, the electrode potential of the anode reaction ½ $O_2 + 2\,H^+ + 2\,e^{-1} \rightarrow H_2O$ is $1.229 - 0.414 = 0.815$ V.

The values are under standard conditions: 298.2 K, [reactants] = 1 M, except $[H^+] = 10^{-7}$. It follows from the table that $E^0_{\text{acceptor}} - E^0_{\text{donor}}$ is positive for all galvanic cells with O_2 as the electron acceptor, that is, O_2 can oxidize all the reduced compounds in Table 9.3. The emf is large – meaning that $(-\Delta G^0_R)$ is large and positive – for all reactions where the electron donor is chosen from

Table 9.3 Half-cell potentials E^0_{acceptor} for electron accepting couples.

Couple	$2\,NO_3^{-}$/ N_2	O_2/ $2\,H_2O$	SO_4^{-2}/ SO_3^{-2}	$2\,H^+$/ H_2	$NADP^+$/ NADPH	NAD^+/ NADH	S/HS^{-1}	FAD^+/ FADH
E^0 mV	824	815	526	415	324	320	278	220
Couple	N_2/ $2\,NH_4^+$	CO_2/ CH_4	SO_4^{-2}/ HS^{-1}	SO_4^{-2}/ S	$2\,HCO_3^{-1}$/ CH_3COO^{-1}	NO_2^{-1}/ NH_4^+	NO_3^{-1}/ NO_2^{-1}	Fe^{+3}/Fe^{+2}
E^0 mV	277	239	218	200	−291	−340	−431	−770

Note: E^0_{donor} for the corresponding electron donating couples are the negative of the tabulated values. The emf for a galvanic cell in which the reaction in one half cell donates electrons to the reaction in the other half cell is emf $= E^0_{\text{acceptor}} - E^0_{\text{donor}}$.

the first entries of the table, but smaller when the electron donor is chosen from the last entries.

Example 9.4 Examples of the use of galvanic cells for the calculation of ΔG_R

Combining the electron acceptor couple $O_2/2\,H_2O$ with the electron donor couple $N_2/2NO_3^{-1}$ yields

$$O_2 + 4\,e^- + 4H^+ \rightarrow 2H_2O \text{ coupled to } N_2 + 6\,H_2O \rightarrow 2\,NO_3^- + 12\,H^+ + 10\,e^-. \text{ The emf} = 815 - (-824) = 1639\,\text{mV}.$$

Balancing the electrons donated and accepted gives the net reaction for 10 e^- transferred:

$$2.5\,O_2 + N_2 + H_2O \rightarrow 2\,NO_3^- + 2H^+ \text{ with } -\Delta G^0_R = 1.639 \cdot (10 \cdot 96.472) = 1580\,\text{kJ(mol } N_2)^{-1}.$$

When reduction of O_2 is coupled to oxidation of Fe^{+2}, $4e^-$ are transferred and

$$O_2 + 4\,Fe^{+2} + 4\,H^+ \rightarrow 4\,Fe^{+3} + 2\,H_2O \text{ with } -\Delta G^0_R = (0.815 - 0.770) \cdot 4 \cdot 96.472 = 17.4\,\text{kJ(mol } O_2)^{-1}.$$

This example shows that combination of a strongly reducing half-cell reaction with another half-cell reaction where reduction is also favored ($Fe^{+3} \rightarrow Fe^{+2}$) will give a small emf and hence a small value of $-\Delta G^0_R$.

If the reduction of SO_4^{-2} to SO_3^{-2} is coupled to oxidation of Fe^{+2} to Fe^{+3}, (ΔG^0_R) is positive = 47.3 kJ (mol SO_3^{-2}), while correspondingly the reaction $SO_3^{-2} + 2Fe^{+3} + H_2O \rightarrow SO_4^{-2} + 2\,Fe^{+2} + 2\,H^+$ proceeds willingly with emf = −525 − (−770) = 245 mV, and $(-\Delta G^0_R) = 2 \cdot 0.245 \cdot 96.472 = 47.3\,\text{kJ(mol } SO_3^{-2})^{-1}$.

The value of ($-\Delta G^0_R$) does not necessarily predict whether a process will proceed willingly under actual physiological conditions. A certain group of iron-oxidizing bacteria that preferentially grows at low pH can gain much energy by the reaction $O_2 + 4\ Fe^{+2} + 4H^+ \rightarrow 4\ Fe^{+3} + 2H_2O$ discussed above. At pH = 3, the concentration of the substrate H^+ is 10^4 times higher than under standard conditions and by (9.10)

$$\begin{aligned}-\Delta G_R &= (-\Delta G_R^0) + RT\ln((10^4)^4)\\ &= 17.4 + 8.314 \cdot 0.2982 \cdot 2.303 \cdot 16 = 108.8\ \text{kJ mol}\ O_2.\end{aligned}$$

At higher pH, for example, at pH = 7, $Fe(OH)_3$ precipitates. Since the solubility product $[Fe^{+3}]\ [OH^-]^3 = 4 \cdot 10^{-38}$, the concentration of $Fe^{+3} = 4 \cdot 10^{-38} \cdot 10^{21} = 10^{-16.40}$ at pH = 7, where $[OH^{-1}] = 10^{-7}$. Consequently, at pH = 7

$$-\Delta G_R = 17.4 - RT\ln(10^{-16.40 \cdot 4}) = 17.4 + 374.5 = 392\ \text{kJ mol}\ O_2^{-1}.$$

The strong oxidizing potential of the Fe^{+3}/Fe^{+2} couple has been used, especially in bioleaching of iron ore. In recent years, oxidation of, for example, acetate to HCO_3^{-1} has been proposed to harvest electrical energy in fuel cells. The conventional $O_2/2H_2O$ electron acceptor couple (815 mV) tends to be kinetically limited, and it is replaced with an almost equally strong Fe^{+3}/Fe^{+2} couple (770 mV). Heine, Hamelers, and Buisman [11] describe a microbial fuel cell shown in Figure 9.1 that can create a sizeable energy current from waste material.

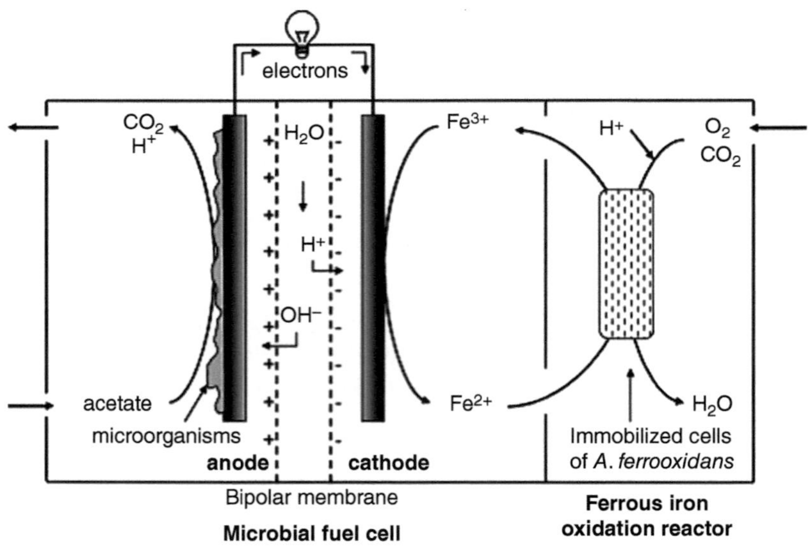

Figure 9.1 Conversion of acetate to CO_2 at the anode by reduction of Fe^{+3} to Fe^{+2}. A bipolar membrane separates H_2O into OH^{-1} that reacts with CO_2 to HCO_3^{-1} and H^+ that migrates to the cathode. Here O_2 is reduced to H_2O with simultaneous reoxidation of Fe^{+2} to Fe^{+3}, catalyzed by immobilized cells of the bacterium *Acidithiobacillus ferrooxidance*. In the cathode half-cell, pH ≤ 2.5, maintained by the supply of H^+, and this prevents precipitation of $Fe(OH)_3$ in the half cell [11].

E_{donor} is determined by pairing the acetate oxidation reaction with reduction of O_2 to the net reaction:

$$\begin{aligned} CH_3COO^{-1} + 2\,O_2 \rightarrow 2HCO_3^{-1} + H^+ \text{ with } (-\Delta G_0) \\ = -2 \cdot 386.7 - (-369.3) \\ = 404.1\text{ kJ mol acetate}^{-1} \text{ by (9.17).} \end{aligned}$$

Since $8 \cdot (E_{acceptor} - E_{donor}) \cdot 96.472 = -\Delta G_0$ for the net reaction, and $E_{acceptor} = 815$ mV, one obtains $E_{donor} = 291$ mV, that is, when $HCO_3{}^{-1}$ is reduced to CH_3 COO^{-1}, $E_0 = -291$ mV, the value entered into Table 9.3.

Since pH ≤ 2.5 in the fuel cell, Figure 9.1, $-\Delta G$ is at least RT ln $10^{4.5} = 26$ kJ smaller than $-\Delta G^0 = 404$ kJ mol acetate^{-1}, but the overall process still has a large positive affinity.

In Section 3.6, several examples with relation to bio-remediation processes were given.

Sulfate reducing bacteria operate according to $SO_4{}^{-2} + 8\ e^- + 9\,H^+ \rightarrow HS^{-1} + 4\,H_2O$. When the electrons are donated by $H_2 \rightarrow 2\,H^+ + 2\ e^-$, one obtains the net stoichiometry as follows:

$$\begin{aligned} 4\,H_2 + SO_4^{-2} + H^+ \rightarrow HS^{-1} + 4\,H_2O \text{ with } -\Delta G_R^0 \\ = 8 \cdot (0.218 - (-0.415)) \cdot 96.472 \\ = 488\text{ kJ(mol } SO_4^{-2})^{-1}. \end{aligned}$$

In a natural habitat, the concentration of dissolved H_2 is, however, far smaller than 1 M. With $[H_2]_{aq} = 10^{-6}$ M and pH = 7, the value of $-G_R^0$ is reduced by RT ln $10^{-24} = 134$ kJ (mol $SO_4{}^{-2})^{-1}$. Also, the pH of the habitat where significant sulfate reduction occurs is likely to be alkaline (pH = 9 to 10) that further reduces $-\Delta G_R$, that is, the affinity of the process is smaller.

Nitrate reduction in wastewater was also discussed in Section 3.6.

$$\begin{aligned} 2\,NO_3^{-1} + 10\,e^- + 12\,H^+ \rightarrow N_2 + 6\,H_2O \text{ and } 2\,NH_4^+ \\ \rightarrow N_2 + 8\,H^+ \\ + 6\,e^-, \text{ i.e., an overall reaction.} \end{aligned}$$

$$\begin{aligned} 3\,NO_3^{-1} + 5\,NH_4^+ \rightarrow 4\,N_2 + 2\,H^+ + 9\,H_2O \text{ with } -G_R^0 \\ = (0.824 - (-0.277)) \cdot 15 \cdot 96.472 \\ = 1593\text{ kJ}(4\,N_2)^{-1}. \end{aligned}$$

In a wastewater treatment situation, $[NO_3{}^{-1}]$ and $[NH_4{}^+]$ are far below 1 M, perhaps 10^{-4} M.

$-\Delta G_R = 1593 - RT\ \ln[10^{3 \cdot 3} \cdot 10^{5 \cdot 3}] = 1593 - 137 = 1456$ kJ $(4\,N_2)^{-1}$, but still the reaction has a high affinity.

The examples given above show that the galvanic cell is a powerful instrument to obtain ΔG_R, and since emf can be measured with an accuracy of at least 1 mV, the values obtained are as good as those obtained from calorimetric measurements for $-\Delta H_R$ coupled with spectroscopic data for ΔS_R. The pioneers of electrochemistry were Walther Nernst (1864–1941) and Johannes Brønsted (1879–1947). In their laboratories, countless experiments were made with different electron acceptor and electron donor pairs, and these are the foundations of extensive tables of electrochemical data. The tables express the same information as do the equally comprehensive data for free energy of formation of compounds that are available in data banks and in tables.

9.3 Conversion of Free Energy Harbored in NADH and $FADH_2$ to ATP in Oxidative Phosphorylation

The mechanism by which free energy packages, ATP, are generated through an energy conversion process from other energy packages, namely, NADH or $FADH_2$ (NADH with an energy content $\Delta G_{NADH} = 219\ \text{kJ mol}^{-1}$), has been studied for at least 75 years. A complete understanding of the mechanism has been one of the key challenges of the biosciences in the twentieth century.

The energy conversion process, oxidative phosphorylation, is different in bacteria (prokaryotes), in eukaryotes such as yeast, filamentous fungi and animal cells, and in plant cells. Here only the process that takes place in the mitochondria of eukaryotes will be discussed – and only semiquantitatively, without inferring all the experimental data and the detailed theoretical investigations that have been devoted to the subject.

Figure 9.2 is a 190 000 times magnification of a mitochondrion organelle from pancreas cells by the 1974 Nobel Prize winner G.E Palade (reproduced by [12]).

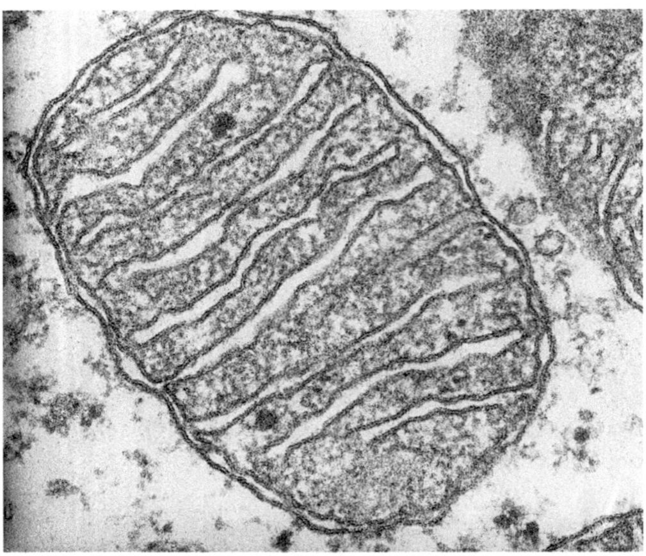

Figure 9.2 Microphotograph of mitochondria in pancreas cells. (Reproduced with permission from Ref. [12].)

The organelle is surrounded by an inner and an outer membrane, and the inner membrane is folded into "*cristae mitochondriales*" (a term coined by Palade) that appear as microtubes extending almost (or wholly) across the mitochondria. Clearly, the area of the inner membrane is vastly extended, and the cristae may even be connected to the outer membrane by a separate membrane "stopper."

Closely spaced along the cristae are the sites where the free energy of the nucleotides is tapped and the energy is transferred via directly connecting ion-channels to the F_0-F_1 ATPase enzyme located just across the membrane. Here, ATP^{-4} is produced from ADP^{-3} (+ an anion) and transported back to the inside of the cristae tube. ADP^{-3} is transported to the cristae and ADP^{-4} transported away (perhaps through as yet undiscovered membrane stoppers?) to be used for reactions in the cytosol.

Whether or not the energy conversion occurs as described above is speculative, but the way in which the energy of NADH and $FADH_2$ is sequestered in the electron transfer chain (ETC) is known in detail.

The ETC consists of a closely cooperating group of four enzyme complexes (complex I–IV) where the electrons produced by oxidation of the nucleotides are finally donated to molecular O_2 that is reduced to H_2O. The electron gradient created from complex I to complex IV permits ions to be transported against a concentration (activity) gradient across the cristae tube membrane and they return through the ATPase, creating ATP from ADP on the way back.

In *complex I* NADH produced in the TCA cycle (Figure 4.6) is oxidized by $NADH \rightarrow NAD^+ + 2\ e^- + H^+$, and the electrons are transferred to the oxidated form of ubiquinol (a derivative of benzoquinol, also called coenzyme Q_{10} or just "Q") by $Q_{ox} + 2\ e^- \rightarrow Q_{red}$ in *complex II* of the ETC. Also contributing to Q_{red} in complex II is $FADH_2$, a product from the TCA cycle obtained in the reaction where succinate is oxidized to fumarate. The third contribution to Q_{red} is from cytosolic NADH that in the form of $FADH_2$ is transported into the mitochondria to complex II. In *complex III*, the total amount of Q_{red} produced in complexes I to II transfers two electrons to the oxidized form of cytochrome c by Q_{red} + cytochrome $c_{ox} \rightarrow Q_{ox}$ + cytochrome c_{red}. Finally, in complex IV, cytochrome c_{red} delivers the 2 e^- to ½ O_2 that is reduced to H_2O.

The total ΔG^0 released per NADH in the processes leading to cytochrome c_{ox} is 69.5 (complex I) + 36.7 ($FADH_2$ to complex II) + 112 (cytosolic NADH to complex II) = 218 kJ (mol cytochrome c)$^{-1}$.

Considering the different free energy of NADH and $FADH_2$, it appears that an ideal P/O ratio for the overall reactions NADH (or $FADH_2$) + ½ $O_2 \rightarrow$ P/O ATP (4.21) should be around 2.5. Since 24 redox equivalents are produced by conversion of $C_6H_{12}O_6$ to CO_2 it is evident that oxidative phosphorylation is by far more effective than substrate-level phosphorylation in glycolysis (30 redox units compared to 2 per glucose molecule).

When ATP is used to energize metabolic reactions, the thermodynamic efficiency is, however, far below 100%. In Chapter 4, it was described that the experimental P/O ratio observed for oxidative fermentation of yeast is only about 1.25

to 1.5 – but still respiration provides much more ATP to living cells than does the glycolytic process.

As a final remark, the way in which the energy is transported across the membrane of the cristae is not definitely settled. Nath and Villadsen [13] discuss the uncertainties that still remain before this question is definitely answered.

9.4 Calculation of Heat of Reaction $Q=(-\Delta H_c)$ and of $(-\Delta G_c)$ Based on Redox Balances

In the design of bioreactions on industrial scale, it may become a significant problem that the heat of reaction cannot be removed by simple means, such as spraying the outer surface of the reactor with water or through the use of cooling coils inside the reactor. If the heat of reaction is not removed, it becomes impossible to keep a desired reaction temperature. Here is an example from the Chilean wine industry: white wine fermentation was performed at 13 °C in order to retain a good bouquet of the final product. A sheet of water (10 °C at the top) was continually drained down the outside of the 80 m^3 fermentation reactor that had no means of internal stirring. Measurements showed that the temperature across the 1.8 m tank diameter varied from 12 °C at the wall to 20 °C at the center of the tank. This had a major influence on the quality of the wine. Installation of a simple recirculation of the medium from the bottom of the tank to a point in the top where the medium was reinjected using a rotary jet head (RJH) – a device described in Chapter 10 – reduced the temperature variation to 2 °C across the tank.

In aerobic fermentation, and especially when Y_{so} is large, removal of the heat of reaction Q becomes a real problem, for example, in production of single-cell protein (Examples 2.2 and 9.5) from methane. $Q=41.6$ M J (kg CH_4 converted)$^{-1}$ or 52.3 M J (kg biomass $CH_{1.8}O_{0.5}N_{0.2}$)$^{-1}$. To remove the heat of reaction, several large hi-tech heat exchangers have to be installed.

The heat of reaction Q is determined from *the heat of combustion* Q_i of the reactants:

$$Q = (-\Delta H_R) = \sum\nolimits_i Q_i Y_{si} - \sum\nolimits_k Q_k Y_{pk}. \tag{9.21}$$

Q is the *loss* of caloric heat from the reacting system with stoichiometry (9.22):

$$0 = \sum\nolimits_i Y_{si} S_i - \sum\nolimits_k Y_{pk} P_k. \tag{9.22}$$

Q is *positive* when the sum of heats of combustion of the substrates, weighted with the stoichiometric coefficients, is larger than the corresponding sum for the products.

Heats of combustion Q_i of compounds are available from thermodynamic tables or data banks. They are similar to $H_{f,I}$, the negative of the enthalpy of formation H_f of compounds used in Eq. (9.9) to define ΔH_R, but there is one significant difference: *Q_I is defined to be 0 for compounds that are the "natural"* *final products of combustion processes, namely, liquid H_2O, CO_2, and N_2.* Thus, while $(-H_{f,H2O}) = 285.8$ kJ (mol H_2O_l)$^{-1}$, $Q_{H2O,l} = 0$ under standard conditions $P = 1$ atm, $T = 298.2$ K.

The choice of standard conditions, here that $Q_i = 0$ for H_2O_l, CO_2, and N_2, does not influence the value of Q, the heat of reaction determined by Eq. (9.21). That the choice of standard state is irrelevant was seen previously for calculation of $(-\Delta G_R)$ and in redox balances by

$$\sum_i Y_{si}\kappa_{si} - \sum_k Y_{pk}\kappa_{pk} = 0. \tag{9.23}$$

Q_i is equal to the HHV (higher heating value) for fuels.

The last column of Table 9.4 lists Q_i (kJ C-mol)$^{-1}/\kappa_i^*$ for a number of compounds that occur in fermentation reactions. κ_i^* is the degree of reduction per C of

Table 9.4 Heat of combustion under standard conditions (pH = 7, liquid water, CO_2 dissolved in H_2O and dissociated to HCO_3^{-1}).

Compound	Formula	Degree of reduction (κ^*) per carbon	Q_i^0 (kJ/mole)	Q_i^0 (kJ/C mole)	Q_i^0/κ^* (kJ/C mole)
Formic acid	CH_2O_2	2	255	255	127.5
Acetic acid	$C_2H_4O_2$	4	875	437	109.5
Propionic acid	$C_3H_6O_2$	4.67	1527	509	109.0
Butyric acid	$C_4H_8O_2$	5.00	2184	546	109.2
Valeric acid	$C_5H_{10}O_2$	5.20	2841	568	109.2
Palmitic acid (s)	$C_{16}H_{32}O_2$	5.75	9978	624	108.5
Lactic acid	$C_3H_6O_3$	4	1367	456	114.0
Gluconic acid	$C_6H_{12}O_7$	3.67	2575	429	117.0
Pyruvic acid	$C_3H_4O_3$	3.33	1082	361	108.3
Oxalic acid	$C_2H_2O_4$	1	246	123	123.0
Succinic acid	$C_4H_6O_4$	3.50	1491	373	106.6
Fumaric acid	$C_4H_4O_4$	3	1335	334	111.3
Malic acid	$C_4H_6O_5$	3	1328	332	110.7
Citric acid	$C_6H_8O_7$	3	1961	327	109.0
Glucose	$C_6H_{12}O_6$	4	2803	467	116.8
Fructose	$C_6H_{12}O_6$	4	2813	469	117.2
Galactose	$C_6H_{12}O_6$	4	2805	468	117.0
Sucrose	$C_{12}H_{22}O_{11}$	4	5644	470	117.5
Lactose	$C_{12}H_{22}O_{11}$	4	5651	471	117.8
Hydrogen	H_2	2	286		(143.0)
Methane (g)	$(CH_4)_g$	8	890	890	111.2
Ethane (g)	$(C_2H_6)_g$	7	1560	780	111.4
Propane (g)	$(C_3H_8)_g$	6.67	2220	740	110.9
Methanol	CH_4O	6	727	727	121.2

(continued)

Table 9.4 (Continued)

Compound	Formula	Degree of reduction (κ^*) per carbon	Q_i^0 (kJ/mole)	Q_i^0 (kJ/C mole)	Q_i^0/κ^* (kJ/C mole)
Ethanol	C_2H_6O	6	1367	683	113.9
iso-Propanol	C_3H_8O	6	2020	673	112.2
n-Butanol	$C_4H_{10}O$	6	2676	669	111.5
Ethylene glycol	$C_2H_6O_2$	5	1179	590	118.0
Glycerol	$C_3H_8O_3$	4.67	1661	554	118.6
Acetone	C_3H_6O	5.33	1790	597	111.9
Formaldehyde	$(CH_2O)_g$	4	571	571	142.7
Acetaldehyde	C_2H_4O	5	1166	583	116.6
Urea	CH_4ON_2	6	632	632	105.3
Ammonia (aq)	NH_3	3	383		(127.7)
Hydrogen sulfide (g)	H_2S	8	909		(113.6)
Biomass (s)	$CH_{1.8}O_{0.5}N_{0.2}$	4.80	560	560	116.7

Note: Otherwise, 1 M concentration of all compounds (dissociated and undissociated). $\kappa^* = \kappa$ of Chapter 3, except for N-containing compounds (N is defined to have $\kappa^* = 0$).

the compound *based on nitrogen having the degree of reduction = 0*. It is seen that many compounds have $Q_i/\kappa_i^* \approx 115$, that is, that $Q_i \approx 115\,\kappa_i^*$ kJ C-mol^{-1}. The outliers are (methanol), formaldehyde, formic acid, oxalic acid, and ammonia that all deviate from $Q_i/\kappa_i^* \approx 115$ by more than +5%.

If the approximate relation $Q_\text{I} \approx 115\,\kappa_i^*$ kJ C-mol^{-1} is accepted, Eq. (9.21) simplifies to

$$Q = (-\Delta H_\text{R}) = 115 \cdot \left(\sum\nolimits_i \kappa_{si}^* Y_{si} - \sum\nolimits_k \kappa_{pk}^* Y_{pk}\right). \tag{9.24}$$

In an anaerobic process, the redox content of the substrates is redistributed to the products, and $Q = 0$ according to Eq. (9.23). In an aerobic process, $\sum_i \kappa_{si}^* Y_{si} - \sum_k \kappa_{pk}^* Y_{pk} = 4\,Y_{so}$, and consequently $Q = 4 \cdot 115\ Y_{sO} = 460\ Y_{so}$. Since Q is independent of the redox level chosen for the elements, the last equation can also be written as $\sum_i \kappa_{si} Y_{si} - \sum_k \kappa_{pk} Y_{pk} = 4\,Y_{so}$, and one does not need to introduce a new set of redox contents for the N-containing compounds.

The result $Q = 460\ Y_{so}$ kJ/C-mol S can also be reformulated to $Q = 460\ Y_{so}$ kJ/(C-mol S converted m^{-3} h^{-1}) or $Q = 460$ kJ/(mol O_2 converted m^{-3} h^{-1}), for example, a heat production of $460 \cdot 200 \cdot 3600^{-1} = 25.6$ kW in a bioreactor where 200 mol O_2 (m^3 h)$^{-1}$ is consumed by the reaction. As discussed in Section 16.1, this represents a considerable burden on the heat exchanger system of the bioreactor.

Example 9.5 Calculation of the heat of reaction Q and extension to approximate calculation of ΔG_R

1) Biomass is produced with CH_4 as carbon source and NH_3 as nitrogen source:

$$CH_4 + 1.454\,O_2 + 0.105\,NH_3 \rightarrow 0.520\,CH_{1.8}O_{0.5}N_{0.2} + 0.48\,CO_2 + 1.688\,H_2O. \qquad (1)$$

Using the κ values defined in Chapter 3 ($\kappa_{NH3} = 0$), one obtains

$$Q = 115 \cdot (8 - 0.52 \cdot 4.20) = 669\,\text{kJ/mol}\,CH_4 \text{ or } 1.454 \cdot 460 = 669\,\text{kJ/mol}\,CH_4 (= 52.3\,\text{MJ/kg biomass}).$$

In Example 10.10, the design of a loop reactor for production of 667 kg biomass h^{-1} from methanol is discussed. The total reactor volume is divided into 11 loop reactors, each of volume 40 m^3. The oxygen uptake is 7.89 mol s^{-1}, or 7.89/40 = 0.197 mol $m^{-3}\,s^{-1}$. Consequently, the heat release is $460 \cdot 0.197$ kJ $m^{-3}\,s^{-1}$ or 90 kW m^{-3}. There is no way in which this large heat release can be removed in internal heat exchangers. The medium has to be sent through highly effective heat exchangers located *outside* the reactor and transported back to the reactor. This is a considerable complication in the reactor design, and the best solution is probably to feed less O_2 per 40 m^3 reactor and install more than 11 parallel reactors.

The heat of combustion $Q_c = Q_c^0$ under standard conditions, and the corresponding free energy of combustion $-G_c = G_c^0$ for a small number of compounds, is shown in the following table:

Compound	H_f^0	G_f^0	S^0	Q_{cl}^0	$-G_c^0$ (units kJ/mol, except for S^0 with unit J (K mol)$^{-1}$)
$(CO_2)_g$	−393.6	−394.5	214	0	0
CH_4	−74.6	−50.5	186	890	818
C_2H_5OH	−277.7	−174.8	161	1367	1326
CH_3COOH	−484.6	−389.9	160	874	933
$C_6H_{12}O_6$	−1273			2803	2872 (glucose)

Combustion of CH_4 follows the stoichiometry $CH_4 + 2\,O_2 \rightarrow CO_2 + 2\,H_2O_l$

$Q_{CH4} = -74.6$ ($= H_f\,(CH_4) - 2\,(H_f\,(H_2O)) - H_f\,(CO_2)) = -74.6 + 2 \cdot 285.8 + 393.6 = 890$ kJ mol^{-1} (see also Table 9.2).

$$-G_c^0(CH_4) = Q_{CH4} + T\left(\sum\nolimits_k Y_{kp}S_{kp}^0 - \sum\nolimits_i Y_{is}S_{is}^0\right)$$
$$= 890 + T(214 + 2 \cdot 70 - 2 \cdot 205 - 186)10^{-3} = 818\,\text{kJ mol}^{-1}$$

In the same way, $-G_c^0$ is calculated for the other entries of the table. For biomass $= CH_{1.8}O_{0.5}\ N_{0.2}$, $-G_c^0 = 536$ kJ(C-mol)$^{-1}$. For reaction (1), $-\Delta G_R^0 = -0.520 \cdot 536 + 818 + 0.104 \cdot 329 = 573$ kJ mol CH_4^{-1}.

Note that H_c and G_c for both H_2O_l and CO_2 are defined to be zero.

2) Next, consider aerobic fermentation of yeast with glucose as carbon substrate:

$$\begin{aligned} CH_2O + 0.1745\,O_2 + 0.0474\,NH_3 \rightarrow\ & 0.279\,CH_{1.83}O_{0.56}N_{0.17} \\ & + 0.355\,CH_3O_{1/2} \\ & + 0.366\,CO_2 + 0.283\,H_2O. \end{aligned} \tag{2}$$

Again, the approximate calculation gives $Q = 0.1745 \cdot 460 = 80.2$ kJ (C-mol glucose)$^{-1}$

Calculation according to the heat of combustion values in Table 9.2 gives:

$$\begin{aligned} Q &= 2803/6 + 0.0474 \cdot 383 - 0.279 \cdot 560 + 0.355 \cdot 1367/2 \\ &= 86.4\ \text{kJ(C-mol glucose)}^{-1}. \end{aligned}$$

Considering the uncertainty of Q_c(biomass) and that the result is obtained by subtraction of terms of almost equal size, the approximate method gives an adequate result – to be used in design of heat exchangers for removal of the heat of reaction.

The value of $-\Delta G_R^0$ is calculated from the tabulated values as shown in the previous example.

Based on a large set of tabulated values for $-\Delta G_c^0$, Roels [14] derived the following approximate relation for the change of free energy by reaction:

$$-\Delta G_R^0 = 98.4\left(\sum\nolimits_i \kappa_{si}^* Y_{si} - \sum\nolimits_k \kappa_{pk}^* Y_{pk}\right) + 86.6\left(\sum\nolimits_i Y_{si} - \sum\nolimits_k Y_{pk}\right). \tag{3}$$

For reaction (2), one obtains: 98.4 $(4 + 3 \cdot 0.0474 - 4.71 \cdot 0.279 - 6 \cdot 0.355) + 86.6\ (1.0474 - 1) = 72.8$ kJ C-mol glucose^{-1}.

3) Finally consider the methanation reaction of (9.7):

$$\begin{aligned} H_2 + 0.260\,CO_2 + 0.0048\,NH_3 \rightarrow\ & 0.02\,CH_{1.68}O_{0.39}N_{0.24} \\ & + 0.240\,CH_4 + 0.511\,H_2O. \end{aligned} \tag{4}$$

By the approximate method, $Q = (2 - 0.02 \cdot 4.33 - 8 \cdot 0.240) \cdot 115 = (-0.0066) \cdot 115 \approx 0$.

This is very far from the large Q measured experimentally (or calculated using the true values of Q_i for the reactants in (4) – see Ref. [3]. The true value of Q is 63.6 kJ (mol H_2)$^{-1}$ reflecting that Q_c^0 for H_2 is 285 kJ/mol when it is combusted to H_2O, and $Q_{CH4} = 890$ kJ/mol.

The simple formulas for calculation of Q and ΔG_R will not work when $115 \cdot \kappa^*$ for Q is appreciably different from the true value of Q_c (as is the case especially for H_2 and to a certain extent also for CH_4).

References

1 Nordkvist, M. (2005) Mixing and mass transfer by rotating jets: fundamentals and applications. PhD thesis, DTU (DK).

2 Geng, M. and Duan, Z. (2010) Prediction of oxygen solubility in pure water and brines up to high temperatures and pressures. *Geochim. Geochosm. Acta*, **74**, 5631–5640.

3 Villadsen, J., Nielsen, J., and Lidén, G. (2011) *Bioreaction Engineering Principles*, 3rd edn, Springer, New York.

4 Lebon, G., Jou, D., and Casaz-Vasques, J. (2008) Understanding non-equilibrium thermodynamics, in *Foundations, Applications and Frontiers*, Springer, Berlin.

5 Alberty, R.A. (2003) *Thermodynamics of Biochemical Reactions*, John Wiley & Sons, Inc., New Jersey.

6 Schill, N.A., Liu, J.S., and von Stockar, U. (1999) Thermodynamic analysis of growth of *Methanobacterium thermoautotrophicum*. *Biotechnol. Bioeng.*, **64**, 74–71.

7 Lehninger, A.E. (1975) *Biochemistry*, 2nd edn, Worth, New York.

8 Bekers, K., van Gulik, W., and Heijnen, J.J. (2015) Determination of the *in vivo* NAD/NADH ratio in *S. cerevisiae* under anaerobic conditions using alcohol dehydrogenase (ADH) as sensor. *Yeast*, June 9, 2015, E pub.

9 Jovanovic, S., Jovanovic, N., and Jovanovic, A. (2006) High glucose concentration protects single beating adult cardiomycetes against bacteria. *Biochem. Biophys. Res. Comm.*, **341**, 57–66.

10 Vojinovic, V. and von Stockar, U. (2009) Influence of uncertainties in pH, p(Mg), activity coefficients, metabolic concentrations, and other factors on the analysis of thermodynamic feasibility of metabolic pathways. *Biotechnol. Bioeng.*, **103**, 780–795.

11 Heine, A.M., Hamelers, H.V.M., and Buisman, C.J.N. (2007) Microbial fuel cell operation with continuous biological ferrous iron oxidation at the catholyte. *Environ. Sci. Technol.*, **41**, 4130–4134.

12 Scheffler, J.E. (2008) *Mitochondria*, 2nd edn, John Wiley & Sons, Inc., New York.

13 Nath, S. and Villadsen, J. (2015) Oxidative phosphorylation revisited. *Biotechnol. Bioeng.*, **112**, 429–437.

14 Roels, J.A. (1983) *Energetics and Kinetics in Biotechnology*, Elsevier, Amsterdam.

Part Two
Bioreactors

Fundamental Bioengineering, First Edition. Edited by John Villadsen.

10
Design of Ideal Bioreactors

John Villadsen

Summary

The ideal bioreactor, both the ideal stirred tank reactor (STR) and the ideal plug flow reactor (PFR), is the subject of this chapter. The focus is on design of different types of industrially important bioreactors. These are generally STRs, first of all the continuous stirred reactor CSTR, working in steady state. The STR is characterized by no spatial variation in any of the variables, concentration of reactants, temperature, and pH. Different ways of operating steady-state CSTRs are described with maximum volumetric productivity as an important objective. The advantage of using biomass recirculation and of operating with several reactors in parallel or in series is described. Next fed-batch operation is discussed, and this mode of operation is compared with steady-state CSTR. New types of reactors, the so-called loop reactors, which are better suited for high mass transfer of, for example, oxygen are introduced and their use is exemplified by the design of an industrial plant for single-cell protein production. Finally, the design of an industrially important process, the production of succinic acid from glucose in several reactor types, is used as an example.

10.0
Introduction

In this chapter, the fundamental disciplines of stoichiometry, biochemical kinetics, and thermodynamics from Chapters 2, 3, 7, and 9 will be combined to design reactors in which the processes can be carried out at any scale. Transport phenomena, particularly mass transport to and from a gas phase, will be included, but the kinetics of the transport process, that is, the mass transport coefficient $k_l a$ will be assumed to be a known parameter.

The reactors are assumed to be "ideal," by which it is meant that mixing the liquid phase is so efficient that no concentration gradients exist. In Chapter 11,

Fundamental Bioengineering, First Edition. Edited by John Villadsen.

metrics for mixing are defined, and the power input needed to obtain a certain degree of mixing will be discussed. Thereafter, empirical relations between the degree of mixing and the rate of transport processes – especially mass transport – will be derived, and the general conclusion will be that transport processes primarily depend on the power input. Combination of the design principles in Chapter 10 and the mostly empirical relations from Chapter 11 will allow a realistic design of a large-scale industrial bioreactor to be made, including estimation and comparison of the production costs in reactors of different designs.

It will be argued that the steady-state CSTR is the reactor that is potentially able to entail the smallest capital + production costs for any bioreaction process, and many variants of steady-state CSTR operation are discussed. In industrial practice, another type of STR operation, namely, fed-batch, is currently preferred to steady-state CSTR operation and consequently fed-batch operation will be given much attention. New reactor designs, loop reactors where the mixing is done by means that are different from the mechanical stirrers used in CSTR designs, will be introduced. The batch reactor and the continuous plug flow reactor are briefly discussed in Example 10.5. The batch reactor is used – usually in small scale – for the production of specialty chemicals, whereas the plug flow reactor is quite impractical as a stand-alone reactor in the bio-industry since it requires a nonsterile feed. As a scavenger of the remainder of the substrate in the effluent from a steady-state CSTR, the plug flow reactor (or an approximation to a plug flow reactor as a series of sequential CSTRs) serves a useful purpose in wastewater treatment, and this application of the plug flow reactor is illustrated in Example 10.5.

Finally, in general only Monod-type kinetic expressions will be used in Chapters 10 and 11. The emphasis is on *reactor* design, and simple kinetic expressions are used to avoid unnecessary complications. In any case, it has been shown in Chapter 7 that with our present knowledge of cell kinetics it is futile to design industrial bioreactors based on anything more complicated than the *steady-state* kinetics, typically represented by Monod kinetics. In control of bioreactors (Chapter 15), transient behavior *must* be considered, but here the solution is to control the state of the process using a data-based model that is continuously updated on the process computer.

In the design of enzymatically catalyzed processes, the mechanistic kinetic models of Chapter 7 can, of course, be used, just as in any other chemical reaction.

10.1 The Design Basis for a Once-Through Steady-State CSTR

Figure 10.1 is a schematic representation of a continuous stirred tank reactor, and a reaction carried out in the reactor is modeled by Eq. (10.1):

$$\frac{d(V\mathbf{c})}{dt} = V(\mathbf{q}^{T} + \mathbf{q}) + v_f\,\mathbf{c_f} - v_e\,\mathbf{c_e} \tag{10.1}$$

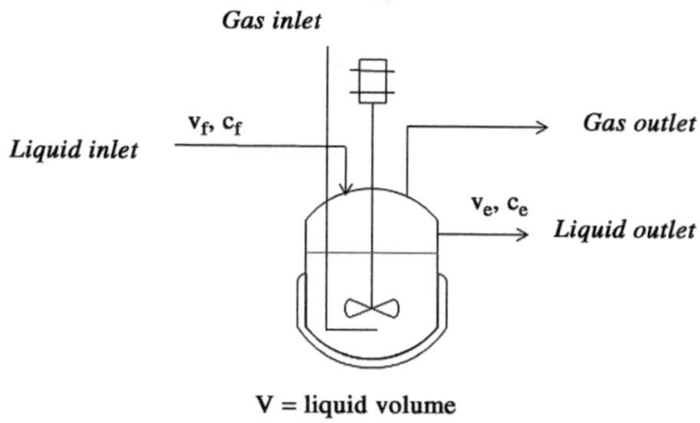

Figure 10.1 A CSTR for a cell reaction with one liquid feed stream v_f in which the concentrations of substrates S_i are given by the vector $\mathbf{c_f}$. The effluent flow is v_e that is not necessarily equal to v_f, and the concentration of effluent components (products and remaining substrates) are given by the vector $\mathbf{c_e}$. The concentrations of substrates and products *in* the reactor are given by the vector $\mathbf{c}$ (not shown on the figure). When the reactor is number 2,3 in a sequence of CSTRs, $\mathbf{c_f}$ will also contain concentrations of biomass and products. The nomenclature for enzymatic reactions is much simpler than for cell reactions.

The mass balance (10.1) holds both for a cell reaction and for an enzymatic reaction where, in all cases $v_f = v_e$, and $\mathbf{c}_f = \mathbf{c}$, the vector of reaction species concentrations. The accumulation term $d(V\mathbf{c})/dt$ is zero for a steady-state process, Sections 10.1–10.4. On the r.h.s. of Eq. (10.1), vectors $\mathbf{q}$ and $\mathbf{q}^T$, $v_f\,\mathbf{c}_f$ and $v_f\,\mathbf{c}_f$ are, respectively, reaction rate terms, which express transfer between a gas phase and the liquid phase, and flow terms that express the difference between input and output of reaction components. The reaction rate terms in $\mathbf{q}$ are functions of $\mathbf{c}$. The mass balance for an enzymatic reaction follows by obvious simplifications of the balances treated below in Sections 10.2–10.5.

In this section, the steady-state CSTR will be treated. This mode of operation of the CSTR was discussed in Chapters 2–4 and in Chapter 7. With a single CSTR $[x_f, p_{if}]$ are always assumed to be zero. $v_f = v_e$, and $\mathbf{c_e} = \mathbf{c}$.

In Section 10.2, combinations of steady-state CSTRs in parallel or in sequence are considered. In sequential operation, the feed to the first reactor has $[x_f, p_{if}] = 0$, but the feed to subsequent reactors contains both products and biomass, $[x_{f,j}\, p_{if,j}] \neq 0, j = 2{,}3, \ldots$.

In Section 10.3, recycling of biomass is included for a single CSTR. Here $v_f = v_e$, but $x_e \neq x$, while for all other reactants $\mathbf{c_e} = \mathbf{c}$.

When fed-batch operation is treated in Section 10.5, $v_e = 0$, and the liquid feed stream v_f contains only substrates, $[x_f, p_{i,f}] = 0$.

The term $\mathbf{q} = [q_x, (-q_s), q_{pi}]$ will be represented by Monod expressions, $q_x = r_x\, x = r_x Y_{sx}^{obs}(s_f - s)$, where r_x is given by one of the expressions (7.15), (7.21)–(7.23), and Y_{sx}^{obs} is the observed yield coefficient of biomass *on the limiting substrate* S. All the other rates in $\mathbf{q}$ are given by $q_i = Y_{x,i}^{obs}\, q_x$.

In terms of the true yield coefficients Y_{ij}^{true}, the mass balances for the case of one metabolic product P, $[x_f, p_f] = 0$, and a simple Monod expression (7.15) for $r_x = \mu$ become:

$$q_x = \mu x = \mu_{\max} \frac{s}{s + K_s} x = Dx. \tag{10.2}$$

$$q_s = -D(s_f - s) = -(Y_{xs}^{\text{true}} D + m_s)x. \tag{10.3}$$

$$q_p = Dp = (Y_{xp}^{\text{true}} D + m_p)x. \tag{10.4}$$

From Eqs. (10.2) to (10.4), the biomass and product concentrations are obtained in Eq. (10.5):

$$D = \mu = \mu_{\max} \frac{s}{s + K_s}; \quad X = \frac{S(1 - S)}{S + b(S + a)}; \quad P = \frac{S + b_1(S + a)}{S + b(S + a)}(1 - S) \tag{10.5}$$

In Eq. (10.5), dimensionless variables [S, P, X] and dimensionless parameter combinations are introduced:

$$X = \frac{x}{s_f Y_{sx}^{\text{true}}}; \quad P = \frac{p}{s_f Y_{sp}^{\text{true}}} \quad \text{and} \quad S = \frac{s}{s_f}; \quad a = \frac{K_s}{s_f}; \quad b = \frac{m_s Y_{sx}^{\text{true}}}{\mu_{\max}};$$

$$b_1 = \frac{m_p Y_{px}^{\text{true}}}{\mu_{\max}}. \tag{10.6}$$

Example 10.1 Calculation of biomass and product concentrations as functions of *D*

In Figure 7.5, data are presented for aerobic fermentation of *Aerobacter aerogenes* on glycerol. Parallel to the main reaction with stoichiometry given by Eq. (1), glycerol is oxidized to CO_2 in a maintenance reaction (2):

$$\begin{aligned} &CH_{8/3}O + 0.4483\, O_2 + 0.1368\, NH_3 \\ &\quad \rightarrow 0.6842\, CH_{1.8}O_{0.5}N_{0.2} + 0.3158\, CO_2 + 0.923\, H_2O. \end{aligned} \tag{1}$$

$$CH_{8/3}O + 1.1667\, O_2 \rightarrow CO_2 + 1.3333\, H_2O. \tag{2}$$

$$\begin{aligned} Y_{sx}^{\text{true}} &= 0.6842 \cdot (24.6/30.67) = 0.549 \text{ g biomass/g glycerol.} \\ Y_{sp}^{\text{true}} &= 0.3158 \cdot (44/30.67) = 0.4531 \text{ g } CO_2\text{/g glycerol.} \\ Y_{px}^{\text{true}} &= Y_{sx}^{\text{true}}/Y_{sp}^{\text{true}} = 1.212 \text{ g biomass/g } CO_2. \end{aligned} \tag{3}$$

The maintenance coefficient m_s is fitted from the data in Figure 7.5 to $m_s = 0.08$ g glycerol (g X h)$^{-1}$, and $\mu_{\max} = 1\ h^{-1}$. According to Eq. (2), one C-mol CO_2 is formed (C-mol S)$^{-1}$ in the maintenance reaction.

$$b = \frac{0.08 \cdot 0.6842 \cdot 24.6}{30.67 \cdot 1} = 0.04392.$$

$$m_p = \frac{0.08 \cdot 44}{30.67} = 0.1147\,\text{g}\,CO_2(\text{g biomass h})^{-1} \text{ and } b_1 = 0.1147 \cdot 1.212 = 0.1391. \tag{4}$$

For $D \to 0\,[X,P] \to \left[0, b_1/b\right] = \left[0, \frac{m_p}{m_s} \cdot \frac{Y_{px}^{true}}{Y_{sx}^{true}}\right] = \left[0, 1 \cdot \frac{0.6842/0.3158}{0.6842/1}\right] = [0, 3.17]$.

Differentiation of the expression for X in Eq. (10.5) with respect to S yields:

$$X_{max} = \frac{\left(\sqrt{1 + a^*} + \sqrt{a^*}\right)^2}{1+b} \quad \text{for } S = -a^* + \sqrt{a^{*2} + a^*},$$

where $a^* = \dfrac{a\,b}{1+b}$.

For $a = (0.01/10) = 10^{-3}$ (see Figure 7.5), one obtains $a^* = 4.21 \cdot 10^{-5}$, $S = 6.44\,10^{-3}$, $X = 0.964$ at $D = 0.866\,\text{h}^{-1}$.

Inserting $S = 6.44\,10^{-3}$ in the formula (10.5) for P gives $P = 1.097$, and $P/X = 1.138$.

From the dimensionless formulas for P and X, one obtains $P/X = Y_{px}^{true}(p/x) = Y_{px}^{true} \cdot Y_{xp}^{obs}$. This product should be exactly 1 if there is no maintenance production of CO_2. The above result shows that even at $D = 0.866\,\text{h}^{-1}$, which is close to the value $D_{washout} = 10/(10 + 0.01) = 0.999$, there is still a considerable influence of the maintenance reaction.

Figure 10.2 shows $P(D)$ and $X(D)$ for $0 \leq D \leq 0.999\,\text{h}^{-1} = D_{washout}$. It adds considerable insight to Figure 7.5, although it is of course based on

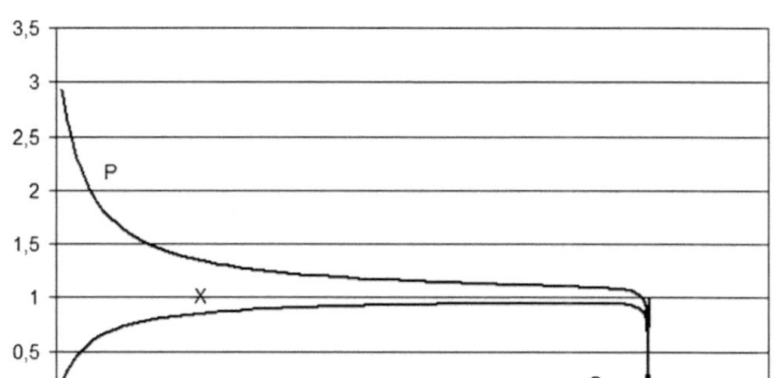

Figure 10.2 Aerobic cultivation of *A. aerogenes* on glycerol with biomass and CO_2 as products. Dimensionless values of glycerol concentration (S), biomass concentration (X), and product ($P = CO_2$) as determined by Eqs. (10.5) and (10.6) are shown. $D_{max} = D_{washout} = 0.999\,\text{h}^{-1}$. For other data, see Figure 7.5.

the same parameters in the kinetic expression, but the maintenance reaction is explicitly stated in the fashion explained in earlier examples, such as Example 3.6.

An important design criterion is the *productivity* of a given reactor. For biomass and metabolic products, one wants to maximize Dx and Dp (with $[x_f, p_{if}] = 0$), and in a bio-remediation process one wants to maximize $D(s_f - s)$. In all cases, one is looking for the maximum value of $[q_x, q_{pi}, (-q_s)]$, the *maximum volumetric productivity of the reactor.*

Since $q_x = r_x\, x$, it is always possible to tabulate q_x as a function of s, since $r_x = f(s,p), x = Y_{sx}^{obs}(s_f - s)$, and $p = Y_{sp}^{obs}(s_f - s)$. In this way, the s value is obtained for which q_x is maximum without recourse to analytical differentiation of q_x. Once the optimal value of s and q_x^{max} has been obtained, the corresponding $(-q_s)^{max}$ and q_p^{max} are found as, for example, $(-q_s)^{max} = Y_{xs}^{obs}\, q_x^{max}$.

The following analytical expression for the optimal value of $S = s/s_f$ and the corresponding optimal value for $D = D_{opt}$ and maximum value for q_x can, however, easily be obtained. The results (10.7) are correct *for a simple Monod rate expression* $r_x = \frac{\mu_{max} s}{s+K_m} = \frac{\mu_{max} S}{S+a}$ *with no maintenance included*:

$$q_x = Dx = \frac{\mu_{max} S}{S+a} Y_{sx} s_f (1-S), \text{ and } Q_x = \frac{q_x}{\mu_{max} Y_{sx} s_f} = \frac{S(1-S)}{S+a}.$$

$$Q_x^{max} = \left(\frac{q_x}{\mu_{max} Y_{sx} s_f}\right)_{max} = \left(\sqrt{1+a} - \sqrt{a}\right)^2 \text{ for } S_{opt}$$
$$= -a + \sqrt{a^2 + a} \text{ and } D = D_{opt} = \frac{S_{opt}}{S_{opt} + a}. \tag{10.7}$$

For substrate-inhibited kinetics, Eq. (7.21), with $S = s/s_f, a = K_m/s_f, Y_{ij} = Y_{ij}^{obs}$, and $b = s_f/K_i$:

$$r_x = \frac{\mu_{max} S}{b S^2 + S + a} \text{ and } Q_x = \frac{q_x}{\mu_{max} Y_{sx} s_f} = \frac{S(1-S)}{S + a + bS^2}.$$

$$S_{opt} = -c + (c^2 + c)^{1/2}, \quad \text{where} \quad c = a/(1+b). \tag{10.8}$$

For product inhibition following (7.22):

$$Q_x = \frac{S(1-S)(1 - d(1-S))}{S+a}, \quad \text{where} \quad d = Y_{sp}\, s_f/p_{max}. \tag{10.9}$$

For maintenance added to the simple Monod model:

$$Q_x = \frac{S^2(1-S)}{(S+a)(S+b(S+a))}, \quad \text{where } b = m_s\, Y_{sx}^{true}/\mu_{max} \text{ as defined in (10.6).} \tag{10.10}$$

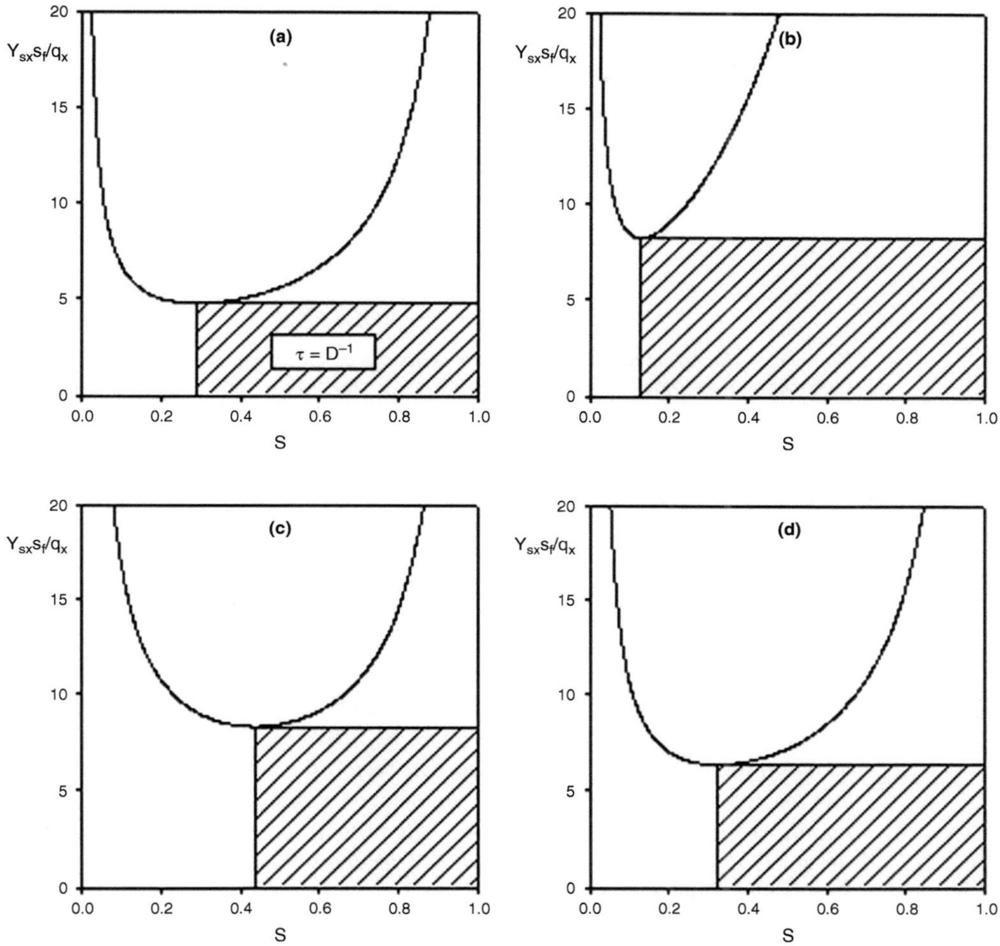

Figure 10.3 Productivity of a steady-state stirred tank reactor. (a) Monod kinetics: $\mu_{max} = 0.5\,h^{-1}$, $K_s = 0.4\,g\,l^{-1}$. $s_f = 2\,g\,l^{-1}$, $Y_{sx} = 0.5\,g\,g^{-1}$, Eq. (10.7). (b) Substrate inhibition: as in (a), and $K_i = 0.25\,g\,l^{-1}$, that is, $b = 8$, Eq. (10.8). (c) Product inhibition: as in (a), and $s_f Y_{sp}/p_{max} = 2/3$, Eq. (10.9). (d) Maintenance: as in (a), and with $b = 0.2$, Eq. (10.10). Hatched area: Holding time τ to achieve maximum productivity (a) 3.37 h; (b) 7.17 h; (c) 4.69 h; (d) 4.29 h.

Figure 10.3 shows $(Y_{sx}s_f/q_x)$ for the four cases (10.7)–(10.10) as functions of $S = s/s_f$. Calculation of Q_x and determination of Q_x^{max} is done by tabulation as described above for Eqs. (10.9) and (10.10), since analytical calculation is far too complicated to be worth the effort.

The reason for using $\frac{1}{Q_x \cdot \mu_{max}} = \frac{Y_{sx}s_f}{q_x}$ as the ordinate (rather than simply Q_x^{-1}) is that for $x_f = 0$ this is related to D^{-1} = holding time τ – or residence time in the language of chemical reaction engineering

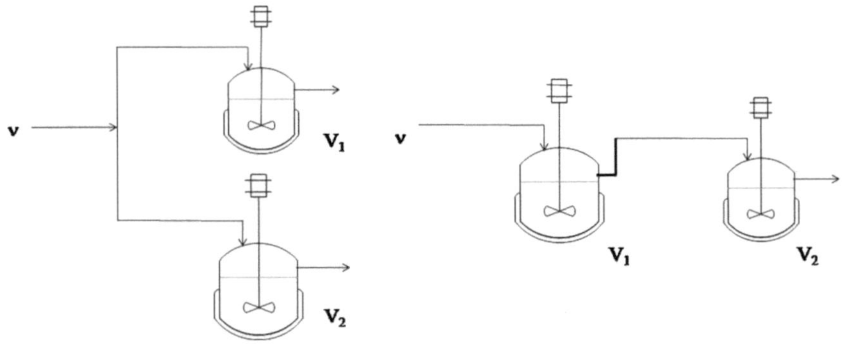

Figure 10.4 Combination of two reactors of volume V_1 and V_2, either in parallel or in sequence. In the example, $V_1 = V_2 = 20\ \mathrm{m}^3$. Different values of v are studied: $v = [20, 30$ and $60\ \mathrm{m}^3\ \mathrm{h}^{-1}]$.

$$q_x = r_x x = Dx = DY_{sx}^{obs}\, s_f(1 - S) \rightarrow D^{-1} = \tau = \frac{Y_{sx}s_f}{q_x} \cdot (1 - S). \qquad (10.11)$$

Hence, the shaded area on the figures represents the holding time needed to obtain a given value of S. It is seen that *the smallest possible holding time for a steady-state CSTR* to reach S_{opt} is obtained from the rectangle that touches the curve for $(Y_{sx}\ s_f/q_x)$ at its minimum. The highest productivity is obtained at this holding time. Using a sequence of steady-state CSTRs down to S_{opt} would give a higher value of τ. On the other side of S_{opt}, that is, for s values, where q_x *increases* with increasing s (as in a normal chemical reaction), it is better to follow the curve to obtain the smallest value of $\tau = (Y_{sx}s_f/q_x) \cdot (1 - S)$. This is the reason why a *plug flow* reactor rather than a CSTR (or a series of CSTRs) gives the smallest value of the *extra* τ needed to reduce S from S_{opt} to a smaller value. This is illustrated in Example 10.5.

One might think that operation at the s value for which r_x, r_p, or $(-r_s)$ is as large as possible would offer another optimization opportunity. This is not so.

For Eqs. (10.7), 10.9 and (10.10), r_x is at a maximum for $s = s_f$, but here $x = 0$, and the productivity is zero. For Eq. (10.8) the maximum value of r_x is obtained by differentiation of

$$r_x = \frac{\mu_{max}S}{bS^2 + S + a} \rightarrow S_{opt} = \sqrt{\frac{a}{b}} \quad \text{or} \quad s = \sqrt{\frac{K_m K_i}{s_f^2}}, \qquad (10.12)$$

unless $a \geq b$, in which case the value S_{opt} at which r_x is maximum $= \frac{\mu_{max}}{b+1+a}$.

Operating at $s = s_f$ will give $x = 0$ $(a/b \geq 1)$, and the productivity is 0. Operating at $S = \sqrt{\frac{a}{b}}$ for $a/b < 1$ undoubtedly leads to an x value>0, but operation at this S value is, however, extremely dangerous since even a small increase in D will lead to washout, because D now becomes larger than r_x^{max}. The operation point for maximum productivity is given by Eq. (10.8), and for $[a, b] = [0.2, 8]$ as in (b) in Figure 10.3, this is at $S = 0.1285$. The S value for which r_x is as large as possible is at $S = 0.1581$, but here q_x is smaller, and operation at this value of S (or at higher S values) is unstable.

10.2 Combination of Several Steady-State CSTRs in Parallel or in Series

In Section 10.2, the design and operation of a single steady-state CSTR was discussed. In the bio-industry, one would design the CSTR to work with the expected feed flow, and one would operate the reactor according to a mixture of objectives: try to obtain the highest overall economic benefit from the operation without violating the rules for maximum effluent concentration of any environmentally unfriendly reactant. In a longer perspective, one might consider to install a second reactor if the demand for the product is expected to increase.

This is not an option in a wastewater treatment plant where the concentration of the pollutant can vary significantly over short time intervals and where especially the incoming flow v_f could vary drastically, for example, after a sudden rainstorm. The overriding concern is to avoid washout of the biomass since regeneration of the often very slow growing consortium of microorganisms (as flocs or immobilized on inert particles) can take months.

The subject of optimizing operation of a wastewater treatment plant can be approached only by means of examples since there are so many variables: the kinetics of the reaction, the relative stability of different organisms in the microbial consortium, and so on. In Example 10.2 and in Figure 10.4, simple Monod kinetics will be used, and the optimal operation of two equal sized bioreactors will be studied for different values of the flow $v_f = v_e$ through the system.

Example 10.2 Optimal combination of two steady-state CSTRs in a wastewater treatment plant

Two reactors, each of volume $V = 20\,\text{m}^3$ is used to remove a pollutant S from a feed stream v_f where the concentration of S is $s_f = 60\,\text{mg l}^{-1}$. The kinetics of the reaction is given by

$$r_x = \frac{\frac{4}{3}\,s(\text{h}^{-1})}{s + 4\left(\frac{\text{mg}}{\text{l}}\right)};\, q_x = r_x\, x,\ \text{and}(-q_s) = Y_{xs}\, q_x = 10\, q_x \cdot x.$$

$$x_f + Y_{sx}(s_f - s), \quad \text{where}\ x_f = 0\ \text{in the feed to the plant.}$$

$$v = \mathbf{20}\, m^3\, h^{-1}.$$

For this small value of v, the plant is overdesigned. From the kinetics and using Eq. (10.7), one obtains that the maximum productivity is obtained at $S = (1/5) = 0.2$, that is, at $s = 12\,\text{mg l}^{-1}$. *In a single steady-*

state CSTR with $V = 20\,\mathrm{m}^3$, a maximum removal of S is obtained at $D = 1\,\mathrm{h}^{-1}$, and $(-q_s)^{\max} = 48\,\mathrm{mg\,l^{-1}\,h^{-1}}$. $x_e = 4.8\,\mathrm{mg\,l^{-1}}$ and $s_e = 12\,\mathrm{mg\,l^{-1}}$.

Thus, to treat $20\,\mathrm{m^3\,h^{-1}}$, one will obtain the maximum productivity when only one of the two parallel tanks is used and the other is at standby. When both tanks are used, the combined effluent concentration is below $12\,\mathrm{mg\,l^{-1}}$, that is, the *quality* of the treatment process is improved. Should the feed be split equally between the two reactors to obtain the maximum increase in quality of the process?

When $10\,\mathrm{m^3\,h^{-1}}$ is sent to each tank, $D = 0.5\,\mathrm{h}^{-1}$ in both tanks, and from the kinetic expression $s_e = 2.4\,\mathrm{mg\,l^{-1}}$, $x_e = 5.76\,\mathrm{mg\,l^{-1}}$, and the productivity is $(-q_s) = 57.6/2 = 28.8\,\mathrm{mg\,l^{-1}\,h^{-1}}$ for the total volume $40\,\mathrm{m}^3$ of the two tanks.

An uneven distribution of $v = 20\,\mathrm{m^3\,h^{-1}}$ results in a *smaller* volumetric productivity, and *a smaller* quality improvement – for example, with $v_1 = 9\,\mathrm{m^3\,h^{-1}}$ to tank 1 and $v_2 = 11\,\mathrm{m^3\,h^{-1}}$ to tank 2:

$$s_1(= s_e \text{ from tank 1}) = 2.037\,\mathrm{mg\,l^{-1}}, \quad s_2 = 2.809\,\mathrm{mg\,l^{-1}} \rightarrow s_{e,\text{average}} = 2.46\,\mathrm{mgl^{-1}}, \quad (-q_s) = 28.77\,\mathrm{mg\,l^{-1}\,h^{-1}}.$$

In the consecutive arrangement of the two tanks $v = 20\,\mathrm{m^3\,h^{-1}}$ is sent to the first tank, and the effluent from tank 1 is further treated in tank 2:

$[s_1, x_1]$ are the substrate and biomass concentrations between the two reactors and they are given by $s_1 = 12\,\mathrm{mg\,l^{-1}}$ and $x_1 = 4.8\,\mathrm{mg\,l^{-1}}$, since tank 1 operates at its optimal productivity ($D = 1\,\mathrm{h}^{-1}$) and $x_f = 0$.

For the second tank, also with $v/V = D = 1\,\mathrm{h}^{-1}$, the mass balance is

$$v/V(x_2 - x_1) = 1 \cdot (x_2 - x_1) = D(s_2)x_2 = r_x(s_2)x_2;\ [s_2, x_2] \text{ are the effluent concentrations from tank 2} \quad (1)$$

Solution of (1): $x_2 - 4.8 = \frac{4}{3}(60 - 10\,x_2)/(64 - 10\,x_2) \cdot x_2$ gives $x_2 = 5.933$ and $s_e = 0.67\,\mathrm{mg\,l^{-1}}$.

Splitting the feed is not optimal since this would lead to suboptimal production in both tanks. Thus, for v_1, respectively, equal to 18 and $16\,\mathrm{m^3\,h^{-1}}$ leads to $x_2 = 5.923$ and $5.900\,\mathrm{mg\,l^{-1}}$, and s_e is larger than $0.67\,\mathrm{mg\,l^{-1}}$.

Clearly, a consecutive arrangement of the two reactors gives a much better quality of the effluent water, and also a slightly better productivity than the parallel arrangement. In an arrangement of reactors where the total reactor capacity is large, the effluent substrate concentration reaches a very low level, and then, as mentioned in the introduction to this chapter, an approximation to plug flow – here a sequence of two reactors – gives the highest productivity and also the maximum conversion of substrate.

$$\boldsymbol{v = 40}\,\mathrm{m}^3\,h^{-1}.$$

For this flow rate, the parallel arrangement of the two reactors with $20\,m^3\,h^{-1}$ fed to each reactor gives the maximal productivity of 48 mg substrate removed $l^{-1}\,h^{-1}$ in each of the reactors, and $s_f = 12\,mg\,l^{-1}$. No uneven distribution of v between the reactors is going to improve on this result.

The consecutive arrangement of the two reactors with $40\,m^3\,h^{-1}$ fed to the first reactor will not work since $v/V_1 = 2\,h^{-1}$, which is larger than $D_{max} = \frac{4}{3}\frac{s_f}{s_f+4} = \frac{5}{4}$.

The consecutive arrangement offers an extra degree of freedom: one may send only $v_1\,m^3\,h^{-1}$ through the first tank, while $v_f - v_1 = v_2$ bypasses the first tank and is reunited with v_1 before the second tank.

Let v_1 be $20\,m^3\,h^{-1}$. Now $D_1 = 1\,h^{-1}$, and the effluent concentrations from tank 1 are $[x_1, s_1] = [4.8,\ 12\,mg\,l^{-1}]$. After mixing of v_1 and v_2, the combined feed $40\,m^3\,h^{-1}$ to tank 2 is $[x_1, s_1]' = [2.4,\ 36\,mg\,l^{-1}]$.

Using (1) one obtains $2(x_2 - 2.4) = \frac{4}{3}(60 - 10\,x_2)/(64 - 10\,x_2) \cdot x_2 \rightarrow x_2 = 4.8\,mgl^{-1}$, exactly the same result as in the parallel arrangement of the two $20\,m^3$ reactors. The reason is that with $x_2 = 4.8\,mg\,l^{-1}$, both tanks operate at the optimal value of the specific growth rate r_x.

Would it be an advantage to use an uneven distribution of v_1 and the bypass v_2?

Let v_1 be $18\,m^3\,h^{-1:}$ $D_1 = 0.9\,h^{-1}$, $[x_1, s_1] = [5.17,\ 8.31\,mg\,l^{-1}]$, $[x_1, s_1]' = [2.327,\ 33.73\,mg\,l^{-1}] \rightarrow x_2 = 4.794\,mg\,l^{-1}$.

For $v_1 = 16\,m^3\,h^{-1}$: $D_1 = 0.8\,h^{-1}$, $[x_1, s_1] = [5.4,\ 6\,mg\,l^{-1}]$, $[x_1, s_1]' = [2.16,\ 38.4\,mg\,l^{-1}] \rightarrow x_2 = 4.532\,mg\,l^{-1}$.

For $v_1 = 22\,m^3\,h^{-1}$: $D_1 = 1.1\,h^{-1}$, $[x_1, s_1] = [4.11,\ 18.85\,mg\,l^{-1}]$, $[x_1, s_1]' = [2.263,\ 37.37\,mg\,l^{-1}] \rightarrow x_2 = 4.651\,mg\,l^{-1}$.

As could be expected, an uneven distribution of v_f is suboptimal compared to $20\,m^3\,h^{-1}$ fed to the first tank, and $20\,m^3\,h^{-1}$ bypassed to the second tank, where it is calculated that r_x attains its optimal value $= 1\,h^{-1}$ in both reactors for $x_2 = 4.8\,mg\,l^{-1}$.

One concludes that the parallel and the consecutive arrangements work equally well when the feed stream is distributed in the optimal way.

$\boldsymbol{v_f = 60\,m^3\,h^{-1}}$.

When $v_f = 60\,m^3\,h^{-1}$, two tanks each of $20\,m^3$ volume will not work at all if they are arranged in parallel since r_x^{max} is $1.25\,h^{-1}$ and by equal distribution of the feed both tanks will wash out when $v/V = 30/20 = 1.5$. The only solution is to let $20\,m^3\,h^{-1}$ feed pass through each reactor and bypass the remaining $20\,m^3\,h^{-1}$. The effluent concentration of substrate in the combined effluent is $(12 \cdot 40 + 60 \cdot 20)/60 = 28\,mg\,l^{-1}$.

In analogy with the case for $v_f = 40\,m^3\,h^{-1}$, one might feed $20\,m^3\,h^{-1}$ to the first reactor and $40\,m^3\,h^{-1}$ between the two reactors:

$$[x_1, s_1] = [4.8, 12\,mg\,l^{-1}], [x_1, s_1]' = [1.6, 44\,mg\,l^{-1}] \rightarrow x_2 = 2.65\,mg\,l^{-1} \text{ and } s_f = 33.5\,mg\,l^{-1}.$$

For $v_1 = 18$ and $22\,\text{m}^3\,\text{h}^{-1}$, the result is $x_2 = 2.575$ and $2.508\,\text{mg l}^{-1}$, respectively. This is intuitively correct: when the consecutive arrangement of the tanks is overloaded, it is best to let the first tank operate at a dilution rate where its productivity is at a maximum.

There is, however, no better solution than totally bypassing $20\,\text{m}^3\,\text{h}^{-1}$ and treating the remaining $40\,\text{m}^3\,\text{h}^{-1}$ as described above for a total flow of $40\,\text{m}^3\,\text{h}^{-1}$, namely, to send $20\,\text{m}^3\,\text{h}^{-1}$ through the first reactor, while the remaining $20\,\text{m}^3\,\text{h}^{-1}$ joins after the first reactor. Thus, $40\,\text{m}^3\,\text{h}^{-1}$ is treated to an s value of $12\,\text{mg l}^{-1}$ and the final effluent contains $(40 \cdot 12 + 20 \cdot 60)/60 = 28\,\text{mg l}^{-1}$, the result obtained in the parallel arrangement.

The result for the severely overloaded system fits with the conclusions from $v_f = 20$ and $40\,\text{m}^3\,\text{h}^{-1}$: for a feed rate that is small compared to the combined capacity of the reactor system, consecutive operation of the two tanks is clearly the best choice, since this system approximates the behavior of a plug flow reactor. When the feed rate increases, the parallel arrangement gradually improves, and when $v_f \geq$ the value where $v_f/2V = 1\,\text{h}^{-1}$ the optimal r_x for each tank, one obtains the same optimal reduction of substrate concentration in both arrangements, the parallel with equal distribution of v_f to the reactors, the consecutive when the total flow of substrate is distributed such that each reactor operates at its optimal value of r_x. In both cases, part of the incoming feed is bypassed in the reactor system.

Obviously, calculations must be done for other kinetic expressions to see if the above-described design criteria can be used in general. It seems, however, that the conclusions of the example are based on fundamental insight into a cell reaction system.

10.3
Recirculation of Biomass in a Single Steady-State CSTR

The volumetric production rate is proportional to the biomass concentration in the bioreactor. Thus, if the biomass can be retained longer in the reactor than the holding time $\tau = D^{-1} = V/v$, then one may expect a higher productivity in a given reactor volume V.

Two different solutions are used by industry to retain the biomass in the reactor, and these are shown schematically in Figure 10.5a and b.

In Figure 10.5a, a settler is inserted *between the reactor outlet and the effluent from the system.* This design is widely used in wastewater treatment plants, and with a sufficiently large settler one can obtain a concentration x_R in the recycle stream Rv that is up to 50% higher than the concentration of cells x in the effluent from the reactor. In mammalian cell cultivations, *cell perfusion* by gentle settling of the cells, returning the major part of the cells and nutrients to the reactor – perhaps after separation of toxic products – is widely used, especially in "single-use bioreactors."

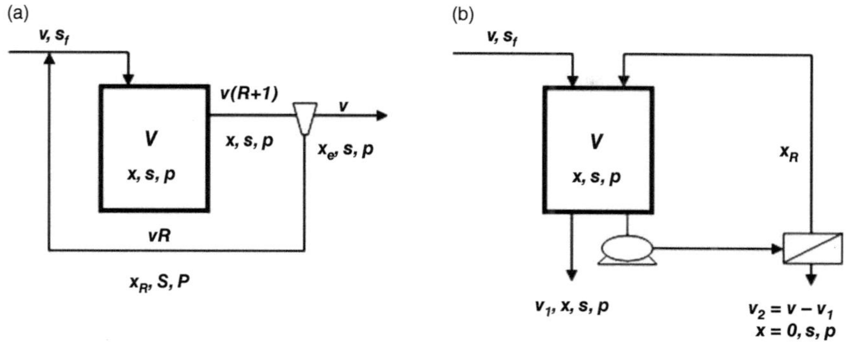

Figure 10.5 Cell recirculation achieved by (a) a settler and (b) by a membrane filter.

With a more efficient cell up-concentration in a hydrocyclone, one avoids the capital and operation costs of the large settler, and this option is sometimes used in the bio-industry. The industry appears to favor this cell separation method over simple settler systems.

In Figure 10.5b, a complete separation of cells from the liquid effluent is achieved by means of a microfilter, an ultrafilter, or even a nanofilter (for virus cultivation). To obtain steady state, the cells produced are purged in a separate effluent stream v_1 on the figure. This design is the most widely used in industry as well as in production of germ-free drinking water. The capital cost for the filter and the pump is quite high compared to the cost for the tank reactor, and so is the operation cost, but the efficiency is high and sterile operation is easier to maintain.

In Figure 10.5a, a cell mass balance around the settler gives:

$$v(R+1)x = v\,R\,x_R + v\,x_e \rightarrow (R+1) = R\frac{x_R}{x} + \frac{x_e}{x} = R\beta + f. \tag{10.13}$$

R is the *recirculation rate* and β is the *cell up-concentration ratio.* For given values of R and β, the ratio f, the up-concentration of cells in the reactor relative to the effluent cell concentration x_e can be calculated. β depends on the efficiency of the settler. $\beta < 1.3$–1.5 unless a very large settler is used. For a hydro-cyclone, $\beta < 1.5$–2.5. R is the ratio between the recirculated medium Rv and the net flow v through the system. Usually, $0.4 < R < 0.8$, and the value of R chosen depends on the pumping costs, which may be high, especially for viscous media.

An overall balance for cells is as follows:

$$D(x_e - x_f) = \frac{v}{V}(x_e - x_f) = q_x = x\,r_x \text{ or } D\left(f - \frac{x_f}{x}\right) = r_x. \tag{10.14}$$

Usually, cell recirculation is used in single reactors, and then $x_f = 0$. r_x is *only* a function of s and p, the concentrations of substrate and product *in the reactor.* Eq. (10.14) expresses that by a suitable choice of f, one may achieve steady-state operation of the system *at any desired value of r_x or D.* This is used either to increase the capacity v of a given reactor volume V or to achieve a given (high) conversion of substrate for given v and V.

As long as the substrate S or the product P does not adhere to X in the circulation loop (it could adsorb on the biomass), both the substrate and the product concentrations are the same in the reactor, in the loop, and in the effluent. Thus, the only effect of cell recirculation is *to increase the productivity of the reactor by increasing the concentration of biomass in the reactor.*

In Figure 10.5b, the feed v to the system is split into v_1, the purge stream, and v_2, the cell free permeate from the membrane filter. All the biomass produced in the reactor leaves through v_1:

$$v_1 x = V r_x x \rightarrow r_x = \frac{v_1}{V} = \frac{v - v_2}{V} = \frac{v}{V}\left(1 - \frac{v_2}{v}\right) = D(1 - \Omega). \tag{10.15}$$

In Eq. (10.15), Ω is the *filter separation factor.* With sufficient energy input to the recirculation loop, one can have virtually the whole net flow through the system leave through the filter, and as in Eq. (10.14) r_x is "artificially" increased by the large factor $\frac{1}{1-\Omega}$.

The improvement of reactor performance obtained by cell recirculation is independent of a possible maintenance substrate consumption that occurs in parallel to the main reaction. The maintenance increases proportional to the production by the main reaction, and if the production increases by a certain factor maintenance increases by the same factor, that is, the maintenance substrate consumption is constant per unit product produced. The operation costs in terms of pumping and capital costs for extra equipment do, however, make recirculation less profitable as seen in Example 10.4. For reactions where the biomass yield is very small, for example, in wastewater treatment, the pumping costs are usually small relative to the gain in capacity.

Example 10.3 Recirculation as a means of improving the capacity of a bio-remediation plant

In Example 10.2, the operation of a plant in which two reactors, each of volume $20\,m^3$, coupled sequentially is used to treat $v_f = 60\,m^3\,h^{-1}$ wastewater with $s_f = 60\,mg\,l^{-1}$ can be significantly improved if recirculation is applied to the second reactor.

If $v_2 = 40\,m^3\,h^{-1}$ raw sewage is joined with $v_1 = 20\,m^3$ treated in the first $20\,m^3$ reactor, the joint feed to the second reactor has $x_1 = 1.6\,mg\,l^{-1}$ and $s_1 = 44\,mg\,l^{-1}$. It is desirable to operate also the second reactor with an apparent dilution rate of $D_{app} = 1$ since this is the dilution rate for which q_x is maximum with the kinetics of the reaction.

$$D(x_{2,e} - x_1) = 3(x_{2,e} - x_1) = \frac{4}{3}\,\frac{60 - 10\,x_{2,e}}{64 - 10\,x_{2,e}}\,x_2$$

$x_{2,e} = 4.8\,mg\,l^{-1}$ and $s_e = 12\,mg\,l^{-1}$

The biomass concentration

x_2 in the second reactor is $x_2 = 2x_{2,e} = 9.6\,mg\,l^{-1}$.

Hence $f = 1/2$ for the second reactor.

This is the result obtained in Example 10.2 for $\nu_f = 40\,m^3\,h^{-1}$ with the same split of the feed. The same result would, of course, be obtained if a single reactor of volume $60\,m^3$ were used, but this is not an option with an existing design of the reactor system.

According to Eq. (10.13), a factor $x_2/x_{2,e} = f = \frac{1}{2}$ will be obtained, for example, with $\beta = 11/8$ and $R = \frac{4}{3}$. The recirculation flow is $\frac{4}{3} \cdot 60 = 80\,m^3\,h^{-1}$, which requires both a large settling tank and some pumping costs. The biomass concentration x_R in the recirculation loop is still very low, $x_R = \beta\, x_2 = 11/8 \cdot 2 \cdot 4.8\,mg\,l^{-1} = 13.2\,mg\,l^{-1}$, and the circulating liquid has a viscosity like water.

Example 10.4 Lactic acid production from lactose and using a membrane filter for biomass recirculation

The wastewater from a cheese factory contains 4.5 wt.% lactose after the soluble proteins have been separated by an ultrafiltration and used for "upgrading" cheap animal food products. The total wastewater outlet is $15\,m^3\,h^{-1}$, and it contains a total of $675\,kg\,h^{-1}$ lactose. It is (marginally) profitable to convert the lactose to lactic acid by the homolactic metabolism of Figure 4.4b.

For a given kinetics, the reactor volume needed to convert 98% of the lactose to lactic acid is V m^3. It is desired to reduce the reactor volume to $V_1 = 0.1\,V$. To this purpose a microfilter is installed as shown in Figure 10.5b.

For $\Omega = 0.9$ in (10.15), $\nu_1 = 1.5\,m^3\,h^{-1}$ and ν_1 contains all the biomass produced. $\nu_2 = 13.5\,m^3\,h^{-1}$ and contains no biomass. Ten per cent of the lactic acid is lost with ν_1 unless another filter is installed to recapture the product from ν_1, but also in a once-through operation the biomass must be filtered from the liquid flow of $15\,m^3\,h^{-1}$.

From brochures of microfilters, one obtains the following data:

The filter permits a flow of $10\,l\,m^{-2}$ filter h^{-1}. This gives a filter area of $1350\,m^2$.

To sustain the flow through, the filter $0.74\,m^3\,h^{-1}$ liquid must be pumped per m^{-2} filter, transverse to the flux direction through the filter. Hence, approximately $1000\,m^3\,h^{-1}$ medium is pumped through the loop.

The pressure loss across the filter is $4\,bar = 4 \cdot 10^5\,kg\,m^{-1}\,s^{-2}$, assuming that the filter is not fouled.

The energy input to be delivered by the pump $= 4 \cdot 10^5\,kg\,m^{-1}\,s^{-2} \cdot 1000/3600\,m^3\,s^{-1} = 111\,kW$.

The reactor volume has been reduced by a factor 10, but the cost (filters and pumping costs) may well cancel out the reduction in the capital cost for the reactor. An economically better solution could be to increase the feed concentration of lactose by evaporating part of the water. Now a membrane filter may have to be installed to remove the lactic acid product, since a high lactic acid concentration severely inhibits cultivation with lactic bacteria. Evaporation of water from the product, lactic acid, must, however, be included in any case, and intensifying the process by providing a higher concentration of lactose in the feed is not an extra cost.

Example 10.5 A plug flow reactor installed after a steady-state CSTR

In the introduction to this chapter, the advantage of using a plug flow reactor *after* a steady-state CSTR is mentioned.

The mass balance for substrate conversion in a steady-state plug flow reactor is as follows:

$$\frac{ds}{dz} = \frac{v}{A} q_s, \tag{10.16}$$

where z is a coordinate along the reactor axis $0 \leq z \leq L$ (the length of the reactor) and $\frac{v}{A}$ is the volumetric liquid flow per area of the free reactor tube, the *superficial linear velocity, u*. The solution of Eq. Eq. (10.16), assuming no axial diffusion in the reactor (i.e., $s = s_f$ at $z = 0$) *and for a simple Monod kinetics* is

$$\begin{aligned} \Theta &= \mu_{max} \frac{V}{v} = \mu_{max} \tau \\ &= \left(1 + \frac{a}{1 + X_f}\right) \ln\left(\frac{X}{X_f}\right) - \frac{a}{1 + X_f} \ln(1 + X_f - X) \\ &= (1 + a) \ln\left(\frac{x}{x_f}\right) - a \ln\left(\frac{s}{s_f}\right). \end{aligned} \tag{10.17}$$

In the first part of Eq. (10.17), the product of the holding time τ, and μ_{max}, the maximum value of r_x in Eq. (7.15), is related to the dimensionless variables and parameters in Eq. (10.6). $X_f = (x_f / s_f Y_{sx})$ and $a = (K_m / s_f)$. Formula (10.17) is derived in several textbooks (e.g., [1], Chapter 9) and it is applied for a stand-alone plug flow reactor. The second expression is used to calculate the holding time for a plug flow reactor placed after a sequence of steady-state CSTRs, and without withdrawal of reactants between the reactors.

Consider a plug flow reactor placed after the second of the two 20 m^3 CSTRs in the consecutive arrangement discussed in Example 10.2 for $v_f = 60\ m^3\ h^{-1}$. $X_2 = 2.65$ mg l^{-1}, and $s_2 = 33.5$ mg l^{-1}.

$$\begin{aligned} \frac{4}{3}\tau &= \left(1 + \frac{1}{15}\right) \ln \frac{0.1(60 - s_e)}{x_2} - \frac{1}{15} \ln\left(\frac{s_e}{s_2}\right) \\ &= \frac{16}{15} \ln \frac{0.1\ (60 - s_e)}{2.65} - \frac{1}{15} \ln\left(\frac{s_e}{33.5}\right). \end{aligned} \tag{1}$$

For $s_e = 3\,\text{mg l}^{-1}$, one obtains $\tau = 0.978\,\text{h}^{-1} \rightarrow V_{\text{plug flow}} = 0.978 \cdot 60 = 58.7\,\text{m}^3$.

With a more modest objective of reducing s to $s_e = 12\,\text{mg l}^{-1}$, $V_{\text{plug flow}} = 42.1\,\text{m}^{-3}$.

This could well prove to be a viable solution compared to installing cell recirculation to bring the substrate concentration down from 33.5 mg l^{-1} in the example with two 20 m^3 reactors in series.

Note: Expression (10.17) is identical to the integrated mass balance for a *batch reactor with simple Monod kinetics.* The quantity $\frac{zA}{v}$ that in Eq. (10.17) represents the holding time in a continuous steady-state plug flow reactor becomes the reaction time t (>0) in the unsteady-state mass balance for a batch reactor.

$$\frac{ds}{dt} = q_s, \quad \text{which integrates to } \Theta = \mu_{\max}\, t$$
$$= \left(1 + \frac{a}{1 + X_0}\right)\ln\left(\frac{X}{X_0}\right) - \frac{a}{1 + X_0}\ln(1 + X_0 - X). \tag{10.18}$$

$X_0 = \frac{x_0}{s_0\, Y_{sx}}$. $[x_0, s_0]$ are, respectively, the concentration of biomass (the inoculum) and of substrate at $t = 0$.

In this chapter, the batch reactor will not be further discussed, mostly because large-scale production of commodity chemicals will become too expensive in a batch reactor. The treatment of rapid transients in Section 7.3 has shown that the time profile of reactant concentrations cannot be represented by any Monod-type kinetics – and at present one lacks fundamental knowledge of how any specific rate can be modeled for design of bioreactors with rapid transients. During a large part of the batch reaction, the cells are, however, in metabolic steady state with substrate concentrations far above the value of the Monod constant K_m. The relative biomass concentration x/x_0 increases exponentially with time ($q_x = \mu_{\max}\, x$). This part of the batch time profile is uninteresting, except for determination of $\mu_{\max}$. The lag phase in the beginning of the batch fermentation cannot be represented by Monod expressions (cybernetic models, Section 7.4, or the equally empirical compartment models discussed in Section 7.3 could be used), and the end of the batch, when substrate concentration decreases below K_m, cannot either be modeled by Monod-type kinetics – although the downgrading of cell machinery is slow and the reaction will come to almost completion before the structure of r_x has changed dramatically.

For process control of industrial bioreactors – for example, to keep certain variables at given set-point values – one must apply a continuously changing, data-driven reactor model.

10.4
A Steady-State CSTR with Uptake of Substrates from a Gas Phase

In previous examples, only the liquid-phase reaction has been considered. There are, however, many reactions in which substrates are supplied from a gas phase, and products are desorbed to the gas phase. In order to obtain the steady-state design of the reactor, one must assure that the rate of consumption of all substrates corresponds to the stoichiometry of the reaction. Gas to liquid transport is determined by mass transfer with rate constant $k_l a$ as described in the early part of Chapter 2, while the reaction in the liquid phase is described by the kinetic expressions developed in Chapter 7.

To illustrate the rather obvious requirement that all rates must be equal in steady-state operation, a single example will be given.

Example 10.6 Uptake of O_2 and reaction with methanol in single-cell biomass production

The production of single-cell protein (SCP) with C1 compounds (CH_4 or CH_3OH) as carbon source has become a valuable option for countries with a large production of natural gas that has recently lost much of its commercial attraction due to the advent of gas production through "fracking." Conversion of the natural gas to methanol is not either attractive since the market for methanol is largely saturated. Hence, countries such as Trinidad and Tobago with a daily production of methanol in excess of 20 000 tons and a weak market for sale of natural gas to the United States of America has considered production of SCP for the Caribbean market, mostly to supply a fish-meal substitute for the large aquaculture industries in the region.

Conversion of methanol to SCP (with the formula $X = CH_{1.8}O_{0.5}N_{0.2}$) is by the stoichiometry given in Eq. (1):

$$\begin{aligned} &CH_3OH + 0.7317\,O_2 + 0.1463\,NH_3 \\ &\quad \rightarrow 0.7317\,X + 0.268\,CO_2 + 1.56\,H_2O. \end{aligned} \tag{1}$$

An excess of CH_3OH severely inhibits the growth of the production organism (*Methylococcus capsulatus* is often used, but other bacteria, for example, *Pseudomonas* sp. oxidize methanol to biomass without excessive production of CO_2).

For a methanol (S) limited cultivation at the optimum temperature 45 °C, the kinetics of Eq. (2) is proposed:

$$q_x = \frac{0.9\,s}{s + 0.0834(\text{kg m}^{-3})}x. \tag{2}$$

With a methanol feed concentration $s_f = 50\,\text{kg m}^{-3}$, $q_x^{max} = 0.9\frac{0.7317 \cdot 24.6}{32} \cdot 50 \cdot 0.921 = 23.33\,\text{kg X m}^{-3}\,\text{h}^{-1}$ is obtained by Eq. (10.7) at a methanol concentration $s = 1.958\,\text{kg m}^{-3}$ and dilution rate $D = 0.863\,\text{h}^{-1}$. The biomass concentration in the effluent is $x_e = 0.5625\,(50 - 1.958) = 27.0\,\text{kg m}^{-3}$.

For production of 50 000 ton biomass year^{-1} = 16.67 ton biomass h^{-1}, one needs a volumetric flow through the reactor of $v = 16667/27.0 = 617\,\text{m}^3\,\text{h}^{-1}$ and a total reactor capacity of $V = 617/0.863 = 715\,\text{m}^3$ (distributed between seven or eight CSTRs, each with a working volume of about $100\,\text{m}^3$).

To maintain a steady-state production of 16.67 ton biomass per hour, the oxygen transfer must be

$$\begin{aligned} q^{\text{T}}_{\text{O}_2} &= \frac{0.7317}{0.7317}\,\text{mol O}_2(\text{C-mol X h})^{-1} = \frac{1000}{24.6} \\ &= 40.65\,\text{mol O}_2(\text{kg X h})^{-1} = 948\,\text{mol O}_2(\text{m}^3\,\text{h})^{-1}. \end{aligned} \quad (3)$$

In Eq. (3), it is assumed that one can neglect the oxygen dissolved in the liquid feed and effluent.

If the reactor is sparged with technically pure (e.g., 99% pure) O_2 at 2 atm total pressure in the reactor and 45 °C reactor temperature, $s^*_{\text{O}} = 1.96\,\text{mmol l}^{-1} = 1.96\,\text{mol m}^{-3}$ by Eq. (9.2). s_{O} is about $0.1\,s^*_{\text{O}}$ and

$$948 = k_1a \cdot 0.9 \cdot 1.96 \rightarrow k_1a = 537\,\text{h}^{-1}. \quad (4)$$

If this mass transfer coefficient cannot be achieved in the large ($\sim 100\,\text{m}^3$) bioreactor – and the discussion in Chapter 11 will show that a large power input is needed to achieve $k_1a = 537\,\text{h}^{-1}$ – then the methanol feed concentration has to be decreased, and this will result in a larger total bioreactor volume.

10.5 Fed-Batch Operation of a Stirred Tank Reactor in the Bio-Industry

A majority of industrial bioreactors operate in the fed-batch mode. The large-scale of production in the commodity chemical industry dictates that simple, easily controlled equipment should be used. The steady-state CSTR will, as discussed in this section, have a higher volumetric production rate than the fed-batch STR, and it is sometimes an advantage that downstream operation can be done in continuous mode and not sequentially after each fed-batch run. Some companies fear that the continuous feed policy may increase the risk of infection, but continuous sterilization of the feed to a CSTR is now routinely done with hardly any risk of infection, and also in the fed-batch mode the reactor is continuously fed with substrates. The greatest worry in continuous operation is that the often heavily engineered production organism should revert to an unproductive variety, in which case the promise of continuous operation for several months is not fulfilled. Still, large productions are now made in steady-state CSTRs, also for valuable and sensitive products such as human insulin. The Novo-Nordisk company uses campaigns of about 4 weeks continuous yeast culture, engineered to make human insulin, in their large (~75 m^3) production reactors.

The classical motivation for operating a cultivation process in fed-batch mode is to secure that the specific growth rate stays at or just below the value where overflow metabolism leads to production of an unwanted by-product, for example, ethanol in baker's yeast production. For simple kinetics, the feed profile in fed-batch cultivation can be exactly predicted to assure that the specific growth rate remains at a fixed value. The fed-batch mode of operation was actually invented in 1917 by the chief engineer Søren Sak (1883–1950) of the Danish Yeast Company, exactly to be able to feed a concentrated sugar solution to make baker's yeast, at a low and constant sugar concentration in the reactor. For many years, the technology was named "Sak's method," but now "fed-batch operation" is universally used.

A fed-batch starts as a batch reaction in a medium volume V_0. Medium is inoculated, and batch operation continues until the substrate concentration has decreased to a low value s_0 and the biomass concentration has increased from, for example, 10 g m^{-3} to 1 kg m^{-3}. s_0 is chosen such that the specific growth rate attains its desired value for $s = s_0$. Thereafter, feed of the limiting substrate is started at flow rate v_f and with very high substrate concentration s_f. For recombinant organisms, an inducer is added to start production of the desired product, and often the O_2 feed is set at a lower value since the oxygen demand in the production phase is smaller (the product may even be formed only at near steady-state biomass concentration – but this is of course not the case for baker's yeast production). The fed-batch stops when a certain volume of medium V has been obtained or when the titer (concentration) of a metabolic product has reached the desired value. The reactor is emptied, its contents are sent to downstream processing, the reactor is sterilized, and a new fermentation starts. Sometimes, a part of the medium is retained in the reactor, and a new fed-batch is immediately started. This is called *repeated fed-batch*, and the productivity becomes much higher since sterilization is avoided, and also the long and unproductive batch phase. Unfortunately, repeated fed-batch leads to a gradually decreasing activity of the biomass, and mutations happen at increasing frequency – an unpleasant fact in, for example, fed-batch production of penicillin where the high producing industrial strain of *Penicillium crysogenum* reverts to a "bald," poor producing "natural" strain.

For one metabolic product, the transient mass balances for biomass X, limiting substrate S and product P are as follows:

$$\frac{d(x\,V)}{dt} = q_x\,V = \mu\,x\,V. \tag{10.19}$$

$$\frac{d(sV)}{dt} = V\frac{ds}{dt} + s\frac{dV}{dt} = -\,Y_{xs}\,\mu\,x\,V + v_f s_f, \quad \text{or} \quad \frac{ds}{dt} = -Y_{xs}\,\mu\,x\,V + D(t)(s_f - s). \tag{10.20}$$

$$\frac{d(pV)}{dt} = Y_{xp}\,\mu\,x\,V \tag{10.21}$$

Neither X, S, nor P leaves the reactor, and X and P are not present in the feed v_f. The occurrence of what looks like a flow term $D(t)$ s in Eq. (10.20) origins in the term $s\,\frac{dV}{dt}$, while $D(t)$ s_f represents the inflow of S with the feed v_f. $D(t)$ is only formally a dilution rate, since both v_f and V are functions of t to be found either analytically or by numerical integration of the mass balances.

An analytical solution of Eq. (10.19) is obtained when μ is a constant $=\mu_0$: $x\,V = x_0\,V_0 \exp(\mu\,t)$. A constant $\mu=\mu_0$ can be maintained only when μ is a function *only* of the limiting substrate concentration s that is kept constant during the fed-batch. If μ depends on the product concentration p, the analytical integration of Eq. (10.19) is not possible.

In the following calculations, it is also assumed that there is no maintenance substrate consumption in parallel to substrate consumption in the main reaction since this would invalidate the assumption that $(-q_s)$ is proportional to q_x in Eq. (10.21), that is, $Y_{xs}^{true} = Y_{xs}^{obs} = Y_{xs}$.

When $\mu=f(s)$, and when the yield coefficients Y_{xs} and Y_{xp} are constants, the solution of Eq. (10.22) is trivial, and $v_f(t)$ is obtained.

$$\frac{ds}{dt} = 0 = -Y_{xs}\,\mu x(t) + \frac{v_f(t)}{V(t)}(s_f - s) \rightarrow v_f(t) = \frac{Y_{xs}}{(s_f - s_0)}\mu_0 x V$$
$$= \frac{Y_{xs}}{(s_f - s_0)}\mu_0 x_0 V_0 \exp(\mu_0\,t). \tag{10.22}$$

Equation (10.22) shows that the feed profile is exponential: $v_f(t) = b\,\mu_0\,x_0\,V_0 \exp(\mu_0\,t)$ with $b = \frac{Y_{xs}}{(s_f - s_0)}$.

Now $V(t)$ is obtained by integration of $\frac{dV}{dt} = v_f = b\mu_0 x_0 V_0 \exp(\mu_0 t)$, with $V = V_0$ for $t=0$:

$$\frac{V}{V_0} = 1 - b + bx_0 \exp(\mu_0 t) = 1 - b - bx_0 y, \quad \text{where } \alpha = \exp(\mu_0\,t). \tag{10.23}$$

Finally, x is determined from $xV = x_0 V_0 \exp(\mu_0 t) \rightarrow \frac{x}{x_0}$

$$= \frac{\exp(\mu_0\,t)}{1 - bx_0 + bx_0 \exp(\mu_0\,t)} = \frac{V_0}{V}\alpha. \tag{10.24}$$

$\frac{x}{x_0}$ increases exponentially for small t, but $\frac{x}{x_0} \rightarrow \frac{1}{bx_0}$ for large t, that is, $x \rightarrow \frac{Y_{xs}}{s_f - s_0}$.

Thus, when $t \rightarrow \infty$ the biomass concentration increases to the value that would be obtained if all the substrate in v_f is converted to X with yield coefficient Y_{sx}.

Consequently, when the conditions, (1) μ is independent of p and (2) no maintenance consumption of substrate, are satisfied, the time profiles for all variables are calculated analytically.

Example 10.7 Design of a fed-batch process for baker's yeast production

A typical industrial strain of *Saccharomyces cerevisiae* converts glucose to biomass (baker's yeast) by the following stoichiometry:

$$CH_2O + 0.3944\,O_2 + 0.115\,NH_3 \rightarrow 0.5768\,X(CH_{1.8}O_{0.5}N_{0.2}) + 0.4232\,CO_2 + 0.65\,H_2O. \quad (1)$$

For glucose as the limiting substrate, the specific growth rate is given by

$$r_x = \mu = \frac{0.4\,(h^{-1})s}{s + 150(\text{mg l}^{-1})}. \quad (2)$$

When $\mu > 0.25\,h^{-1}$ some ethanol will be produced as a byproduct. To avoid this, and to obtain the highest instantaneous production rate $q_x = \mu x$ of biomass during the fed-batch production, one decides to work at $\mu = \mu_0 = 0.25\,h^{-1}$. At this value of μ, the glucose concentration obtained from Eq. (2) is $s = 250\,\text{mg l}^{-1}$, and s should be kept at $s_0 = 250\,\text{mg l}^{-1}$ throughout the fed-batch process. In a preceding batch process, the biomass concentration has reached $x_0 = 1\,\text{g l}^{-1}$, and s is adjusted to $s_0 = 0.25\,\text{g l}^{-1}$.

During the fed-batch phase, glucose (+NH_3 and minerals) is added at a feed rate v_f [l h^{-1}] and at $s_f = 100\,\text{g l}^{-1}$. It is decided to stop the fermentation when V has reached 4 V_0, where V_0 is the medium volume at $t \leq 0$. The reactor is sparged with dry air at the fermentation temperature of 30 °C. Using an efficient Rushton turbine (see Chapter 11), one expects to obtain a mass transfer coefficient up to $k_la = 650\,h^{-1}$.

From Eq. (1), the yield coefficient Y_{xs}^{obs} (assumed to be equal to Y_{xs}^{true}) = $\frac{1 \cdot 30}{0.5768 \cdot 24.6} = 2.114$ g glucose g X^{-1}.

Inserting the parameter values $b = \frac{2.114}{100-0.25} = 0.0212$ and $\mu_0 = 0.25\,h^{-1}$ in Eqs. (10.22)–(10.24), one obtains

$$v_f = 0.0212 \cdot 0.25 \cdot 1 \cdot V_0 \exp(0.25\,t) = 0.00530\,V_0 \exp(0.25\,t) = 0.00530\,V_0\alpha(t).$$

$$V/V_0 = 0.9788 + 0.0212\exp(0.25t) = 0.9788 + 0.0212\,\alpha(t). \quad (3)$$

$$x/x_0 = V_0/V \cdot \alpha(t) = \frac{\alpha(t)}{0.9788 + 0.0212\,\alpha(t)}.$$

The total fed-batch time t_{end} is obtained by solution of $4 = 0.9788 + 0.0212\exp(0.25\,t_{end}) \rightarrow t_{end} = 19.84\,h$.

The oxygen demand can be calculated from Eq. (1) – again assuming that $Y_{xo}^{obs} = Y_{xo}^{true}$:

$$(-q_o) = Y_{xo}^{obs} q_x = Y_{xo}^{obs}\mu_0 x = \frac{0.3944 \cdot 0.25}{0.5768 \cdot 24.6}x = 0.00695\,x\,\text{mol } O_2(\text{l h})^{-1}$$

At the start of the fed-batch, say at $x = 2\,\text{g l}^{-1}$, the oxygen demand is low:

$(-q_o) = 0.00\,695 \cdot 2 = 0.0139$ mol O_2 $(\mathrm{l\,h})^{-1}$. At 30 °C, 1 atm $s^*_{O2} =$ 0.244 mmol l^{-1} when sparging with air. Assuming that $s_{O_2} = 0.1\, s^*_{O_2}$, it is seen that this oxygen demand is satisfied with $k_l a = \frac{0.0139 \cdot 10^3}{0.9 \cdot 0.244} = 69.5\,\mathrm{h}^{-1}$.

Consequently, stirring of the reactor can be much less intensive than the maximum stirring capacity that would give $k_l a_{\,max} = 650\,\mathrm{h}^{-1}$. The stirrer speed (N s^{-1}) can be low, and thereby (with a fixed stirrer diameter d_s) the power input to the Rushton turbine can be significantly reduced. The quantitative relation between power input and $k_l a$ is discussed in Chapter 11.

Repeating the calculation for the highest permitted stirrer speed that will lead to $k_l a = 650\,\mathrm{h}^{-1}$ gives a maximum oxygen transfer rate $(-q_o)^T = 650 \cdot 0.9 \cdot 0.244 = 142.7$ mmol O_2 $(\mathrm{l\,h})^{-1}$.

This corresponds to an x value of $x = \frac{142.7}{6.95} = 20.53\,\mathrm{g l}^{-1}$. *This is the highest x value that can be reached if the constant μ policy is maintained.*

To obtain an x value larger than 20.53 g l^{-1}, one must adopt a different feed policy for $t > t^1 = 14.29$ h, the fermentation time when x has reached 20.53 g l^{-1} in Example 10.7. The feed rate v_f must be lower than that would be calculated by Eq. (3) of the example: $v_f = 0.00\,530\ V_o\ \alpha(t)$.

In industrial practice, different feed policies are pursued after the point has been reached where sufficient oxygen cannot be transferred due to restrictions on the maximum power input allowed. A very conservative feed policy is to maintain the feed rate at the end of the constant μ phase, in Example 10.7, to continue with $v_f = 0.00\,530\ V_0 \exp(0.25 \cdot 14.29) = 0.188\ V_0$ l h^{-1}. A less conservative policy is to continue with a linearly increasing feed rate $v_f = 0.188\ V_0\ (1 + \beta\ (t - t^1))$, where the constant β is the slope of the curve $v_f\ (t)$ at t^1, that is, $\beta = 0.25 \cdot 0.188\ V_0 = 0.0472\ V_0$ in Example 10.7. In both cases, μ will be a decreasing function of t for $t > 14.29$ h.

Villadsen and Patil [2] showed that one may continue with exponential feeding, also for time larger than t^1 where the maximum oxygen transfer capacity $(-q^{max}_{O_2})$ is reached.

The policy is called "constant $(-q_{O2})$ feed policy," and it was found that this allowed feeding at a rate $v_f = v^*_f \exp\left(\frac{(-q^{\,max}_{O_2})}{Y_{so} \cdot s_f}\right)$. In Example 10.7, one obtains $v_f = 0.188\, v^*_f \exp\left(\frac{0.1427 \cdot 30}{0.3944 \cdot 100}(t - t^1)\right) = 0.188\, V_0 \exp(0.1085(t - t^1))$. For $t \gg t^1$, the "constant $(-q_{O2})$ feeding policy" will lead to the goal $V = 4\ V_0$ of Example 10.7 much faster than the "linear feeding policy." Thus, at $t = t^1$, the volume of medium in the tank is $V_1 = (0.9788 + \exp(0.25 \cdot 14.29))\ V_0 = 1.733\ V_0$. The time $t - t^1$ to reach the goal $V = 4\ V_0$ is calculated by integration v_f from t_1 to t_{end}:

$$\frac{4}{1.733} = \exp(0.1085(t_{\mathrm{end}} - t^1)) \rightarrow t_{\mathrm{end}} - t^1 = 7.71\,\mathrm{h} \text{ and } t_{end} = 22.00\,\mathrm{h}.$$

At the end of the fed-batch phase, the glucose converted to biomass is $s_f\ Y_{sx}$ $(3V_0)$ since the glucose concentration is still $s_0 = 1$ g l^{-1}. Hence, the biomass produced is $100 \cdot (1/2.114) \cdot 3V_0$. Thus, the final biomass concentration is $x_{end} = (1 \cdot 1 + 3 \cdot 47.30)/4 = 35.73\,\text{g l}^{-1}$. This result is independent of the way in which V has increased from V_0 to $4\ V_0$. The obvious advantage of the "constant $(-q_{O2})$ feed policy," is that the result is obtained after a shorter fermentation time of 22 h compared to 26.2 h for the "constant feed policy" and 24.1 h for the "linearly increasing feed policy" – both obtained by integration of $v_f\,(t)$ after oxygen limitation has started.

The increase of fermentation time when a suboptimal feeding policy is chosen is less important than the decrease of product quality by reducing μ during the oxygen-limited part of the fermentation as shown in Figure 10.6. When the optimal policy is chosen, $(-q_{O2})$, and therefore also $q_x = \mu\, x$, remains constant. Thus, μ decreases inversely proportional to x, and when x has reached $35.73\,\text{g l}^{-1}$, μ has decreased to $0.25\ (20.53/35.73) = 0.1436\,\text{h}^{-1}$, which means that s has decreased to $84\,\text{mg l}^{-1}$. With the other policies, $s(t_{end})$ is obtained by numerical integration of the mass balances to, respectively, 51 and $37\,\text{mg l}^{-1}$. As reported by Svensson *et al.* [3], fermentation at a low value of s provokes a severe stress response in *Escherichia coli* leading to the production of endotoxins together with the heterogeneous protein that is the main product. The endotoxins are very difficult to separate from the proteins in downstream processing, and the increasing production costs may endanger the economic viability of the process. Svensson *et al.* [3] proposed to work with a decreasing temperature profile during the decreasing μ phase, but the solution proposed above may be simpler to implement.

In the discussion of Example 10.7, the fed-batch operation strategy has been illustrated for a simple case where an analytical solution of the mass balances

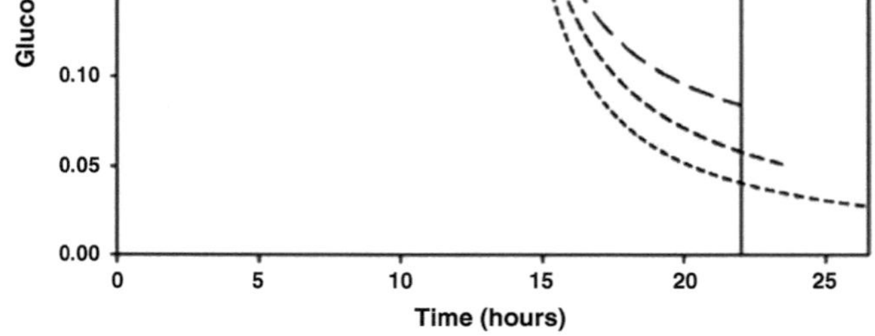

Figure 10.6 Decrease of s(glucose) after switching to different feed profiles at $t = 14.29$ h, see Example 10.7.

can be determined. The same procedure can be followed when O_2 rather than glucose is the limiting substrate. The feed of air is slowly increased at a given value for s_{O2}, for example, $0.1\, s^*_{O_2}$. The feed rate of a glucose solution of concentration s_f can be determined just as the feed rate of O_2 could be determined as a function of the glucose feed rate. Also, the fermentation can be continued until a certain titer of product is reached. In both cases, an analytical solution is (as in Example 10.7) possible only when the specific growth rate is *only* a function of the limiting substrate that is chosen, and if the yield coefficients are equal to Y_{ij}^{true}. Otherwise, as shown in Example 10.8, numerical solution of the set of transient mass balances is necessary.

Example 10.8 Production of succinic acid in a fed-batch process. Comparison with a steady-state CSTR

Example 7.2 considered succinic acid production from glucose by stoichiometries (1) and (2) combined to (3):

$$\begin{aligned}\text{Main reaction}: &\ CH_2O + 0.119\,CO_2 + 0.0238\,NH_3 \\ &\rightarrow 0.119\,CH_{1.8}O_{0.5}N_{0.2} + 1.00\,CH_{1.5}O + 0.179\,H_2O.\end{aligned} \tag{1}$$

$$\text{Maintenance reaction}: CH_2O + \frac{1}{7}CO_2 \rightarrow \frac{8}{7}CH_{1.5}O + \frac{1}{7}H_2O. \tag{2}$$

The combined stoichiometry for Eqs. (1) and (2) was found to be

$$\begin{aligned} CH_2O &+ \frac{D+0.2342}{8.403\,D+1.64}CO_2 + \frac{0.2\,D}{8.403\,D+1.64}NH_3 \rightarrow \frac{D}{8.403\,D+1.64}X \\ &+ \frac{8.403\,D+1.874}{8.403\,D+1.64}SA + \frac{1.504\,D+0.2342}{8.403\,D+1.64}H_2O, \end{aligned} \tag{3}$$

when the specific rates of substrate consumption and succinic acid production are given by

$$\begin{aligned}(-r_s) &= 10.71\,r_x + 2.00, \quad \text{and}\ r_{SA} \\ &= 10.56\,r_x + 2.24, \quad \text{both in units of g(g X h)}^{-1}.\end{aligned} \tag{4}$$

$$\text{The specific growth rate } r_x = \mu \text{ was determined to } r_x = \mu = \frac{0.8(\text{h}^{-1})s}{s + 50\ \text{mg l}^{-1}} \cdot \frac{10\ \text{g l}^{-1}}{\text{SA} + 10\ \text{g l}^{-1}}. \tag{5}$$

The specific rate $r_x = \mu$ depends on both glucose concentration s and succinic acid concentration SA. Also, the observed yield coefficients Y_{ij} are not proportional. Hence, analytic solution of the mass balances (10.22)–(10.24) is not possible.

The fed-batch mass balances (applicable for $t > 6.3$ h) are as follows:

$$\frac{\mathrm{d}(x\,V)}{\mathrm{d}t} = \mu(s,p)x\,V \rightarrow \frac{\mathrm{d}x}{\mathrm{d}t} = \mu\,x - \frac{v_\mathrm{f}\,(t)}{V(t)}x \text{ with } x = x_0 \quad \text{for } t = 0. \tag{6}$$

$$\frac{\mathrm{d}(sV)}{\mathrm{d}t} = -\,Y_\mathrm{xs}^\mathrm{obs}\,\mu x + v_\mathrm{f}(t)s_\mathrm{f} \rightarrow \frac{\mathrm{d}s}{\mathrm{d}t} = -Y_\mathrm{xs}^\mathrm{obs}\,\mu\,x + \frac{v_\mathrm{f}}{V}(s_\mathrm{f} - s) = 0 \text{ with } s = s_0 \quad \text{for } t \geq 0. \tag{7}$$

$$\frac{\mathrm{d}(pV)}{\mathrm{d}t} = Y_\mathrm{xp}^\mathrm{obs}\,x\,V \rightarrow \frac{\mathrm{d}p}{\mathrm{d}t} = Y_\mathrm{xp}^\mathrm{obs}\,\mu\,x - \frac{v_\mathrm{f}\,(t)}{V(t)}p \text{ with } p = p_0 \quad \text{for } t = 0. \tag{8}$$

$$\text{From (7) one obtains } \frac{v_\mathrm{f}}{V} = \frac{\mathrm{d}\left(\ln \frac{V}{V_0}\right)}{\mathrm{d}t} = \frac{\mathrm{d}z}{\mathrm{d}t} = \frac{Y_\mathrm{xs}^\mathrm{obs}\,\mu\,x}{s_\mathrm{f} - s_0}; \quad \frac{V}{V_0} = \exp(z). \tag{9}$$

$$Y_\mathrm{xs}^\mathrm{obs} = \frac{Y_\mathrm{xs}^\mathrm{true}\,\mu + m_\mathrm{s}}{\mu} \quad \text{and} \quad Y_\mathrm{xp}^\mathrm{obs} = \frac{Y_\mathrm{xp}^\mathrm{true}\,\mu + m_\mathrm{p}}{\mu}. \tag{10}$$

The parameters in Eq. (10) are $[Y_\mathrm{xms}^\mathrm{true}, m_\mathrm{s}, Y_\mathrm{xp}^\mathrm{true}, m_\mathrm{p}] = [10.71, 2.00, 10.56, 2.24]$.

The initial values after a batch upstart of the reactor are $[s, x, p, z] = [0.999 = s\ (t > 0), 0.6367, 9.189\ \mathrm{g/l}, 0]$.

The three dependent variables $[x, p, z]$ in Eqs. (6)–(9) are found as functions of t by numerical integration. Clearly, all three variables are connected through the dependence of μ on p (s is constant $= s_0$).

As illustrated in Figure 10.7a the SA concentration (upper curve) increases very slowly, and the biomass concentration (lower curve) rapidly reaches an almost stationary low value. In Figure 10.7b the specific growth rate rapidly decreases by a factor 5-7. Clearly fed-batch operation is unacceptable.

Note: As a matter of academic interest, one may check Eqs. (6)–(10) by simplifying to $\mu = \mu_0 = \text{constant}$, $r_\mathrm{s} = -Y_\mathrm{xs}\mu$ and $r_\mathrm{p} = Y_\mathrm{xp}\,\mu$, that is, the situation treated in Eqs. (10.22)–(10.24).

From Eq. (6), $\mathrm{d}x/\mathrm{d}t = \mu_0\,x - Y_\mathrm{xs}\,\mu_0\,x^2/(s_\mathrm{f} - s) = \mu_0\,x - b\,\mu_0\,x^2$, where b is defined in Eq. (10.22). This is a Bernoulli d.e., which is converted to a linear d.e. with constant coefficients by the substitution $w = x^{-1}$. When w is introduced, one obtains $\mathrm{d}w/\mathrm{d}t = -\mu_0\,w + b\mu_0$ that is solved from $t = 0$, where $w = 1/x_0$.

The solution is $w = A\,\exp(-\mu_0 t) + b$, where the arbitrary constant $A = (x_0^{-1} - b)$.

$w = x^{-1} = (x_0^{-1} - b)\exp(-\mu_0 t) + b$ or $x/x_0 = \alpha\ (1 - bx_0 + bx_0\alpha)^{-1}$ with $\alpha = \exp(\mu_0 t)$, and this is Eq. (10.24). Similarly, with $\mu = \mu_0$ Eq. (6) integrates to the solution of Eq. (10.19): $x\ V = x_0 V_0\ \exp(\mu_0 t)$, and when this is inserted in v_f/V_0 one obtains $v_\mathrm{f} = x_0 V_0\ \exp(\mu_0 t) Y_\mathrm{xs}\mu_0/s_\mathrm{f} - s) = x_0 V_0\ \exp(\mu_0 t) b\mu_0$, which is the result (10.22).

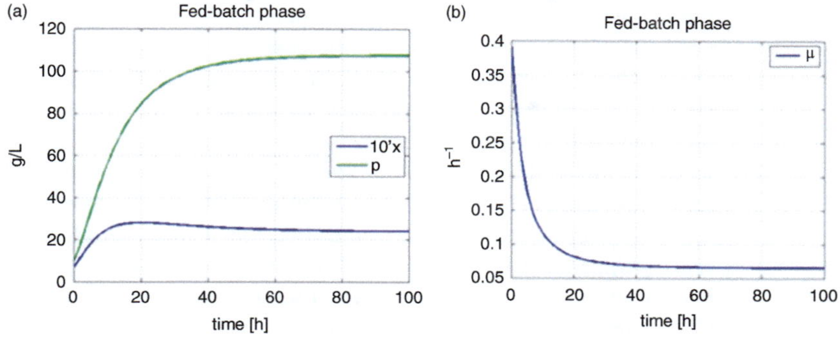

Figure 10.7 (a) Concentration of respectively x and SA during a fed-batch in which $s_f = 80\,g\,l^{-1}$ started at $t = 0$ with the initial values shown in the text. The SA concentration increases to above $100\,g\,l^{-1}$, but only when V/V_0 has reached an unacceptably high value of 100. The highest productivity q_p (g SA (l h)$^{-1}$) = 12.43 is obtained at $[t, x, V/V_0] = [18\,h, 2.78\,g\,l^{-1}, 3.77]$. The reason for the slow increase of SA concentration is seen in Figure 10.7b. The specific growth rate drops precipitously due to the inhibition by the produced SA.

Operation in a steady-state CSTR is also very difficult as seen in Table 10.1.

The values in Table 10.1 have been obtained by solving the steady-state mass balances at different values of the dilution rate D. This is done for three different values of the feed concentration s_f of glucose.

At each value of s_f, the value of D that gives the maximum value of q_{SA} is determined.

For a simple Monod rate expression, the value of D for which q_p is maximum is independent of the feed concentration of substrate. Hence, the productivity increases (almost) proportional to s_f. This is obviously not the case for a kinetic rate expression with product inhibition. The productivity does increase with s_f but far from proportional to s_f.

The most serious difficulty with product-inhibited kinetics is, however, that the maximum value of q_p is obtained at a D value that decreases

Table 10.1 The maximum productivity of SA = q_p^{max} obtained at different values of the glucose feed concentration s_f.

s_f (g l^{-1})	D (h^{-1})	s (g l^{-1})	p (g l^{-1})	x (g l^{-1})	q_p (g $l^{-1}\,h^{-1}$)
20	0.26	1.225	19.56	1.02	5.09
40	0.16	2.979	39.17	1.595	6.27
80	0.09	7.275	78.28	2.208	7.05

Note; For each s_f, the value of D where q_p^{max} is obtained is also shown.

rapidly with s_f. This means that the capital costs needed to obtain a given production of the desired product increase dramatically with increasing s_f.

The CSTR operation does, however, offer one possible solution to the deleterious influence of the product. By employing a suitable membrane filtration technology, one may separate the product from the reaction medium, and cell retention as in Section 10.4 can be implemented in the same operation. The cost of the membrane filtration system is, as illustrated in Example 10.4, quite high, but for strongly product-inhibited reactions, product removal and cell retention offers a much better prospect for economic viability of the process than a simple once-through CSTR. In a fed-batch reactor, it appears to be much more difficult to include a membrane filtration system.

In the case of succinic acid production on commercial scale (1500 kg h^{-1}), an electrochemically enhanced filtration system is proposed by the Danish company Jurag (private communication). The dilute (25 g l^{-1}) solution of Na_2SA that is the product obtained in a bacterial cultivation at pH 7 is passed rapidly past a battery of anion membranes and a filtrate of Na_2SA with a molarity as high as 1–1.5 (~150 g l^{-1}) is obtained by electroosmosis. In a subsequent unit, the Na^+ ions are exchanged with H^+ by the use of bipolar membranes. The resulting 150 g l^{-1} solution of SA can be crystallized at about 2 °C, and succinic acid of a quality demanded to make high-quality polyesters is obtained. The option of using electrodialysis to up-concentrate the Na_2SA solution and finally to obtain pure SA is further discussed in Chapter 13.

Other solutions to the SA production problems are available. One particularly elegant solution is that of the Dutch company DSM. In their industrial process, they use yeast as production organism rather than bacteria, and thereby they can work at a pH that is sufficiently low to produce mostly undissociated SA rather than an SA salt. Possibly, in the downstream processing some H_2SO_4 has to be added to the product, but the amount of gypsum produced when the SO_4^{-2} is removed appears to be quite small. Electrodialysis is further discussed in Section 13.8. Processes for economic evaluation of different SA production processes are discussed in Chapter 17.

In Example 10.8, the advantages of steady-state CSTR production over other modes of operation were quite obvious. The possible danger of steady-state CSTR operation, particularly the risk of mutation of the production organism, when continuous operation over many weeks is needed to ensure the advantage of the CSTR, has been pointed out previously.

To summarize the discussion of continuous CSTR operation versus a fed-batch STR, the simple mass balances offer an opportunity for a strict quantitative comparison. In Example 10.9, the baker's yeast production by fed-batch is compared with a steady-state CSTR design. As a result in the steady-state CSTR, three times higher productivity than in the fed-batch can be obtained. It is suggested that rather than persisting to use fed-batch production, one should by a targeted research effort find means to avoid an undesired reversion of the highly effective production strain to a less efficient parental strain.

Example 10.9 Baker's yeast production in continuous CSTR compared to fed-batch production

In Example 10.7, an increase of biomass concentration to $x_{end} - x_0 = 35.73 - 1 = 34.73\,g\,l^{-1}$ was obtained after 22 h of fed-batch operation. This biomass was contained in a volume 4 V_0, where V_0 is the medium volume at the start, $t = 0$ of the fed-batch.

Consider an aerated, steady-state CSTR of volume V_0 operated at $x_f = 0$, and with $x_e = 34.73\,g\,l^{-1}$. This gives a feed concentration $s_f = 74.77\,g\,l^{-1}$ for a yield coefficient of 2.124 g glucose g X^{-1} and an effluent concentration of glucose $s = 1\,g\,l^{-1}$, the value used in the fed-batch cultivation. The D value to be used to obtain $s = 1\,g\,l^{-1}$ is $D = \frac{0.4}{1+0.15} = 0.3478\,h^{-1}$. The productivity is $q_x = 0.3478 \cdot 34.73 = 12.07\,g\,X\,(l\,h)^{-1}$.

Assuming that the mass transfer coefficient is $650\,h^{-1}$, the value used in Example 10.7, then the oxygen transfer is $(-q_{O2}) = 142.7\,mmol\,l^{-1}$. This gives a biomass production rate $q_x = 20.53 \cdot \mu_0 = 5.13\,g\,X\,(l\,h)^{-1}$. One might either reduce s_f to $32.29\,g\,l^{-1}$ in which case x_e is reduced to $14.73\,g\,l^{-1}$ or use $D = 0.147\,h^{-1}$ (giving $s = 0.059\,g\,l^{-1}$ and $x_e = 34.88\,g\,l^{-1}$). In both cases, one obtains $q_x = 5.13\,g\,l^{-1}\,h^{-1}$.

If s_f is reduced to $32.29\,g\,l^{-1}$, the hourly production of biomass is $D\,x = 5.12\,g\,(l \cdot h)^{-1}$.

After 22 h fed-batch operation, one has produced 4 $V_0 \cdot 34.73$ g l biomass or 6.31 g $(l\,h)^{-1}$ biomass in a reactor that has the volume 4 V_0.

The steady state CSTR has a productivity $\frac{4 \cdot 5.12}{6.31} = 3.24$ times higher than the fed-batch.

10.6 Loop Reactors: a Modern Version of Airlift Reactors

Bioreactors with circulation of medium in an exterior loop have been used for more than 60 years. The biomass recirculation reactor is one example, but circulation can also be done with the purpose of cooling the reaction medium. Aerobic reactions with a large O_2 yield coefficient will, as discussed in Section 9.4, also have a high heat production per C-mole substrate converted. For large reactor volumes, it becomes difficult to remove the heat produced through internal cooling coils since the heat produced increases with scale λ as λ^3, while the heat transfer area increases only with λ^2. Here recirculation of the medium through efficient heat exchangers, for example, Alpha-Laval plate-and-frame heat exchangers, can effectively control the reactor temperature at the set-point value.

A reactor with *internal* circulation of the medium is the *airlift reactor.* In its typical form, the airlift reactor consists of an inner tube through which the gas rises in the medium. At the top of the reactor, the gas is separated from the liquid that descends through the space between the inner tube and the reactor wall. This gravity-driven circulation is supported only by the kinetic energy delivered to the gas phase when it is pumped through the sparger into the bottom of the reactor. The reactor is quite simple in its construction with no mechanical stirrer and few internals. The power input to an airlift reactor is not nearly as high as that admitted to an STR, but still the natural draft airlift reactor (as well as the even poorer mixed bubble column) can be used for aerobic processes, as long as the O_2 requirement is not large. The impressive reactor shown on Figure 10.8 – probably the largest fermentor ever built – was used by the ICI company in its single-cell protein cultivation of methanotrophs on methanol. Mixing of the (inhibitory) substrate through hundreds of injectors along the

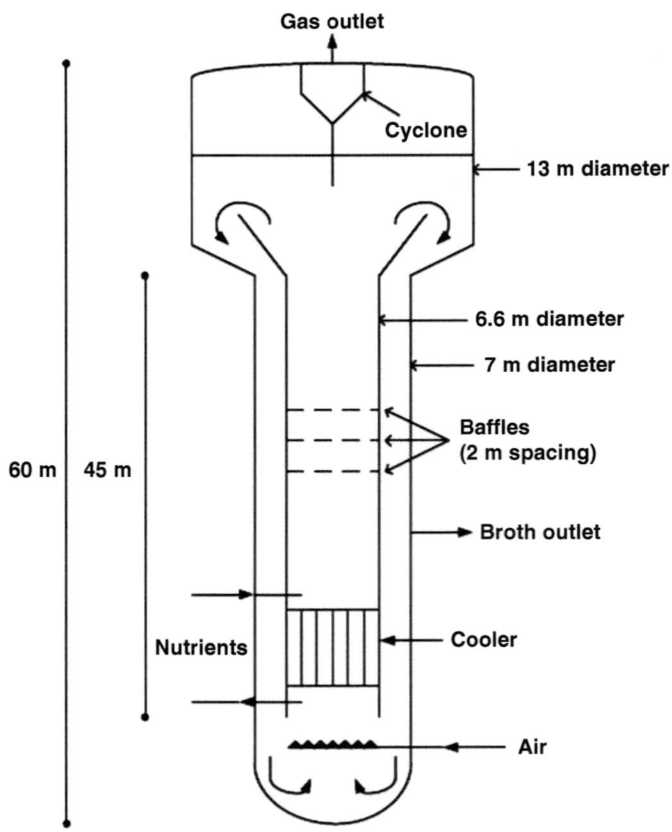

Figure 10.8 Airlift – reactor commissioned around 1970 by ICI at their Billingham (UK) plant for production of single-cell protein from methanol. The reactor volume is 1900 m^3 and the height is 60 m.

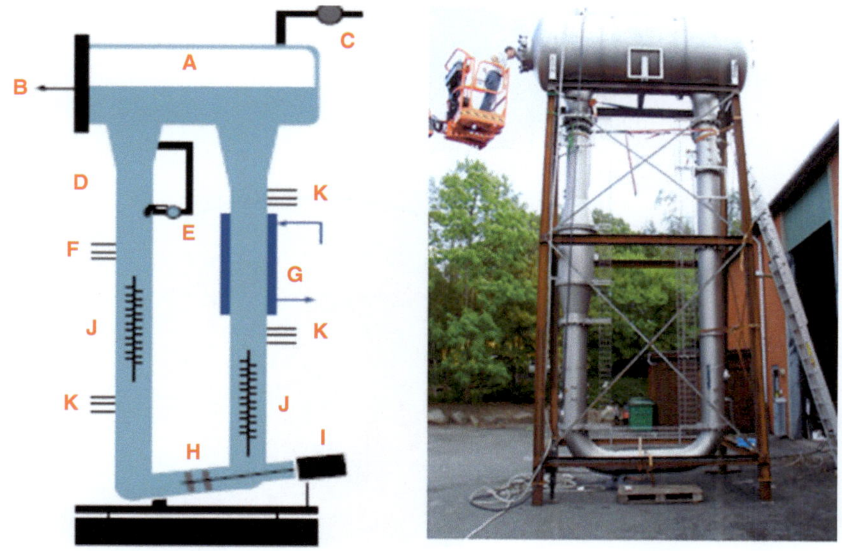

(a) (b)

Figure 10.9 Forced circulation loop reactors. (a) Schematic representation of the 0.6 m^3 active volume reactor built and tested for SCP production at DTU (DK) between 2000 and 2004. A: a large headspace gives effective separation of gas from liquid; B: Liquid effluent valve (feed is through the gas liquid ejector (E)); J: the two Sulzer static mixers used; H pump (75 kW); G: heat exchanger; K and F: sensors for O_2, biomass concentration, and temperature. (b) Demonstration reactor built by Unibio A/S (DK) at Point Lisas, Trinidad, in 2008–2009 as an exact linear scale-up of the DTU pilot plant reactor. In the 1.5 m^3 working volume, 90–100 kg SCP is produced per day.

60 m high reactor proved insufficient, and the mass transfer coefficients was far too low.

A loop reactor with *forced circulation* can provide excellent liquid-phase mixing and also a much higher mass transfer rate than the airlift reactor, but at the cost of a massive power input, usually provided by *static mixers*. Figure 10.9 shows a reactor of this type. It is also used for single-cell protein production, and the power input is so high that one can obtain a mass transfer coefficient as high as 700–900 h^{-1} by the use of static mixers of the Sulzer type. These mixers consist of porous plugs of corrugated iron or polymers through which the gas–liquid mixture is forced at high linear velocity. A sufficient number of these plugs are inserted between the gas input site and the gas exit to ensure that 90–95% of the (technically pure) O_2 substrate is consumed. The liquid medium has a holding time τ of about 5 h (dilution rate 0.2 h^{-1}), and the liquid phase reactants are fully mixed. The gas flows approximately in plug flow from gas inlet through the loop to the large headspace in about 30 s for the demonstration reactor in Figure 10.9b, and as shown in Example 10.10, the reaction is transport limited in about half the loop volume.

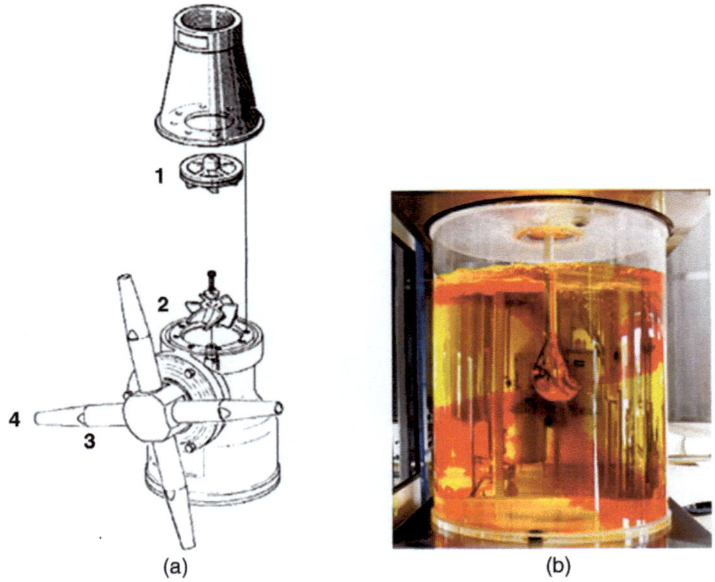

Figure 10.10 Rotary jet head is a recirculation reactor. (a) Rotary jet head. The liquid enters at 1 and activates a turbine that turns the unit 2 in a horizontal rotation, and the ejector (3) with four nozzles in a vertical rotation. The machine shown has a total weight of 12.2 kg with nozzle diameters (4) of 8, 9, or 10 mm (interchangeable). $N = 3.5$ revolutions min^{-1} in the horizontal plane at 10 bar gauge pressure and liquid flow $v_l = 46\,m^3\,h^{-1}$. The RJH in (a) effectively mixes an approximately spherical volume of about 6 m^3.

As a final example of modern loop reactors, Figure 10.10a and b shows the *rotary jet head* (*RJH*) reactor developed by Isomix A/S (DK) – now a part of Alpha-Laval – during 2002–2005. The medium is transported from the bottom of the reactor and is reinjected into the medium through one of the several RJHs. The jets of liquid sweep a given reactor volume very evenly, and mixing of liquid (or dispersion of gas injected with the liquid) through the nozzles of the jet head into the reactor medium is quite effective. One advantage of the RJH concept is that no internals (baffles, etc.) are needed in the reactor and that heat exchange is done in the external loop. The jet heads are hydrodynamically very well balanced, and the vibrations of, for example, Rushton turbines that over time leads to development of cracks in the reactor are avoided.

Figure 10.10b shows the visualization of the mixing action of small RJH in a 400 l tank. A base pulse is injected through the RJH and the color of the base indicator (methyl orange) changes from yellow to red.

The modern bioreactors are slowly entering the very conservative market for bioreactors. Thus, the RJH is extensively used in the brewing industry in the production of 4.5 vol% pilsner beer from the primary 11–12 vol% brew by dilution with carbonated water from which at the same time the O_2 is stripped to a level of 1–5 ppb (to improve the stability of the beer). An even larger application is in the cooling of the brew from 20 °C to the maturation temperature of 2–3 °C, an operation that lasts many hours in a 400 m^3 brewing tank. This cooling down time is drastically reduced by the gentle mixing offered by the RJH machine. In the "real" bio-industry (e.g., for production of industrial enzymes), the RJH has until now not been accepted.

Example 10.10 Design of a single-cell biomass production in a loop reactor

In Example 10.6, the mass transfer coefficient needed to match the liquid-phase volumetric productivity of biomass in a steady-state CSTR was calculated. The stoichiometry is

$$CH_3OH + 0.7317\,O_2 + 0.1463\,NH_3 \rightarrow 0.7317\,X + 0.268\,CO_2 + 1.56\,H_2O. \tag{1}$$

The volumetric rate of production with methanol (feed concentration 50 kg m^{-3}) as the limiting substrate is

$$q_x = \frac{0.9\,s}{s + 0.0834(\text{kg m}^{-3})}x. \tag{2}$$

The methanol concentration is kept at its optimal value 1.958 kg m^{-3} (see Example 10.6) at which the maximum volumetric production rate q_x^{max} = 23.33 kgm^{-3} h^{-1} is obtained for $D = 0.863\,h^{-1}$ and a biomass concentration $x = q_x^{max}/D = 27.03$ kgm^{-3}.

At the gas injection point (E on Figure 10.9a), technically pure O_2 is injected with fresh feed. The total pressure in the loop is 2 atm and $s^*_{O_2} = 1.96$ mM. To avoid any risk of harming the microorganism the minimum value of s_{O2} is set to 100 μM. With $k_l a = 790\,h^{-1}$, a mass transfer coefficient that can easily be obtained with static mixers, $(-q_{O2}) = 790 \cdot 1.86 = 1.469$ mol O_2 (l h)$^{-1}$. This corresponds to a biomass production of $1.469 \cdot (0.7317/0.7317) \cdot 24.6 = 36.13$ kg m^{-3} h^{-1}. At the point of injection, the oxygen transfer capacity is clearly larger than the biomass production rate 23.33 kg m^{-3} h^{-1} obtained at the optimal D by Eq. (2).

The fermentation is limited by the rate of (2). This situation persists until $s^*_{O_2}$ has decreased to 1.304 mM when $790 \cdot (1.304 - 0.1) \cdot 24.6 \sim 23.33$ kg X m^{-3} h^{-1}.

Let z be the volume of O_2 that remains at a certain point in the loop per m^3 O_2 injected. $1 - z$ is *the degree of conversion of oxygen*. Since $s^*_{O_2}$ is proportional to the partial pressure of O_2 in the gas, one obtains

$$y = \frac{O_2}{O_2 + CO_2} = \frac{z}{z + (1-z)\left(\frac{0.2683}{0.7317}\right)} = \frac{z}{z + (1-z) \cdot 0.3667}. \tag{3}$$

In Eq. (3), it is assumed that no water is taken up or deposited from the gas phase.

At the point where the two rates are equal $s^*_{O_2} = 1.304\,\text{mM}$, $y = \frac{1.304}{1.96} = 0.6653$, and $z = 0.4217$ $(1-z)\cdot 100$ ~58% of the oxygen injected has been consumed at the point of equal rates. After this point, the rate becomes oxygen limited, and the rate of biomass consumption becomes limited by the oxygen transfer, that is, by a first-order process with rate constant $k_l a = 790\,\text{h}^{-1}$.

Let the inlet feed of oxygen be G_0 mol s^{-1} and let the total reactor volume be $V\,\text{m}^3$.

In the first part V_1 of the reactor O_2 is consumed at a rate $(-q_{O_2}) = \frac{23.33\cdot 1000}{24.6\cdot 3600} = 0.263\,\text{mol}\,O_2\,\text{m}^{-3}\,\text{s}^{-1}$. The volume V_1 in which the consumption rate of O_2 is constant is $V_1 = G_0\,(1-z)/0.263\,\text{mol}\,(\text{m}^3\,\text{V}_1\,\text{s})^{-1}$. If the constant oxygen uptake rate persists until $V_1 = V$ (i.e., $z = 0$ at the outlet), then $3.80\ G_0 = V$.

To continue the calculation, one must now choose a reactor volume V. *Let the choice be* $V = 30\,\text{m}^3$. If the uptake rate were constant, then $G_0 = 30/3.80 = 7.89\,\text{mol}\,O_2\,\text{s}^{-1}$ would be converted in the $30\,\text{m}^3$ reactor.

In the volume $V_1 = 7.89\cdot(1-0.4217)/0.263 = 17.34\,\text{m}^3$, the first part of the reactor of volume $V = 30\,\text{m}^3$, the uptake rate of oxygen is $7.89\,\text{mol}\,\text{s}^{-1}$. In the last $12.66\,\text{m}^3$, the uptake rate is proportional to $s^*_{O2} - 0.1\,\text{mM}$

$$G_0 \frac{\text{d}z}{\text{d}V} = -\frac{790}{3600}(1.96\,y - 0.1)\text{mol}\,O_2\,\text{m}^{-3}\,\text{s}^{-1},$$

$$\text{with } G_0 = 7.89\,\text{mols}^{-1}. \qquad (4)$$

The differential equation is integrated from $y = 0.6653$ at $V = 17.34\,\text{m}^3$ until $V = 30\,\text{m}^3$, and the value of y at this point determines the conversion $1-z$.

When $y(z)$ is inserted from Eq. (3), one obtains directly a differential equation in z and V:

$$35.95\frac{\text{d}z}{\text{d}V} = -\left(1.96\frac{z}{0.6333\,z + 0.3667} - 0.1\right)$$

$$\rightarrow -35.95\frac{0.6333\,z + 0.3667}{1.8967\,z - 0.03667}\text{d}z = \text{d}V$$

$$\rightarrow -\left(1 + \frac{1.1349}{1.897\,z - 0.0367}\right)\text{d}z$$

$$= \frac{\text{d}V}{12.00} \rightarrow -\left(1 + 0.5984\ \frac{1}{z - 0.01933}\right)\text{d}z$$

$$= 0.08333\,\text{d}V. \qquad (5)$$

Integration of Eq. (5) is easy, and when $\Delta V = 30 - 17.34 = 12.66\,\text{m}^3$ is inserted, one obtains the conversion z_f after the gas has passed the $30\,\text{m}^3$ as the solution of:

$$-z_f - 0.5984\ln(z_f - 0.01933) = 1.1778 \rightarrow z_f = 0.13,$$

i.e., 87% of the injected gas 7.89 $m^2\ s^{-1}$ is converted.

If 13% loss of O_2 is unacceptable, one may increase V, for example, to 40 m^3, where $\Delta V = 22.66\ m^3$.

Now $z_f = 0.05$, and the conversion of the injected 7.89 m^3 O_2 is 0.95.

In a reactor of 40 m^3 volume $7.89 \cdot 0.95 = 7.50\ m^3$ O_2 is converted per second. With the stoichiometry of Eq. (1), the production rate of biomass X is $24.6 \cdot 10^{-3}$ kg X mol O_2^{-1} and consequently the biomass production is 664 kg h^{-1}.

If a production of 50 000 t X $year^{-1}$ is desired, then for 300 days of operation, an hourly production of 6944 kg is needed. Eleven reactors, each of 40 m^3 working volume, are needed to achieve this production.

It should be noted that a steady-state CSTR reactor needs to have a much higher total volume to achieve the production calculated above. The reason is that in the loop reactor only the first part of the volume operates with a rate independent of the oxygen partial pressure in the gas. The last part operates by a first-order reaction in the limiting substrate, and here the plug flow behavior of the loop reactor is a great benefit. It would be virtually impossible to have a conversion of 95% of oxygen in the CSTR reactor, since a huge mixing effort must be made to ensure that all the gas bubbles that pass the reactor have a conversion of 95%.

References

1 Villadsen, J., Nielsen, J., and Lidén, G. (2011) *Bioreaction Engineering Principles*, 3rd edn, Springer, New York.

2 Villadsen, J. and Patil, K.R. (2007) Optimal fed batch cultivation when mass transfer becomes limiting. *Biotechnol. Bioeng.*, **98**, 706–710.

3 Svensson, M., Silfwersparre, G., Häggström, L., and Enfors, S.O. (2005) Control of endotoxin release in *E. coli* fed batch cultures. *Bioprocess Biosyst. Eng.*, **27**, 135–152.

11
Mixing and Mass Transfer in Industrial Bioreactors

John Villadsen

Summary

Design of a *real* reactor for a *real* process in industrial scale requires much more than the design of the "ideal" reactors, the subject of Chapter 10. Mass balances and mathematical solution of mass balances must be supplemented by insight derived from construction of and operation of different reactor types in many scales and for many different purposes. This insight is formulated in empirical relations between key process parameters, such as mass and heat transfer coefficients, and the power input to the process. *Mixing* becomes an important issue, and from mixing time as a function of the geometry of the reactor and the power input, one derives expressions for the transfer parameters. With the new types of reactor that were briefly introduced in Chapter 10, new design expressions have to be developed since the present experimental background is mostly derived from the use of mixing aggregates that have become standard through industrial application for many decades. This chapter collects the most important empirical knowledge about design of mixers to achieve a particular goal. The resulting design formulas are not in any way quantitatively correct, but based on dimensional analysis one is able to extrapolate from small- to large-scale operation, and it is shown that linear scale-up may not give the smallest power input for a given mixing objective. The introduction given here is the basis for the visionary scale-up/scale-down design principles discussed in Chapter 16.

11.0
Introduction

In Chapter 10, we dealt with design of the most important types of reactors used by the bio-industry. Sometimes, the design of a reactor *with no spatial variation in the culture variables* can be done quite simply, and in other cases numerical simulation must be used. The mathematical framework for modeling and computation of the operation of these reactors is, however, well described in the literature, and with an increasing understanding of models for cell reactions

Fundamental Bioengineering, First Edition. Edited by John Villadsen.

quite precise design and guidelines for optimal production strategies can be expected.

Beyond the so-called "ideal" models for reactor design, there are, however, fundamental problems for which solutions are as yet only of a semiquantitative nature. *Mixing* is the key source of the problems. Industry claims that more than a third of the operating problems in bioreactions stem from inadequate mixing. Numerous technologies for mixing are available, but the outstanding questions are: which equipment is best, what is the cost of mixing, and how does imperfect mixing influence the quality of the product?

When feed is mixed into the volume of a steady-state CSTR, the ratio between feed concentration and effluent concentration can easily be larger than 10^4. This will frequently result in problems with product quality. Undesired side reactions occur, for example, formation of ethanol in aerobic cultivation of yeast caused by the Crabtree effect and overflow of carbon at the branch point of pyruvate, when the sugar concentration is too high. Sometimes, this causes even more serious problems: the formation of a heterologous protein can stop. In single-cell protein (SCP) production from C1 substrates, methanol is a better carbon source than CH_4 (mass transfer of O_2 is reduced by simultaneous transfer of CH_4 from the gas phase, and the O_2 yield coefficient with methanol as substrate is much lower). Still, the difficulty of adequate mixing of methanol in large-scale reactors presents formidable problems, for example, methanol is oxidized to strongly inhibiting formic acid rather than to the stage of formaldehyde, which is the backbone of SCP biomass.

Mass transfer must be adequate to secure that the right product is formed. An introductory example will demonstrate that the cells rapidly suffer if the oxygen supply is not sufficient:

In respiratory cultivation of yeast, Y_{so} is about 0.42 mol O_2 C-mol glucose^{-1} and at a dilution rate of 0.2 h^{-1}, a feed of 28 g glucose l^{-1} and no glucose in the effluent (see Section 3.3), the substrate consumption rate $(-q_s) = 0.2 \cdot 28 = 5.6$ g l^{-1} h^{-1} = 0.187 C-mol l^{-1} h^{-1}. The oxygen uptake is $(-q_o) = 0.42\,(-q_s) = 78$ mmol l h^{-1}. If sparging is discontinued, the 0.26 mmol O_2 dissolved in the liquid *is used up in 12 s.*

In Section 11.1, the metrics of mixing will be defined and some empirical variables that characterize a mixing process are introduced. Next, in Section 11.2, relations used to calculate the power input needed for a certain degree of mixing are derived. Finally, in Section 11.3, mass transfer will be related to power input and mixing time, and some industrial examples are given to illustrate the problems of mixing.

11.1 Definitions of Mixing Processes

The efficiency of a given type of mixing equipment is defined by the *degree of mixing E* that is determined in tracer experiments:

$$E(t) = \frac{c(t) - c(t = 0)}{c(t \to \infty) - c(t = 0)}. \tag{11.1}$$

Here, $c(t)$ is the tracer concentration measured at time t after the start of the experiment – preferably at several locations of the reactor to allow an average value $E^{av}(t)$ for the reactor to be calculated.

The tracer time profile depends on the geometry of the reactor (e.g., "dead volumes" under the stirrer) and on the construction of the mixer. There may be several time constants that give rise to an oscillating but damped tracer response; but very often a mixing process can be modeled by one time constant and the response is modeled as an exponential function:

$$1 - E = \left| \frac{c - c_0}{c_{\infty - c_0}} \right| \approx \exp(-k_{\text{mix}}\, t). \tag{11.2}$$

Here, k_{mix}^{-1} is the time constant for mixing, and for a given k_{mix}, the time t_{mix} to achieve a given value E of the degree of mixing is obtained by

$$t_{\text{mix},E} = \frac{1}{k_{\text{mix}}} \ln(1 - E). \tag{11.3}$$

The tracer experiment can be carried out based on many concepts:

1) *Conductivity of the liquid*: A pulse of a concentrated salt solution is added to the tank, and the increase of conductivity is recorded as a function of time. The response time of conductivity probes is very small, and a fast transient can be accurately determined. A definite drawback of the method is that after each experiment the tank should be emptied and refilled with water since the sensitivity of conductivity measurements rapidly decreases with increasing salt concentration. Also, gas bubbles between the electrodes severely distort the signal due to their low conductivity.
2) *pH method*: A pulse of base or acid is added, and the development of the pH in the liquid is followed. Gas bubbles do not interfere in measurement with pH probes, and the tracer experiment can be carried out during fermentation if ΔpH is reasonably small. The response time for pH probes is much higher than for conductivity probes, and the method is unsuited for laboratory-scale tanks. With tank volumes above 100–200 l, the measurement of pH after, say a pulse Δ pH of 0.5–2, is by far the best quantitative method. pH is a nonlinear function of the proton concentration $[H^+]$, and Eq. (11.2) does not strictly follow the same $c(t)$ profile in an experiment with a positive and a negative pulse ΔpH. Nordkvist *et al.* [1] showed that when $t_{\text{mix},0.95}$ for a step-up of pH experiment (from 3.50 to 3.80) was averaged with (the smaller) $t_{\text{mix},0.95}$ for a step-down experiment (from 3.80 to 3.50), one would get virtually the same $t_{\text{mix},0.95}$ as in an experiment where $[H^+]$ was changed between $10^{-3.50}$ and $10^{-3.80}$ or vice versa.
3) *Fluorophore method*: This method is very attractive for qualitative studies of how a mixing system operates. Figure 10.10b is a good example: one immediately gets an impression of how the jets from the RJH sweep through the tank volume. One observes that mixing close to the RJH is less effective than even at the end of the jets. This means that in contrast

to a Rushton turbine where very effective radial mixing is achieved close to the turbine, while the axial effect of the turbine is small, the RJH disperses the kinetic energy in the jets very evenly in the tank. Quantitative methods based on fluorophore tracers are being developed, but the pH method is still much more accurate.

4) *Other types of tracer experiments*: A pulse of hot water can be added to the tank, and the mixing of the hot water into the cold water of the tank can be measured by temperature probes. Since the viscosity η of hot water is much lower than that of room temperature water, and the mixing time decreases appreciably with decreasing liquid viscosity, the method is not recommendable unless a serious modeling work of $t_{\text{mix}}(\eta)$ is added to Eq. (11.2).

Very accurate measurements of t_{mix} have been obtained based on isotope measurements. Pedersen *et al.* (1994), [14] demonstrated this in experiments where an indium isotope was added at $t = 0$. Isotope measurements with rather stable isotopes are not at all welcome in the bio-industry environment, and the method has mostly academic interest.

Mixing by any type of mixing equipment is a *macroscopic* process, whereas mixing by molecular diffusion is a *microscopic* process. The classical equipment for macromixing is either a propeller (hydrofoil) or a turbine. The hydrofoil mainly causes axial circulation, while the turbine effectively mixes the liquid radially in an axial compartment. In macromixing, the liquid is churned down to smaller liquid elements by the action of the mixer. Finally, recombination of liquid elements occurs at the same rate as breakup of the elements, and independent of the mixing machinery, a minimum size liquid element with length scale λ is obtained. Using dimensional analysis, the Russian mathematician Andrei Kolmogorov in 1941 derived the two expressions in Eq. (11.4). With a given power input $P(W)$, the liquid is over time divided into small liquid volumes, characterized by the length scale λ, and the time constant of the turbulent transfer of kinetic energy to the liquid is τ. When the Kolmogorov length scale λ is reached, the rates of coalescence and breakup of fluid elements are equal. Further input of power is entirely converted to heat.

$$\lambda = \left(\frac{M\nu^3}{P}\right)^{1/4} \qquad \tau = \left(\frac{\nu M}{P}\right)^{1/2}. \tag{11.4}$$

In Eq. (11.4), P/M is the power input per mass of the liquid ($\text{m}^2\ \text{s}^{-3}$) and ν is the kinematic viscosity of the liquid, η/ρ_1 ($\text{m}^2\ \text{s}^{-1}$). In SI units, λ consequently has the unit m and the unit for τ is s.

For water, the dynamic viscosity η is 10^{-3} kg $(\text{m s})^{-1}$ and $\nu = 10^{-6}\,\text{m}^2\,\text{s}^{-1}$. For a power input $1\,\text{kW}\,\text{m}^{-3}$ $\lambda = 10^{-4.5}\,\text{m} = 3.16\,10^{-5}\,\text{m} = 31\,\mu\text{m}$. For a liquid with viscosity 500 times that of water (the fermentation medium for filamentous fungi with about $5\,\text{g}\,\text{l}^{-1}$ biomass), $\lambda = 500^{0.75} \cdot 31\,\mu\text{m} = 3.28\,\text{mm}$.

The dispersion of, for example, glucose in the minimum size liquid elements occurs *only* by molecular diffusion, and the time constant for molecular diffusion is as follows [13]:

$$t_D = \frac{\lambda^2}{4\,\pi^2 D_{\text{glc}}}.$$

With an aqueous medium, the time to reach perfect mixing at the Kolmogorov length scale is hardly perceptible, since D_{glucose} is 8.5 $10^{-10}\,\text{m}^2\,\text{s}^{-1}$ and $t_{\text{D}} = 31^2\ 10^{-12}/(4 \cdot 9.86 \cdot 8.5 \cdot 10^{-10}) = 2.9\ 10^{-2}\,\text{s}^{-1}$. For the medium with $\nu = 500\ 10^{-6}\,\text{m}^2\,\text{s}^{-1}$, $t_{\text{D}} = 500^{1.5}\ 2.9 \cdot 10^{-2} = 324\,\text{s}$.

Thus, while mixing an aqueous solution with a power input of $1\,\text{kW}\,\text{m}^{-3}$, a modest power input in industrial practice where up to $5\,\text{kW}\,\text{m}^{-3}$ is acceptable, leads to complete mixing by the turbulent mixing mechanism, a severe time delay of more than 5 min is expectable in the cultivation of, for example, *Aspergillus oryzae*, a standard production organism for detergent enzymes. Consequently, large substrate gradients in the medium can be expected.

The Kolmogorov time constant for mixing by turbulent power input is $\tau = 10^{-3}\,\text{s}$ for the aqueous medium and 0.022 s for the viscous liquid when $P/M = 1\,\text{W}$. Macromixing and micromiximg take about the same time for aqueous systems, while micromiximg completely dominates in case of high-viscosity liquids. These time constants are, of course, very small compared to the large residence time V/v for cell reactions in a steady-state CSTR. To reduce the feed concentration by a factor 10^4, one needs a turbulent mixing time of $t_{\text{mix,turbulent}} = \ln\,(10^4)\ \tau = 0.2\,\text{s}$ for the viscous liquid. Intuitively, this mixing time appears to be too small, but both formulas in Eq. (11.4) require an experimentally determined dimensionless parameter, and their primary use is to compare the behavior of liquids with different viscosities, the dominant process parameter in turbulent mixing. When the mechanical mixing is inhomogeneous, for example, if there are "dead zones" in the reactor, $t_{\text{mix, turbulent}}$ can be much larger. In this case, larger concentration gradients can also be expected in more amenable media, for example, in an aerobic *Saccharomyces cerevisiae* cultivation to produce baker's yeast.

The formulas in Eq. (11.4) are supplemented by a number of equally semi-empirical "rules for mixing." The most important of these dimensionally correct formulas are given in Eq. (11.5):

$$t_{\text{mix}}(= t_{\text{mix,turbulent}}) = kN^{-1}, \quad v_{\text{pump}} = N_{\text{f}}(Nd_{\text{s}}^3), \quad t_{\text{c}} = \frac{V}{v_{\text{pump}}} = \frac{V}{N_{\text{f}} d_{\text{s}}^3}\frac{1}{N} \text{ and } t_{\text{mix}} = 4\,t_{\text{c}}. \tag{11.5}$$

For turbine mixing, the first formula states that the mixing time is inversely proportional to the stirrer speed N. For "normal" mechanical mixers $t_{\text{mix, 0.95}} \sim 20\,N^{-1}$, but a scalar constant k between 10 and 15 is often cited in literature –see also the derivation in Eq. (11.14) where a value of the proportionality constant is calculated as a function of the geometry. The second formula expresses that the liquid pumped by a turbine or hydrofoil into the bulk volume of the tank is proportional to $N \cdot d_s^3$, where d_s is the stirrer diameter. N_f is a parameter that depends on the stirrer type, for example, $N_f = 0.72$ for a Rushton turbine. t_c is the circulation time of the liquid by the action of the stirrer, and t_c is defined as the total liquid volume divided by the liquid flow rate delivered by the mixing equipment. Due to the slow mixing provided by turbines between axial compartments, total mixing can be slow for turbines, but for hydrofoils $t_{mix,95\%}$ is approximately $4t_c$ as shown in the last formula.

The empirical formulas of Eq. (11.5) are collected into a Reynolds number (11.6) for mixing where $\nu = \eta/\rho_l$ is incorporated to distinguish between liquids with different physical properties:

$$\mathrm{Re} = \mathrm{Re}_{mix} = \frac{N d_s^2}{\nu}. \tag{11.6}$$

11.2
The Power Input *P* Delivered by Mechanical Stirring

The classical mixing equipment is either a propeller (hydrofoil) or a turbine. The best known turbine is the Rushton turbine [2,3]. Figure 11.1 is a schematic representation of a Rushton turbine (adapted from [4]) installed in a baffled stirred tank bioreactor. The baffles are necessary to avoid a swirling rotation of the liquid without any axial mixing. A typical Rushton turbine has six blades, welded to a plate. The recommended diameter of the turbine is $d_s \sim 0.35T$, where T is the tank diameter. In tall, slender tanks, $H > 2\ T$, several Rushton turbines are mounted on the shaft at a distance indicated on the figure. The motor is connected to the shaft via bearings through a sterile seal, and is supported by a bearing at the bottom to avoid twisting of the shaft. The motor needed to deliver sufficient power input to a large bioreactor must be large, for example, a 500 kW motor for a 75 m^3 tank.

The Rushton turbine with its very high power number N_p is undoubtedly the best machinery to deliver power input to the usual aqueous media of fermentation processes. Its weakness, however, is that while the radial mixing is excellent, there is little mixing axially between the compartments defined by the individual Rushton turbines, and therefore mixers of the hydrofoil type that can pump axially down – for example, a six-blade Prochem Maxflow™, or axially up (e.g., APV B-2™, which is also good for air dispersion) are installed. The hydrofoils have a much smaller power number than Rushton turbines (N_p from 0.8 to 1.5), and they deliver less kinetic energy to the mixing process.

When mixing pseudo-plastic liquids with very high apparent viscosity at low shear rate, the Rushton turbine gives excellent mixing in a core of liquid with diameter $\sim d_s$, while the liquid in the shell between T and d_s is stagnant. No stirrer construction is really satisfactory in this case. In such untypical situations,

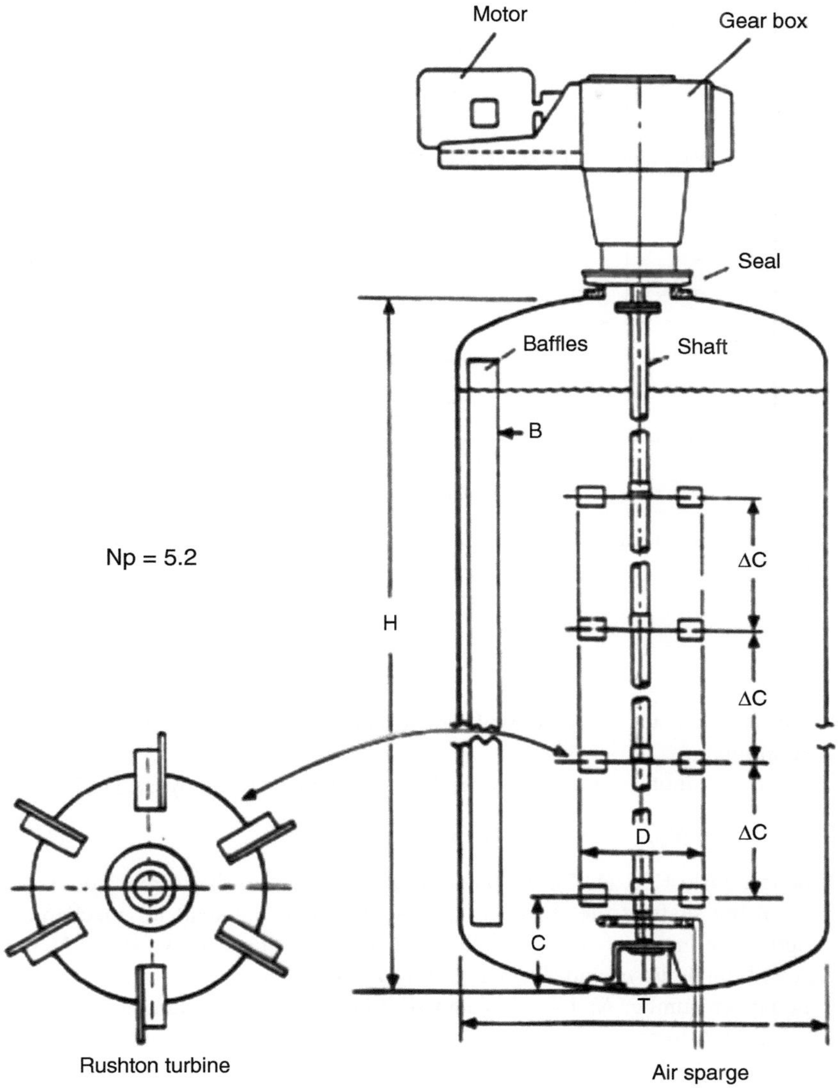

Figure 11.1 A stirred tank bioreactor with 4 six-bladed Rushton turbines. N_p is the *power number* for the mixer, and $N_p = 5.2$ for Rushton turbines. ΔC is the distance between the Rushton turbines, ~0.7 T, where T is tank diameter. The first turbine is placed ~0.3 T above the bottom of the tank, and gas is sparged right under the first turbine. If possible, feed should be delivered right at the turbines. To provide sufficient axial mixing, the Rushton turbines are often supplemented by hydrofoils for upward or downward pumping.

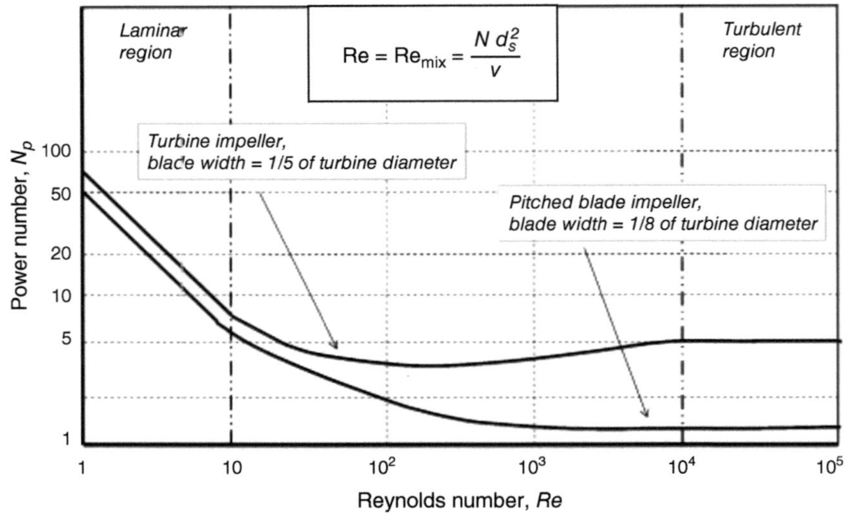

Figure 11.2 Power number N_p as a function of Re_{mix} for a Rushton turbine and for a pitched blade impeller.

one needs to consult "Handbook of Industrial Mixing" [5], the ultimate encyclopedia on mixing.

The power input for *any* type of mechanical mixing is given by Eq. (11.7):

$$P(W) = N_p \rho_l N^3 d_s^5 = 2\pi N \mathrm{To}. \tag{11.7}$$

In the last part of Eq. (11.7), the power input is related to the torque To ($W\,s = N\,m$) on the shaft of the stirrer. It is used to design the shaft for mechanical strength. The first part of Eq. (11.7) relates P to the operating variables N, the density of the liquid ρ_l, and the diameter d_s of the stirrer.

The power number N_p for an unaerated liquid is given as a function of Re_{mix} in Figure 11.2. In the laminar region, $Re_{mix} < 10$, the power number is inversely proportional to Re_{mix}, and in the fully turbulent region $N_p \rightarrow$ a constant. $N_p = 5.2$ for a Rushton turbine, and a smaller value is obtained for hydrofoils. Figure 11.2 is completely analogous to a figure that shows the friction factor for tubular flow of liquids as a function of $Re = (u_l d_h/\nu)$ (u_l = linear velocity; d_h = hydraulic diameter of the tube).

For aerated flow, N_p decreases with the *aeration number* $N_A \equiv \frac{v_g}{N d_s^3}$

$$= \frac{v_g}{v_{pump}} N_f. \tag{11.8}$$

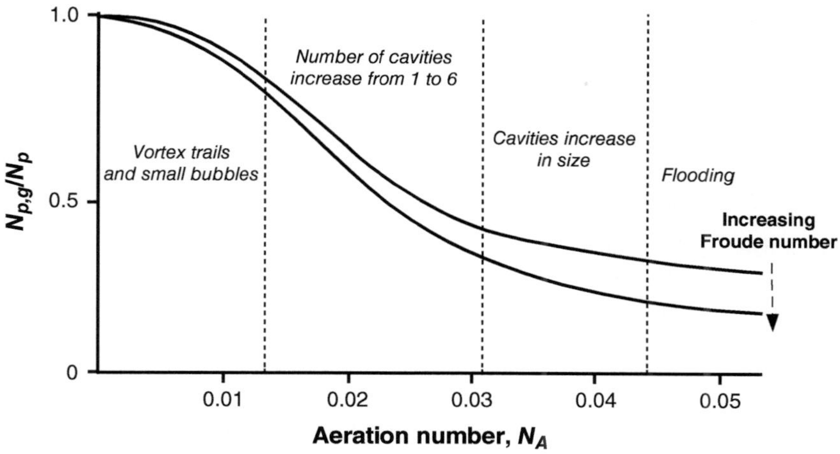

Figure 11.3 Ratio between gassed and ungassed power number as a function of N_A and Fr for a six-bladed Rushton turbine. (Adapted from the *Ekato Hand book of Mixing Technology* (www.EKATO.com).)

Figure 11.3 shows that at constant value of Re_{mix} the ratio $N_{p,g}/N_p$ decreases from 1 at $N_A = 0$ (i.e., volumetric gas flow; $v_g = 0$) to less than 0.2 for large values of v_g or small values of N or d_s. The stirrer diameter has a dominating influence on N_A due to the power 3 of d_s. The figure shows different regions of the gas flow in the tank, from well-dispersed small bubbles at small N_A to flooding of the impeller at large N_A. When flooding starts, very little power is drawn on the mixer, and the time for mixing t_{mix} is large. Since the mass transfer coefficient varies with the inverse of t_{mix}, it is definitely undesired to have a small value of $N_{p,g}/N_p$.

The function $(N_{p,g}/N_p) = f(N_A)$ also depends on the *Froude number* $Fr \equiv (N^2 d_s/g)$, where g is the acceleration of gravity ~9.82 m s^{-2}. For increasing value of Fr and constant N_A, $N_{p,g}/N_p$ is smaller.

Nienow [4] determined two values of N from Eqs. (11.9) and (11.10). If N is smaller than the value determined from Eq. (11.9), flooding of the stirrer results. If N is larger than the value determined from Eq. (11.10), the gas flow v_g is well dispersed by the stirrer, and $N_{p,g}/N_p$ is not much below 1.

$$\text{The aeration number } N_A = \left(\frac{v_g}{N d_s^3}\right)_{\text{flooding}} = 30\left(\frac{d_s}{T}\right)^{3.5} \text{Fr}. \tag{11.9}$$

$$\text{The aeration number } N_A = \left(\frac{v_g}{N d_s^3}\right)_{\text{fully dispersed}} = 0.2\left(\frac{d_s}{T}\right)^{0.5} \text{Fr}^{0.5}. \tag{11.10}$$

Inserting $Fr = (N^2 d_s/g)$ in Eqs. (11.9) and (11.10) allows the value of N to be found for which, respectively, flooding and fully dispersed gas flow can be expected:

$$\text{Flooding occurs when} \quad N < \left(\frac{g}{30} v_g\right)^{1/3} T^{7/6} d_s^{-2.5}. \tag{11.11}$$

$$\text{Good dispersion is expected when} \, N > \left(5 g^{0.5} v_g\right)^{1/2} T^{1/4} d_s^{-2}. \tag{11.12}$$

For a fixed tank diameter T and a fixed gas flow, Eqs. (11.11) and (11.12) imply that the value of N for which good dispersion (and certainly no flooding) occurs decreases very fast with increasing d_s. This of course also means that more gas can be dispersed for a given N. Note that the unit of N in the formulas is s when SI units are used, for example, $g = 9.82\,\text{m s}^{-2}$.

The maximum gas flow that can be dispersed at a given N without risk of flooding depends on the type of impeller. In this respect, the Rushton turbíne is not good due to its high N_p value – see Figure 11.2. A low N_p mixer such as the Scaba™ hollow blade mixer with $N_p = 1.5$ is better, but again the Rushton turbine is, as mentioned, superior if a high-energy input is needed.

It is noted that especially the value of N at flooding is strongly scale dependent ($T^{7/6}$). For a given volume V, tall but slim reactors are better, although axial mixing is a problem. For each $\Delta H = 1.1 - 1.5\,T$, an extra Rushton turbine has to be used, and the total power input needed increases almost proportional to the number of turbines.

Finally, the power input to *bubble columns* is given by Eq. (11.13), which simply tells that energy in the pressure (p)-driven delivery of gas through the sparger is converted to kinetic energy:

$$P(W) = v_g\,\Delta p = u_g\,\rho_l\,g\,V_{\text{liquid}}. \tag{11.13}$$

Δp is the pressure drop over the sparger + the hydrostatic pressure at the bottom of the reactor, and u_g is the superficial gas velocity $= \frac{v_g}{(\pi/4)\,T^2}$.

The power input by Eq. (11.13) for a bubble column is only a fraction of what the mechanical stirring is capable to deliver. This is the reason why mass transfer is low in bubble columns.

Example 11.1 Analysis of mixing in an aerated stirred tank of volume 600 l

The tank diameter T is chosen to be 0.9 m, and for 600 l liquid volume $H = 0.94$ m, giving a suitable value of $H/T = 1.04$. d_s is chosen to $0.35\,\text{m} = 0.39\,T$, and $N = 2\,\text{s}^{-1}$. According to Eq. (11.5), this gives a circulation time of $t_c = \frac{0.6}{2 \cdot 0.72 \cdot 0.35^3} = 9.7$ s for a Rushton turbine with

$$N_f = 0.72, \text{ and } t_{mix,0.95} \sim 4t_c = 39\,\text{s}. \tag{1}$$

The power input to the *unaerated* liquid is $P = 5.2 \cdot 1000 \cdot 2^3 \cdot 0.35^5 = 218\,\text{W} = 218/0.6 = 0.364\,\text{kW per m}^3$ reactor or $= 0.364$ W per kg medium.

Let 100 l air min^{-1} be sparged to the liquid in the tank.

$u_g = \frac{0.1}{60 \cdot 0.785 \cdot 0.9^2} = 2.62 \cdot 10^{-3}\ \text{m s}^{-1}$. The power input to sparge the liquid is $u_g\, \rho_l\, g\, V = 15.4\,\text{W}$.

The total power input is 233 W of which the power input to the gas is a small fraction.

Using Eq. (11.12), one obtains the *minimal* value of N to ensure good dispersion:

$N > \left(5 \cdot 9.82^{0.5} \cdot 0.00167\right)^{0.5} \cdot 0.90.25 \cdot 0.35^{-2} = 1.28\,\text{s}^{-1}$, and with $N = 2$ one finds that $N_A = N_A = \frac{0.00167}{2 \cdot 0.35^3} = 0.0194$, and from Figure 11.3 $\frac{N_{P,g}}{N_P} \cong 0.5$.

Hence, the true power input to the *aerated* tank is approximately $233 \cdot 0.5\,\text{W} = 0.117\,\text{kW m}^{-3} = 0.12\,\text{W kg}^{-1}$. This is a modest power input, and one would expect a low value of $k_l a$.

11.3 Mixing and Mass Transfer in Industrial Reactors

A large number of experimental results in water-like media and at many length scales have been reviewed by Nienow [6]. The results *for* $H \sim T$ (and not for $H = 2\ T$!) are collected in Eq. (11.14), where N_p is the effective power number, that is, $N_{p,g} = N_p \cdot N_{p,g}/N_p$ for aerated liquids:

$$t_{mix,0.95} = 5.3\, N_p^{-1/3}\, N^{-1} (d_s/T)^{-2} = 5.3 \cdot (P/(N_p \cdot \rho_l N^3 d_s^5)^{-1/3} \cdot N^{-1} \cdot (d_s/T)^{-2}$$
$$= 5.74 \cdot (P/M)^{-1/3} (d_s/T)^{-1/3} (H/T)^{-1/3}\, T^{2/3}. \qquad (11.14)$$

Both the first and the last expression for $t_{mix,\,0.95}$ calculates t in units of s when the other quantities are inserted in SI units. The formulas are used for different purposes. The first formula confirms that $t_{mix,\,0.95}$ is inversely proportional to N, and the proportionality constant k in Eq. (11.5) is seen to equal $5.3 \cdot 5.2^{-1/3} \cdot 0.35^{-2} = 25.0$ for a Rushton turbine with the "normal" $d_s/T = 0.35$. The value of the constant k decreases rapidly with increasing d_s/T. Thus, with $d_s = 0.48\ T$, the constant is equal to 13, a value often cited in literature.

Example 11.2 Mixing time in a 600 l unaerated tank

In Example 11.1 and for the unaerated tank, one would calculate $t_{mix,\,0.95} = 5.3 \cdot 5.2^{-1/3} \cdot (0.35/0.9)^{-1}\ N^{-1} = 20.22\ N^{-1} = 10.1\,\text{s}$ for $N = 2\,\text{s}^{-1}$ by the first expression in Eq. (11.14). This is quite different from the result 39 s obtained in Example 11.1, but Eq. (11.14) is much better founded in the large literature study, and the value 10.1 s is more likely to be true than the value 39 s that does not take the stirrer geometry into account. Note that the same result for $t_{mix,0.95}$ is obtained by the second expression in Eq. (11.14), which is just a reformulation of the first expression:

$$t_{mix,0.95} = 5.74 \cdot (218/600)^{-1/3} (0.35/0.9)^{-1/3} \cdot (0.94/0.9)^{-1/3} \cdot 0.9^{2/3}$$
$$= 10.1\ \text{s}. \qquad (2)$$

In the aerated tank with $(N_{p,g}/N_p) \sim 0.5$ that is discussed in Example 11.1:

$$t_{mix,0.95} = 5.74 \cdot (218/600)^{-1/3} \cdot 0.5^{-1/3} \cdot (0.35/0.9)^{-1/3} \cdot (0.94/0.9)^{-1/3} \cdot 0.9^{2/3} = 12.8\ \text{s}.$$

This is not much different from the result for the unaerated tank, but $(N_{p,g}/N_p)$ is also rather high. Note that again the same result is obtained by both formulas in Eq. (11.14), which are just reformulations of one another.

The second expression in Eq. (11.14) relates $t_{mix,0.95}$ to the power input. It is seen that the mixing time decreases quite slowly with the power input per mass of medium. Since the mass transfer coefficient is likely to vary inversely proportional to t_{mix}, one would expect that

$$k_l a \text{ is proportional to } (P/M)^{1/3}. \tag{11.15}$$

It should also be noted that the scale of the process influences $t_{mix,0.95}$. In the second expression of Eq. (11.14), the effect of scale is seen to be $T^{2/3}$, which means that a scale-up from 1 to 100 m^3 will increase $t_{mix,0.95}$ by a factor $(100^{2/3})^{1/3} = 2.78$ and correspondingly reduce $k_l a$ by the same factor. Since $k_l a$ by the first relation in Eq. (11.14) is proportional to $(P/M)^{1/3}$, this means *that scale-up will unavoidably be very expensive in terms of power input* – or alternatively that a lower volumetric productivity must be accepted in the large scale.

Example 11.3 Relations between power input and t_{mix} for other mixing systems

The results of Eq. (11.14) were derived by collecting experimental data from many systems *where Rushton turbines or hydrofoils were used.* The relationships have the same structure when quite different mixing systems are used. As an example, the rotary jet head (RJH) mixing system (Figure 10.10) will be used. As described in Section 10.7, the RJH system functions by conversion of the pumping energy input into kinetic energy in the jets that are sent into the liquid of the reactor through the nozzles in Figure 10.11a.

Nordkvist *et al.* [1] found that the power input used to inject a volume v_l $m^3\ h^{-1}$ through the four nozzles (all of diameter d (mm)) of N RJH machines was given by

$$P(W) = \frac{v_l\ \Delta p}{0.036} = 1086\, v_l^3\, d^{-4}\, N_{RJH}^{-2}. \tag{1}$$

The numerical value on the r.h.s. of Eq. (1) was found by measuring the pressure loss and the corresponding v_l for a number of nozzle diameters, and with $N_{RJH} = 1$ and 2 RJH machines in the tank. The volume v_l is distributed evenly between the RJHs if $N_{RJH} > 1$.

Expression (11.7) can be rewritten by the use of the formula for v_{pump} in Eq. (11.5):

$$P = N_{\text{p}}\,\rho_{\text{l}}N^3d_{\text{s}}^5 = N_{\text{p}}\,\rho_{\text{l}}\left(\frac{v_{\text{pump}}}{N_{\text{f}}d_{\text{s}}^3}\right)^3 d_{\text{s}}^5 = N_{\text{p}}\,N_{\text{f}}^{-3}\,\rho_{\text{l}}\,d_{\text{s}}^4\,v_{\text{pump}}^3. \tag{2}$$

Hence, the action of the RJH and of a stirrer is in principle the same if the reciprocal nozzle diameter is interpreted as the diameter d_{s} of a mechanical stirrer and the circulated liquid volume v_{l} represents the volume v_{pump} ejected from the stirrer.

Expression (18) for t_{mix} in Nordkvist *et al.* [1] is also very similar to those of Eq. (11.14):

$$\begin{aligned}\text{For } N_{\text{RJH}} &= 1 \text{ and water}: \; t_{\text{mix}}(s)\\ &= 9486(P/V\ \text{W m}^{-3})^{-0.308}\,d(\text{mm})^{-1.23}.\end{aligned} \tag{3}$$

Expression (3) closely resembles the second expression of Eq. (11.14) in the power of (P/V), and since the speed of rotation N of the stirrer is inversely proportional to v_{pump} (which is v_{l} in [1]), also the first expression in Eq. (11.14) has the same functionality of the operating variables as is found for RJHs.

Since Nordkvist *et al.* also studied liquids with much higher viscosity than water, they could find an expression for the variation of t_{mix} with the viscosity: besides water they used a thin CMC solution with $\eta = 13.7 \cdot 10^{-3}$ kg m^{-2} s^{-1}. They found (in their equation (14)) that t_{mix} $(s) = 12\ 607$ v_{l} (m^3 h^{-1})$^{-1.39}$ for experiments in a 3.4 m^3 tank; the same as was used in Eq. (3) above. The effect of viscosity on the numerical constant is smaller than that predicted by the Kolmogorov formula for τ_{mix} (η/ρ_{l}), namely, by a factor $\sqrt{13.7} = 3.70$, and the CMC solution also appears to have a shear-thinning effect, since the decrease of t_{mix} with v_{l} is stronger than what would be predicted for an aqueous medium.

In the literature, one finds many, very different expressions for the influence of viscosity on mixing time – and therefore also on $k_{\text{l}}a$. Usually, an extra factor η^{δ} with δ anywhere from −0.4 to −1 is used to correlate data for Newtonian fluids with different viscosity under otherwise the same operating conditions.

Data for $k_{\text{l}}a$ as a function of the operating variables (such as $u_{\text{g}} = v_{\text{g}}/A$, where A is the cross-section area of the tank) have been obtained for many different types of mixing equipment and at different operating scale. Table 11.1 collects several frequently applied correlations.

A very popular model, backed up by a large amount of experimental data, is the van't Riets model:

$$k_{\text{l}}a(\text{s}^{-1}) = 0.026\,u_{\text{g}}^{0.5}(P/V)^{0.4}. \tag{11.16}$$

Table 11.1 Parameters in power law expressions $k_la = k\, u_g^{\alpha}(P/V)^{\beta}$ for different literature correlations.

Medium	k	α	β	Agitator	Reference
Coalescing	0.025	0.5	0.4	Six-bladed Rushton turbine	Moo-Young and Blanch [7]
	0.00495	0.4	0.593	Six-bladed Rushton turbine	Linek *et al.* [8]
	0.01	0.4	0.475	Various agitators	Moo-Young and Blanch [7]
	0.026	0.5	0.4	Not specified	van't Riet [9]
Non-coalescing	0.0018	0.3	0.7	Six-bladed Rushton turbine	Moo-Young and Blanch [7]
	0.02	0.4	0.475	Various agitators	Moo-Young and Blanch [7]
	0.002	0.2	0.7	Not specified	van't Riet [9]

Note: SI units must be used for P/V (W m^{-3}) and superficial gas velocity u_g (m s^{-1}) to obtain k_la in units of s^{-1}. The table shows data both for air/water systems (noncoalescent) and for coalescent media such as fermentation broths, salt solutions, or solutions that contain, for example, methanol or ethanol.

As mentioned above, a viscosity effect for liquids different from water is included as $(\eta/\eta_{H2O})^{-\delta}$, where $\delta = 0.4 - 1$. Equation (11.16) should work from small pilot scale tanks to industrial reactors.

Similar relations for k_la are obtained for mixing systems that are different from those based on mechanical stirring with Rushton turbines and hydrofoils.

For mixing with RJHs, Nordkvist *et al.* [1] obtained:

$$k_la(\mathrm{s}^{-1}) = 0.0958\, u_g^{0.764}\, v_l^{0.700}(\mathrm{SI-units}). \tag{11.17}$$

Assume that almost all the power input goes to circulate the liquid. From Example 11.3, $t_{mix,0.95}$ is proportional to $(P/V)^{-0.308}$, that is, proportional to $v_l^{-0.925}$ for a given nozzle diameter and an unaerated liquid. From Eq. (11.17), k_la is proportional to $v_l^{0.700}$, that is, proportional to $(P/V)^{0.7/3} = (P/V)^{0.23}$ for water-like systems. The dependence of k_la on P/V is smaller than for mechanically stirred tanks as seen in Table 11.1, but the dependence on u_g is almost the same as that found in Heijnen and van't Riet [10], namely, $k_la(\mathrm{s}^{-1}) = 0.32\, u_g^{0.7}$. Comparison of the RJH system with systems based on mechanical stirrers (Figure 10 of Nordkvist *et al.* [1]) shows that the RJH is not as efficient for mass transfer as mechanically stirred systems for large power inputs ($P/V > 1 - 1.5\,\mathrm{kW\,m^{-3}}$), but more efficient for small values of P/V. With a large value of v_l, it appears that the gas is not efficiently dispersed by the jets. To operate in the advantageous range of P/V, the liquid circulated in the tank should be split (equally) to several RJHs, since now a large value of nozzle diameter d can be used (P/V depends on d^{-4}) for the same total v_l.

Static mixers have a very high specific O_2 transfer – up to $4.2\,kg\,O_2\,kWh^{-1}$, that is, an uptake rate of $0.0364\,mol\,m^{-3}\,O_2\,s^{-1}/(kW\,m^{-3})$ (Al-Taweel *et al.* [11]). This is the reason for using static mixers in situations where a very high oxygen transfer rate is required, for example, in SCP production (Example 10.10). With a partial pressure of O_2 of 2 atm (as used in Example 10.10 at the point of injection), one obtains a mass transfer coefficient $k_l a = 364/(1.86\ 10^3) = 0.195\,s^{-1} = 705\,h^{-1}$ at a very modest power input of $1\,kW\,m^{-3}$. Experiments at $P = 2$–5 atm with static mixers and feeding air to a water-like medium gave the relation (11.18) between $k_l a$ and power input P/M (DuPont, Private communication, 2014):

$$k_l a(h^{-1}) = 301(P/M\,kW\,m^{-3})^{0.42}, \quad \text{for } 0 < P/M < 10\,kW\,m^{-3}. \tag{11.18}$$

Equation (11.18) gives approximately the same value at $\pi_{O2} = 2$ atm for $k_l a$ as that found by Al-Taweel *et al.* The exponent of P/M corresponds to that in the van't Tiet expression of Table 11.1. An influence of u_g (not stated in the reference) is hidden in the factor 301.

The best correlations for newer types of reactors and new mixing equipment are probably to be found in Paul, Atieno-Oberg, and Kresta [5], supplied with experimental data from the newest literature. There is no doubt that the cost of mixing to obtain, for example, a specific value of the mass transfer coefficient will decrease when newer reactor designs are accepted by industry.

All the experimental data collected to construct the correlations for $k_l a$ are obtained by one or more of the following methods:

1) *Transient methods*; These are based on the mass balance with O_2 as the dissolved gas:

$$\frac{ds_o}{dt} = k_l a(s_o^* - s_o) - (-q_o). \tag{11.19}$$

The liquid is saturated with oxygen at a certain s_o^* value and with $q_O = 0$. Stripping of O_2 at a fixed rate of N_2 feed is started. Now, $s_o^* = 0$, and the "gassing out" experiment follows an exponential decline $s_o/s_o^* = \exp(-k_l a\,t)$. The transient is likely to be very short since $s_o^*\,V$ is small, and consequently the results are inaccurate.

When the experiment is carried out with SO_3^{-2} as the O_2 acceptor s_{SO3-} $(t = 0)$ is known, and when air or O_2 is sparged to the tank $d\,(s_{SO3\text{-}2})/dt$ is proportional to $(s_{SO3\text{-}2}\ (t = 0) - s_{SO3\text{-}2}\ (t))$. Again an exponential decline is observed and $k_l a$ can be determined by titration of $s_{SO3\text{-}2}\ (t)$. A liquid that contains enough Na_2SO_3 to obtain reliable results of the titration is strongly coalescent. Hence, the results for $k_l a$ will be higher than would be the case if the experiment were carried out with tap water. This leads to no serious error if the experiment is used to evaluate $k_l a$ during an actual fermentation, since fermentation media are also coalescent. The method is, however, quite unpleasant to work with if $k_l a$ is to be found for a large tank. Sacks of Na_2SO_3 must be used, and after an experiment the reactor has to be filled with fresh medium, since otherwise the $k_l a$ value found in the next experiment will be higher than in the first experiment at the same power input.

2) A *steady state* experiment for determination of $k_l a$ is much to be preferred, since the accuracy can be much higher. If Eq. (11.19) is applied to a fermentation reaction at steady state, the terms on the r.h.s. are equal, and when the rate of the fermentation reaction is measured (it could be an aerobic yeast cultivation) by the rate of biomass formation, $k_l a$ is determined for each value of the power input.

A very satisfactory experimental method for the determination of the mass transfer coefficient is the so-called "H_2O_2 method." This was first proposed by Hickman [12] but has later been refined and adapted by Novozymes to $k_l a$ measurements in industrial reactors of size up to 100 m^3.

The medium is continuously sparged with air, and at $t = 0$ a feed of H_2O_2 is started at a rate of q_{H2O2} mol h^{-1}. The medium contains catalase, and after a while the rate of oxygen produced by splitting of H_2O_2 becomes equal to the rate of O_2 transfer to the gas phase. At this point, $s_o > s_o^*$, the saturation concentration of O_2 that would be obtained if O_2 was not continuously produced:

$$\text{Since } H_2O_2 \text{ reacts according to } \quad H_2O_2 \rightarrow \tfrac{1}{2}\, O_2 + H_2O. \tag{11.20}$$

$$k_l a = \frac{q_{H_2O_2}}{2 \cdot \left(s_{O_2}^* - s_{O_2}\right) V} = -\frac{v_{f,H2O2} \cdot c_{f,H2O2}}{2 \cdot \left(s_{O_2}^* - s_{O_2}\right) V}. \tag{11.21}$$

$s_{O_2}^*$ is smaller than s_{O2}, but a little higher than the value for sparging with air, since the gas that leaves the tank includes O_2 stripped from the liquid. The value $s_{O_2}^*$ $(s_o/s_{O_2}^* - 1) = s_{O_2}^*(\text{DOT} - 1)$ is measured for different values of the power input P/V, and in this way the relationship between v_g, P/V, and $k_l a$ is obtained. After each experiment, all that remains of the added H_2O_2 is water, and a new experiment can immediately be started.

Figure 11.4 shows the results of an experiment with the RJH mixing system, and the result (11.16) was found on the basis of many experiments at different u_g and P/V.

11.4 Conclusions

In the previous paragraphs, a small number of empirical methods have been advocated for the calculation of power input to achieve a desired mixing time or mass transfer coefficient at a given scale of the process. The resulting formulas are valid for typical conventional mixing aggregates, and most of the experimental data to back up the formulas has been collected using Rushton turbines and

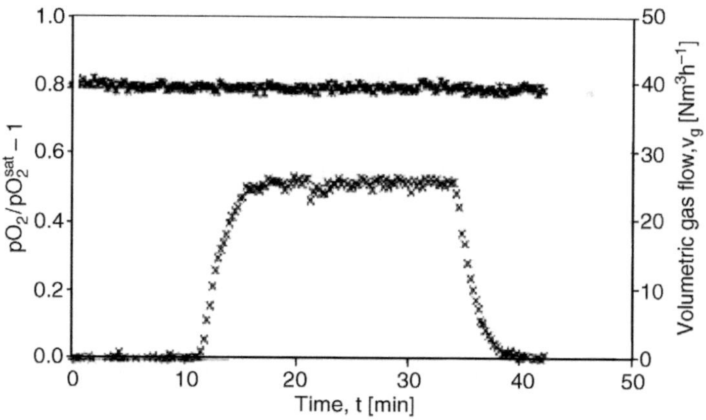

Figure 11.4 Typical experiment using the H_2O_2 steady-state method to determine k_la at 25 °C in a tank of volume 3.4 m^3 sparged with air and using one RJH mixer (see Figure 8 in Nordkvist *et al.* [1]). The feed rate of H_2O_2 is 93.2 mol h^{-1} and k_la is determined: $k_la = \frac{93.2}{2 \cdot 3.4 \cdot (0.5 \cdot 0.21 \cdot 1.26 \cdot 1.10)} = 93.2\ \text{h}^{-1}$.

other mechanical stirrers. Comparison with other types of mixers, in particular rotary jet heads and static mixers, shows that to a certain extent, the results carry over to quite different mixing concepts. One should, however, not conclude that the calculations guarantee that the scale-up of a process by a large factor gives the desired result. There are so many, especially quality, criteria that cannot be predicted in the larger scale, based on simple scale-up formulas. One needs to set the standards that are desired in the industrial scale, and by experiments in smaller scale see how the specific quality criteria change with process conditions. When more solid background information has been derived in small scale, one might scale-up, *perhaps using operation conditions that deviate from those that are found best in small scale.*

This is the concept behind the *scale-down/scale-up* design concept that has won acceptance in recent years. The subject is treated in Chapter 16 of this book.

In the final Example 11.4, all the basic scale-up information treated in this chapter will be illustrated. When supplemented with the concepts of Chapter 16 one might hope to achieve a successful scale-up in practice, avoiding the pitfalls that are not immediately obvious when working only with design criteria such as power input and its relations with process parameters.

Example 11.4 Scale-up of a 600 l pilot plant to an industrial reactor of volume 60 m^3

In Example 11.2, mixing in a tank of volume 600 l at $H/T \sim 1.04$ was considered.

With tank diameter $T = 0.9$ m, stirrer diameter $d_s = 0.35$ m, and with $N = 2\ \text{s}^{-1}$ and a power input of 218 W, a mixing time $t_{mix,0.95} = 10.1$ s was determined for the *unaerated liquid* and 12.8 s for the *aerated liquid.*

The tank volume is increased by a factor 100 to 60 m^3 by *linear scale-up*. In the industrial scale reactor $T = 100^{1/3} \cdot 0.9 = 4.178$ m, $H = 4.363$ m, and the stirrer diameter of the Rushton turbine $d_s = 1.624$ m.

$$\text{If } N = 2\,\text{s}^{-2} \text{ is still used, then } P = 5.2 \cdot 1000 \cdot 2^3 \cdot 0.355 \cdot 100^{5/3}$$
$$= 218 \cdot 2154 = 470\,\text{kW, or } 7.83\,\text{kWm}^{-3}. \tag{1}$$

With this very large power input, one will still get a mixing time of 10.1 s, but in industrial practice one will usually not accept a power input >5 kW m^{-3}. One will have to accept that $t_{mix,0.95} > 10.1$ s. First, it will, however, be seen if a change in the geometry on scale-up will give a smaller power input.

There are two ways in which P can be reduced: (1) a smaller N can be used or (2) d_s can be increased. If N is decreased by a factor 2 or d_s is increased by a factor $\sqrt{2}$ in the basic design of the 60 m^3 tank, then by the first expression in Eq. (11.14) t_{mix} remains the same, t_{mix} is proportional to d_s^{-2} and proportional to N^{-1}. The effect on the power input is, however dramatic: with $d_s = \sqrt{2} \cdot 1.624 = 2.296$ m $(d_s/T) = 0.55$) and $N = 1\,\text{s}^{-1}$, the power input P is a factor $2^{-3} \cdot 2^{5/2} = 2^{1/2}$ smaller than that of the basic design, that is, $P = 332$ kW for $t_{mix,0.95} = 10.1$ s.

In any scale-up process, the same quality of the mixing is obtained when the stirrer diameter is increased by a factor $\sqrt{a}$ *and the stirrer speed reduced by a factor* a. *The power input is, however, reduced by a factor* $\sqrt{a}$. With the above change in the basic design, that is, $a = 2$, the power input *P/V*, is reduced to the more amenable 5.53 kW m^3 without sacrificing the low value 10.1 s^{-1}.

Let us next consider a power input that is the same, 0.364 kW m^3, as in the pilot plant reactor. This gives a power input of 21.84 kW to the 60 m^{-3} tank, that is, a factor 21.52 smaller than that for the basic design.

According to Eq. (11.14), $t_{mix,0.95}$ is inversely proportional to $(P/M)^{1/3}$, and consequently the mixing time is

$$21.52^{1/3} \cdot 10.1 = 28.1\,\text{s when } P/M \text{ is } 0.364\,\text{W kg}^{-1}. \tag{2}$$

One could also use a mixing system different from that of the Rushton turbine. Consider the use of 10 RJH, each of the type with a nozzle diameter 15 mm. These would cover the total volume of the 60 m^3 tank, each servicing about 6 m^3 (with some overlap). This will be explored for $t_{mix,\,0.95} = 28.1$ s.

For a *single* RJH machine, Eq. (3) in Example 11.3 would give a $t_{mix,\,0.95} = 28.1$ s for a power input of $P/V = (9486 \cdot 15^{-1.24}/28.1)^{1/0.308} = 2.97$ kW m^{-3} or 0.297 W kg^{-1}. This is a smaller specific power input than the value 0.364 W kg^{-1} obtained for the Rushton turbine. It is, however, not realistic to hope for one machine to give adequate mixing to a reactor of volume 60 m^3, since good mixing can be guaranteed only in a spherical volume of about 6 m^3 around the RJH. Instead, the power input 2.97 kW m^{-3} is delivered by 10 RJH machines – the same total liquid circulation v_l is used, but divided equally among $N_{RJH}=10$ machines. This is discussed in Villadsen *et al.* (2011), [15] (p. 523), and the result is that the power input, and the same $t_{mix,0.95}$, is obtained with N_{RJH} RJHs:

$$P/V(\mathrm{W\,m^{-3}}) = 8.05\;10^{12}\cdot N_{RJH}^{-2} \cdot d^{-4} \cdot t_{mix,0.95}^{-3.25}. \tag{3}$$

Using Eq. (3) for $t_{mix,0.95}=28.1$ s gives $P/V=0.126$ kW m^{-3} or a total power input $P=60\cdot 0.126=7.56$ kW.

From these calculations, it appears that the total power input can be reduced from 21.84 kW when a *single* Rushton turbine is used to 7.56 kW when 10 RJHs are used. Intuitively, it appears to be right that distributing the power input between several mixers in the same volume gives a better result than expending all the power input in one mixer. Unfortunately, at present the RJH mixing machines have not been adopted by the large bio-industry (but only to large-scale application in the brewery industry and for moderate mixing duties). Further research on the possible advantages of these new mixing aggregates is needed.

Next, the operation of the aerated 60 m^3 tank is considered.

When the reactor volume is scaled up by a factor 100, the superficial gas velocity is scaled up by a factor $100^{1/3}$, and from Example 11.1 one obtains $u_g=2.62\;10^{-2}\cdot 100^{1/3}=0.122\;\mathrm{m\,s^{-1}}$. According to van't Riet's formula (Table 11.1), the mass transfer coefficient is determined by Eq. (4) for the 600 l tank:

$$k_l a = 0.026\cdot 3600\cdot (2.62\,10^{-2})0.5\cdot (117)^{0.4} = 102\;\mathrm{h^{-1}}. \tag{3}$$

When the tank is scaled up to 60 m^3, the value of $k_l a$ increases by the factor $100^{1/3}$ to 472 h^{-1}. The conclusion is that due to the increase of u_g with increasing scale the mass transfer coefficient $k_l a$ increases with scale for the same power input per unit volume P/V or power input per unit mass P/M.

Next, the influence of scale-up to 60 m^3 reactor size on the stability of the air dispersion is considered. Now, $v_g = (0.1/60)\cdot 100 = \frac{1}{6}\;\mathrm{m^3\,s^{-1}}$, and for the basic design with $d_s=1.624$ m, application of Eq. (11.11) gives:

Flooding will occur for $N<(9.82/180)^{1/3}\cdot 4.178^{7/6}\cdot 1.624^{-2.5}=0.60\;\mathrm{s^{-1}}$.

Similarly, to be into the well-dispersed region:

$$N > (5\cdot 9.82^{0.5}/6)^{1/2}\cdot 4.178^{1/4}\cdot 1.624^{-2} = 0.88\;\mathrm{s^{-1}}.$$

For a constant value of $N=2\;\mathrm{s^{-1}}$, $N_A = \frac{1/6}{2\cdot 1.628^3} = 0.0193$ according to Eq. (11.8) $=N_A$ at the 600 l scale since v_g and d_s^3 have both been scaled up by the same factor 100. In the 600 l scale, flooding occurs for $N<1.00\;\mathrm{s^{-1}}$, and good dispersion is obtained for $N>1.28\;\mathrm{s^{-1}}$. Consequently, it might seem that at constant power input P/V or P/M, a much greater gas flow can be handled in large scale than in the smaller scale.

One must, however, also consider that the Froude number $Fr = (N^2 d_s/g)$ given by Eq. (11.9) increases with scale for the same value of N. This is best seen in Figure 16.1. When the scale increases for the same power input per unit mass, the Froude number decreases while N_A (called Fl^G on the figure) increases. If N is kept constant ($= 2\,s^{-1}$) in both scales, Fr has increased by a factor $100^{1/3} = 4.64$, and from Figure 11.3, it is likely that $(N_{p,g}/N_p)$ is much smaller than 0.5 in the large scale.

Consider scale-up of volume V of the reactor and of the volumetric gas flow v_g by a scalar factor 10^a, *at constant effective* $P/M = (N^3 d_s^5/d_s^3)$. When d_s is scaled up by a factor $10^{a/3}$ at constant $N^3 d_s^5/d_s^3$ according to Eq. (11.7), one needs to scale down N by a factor $10^{-2a/9}$ to keep P/M constant. Hence, Fr is scaled down by a factor $(10^{-2a/9})^2\ 10^{a/3} = 10^{-a/9}$. At the same time, the aeration number $N_A = (v_g/N d_s^3)$ is scaled up by a factor $10^{2a/9}$.

If the $60\,m^3$ reactor is operated at the same aerated $P/M = 0.177\,W\,kg^{-1}$ as the 600 l reactor, one needs to scale N down to $2 \cdot 10^{-2a/9} = 2 \cdot 10^{-4/9} = 0.72\,s^{-1}$ when d_s is scaled up by the factor $10^{a/3} = 4.64$ to 1.624 m. The value of Fr decreases $(4 \cdot 0.35/9.82\,) = 0.143$ to $0.143 \cdot 10^{-2/9} = 0.085$. N_A increases to $0.0193 \cdot 10^{2a/9} = 0.0553$. According to Figure 16.1, the large-scale reactor works well into the region of a well-mixed gas phase. [*Note*: As the text in Figure 16.1 explains the three scale-up examples ($a = 1, 2 \ldots 5$), also take into account that the gas volume decreases with a due to the higher hydrostatic pressure at $a = 4$ and 5. Hence, for $a = 4$ and 5, the increase of N_A with a does not quite match the result obtained above.]

As a final comment to the above, one might install a stirrer with a smaller N_p than $N_p = 5.2$ for the Rushton turbine. A four- bladed Lightning™ hydrofoil is better suited for dispersion of gas *at the same N and P*. Since $N_p = 0.8$ for the Lightning mixer, one has to change d_s by a factor $(5.2/0.8)^{1/5} = 1.45$ from $d_s = 1.624$ to $d_s = 2.361$ m. Now, $N_A = 0.019 \cdot 1.45^{-3} = 0.0062$, a very small value, and $N_{p,g}/N_p$ is certainly close to 1. N could be reduced to $1\,s^{-1}$ without seriously decreasing $N_{p,g}/N_p$ and one will still obtain the advantage of a smaller power input by decreasing N and increasing d_s.

References

1 Nordkvist, M., Grotkjær, T., Hummer, J.S., and Villadsen, J. (2003) Applying rotary jet heads for mixing and mass transfer in a forced recirculation tank reactor system. *Chem. Eng. Sci.*, **58**, 3877–3890.

2 Rushton, J.H., Costich, E.W., and Everett, H.J. (1950a) Power characteristics of mixing impellers – Part I. *Chem. Eng. Prog.*, **46**, 395–404.

3 Rushton, J.H., Costich, E.W., and Everett, H.J. (1950b) Power characteristics of mixing impellers – Part II. *Chem. Eng. Prog.*, **46**, 467–476.

4 Nienow, A.W. (2010) Stirred tank bioreactors, and scale-up and scale down

of stirred bioreactors, in *Comprehensive Bioprocess Engineering* (ed. M. Berovic and S.O. Enfors), University of Ljubljana, Slovenia, pp. 135–159 and 179–199.

5 Paul, E.D., Atieno-Oberg, V.A., and Kresta, S.M. (eds) (2004) *Handbook of Industrial Mixing*, 2nd edn, John Wiley & Sons, Inc., Hobeken, NJ.

6 Nienow, A.W. (1997) On impeller circulation and mixing effectiveness in the turbulent flow regime. *Chem. Eng. Sci.*, **15**, 2557–2565.

7 Moo-Young, M. and Blanch, H.W. (1981) Design of biochemical reactors. Mass transfer criteria for simple and complex systems. *Adv. Biochem. Eng.*, **19**, 1–69.

8 Linek, V., Vacek, V., and Benes, P. (1987) A critical review and experimental verification of the correct use of the dynamic method for the determination of oxygen transfer in aerated agitated vessels to water, electrolyte solutions and viscous liquids. *Chem. Eng. J.*, **34**, 11–34.

9 van't Riet, K. (1979) Review of measuring methods and results in non-viscous gas liquid mass transfer in stirred vessels. *Ind. Eng. Chem. Process Dev.*, **18**, 357–364.

10 Heijnen, J.J. and van't Riet, K. (1984) Mass transfer, mixing and heat transfer phenomena in low viscosity bubble column reactors. *Chem. Eng. J.*, **28**, B21–B42.

11 Al-Taweel, A.M., Yan, J., Azizi, F., Odedra, D., and Gomaa, H.G. (2005) Using in-line static mixers to intensify gas–liquid mass transfer processes. *Chem. Eng. Sci.*, **60**, 6378–6390.

12 Hickman, A.D. (1988) Gas–liquid oxygen transfer. A novel experimental technique with results for mass transfer in aerated agitated vessels. Proceedings of 6th European Conference on Mixing, Pavia, pp. 369–374.

13 Crank, J. (1957) *Mathematics of Diffusion*, Oxford University Press, Oxford.

14 Pedersen, A.G., Bundgaard-Nielsen, M., Nielsen, J., and Villadsen, J. (1994) Characterization of mixing in stirred tank bioreactors equipped with Rushton turbines. *Biotechnol. Bioeng.*, **44**, 1013–1017.

15 Villadsen, J., Nielsen, J., and Lidén, G. (2011) Bioreaction Engineering Principles, 3rd edn, Springer, NY.

Part Three
Downstream Processing

Fundamental Bioengineering, First Edition. Edited by John Villadsen.

12
Product Recovery from the Cultures

Sunil Nath

Summary

The vast field of bioproduct recovery is surveyed. The various unit operations and techniques available in the field are classified based on the underlying mechanism of separation. A large number of approaches for recovery of cell products are described for specific processes and products. The products include cells (baker's yeast), polysaccharides (xanthan gum), antibiotics (penicillin), and proteins (interferon and insulin). A detailed case study of the purification operations for the recovery of α-A interferon is presented along with information on the yields and purification factors attained after each step that help in understanding the complexities involved in the design of these processes. Taken together, the approaches and case study illustrate the principles that may be considered as the common elements of various bioproduct recovery processes.

12.0
Introduction

The goal of product recovery operations in downstream processing is to recover and purify the product to the desired specification levels while maximizing yield and minimizing costs. The nature of cell culture media has major implications for all downstream operations. While the use of crude fermentation materials may reduce upstream processing costs, they tend to raise substantially the costs of downstream processing and make it more complex. When the products are cells or are intracellular in nature, the presence of insoluble media materials will increase the amount of contaminants that will have to be separated from the product at some further stage. Hence, the use of raw materials must take into consideration their economic impact on culture productivity and purification. Therefore, fermentation processes in which costs of purification are dominant (which is the norm) tend to use purer raw materials in order to minimize the costs of product recovery.

Cell cultures often produce impurities that are closely related to structure and properties with the desired product, and therefore separation in subsequent

Fundamental Bioengineering, First Edition. Edited by John Villadsen.

downstream processing steps is very exacting. During the production of primary and secondary metabolites by fermentation or biotransformation, related impurities are produced owing to the presence of side reactions or due to product degradation. The presence of these by-products may be minimized by changing the fermentation medium or by altering fermentation conditions. Alternatively, enzyme inhibitors can be used that target the side reactions. In batch fermentation, the harvesting time may be crucial and for cells themselves as products, cell viability can vary over the growth curve. For intracellular products, it is important to harvest the cells before cell lysis in order to ensure retention of the product in the separated solid phase. In the case of extracellular products, cell lysis decreases the specific activity of product in the clarified liquid phase. Furthermore, release of intracellular proteases by cell lysis can also degrade products. The product concentration per unit volume or per unit cell mass at the end of the cell culture step plays an important role in the nature and type of the downstream processing steps required. The number of downstream processing steps impacts process economics in two ways. First, the cost of each purification step in the train of steps has to be taken into account and second, the yield of the product at the end of each step has to be kept track of. Multistep purifications lead to cumulative losses in product yield even when the average yield per step is high. Obviously, it is an ideal situation if the number of steps in a purification strategy can be minimized and yet the desired product specifications can be achieved [1–3].

Table 12.1 shows the variety of separation techniques that are available for biological product recovery and purification. The unit operations involved in recovery and purification are classified based on the underlying mechanism of the separation process. These unit operations can be employed in diverse ways with widely different selectivities.

Table 12.1 Classification of unit operations and product recovery and purification processes based on their underlying mechanism of separation.

Physical basis of separation	Unit operations/bioseparation processes
Solubility	Extraction processes (extraction and precipitation)
Density	Centrifugal processes (differential centrifugation)
Rate of sedimentation	Centrifugal processes (nonequilibrium settling)
Molecular size	Chromatographic processes (gel filtration), membrane processes (filtration, ultrafiltration, and variants)
Net ionic charge	Chromatographic processes (Ion-exchange), electrokinetic processes (continuous electrophoresis, isoelectric focusing)
Hydrophobicity	Chromatographic processes (hydrophobic interaction and reversed phase chromatography)
Binding specificity	Chromatographic processes (affinity chromatography)

12.1
Steps in Downstream Processing and Product Recovery

In some cultivation processes, the fermentation solids are directly utilized or are utilized after evaporation or drying, for example, in breads and cheese products. In other processes, the products are obtained after extraction, distillation, or adsorption operations. The most common first step of purification for products that are intracellular is the solid–liquid separation of cells and other solids from the culture fluid. Such solid–liquid separation is commonly achieved by the use of mechanical operations such as sedimentation, filtration, microfiltration, or centrifugation. After such separation, the cells are lyophilized, solvent-extracted, or mechanically disrupted to release intracellular content, depending on the purpose. The disintegrated cells are generally processed further using solid–liquid separation methods again in order to obtain the solids or soluble fraction. A large number of products of cell bioconversions are present in the soluble intracellular fraction or in the clarified extracellular culture fluid. These products need to be further processed and purified using a variety of bioseparation techniques that are sensitive to differences in physicochemical properties such as charge, size, solubility, volatility, biological affinity, and so on. A most commonly employed train for separation of intracellular protein products from fermentation or cell culture is depicted in Figure 12.1. This methodology may be suitably modified for specific applications. Several downstream processing approaches for recovery of cell products will be illustrated in this chapter for specific processes and products. These products include whole cells, polysaccharides, antibiotics, and proteins.

12.2
Baker's Yeast

The fermentation broth containing 40–60 kg (dry biomass) m^{-3} after yeast fermentation is passed through a continuous nozzle centrifuge to produce yeast cream containing 20 wt.% solids. It is concentrated to 27–28 wt.% by plate-and-frame filtration or more generally by rotary vacuum filtration. Starch is often used as a filter aid in these filtrations. The yeast cake containing 27–28 wt.% solids is sold as fresh baker's yeast [4]. Active dry yeast can also be prepared for which suitable emulsifiers are added to the yeast cake in order to assist subsequent extrusion processes. Dried baker's yeast is then prepared by spray or fluidized bed air drying. Air temperatures of up to 150 °C are used and the yeast temperature is kept in the 25–40 °C range. The final moisture content in the dried baker's yeast is typically around 7%. Yeast is used as a flavoring agent in the food industry and as a nutrient source in various health foods. A continuous process for the manufacture of fresh and dried baker's yeast is shown in Figure 12.2.

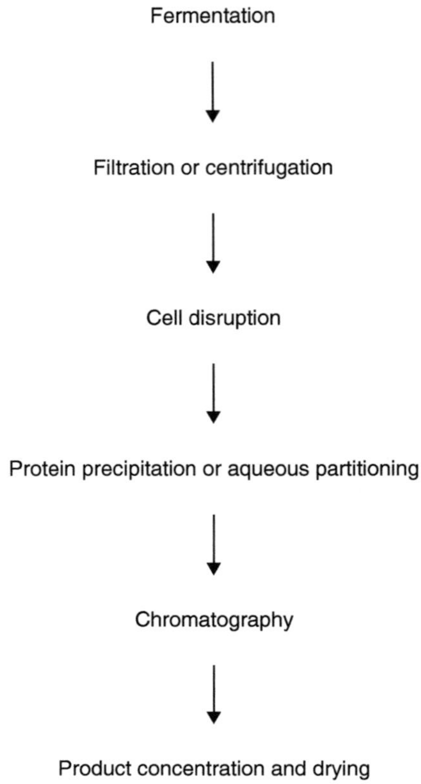

Figure 12.1 Commonly employed sequential steps in intracellular protein purification.

12.3 Xanthan Gum

A process for the recovery of the microbial polysaccharide xanthan from the fermentation broth is shown in Figure 12.3. The high viscosity in the whole broth makes it difficult to carry out filtration or centrifugation operations without greatly diluting the contents. For xanthan gum, the fluid is heated to a temperature greater than 100 °C in order to lower the viscosity of the broth as such treatment does not degrade the gum product. The clarified xanthan is then precipitated using salts or by organic solvents such as alcohol. The wet precipitate is dewatered by pressing or centrifugation and then dried using forced air, vacuum, or spray dryers. The dried product is then milled to a particle size that gives it both good solubility and good dispersion properties.

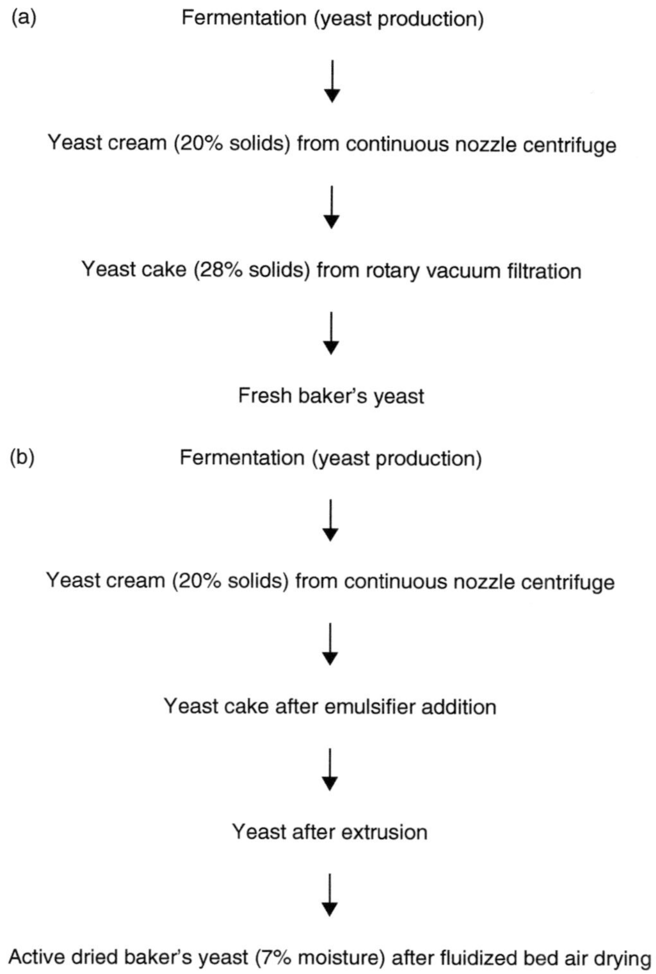

Figure 12.2 Diagram of continuous process for the manufacture of fresh (a) and dried baker's yeast (b).

12.4 Penicillin

After fermentation, penicillin is separated from the whole broth by rotary vacuum filtration, often using filter aids. The filtrate is cooled to 0–4 °C before the solvent extraction step. Penicillin is a strong acid and it is readily soluble in organic solvents and is generally extracted into butyl acetate at pH 3.0 by the use of continuous countercurrent extraction (Figure 12.4). The penicillin in solvent is treated with activated carbon and drum filtered to remove pigments and other impurities. Addition of alkali to the filtrate in a stirred tank makes the

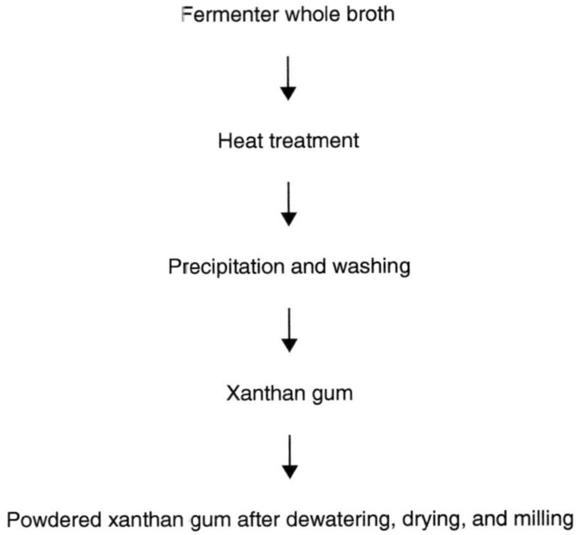

Figure 12.3 Process for the recovery of xanthan gum.

penicillin insoluble and crystals are collected using rotary vacuum filters or basket centrifuges. The crystals are washed with solvents such as butanol to remove impurities and the washed crystals are dried. The dried crystals are purified by recrystallization and filtered in water under sterile conditions to produce pharma-grade penicillin G. The aqueous penicillin G is diluted with butanol, distilled under vacuum to yield a final crystalline penicillin product that can be injected. The outline of a process for recovery of penicillin is illustrated in Figure 12.4 [5].

12.5
α-A Interferon

The purification operations for the recovery of α-A interferon from *Escherichia coli* is schematically illustrated in Figure 12.5. The process essentially consists of product recovery from cells followed by three column steps, that is, immunoaffinity chromatography, cation exchange chromatography, and gel filtration, performed at 4 °C [6]. The process will now be described in greater detail and will serve as a case study for this chapter and an illustration of the intricacies involved in product recovery operations. The product recovery by the procedure in [6] started with disruption of frozen cell paste (6 kg) in a bead mill grinder. To the disrupted cells was added an extraction buffer (2 M guanidine, 2% Triton X-100 v/v, 0.1 M Tris pH 7.5, 24 l) and the pH adjusted to 7.5 with 5 N NaOH. After stirring for 2 h, the mixture was diluted with 150 l of distilled water. This led to an enhanced flocculation of debris and simultaneously to a drastic

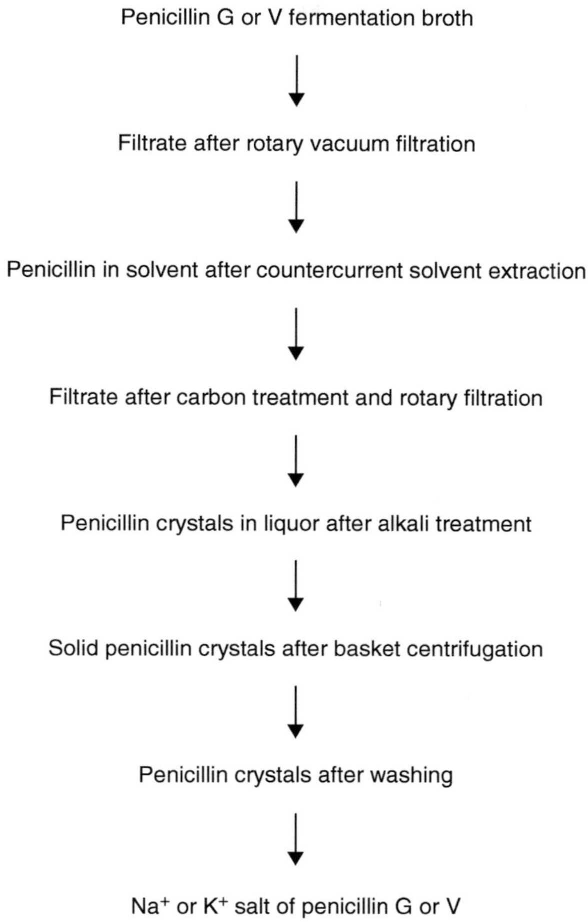

Figure 12.4 Process diagram for the recovery of penicillin.

lowering of the guanidine concentration. Hyflow filter aid (40 g/l) was added to the diluted extraction slurry and filtered using a rotary drum vacuum filter precoated with 10 kg of coating agent (Standard Super Cel). 2-Octanol was added to the clarified solution as an antifoaming agent, and the solution was filtered through a 0.3 μm cartridge filter and stored at 4 °C. The solution was concentrated threefold by tangential flow ultrafiltration using a 10 000 molecular weight cutoff membrane and filtered through a 1.2 μm cartridge.

The next step is an important one involving the use of affinity chromatography consisting of monoclonal antibody coupled to *N*-hydroxysuccinimide ester-activated Sepharose 4B (see the Chapter 13) of density 12.7 mg/ml and a total volume of 720 ml. The column was equilibrated with 2.5 column volumes of equilibration buffer (0.3 M guanidine, 0.3% Triton X-100, 0.1 M Tris pH 7.5) and

Fermentation

↓

Centrifugation

↓

Bead milling

↓

Extraction, dilution, microfiltration, concentration, and filtration

↓

Immunoaffinity chromatography

↓

Concentration and diafiltration

↓

Cation-exchange chromatography and concentration

↓

Gel filtration chromatography

Figure 12.5 Process for the purification of α-interferon.

the sample was loaded at a linear velocity of 38 cm/h. The column was washed with five column volumes of the equilibration buffer and with each of the following buffers: 0.5 M guanidine, 0.2% Triton X-100, 25 mM Tris, pH 7.5; 1 M NaSCN, 0.1% Triton X-100, 25 mM Tris, pH 7.5; 150 mM NaCl, 0.1% Triton X-100. The material eluting with these buffers was thrown away. The product was eluted from the column with elution buffer (0.2 M acetic acid, 0.1% Triton X-100, 150 mM NaCl) with continuous measurement of both pH and UV absorbance at 280 nm. The peak was collected from the instant the effluent decreased below pH 5 and until the time 1.5 column volumes had passed. The column was then reequilibrated with the equilibration buffer and another cycle of the operation was thus begun. The collected pool was adjusted to a pH of 3.0 in order to prevent proteolytic degradation during storage.

The collected pool was concentrated by ultrafiltration, diafiltered with nine volumes of a new equilibration buffer (25 mM NH_4OAc, pH 4.5, 25 mM NaCl) and further concentrated. The ultrafiltration membranes were washed with the same equilibration buffer and the wash was added to the concentrated solution. The pool was loaded on to a cation exchange column that had been preequilibrated with the equilibration buffer used for the diafiltration. Under these conditions, the protein product bound to the column. The column was washed with equilibration buffer until the absorbance of the effluent at 280 nm reached the baseline, indicating that the Triton X-100 had also been removed from the product. The adsorbed protein was eluted with a new eluting buffer (25 mM NH_4OAc, pH 5.0, 120 mM NaCl) and now the peak was collected as a single fraction that was concentrated by ultrafiltration and then filtered through a 0.45 μm filter. The filtered sample was run on a gel filtration column with the elution buffer used during cation exchange chromatography. Fractions representing the monomeric form of the protein were pooled and stored at −20 °C. The activity of the monomeric form was about three times that of the other peak that comprised of combined oligomeric forms of the protein. The yields of the process during the operation are shown in Table 12.2 [6].

It was interesting to note that centrifugation between 10 000–14 000 *g* did not serve to obtain a clarity equivalent to use of the rotary drum filter. The use of immunoadsorbent was the most costly step in the entire train of operations, hence its cycle life under industrial conditions is an important factor for commercialization of the process. The authors of Ref. [6] could use their column over hundreds of cycles. A crucial determinant of column life was the use of well-clarified samples before loading on to the column. The large protein loss on the cation exchange column was due to precipitation of oligomeric forms of the product. The cation exchange chromatography was found to remove both the nonionic detergent Triton X-10 and the enzymes responsible for proteolytic attack. The gel filtration step was effective only at pH 5.0 or below. At higher pH values, the protein polymerized and eluted from the column in the void

Table 12.2 Yields and purification factors for the various steps in the purification of α-interferon.

Step	Volume (l)	Protein (g)	Activity (× 10^{-12} units)	Specific activity (× 10^6 units/mg)	Yield (%)	Purification factor
Crude extract	24	394	2.6	7.0	100	1.0
Drum filter concentrate	52	192	3.1	16	100	2.3
Affinity eluate	16	8.7	1.9	220	61	13.8
Cation exchange eluate	1.0	4.9	1.2	250	63	1.1
Gel filtration	0.5	2.4	0.6	250	50	1.0
Overall					23	36

volume. This case study demonstrates that the most difficult separations in biotechnology are those involving different forms of the same protein.

12.6 Insulin

Insulin was among the first recombinant proteins in biotechnology, and its requirement was in several thousands of kilograms per year. Producers include Eli Lilly, Novo Nordisk, and Roche. In the process of Eli Lilly and Co., the insulin is produced in *E. coli* [7]. After fermentation, cells are chilled, disrupted by bead milling, and inclusion bodies containing the product are isolated by centrifugation (Figure 12.6). The mixture is treated with CNBr to cleave proteins into fragments, with proinsulin remaining as an uncleaved molecule. The *S*-sulfonate derivative of proinsulin is chromatographically purified by steps that are patented and unspecified. The *S*-sulfonate is folded to form proinsulin that is again purified by unspecified steps and transformed enzymatically to human insulin. The material is purified by multiple high-resolution steps to remove impurities that comprise chemical forms of insulin.

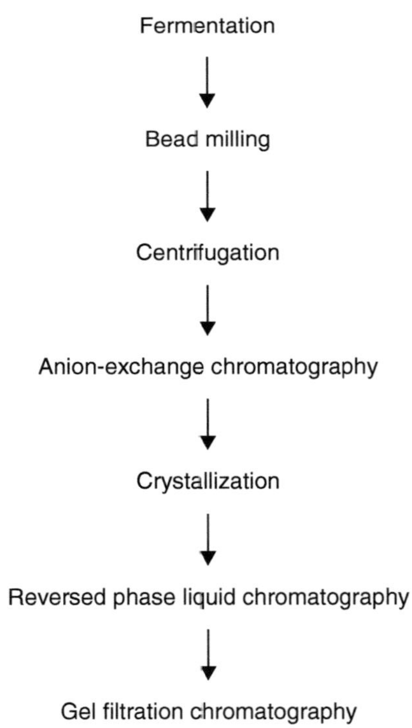

Figure 12.6 Process for the manufacture of insulin.

Anion-exchange chromatography is the first of these important high-resolution steps, followed by crystallization. The crystals are redissolved for the second key step of reversed phase liquid chromatography. The matrix was revealed to be Zorbax Process grade C_8. The insulin was loaded on to the column in a water-rich mobile phase. It was eluted from the column with a linear gradient of 15–30% acetonitrile. The purity was over 97% and the recovery more than 75%. A step gradient that could yield the same purity and recovery could not be found. This could be due to the nonlinear adsorption isotherm exhibited by the product in the overloaded conditions used at which the column was operated. The column was regenerated with buffer (50 mM ammonium phosphate, pH 7.4, 60% acetonitrile). The insulin was recovered from the elution buffer by precipitation with zinc chloride and final purification was carried out by gel filtration chromatography (Figure 12.6). The column dimensions of the largest column in which reversed phase liquid chromatography was successfully carried out measured approximately 50×50 cm with an internal volume of more than 80 l and an insulin load of 1.2 kg [7].

12.7 Conclusions

The above examples illustrate key features of purification operations for materials of biological origin – cells, polysaccharides, antibiotics, and proteins. It is also clear that each example or case study is different from the others and requires its own practical knowledge and experience and that only certain principles may be considered as the common elements of the separations (Figures 12.1–12.6). In particular, fundamental principles of some of the widely used high-resolution column chromatographic techniques mentioned above, along with practical information on operation and further examples of preparative-scale separation of biological molecules, are covered in Chapter 13. A concise review of new approaches and recent advances in bioproduct recovery [8] and novel applications in the interface areas of nanobioproduct recovery [9,10], therapeutics [11], and energy technologies [12,13] are suggested for further reading.

References

1 Endo, I., Nagamune, T., Katoh, S., and Yonemoto, T. (eds.) (2000) *Bioseparation Engineering*, Elsevier Science, Amsterdam.
2 Ghosh, R. (2006) *Principles of Bioseparations Engineering*, World Scientific, Singapore.
3 Shukla, A.A., Etzel, M.R., and Gadam, S. (eds.) (2007) *Process Scale Bioseparations for the Biopharmaceutical Industry*, CRC Press, Boca Raton, FL, USA.
4 Chen, S.L. and Chiger, M. (1985) Production of baker's yeast, in *Comprehensive Biotechnology*, vol. 3 (ed. M. Moo-Young), Pergamon, Oxford, pp. 429–461.
5 Sylvester, J.C. and Coghill, R.D. (1954) The penicillin fermentation, in *Industrial Fermentations*, vol. 2 (eds. C.A. Underkofler and R.J. Hickey), Chemical Publishing Co., New York, pp. 219–263.

6 Tarnowski, S.J., Roy, S.K., Liptak, R.A., Lee, D.K., and Ning, R.Y. (1986) Large-scale purification of recombinant human leukocyte interferons. *Methods Enzymol.*, **119**, 153–165.

7 Kroeff, E.P., Owens, R.A., Campbell, E.L., Johnson, R.D., and Marks, H.I. (1989) Production scale purification of biosynthetic human insulin by reversed phase high performance liquid chromatography. *J. Chromatogr.*, **461**, 45–61.

8 Cramer, S.M. and Holstein, M.A. (2011) Downstream bioprocessing: recent advances and future promise. *Curr. Opin. Chem. Eng.*, **1**, 27–37.

9 Jahanshahi, M., Zhang, Z., and Lyddiatt, A. (2005) Subtractive chromatography for purification and recovery of nano-bioproducts. *IEE Proc. Nanobiotechnol.*, **152**, 121–126.

10 Benavides, J., Aguilar, O., Lapizco-Encinas, B.H., and Rito-Palomares, M. (2008) Extraction and purification of bioproducts and nanoparticles using aqueous two-phase systems strategies. *Chem. Eng. Technol.*, **31**, 838–845.

11 Ho, C.W., Tan, W.S., Kamaruddin, S., Ling, T.C., and Tey, B.T. (2008) The direct recovery of recombinant hepatitis B core antigen from disruptate derived from continuous-flow bead milling. *Biotechnol. Appl. Biochem.*, **50**, 49–59.

12 Balasundaram, B., Skill, S.C., and Llewellyn, C.A. (2012) A low energy process for the recovery of bioproducts from cyanobacteria using a ball mill. *Biochem. Eng. J.*, **69**, 48–56.

13 Ezeji, T.C. and Li, Y. (2010) Advanced product recovery technologies, in *Biomass to Biofuels: Strategies for Global Industries* (eds. A. Vertes, N. Qureshi, H. Yukawa, and H.P. Blaschek), John Wiley & Sons, Inc., Hoboken, NJ, USA, pp. 331–345.

13
Purification of Bio-Products

Sunil Nath

Summary

The objective of all recovery processes (described in Chapter 12) is to obtain the desired purity and yield for a product, and techniques for purification are aimed to remove impurities closely related to the properties of the product. Column chromatography processes constitute the workhorse for purification of biochemical products at an analytical and even a preparative scale. The name originates from the first applications of the technique a century ago when solutions of dyes or vegetable pigments were resolved into bands of various colors by passing them through a column of fine particles (the solid or stationary phase). Molecular sieves, gels, or nonionic resins may be readily used as the stationary phase in column chromatographic processes. The column is packed with beads of solids and a feed solution (the mobile phase) is made to flow through the column and eluted with a suitable solvent, typically water for fermentation products. The separation media selectively slows certain solute components in the feed solution, while other components rapidly move down the column. This leads to a separation as the solutes are transported through the column, and the purity of the respective solutes gets enhanced. Ultimately, the feed components appear ("elute") at different times at the outlet of the chromatography column and can be collected separately. This chapter attempts to summarize aspects of the application of chromatographic processes for carrying out separations on a preparative and industrial scale.

Numerous technical advances have made column chromatography into a sophisticated tool for purification of biological materials. These include (a) incremental developments over the years in the various chromatographic equipment and columns with integrated recycle and control systems and (b) massive developments in the manufacture of resins with narrow and finely controlled particle/bead size distributions and improved resin stability and performance.

Fundamental Bioengineering, First Edition. Edited by John Villadsen.

13.1
Methods and Types of Separations in Chromatography

A plethora of liquid chromatographic methods has been described in the vast literature on bioseparation. These include ion-exchange chromatography, gel filtration chromatography, hydrophobic interaction and reversed phase liquid chromatography, and affinity chromatography [1–5]. In ion exchange chromatography, the stationary phase may consist of either a hydrophilic or a hydrophobic matrix functionalized with different ionizable groups. Reversible polar bonds can be readily formed between the ionic components in the mobile and stationary phases. The eluent is typically an aqueous buffer or water in the case of fermentation products. Anion-exchange and cation-exchange constitute the two key types of ion-exchange. In gel filtration/permeation chromatography, the stationary phase functions as a sieve that separates solutes on the basis of molecular size. Typically, in gel permeation chromatography, cross-linked polystyrene gels swollen in organic solvents are used to separate macromolecules based on size differences. In gel filtration chromatography, columns loaded with cross-linked polydextran gels swollen in aqueous solution are generally used to purify water-soluble proteins and macromolecules on the basis of size. This technique is also called size exclusion or molecular sieve chromatography in the literature. Reversed phase liquid chromatography usually employs columns packed with silica to which a suitable length hydrocarbon is covalently attached. This leads to partitioning of a solute between the polar aqueous mobile phase and the hydrophobic stationary phase. Feed components then elute in order of decreasing polarity because the strength of an eluting solvent increases with decrease in polarity. Charged and highly polar fermentation products are ideal candidates for purification using this technique. Affinity chromatography relies on the highly selective interaction of biological macromolecules with other molecules.

The use of a certain chromatographic technique depends on the type of biological material/feed and the specific physicochemical property between impurities and product selected as the target separation parameter. There also exist differences between chromatographic operations depending on the scale of purification. A typical size of particles and materials used for ion-exchange chromatography is 90 μm. The packing beads used in chromatographic applications on an analytical scale conventionally had a low degree of cross-linking and mean diameters of approximately 100 μm. Such media could be used only at low pressure (typically less than 20 atm) and fairly low flow rates. The beads do not have the stability to withstand higher pressures, and hence they cannot be used for preparative separations at high flow rates. A number of modern advances in resin technology have led to the development of minibeads that are as small as 3 μm and can withstand pressure drops as high as 18 MPa. However, industrial chromatographic applications are forced to use larger diameter particles (typically 250–500 μm) with high degrees of cross-linking.

Separations in column chromatography processes can also be classified as separations based on exclusion by charge, exclusion by size, or affinity differences.

These may be readily described based on the partitioning of a solute between two liquid phases in a three-phase system and quantified by its distribution (or partitioning) coefficient, K_d defined as

$$K_\mathrm{d} = \frac{C_\mathrm{r}}{C_\mathrm{l}}, \tag{13.1}$$

where C_r is the concentration of a solute in the liquid phase inside the resin bead and C_l is the solute concentration in the liquid phase between resin beads, that is, in the interstitial space. The partitioning of the solute depends on the structure and composition of the solute, the type, and properties of the resin and the other components in the solution. The ratio of the distribution coefficients of two solutes is called the separation factor, α defined as

$$\alpha = \frac{K_\mathrm{d1}}{K_{d2}}. \tag{13.2}$$

For $\alpha > 1$, solute 2 will elute the column at an earlier time than solute 1.

In chromatography based on charge exclusion, the ionic components in the feed solution are excluded from the resin by electrostatic repulsion; the nonionic components freely distribute into the intraresin liquid phase. Thus, compared to the ionic components that move only through the interstitial spaces among beads, the noniionic solute takes a much longer path through the column and elutes later than an ionic solute. An important example of industrial chromatography is the separation of sucrose from various ionic solutes in molasses.

In exclusion based on size, the sieve-like resin excludes the larger size molecules but permits the small molecules to enter the bead volume. Thus, smaller molecular weight solutes elute later at the end of the column, that is, they require a larger elution volume than higher molecular weight solutes. Molecules of similar size and charge can be separated based on differential affinities or distribution coefficients for the resin beads. A large-scale example of this type of separation is the purification of fructose from glucose in the corn sweetener industry.

Affinity chromatography has as its basis in the biospecific recognition between a solute and a ligand that is covalently attached to a solid matrix. In the affinity purification process, an enzyme is adsorbed with very high selectivity from a mixture of proteins. After the unadsorbed proteins are washed from the affinity column, the enzyme can be eluted from the ligand and recovered. This technique has been used in several one-step purifications for enzymes, proteins, nucleic acids, and antibodies. Today, affinity chromatographic processes are routinely employed for the purification of monoclonal antibodies, accounting for almost 40% of all biopharmaceutical products. A generalized protein purification scheme comparing conventional chromatographic techniques with affinity chromatography is presented in Figure 13.1. Conventional chromatographic techniques separate proteins based on charge, size, or isoelectric point differences, but require many additional steps also. This leads to lower overall yields of the purified protein compared to the use of affinity chromatography (Figure 13.1).

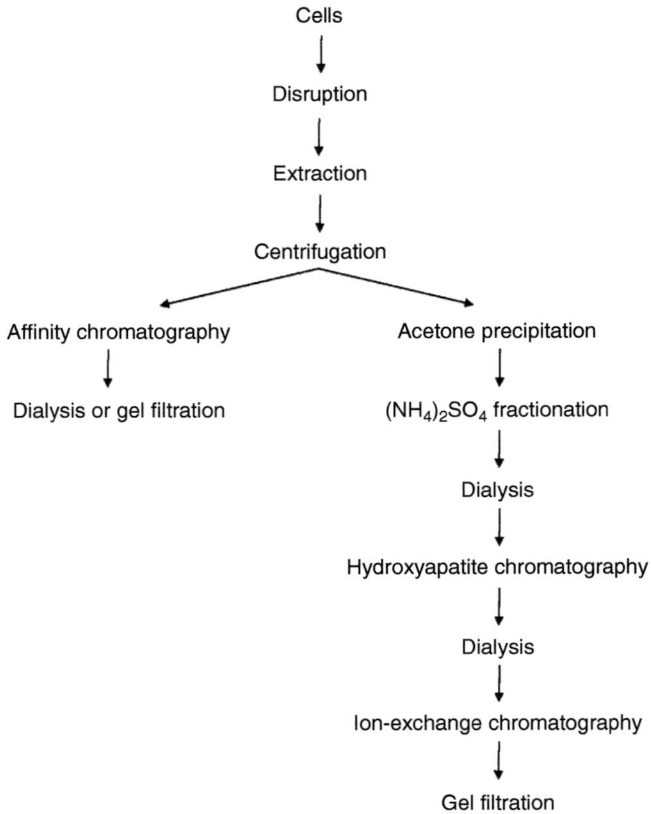

Figure 13.1 Comparison of affinity chromatography with conventional chromatographic techniques.

13.2
Materials Used in Chromatographic Separations

Various ion-exchange resins and hydrophilic and hydrophobic matrices and gels are commonly used as the stationary phase in column chromatographic applications. Both porous and nonporous beads can be used depending on the type of application. Hydrophobic matrices are generally polystyrene-based, while cross-linked dextrans, agarose, and polyacrylamide serve as hydrophilic gels. Dextran gels ("Sephadex") are made of linear polyglucose chains of dextran cross-linked with epichlorohydrin, while polyacrylamide gels ("Biogel") are polymers of acrylamide with methylene bisacrylamide as the cross-linker. The structure of agarose gels ("Sepharose") is maintained by hydrogen bonds, not by chemical cross-linkers, which places limitations on their range of operation for several applications (typically below 40 °C and between pH 5 and 8).

Table 13.1 Range of popular commercially available conventional hydrophilic materials for chromatographic applications.

Gel	Molecular weight range	Hydrated volume (ml/g dry gel)
Sephadex G-10–G-200	700–800 000	2.5–30
Bio-Gel P-2–P-300	200–400 000	4.0–70
Agarose 10%–4%	10 000–10 000 000	

Dextran and polyacrylamide gels are water-insoluble owing to their high degree of cross-linking, yet they are very hydrophilic and swell in aqueous systems. Hydrated gels are available in well-controlled particle size ranges (50–80, 80–150, and 150–300 μm). The molecular weight and hydrated bed volume range that are covered by commonly used commercially available hydrophilic gels are given in Table 13.1.

Also widely used are the very rigid hydrophobic matrices based on styrene polymers ("Stryagel") with divinylbenzene as the cross-linking agent. Such matrices require an organic medium such as carbon tetrachloride for operation and cannot be used in aqueous media. Such materials are commercially available and possess a very high molecular weight fraction range, typically from less than 1000 to 20×10^6 and with porosities as low as 160 nm at the lower molecular weight exclusion limit to one as high as 4 μm.

Ideal ligands for affinity chromatography bind strongly and specifically with the solute to be purified. The binding should also be reversible so that the purified solute can be readily eluted from the ligand. The binding affinity, selectivity, and elution behavior constitute the key factors in a quantitative analysis of the affinity chromatography process. The binding strength and selectivity are generally described mathematically by the association constant, K_a, while the adsorption process is represented by a Langmuir-type adsorption isotherm. High values of K_a are desirable as they represent highly stable complexes. K_a values as high as 10^{15} M are obtained for the biospecific avidin–biotin interaction. Values of the association constant vary in a range between 10^6 and 10^{12} M for antigen–antibody interaction and between 10^4–10^6 M for lectin–carbohydrate and enzyme–substrate interactions.

The most commonly employed matrices for affinity chromatography are polysaccharide, agarose, polyacrylamide, and microporous glass beads. Several other supports such as cellulose, cross-linked dextran, and polystyrene have also been used. A major advantage of support materials such as cellulose is their ultralow cost.

Further data on the operation limits, including the maximum temperature and the maximum pressure drop until which some of these popular materials can be used, are provided in Table 13.2. A large number of innovations in packing materials occur in the market at a fast pace. Notable among these is the use of suitably functionalized and surface-modified porous glass or silica, as also the

Table 13.2 Limits sustainable by some common and popularly used conventional chromatographic packing materials.

Material	Maximum temperature (°C)	Operational pH range	Maximum linear flow or range (m/h)	Maximum pressure drop or range (m water)
Cross-linked dextran	120	>2	0.80	0.15–1.5
Dextran, bisacrylamide	120	3–11	0.25–0.40	0.7–3.0
Polyacrylamide	120	2–10	0.03	0.20–1.0
Agarose	40	4–9	0.03	0.30–1.0
Cross-linked agarose	120	3–14	0.15–0.30	0.5–1.5
Cross-linked polyether	120	1–14		
Surface-modified glass or silica	120	<9		

development of novel polymer-coated silica gels. However, the cost of several designed and tailor-made materials can be exorbitantly high, limiting their general applicability.

13.3 Chromatographic Theory

The theory of chromatography is too vast to cover in any single review [1–5] and only the most basic concepts will be touched upon here. These are based on the original column chromatographic theory of Martin and Synge [6], which can be applied to preparative-scale separations also. The solvent volume required to elute the peak solute concentration can be expressed mathematically by the equation

$$V_{\max} = K_{\mathrm{d}} V_{\mathrm{r}} + V_{\mathrm{l}}. \tag{13.3}$$

Here, K_{d} is the solute distribution coefficient (Eq. (13.1)) and $V_{\max}$ is the liquid volume that has passed through the column at the maximum solute concentration (i.e., at the midpoint of the solute peak). V_{r} refers to the liquid volumes inside the resin and V_{l} to the interstitial volume. This calculation requires the equilibrium distribution data and data on the resin and column dimensions. The assumption of equilibrium implies that the above equation is valid only for small diameter particles and for operation at very low flow rates, that is, for analytical applications. Hence, Eq. (13.3) is only a first approximation for preparative- or industrial-scale column chromatographic applications.

The number of theoretical plates, n, is a measure of the quality of a separation because it compares the column length required for equilibration between a solute in the liquid in the resin versus the solute in the liquid in the interstitial spaces. A simple way to calculate n from chromatographic data is the use of Eq. (13.4), that is,

$$n = \frac{2V_{\max}(V_{\max} - V_1)}{W^2}, \tag{13.4}$$

where W is the half-width of the elution curve at a y-axis value of 1/e (= 0.368) times the maximal solute concentration. For a Gaussian elution curve, W is also related to the standard deviation of the distribution. The height equivalent of a theoretical plate, h, can therefore be readily estimated by dividing the column height by n. There exists an optimal mobile-phase velocity at which a minimum value of h (i.e., maximum n) is obtained [7].

Another way of specifying the required bed height is by calculating the resolution of a separation, R, which is a measure of the peak separation to the average peak width. It is given by the formula

$$R = \frac{V_{\max 1} - V_{\max 2}}{0.5(W_1 + W_2)}. \tag{13.5}$$

The larger the number of theoretical plates, the superior is the resolution. However, column heights cannot extend to infinity and there are practical limits imposed on them in an industrial scenario or on preparative scales. In fact, the value of n can also be changed by varying the flow rate and the mean resin particle size or particle size distribution. For a value of R of 1.0, the contamination in the product is less than 3%. In industrial applications, for solutes of approximately the same concentration, even $R = 0.8$ is considered an acceptable level of recovery and purity.

There is a lack of theoretical work on affinity chromatography. In general, the plate theory described above that is routinely used to model column chromatography is inapplicable to affinity chromatographic processes. A chief reason for this is that the adsorption isotherms in affinity processes are highly curved and cannot be assumed to be linear. However, the theories for analyzing conventional chromatographic systems require linear isotherms and linear adsorption–desorption kinetics. More work needs to be done in future to extend these theories to affinity processes.

13.4 Practical Considerations in Column Chromatographic Applications

Both process operating parameters and resin properties determine the number of theoretical plates and determine the resolution of a chromatographic separation, as discussed above. The mean bead size, the bead size distribution, and the type and degree of cross-linking are key resin parameters in chromatographic applications. Generally, the mean particle size of the resin beads must be selected so that (a) solute diffusion into and from the beads does not become the rate-determining step and (b) the pressure drop across the column does not exceed a preset limit. Small particles both adsorb and release solute rapidly and produce sharp elution profiles. On the other hand, larger bead sizes take longer in solute adsorption and release and lead to a broader, less distinct elution. Obviously, the use of the smallest possible particle sizes achieves the largest number of theoretical plates; however, if the separation ratio for the solutes is sufficiently large, a resin with larger bead size can be readily used. The lower limit on particle size is determined by what constitutes an acceptable pressure drop across the chromatographic column.

Particle size distributions of the resin also play a critical role. With a very polydisperse distribution of bead sizes for a resin of a certain mean particle size, acceptable elution profiles are not always possible to be achieved. Particles smaller than the mean size will elute the solute more rapidly while particles larger than the mean will do so with less rapidity, a fact that will lead to undesirable zone broadening and tailing of the elution profiles. Today, separation resins are available in which beads are all of almost the same size; for example, 90% of the particles are within 10% of the mean particle size with state-of-the-art resin manufacturing technology.

The degree of cross-linking of resin beads also plays a role in determining the location of the elution peak and profile. An increased degree of cross-linking has the effect of decreasing the volume of liquid inside a bead and serve to exclude charged species. For neutral solutes, the rate of diffusion inside the bead decreases with increase in the extent of cross-linking. Thus, the elution peak and elution curve will elute at progressively lower elution volumes as the degree of cross-linking of the resin is increased. In general, both the rigidity/mechanical strength and the extent of swelling of the polymeric beads are enhanced upon increasing the degree of resin cross-linking. Low degrees of cross-linking of a resin may also lead to an increased pressure drop across the column. Thus, porosity and the degree of resin cross-linking have an important role in dictating the performance of chromatographic systems.

Finally, operating process parameters such as flow velocity, solute concentration, and temperature in the column all need to be optimized in order to obtain the required degree of separation. The sharpness of the elution curve depends directly on the flow rate and increases as the flow rate decreases. However, throughput is compromised as flow rate decreases. Increased solute concentration shifts the peak of the elution profile to higher values of elution volume, without affecting the onset of the profile when the solute first emerges out from the column. Increase of temperature is generally found to have an adverse effect on separation factors. However, operation at increased column temperature affords other advantages: it decreases solution viscosity, making further processing easier, and it also inhibits microbial growth on the resin beads and minimizes the effects of fouling. These points illustrate the complex interplay of a large number of factors in determining the efficiency of a chromatographic separation process.

Typically, optimization and lab trials are carried out on chromatographic beds with $L/D = 10$. It is very important to test the stability of the target biomolecule in different equilibration buffers. The stability should be tested at least for a time duration corresponding to the residence time of the biomolecule in the column. In particular, it is necessary to give great importance to the buffer pH, which is a key governing factor for the stability of proteins in solution. The resin/gel should also be allowed to completely swell in an excess of equilibration buffer for a sufficient length of time. The process needs to be repeated with fresh buffer several times. Several modern chromatographic materials that are tailor-made for the purification of biopharmaceuticals are sold preswollen; hence, there is no need for swelling of the gel in such cases. The feed should be dissolved in the same equilibration buffer and injected into the column at constant low flow rates such that equilibrium is attained for the solutes during adsorption. Typically, cationic elution buffers (e.g., tris-HCl) should be used with anionic-exchange resins and anionic buffers (e.g., phosphate) with cationic-exchange resins. Isocratic elution is commonly used in analytical chromatography. Gradient elution that is either step-wise or linear in the ionic strength of eluents is also commonly used for proteins. The elution volume as a function of log (molecular weight) of solutes for most gels/resins displays two regions: a linear region and an asymptotic region, with the former corresponding to the range in which the resin should be used for optimal results.

13.5 Scale-Up

By the procedures discussed above, chromatographic purifications on an analytical scale can be readily optimized for a particular target biomolecule or product. However, separations on a preparative and industrial scale also need to consider (and preferably maximize) throughput, which is not a consideration on the analytical scale. This maximization has to be achieved without sacrificing resolution, which is the most difficult challenge for the scale-up process. This is because the operating parameters in a chromatographic separation generally have opposing effects on resolution and throughput. For instance, resolution varies directly with column length L, but throughput varies inversely as $1/L$. Similarly, wall effects lead to a deleterious effect on resolution as column diameter is increased, especially at small values of the diameter, while throughput scales as D^2. Finally, sample volume and flow rate affect resolution in an inverse fashion, while throughput is linearly related to these process parameters. Hence, scale-up of chromatography columns involves selection of an optimal process, selection of an optimal bed height and diameter, and selection of the detailed column design for the given constraints of space, materials, cost, and so on.

For gel filtration and certain other chromatographic techniques, it is well established that an appropriate criterion for upscaling (as opposed to scale-up; see below) is the maintenance of dynamic similarity. This implies that both the small and the large columns must maintain the same values of both L/D and the Reynolds number for tubular flow, $\mathrm{Re} = \rho v D/\mu$. An additional thumb rule is that the ratio of the volume of protein solution to that of the volume of the gel or adsorbent be maintained the same in both the small and the upscaled column. These criteria permit an unequivocal choice to be made for the dimensions of the column and the flow rate to be used in the process. Such procedures ensure that the same or similar elution profiles are obtained in the upscaling process. In practice, for scale-up, the thumb rule for column dimensions is to increase the column diameter and maintain the bed height. Increasing the diameter of the column during scale-up will, of course, affect process parameters, such as volumetric and linear flow rates. Such a strategy of increasing D while maintaining L during scale-up (as opposed to the maintenance of L/D) has the major advantage that it avoids problems of resolution and pressure drop.

13.6 Industrial Applications

A large number of industrial chromatographic applications have been described in various patents, and it is beyond the scope of this chapter that has primarily addressed fundamental issues to cover this vast subject. An example is the purification of molasses to obtain sugar. Molasses is a by-product in the processing of cane or beet sugar and possesses high sugar content (approximately 50% of the total dissolved solids). It has proved very difficult to reduce this sugar content in molasses by various treatment processes without simultaneously also obtaining unmanageably high waste. The nonsugar portion of molasses contains both inorganic and organic compounds and these impurities are different depending on the source of the molasses (e.g., cane or beet). The basic principle of the chromatographic separation is that upon pumping of the molasses into the column, the ionized ash is readily excluded from the resin and the sugar is adsorbed on to the beads. The distribution coefficients vary depending on whether the Na^+, K^+, or Ca^{2+} forms of the resin are used. Such ion exclusion processes for recovering sucrose have been documented and engineering analysis has been made. Ion exclusion under optimum conditions has been shown to lead to a sucrose recovery of over 90% and a purity of 90% [8–10].

Various strategies for the above purification have been examined. These include the separation of molasses into a nonsugar fraction and a total sugar fraction that contains glucose, fructose, and sucrose, as well as the first inversion of the sucrose in molasses into glucose and fructose followed by the purification of molasses into a nonsugar fraction and a fraction containing only glucose and fructose. It has been shown that inverting the molasses before separation leads to increased capacity of the chromatographic column without corresponding decrease in purity of the product. By continuous chromatography and other innovations, a sucrose recovery of over 90% and a purity of 99% have been routinely obtained and several tonnes of molasses per m^3 resin per day have been treated on the industrial scale.

Several other examples with flowsheets have been covered in Chapter 12 on product recovery.

13.7
Some Alternatives to Column Chromatographic Techniques

In a substantial amount of research, alternatives to chromatography have also been explored for the purification of biological macromolecules. Since fermentation broths are very dilute systems, with typical product concentrations as low as 1–100 g/l, it has been especially important to eliminate energy-intensive thermal separation processes. Electrodialysis and electrophoresis have been offered as promising alternatives, although it must be emphasized that these techniques are still in the research stage and have not been widely commercialized for industrial-scale operation unlike the column chromatographic operations discussed above. One example of the use of electrochemically enhanced filtration is shown in Section 10.5, p. 348.

13.8
Electrodialysis

In electrodialysis, a diffusive transfer of an ionic solute across the membrane is driven by an applied electric field. The membranes are arranged parallel to each other in a typical plate-and-frame type of set up [11,12]. This geometry allows the application of the electrical driving force in an efficient way. The ability of the solute to undergo ionization is a requirement for the successful application of this technique. Basically, cation and anion membranes are arranged alternately between two electrodes and an electrolyte solution is fed into alternate chambers. The application of an electrical field causes the cations to move toward the cathode and the anions to migrate toward the anode. However, the cation/anion cannot diffuse beyond one chamber because its diffusion is impeded by a membrane that is impermeable to that ion. Therefore, one flow stream is depleted of ionic solute, while an adjacent stream is enriched in that solute. The membranes separating the chambers need to have a large enough ion-exchange capacity and pores of small size (approximately 0.3 nm) that can repel oppositely charged ions. The separation of the ionic solute from the feed stream depends on the solute ionization potential, the feed flow rate, and the electrical field strength. The most important parameters that determine the separation efficiency of the electrodialysis process are the electrical resistance of the ion-exchange membranes and its selectivity. Ideally, the resistance of the membranes per unit area to saline water should be as low as possible (typically less than 30 ohm/cm^2). The selectivity is measured by the percent increase in flux of an ion due to the selective membrane compared to the flux in its absence and should be as large as possible.

However, in practice for separation applications from fermentation broths, the separation efficiency is limited by the low solute ionization potential and the low concentration of the solute. Even a high ionization potential salt such as NaCl should be present at least at a concentration of 300–400 ppm for electrodialysis operations. For more dilute systems, an excessive voltage gradient and high-power input does not permit satisfactory operation of the electrodialysis system.

In Chapter 10, up-concentration of a weak solution of sodium succinate illustrates application of electrodialysis on an industrial scale. Finally, undissociated succinic acid is produced using a bipolar electrodialysis membrane to exchange Na^+ with H^+.

13.9 Electrophoresis

In electrophoresis, charged particles migrate due to the application of an electrical field. Various media have been used for biochemical electrophoretic separations ranging from free-flow aqueous solutions or suspensions to gels of varying but controlled pore size. In electrophoresis, separations occur due to the differential migration of various component solutes under the influence of electrical forces. In addition, pH or ionic strength gradients, cellulose acetate membranes, or sieving effects of gels add additional elements that enhance the power of the electrophoresis technique. These have led to the development of several techniques such zone electrophoresis, electrophoretic isoelectric focusing, isotachophoresis, field-step electrophoresis, and capillary electrophoresis [1,13–15]. Electrodialysis can also be viewed as a special electrophoretic technique in which charged membranes constitute the additional discontinuity elements. For proteins, the ionization of various functional groups cause the macromolecule to possess a net negative charge at basic pH values above 7 and a net positive charge at acidic pH values. These differences are translated into differences in electrophoretic mobility (i.e., the velocity of solute migration per unit electrical field strength) during the separation. At the isoelectric point of the protein, the molecule has zero electrophoretic mobility.

The concept of the electrical double layer is important for a fundamental understanding of all electrophoretic techniques. The electrical double layer structure arises from the fixed charges on the surface of the macromolecule along with the relatively mobile counterions in the liquid environment around the molecule. Even though the net charge of the double layer is zero, there exists a nonuniform spatial distribution of charges in a field that gives rise to an electrical potential. The thickness of the electrical double layer is normally characterized by the inverse of the Debye–Hückel constant, κ. The mobile counterions in the immediate vicinity of the macromolecule are bound tightly to it and are able to migrate with it during electrophoresis. This generates a surface of shear that is larger than the size of the macromolecule itself. The electrical potential at this shear plane/boundary determines the electrophoretic mobility of the molecule and is termed the zeta potential. The zeta potential is an elusive experimental variable that cannot be measured simply and directly but has to be calculated based on theories of the electrical double layer. Such theories can be complex, having to take into account electrophoretic migration and attraction, electrophoretic retardation effects, frictional effects in the solvent, and finally relaxation effects.

It is well known that several electrophoretic techniques with and without gels are available for analytical applications; however, it is a lesser known fact that electrophoretic techniques have also been employed for carrying out separations on a preparative scale. We wish to emphasize this point in this chapter, that is, these techniques are not for use on an analytical scale only. Thus, zone electrophoresis and isoelectric focusing are two methods that are also extremely useful for preparative-scale separations of proteins, cells, and cell organelles. Zone electrophoresis is characterized by the complete separation of charged particles and solutes into separate zones. The technique can be readily adapted to carry out separations in free solutions, and it is possible to operate it in a continuous electrophoresis mode, a major advantage over other separation techniques available for preparative-scale purification. However, earlier separations with this technique were performed using filter paper in an electrophoretic cell. Cellulose acetate membranes are now commonly used in such cells as the support medium. High density gels with specific pore sizes are also commonly used as the resolving medium to separate proteins. The technique offers very good resolution owing to the uniform (or controlled gradientl) porosity of the medium.

In isoelectric focusing, a pH gradient is superimposed on a medium that exhibits a density gradient along the column. Approximately, 50% sucrose is generally the preferred medium for such a density gradient. After the density gradient has been established, a mixture of ampholytes that have a wide range of isoelectric points is injected into the chamber and an electrical field is applied, leading to the formation of a pH gradient. The most acidic component of the ampholytes stays near the anodic end of the column, while the most basic among the ampholytes migrates to the cathode, and the other components cover the range in between, depending on the value of their isoelectric points. The protein mixture to be separated is applied to the electrophoresis column and undergoes fractionation after which the column is emptied into separate fractions. The protein fractions are then separated from the sucrose and the ampholytes by dialysis or ultrafiltration. Mixtures of various ampholytes covering various pH ranges (typically 3–10) are available commercially in a low molecular weight range (typically 300–600) such that separation of proteins from them is easily achieved after the fractionation step.

It is possible to establish an overall unity in the mechanism of protein separation on a preparative scale by free-flow zone electrophoresis and ion-exchange chromatography. This can be done by comparing the apparent electrophoretic

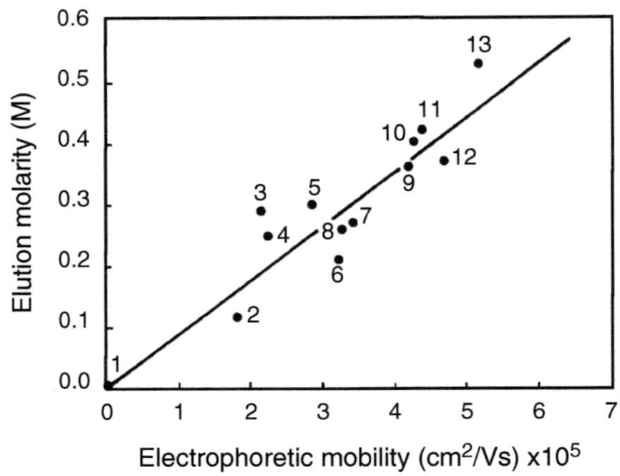

Figure 13.2 Correlation between the electrophoretic mobility in preparative continuous free-flow zone electrophoresis and the elution molarity in ion-exchange chromatography for various proteins. The electrophoresis was carried out in tris-acetate buffer; pH varied from 6.8 to 7.9 depending on the protein system. During chromatography, the pH of the tris-acetate elution buffer was 7.5. (1) Myoglobin from horse heart. (2) Formate dehydrogenase from *Candida boidinii.* (3) Catalase from bovine liver. (4) D-Hic dehydrogenase. (5) Catalase. (6) Formaldehyde dehydrogenase from *C. boidinii.* (7) Bovine serum albumin. (8) Ovalbumin from chicken egg. (9) Ferritin from horse spleen. (10) β-galactosidase from *Escherichia coli.* (11) Bovine thyroglobulin. (12) Methanol oxidase from *C. boidinii.* (13) L-Hic dehydrogenase.

mobility measured in a continuous free-flow zone electrophoresis experiment with the elution molarity measured in ion-exchange chromatography for various proteins. The results of such an experiment are depicted in Figure 13.2. The linear correlation observed between the apparent electrophoretic mobility and the elution molarity confirms the primary role of net surface charge in separations by the two techniques. The deviations from perfect linearity in Figure 13.2 occur chiefly due to the effect of the buffer pH, which varied between 6.8 and 7.9 in the continuous electrophoresis experiments but was kept fixed at a pH value of 7.5 in the ion-exchange chromatography experiments. However, the unity may not hold over the entire range of ionic strength (from millimolar to molar salt concentrations). At low salt concentrations, the electrophoretic mobility is generally very sensitive to variations in salt concentration, while the elution molarity is found to vary only slightly. By contrast, the high salt limit displays the opposite trend. Hence, continuous free-flow electrophoresis techniques are in many cases superior, more sensitive, and higher in resolution than ion-exchange chromatographic techniques in the low salt range. Proteins with differences greater than an elution molarity of 0.05 M in ion-exchange chromatography can be readily separated by continuous zone electrophoresis on a preparative scale. This corresponds to a preparative separation of proteins that differ in their apparent electrophoretic mobilities by only 0.70×10^{-5} $cm^2/(Vs)$ [16].

13.10 Conclusions

This chapter has focused on certain fundamental aspects of purification of biomolecules by column chromatography and electrophoresis. Several other important purification techniques, both old and new, could not be covered due to the vastness of the topic. Rather than deal with each of these purification tools and methods, which was an impossible task, it was decided to summarize aspects of the most widely used technique in the field for separations on a preparative and industrial scale (column chromatography) and the adaptation of the most prevalent method for separations on an analytical scale (electrophoresis) in the biological sciences and engineering to the preparative scale. Certain interesting correlations between the two techniques on a preparative scale have also been discussed. The latest advances in chromatographic techniques are covered in several other sources [17–26]. However, the choice of the techniques does not mean that other purification tools and methods are any less important. Advances continue to be made in every subfield of bioseparation, and each technology continues to have its own applications, to offer its own unique advantages in purification of various biological systems, and to present its own challenges.

References

1 Jansson, J.-C. (2011) *Protein Purification: Principles, High Resolution Methods, and Applications*, John Wiley & Sons, Inc., Hoboken, NJ, USA.

2 Scopes, R.K. (1994) *Protein Purification: Principles and Practice*, 3rd edn, Springer, New York.

3 Endo, I., Nagamune, T., Katoh, S., and Yonemoto, T. (eds.) (2000) *Bioseparation Engineering*, Elsevier Science, Amsterdam.

4 Ghosh, R. (2006) *Principles of Bioseparations Engineering*, World Scientific, Singapore.

5 Shukla, A.A., Etzel, M.R., and Gadam, S. (eds.) (2007) *Process Scale Bioseparations for the Biopharmaceutical Industry*, CRC Press, Boca Raton, FL, USA.

6 Martin, A.J.P. and Synge, R.L.M. (1941) A new form of chromatogram employing two liquid phases. 1. A theory of chromatography. 2. Application to the micro determination of the higher mono amino acids in proteins. *Biochem. J.*, **35**, 1358–1368.

7 Dean, J.A. (1969) *Chemical Separation Methods*, Van Nostrand, New York, p. 60.

8 Saska, M. and Lancrenon, X. (1994) Applications of continuous chromatographic separation in the sugar industry. III. Desugarization of canemolasses. *Int. Sugar J.*, **102**, 403–410.

9 Silva, V.M.T.M., Gomes, P.S., and Rodrigues, A.E. (2012) Use of ion exchange resins in continuous chromatography for sugar processing, in *Ion Exchange Technology II: Applications* (ed. M. Luqman Inamuddin), Springer, New York, pp. 109–135.

10 Charton, F. and Nicoud, R.M. (1995) Complete design of a simulated moving bed. *J. Chromatogr. A*, **702**, 97–112.

11 Lacey, R.E. (1979) Dialysis and electrodialysis, in *Handbook of Separation Techniques for Chemical Engineers* (ed. P.A. Schweitzer), McGraw-Hill, New York, section 1.14, pp. 1–449.

12 Tanaka, Y. (2013) *Ion Exchange Membrane Electrodialysis: Fundamentals, Desalination, Separation*, Nova Science, Hauppauge, NY, USA.

13 Bier, M. (1959) *Electrophoresis*, vol. **I**, Academic, New York.

14 Shaw, D.J. (1969) *Electrophoresis*, Academic Press, New York.

15 Guttman, A. (2014) *Capillary Gel Electrophoresis and Related Microseparation Techniques*, Elsevier Science, Amsterdam.

16 Nath, S., Schütte, H., Hustedt, H., and Deckwer, W.-D. (1993) Application of continuous zone electrophoresis to preparative separation of proteins. *Biotechnol. Bioeng.*, **42**, 829–835.

17 Hanke, A.T. and Ottens, M. (2014) Purifying biopharmaceuticals. Knowledge-based chromatographic process development. *Trends Biotechnol.*, **32**, 210–220.

18 Nfor, B.K., Verhaert, P.D.E.M., van der Wielen, L.A.M., Hubbuch, J., and Ottens, M. (2009) Rational and systematic protein purification process development: the next generation. *Trends Biotechnol.*, **27**, 673–679.

19 Titchener-Hooker, N.J., Dunnill, P., and Hoare, M. (2008) Micro biochemical engineering to accelerate the design of industrial-scale downstream processes for biopharmaceutical proteins. *Biotechnol. Bioeng.*, **100**, 473–487.

20 Jungbauer, A. (2013) Continuous downstream processing of biopharmaceuticals. *Trends Biotechnol.*, **31**, 479–492.

21 Vogel, J.H., Nguyen, H., Giovannini, R., Ignowski, J., Garger, S., Salgotra, A., and Tom, J. (2012) A new large-scale manufacturing platform for complex biopharmaceuticals. *Biotechnol. Bioeng.*, **109**, 3049–3058.

22 Fekete, S., Beck, A., Veuthey, J.-L., and Guillarme, D. (2014) Theory and practice of size exclusion chromatography for the analysis of protein aggregates. *J. Pharm. Biomed. Anal.*, **101**, 161–173.

23 Sandra, K., Vandenheede, I., and Sandra, P. (2014) Modern chromatographic and mass spectrometric techniques for protein biopharmaceutical characterization. *J. Chromatogr. A*, **1335**, 81–103.

24 Marichal-Gallardo, P.A. and Álvarez, M.M. (2012) State-of-the-art in downstream processing of monoclonal antibodies: process trends in design and validation. *Biotechnol. Prog.*, **28**, 899–916.

25 Anspach, F.B., Curbelo, D., Hartmann, R., Garke, G., and Deckwer, W.-D. (1999) Expanded-bed chromatography in primary protein purification. *J. Chromatogr. A*, **865**, 129–144.

26 Guiochon, G. and Beaver, L.A. (2011) Separation science is the key to successful biopharmaceuticals. *J. Chromatogr. A*, **1218**, 8836–8858.

Part Four
Process Development, Management and Control

Fundamental Bioengineering, First Edition. Edited by John Villadsen.

14
Real-Time Measurement and Monitoring of Bioprocesses

Bernhard Sonnleitner

Summary

We need to substantially improve our understanding of production biosystems and processes in order to ensure a desired quality performance of (bio-)products. An absolute prerequisite for a better understanding is information and knowledge. This must be created and extended by timely quantitative analysis of the processes, and all this requires process analytical tools, data evaluation algorithms, and adequate models as well as clever process control. The state-of-routine monitoring items established in the bioprocess industry and some important state-of-the-art methods are described and potential pitfalls discussed. Physical and chemical variables such as temperature, pressure, weight, volume, flow rates, pH, redox potential, gas partial pressures in the liquid and molar fractions in the gas phase, infrared spectral analysis of the liquid phase, and calorimetry over an entire reactor are important aspects. Both classical and optical sensor versions are addressed. Biomass and bioactivity monitoring by turbidity, permittivity, *in situ* microscopy, and fluorescence is analyzed. Instrumental analytical tools and interfaces to bioprocesses are explained, such as chromatographic methods, mass spectrometry, flow and sequential injection analyses, field-flow fractionation, and flow cytometry.

14.1
Introduction

What we can measure or monitor in bioprocesses are the variables (or state variables), such as concentrations of reactants, fluxes of fluids, purities of products, or, simply, the pH value. What we mostly desire are the parameters of the biological system in the process. These are the properties of the system that determine the timely development of the state variables. So, besides *m*easurement and *m*onitoring, we need to include a considerable *m*odeling and *c*alulation effort in order to transform the raw data into process information.

Fundamental Bioengineering, First Edition. Edited by John Villadsen.

Biological systems, as they are used in industrial processes, are quite complex, especially from an analytical perspective: there are many (probably not only essential) components of varying molecular sizes and concentrations in a dynamically changing matrix. We usually need to apply appropriate interfaces between the process and the monitoring devices in order to avoid compromising the sterile barrier between process and environment. Also, the critical quality attributes of a process may change quite rapidly. Therefore, we should aim at substituting sampling and transport of the samples to a dedicated laboratory where analysis takes place with *in situ* sensors and automated instrumental analytical methods and instruments – the so-called process analyzers – in order to produce process-relevant data almost in real time and, if possible, continuously and noninvasively. Then, we also need the models and algorithms to work in real time for transformation of the raw data into process information we should exploit for closed loop control of the processes. This must include real-time event detection that can be realized only if the monitoring is performed in the necessary high-frequency domain. Further, we must collect, analyze, structure, and organize the process information (ideally on exactly one platform, namely, the process computer, which is beneficial) in such a way that we can develop expert knowledge. Among others, this idea is strongly supported by authorities, for instance, in the guidances for industry: process analytical technology (PAT) or process validation [1–6].

It is a standard to monitor "integral variables," which means that we try to quantify a process property as the mean value over all members of a cell population, for instance, the cell mass concentration. However, there are a few techniques at hand that permit the analysis on a single-cell level, for instance, flow cytometry or *in situ* microscopy. With these techniques, we can characterize the distribution of a population. A very heterogeneous distribution may not be desirable for a production process, for instance, when only part of the population is productive and the rest is not. Sometimes, we want to acquire direct measures, for instance, the concentration of the limiting substrate in a fed-batch process. However, there are also other indirect useful measures, for instance, the optical density or permittivity of a suspension. Yet, such measurements require model(s) to derive what we are really interested in, in this particular case, the (live) biomass concentration or activity. Models for this purpose are usually quite simple, for example, predicting a direct proportionality, but they must be verified and parameterized experimentally from case to case.

At the small laboratory scale, we usually ignore the nonideal mixing behavior of a bioreactor, that is, the monitored properties of a biosuspension do not vary with the space coordinates. Yet, if the flow behavior becomes non-Newtonian and, certainly, on a larger scale, nonhomogeneity in the reactor assumes critical importance, this should be accounted for by installing more sensors in different parts of the reactor. However, the installation of many (more) sensors is typically interpreted as an increasing risk for contamination and therefore not very likely to be favored at the production scale.

Example 14.1 Performance of a temperature controller in a high-density mycelial culture

Fungal cultures can form mycelial or pellet-type morphology. The latter is preferred in industrial processes because the biosuspension remains less viscous and oxygen transfer is simpler to achieve. However, sometimes mycelial morphology is desired, and the example shows the deterioration of the temperature controller over 100 h of fermentation. The suspension became highly non-Newtonian during this time and the single temperature sensor was trapped in a slowly or poorly mixed space, although the 16 l reactor with an *H*/*D* ratio of approximately 3, equipped with three Rushton turbines powered by a 1 kW electromotor "tried its best." This can be visually estimated from the photos of the viewing glass that were taken at approximately 5 min time intervals showing more or less stationary hydrodynamic behavior. The controller, which worked perfectly at the beginning, was not reparametrized and it was unable to cope with the fact that a single sensor in such an environment can deliver only space-restricted information; the controller, however, is thought to control the entire reactor. The space of measurement (side port through wall in lower third of reactor) and the space of effective heat transfer (entire bottom of reactor) are hardly impacted by the heat exchange.

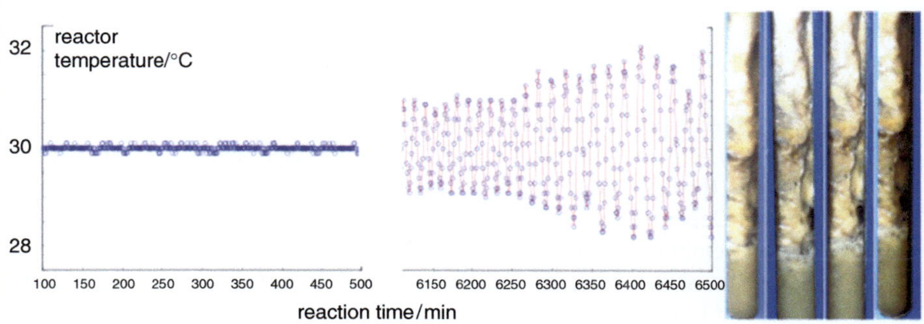

Deteriorating performance of a temperature control loop.

At the same time, we also need to consider the representativity of sample aliquots removed and the timing of sample preparation. It has been shown that the pattern of concentration gradients varies, for instance, with the point of substrate addition [7,8]. Reversely, this holds true also for the position of a sampling port. If we remove a sample aliquot, independent of whether this is done manually or automatically, the cells will not stop their activities per se and continue to convert substrates to products and by-products. We must pay attention to this and implement elements to slow down the metabolism or even inactivate cells appropriately and quickly because, otherwise, even an error-free analysis could be prone to dramatic falsification: this is really important in the case of limiting medium components or metabolites with a high turnover rate.

Example 14.2 Determination of glucose concentration in a glucose-limited culture (fed-batch or chemostat)

A miniaturized microfiltration module is used to separate the cells from the supernatant. In this way, a few μl of cell free permeate per minute is continuously prepared for automatic glucose analysis. The mean residence time of the suspension in the tube between sampling port and filter may be 4 s. What is the concentration of glucose at the entry point of the tube if the (error-free) analyzer displays 50 mg l^{-1}, given that the cell density is 40 g l^{-1} as determined by other measurements and the specific sugar uptake rate r_S equals −3.0 g $g^{-1}\,h^{-1}$ (and is constant, that is, zero-order reaction with respect to glucose)?

$$q_S = r_S \cdot x = -3 \cdot 40 = -120\,\text{g}\,\text{l}^{-1}\,\text{h}^{-1} = -33.3\,\text{mg}\,\text{l}^{-1}\,\text{s}^{-1}.$$

Assumption: steady state in the tube and the biomass concentration does not change significantly, then:

$$c_t = c_0 + \int q_S\,\text{d}t \text{ that simplifies to}$$

$$c_t = c_0 + q_S \cdot t, \text{ because } q_S \text{ remains constant.}$$

After rearranging and solving for the initial condition, c_0, we get

$$c_0 = c_t - q_S \cdot t = 50 + 33.3 \cdot 4 = 183\,\text{mg}\,\text{l}^{-1}.$$

The error due to neglecting the sample preparation is, therefore, 133 mg l^{-1} or 266% with respect to the display of the glucose analyzer!

Other aspects to consider are the response times, dead times, and fouling of *in situ* sensors. In several cases, sensors are used that are covered with a membrane through which the analyte must diffuse (prior to analysis within the sensor). Depending on the nature of the membrane material and the thickness of the membrane, we must expect a relevant response delay.

Exhaust gas analyzers are placed along the exit line of the exhaust gas and should have some equipment installed that allows removal of all kinds of liquid. It takes some time for the gas to "travel" to the actual measuring device that causes a dead time.

Membranes and diaphragms are not "perfect" materials, and during fermentation deposits of precipitates or biofilms can build up on the membrane surface, thus separating the membrane from the biosuspension (i.e., the measuring solution). This may just delay the response time, but it may also create new selectivity.

Altogether, there are many pitfalls to be considered carefully.

14.2 Variables that should be Known

What does a process engineer need to know? The active players in bioprocesses are certainly the biocatalysts, most often modeled as the cell mass concentraion,

x, the feed concentrations, of which most often only the so-called limiting substrate concentration, s_i, is of importance, and there are the product concentrations, p_j, including concentrations of by-products. Yet, dead cells do not effectively contribute to a productive process. It is obviously important to know the viability and the activity of the biocatalyst. The problem is there are no good (innate) human senses for those variables!

Many industrial productions use aerobic organisms; if they are strictly aerobic, the oxygen supply rate will probably limit the entire reaction rate and if they are facultative aerobes, they have the possibility to switch their metabolism to another type that is less productive or even unproductive with respect to the desired product. Consequently, we need to know the actual availability of oxygen for the populations and – also – actively manipulate the oxygen transfer (rate) from the gas into the liquid phase accordingly. Strictly, anaerobic organisms, on the other hand, will not grow and/or not produce if oxygen is not excluded or, at least, not sufficiently scavenged. Carbon dioxide is not only a "loss of carbon" but it certainly may affect the activity of the cells; we need to know its presence quantitatively and remove it by appropriate engineering means.

The activity of biocatalysts depends also on environmental variables such as temperature and pH or ionic strength. Therefore, they must be known and be controlled, as well. In additon, operational variables such as feed rate, dilution rate, or recirculation rate have a decisive effect on the productivity of a process. Therefore, also these operation variables must be known and be controlled.

Part of the addressed variables can be relatively easily acquired because there are plenty of well-established and reliable sensors available on the market. The variables are, since long, regarded as "necessary," and therefore implemented as an industrial standard. The "rest of" the variables are (much) more tedious or difficult to monitor or estimate. There are no plug-and-play solutions available on the market and they are mostly omitted in industrial processes. The following sections are structured according to this situation. More details can be found in [9–15].

14.3 Variables Easily Accessible and Standard

Among the physical variables *always* measured (somewhere) in the bioreactor are only *temperature* and *pressure*. The devices commercially available are generally very robust and reliable but not simple to (re-)calibrate accurately, although they are generally very precise. Proper temperature compensation of pressure sensors and proper compensation for long cabling distances between temperature sensor and amplifier is definitely only a question of costs. The sensors are mounted on housings made of stainless steel and usually do not cause contamination problems; however, their static seals (mostly O-rings) to the reactor ports deserve repetitive and, desirably, preventive maintenance work.

Other physical variables such as *flow rates of liquids* (e.g., ultrasound, Coriolis, magneto-inductive principles) are mounted either on pipes or tubes connecting the reactor with some other (supply or buffer) vessels, yet within the axenic space. Probable contamination problems are not the sensors themselves but, as above, the seals between pipes and sensors.

Flow rates of gas(es) are usually measured in the supply line because the fresh gas is least contaminated (chemically and hygienically, mostly dry and free of oil) and the devices (often thermoanemometers) mounted before the sterile filter. Therefore, they cannot create a contamination risk.

Generally, there are many more measuring principles implemented as commercially available products that deliver a volume-flux signal, rather than those delivering a mass-flux signal. The latter type is preferred because engineers often calculate mass balances and the signals are directly useful in such cases. It is also important to be aware of the interferences such as electrical conductivity of the medium, gas or foam contents of the medium, particles in the medium, for which the different principles are differently sensitive. Professional calibration is critical in all cases. The user should check carefully for what he needs or wants (in the design phase) or what he has installed (in an existing plant).

Mass fluxes can also be derived from mass measurements: the weight gain or loss of a vessel per time unit is the mass flux in or out, provided there is only one flux active at a time. Vessels can be easily mounted on scales or balances, which are outside the axenic space, but the challenging problem is that all connections to the surroundings (piping, tubing, electrical cables, or other accessories as well as wind exposure) must be strictly force-compensated. The accuracy of the differentiated weight signal depends on the time intervals and the resolution of the scale/balance, and this signal may be quite noisy. However, we often need/use the integrals of fluxes and, in this case, this is what we measure directly. This mass signal also gives the mass-filling degree of a vessel and represents its "working volume" totally independent of swirling surfaces, gas hold-up, or foam layers on top of solutions or suspensions. Other measuring principles suffer extensively from such interferences.

What remains among the interesting physical variables is the *power input* into a fermenter. The most prominent information would be the local energy dissipation gradients in the biosuspension (see Chapter 11.2) because this would tell more about mechanical or shear stress of the cells. However, this is usually substituted by measuring the electrical power consumption of the driving motor (for economic reasons) or the revolution speed of the stirrer shaft (for simplicity and cost reasons) giving only a very nebulous estimate of what we are really interested in. These measurements are all outside the axenic space and can never compromise the sterile barrier.

The performance of biological agents does depend not only on the physical but certainly also on the chemical environment. Chemistry poses more analytical challenges and this is most probably why there are not many "simple" sensors for the chemical environment available. There are only a few commercialized types available:

- *Sensors selective for ions*: most prominent for protons – the *pH electrodes.*
- *Sensors for dissolved gases*: most prominent for the partial pressures of O_2 and CO_2 over a solution – the *p*O_2 and *p*CO_2 *electrodes.*
- *Sensors for dissolved redox couples*: the *redox electrodes.*

The other, probably much more decisive, chemical species require more effort than just an *in situ* electrode.

The classical versions of those electrodes are electrochemical types, part of them exploiting the potentiometric principle, part of them the amperometric principle. The newer electrodes employ optical principles, do not consume their analytes, and are not prone to interferences by electromagnetic noise, an aspect that is very important in an industrial production environment. Only recently, electrochemical sensors are commercially available in which all the critical electronics are moved into the head of the single-rod electrode. This minimizes or even eliminates electromagnetic influences from the closer environment, and the communication takes place via digital (ethernet, other buses, and even wireless) rather than by analog signals that are tedious to cable.

pH and redox potential are potentiometrically measured, that is, a minute potential difference (voltage) between glass membrane or noble metal, exposed to the measuring solution, and a reference electrode is determined, therefore: "potentiometric type." This requires both a (very) high input impedance of the amplifiers and a reference potential of the reference electrode with respect to the measuring solution. The latter is accomplished in single-rod electrodes by electrically connecting the reference electrolyte with the measuring solution via a diaphragm. Such a diaphragm can quickly deteriorate and is, therefore, left out in newer electrodes that require solidification of the reference electrolyte with a hydrogel. Instead of the glass membrane, the gate electrode of a field effect transistor (FET) can be exposed to the measuring solution. Its potential will then steer the current between source and gate electrode of the FET. However, the reference electrode and diaphragm are still needed. The advantage of the pH-FETs is that they can be made of unbreakable material – an immense gain for the food industry. A great disadvantage, however, is the substantial drift of the electrodes during steam sterilization.

Electrodes for the partial pressures of dissolved gases must be made selective for gases – actually for almost all volatiles – by separating them from the measuring solution, using a gas-permeable membrane. The gas must dissolve in the membrane, diffuse through the membrane, and then can be analyzed simply. This means that these electrodes work under diffusion-controlled condition, and the membrane materials (usually silicon and/or Teflon) must be appropriately selected. The pO_2 electrode "finds" only the oxygen by its construction as a polarographic electrode operated at a polarization voltage of around 650 mV: the O_2 is then reduced at the Pt-cathode whereas the Ag-anode dissolves, thus liberating four electrons per O_2 molecule. It is exactly this current that is (amplified and) measured. Therefore, the sensor is of the "amperometric type." Under these "inner conditions," CO_2 is inert and does not react, although it also enters the inner electrode through the membrane. Similarly, the pCO_2 electrode "finds" its CO_2 as an acidifying agent in the inner electrolyte, and the pH change can be simply detected using a pH electrode for which O_2 is inert. However, any other acidic or alkaline volatiles that can pass through the membrane will, of course, also contribute to a pH change and so interfere with the pCO_2 determination. Both electrodes are membrane covered and consume their analyte and internals (Ag anode and pH buffer, respectively) that requires frequent maintenance work and calibration.

Alternative electrodes based on optical principles are commercially available now. They have the advantage that they do not consume analyte or internals and do not require inner electrolytes. The optical sensors generally respond more quickly than the electrochemical ones. A dye or fluorophore is immobilized in a very thin patch of hydrophobic carrier material on top of an optical window and directly exposed to the measuring solution. The other side of the window is connected to optical fibers coming from an illumination source or going back to an optical detector. If the dye is a pH indicator (or a mixture of more than one), the electrode can sense the pH. If the fluorophore is oxygen sensitive, the electrode can sense pO_2. Metal organic complexes of Ru or Pt, for instance, fluoresce in the absence of O_2, while their fluorescence is quenched by O_2 [12]. Consequently, the active signal is maximal at minimal pO_2, which is opposite to the amperometric electrode. Hence, both sensor types are complementary and have their best signal/noise ratio at the opposite ends of the measuring ranges. Using both sensor types at the same time is definitely a performance gain!

The oxygen needed (and, of course, other gases to be converted) can be consumed by the cells only if dissolved in the aqueous liquid phase; however, we provide oxygen via the gas phase. A major function of bioreactors is to facilitate the transfer from the gas into the liquid phase. The reverse holds true for all produced gases such as CO_2: they need finally to be removed via the gas phase. One aim of a process engineer is to provide just the right transfer capacity. This can be a considerable challenge because gases are generally very poorly soluble in water. According to the mass balances of individual gas components over the gaseous phase, one can calculate the transfer rate if the gas-mass flow rate and the gas-phase composition is known (see Chapter 2, Eq. 2.5). Therefore, it is wise to analyze the gas composition. Many industrially relevant biosystems need O_2 and produce CO_2, while a few others may consume alkanes and H_2 or produce CH_4. There are many gas analyzers commercially available, based on various measuring principles. O_2, for instance, is a paramagnetic substance and this property can be quantified. Metal oxide sensors need to be heated and are, therefore, not favored. CO_2 and CH_4 are green-house gases and their infrared absorption can be quantified. The gas composition also affects the sound velocity in a gas and can be determined with electro- or magnetoacoustic means [16]. Any gas can be determined using a mass spectrometer [17,18]. Care must be taken to protect the analyzers from foam or cells and to exclude water (vapor), especially if exhaust gas is measured. It is also important to understand that a molecule density per volume, n_i/V, is the primary result, and not, for example, a volume fraction y_i, though most instrument displays are labeled with "%"; n = total number of molecules, n_i = number of molecules of interest (e.g., of O_2), y_i = molar fraction of the molecules of interest, and V = volume of measuring cuvette. Further, $n/V = p/(RT)$, and $n_i/V = y_i p/(RT)$, which means that the primary result depends on total pressure (p) and temperature (T). Many instruments are temperature controlled but ventilated toward ambient atmosphere, which makes the result dependent on atmospheric pressure changes.

Caveat: Probably, some calculations may be necessary to derive the percentage or molar fractions of the individual components in the gas phase from the primary n_i/V results.

Another caveat is important in case of very low gas flow rates: the analyzers may consume more gas than what is provided and will, therefore, deliver false results (due to "dilution" of exhaust gas with fresh ambient air).

To know the gas composition is, by itself, not very informative, but one can calculate important variables from these data, that is, convert raw data into information – they are the so-called soft(-ware) sensors: all the transfer rates and ratios between gases produced and consumed. These calculations are so simple that they could/should be implemented on every automatic process control unit. Soft sensors are, generally, calculation methods using a (simple) model to compute nonmeasurable variables such as rates or specific rates or nonmeasured variables from two or more variables derived from respective hardware sensors [19].

Example 14.3 Oxygen transfer rate (OTR): the engineer's domain

The mass balances for oxygen are as follows (see Section 2.1):

$$\frac{d(V_G \cdot y_{O_2})}{dt} = \underbrace{\text{GMFR}^{\text{fresh}} \cdot y_{O_2}^{\text{fresh}} - \text{GMFR}^{\text{exhaust}} \cdot y_{O_2}}_{\text{in- and outflow}} - \underbrace{\{\text{transfer rate into the liquid phase}\}}_{\text{physical reaction}}$$

$$d\frac{(V_L \cdot c_{O_2})}{dt} = \underbrace{F \cdot c_{O_2}^{\text{in}} - F \cdot c_{O_2}}_{\text{convection through liquid phase}} - \underbrace{\text{OUR}}_{\substack{\text{biological}\\ \text{reaction}}} + \underbrace{\{\text{transfer rate into the liquid phase}\}}_{\text{physical reaction}},$$

with

V_G = total gaseous volume in the reactor: sum of all bubbles + headspace: assumed as ideally mixed !

V_L = total volume of the liquid phase in the reactor: assumed as ideally mixed

y_{O2}	=	molar fraction of O_2 either in the fresh gas (superscript ^{fresh}) or in the ideally mixed gaseous phase identical with molar fraction in the exhaust gas = %/100
c_{O2}	=	concentration of dissolved oxygen in the inflowing medium (superscript ^{in}) or in the ideally mixed liquid phase identical with concentration in the outflow
GMFR	=	gas mass-flow rate
F	=	flow rate of fresh medium or biosuspension (equal as long as V_L is constant)
OUR	=	oxygen uptake rate (of the entire biomass in the reactor); dimension = *(mass of oxygen taken up by entire cell mass) time*$^{-1}$

The left-hand parts of these equations, that is, the differentials, are very small compared to the right-hand terms, even in a batch cultivation. If both can be neglected, that is, under (pseudo-)stationary conditions, one can, as discussed in Chapter 2.1, substitute the term "transfer rate" in one equation with the expression defined in the other equation. In other words, only what was transferred into the liquid phase can be taken up by the cells and is consumed. Since the concentrations of dissolved oxygen are very small, one can also neglect the convection term through the liquid phase (which vanishes anyhow in a batch, since $F = 0$). The combination of the equations then reads as follows:

$$\mathrm{OUR} = \mathrm{GMFR}^{\mathrm{fresh}} \cdot y_{\mathrm{O_2}}^{\mathrm{fresh}} - \mathrm{GMFR}^{\mathrm{exhaust}} \cdot y_{\mathrm{O_2}}.$$

This variable OUR describes the demand for oxygen of the entire reactor and will usually increase (exponentially) during a batch; its maximal value is important for the plant design.

The volume-specific oxygen demand rate is called oxygen transfer rate (OTR) and defined as follows:

$$\mathrm{OTR} = \frac{\mathrm{OUR}}{V_{\mathrm{L}}}.$$

The dimension of OTR ($= -q_o$ in the nomenclature of Chapter 2.1) is consequently *(mass of oxygen transferred) (liquid volume)*$^{-1}$ *time*$^{-1}$. Its value will, of course, also increase during a batch and its maximal value is important for the reactor design or reactor selection. Sometimes, the biomass can become so high that the demand for oxygen exceeds the maximal transfer capacity of the reactor: then, the cultivation runs notoriously into oxygen (transfer) limitation – as discussed in Chapter 10.

Example 14.4 Respiratory quotient (RQ): the physiologist's domain

Biologists go even further and define the specific oxygen uptake rate, r_{O2}, as the uptake rate divided by the entire cell mass; similarly, there is a specific carbon dioxide production rate defined, r_{CO2}:

$$r_{O_2} = \frac{\text{OUR}}{x \cdot V_L} \quad \text{and} \quad r_{CO_2} = \frac{\text{CPR}}{x \cdot V_L},$$

where x is the biomass concentration and CPR is the carbon dioxide production rate of the entire mass of cells in the reactor. It can be calculated using the mass balances for CO_2 analogous to OUR as shown above.

Then, the respiratory quotient (RQ) is defined as the ratio of those specific rates:

RQ = $\left|\frac{r_{CO_2}}{r_{O_2}}\right|$; however, this is – per definition – a molar ratio (and always positive):

$$\text{dimension} = (\text{mol of } CO_2 \text{ produced})\,(\text{mol of } O_2 \text{ consumed})^{-1}.$$

Note that the value of RQ is not necessarily unity; this would be so only if the general degree of reduction of the substrate is identical with that of the cells and growth is balanced and no products other than CO_2 and H_2O are released. If this is not the case – which is the rule rather than an exception – the gas mass-flow rates into and out of the reactor cannot be equal. Instead of measuring both gas flow rates individually, one can derive one flow rate from the other one by applying an inert gas balance assuming only O_2 and CO_2 are reactive components in the gaseous phase (but there must be at least one inert component in the gaseous phase such as largely nitrogen in the air); as discussed above, one would prefer to measure the fresh gas flow rate:

$$\text{GMFR}^{\text{exhaust}} = \text{GMFR}^{\text{fresh}} \cdot \frac{1 - y_{O_2}^{\text{fresh}} - y_{CO_2}^{\text{fresh}}}{1 - y_{O_2} - y_{CO_2}}.$$

Although defined as ratio of (specific) rates, the calculation of RQ simplifies because volume, biomass concentration, and gas flow rate cancel out; one needs only the gas composition data:

$$\text{RQ} = \frac{y_{CO_2} - y_{CO_2}^{\text{fresh}} - y_{O_2}^{\text{fresh}} \cdot y_{CO_2} + y_{O_2} \cdot y_{CO_2}^{\text{fresh}}}{y_{O_2}^{\text{fresh}} - y_{O_2} - y_{O_2}^{\text{fresh}} \cdot y_{CO_2} + y_{O_2} \cdot y_{CO_2}^{\text{fresh}}}.$$

Caveat 1: the quadratic terms in this equation derive from the inert gas balance. If the calibration of the gas analyzer is not close to perfect, those terms can contribute substantially to error propagation and it may be "safer" (though systematically false) to approximate the RQ with the reduced formula (which would be definitely correct only in case of RQ = 1.00):

$$\text{RQ} = \frac{y_{CO_2} - y_{CO_2}^{\text{fresh}}}{y_{O_2}^{\text{fresh}} - y_{O_2}}.$$

Caveat 2: The mass balance for CO_2 over the liquid phase, if setup like for O_2, holds true only for quite acidic conditions: if the pH is high, CO_2

undergoes chemisorption: HCO_3^- or CO_3^{2-} are much better soluble in water than CO_2 and, hence, these species must be taken into account in the mass balances as well: exhaust gas analysis alone would give incorrect RQ values.

14.4 Variables Requiring More Monitoring Effort and Not Yet Standard

14.4.1 Biomass

Although biomass concentration is believed to be the (most) crucial variable in bioprocesses, online monitoring does not seem to be widely accepted. The problem is posed by the many interferences that are captured with most online techniques. The respective results are often not trusted in spite of a good coincidence between actual relative data and those of approved reference cultivations, though this should be enough indication for a good reproducibility (of the process). Obviously, many people rely on offline reference methods only. The commercially available sensors and instruments useful to quantify biomass are either indirect methods – and deliver estimates – or methods implementing counting of suspended particles, and these deliver a cell number concentration rather than a cell mass concentration.

14.4.1.1 Mass Concentration or Number Concentration: a Scalar Variable

A number of *in situ* sensors determines the "optical density" of biosuspensions, some in forward direction (like in a lab photometer, also called turbidimetry), some at a 90° angle (also called nephelometry), and some others measure the backward reflected light. Unfortunately, they all have disadvantages in common: turbidity is created by suspended cells, this is what we want to measure, but it is also created by particulate matter other than cells (e.g., solid medium components or precipitates), by gas bubbles, and by droplets (of water-immiscible liquids such as many antifoam agents or of a water-immiscible liquid partner phase). The greatest impact of those interferences certainly comes from the gas bubbles and may change considerably during a fermentation. Some sensor suppliers try to reduce this effect by attempting to exclude gas bubbles from the sensors viewing field. This can be achieved with a protective mesh around the sensor or by mimicking the manual laboratory method: fill the optical path with suspension, close the inlet, and wait for degassing, then measure and repeat this sequence. The medium itself may be colored and this can also mimic turbidity simply by absorption of light. Therefore, most of the sensors use light in the near-infrared (800–1100 nm) to reduce this effect. Nowadays, lasers or laser diodes provide sufficient power to monitor even high-density cultivations. None of those sensors can, however, discriminate between live or dead or inactive cells or abiotic particles.

Another type of sensors measure the volume of intact cells: cells are enclosed by the cytoplasmic membrane that functions somehow as an electrical insulator.

If the cells are exposed to an alternating electrical field, the charged species inside the cells are attracted to one or the other side of the field, but they cannot leave the cell membrane. This mirrors an electrical capacitor, yet the capacitance depends on the size and the contents of the membrane-enclosed volume and the frequency of the electrical field. Further, the capacitance can be "short-circuited" by the conductivity (a quasi-parallel resistor) of the medium (i.e., the freely moving charged species not enclosed by a membrane). The breakdown of the impedance ΔC (also known as β-dispersion) is an estimate for the cellular volume, and the critical frequency at which this breakdown occurs is an estimate for the size of the cells. Nowadays, there are instruments on the market that can resolve (or scan) a range of frequencies, approximately from 100 kHz to 20 MHz, and so deliver dielectric spectroscopy information. This method usually works better with large cells, for instance, CHO cell lines. Common to all these sensor types is that they need to be specifically calibrated for each biological system [20,21].

An alternative to the cell-mass and cell-volume estimation is to count the cells per unit volume resulting in a cell number concentration. Both *in situ* microscopy (ISM) and online flow cytometry (FCM) are reported to be useful – but they can afford more than just count cells as discussed later in the section particles and particle-associated compounds.

An immersed microscopic objective produces a two-dimensional image that must be evaluated; however, in a flow cytometer the cells pass the objective singly and these events are counted over time. Appropriate software for image analysis must be available to derive an estimate for the cell number concentration (and probably also get information about the size distribution and morphology).

Flow cytometry, however, needs an *ex situ* set up and the protection of the sterile barrier of the process: first, a degassing unit in bypass to the reactor and, further, either a flow injection (FIA) or a sequential injection analysis (SIA) type interface is necessary, perhaps to stain the cells, but most probably to dilute the cells, and to wash them in a membrane-covered microchamber (most probably the two are used in combination) and, eventually, feed them to the flow cytometer. A cell density of approximately 10^6 ml^{-1} (or less) is the best for the FCM measurement. Around 10^5 events can then be evaluated at an event rate of approximately 10^3 s^{-1}. FCM is the only technique that permits analysis of populations from a bioprocess on the single-cell level including individual cell properties. Depending on the staining method applied, one can, therefore, evaluate what percentage of a population is, for example, highly active, what percentage is dormant but functional, and what percentage is dead. The first online application was simply the control of cell number concentration in the so-called cytostat [22]. Enumeration of cells in drinking-water preparation processes can be performed directly [23].

Fluorescence monitoring at a distinct wavelength (couple: 1 specific excitation and 1 specific emission) or with scanning extensions (the so-called two-dimensional fluorescence monitoring) is, of course, less expensive than FCM but yields a population average only. For many purposes, this may be sufficient and desirable as one can follow growth or product formation or physiological changes of a population extremely rapidly and noninvasively. The interpretation of the resulting signals deserves some care, however, as the history of one pioneer fluorescence sensor demonstrates. This sensor was built specifically for NAD(P)H components that are both essential for every living cell. The conclusion that such a sensor would monitor the biomass concentration – the marketing argument for the sensor – was, however, false. Indeed, every living cell contains NADH and NADPH, but depending on the metabolic state of the cell, the ratio between reduced and oxidized forms can change very rapidly, in a matter of seconds to minutes for the entire population in a bioreactor. Since only the reduced forms do fluoresce, the sensor signal reflects the (inner) redox state of a population rather than the biomass concentration. If this fact is not made absolutely clear, the user of this sensor will eventually be completely misled. The commercial solution of this dilemma was withdrawal of the sensor from the market.

Fluorescent reporter molecules have become well established over the years in developmental and cell biology. They are, unfortunately, scarcely used for quantitative indication of product formation in production processes but could be an excellent means to move some delicate lab analyses to a simple real-time monitoring task. Since the *in situ* sensors always pick up the excitation light as scattered (or reflected) light it is worthwhile considering the use of fluorescent reporter molecules that have a maximal distance between excitation and emission wavelength, for instance, wild-type green fluorescent protein (GFP: 395/509 nm) rather than the enhanced variant (eGFP: 484/507 nm) in order to achieve a reasonable separation of the desired fluoresence signal from the (high-intensity) stray light [24–26].

Yet, the decisive fact for a productive process is not some available mass of a potential biocatalyst, it is the overall activity of the biocatalyst. And this can be estimated via indirect methods such as gas balancing or energy balancing: aerobic organisms – if active – will consume oxygen and this can be estimated simply via the above discussed approach. Almost all organisms – if active – will release enthalpy, thus raising the temperature of the medium. In order to keep the temperature in a permissive range, if possible close to the optimal temperature, most technical cultivation systems are equipped with a temperature controller. Its major target is to remove the biogenic heat from the cultivation system. If this heat flux can be determined, one gets a good measure of the activity of the entire population inside a bioreactor. Many temperature controllers are designed as cascade controllers and these require the input information from the coolant circuit as well; that is, this information is usually collected but used for technical reasons only – and not to create information on bioactivity. If only the flow rate of the coolant is known (or at least kept constant), one can simply derive the heat flux from these data. The larger the scale of a reactor the less important are errors added due to heat transfer to the ambient (the "more adiabatic" is a reactor), but the technique is also reported for bench-scale bioreactors with a sensitivity as low as 50 mW l^{-1} [27]. This provides a simple, heat-based biomass and specific growth rate estimator for microbial cultures. Of course, there are some systematic disturbances included in this result such as mechanical power dissipation to the liquid, friction converted to heat in bearings and sealing, or the evaporation loss of heat when dry air is used to aerate a fermenter. However, all those effects can be quantified in abiotic control experiments, that is, "calibrated", and, later, accounted for in the real cultivation processes. Since this technique balances the entire reactor, one gets a direct estimate for the performance of a production lot. Interesting enough, the ratio between biogenic heat production rate and oxygen consumption rate (of aerobic organisms) is relatively constant, that is, independent of the type of organism, and has a value on the order of 460 kJ per mole of oxygen [28] – as also shown in Chapter 9, Eq. (9.24). This ratio is known as the oxocaloric yield. So, by knowing only one rate, one can estimate the other rate; however, determining both rates in real time is a reliable and redundant means to estimate bioactivity.

14.4.1.2 Population Distribution of Particles and Particle-Associated Compounds – a Multidimensional Variable: FCM, FFF, and ISM

Undoubtedly, the most important particles in bioprocesses are intact, active cells. Often, they grow, but this is not necessarily mandatory for all biotransformation processes since also nongrowing cells can be good active biocatalysts. In many cases, the number concentration of cells is of importance and, hence, counting cells per volume or sizing the cells can give valuable information on the process performance.

This goal can be achieved by a number of different methods, for instance, by using an appropriate dilution of the cell suspension and detecting the cells either with a flow cytometer (FCM), a field-flow fractionation apparatus (FFF), or using an *in situ* microscope (ISM). But all these techniques can also provide more, additional information on the population distribution: the individual cells can be classified, for instance, according to their size or whether they are aggregated or single. This opens the possibility to judge the homo- or heterogeneity of a population and can be very useful for (automatically) deciding whether to tune (control), to continue, or to stop an actual bioprocess.

In an FCM, the cells are usually hydrodynamically focussed using a sheath fluid into which the sample stream is concentrically injected or aspirated: the individual cells align like a string of pearls and pass a measuring cuvette in which the individual events are optically detected: the stream line is excited with one or more focused laser beams; cells scatter the light in forward and sideward direction, and the scattered light is detected using photo multipliers or photodiodes. The forward scatter is associated with the particle size, while the side scatter bears information on the granularity of the individual particle. More sensitive than the forward scatter, the slit scan (or time-of-flow) method can estimate the long axis of particles. If the particle contains fluorophores, either by itself or after selective staining of nonfluorescent components, the emitted fluorescence intensity can also be measured and may provide quantitative information on a specific quality of a cell. Nowadays, there are many different dyes available: some of them are membrane permeant and some are membrane nonpermeant. Hence, one can differentiate between cells with intact membrane and cells with a compromised membrane (e.g., discriminate live/dead) [29,30]. Among other characteristics, the possibility to derive DNA fingerprints from flow cytograms definitely allows the process engineer to detect contaminations very early. Or, heterologous proteins could be tagged with a fluorescent reporter (e.g., a fluorescent protein such as GFP) and, then, the fluorescence signal bears information on the expression quantity per individual cell. Flow cytometers can easily analyze thousands of cells per second and, so, one can acquire statistically relevant data in very short time. In other words, one directly obtains, almost in real time, a population distribution based on the single-cell level. Example 14.6 illustrates the monitoring of intracellular products.

Example 14.5 An approach to prepare a cell suspension for online flow cytometric monitoring

This particular example shows the principal setup to analyze polyhydroxyalkanoates that exist as particulate solid granules within the cells. These granules can be stained with nile red resulting in fluorescent "inclusions bodies" easily detectable with a flow cytometer (FCM).

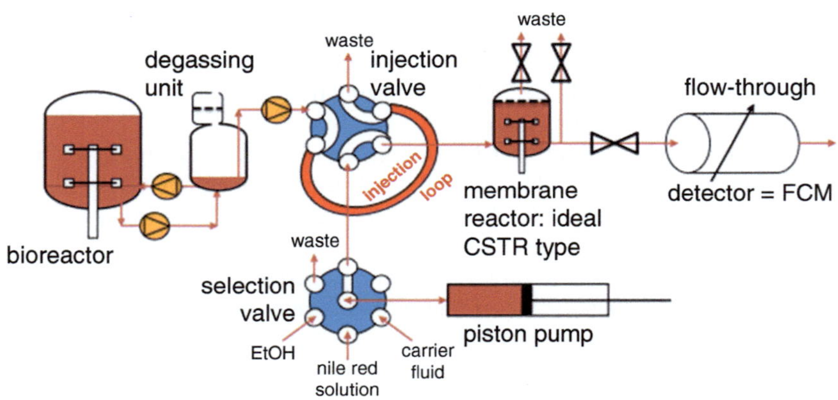

The biosuspension is discontinuously withdrawn from the bioreactor into a degassing unit from which an injection loop of well-known volume is then loaded while the piston pump aspirates ethanol and alcoholic nile red solution. After switching the injection valve, the sample is then pushed out of the loop into the membrane reactor and there conditioned with the alcoholic staining solution; the valve to the waste exit must be open, the valve to the FCM closed. Staining is completed within less than 2 min. Excess nile red is then washed out with ethanol, the suspension of stained cells can be further diluted with aqueous carrier fluid to a density appropriate for the FCM and, only then, a defined volume is routed through the FCM by switching the valves appropriately.

FFF works differently to separate particles (and/or larger dissolved molecules): a suspension aliquot is injected into a shallow flow channel through which it flows with a (laminar) parabolic flow profile. The separation is achieved by applying a field perpendicular to the flow; the field can be a gravitational, a thermal, an electrical, or yet another flow field that forces the individual particles more or less strongly toward one wall of the flow channel, the so-called accumulation wall. The more the cells are forced to this accumulation wall the smaller is their traveling speed compared to the average linear fow rate of the liquid flow in the channel (the so-called "steric mode" that applies for particles larger than approximately 1 μm). Hence, this fraction of the population is retarded that is detected at the outlet with an appropriate (optical or electrical) detector. This also informs on the population distribution, but not on the single-cell level. On the other hand, also dissolved molecules can be forced toward the accumulation wall by the field and are then separated according to their mobility: the smaller the molecules or aggregates are, the higher is their diffusivity and, hence, they will spend more time in flow regions with higher speed and be retarded less than their larger partner molecules. Several different FFF instruments have entered the commercial market in recent years.

ISM resembles desk-top microscopy very much; however, the microscope objective, the illumination source (usually a powerful LED) and probaly some mechanical actors are inside the sterile barrier and must be sterilized *in situ*, the rest including camera and computer is, of course, outside. The "sample zone" directly in front of the objective is usually on the order of 50 μm for the monitoring of larger (animal) cells. The sample zone can be open and thus permits the acquisition of sequences of images with exposure times of 1 ms or less. Focusing may become a significant problem at higher cell densities and the software may no longer be able to distinguish the individual cells. Microbubbles can be quite easily detected by image evaluation software due to their size and almost perfectly spherical shape. Since they occupy part of the measured volume, this must be taken into account during calculation of the cell number concentration. Yet, particles other than cells such as crystals (of product) and even cell aggregates such as dense mycelia or pellets can be additionally analyzed using this technique [31–34].

14.4.2
Products and Substrates

Substrates may be solid, liquid, or gaseous as pure compounds; however, they must eventually be dissolved in the medium in order to be consumed by the cells. In batch processes, dissolved substrate concentrations can be (very) high. Yet, in many technical processes, a substrate limitation is desired and then the substrate concentrations are quite low – and this could be a challenge for the process analytical techniques applied. In general, substrate monitoring techniques should span a considerably large dynamic range. This may not so strictly apply for products since industrial processes very often aim at maximizing the product titer. However, products are not necessarily excreted from the cells, that is, they may be intracellular, or they could precipitate in the medium or be so volatile that they are removed continuously via the exhaust gas. In other words, product analysis faces other challenges, namely "looking into the cells," opening cells, or dealing with suspended solid or evaporated gaseous substances. Except for volatile products and for those materials where *in situ* sensors exist, we need an axenic interface to prepare a representative sample ready for analysis.

14.4.2.1 **Interfacing Process Analyzers Axenically to the Processes**

The simplest way to acquire a (continuous or intermittent) sample stream from the reactor is using a one-way pump. If a sample shall be taken at discrete time intervals, a sampling valve will be the right choice; this can work without a pump if the reactor is (slightly) pressurized. If a valve is employed, this should (1) have a minimized dead volume, (2) seal at the reactor wall or reach into a turbulent reactor zone rather than sitting at the bottom, (3) have a connection to a steam line for cleaning and hygienizing the valve's dead volume before and/or after sampling, and (4) probably have a connection to a sterile air line for drying after cleaning; these extra efforts help avoiding carryover and dilution of samples. Cells are not affected by this type of sample acquisition as long as the time required is (very) short, for example, between a few 100 ms and a few seconds, depending on the cell density and specific substrate consumption rates. If it takes longer, substrates and, most probably first, oxygen will become limiting.

If, however, one is not interested in cell properties, but in concentrations of dissolved reactants only, the option is either to remove or inactivate the cells during sample acquisition: (micro- or ultra-)filtration, poisoning, heating, or cooling right at the exit of the reactor can be technical solutions. Filters can be mounted *in situ* or in bypass. The *in situ* version needs a proper mount point in a turbulent space of the reactor in order to avoid fouling or even clogging of the filter. The bypass version has the advantage that the filter can be changed during cultivation, if necessary, but it has the disadvantage that the mean residence time of cells in contact with the liquid outside the (well mixed) reactor is nonzero, thus giving rise to falsification of the subsequent analyses as discussed in Example 14.2. A promising alternative could be the dielectrophoretic cell retention or using spin filters.

Poisoning the removed biosuspension with cyanide or azide or any other appropriate agent in a double-lumen catheter does stop only a portion of metabolic actions of the cells. Heating will destroy all thermolabile molecules that are intracellular biomolecules (certainly most enzymes) and probably also extracellular components one might be interested in. Cooling is usually restricted by the solidification of the suspension at around 0 °C, which might not be low enough to arrest significant metabolic reactions that could introduce systematic errors. If the suspension is mixed with prechilled solvent, such as methanol at −40 °C, one can reach much lower temperature in the liquid state, however, at the expense of dilution and probably extraction of intracellular components into the liquid.

Probably the most versatile tools to automate (bio-)chemical analyses are flow injection analysis (FIA) or sequential injection analysis. Basic components of any FIA equipment are a transport system consisting of tubing, pumps, valves, and a continuously flowing carrier stream into which a technical system, for instance, an injection valve hosting an injection loop of defined, constant volume injects a sample from time to time. Contrarily, an SIA system aspirates the necessary components sequentially and ejects them thereafter, and the flow may be segmented; however, this setup can save reagents and reduce waste. In both cases, a suitable detector must be placed at the end of this line; the detector could be anything in the range from a simple thermosensor to a complex flow cytometer. While flowing from injector to detector, any physical or chemical reaction or combinations thereof may occur.

In this sense, (liquid or gas) chromatographs or electrophoreses are also special cases of an FIA. There are no chemical reactions performed but mixtures are separated; however, pre- or postcolumn derivatization can be implemented. Specifically, high-performance liquid chromatography (HPLC and increasingly ultra-performance liquid chromatography, UPLC) is widely accepted and established, also in downstream processing [35]. Although this equipment is not inexpensive, the cost per analysis has to be divided by the number of components that can be analyzed, which is usually more than a dozen in one sample.

Further important features are that sample dilution can be easily performed in an FIA and that these setups can work in a mini- or microfluidic system. The automation may reproduce systematic errors if calibration is done improperly and/or recalibration is omitted. Frequent recalibration may be necessary. Whenever injection loops are implemented, it is mandatory to avoid the occurrence of gas bubbles in the sample stream since they falsify the volumetric dosage. This error would, of course, be propagated by the dilution factor. Degassing vessels (i.e., sedimenters for the liquid phase) are simple to construct but add a mixing device upstream the dilution and/or analysis whereas degassing membranes may not be as effective but they do not mix (they just disperse axially).

Example 14.6 One possible FIA format for automated sucrose analysis

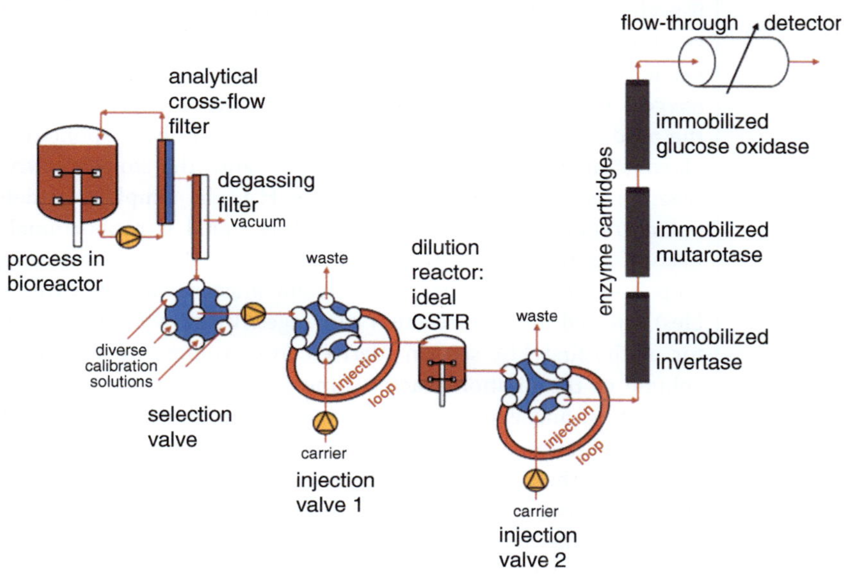

The component of interest in this specific example is the substrate sucrose in the bioreactor.

The external analytical cross-flow microfilter serves as a device to remove cell free permeate from the reactor and it represents (and does not compromise) the sterile barrier. Cells are returned to the reactor and not sacrificed. It may be necessary to take the cell retention into account when calculating mass balances. Further downstream, the permeate is degassed via a vaccuum filter and then, in the valve position shown in the figure, transported into the injection loop of the first injection valve. The selection valve in between would allow to feed various calibration solutions, if desired, alternatively. After switching injection valve 1, the content of the injection loop 1 is pushed into the dilution reactor by the carrier stream (while sample goes directly to waste). The outflow of the dilution reactor, which should be a well-mixed continuously stirred tank reactor (CSTR) of a volume as small as possible, is further fed into the loop of injection valve 2. So far, sucrose is diluted only with the carrier fluid. After a distinct time gap, which must be calculated by the FIA control algorithm in real time, injection valve 2 is also switched thus feeding a defined volume of the appropriately diluted sucrose solution into the first enzyme cartridge. This contains immobilized invertase, and sucrose is now hydrolyzed into α-D-glucose and fructose. Even though glucose mutarotates ($\alpha \leftrightarrow \beta$) spontaneously in an aqueous salt buffer, this would be too slow in order to reach equilibrium. Therefore, the next cartridge contains immobilized mutarotase in order to speed up equilibration. In the last cartridge, immobilized glucose oxidase will convert (exclusively) the β-D-glucose. The molecular oxygen necessary for this stoichiometric reaction into H_2O_2 and gluconolactone must be provided dissolved in carrier fluid 2. It is decisive that glucose limits this reaction and not oxygen! Finally, the flow-through detector may determine either one of the products (H_2O_2 or gluconolactone that hydrolyzes spontaneously to gluconic acid) or the cosubstrate O_2 because the stoichiometry of the reaction is well known. Usually, the H_2O_2 is simplest to determine polarographically; the integral over this signal is proportional to the mass of sucrose (concentration of diluted sample × volume of injection loop 2 – this is why the entire system must be operated free of bubbles) injected into the reaction cartridges. By knowing the dilution factor of the first FIA section (derived from volume of injection loop 1, volume of the dilution reactor, time gap between switching the injection valves, and exponential washout profile of the (ideal) CSTR), one can straightaway calculate the sucrose concentration in the bioreactor. As soon as the concentration of sucrose in the dilution reactor is negligibly small, the injection valves can be returned into their starting positions and the next analytical cycle can be started. Results are so produced discontinuously but with high frequency.

Of course, this setup is not the only one thinkable for sucrose. One could, for example, first mix the permeate with alkali and then inject the so prepared sample onto an HPLC column in order to separate the components of the mixture and quantify the respective sucrose peak either with an inexpensive refraction index detector or with a more expensive mass spectrometer as the detector.

14.4.2.2 IR Active Molecules in Solution or Suspension

A very powerful technique to analyze organic compounds is the infrared (IR) spectrometry. This technique can be applied *in situ*, that is, it does not require an interface, works in real time, and is neither invasive nor destructive. Various different functional groups (keto, hydroxy, thiol groups, ether bonds, etc.) in organic molecules absorb distinct energy bands when excited with IR radiation and herewith power typical molecular vibrations. This represents a "functionality fingerprint" of the monitored molecules.

Water, the solvent for most bioprocesses, absorbs massively in the far IR; this domain is not really informative. The mid-IR (MIR) shows the most fundamental molecular vibrations. This domain spans the wave number range from 200 to 4000 cm^{-1}. The respective absorption bands are relatively narrow and can usually (i.e., certainly in the case of pure compounds) be attributed (qualitatively) to typical functional groups. The near-IR domain (NIR, wave number range from 4000 to 13000 cm^{-1}) shows typically vibrational combinations or overtones and is characterized by (much) broader absorption bands. This is also why components in aqueous solution can be determined at much lower concentrations and with a higher degree of spectral resolution in the MIR domain. NIR instruments need a significantly better signal to noise ratio [12,36–39].

Unfortunately, bioprocess suspensions are multicomponent mixtures of many different compounds. Hence, the IR signals of many functional groups distributed over many different compounds present, probably also in highly different concentrations, are convoluted. This makes the qualitative and quantitative interpretation of the spectral "information cocktail" much more difficult than in case of chemical processes that involve a minimum number of compounds at usually much higher concentrations. Multivariate evaluation and calibration algorithms, typically chemometric procedures, must be applied to retrieve information from the raw spectral data [40–42]. The chemometric calibration models are critical and should be constructed and verified according to an elaborated design of experiments. One usually builds them from spectrometric analyses of solutions of either pure compounds or mixtures of a few compounds in well-known concentrations; this is preferred rather than spiking spent culture supernatants with some pure compounds. The danger herewith is that the model "learns" a biological rather than an analytical correlation, for instance, when the substrate concentration is high the product concentration is low (or vice versa). An exciting and very successful verification method is spiking individual pure components (in well-known amounts) directly into an active process. Such disturbances can usually be tolerated or are even part of the protocol (e.g., repetitive, discontinuous feed additions) but the response is positive: when the calibration model used to deconvolute the raw data predicts an appropriate concentration increase of only the spiked species but no changes for all other species, the model is obviously useful. If this is so, plenty of relevant compounds can be quantified, such as sugars, alcohols, organic and amino acids, proteins, or biomass itself [43].

Today, the combination of the Fourier transform (FT) method with the use of attenuated total reflection (ATR) probes is believed to be ideal for analyzing complex mixtures in aqueous solution or suspension. An ATR probe is in direct contact with the biosuspension; the material is preferentially diamond because of its high refractive index. It is the evanescent field (or wave) that interacts with the dissolved analytes [36]. Optical fibers are available today and rapidly improving. This will render this monitoring technique really attractive (the so far used "conduits" between process probe and IR analyzer, consisting of blackened tubes connected via mirrors and junctions, are definitely not very suitable in an industrial production environment).

Raman spectroscopy is an emerging technique in bioprocess monitoring, but is not widely applied. Besides following substrate and product concentrations, the spectra can be correlated (via partial least square (PLS) models) with total cell mass and viable cell mass concentrations, thus probably allowing a simple noninvasive real-time discrimination of live and dead cells [44].

14.4.2.3 Volatile Components

Sometimes, volatile organic compounds (VOC) are the substrates (e.g., methanol) or products (e.g., all kinds of "biofuel" and "biosolvent," "ABE fermentation," etc.) of interest: among those are alcohols, aldehydes, minor fermentation byproducts that are interesting in the flavor and fragrance industry, as well as gasses other than O_2, CH_4, and CO_2.

Mass spectrometry (MS) seems to be the method of choice to monitor these compounds: they can enter the spectrometer directly from the gas phase or enter via a pressure-drop capillary, or they can be sampled from the liquid phase via a solution-diffusion membrane. In the first case, sampling occurs from the exhaust gas, that is, outside the sterile barrier, and does not require special hygienic precautions; in the second case, however, the membrane is not only the part responsible for selectivity but also plays the roles of sterile and pressure barrier making its use quite delicate (if it should leak, the entire high-vacuum interior of the MS would be flooded instantaneously).

Many gas-inlet MS operate interfaced via multichannel selection valves and monitor the exhaust gases of more than one reactor in a round-robin schedule (multiplexed, that is, discontinuously, but since the turn-around time can be as small as a few seconds and the measurement of a mass spectrum takes only a few seconds or even less, this is called quasi-continuous). This approach reduces the investment costs, but the reactors need to be placed in close proximity.

Besides the classical instrument types is now the upcoming proton transfer reaction (PTR)-MS. Their highly sensitive chemical ionization technique – a proton is "softly" transferred from H_3O^+ to the analyte that is then detected at

M^{+1} – results in no fragmentation of the analyte molecules, extremely low detection limits (down to 10^{-12} vol/vol), and very high speed (dynamic resolution in the subsecond domain). No sample preparation is needed [18,45].

Another promising application of mass spectrometric monitoring (in academic research) is the isotopic discrimination of, for example, $^{12}CO_2$ and $^{13}CO_2$ in omics studies: the fates of ^{13}C-atoms from specifically ^{13}C-labeled substrates could so be tracked [46,47].

In the past, the so-called electronic noses had been used frequently, but they appear to have faded out. Electronic noses are arrays of individual gas sensor elements, probably up to 1024, and these arrays are repeatedly and alternatively exposed to exhaust gas ("excited") and an inert reference gas ("relaxed"). The array elements translate chemical information into an electrical signal and they are either metal oxide or polymer sensors that change their electrical resistance while volatile molecules are adsorbed or they are quartz microbalances that change their resonance frequency while their mass changes due to some adsorption. The adsorption characteristics are (even qualitatively) mostly unknown. The signal trajectories of the elements are followed both during excitation and during relaxation [48]. Again, multivariate regression models are required to extract useful information from the huge set of raw data and appropriate calibration (or "training") of the sensors is an absolute prerequisite.

14.5 Data Evaluation

Many of the above-discussed measurement and monitoring techniques, and certainly all of the instrumental analytical techniques, produce large amounts of raw data. The raw data as such are not really informative, although they contain the desired information: the data need to be evaluated. Most of the suppliers of analytical instruments also provide some (proprietary) software for data evaluation according to the needs in their most important market; and this, as a rule, is an analytical laboratory, whereas the application for process monitoring still remains the exception. The available software tools are often tailored to good laboratory or clinical practice constraints that can make them cumbersome to work with. The software tools are often an isolated piece that cannot communicate simply and easily with other software, for example, running on dedicated process computers thought to control the production processes. In an academic environment, the straightforward solution is to write ad hoc algorithms trying to circumvent (and not to solve) those problems. Also in our lab, we have produced too many of such stupid work-arounds and wasted time. This collection and synchronization of individual data from different analyzers onto one single platform (process computer) including reformatting the data into one common format is sometimes called data alignment. In the era of PAT, QbD, and process validation, we urgently and rapidly need more flexible and open commercial software solutions to follow and foster the respective goals!

Though this might become a reality soon, we will have some information on or estimates for concentrations at hand in due time. This may, for example, be the traget to control a feed so that the concentration of a substrate or a starting material remains constant. But it may well be that the decisive variable to be controlled is a ratio of concentrations (such as the RQ value or various yield coefficients), a volumetric or a specific reaction rate. In the case of ratios, one should not work with single, noisy values but, instead, regress over a usefully long historic time horizon. In geometric terms, this means to find the slopes in phase plane plots. Since the basic variables are all somehow error-prone, the regression must minimize the orthogonal distances to the individual raw data points. In the case of reaction rates, we face a simple and straightforward situation only if the process runs at steady state: then, the accumulation terms in the mass balance equations vanish and the rates are determined (as discussed in Chapters 2 and 3) by concentration differences and the independent operational variable (e.g., the dilution rate or, its inverse, the mean hydraulic residence time). Unfortunately, continuous bioprocesses operated at steady state are the exception, not the rule. In nonsteady-state processes, however, we need to find the transient rates from the time derivatives of the individual variables; since these are measured or monitored with a given uncertainty, the noise of the transient data will be propagated. In order to mitigate this problem, we need the concentration data at a high frequency.

Sometimes, the raw data sets are very complex and multidimensional (e.g., from spectral analyzers, electronic noses, or flow cytometers); it is likely that part of such a data set contains valuable information and part of it no useful information at all. Therefore, multivariate (chemometric) models are often used to extract a reasonably big portion of valuable information from the raw data sets. Care must be taken that the models are not be trained to reproduce biological effetcs (such as more or less constant yield coefficients) but analytically based and validated correlations instead. This could be proven by clever spiking experiments as discussed above [49]. Anyhow, these methods are not plug-and-play and we need to ascertain that users are well educated and trained, and understand what they do! *Caveat*: Any extrapolation using such data-driven models over the horizon of the training data set may be very dangerous because the domain of validity is basically unknown.

References

1 FDA (2004) Guidance for Industry PAT: A Framework for Innovative Pharmaceutical Development, Manufacturing, and Quality Assurance, www.fda.gov/ (accessed June 29, 2015).

2 FDA (2011) Guidance for Industry – Process Validation: General Principles and Practices, www.fda.gov/ (accessed June 29, 2015).

3 FDA (2011) Implementation of Quality by Design (QbD): Current Perspectives on Opportunities and Challenges, www.fda.gov/ (accessed June 29, 2015).

4 Duennebier, G. and Tups, H. (2007) FDA PAT initiative: eine anwendersicht zu technischen möglichkeiten und aktueller industrieller umsetzung. *Chem. Ing. Tech.*, **79** (12), 2019–2028.

5 Junker, B.H. and Wang, H.Y. (2006) Bioprocess monitoring and computer control: key roots of the current PAT initiative. *Biotechnol. Bioeng.*, **95** (2), 226–261.

6 Glassey, J., Gernaey, K.V., Clemens, C., Schulz, T.W., Oliveira, R., Striedner, G., and Mandenius, C.F. (2011) Process analytical technology (PAT) for biopharmaceuticals. *Biotechnol. J.*, **6**, 369–377.

7 Enfors, S.O., Jahic, M., Rozkov, A., Xu, B., Hecker, M., Juergen, B., Krueger, E., Schweder, T., Hamer, G., O'Beirne, D., Noisommit-Rizzi, N., Reuss, M., Boone, L., Hewitt, C., McFarlane, C., Nienow, A., Kovacs, T., Trägårdh, C., Fuchs, L., Revstedt, J., Friberg, P.C., Hjertager, B., Blomsten, G., Skogman, H., Hjort, S., Hoeks, F., Lin, H.Y., Neubauer, P., van der Lans, R., Luyben, K., Vrabel, P., and Manelius, A. (2001) Physiological responses to mixing in large scale bioreactors. *J. Biotechnol.*, **85** (2), 175–185.

8 Schmalzriedt, S., Jenne, M., Mauch, K., and Reuss, M. (2003) Integration of physiology and fluid dynamics. *Adv. Biochem. Eng. Biotechnol.*, **80** 19–68.

9 Sonnleitner, B. (2000) Instrumentation of biotechnological processes. *Adv. Biochem. Eng. Biotechnol.*, **66**, 1–64.

10 Clementschitsch, F. and Bayer, K. (2006) Improvement of bioprocess monitoring: development of novel concepts. *Microb. Cell Fact.*, **5**, 19.

11 Julien, C. and Whitford, W. (2007) Bioreactor monitoring, modeling, and simulation. *BioProc. Int.*, **5** (1), 10–17.

12 Beutel, S. and Henkel, S. (2011) *In situ* sensor techniques in modern bioprocess monitoring. *Appl. Microbiol. Biotechnol.*, **91** (6), 1493–1505.

13 Carloni, A. and Turner, A.P.F. (2011) Bioprocess monitoring. *Encycl. Ind. Biotechnol.*, **2**, 766–783.

14 Sonnleitner, B. (2013) Automated measurement and monitoring of bioprocesses: key elements of the M^3C strategy. *Adv. Biochem. Eng. Biotechnol.*, **132**, 1–33.

15 Havlik, I., Lindner, P., Scheper, T., and Reardon, K.F. (2013) On-line monitoring of large cultivations of microalgae and cyanobacteria. *Trends Biotechnol.*, **31** (7), 406–414.

16 Pollard, D. and Christensen, J. (2010) Vent gas analysis. *Encycl. Ind. Biotechnol.*, **7**, 4759–4773.

17 Aehle, M., Kuprijanov, A., Schaepe, S., Simutis, R., and Luebbert, A. (2011) Simplified off-gas analyses in animal cell cultures for process monitoring and control purposes. *Biotechnol. Lett.*, **33** (11), 2103–2110.

18 Luchner, M., Gutmann, R., Bayer, K., Dunkl, J., Hansel, A., Herbig, J., Singer, W., Strobl, F., Winkler, K., and Striedner, G. (2012) Implementation of proton transfer reaction-mass spectrometry (PTR-MS) for advanced bioprocess monitoring. *Biotechnol. Bioeng.*, **109** (12), 3059–3069.

19 Wechselberger, P., Sagmeister, P., and Herwig, C. (2013) Real-time estimation of biomass and specific growth rate in physiologically variable recombinant fed-batch processes. *Bioprocess Biosyst. Eng.*, **36** (9), 1205–1218.

20 Maskow, T., Roellich, A., Fetzer, I., Ackermann, J.U., and Harms, H. (2008) On-line monitoring of lipid storage in yeasts using impedance spectroscopy. *J. Biotechnol.*, **135** (1), 64–70.

21 Ansorge, S., Esteban, G., and Schmid, G. (2010) On-line monitoring of responses to nutrient feed additions by multi-frequency permittivity measurements in fed-batch cultivations of CHO cells. *Cytotechnol*, **62** (2), 121–132.

22 Gilbert, A. and Srienc, F. (2009) Optimized evolution in the cytostat: a Monte Carlo simulation. *Biotechnol. Bioeng.*, **102** (1), 221–231.

23 Arnoldini, M., Heck, T., Blanco-Fernández, A., and Hammes, F. (2013) Monitoring of dynamic microbiological processes using real-time flow cytometry. *PloS One*, **8** (11), e80117.

24 Reischer, H., Schotola, I., Striedner, G., Poetschacher, F., and Bayer, K. (2004) Evaluation of the GFP signal and its aptitude for novel on-line monitoring strategies of recombinant fermentation processes. *J. Biotechnol.*, **108** (2), 115–125.

25 Broger, T., Odermatt, R.P., Ledergerber, P., and Sonnleitner, B. (2009) Exploiting

fluorescent reporter molecules for Process Analytical Technology (PAT). *Chimia*, **63** (3), 171–173.

26 Hewitt, B.M., Singhal, N., Elliot, R.G., Chen, A.Y.H., Kuo, J.Y.C., Vanholsbeeck, F., and Swift, S. (2012) Novel fiber optic detection method for *in situ* analysis of fluorescently labeled biosensor organisms. *Environ. Sci. Technol.*, **46** (10), 5414–5421.

27 García-Payo, M.C., Ampuero, S., Liu, J.S., Marison, I.W., and von Stockar, U. (2002) The development and characterization of a high resolution bio-reaction calorimeter for weakly exothermic cultures. *Thermochim. Acta*, **391** (1–2), 25–39.

28 Maskow, T. (2013) Miniaturization of calorimetry: strengths and weaknesses for bioprocess monitoring and control, in *Biothermodynamics: The Role of Thermodynamics in Biochemical Engineering* (ed. U. von Stockar), EPFL Press, pp. 423–442.

29 Hewitt, C.J. and Nebe-von-Caron, G. (2004) The application of multi-parameter flow cytometry to monitor individual microbial cell physiological state. *Adv. Biochem. Eng. Biotechnol.*, **89**, 197–223.

30 Broger, T., Odermatt, R.P., Huber, P., and Sonnleitner, B. (2011) Real-time on-line flow cytometry for bioprocess monitoring. *J. Biotechnol.*, **154**, 240–247.

31 Pons, M.N. and Vivier, H. (2000) Biomass quantification by image analysis. *Adv. Biochem. Eng. Biotechnol.*, **66**, 133–184.

32 Hoepfner, T., Bluma, A., Rudolph, G., Lindner, P., and Scheper, T. (2010) A review of non-invasive optical-based image analysis systems for continuous bioprocess monitoring. *Bioproc. Biosys. Engin.*, **33** (2), 247–256.

33 Akin, M., Prediger, A., Yuksel, M., Hoepfner, T., Demirkol, D.O., Beutel, S., Timur, S., and Scheper, T. (2011) A new set up for multi-analyte sensing: at-line bio-process monitoring. *Biosens Bioelectron.*, **26** (11), 4532–4537.

34 Prediger, A., Bluma, A., Hoepfner, T., Lindner, P., Beutel, S., Mueller, J., Hilterhaus, L., Liese, A., and Scheper, T. (2011) *In situ*-mikroskopie zur online-ueberwachung von enzymtraegern und zweiphasenprozessen. *Chem. Ing. Tech.*, **83** (6), 884–887.

35 Rathore, A.S., Wood, R., Sharma, A., and Dermawan, S. (2008) Case study and application of process analytical technology (PAT) towards bioprocessing: II. Use of ultra-performance liquid chromatography (UPLC) for making real-time pooling decisions for process chromatography. *Biotechnol. Bioeng.*, **101** (6), 1366–1374.

36 Roychoudhury, P., Harvey, L.M., and McNeil, B. (2006) The potential of mid infrared spectroscopy (MIRS) for real time bioprocess monitoring. *Anal. Chim. Acta*, **571** (2), 159–166.

37 Roychoudhury, P., McNeil, B., and Harvey, L.M. (2007) Simultaneous determination of glycerol and clavulanic acid in an antibiotic bioprocess using attenuated total reflectance mid infrared spectroscopy. *Anal. Chim. Acta*, **585** (2), 246–252.

38 Roychoudhury, P., O'Kennedy, R., McNeil, B., and Harvey, L.M. (2007) Multiplexing fibre optic near infrared (NIR) spectroscopy as an emerging technology to monitor industrial bioprocesses. *Anal. Chim. Acta*, **590** (1), 110–117.

39 Tiwari, S., Suraishkumar, G.K., and Chandavarkar, A. (2013) Robust near-infra-red spectroscopic probe for dynamic monitoring of critical nutrient ratio in microbial fermentation processes. *Biochem. Eng. J.*, **71**, 47–56.

40 Kornmann, H., Marison, I., and von Stockar, U. (2002) Real-time adaptation of partial least squares calibration model used in on-line monitoring of bioprocess by mid-infrared spectroscopy. *Chem. Ing. Tech.*, **74** (5), 703.

41 Lourenço, N.D., Lopes, J.A., Almeida, C.F., Sarraguça, M.C., and Pinheiro, H.M. (2012) Bioreactor monitoring with spectroscopy and chemometrics: a review. *Anal. Bioanal. Chem.*, **404** (4), 1211–1237.

42 Glassey, J. (2013) Multivariate data analysis for advancing the interpretation of bioprocess measurement and monitoring data. *Adv. Biochem. Eng. Biotechnol.*, **132**, 167–191.

43 Landgrebe, D., Haake, C., Hoepfner, T., Beutel, S., Hitzmann, B., Scheper, T., Rhiel, M., Reardon, K.F. (2010) On-line infrared spectroscopy for bioprocess monitoring. *Appl. Microbiol. Biotechnol*, **88** (1), 11–22.

44 Abu-Absi, N.R., Kenty, B.M., Cuellar, M.E., Borys, M.C., Sakhamuri, S., Strachan, D.J., Hausladen, M.C., Li, Z.J. (2011) Real time monitoring of multiple parameters in mammalian cell culture bioreactors using an in-line Raman spectroscopy probe. *Biotechnol. Bioeng*, **108** (5), 1215–1221.

45 Biasioli, F., Gasperi, F., Yeretzian, C., Maerk, T.D. (2011) PTR-MS monitoring of VOCs and BVOCs in food science and technology. *Trends Anal. Chem*, **30** (7), 968–977.

46 Yang, T.H., Wittmann, C., and Heinzle, E. (2006a) Respirometric ^{13}C flux analysis, Part I: design, construction and validation of a novel multiple reactor system using on-line membrane inlet mass spectrometry. *Metabol. Eng.*, **8** (5), 417–431.

47 Yang, T.H., Wittmann, C., and Heinzle, E. (2006b) Respirometric ^{13}C flux analysis, Part II: *in vivo* flux estimation of lysine-producing *Corynebacterium glutamicum*. *Metabol. Eng.*, **8** (5), 432–446.

48 Bachinger, T., Mandenius, C.-F., Striedner, G., Clementschitsch, F., Durrschmid, E., Cserjan-Puschmann, M., Doblhoff-Dier, O., and Bayer, K. (2001) Non-invasive detection of the metabolic burden on recombinant microorganisms during fermentation processes. *J. Chem. Technol. Biotechnol*, **76** (8), 885–889.

49 Schenk, J., Marison, I.W., and von Stockar, U. (2007) Simplified Fourier-transform mid-infrared spectroscopy calibration based on a spectra library for the on-line monitoring of bioprocesses. *Anal. Chim. Acta*, **591** (1), 132–140.

15
Control of Bioprocesses

Jakob Kjøbsted Huusom

Summary

The fundamental principles for control of biochemical processes are presented. Through examples, the selection of manipulated and controlled variables in the classical reactor configurations is discussed, so are control objectives and the challenges in obtaining good control of the bioreactor. A number of typical control loops are given for different objectives, for example, for control of the growth of the organism or the environmental condition that surrounds it. Then, follows a brief introduction to the general principles of process control, the PID control algorithm is discussed, and the design and effect of tuning are shown in an example. Finally, a discussion of novel, model-free control approaches for bioreactors is provided.

15.1
Introduction

As in any other field of control and operation, the purpose of bioprocess control is to ensure that the plant operates as designed, that is, produces the demanded amount of product within specifications while minimizing consumption of energy, process water, and other resources. The process should operate safely and produce only a minimum of waste streams.

Hence, the aim is to achieve the desired performance of the process despite the presence of process disturbances. This implies monitoring the state of the bioreactor using direct and/or indirect measurements (the so-called soft sensors) as described in Chapter 14. To impose a manual or an automatic actuation on the process, based on the input variables (the available degrees of freedom), either feedback or feed-forward control is used. Basically, it is implied that the control action ensures that the fundamental needs of the microorganism are always satisfied.

For most bioprocesses, the main challenge for the process control system is to ensure the production of a product with a given set of specifications, while respecting the safety and environmental constraints of operation. Jeopardizing

Fundamental Bioengineering, First Edition. Edited by John Villadsen.

product quality can be so costly that for some processes this is the only important objective. Any potential economic benefit gained from a quantitative optimization of the operation has little value compared to the cost of not meeting the basic product specifications, especially concerning the product quality.

The industrial bioprocess consists of a number of subprocesses from sterilization, inoculation in small-scale batch reactors, further growth of the culture in larger scale bioreactors using batch or fed-batch mode, to the production phase in fed-batch or in continuous mode, and finally to downstream processing. The objective of this chapter is to discuss the *bioreactor control problems* and to highlight some general traits that distinguish operation of bioprocesses from operation of processes in the conventional chemical process industries. A brief (and certainly incomplete) introduction of the concepts of process control in general will be given before discussing specific strategies for bioprocess control. For a more comprehensive treatment of the general subject of bioreactor control, the reader is referred to Van Impe, Vanrolleghem, and Iserentant [1], Dochain *et al.* [2] and Stanke and Hitzmann (2013).

15.2 Bioprocess Control

Regardless of the operational mode of the reactor (batch, fed-batch, or continuous), the purpose of the reactor control is to maintain stable operation within an acceptable operating window. This is to be achieved in the presence of process disturbances by actuation (changes in manipulated variables). If possible, the control should also optimize the operation if this secondary objective is not jeopardizing the primary objective. It is, therefore, important to know what the terms stability, optimality, actuation, and disturbances mean in the context of a bioreactor.

Example 15.1 Bioreactor operation

Given a simple bioprocess in a stirred tank where a substrate S is converted into cells. Simple unstructured kinetics with substrate inhibition (Chapter 7, Eq. (7.21)) is used. This simplistic model does not include the need for nutrients other than the substrate S in the reactor environment. It is assumed that the operating conditions have been optimized and are fixed during operation. The simple mass balance expressed in concentrations is

$$\frac{\mathrm{d}x}{\mathrm{d}t} = \mu x - \frac{v}{V}x. \tag{15.1}$$

$$\frac{\mathrm{d}s}{\mathrm{d}t} = \frac{v_0}{V}s_0 - \frac{v}{V}s - \frac{\mu x}{Y_{sx}}. \tag{15.2}$$

$$\frac{\mathrm{d}V}{\mathrm{d}t} = v_0 - v. \tag{15.3}$$

Here, V is the active reactor volume and $[s, x]$ are the concentration of substrate and biomass, respectively. v_0 and v are the volumetric flow rates in and out of the reactor, and the inflow has the substrate concentration s_0. Y_{sx} is yield coefficient of biomass on the substrate and μ is the specific growth rate of the biomass. The specific growth rate with substrate inhibition as a function of the process variable s is given by

$$\mu = \mu_{max}\left(\frac{s}{K_s + s + \frac{s^2}{K_I}}\right), \tag{15.4}$$

where K_S is the Monod constant and K_I is the inhibition constant. μ_{max} is the maximum possible specific growth rate for a given reaction temperature as will be shown in Example 15.2.

This model covers all three reactor types, the batch reactor when $v_0 = v = 0$, the fed-batch reactor $v_0 \neq v = 0$, and the continuous reactor $v_0 = v \neq 0$. For the continuous reactor, the dilution rate $D = v/V$ is a given *constant* operating variable.

For the process described, it is reasonable in many production facilities to regard the substrate concentration s_0 in the feed to the fed-batch or the continuous process or the initial concentration s_0 for the batch as a *disturbance variable* that may fluctuate. In some reactor configurations, it may be a *manipulated* variable, available for control. The value of s_0 may be measured or not. In any case it is the operation of the reactor that determines if the desired biomass concentration can be achieved after ending the batch or fed-batch phase or in the effluent from a steady-state continuous process, that is, in this example the *performance* of the process is directly related to the biomass concentration of the product. The available means to influence the process through actuation depends on the mode of operation.

In the batch mode, the reaction time is the only available degree of freedom for the operator. To predict the fermentation time, either the initial concentration $s = s_0$ at $t = 0$ must be known (based on what is charged to the reactor) to predict the reaction time, or the biomass concentration in the reactor needs to be monitored online.

For the fed-batch, both the feed rate and the total reaction time can be manipulated, which makes it possible to *optimize* the operation, that is, to find how the feed to the reactor should be manipulated to achieve the desired volume and concentration as fast as possible. This strategy can be implemented offline based on a given value for s_0 in the feed, and it results in a feeding strategy (recipe) for the operation. In case of a disturbance, changes in this strategy are needed.

For continuous operation, the feed rate determines the dilution rate of the process, that is, the reciprocal of the residence time for the reactor. This variable affects the steady-state concentration profile of both the biomass and the substrate within the reactor and hence the conversion of substrate into biomass. The dilution rate is the only variable that can be manipulated to control the product concentration.

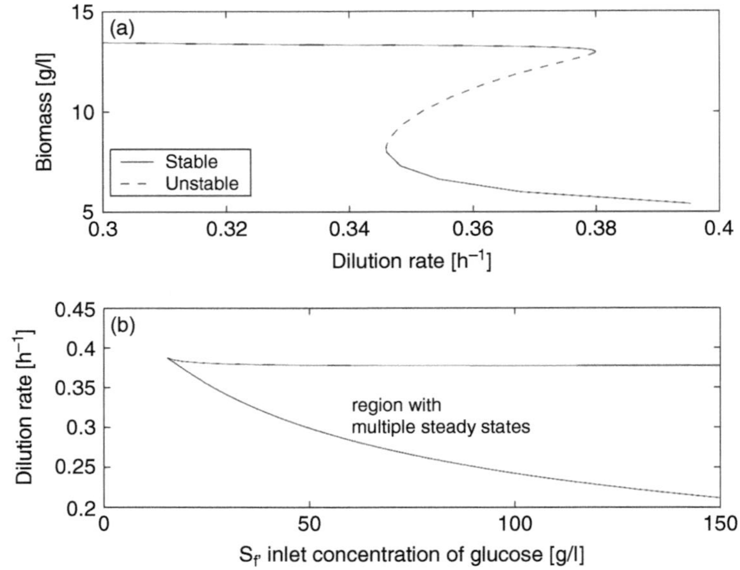

Figure 15.1 One- and two-parameter bifurcation diagram for continuous cultivation of *S. cerevisiae* [3]. In the top graph (a) stable and unstable steady-state solution for the biomass concentration is shown for a fixed dilution rate. In graph (b) the region where multiple steady states are possible is shown as a function of dilution rate and feed concentration of glucose.

Stable operation is paramount in any reactor configuration. It is known that in continuous fermentation processes and with certain forms of the kinetic expression, one may run into operational conditions, for example, in a range of D values, where the steady-state mass balance has more than one solution $s(D)$. A detailed study of continuous cultivation of *Saccharomyces cerevisiae* has shown this behavior, and in Figure 15.1 a bifurcation diagram is shown for this system from Lei [3]. The bifurcation diagram is constructed based on static models (much more complex than Figure 15.4) and a semiquantitative explanation is given by Villadsen and Jørgensen [4]. This behavior has very important implication in both the design of the reactor and the choice of operation strategy.

Besides the batch, fed-batch, and continuous operating mode, a bioreactor can also be operated as a sequential batch reactor (SBR) with cycles of different distinctive phases [5]. Typical phases of SBR operation are as follows: filling, reaction in batch or fed-batch mode, and settling of solids. The individual SBR phases can be operated in aerobic or anaerobic mode. A control strategy for the SBR is formulated for each of the phases, and a policy is devised for choosing the time of shift from one phase to the next. Since the transition time between different phases is often short compared to the duration of each phase, the control of the SBR can be designed based on the same principles as are used for the individual operation phases. The shifting criteria will have to be based either on a fixed operation time for each phase or on online measurements of process variables that determine when to shift. The performance of the SBR is highly dependent on the strategy for shifting between the different phases of the total process.

15.2.1
Design Variables in Bioreactor Control

As seen from Section 14.4, it is far more difficult to model and measure the state of the microorganism than the environment that surrounds it. The state vector for the microorganism includes concentrations of biomass, substrate, products, and variables that determine the redox state of the culture ($NADH/NAD^+$), as well as other metabolic fluxes that may indicate a change in the metabolism of the organism. The environmental variables are temperature, pH, dissolved oxygen concentration, and so on. Often, the environmental variables can be used to predict the state of the microorganism.

The inhibitory effect of high dissolved oxygen concentrations may be expressed by the same type of unstructured kinetic expression as given in Eq. (15.4). Inhibition due to products or to nonlimiting substrates such as NO_3^- can be modeled as shown in Chapter 7. The effect of temperature and pH is directly seen in the value of μ_{max} in the kinetic expressions, and an optimum T and pH exists. The reason for an optimum cultivation temperature T is easily understood. For low values of T, the rate constant μ_{max} increases with temperature by the Arrhenius expression. Above the optimum T, degradation of proteins sets in, and μ_{max} decreases rapidly with T. The temperature effect on μ_{max} is modeled by Eq. (15.5), an expression used by Roels [6] and by Villadsen Nielsen, and Lidén [7].

$$\mu_{max} = A\exp\left(\frac{-E_g}{RT}\right) / \left(1 + B\exp\left(\frac{-\Delta G_d}{RT}\right)\right). \tag{15.5}$$

Here, A and B are two preexponential constants, E_g is the activation energy of the growth process, and ΔG_d is Gibbs free energy for the denaturation process that leaves the proteins inactive. R is the universal gas constant and T is the temperature in Kelvin.

Example 15.2 Control of the optimal growth rate of biomass

Consider the bioprocess described by the model (15.1)–(15.4) in Example 15.1, it is desired to optimize the volumetric productivity q_x of biomass. The kinetic parameters have the following values: $K_S = 10$ g/l, $K_I = 0.1$ g/l, and μ_{max} is given by model (15.5). The yield coefficient for biomass is $Y_{sx} = 0.1$ g X/g S and the feed concentration of limiting substrate S is $s_0 = 4$ g l^{-1}. In Figure 15.2, the temperature dependence for μ_{max} is shown using the following set of parameters that are reported for *Escherichia coli* in glucose-rich medium by [7]: $E_g = 58$ KJ/mol, $\Delta G_d = 550$ KJ/mol, $A = 1.0\ 10^{10}$ h^{-1}, and $B = 3.0\ 10^{90}$.

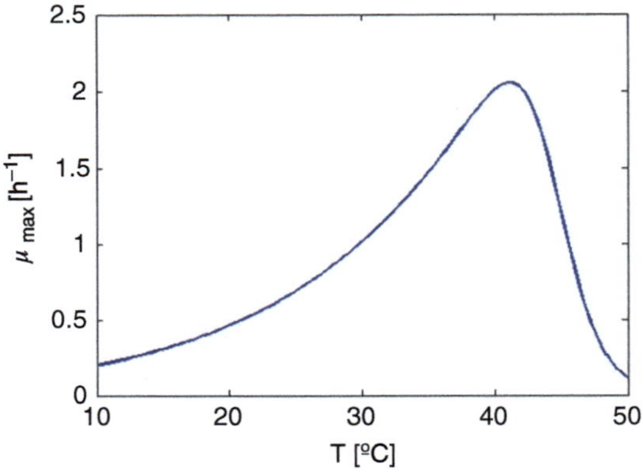

Figure 15.2 Temperature effect on the specific growth rate. The largest value is approximately 2 h^{-1} for μ_{max}, obtained at 42 °C.

It is clearly seen that there is need to keep the reactor temperature in a narrow band around the optimal value in order to secure optimal growth in the reactor. Hence, a bioreaction producing heat cannot be operated optimally if the cooling is not sufficiently effective. Such effects can be experienced locally in the reactor if the mixing is inadequate.

The maximum theoretical value of μ in the rector is obtained by optimizing the temperature and substrate concentration in relation to Eqs. (15.4) and (15.5). Inserting the s value $s = \sqrt{K_s K_I} = 1\,\mathrm{g\,l^{-1}}$ (obtained by differentiation of (15.4) as in Eq. (10.12)) into (15.4) gives $\mu = 2 \cdot \frac{1}{10+1+10} = \frac{2}{21} = 0.095\,\mathrm{h^{-1}}$. In steady state, $\mu = D$, but operation at $D = 0.095$ h^{-1} is not possible, since even the slightest increase of D will lead to washout. The maximum volumetric productivity is obtained as described in Chapter 10, Eq. (10.8). $S_{opt} = -c + \sqrt{c^2 + c}$, where $c = a/(1+b)$ and $[a, b] = [K_s/s_0, s_0/K_i]$. For $[K_s, K_i, s_0] = [10, 0.1, 4\,\mathrm{g\,l^{-1}}]$, one obtains $S_{opt} = 0.19\,337$ ($\rightarrow s_{opt} = 0.7735\,\mathrm{g\,l^{-1}}$). $D_{opt} = 0.09\,232\,\mathrm{h^{-1}}$, and finally $q_x{}^{max} = D_{opt}\, Y_{sx}(s_0 - s_{opt}) = 0.09\,232 \cdot 0.1 \cdot (4 - 0.7735) = 0.0298\,\mathrm{g\,X\,l^{-1}\,h^{-1}}$.

Operating at $D_{opt} = 0.09\,232\,\mathrm{h^{-1}}$ is not very safe since D_{opt} is close to the washout dilution rate $0.095\,\mathrm{h^{-1}}$. One must accept a lower q_x by operating at a lower D value. For $S = 0.12$ ($s = 0.48\,\mathrm{g\,l^{-1}}$), $D = 0.0744\,\mathrm{h^{-1}}$. Here, $q_x = 0.0744 \cdot 0.1 \cdot (4 - 0.48) = 0.0262\,\mathrm{g\,X\,l^{-1}\,h^{-1}}$. Figure 15.3 shows μ and q_x as functions of substrate concentration.

The example shows, however, that particular attention must be paid to a robust control of the dilution rate. The same disastrous effect of a poor dilution rate control is not found in product-inhibited kinetics, but clearly every effort must be made to control D, especially when s_0 fluctuates since the maximum value of q_x is strongly dependent on feed concentration s_0 in the nonlinear rate expressions.

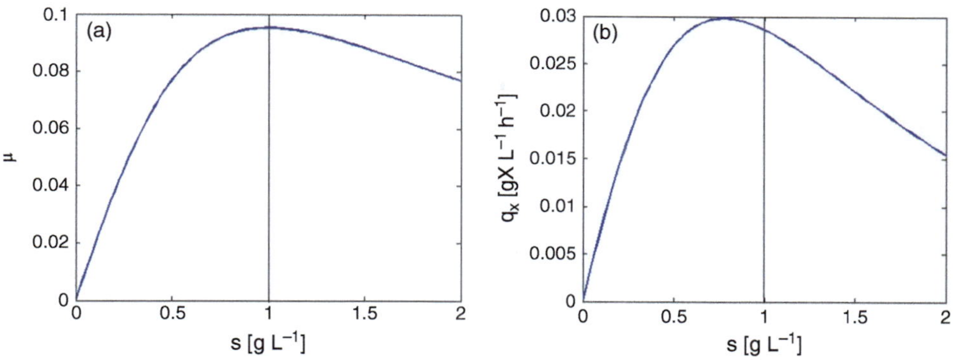

Figure 15.3 The specific growth rate μ (a) and the volumetric productivity q_x (b) as functions of the steady-state substrate concentration in the bioreactor. The vertical lines indicate the upper limit for washout conditions in the reactor.

Example 15.2 shows how the growth rate and thereby the performance of the process is affected by temperature and by substrate inhibition. If the growth is inhibited by products, by dissolved oxygen, or by a nonlimiting substrate such as NO_3^{-1}, all of these variables will have to be controlled. This requires that the operator can make changes to an input variable that subsequently influences the value of the process variable that needs to be controlled. This very general problem can be investigated by looking at the sensitivities of the process variables to all the available manipulated variables. Likewise, a sensitivity analysis between the process variables and the disturbance variables gives information on how difficult the process may be to keep under the desired operation condition.

An alternative approach is to derive a model that explicitly describes the dynamic relation between a given combination of a process variable and an input variable. This type of model is called a *transfer function*, and the transfer function can be derived from a linear state space description. Often, such transfer function models form the basis for design of control loops.

Example 15.3 Sensitivities to operational variables on continuous fermentations

In the example there are two substrates, the carbon source S and oxygen. The kinetics is given by Eq. (15.9), and the transient mass balances for X and the two substrates by Eqs. (15.6)–(15.8). Investigations of the sensitivity for the state variables in a continuous fermentation process on some of the operational variables will be shown. This investigation can show if an actuator variable can be used to affect the value of a controlled variable. For this example, the following simple CSTR model is used, that is, $v_0 = v$ and $v/V = D$, which is the dilution rate of the reactor. The model describes the concentration of biomass, x, substrate, s, and dissolved oxygen in the reactor assuming double-Monod kinetics.

$$\frac{dx}{dt} = -Dx + \mu x. \tag{15.6}$$

$$\frac{ds}{dt} = D(s_0 - s) - \frac{\mu x}{Y_{sx}}. \tag{15.7}$$

$$\frac{do}{dt} = D(o_0 - o) - Y_{xo}\mu x + k_l a(o^{sat} - o). \tag{15.8}$$

$$\mu = \mu_{max}\left(\frac{s}{K_s + s}\right)\left(\frac{o}{K_o + o}\right). \tag{15.9}$$

The parameters of the model is $K_S = 0.1$ g/l, $K_O = 0.2$ mg/l, $\mu_{max} = 0.689$ h^{-1}. $Y_{sx} = 0.4\,g\,X\,(g\,S)^{-1}$, and $Y_{xo} = 1.5\,g\,O_2\,(g\,X)^{-1}$. We shall examine the biomass concentration as a function of D in the interval $0.2 \leq D \leq 0.5$ for discrete values of the two variables s_0, and $k_l a$. A set of nominal operational values are chosen: $s_0 = 10\,g\,l^{-1}$, $k_l a = 250\,h^{-1}$. The concentration of oxygen o_o in the liquid feed is assumed to be equal to 0. The medium is sparged with air, and at a reaction temperature of 25 °C, the saturation concentration of O_2 is 8.36 mg $O_2\,l^{-1}$ based on Eq. (9.2) in Chapter 9.

Solving the steady-state mass balances as functions of D for specified values of s_0 and $k_l a$ gives the curves shown in Figures 15.4 and 15.5. Besides the results for the nominal values of the two control variables s_0 and $k_l a$ (which can be changed by manipulating the stirring rate), results are also shown for variations of 10% up and down for these control variables. It is seen that for low dilution rates the biomass can effectively use all the substrate fed to the reactor, and the concentration of biomass is unaffected by the dilution rate, but can be increased by increasing the substrate feed concentration. For high dilution rates, the biomass concentration is not sensitive to the substrate feed concentration but depends only on the dilution rate. The transition between the two regimes happens for a value of the dilution rate between 0.3 and 0.35 h^{-1}. For $D > 0.35$ h^{-1}, the biomass concentration x increases with increasing value of $k_l a$, while no effect of increasing mass transfer is seen at low D.

As shown in Example 15.3, it is necessary to identify which input variables through actuation can keep the controlled variables at the desired point of operation. Table 15.1 lists a number of classical control loops used in industrial fermentation processes to control the environmental variables. Even in the batch reactor, several of these loops in the table are active [8]. For fed-batch operation, the rate of substrate addition v_0 and the feed concentration s_0 substrate are included, and for the continuous reactor, one may also include a loop on the effluent flow if biomass recirculation is used.

For CSTR operation of the bioreactor, a number of specific operational modes have been defined that are widely used in industry. All these modes aim at controlling the operation such that some property remains constant for the duration

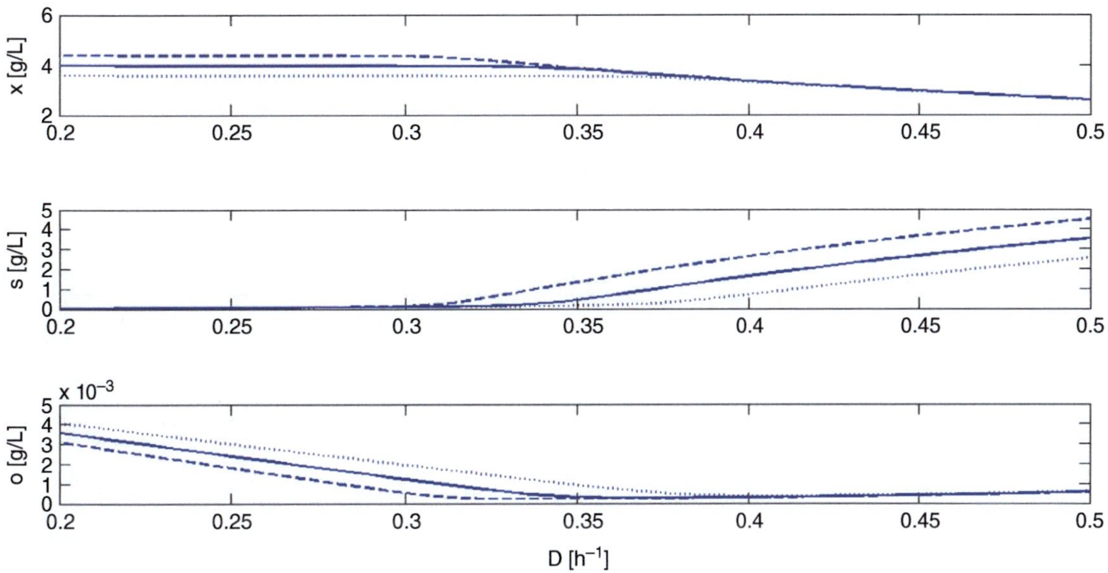

Figure 15.4 Steady-state profiles of biomass, substrate, and dissolved oxygen in a continuous fermentation with double-Monod kinetics for substrate and oxygen. The dashed line represents 10% increase in the feed substrate concentration and the dotted line represents a 10% decrease in this variable.

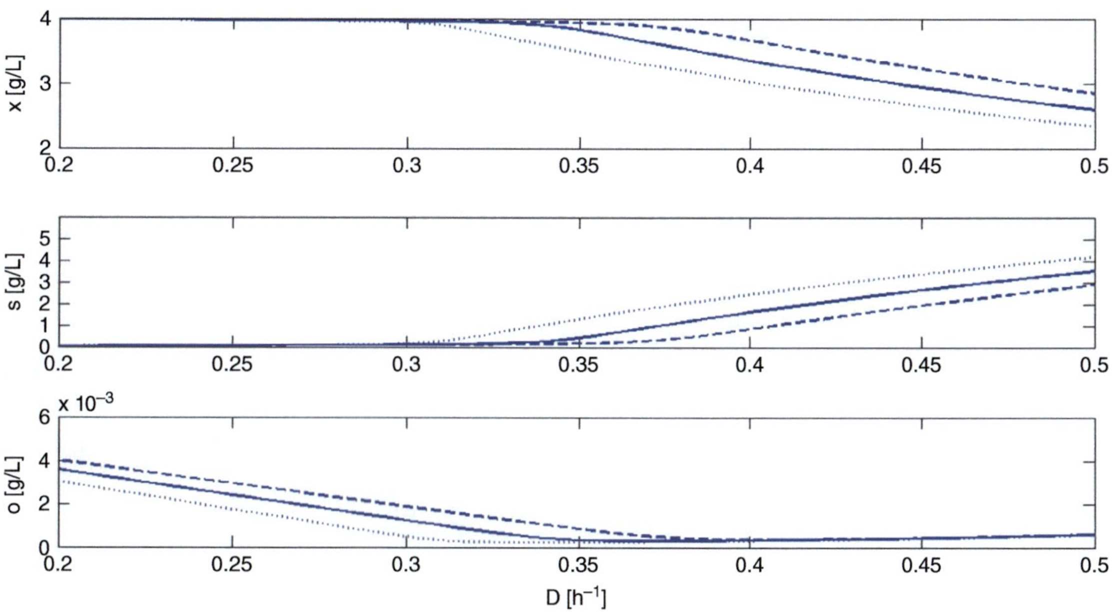

Figure 15.5 Steady-state profiles of biomass, substrate, and dissolved oxygen in a continuous fermentation with double-Monod kinetics for substrate and oxygen. The dashed line represents a 10% increase in $k_l a$ and the dotted line a 10% decrease in this variable. This is achieved by changing the mixing intensity of the reactor.

Table 15.1 Examples of single-loop control strategies for a number of environmental variables in industrial fermentation.

Controlled variables	Actuator variable	Measurement	Control algorithm
Temperature	Heating or cooling medium flow rate	Reactor temperature or jacket temperature	PI(D), possibly in cascade with jacket temperature
Agitation	Stirrer speed	Tachometer	PI(D)
Pressure	Effluent back-pressure control valve	P sensor	PI
pH	Acid/base flow rate	pH electrode	PI with dead band
Dissolved oxygen	Air or gas flow rate	DO electrode	PI
Substrate supply	Substrate feed rate	Flow meter	Recipe
Nutrient supply	Nutrient feed rate	Flow meter	Ratio to substrate
Effluent flow	Harvest rate	Flow meter	
Weight or level	Substrate feed rate	Load cell or differential pressure.	PID cascade from recipe
Foam	Antifoam supply	Conductance/capacitance	On–off control

of the reaction. For the CSTR, only the dilution rate is an available degree of freedom in terms of controlling the reaction media, besides separate nutrient feeds, aeration, cooling, and agitation that is not considered here. These operational modes as mentioned in Chapter 2 are as follows:

Chemostat: Keeps the weight or mass of medium constant by manipulating the inflow of medium v_f. The constant dilution rate is used to control the growth rate.

Turbidostat: Works similar to the chemostat but uses feedback between the turbidity (biomass effluent concentration) and the dilution rate instead of the mass of medium in the reactor to give constant D.

Productstat: Similar to chemostat but using the product concentration to control the dilution rate.

Auxostat: Controls D such that one measured property of the bioreactor remains constant, for example, pH or dissolved oxygen. In this way, it is a generalization of the chemostat, turbidostat, and productstat control strategies.

Retentostat: Liquid is removed continuously but biomass is retained in the reactor so that the biomass concentration increases until it is limited by the rate of added nutrients.

In batch operation, it is important to compensate for variation in the feed stock, and control algorithms have been developed to maintain the *final* result of successive batches on the given specifications (see Bonné and Jørgensen [9] and Jørgensen and Bonné [10]). Reproducibility in biochemical batch operation is often of higher priority than optimality since reproducible results from fermentation processes directly translate to safe operation in financial terms. The risks of having a production process where specifications are not met is considerable, since the living cells are complex organisms that can well change metabolism and undergo unexpected disturbances.

15.2.2 Challenges with Respect to Control of a Bioreactor

It is not easy to run a bioreaction in the optimal way. Besides obtaining online high-quality measurements that are representative for the whole reactor volume – a problem well known also for ordinary chemical reactions – the development of a dynamic process model of sufficient quality remains an obstacle for advanced model based control. The fact that most industrial production processes rely primarily on a few secondary measurements remains as one of the key obstacles to implementation of efficient control schemes. While significant advances in development of sensors has been realized in recent years, implementing process analytics is expensive and requires maintenance that constitutes as barrier for a more widespread use. Another main problem is that the simple unstructured models as presented in Examples 15.1 and 15.3 are based on the assumption of metabolic steady-state behavior for the organism, an issue already discussed in Chapter 7. In reality, the dynamic response of the microbial population may be dominated by overflow metabolism or other phenomena that are not at all described by the models conventionally used for control design. Also, the metabolism in the transient state can be dominated by highly nonlinear dynamics, and most control design methods are based on approximate linear input/output relations. This means that care must be exercised when an unstructured model (usually all that is available) is used for design and validation of suitable control structures. Nevertheless, the unstructured model does contain a great deal of information that should be utilized when designing the bioreactor operation.

15.3 Principles and Basic Algorithms in Process Control

15.3.1 Open Loop Control

An open loop control strategy is based on a predetermined actuation sequence for operation of the process. The strategy may be constructed based on an extensive experimental optimization procedure or from a mathematical optimization based on a process model. In any case, the bioreactor is operated based on a recipe that is repeated for every batch or fed-batch. This could easily be the applied strategy in terms of reaction time, initial loading of the reactor and the feeding profile of substrates, and other nutrients while pH and temperature are operated in closed loop with feedback control.

15.3.2 Feed-forward and Feedback Control

There are two fundamental types of control strategies, namely, feedback and feed-forward control. The feed-forward control assumes that process disturbances can be measured before they affect the process, and action is taken to counteract the disturbance. In the ideal case, the process will not be affected by the disturbance at all. This is the case where a feed to a process is a blend from two different streams. A change in concentration in one of these could be counteracted by changing the ratio between the two. Perfect feed-forward requires perfect process information in terms of a model of how both disturbances and the manipulated inputs affect both the process variables and the perfect measurements of the disturbance. This idealized scenario is an abstraction far from reality in most processing plants. Nevertheless, feed-forward control can be very effective in case of reasonable process insight and good process measurements. It should be utilized when the opportunity is there.

Feedback control is using measurements of the controlled process variables and compares these against their target values (set points). In case of an offset between the controlled variables and their set points, the values of the errors are used in an algorithm to adapt the manipulated variables with the purpose of minimizing this error. Figure 15.6 shows the conceptual structure of a feedback and feed-forward control loop. More information on these basic control principles can be found in Bequette [11] and Seborg, Edgar, and Mellichamp [12].

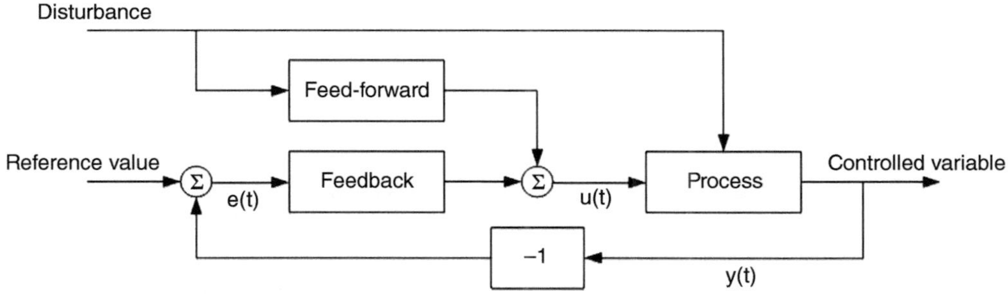

Figure 15.6 Conceptual structure of feed-forward and feedback control loops.

15.3.3
Single-Loop PID Control

The dominating algorithm for process control in industry is the single-loop PID controller. The acronym indicates that the control action can include *P*roportional, *I*ntegral, and *D*erivative action when tracking the error between the controlled variable and its set point(s). Not all three types of actions need to be applied, so it is common to see P, PI, and PD controllers as well as the full PID algorithm. Several special types of industrial implementations exist, and the reader is referred to Åström *et al.* (1995) and Visioli [13] for more details. In the following, a simple introduction to the PID algorithm is presented. Here, the standard nomenclature for control systems will be adapted where u is the manipulated variable, y is the controlled variable with set point y^{sp}, and e is the error signal between the measured value of the controlled variable and its set point.

The feedback control law is an algorithm to calculate the value of the manipulated variable based on the error signal:

$$u(t) = f(e(t)). \tag{15.10}$$

$$e(t) = y^{\text{sp}}(t) - y(t). \tag{15.11}$$

Measurements of the system are taken, the error on the controlled variable is evaluated, and the actuation on the system is determined through the control law (15.10) that gives the new value for the manipulated variable.

For the PID controller, the feedback law takes the following forms:

Proportional action:

$$u(t) = u^{\text{nom}} + K_{\text{c}}e(t) \tag{15.12}$$

Proportional and integral action:

$$u(t) = u^{\text{nom}} + K_{\text{c}}\left(e(t) + \frac{1}{\tau_I}\int e(t)\text{d}t\right) \tag{15.13}$$

Proportional and derivative action:

$$u(t) = u^{\text{nom}} + K_{\text{c}}\left(e(t) + \tau_{\text{D}}\frac{\text{d}e(t)}{\text{d}t}\right). \tag{15.14}$$

Proportional, integral, and derivative action:

$$u(t) = u^{\text{nom}} + K_{\text{c}}\left(e(t) + \frac{1}{\tau_I}\int e(t)\text{d}t + \tau_{\text{D}}\frac{\text{d}e(t)}{\text{d}t}\right). \tag{15.15}$$

In all these algorithms, u^{nom} is the nominal value of the manipulated variable that should be used when the system is in steady state with the controlled variable at its set point. The parameters K_{c}, τ_{I}, and τ_{D} are tuning parameters for the control algorithm that are used to shape the dynamic response of the closed loop, that is, the observed performance of the controlled process.

The interpretation of the proportional action is straightforward as the control law is affine in the error signal with the proportionality constant K_c. This parameter is referred to as the controller gain. This means that large errors give large deviations in the manipulated variable compared to the nominal value u^{nom}, that is, a high degree of actuation on the process. The sign of the gain, K_c, is important and needs to be selected based on the sign of the gain for the process in the transfer function between y and u, that is, a positive gain in the process transfer function means that an increase of u will lead to an increase of y when the system has reached a new stationary point. Therefore, if the measured value of the process variable is too low compared to the set point, the error is positive, the control law should produce an increase in u compared to u^{nom}, and the gain K_c should therefore also be positive. The unit of controller gain, K_c, takes the inverse of unit of the process gain for the process transfer function. The integral action computes the integral of the error from $t=0$ to the present time t and this value is multiplied by K_c/τ_I, which can be regarded as the integral gain. τ_I is a positive constant and is labeled the integral time constant. It is used to balance the integral action versus the proportional action. The advantage of using the integral part of the controller is that static offset can be eliminated in the controlled variable, which is not possible with only proportional action. The contribution from the derivative part adds the derivative of the error with the derivative gain $K_c\tau_D$, where τ_D is labeled the derivative time constant. It can be an advantage to use derivative action since this part has information on the direction in which the error signal is moving. This can lead to significantly faster dynamic responses, but it also means that a noisy measurement can be amplified through the control law and hence lead to a very noisy behavior of the control signal in the manipulated variable. For this reason, the derivative action is often not implemented.

A general interpretation of the PID control law is that the proportional action gives a contribution based on the present error, the integral part integrates the past of the error and the derivative part forecasts the future movement of the error. There exist numerous sets of guidelines for tuning of these parameters. Most of these guidelines are based on simple transfer function-type models and an overview can be seen in O'Dwyer [14] and Seborg, Edgar, and Mellichamp [12]. In general, the method in Skogestad [15] can be recommended. Methods such as autotuning algorithms also exist for tuning in case one has no accurate model information, but these methods require that data are obtained directly from the process [16–18].

Example 15.4 Bioreactor operation with single-loop PID control

Given the process model in Example 15.3, it is desired to control the biomass concentration in a continuous bioreactor through manipulation of the dilution rate. Two scenarios will be considered. In scenario (a) it is desired to move the operational point from one biomass concentration to another and in scenario (b) it is desired to maintain the biomass concentration, but a disturbance is affecting the feed concentration of the substrate. This disturbance is considered to be unknown and therefore has to be mitigated by feedback control only.

In order to derive an appropriate control law, it is advantageous to know the transfer function that gives the dynamic relation between the manipulated variable, the dilution rate, and the controlled variable, the substrate concentration in the bioreactor. This transfer function can be derived based on a linear-state space model that one would obtain from Taylor series expansion of the nonlinear model (15.6)–(15.9) based on a nominal point of operation. From Example 15.3, $D = 0.3\,\text{h}^{-1}$ is chosen that gives the concentrations $x = 3.96\,\text{g/l}$, $s = 0.102\,\text{g/l}$, and $o = 1.23\,\text{mg/l}$. If the dilution rate, D, and the aeration through $k_l a$ are chosen as manipulated variables and the feed concentration of substrate a disturbance, the linear-state space model in deviation variables is as follows:

$$\frac{d}{dt}\begin{bmatrix} x \\ s \\ o \end{bmatrix} = \begin{bmatrix} 0 & 5.7 & 135 \\ -0.75 & -15 & -336 \\ -0.45 & -2.2 & -452 \end{bmatrix}\begin{bmatrix} x \\ s \\ o \end{bmatrix} + \begin{bmatrix} -4.0 & 0 \\ 9.9 & 0 \\ -0.0012 & 0.0071 \end{bmatrix}\begin{bmatrix} D \\ K_l a \end{bmatrix} + \begin{bmatrix} 0 \\ 0.30 \\ 0 \end{bmatrix} s_0. \tag{15.16}$$

If the controlled variable is the biomass concentration in the reactor and the manipulated variable the dilution rate, then the transfer function between this input–output pairing is

$$\frac{X(s)}{D(s)} = \frac{-3.96s - 1790}{s^2 + 467s + 5810} = \frac{-3.96(s + 452)}{(s + 454)(s + 12.8)} \approx \frac{-3.96}{(s + 12.8)} = \frac{-0.309}{0.0781s + 1}. \tag{15.17}$$

This is approximately a first-order system with a negative gain of $-0.309\,(\text{g h})/\text{l}$ and a time constant of 0.0781 h. That means that the model predicts that if the dilution rate is increased by $1\,\text{h}^{-1}$, then the biomass concentration will be reduced by 0.309 g/l, and 63% of the change has occurred after 0.0781 h. The sign of the gain is in correspondence with Figure 15.4 in Example 15.3. Equation (15.17) is expressed in the Laplace domain, which is indicated by using $X(s)$ as the transformed variable of the biomass concentration $x(t)$. s refers to an independent complex variable.

Based on this information in the transfer function (15.17), a controller can be designed that uses the dilution rate to control the biomass concentration. It is seen that the controller gain needs to be negative. In Figure 15.7, a set of closed loop simulations are presented using only proportional control.

It is seen in the figure that none of the closed loop responses is able to eliminate the offset between the biomass concentration and its set point. It is also seen that the controller gain affects the size of the offset and how fast the closed loop system settles at a new steady state. For the case where a disturbance affects the system, the system has been simulated

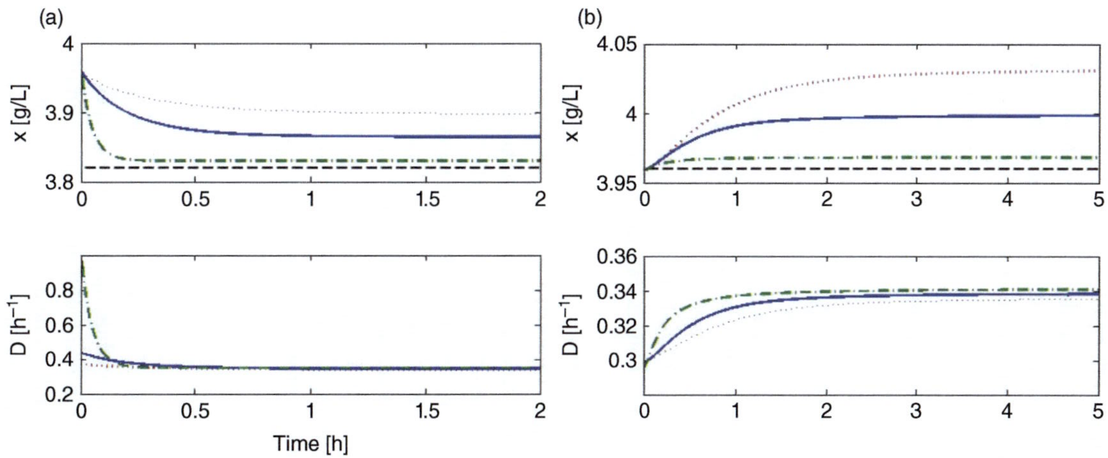

Figure 15.7 Closed loop response for a bioreactor with proportional control. The gain of the controller, K_c [l/(g h)], is varied with −0.5 for the red dotted curve, −1 for the full blue curve, and −5 for dash dotted green curve. The dashed black line is the set point for the biomass concentration. In (a) the set point is changed from 3.95 g/l to 3.82 g/l and in (b) the feed concentration of substrate is changed from 10 g/l to 11 g/l.

with an increase of the substrate concentration. In the opposite case with a step in the feed concentration of substrate from 10 to 9 g/l, the control problem becomes infeasible for this set point and controller tuning and the controller force the dilution rate toward zero.

In Figure 15.8, the same two closed loop scenarios are simulated for the system, but now the controller is a PI controller using the recommended tuning rules by Skogestad [15] with a desired closed loop time constant τ_C of 0.05 or 0.10 h, which is either slightly faster or slower than the open loop system. This gives a controller gain K_c of -5.06 or $-2.03\,l/(g\,h)$ and an integral time $\tau_I = \tau = 0.0781$ h in both cases.

It is seen in Figure 15.8 that the PI controller eliminates the steady-state offset that was observed using the P controller. The signal for the controlled variable exhibits an overshoot before stabilizing at the set point. When the closed loop response is tuned to be faster, the numerical value of the gain is larger and the control signal reacts more aggressively to an error. In this case, the value of $\tau_C = 0.05$ h gives a very reasonable response.

Example 15.4 shows how a simple PID control structure can be developed for the bioreactor based on the unstructured process model from linearization by Taylor series expansion and development of transfer functions. In case the linear model is used for design of more than one loop, it is necessary first to determine the available degrees of freedom for control based on the nonlinear model as this information is lost in the linearization [12,19]. For example, it is not possible to control both biomass and substrate concentrations in the bioreactor independently.

The great success of the PID controller in the process industry in general and its widespread use today in the biochemical industry is due to its simple structure, well-documented theory, and its versatile application range across the field of automation. It does not require a process model but can utilize the information if a model is available. Further, it can be easily tuned to be robust over a large working range. The main drawback of the PID controller is that it is a decentralized single-loop control strategy that may be difficult to tune in case of strong interaction between individual control loops. It also lacks the predictive capabilities of a model-based control strategy, which therefore is better to incorporate loop interactions and feed forward information into the control law.

15.3.4
Diagnostic Control Strategies

As the true dynamics of the system is often not accurately known, it can be advantageous to perform system diagnostics while operating the process. Based on knowledge of the operational variable in combination with, for example, off-gas analysis measurements it may be possible to deduce more accurate information on the state of the organism that can be used to make better decisions

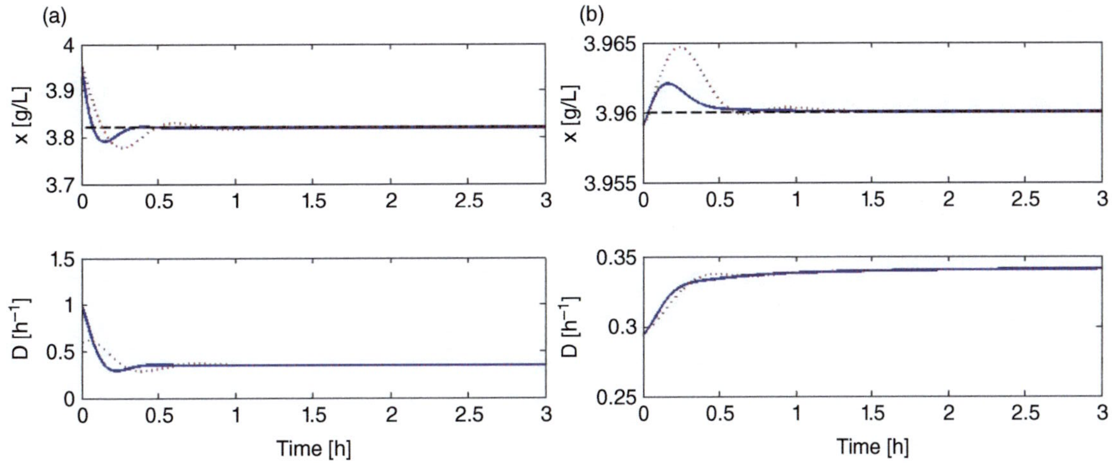

Figure 15.8 Closed loop response for a bioreactor with proportional and integral control. The desired closed loop time constant, τ_C, is varied with 0.10 h for the red dotted curve and 0.05 h for the full blue curve. The dashed black line is the set point for the biomass concentration. In (a) the set point is changed from 3.95 g/l to 3.82 g/l and in (b) the feed concentration of substrate is changed from 10 g/l to 11 g/l.

regarding future control moves. To get data with sufficiently high information content in comparison to the measurement noise, it is necessary to excite the system through the input variables. Operation with this type of online diagnostics and adaptation of the control strategy is used in a number of different algorithms; probing control or extremum seeking control are some of these.

15.3.4.1 Probing Control

This method was originally developed for preventing accumulation of acetate in cultivations of *E. coli*, but has also been used for other organism with overflow metabolism and for the control of aeration in large-scale fermentation processes close to oxygen transfer limitations [20,21]. The principle is to optimize productivity by feedback and identification of reactor behaviors related to the optimal conditions. The basic idea is that the feeding in fed-batch mode is controlled by monitoring of, for example, dissolved oxygen. Pulses in the feed rate are imposed on the reactor, and the response in dissolved oxygen is recorded. If the response in the dissolved oxygen is sufficiently large, the feed rate can be increased proportionally to the size of the pulse response. If there is no detectable response in the dissolved oxygen, it is an indication that the maximum oxygen transfer capacity of the reactor has been reached and the feed rate needs to be reduced in order to maintain the aerobic conditions in the reactor [22].

15.3.4.2 Extremum Seeking Control

The purpose of extremum seeking control is to find a set of parameters or an operation point that minimizes some objective function. The algorithm can be used for online optimization when the input/output dynamics are unknown. The requirement is that an objective function can be constructed from the available measurements on the system. The idea is to use a continuous perturbation, typically in the form of a sine wave on the input and recording the response in the output. The output signal is passed through a filter and multiplied by the perturbation signal that then produces an estimate of the derivative of the objective function with respect to the decision variables. These can then be updated ensuring that the gradient is forced toward zero, which implies that the process is being optimized [23].

15.3.5 Plant-Wide Control Design

A plant-wide control design refers to the structure of the controller rather than the control algorithm or control law itself. In most processing plants, there are several operational degrees of freedom available for control of a number of different process variables. These actuator variables are typically flow rates of material, heating, or cooling as well as agitation rate as seen in Table 15.1. As each of these process inputs may influence several process variables, it is important to make a proper selection of controlled variables based on the available degrees of freedom, so is making an appropriate pairing between these for a single-loop control algorithm such as a PID controller. The proper pairing should ensure that each of the controlled variables is sensitive to its corresponding actuator variable as discussed in Example 15.3. Further, it is desired that cross sensitivities between different pairs are as weak as possible. If strong coupling exists, it is typically necessary to detune at least one of the loops. Part of the plant-wide control design is also to evaluate the potential enhanced closed performance by combined feedback and feed-forward control as well as more advanced control structures such as cascade control, decoupling strategies, or multivariable control. A more in-depth introduction to plant-wide control is provided by Larsson and Skogestad [19] and in Stephanopoulus and Ng [24]. A number of industrial examples are provided in Rangaiah and Kariwala [25].

15.3.5.1 Cascade Control

Cascade control means that two single-loop controls are nested such that the outer loop uses the set point of the inner loop as a manipulated variable. It is a master (primary loop) and slave (secondary loop) structure where the inner loop is the slave of the outer loop. Typically, the inner loop is executed much faster than the outer loop and both loops are of the PID type [11,12]. An example of a cascade control structure for a bioreactor can be for the control of the reactor temperature. The primary loop then use the reactor temperature as the controlled variable and this is measured. This measurement is compared to its set point, and the control action of the primary loop could then be the set point for the temperature in the cooling jacked or coils for the reactor. The secondary loop uses this set point and a temperature measurement to determine if the flow of cooling water should increase or decrease. The advantage is that the desired control variable can be fixed in the primary loop, here the driving force for heat transfer, while the secondary loop translates this desired objective into a control signal that can be manipulated directly.

When tuning cascade loops, the inner loop is tuned first and then closed. Then, the outer loop is tuned based on the dynamics of the resulting closed inner loop.

15.3.5.2 Multiloop Control Strategies

A multiloop control strategy involves more than one manipulated variable in the loop to control one or more controlled variables. It is implied that the control calculations treat all the manipulated variables in one solution strategy in contrast to decentralized single-loop control, which typically consists of several PID loops on the same process. Multiloop control strategies are model based, and coordinated action can therefore be implemented. For this to be effective, good process models of the process dynamics are needed, and these advanced techniques are seldom seen applied for bioreactor control in the industry. The advantage of multiloop and a model-based control strategy is obvious, since the different control loops will not be in conflict, and optimal action can be implemented based on predicted process behavior. Nevertheless, these more advanced control technologies are almost exclusively used in production systems where minimizing production cost is the primary competitive factor, and they are preferred in continuous production systems where the dynamics can be well approximated by linear models.

References

1 Van Impe, J., Vanrolleghem, P., and Iserentant, D. (1998) *Advanced Instrumentation, Data Interpretation and Control of Biotechnological Processes*, Kluwer, Amsterdam.

2 Dochain, Denis, ed. (2008) *Bioprocess Control*, ISTE Ltd. and John Wiley & Sons, Inc.

3 Lei, F. (2001) Dynamics and nonlinear phenomena in continuous cultivation of *Saccharomyces cerevisiae*. PhD thesis. Dept. Chemical Engineering, Technical University of Denmark, Denmark.

4 Villadsen, J. and Jørgensen, S.B. (2014) Reflections on the aerobic fermentation stoichiometry of Crabtree positive yeasts. *Biotechnol. Bioeng.*, **111**, 632–637.

5 Wang, L.K. and Li, Y. (2009) Sequencing Batch Reactors, in *Biological treatment processes*, Handbook of Environmental Engineering, vol. **8** (eds L.K. Wang, N.C. Pereira, Y.-T. Hung, and N.K. Shammas), Humana Press, New York.

6 Roels, J.A. (1983) *Energetics and Kinetics in Biotechnology*, Elsevier Biomedical Press, Amsterdam.

7 Villadsen, J., Nielsen, J., and Lidén, G. (2011) *Bioreaction Engineering Principles*, 3rd edn, Springer, New York Dordrecht Heidelberg London.

8 Caccavale, F., Inmarino, M., Pierri, F., and Tufano, V. (2011) *Control and Monitoring of Chemical Batch Reactors*, Springer London.

9 Bonné, D. and Jørgensen, S.B. (2001) Development of learning control for reproducible and high quality operation of batch processes, in *Proceedings of the 6th IFAC Symposium on Dynamics and Control of Process Systems*, 449–454.

10 Jørgensen, S.B. and Bonné, D. (2006) Systematic methodology for reproducible and optimizing batch operation. *Computer-Aided Chemical Engineering, Elsevier*, **21**, 1275–1280.

11 Bequette, W.B. (2003) *Process Control: Modeling, Design and Simulation*, Prentice Hall.

12 Seborg, D.E., Edgar, T.F., and Mellichamp, D.A. (2004) *Process Dynamics and Control*, 2nd edn, John Wiley & Sons Inc.

13 Visioli, A. (2006) *Practical PID Control*, Springer London.

14 O'Dwyer, A. (2000) A summary of PI and PID controller tuning rules for processes with time delay. Part 1: PI controller tuning rules. In Proceedings of PID 00: IFAC Workshop on Digital Control, pp. 175–180.

15 Skogestad, S. (2003) Simple analytical rules for model reduction and PID controller tuning. *J. Process Contr.*, **13** (4), 733–764.

16 McMillan, G.K. (1994) *Tuning and Control-Loop Performance*, 3rd edn, Instrument Society of America, Research Triangle Park, NC.

17 Shinskey, F.G. (1994) *Feedback Controllers for the Process Industry*, McGraw-Hill Inc., US.

18 Yu, C.C. (1999) *Autotuning of PID Controllers*, Springer London.

19 Larsson, T. and Skogestad, S. (2000) Plantwide control: a review and a new design procedure. *Model. Ident. Control*, **21** (4), 209–240.

20 Åkesson, M. (1999) Probing control of glucose feeding in *Escherichia coli* cultivations. PhD these ISRN LUTFD2/TFRT-1057-SE Department of Automatic Control, Lund Institute of Technology, Sweden.

21 Velut, S., de Maré, L., and Hagander, P. (2007) Bioreactor control using a probing feeding strategy and mid-range control. *Control Eng. Pract.*, **15**, 135–147.

22 De Maré, L., Velut, S., Ledung, E., Cimander, C., Norrman, B., Karlsson, E.N., Holst, O., and Hagander, P. (2005) A cultivation technique for *E. coli* fed-batch

cultivations operating close to the maximum oxygen transfer capacity of the reactor. *Biotechnol. Lett.*, **27**, 983–990.

23 Dochain, D., Perrier, M., and Guay, M. (2011) Extremum seeking control and its applications to process and reaction system: a survey. *Math. Comp. Simul.*, **82**, 369–380.

24 Stephanopoulos, G. and Ng, C. (2000) Perspectives on the synthesis of plant-wide control structures. *J. Process Contr.*, **10**, 97–111.

25 Rangaiah, G.P. and Kariwala, V. (2012) *Plantwide Control: Recent Developments and Applications*, John Wiley & Sons.

26 Åström, K.J. and Hägglund, T. (1995) *PID Controllers: Theory, Design and Tuning*, 2nd edn. Instrument Society of America, Research Triangle Park, NC.

27 Stanke, M. and Hitzmann, B. (2013) Automatic control of bioprocesses. *Adv. Biochem. Eng. Biotechnol.*, **132**, 35–63.

16
Scale-Up and Scale-Down

Henk Noorman

Summary

As described in Chapter 11, the scale-up of a process from laboratory scale, through pilot scale, and finally to the large industrial scale is far from easy. Mixing of the medium is necessary to minimize gradients of substrates, but intensive mixing may become very costly. Sparging of the medium with, for example, oxygen may lead to operation in flow and reaction regimes that change drastically with power input and scale, and a gas–liquid contact pattern that was found to be excellent in small scale may prove to be inadequate in large, industrial scale. To gain a qualitative or semiquantitative insight into the major effects of scale-up, a design philosophy based on scale-down of the large-scale operation to a smaller scale has been developed. One starts by defining the desirable large-scale process conditions to give a product that has certain characteristics, such as quality, yield, ease of downstream processing, and so on. Then one constructs a smaller scale version in which these characteristic properties can be examined as a function of the process conditions. This chapter describes how the proper large-scale performance can be achieved through well-designed scale-down experiments, using regime analysis, compartment models, and other engineering models by which the desired qualities of the product in large scale can be studied.

16.1
Introduction

It is the ultimate purpose of a bioengineer to design and implement bioprocesses in large-scale factories, to deliver bioproducts that are needed by mankind, at affordable cost levels and with minimal environmental burden. Currently, thousands of such factories are producing a wide range of products for various markets. The most important are as follows:

- Biofuels, that is, bioethanol, biobutanol, isoprenoids, and biogas
- Amino acids, notably glutamate, lysine, and threonine

Fundamental Bioengineering, First Edition. Edited by John Villadsen.

- Building blocks for bioplastics, such as succinic acid, 1,4-butanediol, 1,3-propanediol, and lactic acid, or bioplastics like polyhydroxyalkanoates
- Food ingredients, such as enzymes, sweeteners, baker's yeast, lactic acid bacteria, xanthan, and citric acid
- Nutritional ingredients, for example, vitamins, carotenoids, and polyunsaturated fatty acids
- Pharma products, such as anti-infectives, pharmacological agents, and therapeutic proteins.

The processes in the factories represent a value chain, where products are formed from renewable feedstocks in a sequence of operations, usually alternating conversion and separation steps. The feedstocks are being produced from atmospheric CO_2 by plants or with increasing importance from marine species such as seaweed and microalgae. The subsequent bioconversions are done with the help of microorganisms, mammalian cells, or enzymes. There are also hybrid concepts with chemical conversion steps included, like the postfermentation tailoring of active pharmaceutical products. And finally, also wastewater treatment is of importance. The biological oxygen demand of the wastewater is reduced and water is cleaned.

The large-scale operations are usually done with great care to minimize safety risks and contamination, and with emphasis on economy and product quality. Further development of the bioprocess as such at the large-scale level is, therefore, limited to the bare minimum.

Historically, the route to design and improvement of industrial-scale processes has followed a trial-and-error approach, starting at the lab scale. Interesting findings such as an improved product titer were embraced and transferred to the plant, but with mixed success. A classic example is the higher titers that can be easily achieved in the small scale if the bioreactor is equipped with a larger motor to drive the impeller, or if the process time is extended, or if the air flow rate is increased – resulting in a larger evaporation rate. High titers achieved in this way do not represent improvements of the volumetric productivity or the conversion yield, and generally cannot be scaled up and reproduced in the plant. Another example is to keep certain factors the same upon scale-up, such as power input per volume, mixing time, or impeller tip speed (see Chapter 11). However, it can be easily shown that while some factors are kept constant, other factors will change, sometimes with serious economic repercussions. For example, keeping the mixing time at large scale the same as in the lab will lead to an excessive power input for mixing. The complications inherent to scale-up are particularly visible for high-viscosity fermentation media when working with filamentous bacteria and fungi, or in cultures where biopolymers (e.g., xanthan) are produced outside the cell.

Over the past few decades, the *scale-down* approach has developed into a better alternative, and it is now established as a good engineering practice. The following sequence of events can be recognized:

1) First a description of the large-scale bioprocess is made, with a detailed analysis of the process operating regimes and rate-determining steps.

2) Second, the large-scale environment of the cells is mimicked in the lab (scaled-down) with the best possible approximation, that is, keeping the process within the same operating regimes and the same rate-determining mechanism.
3) Third, in the lab, research and development work is done at medium- or high-throughput rate at limited risk and cost.
4) Finally, improvements obtained in the small scale are transferred to the large scale and implemented with drastically reduced chance of failure.

The strength of this four-step approach is that the scale-up work is started from the correct large-scale perspective and not from the small scale from which solutions may not be transferrable to the factory. The following sections of this chapter will elaborate on each of the four steps.

16.2 Description of the Large Scale

The bioreactor, or fermenter, is the central unit of operation in the bioprocess value chain. Various sizes, shapes, and operation drivers have evolved in the industry. As described in Chapters 10 and 11, the most important workhorses for microbial and cell cultures without solids are *stirred tank reactors, bubble columns,* and *air-lift loop reactors* (with or without mechanical power input through mixers). A particular category applies when solids are present. Examples are carriers for immobilized cells or biofilms, or biomass feedstock material in packed beds or solid-state reactors, and slurry reactors where agglomerates of microbes and insoluble substrates need to be kept in a fluidized mode. These special cases will not be treated in this chapter.

The maximum size of stirred tanks with gas supply is around 700 m^3 gross volume, while for anaerobic and ungassed systems tank volumes of more than 3000 m^3 have been constructed and operated. In anaerobic biogas reactors and in wastewater treatment reactors, volumes as high as 20 000 m^3 are possible. Usually, the total volume is contained in a single compartment, but there are also multistage reactors with several stacked compartments, or trays, separated by perforated plates. Although systems with solids and anaerobic processes are important, this chapter will mostly deal with the *gassed* bioreactors, especially aerated, since sparged reactors are dominant in the bio-industry. Note that the same principles apply to bioprocesses using gaseous hydrogen, carbon monoxide, and carbon dioxide. Typically, the industrial reactors have a high aspect ratio (height over diameter, H/T), in the range, 3–10, to maximize gas residence time and utilize the gas in the best possible way. Because of the large volumes, also the length scales are large, with tank diameters of 5 m or more, and heights up to 20 m, which appears to be the maximum for mechanical strength reasons and to comply with local building regulations. Inside the reactors, various internals are installed, notably impellers, baffles, cooling coils, draft tubes, a shaft on which the impellers are mounted, a bearing for the shaft, and a gas sparger for supply and distribution of the gas (see Figure 11.1). The size, geometry, and internals all have a profound impact on the detailed fluid flow patterns, which is further determined by the power input of the gas and/or the impellers, and the properties of the broth, especially when the rheology is non-Newtonian and at high apparent viscosities. Such properties will severely restrict the free flow of fluid.

Similar vessels can be in use in other sections of the process value chain: upstream in enzyme hydrolysis reactors and downstream in biocatalysis and biotransformation steps with enzymes and active cells, respectively, to further convert intermediate fermentation products into tailored end products.

The necessity to operate at such large scales is dictated by the economy: large equipment brings advantages in economy of scale. As a rule of thumb, fixed costs per kg product finally produced decreases proportionally to $[\text{volume}]^{-2/3}$, or products will get roughly 5 times cheaper at 10-fold higher production rate. It is noted that this rule applies to specialty products, where fixed costs constitute the major share of the total costs. For bulk molecules produced at a scale of more than 50 000 tonnes per year, such as is the case for biofuels and bio-based monomers, the relative contribution of the variable costs, that is, feedstock, is dominant and can be 70% or higher. Here a further increase of scale will bring less benefits.

Understanding the processes inside a large bioreactor starts with the study of fluid flow phenomenon and different flow regimes. Flow regime extremes are as follows:

- Ideally mixed. Here mixing is very fast compared to other mechanisms.
- Ideal plug flow. Here there is no mixing between fluid elements, but only bulk liquid convection.

In reality, these extremes are not encountered, but rather some dispersion, or backmixing, and inhomogeneities are observed. The mixing patterns in stirred tanks, bubble columns, and air-lift reactors are different and should be distinguished.

In gassed stirred tanks, gross circulation is determined by interplay of the impeller pumping and gas-driven circulation. There are two main mechanisms:

1) If the gas energy is low compared to the impeller energy, then gas will be effectively dispersed and recirculated through the whole vessel – the total recirculation regime. When the energy input via the gas is high, the impeller(s) cannot disperse the gas and as a consequence bubbles will rise close to the center of the vessel, dragging along fluid packages in their wake, thus creating long axial circulation loops. Here, the reactor operates in the flooding flow regime, which is not desired. Between the two regimes is the partial gas recirculation regime, which is preferred and most dominant in large, stirred bioreactors. A special variant of this flow regime is the loading regime. Here, the region below the bottom impeller contains hardly any gas.

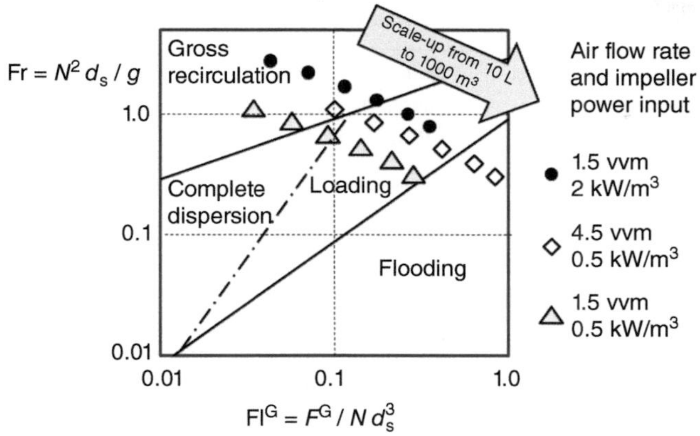

Figure 16.1 Flow map, useful for determining the flow regime at a certain geometry and operation. Regime boundaries are taken from [1,2]. Scale-up from 10 l to 1000 m^3 in steps of a factor 10 is indicated for three combinations of gas flow rate (in vvm, recalculated for standard conditions, 1 bar and 0 °C) and impeller power input. The real volumetric gas flow rate in the reactor is lower as it is compressed proportional to the total pressure in the tank. In all cases, a headspace pressure of 2 bar is assumed, an aspect ratio $H/T = 3.0$, and $d_s/T = 0.4$ (see also Table 16.1). The example demonstrates that regime changes can occur upon scale-up at constant power input and gas flow rate per volume, and also shift with operating conditions.

A valuable representation of all flow regimes can be provided by plotting the dimensionless Froude number $\mathrm{Fr} = N^2\ d_s/g$ (N is impeller rotational speed, d_s is impeller diameter, and g the acceleration of gravity, as defined in Section 11.2, p. 365) against the dimensionless gas flow number $\mathrm{Fl}^G = F^G/Nd_s^3$ (F^G is the gas flow rate) – called "aeration number" N_A in Eq. (11.8). Fl^G is shown in "flow maps" such as Figure 16.1.

2) The gas phase forms cavities behind impeller blades that can have a strong effect on the power transfer into the liquid, and liquid velocities in the impeller discharge stream: at high gas flow rates, large cavities develop, either "ragged" at the highest gas flow rates, or in an alternating configuration with three larger and three smaller forms (the "3-3" structure), and power input can be reduced to ~40% of the ungassed situation. At low gassing rates, small vortex or clinging cavities are found, and the power input is less reduced. Generally, the changes are most prominent with Rushton turbine impellers. When hollow blade impellers are used, or with axially pumping agitators, large cavities hardly develop and the drop in power input is much smaller or even absent.

For bubble columns, chaotic flow patterns with large circulation loops, oriented on top of each other, have been reported at high gas energy [3]. Such a

Table 16.1 Geometry details for differently sized bioreactors.

Reactor liquid volume, V (m^3)	Mean broth pressure, p (bar)	Ungassed liquid height, H (m)	Tank diameter, T (m)	Internal tank wall area, A (m^2)	Tank wall area/volume A/V (m^2/m^3)
0.01	2.0	0.48	0.16	0.2	25.0
0.10	2.1	1.05	0.35	1.2	11.4
1.0	2.1	2.25	0.75	5.3	5.3
10	2.2	4.8	1.6	24.1	2.5
100	2.5	10.5	3.5	115	1.1
1000	3.1	22.5	7.5	530	0.5

The aspect ratio $H/T = 3$, and the headspace pressure is 2 bar in all cases.

heterogeneous flow regime is observed when the superficial gas velocity, calculated as m^3 gas flow rates^{-1} (m^2 cross section of the reactor)$^{-1}$ is higher than about 0.04 m s^{-1}. Below this level, the flow is homogeneous and described by a more orderly rising gas phase, or bubbly flow, throughout the reactor. In reality, even for moderate gassing rates, large bubble columns operate in the heterogeneous flow regime [4].

Air-lift loop reactors are not commonly used, except when (as discussed in Chapter 11) controlled flow with intense recirculation is desired [5]. Less attention has been devoted to this setup, and the literature has only few practical guidelines on their operation [6].

In large bioreactors, it is inevitable that transport phenomena, and not the kinetics of bioconversion, limit the overall process rate. Usually, mixing, mass transfer, CO_2 removal, cooling, or gas–liquid separation in the headspace of the bioreactor is identified as the rate-limiting step, and in mature operations there will be more than one rate-limiting transport mechanism under each operating condition. Each of these five process steps will now be further elaborated. For illustration, example calculations are provided for a hypothetical scale design case, with reactor volumes moving between 10 l and 1000 m^3 in steps of a factor 10. In Table 16.1, key geometry details are provided per volume.

16.2.1 Mixing

Mixing may be the most prominent limiting factor in large-scale bioreactors. A subdivision is to be made in macromixing and micromixing.

Macromixing is primarily caused by bulk liquid flow and dispersion. The length of circulation loops dramatically increases upon scale-up, and can increase by a factor 50 when going from 10 l to 1000 m^3 scale. As a consequence, cells can reside anywhere from tens of seconds to a few minutes in regions where there is no fresh supply of limiting substrate, like in a fed-batch or continuous fermentation. Mixing is often expressed as a certain degree of homogeneity, usually 63% (=1−1/e), 90% or 95% of the final concentration of an inert tracer that is injected into the vessel at one location (see Section 11.1). As a rule of thumb, 95% mixing times are roughly equivalent to 4× the main fluid circulation time. From Eq. (11.14), at constant power input per volume, one obtains that mixing time t_m scales with [tank diameter]$^{2/3}$ and empirically it has been found in ungassed stirred tanks, with a single impeller that $N\, t_m$ = constant. Both results convert to the rule that for a 10-fold bigger vessel, with the same power input per volume, the mixing time t_m increases by about 70%, while the increase of t_m for a reactor volume increase by a factor 10^3 amounts to a factor 5. Absolute values of the mixing time are then on the order of 10 s at 100 l scale and 50 s in a 100 m^3 tank equipped with one Rushton impeller ($d_s/T = 0.4$), when the power input per volume P/V is 2 kW m^{-3} and $H/T \sim 1$. Even at lab scale the mixing times are at best on the order of seconds, which is often neglected. The assumption of ideal mixing may then not be correct.

Such results hold both for stirred tanks equipped with one impeller and also for small bubble columns at low viscosity. For large reactors with high aspect ratio and equipped with multiple impellers, "zoning" is observed, that is, the liquid exchange between zones around the impellers is limited. This is especially true for radially pumping impellers, but it also has been observed for axially pumping agitators. At high viscosity, cavities (the so-called "caverns") are formed around the impellers. Inside the cavern, the fluid is in motion, but outside there is hardly any movement of liquid. All of this results in mixing times that are much higher (10-fold going from full turbulence to the transition region with Reynolds numbers less than 1000). Complete mixing can take very long, and may even be impossible if stagnant regions are formed.

Under gassed conditions, the bulk flow and dispersion is greatly changed compared to ungassed conditions, and the mixing efficiency is different. A powerful, general approach for the prediction of mixing times in bioreactors with multiple impellers under gassed and ungassed conditions, also uniting stirred tanks and large-scale bubble columns with different flow regimes, has been presented by Van't Riet and Van der Lans [7]. A dimensionless mixing number, N_{mix}, was defined:

$$N_{mix}[-] = t_m(P/M)^{1/3}\, T^{-2/3} \quad [\mathrm{s}][\mathrm{m}^{2/3}\,\mathrm{s}^{-1}][\mathrm{m}^{-2/3}]. \tag{16.1}$$

T is a characteristic dimension of the tank, typically defined as the tank diameter, and P/M is the power input to the mass of medium in the tank, with unit W/kg = $m^2\ s^{-3}$. From the definition of a characteristic path for the gross liquid recirculation, with axial dispersion superimposed, and by applying isotropic turbulence theory, one obtains Eq. (16.2) that relates N_{mix} to the design of the reactor:

$$N_{mix} = t_m(P/M)^{1/3}\, T^{-2/3} = 11(T/d_s)^{1/3}(H/T)^2. \tag{16.2}$$

Now, one can calculate t_m as a combined function of the reactor geometry and the power input:

$$t_m = 11(P/M)^{-1/3}(T/d_s)^{1/3}(H/T)^2 = 10.1\,N_{p,g}^{-1/3}\,N^{-1}(T/d_s)^2(H/T)^{7/3}. \tag{16.3}$$

Equation (16.3) is a perfect analog to Eq. (11.14), both when t_m is expressed in terms of (P/M) (the first expression) and when t_m is expressed in terms of the speed of rotation of the stirrer N. $N_{p,g}$ is defined in Eq. (11.8) and Figure 11.2, and $N_{p,g} = N_p$ for unaerated mixing.

The difference between Eqs. (16.3) and (11.14) is the huge influence of the aspect ratio H/T. Whereas Eq. (11.14) is correct *only* for $H \sim T$, Eq. (16.3) is correct for $H/T \gg 1$, that is, for tall, slim reactors. For intermediate values of H/T, the exponent on (H/T) increases from zero when t_m is expressed in terms of N in Eq. (11.14) to 7/3 – and from 1/3 to 2 when t_m is expressed in terms of P/M. The conclusion is the same as that suggested in Chapter 11: a single mechanical mixer performs very poorly when $H/T \gg 1$ and to bring down t_m to an acceptable value, several mixers must be used, one for each segment of the reactor $\Delta H = T$.

Equations hold for both gassed and ungassed conditions, although depending on the flow regime a factor 2 or 0.5 of variation is possible. Operating in the flooding regime will give better axial mixing and lower mixing times, while the full gas recirculation regime will slow down mixing. In case a 3-3 cavity structure is developed behind the impeller blades, then the same mixing time is found for gassed and ungassed media.

For *bubble columns*, the power input per mass P/M is determined as $g\,v^C$ ($m^2\,s^{-3}$) where v^C is the superficial gas velocity, calculated at the actual pressure. For $H/T > 3$, Eq. (16.2) still applies (obviously without the factor $(T/d_s)^{1/3}$) but with a much smaller proportionality constant:

$$N_{mix} = t_m(g\,v^C)^{1/3}\,T^{-2/3} = 1.6(H/T)^2 \tag{16.4}$$

equivalent to

$$t_m = 1.6\,T^{2/3}(g\,v^C)^{-1/3}(H/T)^2. \tag{16.5}$$

However, if $H/T < 3$, experiments show that N_{mix} is *independent* of H/T:

$$N_{mix} = 16 \rightarrow t_m = 16\,T^{2/3}(g\,v^C)^{-1/3}. \tag{16.6}$$

Finally, for air-lift loop reactors, the mixing time is about 4–7× the average circulation time t_C [6]. The circulation velocity is very much dependent on the scale, gas flow rate, and precise configuration. Especially, the friction forces on the liquid when the flow is reversed between the riser and the down-comer sections and the gas hold-up difference in both compartments will determine the liquid velocity and circulation flow rate. From the example in Van't Riet and Van der Lans [7], with $g\,v^C = 2\,W\,kg^{-1}$, a rough estimation gives $N_{mix} = 80$ (–).

Table 16.2 Comparison of mean liquid circulation time $t_C(s)$ in various reactor types and sizes, with aspect ratio $H/T = 2$, all operated at 2 W/kg total power input and with $0.10\,m\,s^{-1}$ superficial gas velocity (calculated from [7] assuming circulation time = mixing time/4).

	0.01 m³	10 m³	100 m³	1000 m³
Airlift	5	18	35	60
Bubble column	1	6	10	17
Stirred tank	4	18	30	50

A comparison between reactors reveals the following:

- *Low viscosity* (i.e., lower than about 100 mPa s, 100 times the viscosity of water): Only for very high aspect ratios ($H/T > 5$) will air lift reactors give the best mixing. In other cases, the bubble column is best. Mixing is 2–5× faster than in a stirred tank with multiple impellers, when $H/T = 2–3$. For aspect ratios below 2, the bubble column and stirred tank performance become similar (see Table 16.2 and Figure 16.2). It is noted that this holds under the assumption that the total power inputs are identical (i.e., the power input by the gas in bubble columns and air-lift reactors equals the value of gas plus impellers in stirred reactors. This, of course, requires a higher gas velocity in the unagitated reactors).

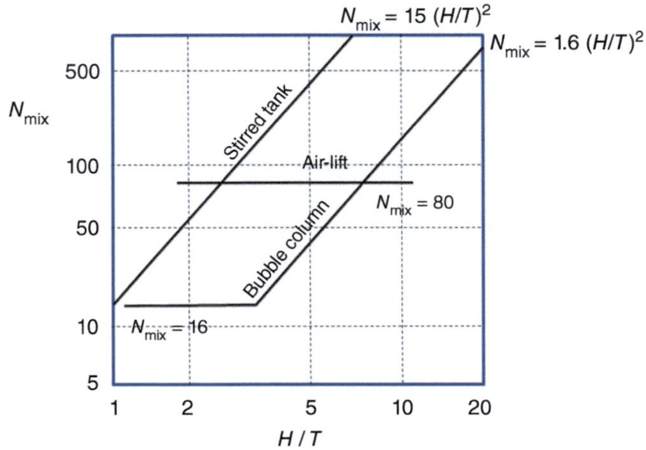

Figure 16.2 Approximate comparison of N_{mix} for the three reactor types, using Eqs. (16.2) (stirred tank), (16.4) (bubble column $H/T > 3$), and (16.6) (bubble column $H/T < 3$), and $N_{mix} = 80$ (air-lift). In all cases $P/V = 2\,kW/m^3$, whereas for the stirred tank $d_s/T = 0.4$.

- *High viscosity* (i.e., higher than about 100 mPa·s): Mixing is better in stirred tanks, on all scales. However, there are mechanical and construction limits to the scale of stirred tank reactors, due to the large weight of the impeller motor mounted on top of the reactor and the presence of moving parts. This is an important reason why in general industrial stirred tanks are smaller in size than nonstirred reactors.

Micromixing is mixing on the molecular scale. Although micromixing is really important only for fast chemical reactions and crystallization, it may sometimes also influence slower bioprocesses. Local variations of concentrations and shear rates may occur in the smallest eddies of the turbulent flow, or in the immediate vicinity of the feed inlet points, and these stochastic variations occur on a scale similar to the size of the cells. On this scale, at one particular location there may be rapid fluctuations of the limiting nutrient concentrations, with a characteristic time of seconds or less, and amplitude of two–four times the average limiting substrate concentration. This aspect of mixing is less understood but could also have an impact on the performance of microorganisms. The report by Enfors *et al.* [8] with computations of the variation of glucose concentration close to the feed inlet will give some first insight. Another report from Larsson *et al.* [9] reveals similar experimental results, measured at three heights in a large bioreactor. The question remains to be answered whether the cells observe such rapid fluctuations or not, as the cells move along with a fluid package that may have more constant composition.

16.2.2
Mass Transfer

The mass transfer from the gas bubbles to the cells takes place in a number of sequential steps, of which the transfer from the liquid interface with the gas phase to the bulk liquid phase is most likely to be rate-limiting. Several key factors determine together the overall mass transfer rate, that is, the mass transfer coefficient, k_l, and the bubble-specific interface area, a. The solubility C^* of the gas in the liquid phase at the actual temperature and pressure is found as described in Chapter 9.

$$\text{Mass transfer rate}\,[\text{mol m}^{-3}\,\text{s}^{-1}] = k_l\, a(C^* - C)[\text{m/s}][\text{m}^2/\text{m}^3][\text{mol/m}^3]. \tag{16.7}$$

Because of difficulties to separate estimation of k_l and a, while the product $k_l \times a$ can be easily measured, usually k_la is taken as one parameter, the *volumetric gas transfer coefficient*. $(C^* - C)$ is the driving force for mass transfer.

The k_la is dependent on four factors: the power input to the bioreactor (P/V), the superficial gas velocity v^C (calculated at the actual pressure, deviating from reference pressure of 1 bar, p_0), the coalescence behavior of the liquid, and the apparent viscosity. For pure water, bubbles easily coalesce, and for stirred tanks

Eq. (16.8) represents experimental data for widely different scales and operating conditions [10]:

$$k_l a = 0.026(P/V)^{0.4}(v^C)^{0.5}. \tag{16.8}$$

In the presence of soluble species such as dissociated salts or interface active agents, coalescence may be inhibited and the mass transfer coefficient can be two–three times higher. In this case, expression (16.8) changes into

$$k_l a = 0.002(P/V)^{0.7}(v^C)^{0.2}. \tag{16.9}$$

These expressions hold for a wide range of reactor volumes and power input, and are independent of the type of impeller used.

For bubble columns with coalescing media and low viscosity, $k_l a$ is represented by

$$k_l a = 0.32(v^C)^{0.7}. \tag{16.10}$$

Air-lift loop reactors, with internal riser and downcomer sections, usually have a higher gas hold-up compared to bubble columns due to interaction with the liquid flow. As a consequence, $k_l a$ can be a factor 2–4 higher compared to the values observed in bubble columns [5].

A comparison of the mass transfer performance of large bioreactors, with some typical values for the operation variables, is given in Table 16.3.

Table 16.3 Volume-specific mass transfer coefficient for several types and scales of bioreactors.

	Power input ($W\,m^{-3}$)		$k_l a$ (s^{-1})			
Reactor volume (m^3)	**Gas**	**Impeller**	**Stirred tank, non coalescing**	**Stirred tank, coalescing**	**Bubble column**	**Air-lift reactor**
0.01	58	1942	**0.144**	**0.047**	**0.009**	**0.013**
0.10	125	1875	**0.163**	**0.053**	**0.015**	**0.023**
1.0	261	1739	**0.180**	**0.074**	**0.025**	**0.038**
10	526	1474	*0.184*	*0.099*	**0.041**	**0.062**
100	1020[a)]	980	*0.158*	*0.117*	**0.066**	*0.098*
1000	1766[a)]	234	*0.065*	*0.087*	*0.096*	*0.145*

Aeration rate is 1.5 vvm (m^3 gas at standard conditions/m^3 liquid/min), the power input of gas plus impeller is taken to be 2 kW m^{-3}, and gas-impeller split is indicated. For stirred tanks, the results for coalescing and noncoalescing liquids are listed; for bubble columns, the expression for coalescing liquids is used; and for air-lift reactors, it is assumed that $k_l a$ is 1.5× the bubble column value. The *italic* numbers denote conditions where the circulation time (from Table 16.2) is higher than the mass transfer time ($1/k_l a$), and poor mixing is a more dominant rate-limiting factor than mass transfer. For numbers in **bold**, the reverse is true.

a) Impeller flooding conditions.

Table 16.4 Scale dependency of oxygen transfer capacity (OTC, equal to $k_l a\ C^*$), assuming headspace pressure 2 bar, aspect ratio 3, gas flow inlet rate 1.5 vvm (m^3 gas under standard conditions/m^3 liquid/min), gas depletion 0.5% per meter height [11], and the same impeller power input per volume (2 kW m^{-3}).

	Relative oxygen transfer capacity			
Reactor liquid volume (m^3)	**OTC ~ $[v^C]^{0.2}$**	**OTC ~ $[v^C]^{0.4}$**	**OTC ~ $[v^C]^{0.5}$**	**OTC ~ $[v^C]^{0.7}$**
0.01	1.0	1.0	1.0	1.0
0.10	1.1	1.3	1.5	1.7
1.0	1.3	1.8	2.0	2.7
10	1.4	2.2	2.7	4.1
100	1.3	2.4	3.1	5.5
1000	0.9	1.7	2.5	4.8

Four different dependencies of the OTC on the pressure-corrected superficial gas velocity, v^C are provided, that is, OTC ~ $[v^C]^a$ with $a = 0.2$, usually found in stirred tanks with noncoalescing broth, $a = 0.4$ or 0.5, which applies to stirred tanks with coalescing broth, and $a = 0.7$ as reported for bubble columns.

In large bioreactors, the transfer of oxygen from the bubbles to the liquid phase is usually better than at lab scale, even when the total power input per volume is kept constant. This is due to a larger contribution of the gas phase, via the superficial gas velocity, and a larger driving force, related to the higher head-space pressure plus the hydrostatic head, often the average of 3 bar, resulting in a higher gas solubility. The larger driving force may be somewhat counteracted by partial depletion of the gas phase: in the gas outlet, the oxygen content of the bubbles may have dropped from 21 to 10%. In total, the oxygen transfer rates then can be two–five times higher than at lab scale (see Table 16.4).

A further counteracting factor may be the actual dissolved gas concentration, which may vary from place to place in the bioreactor. For reasons of operational stability, it could well be that the actual concentration at the location, where the gas concentration is lowest, is kept at the critical level but then elsewhere in the reactor it is higher [6]. As a consequence, the driving force for mass transfer will be reduced. As a ballpark value, the oxygen transfer rate in large fermenters can be as high as 200 mol $m^{-3}\ h^{-1}$. Apart from the gas, impellers also contribute to such high oxygen transfer rates. The upper limit of the impeller power input is constrained to around 4–5 kW m^{-3} because of the following reasons:

- Higher values could result in locally too high shear stresses for the microbes, with fragmentation of hyphae and pellets, cell lysis, and loss of productivity as a result.
- Operational stability problems like resonance phenomena can occur.
- Costs of electricity, motor investment costs, and added costs for construction of the bioreactor and civil engineering strongly influence this upper limit.

A special mention should be devoted to non-Newtonian, high-viscosity systems. Such systems typically occur at high biomass densities when filamentous microorganisms are used or when biopolymers such as polysaccharides are produced. There the mass transfer coefficient will be much smaller because the free flow of fluid is constrained. It is generally assumed that the mass transfer coefficient reduces with [apparent viscosity]$^{-b}$, for viscosities higher than 50 mPa·s, and with b ranging from 0.5 to 1.0 (0.7 is an adequate average). As a result, the OTC will decrease by a factor 5 when the viscosity increases from 50 to 500 mPa·s and by a factor 25 at 5000 mPa·s.

In case other feed gases are used, such as synthesis gas, that is, a mixture of H_2 and CO, then the solubility and diffusivity of these compounds should be accommodated in the mass transfer equations. For example, for H_2 at ambient temperatures, the solubility is 60% of the pure oxygen value. This reduces the driving force for mass transfer by a factor 1.4 (compared to pure oxygen). On the other hand, the diffusion coefficient is 1.6× that of oxygen, and this results in a proportionally higher volumetric mass transfer coefficient.

16.2.3 CO_2 Removal

CO_2 removal goes in the opposite direction to oxygen supply, and similar reasoning applies, although the solubility of CO_2 is 30–40 times higher than of oxygen. If the rate of removal is insufficient, then the cells may suffer from inhibition with a problematic reduction of the fermentation performance. For tall reactors, the inhibition effect usually will be strongest in the bottom section where pressure is the highest. But in the top section also, the gas can be so much enriched in CO_2 that back transfer to the liquid phase will occur, causing extra inhibition.

16.2.4 Cooling

Cooling can be limiting if insufficient cooling area is installed, if the temperature difference between broth and coolant is too small, or if the overall heat transfer coefficient from broth to the coolant phase is too small. We can recognize three system-dependent and two fouling-related steps-in-series for heat transfer:

- A relatively stagnant liquid boundary layer on the interface with the cooling surface.
- A fouling layer between the film and the cooling surface.
- A conductivity barrier formed by the material of the bioreactor, notably the cooling coils and the wall – especially plastic tubes, like those used in disposable bioreactors – but stainless steel has also a relatively low heat conductivity.

- A fouling layer on the side of the coolant – this can be a notorious problem and regular maintenance and cleaning should be done to minimize its influence.
- A relatively stagnant liquid boundary layer on the coolant side.

All of these steps usually have a similar contribution to the overall heat transfer resistance. For each step, heat transfer coefficients are estimated between 3000 and 10 000 $W\,m^{-2}\,K^{-1}$. Overall heat transfer coefficients, U, for all five steps together have been reported in the range 500–1500 $W{\cdot}m^{-2}{\cdot}K^{-1}$. In Chapter 9 (Eq. (9.24), it is shown that the heat evolution is $Q = 460\ Y_{so}$ kJ (C-mole substrate S converted)$^{-1}$. Assuming again an OTC of 200 mol m^{-3} h^{-1}, the heat production rate, Q, will be around 460 $Y_{so}\ (-q_s) = 460\ (-q_o)$ kJ/(mol O_2 h)$^{-1}$]. Let $(-q_o) = 200$ [mol $m^{-3}\,h^{-1}$], and one obtains $Q = 460 \times 200 \times 1/3600 = 25.5$ kW m^{-3}. Assuming further a temperature difference $\Delta T = 25$ K and $U = 1000\ W\,m^{-2}\,K^{-1}$, then it follows that $A = Q/(U \times \Delta T) = 1.0\ m^2/m^3$ cooling area is needed. The total heat production may be further increased with 10–30% by relatively hot gas and liquid feeds into the reactor and the dissipation of mechanical power input to heat by the impellers in stirred tanks. With the standard bioreactor geometry of $H/T = 3$, this means that at volumes above 10–100 m^3 there will be not enough wall area available for cooling, and additional structures should be added, such as a cooling coil, cooling baffles, or an external loop with a pump and heat exchanger. Also for such structures, the same values of U and ΔT apply and cooling rate will be determined by the total cooling area. The sensitivity of cooling on viscosity is not as dramatic as observed for mass transfer, but still the heat transfer coefficient at the broth side can be reduced proportional to [apparent viscosity]$^{-0.3}$, that is, a reduction by a factor 10 if the apparent viscosity is 1000× higher.

16.2.5
Gas–Liquid Separation

Sometimes, it is observed that the escape of the gas phase from a large-scale bioreactor is problematic. This happens with extremely noncoalescing liquids, like with certain proteins and surface-active compounds. It certainly occurs with viscous systems, where small bubbles with a diameter below 0.5 mm do not coalesce and keep circulating in the broth for prolonged times, that is, longer than 15 min. This will give a steady increase of the total gas hold-up. This gas fraction will become depleted and does not contribute anymore to the mass transfer. A second phenomenon is foaming, which can become severe and result in foaming out of a large part of the broth, even in the presence of antifoam agents, unless the gas flow is reduced. Avoiding foaming may result in operation with low gas flow rates, onset of transport limitation, or further reduction of the limiting rates of mass transfer and mixing.

If the characteristic transport time for at least one out of these five transport steps is higher than the characteristic consumption time of the limiting substrate, of mixing of added titrant, or for refreshment of boundary layers at the cooling surfaces, then gradients of substrate concentrations and temperature will be present. These so-called "hot spots and cold spots" can consist of pockets of high and low oxygen and sugar and precursor concentrations, of high and low pH around the titrant addition point, or of high and low shear rates. Further, high temperatures at the inlets of warm gas and liquid feeds and low temperatures at the cooling interface. A powerful way to estimate which of the steps contributes most to overall rate limitation is done via a regime analysis. This entails a comparison of the various characteristic times of the subprocesses (transport and reaction). Reaction time can be estimated from the ratio of concentration to consumption rate, or $C_s/(-q_s)$ [s] for the limiting substrate. This is compared with the gas transfer time, $1/k_l a$ [s], the gas residence time [s], the mixing or liquid circulation time [s], the heat removal time or the residence time in a cold or hot zone.

In the following section, some of the most important mechanisms will be illustrated and compared.

The gas residence time can be estimated from the gas hold-up, which can be 10–20% or higher in large bioreactors. In a 100 m^3 reactor, assuming 10% gas hold-up (i.e., 10 m^3 total gas phase volume), then for a 1 vvm gas feed (100 m^3/min), compressed to an average pressure of 2 bar in the tank, the average residence time is 10 m^3 gas/(50 m^3 gas min^{-1}) = 12 s. If the transfer time of gas from the bubbles is much shorter than this residence time, as is the case for a fast bioreaction, then the gas will become depleted before it leaves the reactor via the headspace.

Quantitative estimations for the gas hold-up can be given for stirred tanks and bubble columns, respectively, via

$$\epsilon = 0.13(P/V)^{0.33}(v^{C})^{0.67} \tag{16.11}$$

and

$$\epsilon = 0.6(v^{C})^{0.7}. \tag{16.12}$$

Both expressions (16.11) and (16.12) apply to coalescing liquids. For noncoalescing liquids, the value of ε can be considerably higher, ranging from 50% to more than 200% higher, depending on the sparger type used. Table 16.5 shows some estimates for the gas residence time, based on the above expressions for gas hold-up, and comparison with the mass transfer time.

The characteristic production times of metabolic products and biomass $[C_p/q_p, C_x/q_x]$, as well as the consumption time of nonlimiting substrates, are usually much larger than the consumption time of the limiting substrates, and they do not contribute to an overall rate limitation. Time limitation related to the limiting substrates can be best illustrated by an example: assume the conversion of glucose and oxygen into 1,3-propanediol (PDO), according to the following stoichiometry:

$$C_6H_{12}O_6 + 2\,O_2 \rightarrow C_3H_8O_2 + 3\,CO_2 + 2\,H_2O. \tag{16.13}$$

The product yield on glucose is set at 1 mol/mol = 0.42 g g^{-1}.

Table 16.5 Comparison of gas residence time with mass transfer time ($1/k_la$), taken from Table 16.3, for different reactor types across the scales.

	Gas residence time (s)			
Reactor volume (m^3)	**Stirred tank, non coalescing**	**Stirred tank, coalescing**	**Bubble column**	**Air-lift reactor**
0.01	**6**	**4**	**1**	**2**
0.10	*10*	**7**	**2**	**3**
1.0	*17*	**11**	**4**	**6**
10	*27*	*18*	**7**	**10**
100	*42*	*28*	*12*	*19*
1000	*47*	*31*	*23*	*34*

The gas hold-up in stirred tanks with noncoalescing liquid is taken to be 50% higher than for coalescing liquid. For the air-lift reactor, a 50% increase is chosen compared to the bubble column (see [5] for details). Other assumptions are as in Tables 16.1 and 16.2. Numbers in *italics* indicate conditions where the gas residence time is longer than the mass transfer time, and gas depletion is likely to occur. The numbers in **bold** indicate no serious depletion.

If the oxygen uptake rate is $100\,\mathrm{mol\,m^{-3}\,h^{-1}}$, then the rate of glucose consumption is $50\,\mathrm{mol\,m^{-3}\,h^{-1}}$. Assuming an affinity, K_m, for glucose uptake in the $0.1\,\mathrm{mol\,m^{-3}}$ range, a fed-batch process with low specific growth rate will operate at glucose levels below the K_m value. For example, if the glucose concentration is $0.01\,\mathrm{mol\,m^{-3}}$, the uptake time of glucose is less than $0.01 \times 3600/50 = 1\,\mathrm{s}$. If the dissolved oxygen concentration is controlled at $0.1\,\mathrm{mol\,m^{-3}}$ (30% of air saturation at 1 bar), the characteristic time is $0.1 \times 3600/100 = 4\,\mathrm{s}$. Both are smaller than the mixing times and mass transfer times ($1/k_la$) estimated before (see Tables 16.2 and 16.3), which means that transport cannot cope with the fast reaction. This also is clear already at $0.01\,\mathrm{m^3}$ scale under the assumptions applied, which means that such a high-intensity process cannot be run in the lab. Measures should be taken to relieve the transport limitation or reduce the oxygen demand and sugar feed rate.

Another feature of large bioreactors is the shear rate gradients. Close to the impeller and at breaking or bursting bubbles, there can be high peaks in the shear rate, while in more distant locations, the shear rates can be very low. If on average, the shear rate, γ, is equal to [12] about 10 N [s^{-1}] then close to the impellers it could be 30 times higher [13], or $300 \times N$, and very locally close to bursting bubbles, even higher. This approximation has been widely used, although it strictly applies to laminar flows. Moving up in scale, lower values of N are used (to avoid excessive power input) and the shear rate becomes smaller. Sanchez Perez *et al.* [14] derived the following alternative for pseudo-plastic fluids under turbulent conditions:

$$\gamma = m\,N^{3/1+n} \tag{16.14}$$

Table 16.6 Estimation of average and peak shear rates, using the different expressions and comparing Newtonian ($n = 1$, $K = 1$) and non-Newtonian ($n = 0.3$, $K = 10$) high viscosity broths.

	Shear rate, γ [s^{-1}]					
	Average laminar	Peak laminar	Average turbulent, Newtonian	Peak turbulent, Newtonian	Average, turbulent, non-Newtonian	Peak turbulent, non-Newtonian
Reactor volume (m^3)	$\gamma = 10\,N$	$\gamma = 100\,N$	$\gamma = m\,N^{3/2}$	$\gamma = 30\,m\,N^{3/2}$	$\gamma = m\,N^{3/1.3}$	$\gamma = 30\,m\,N^{3/1.3}$
0.01	207	2070	66	2000	107	3200
0.10	124	1240				
1.0	74	740				
10	45	450				
100	27	270				
1000	16	160				

$N_p = 2.5$.

with

$$m = (4\,N_p\,\rho\,d_s^2/27\,\pi\,K)^{1/1+n}, \tag{16.15}$$

where n is the flow behavior index, K the consistency index, both parameters in a power law equation representation of pseudo-plastic rheology. Usually in fermentation broths, n will drop from 1 (Newtonian) to about 0.3 [–], while K is strongly dependent on the biomass concentration, and can be as high as 10 [N s^n m^{-2}]. At fixed aspect ratio, scale-up using this expression predicts a constant γ at different scales. Some comparative calculations are provided in Table 16.6.

In fermentations with filamentous organisms, it has been frequently reported that cells are damaged by high shear stresses. Although there is a lot of uncertainty concerning the precise mechanisms and scale dependency, a valuable correlation for different impellers and scales is the energy dissipation–circulation function, EDC [15]. Here, morphological damage was shown dependent on the energy dissipation rate close to the impeller ($\sim P/d_s^3$) and on how frequent cells pass through ($\sim N\,d_s^3/V \sim$circulation time^{-1}). The length scale for this type of cells is larger than 100 μm, which for modest viscosity of the medium is (as explained in Chapter 11) larger than the Kolmogorov microscale of turbulence, ranging from 30 μm for local P/V of 1 kW/m^3 to 13 μm when local P/V is 30 kW/m^3. Hence, it is seen that fluid mechanical stress is causing the problem. It is further noted that bacteria and yeasts are not damaged at all under the same conditions, because they are smaller than the Kolmogorov eddy size. In the literature, poor biological performance is often attributed to shear damage, but upon scrutiny of the data more likely effects seem to have been overlooked, such as

CO_2 inhibition, osmotic stress, or poor mixing of titrants and nutrients. Also, mammalian cells are often stated to be extremely fragile because they do not have a cell wall. However, systematic studies have made clear that this is not true (13]; Godoy-Silva *et al.*, 2009 [16]) provided that Pluronic F68 is added to keep cells away from gas–liquid interfaces.

Given the sensitivity of companies to share commercial data from large industrial bioreactors, only limited data sets of sufficient quality are available to develop a generic picture. One important exception is the many publications made in a Nordic consortium for studies of *Saccharomyces cerevisiae* and *Escherichia coli* using the 30 m^3 bioreactor in Stavanger, Norway. Part of the conclusions above is based on publications from this consortium (e.g., [8,9,17–24). Another good source of information is the early work done by Oosterhuis and Kossen [25] applied to a 25 m^3 gluconic acid fermentation and Sweere *et al.* [26] dealing with baker's yeast. Later studies on baker's yeast fermentation at 215 m^3 scale [27], and *Aspergillus oryzae* in 80 m^3 fermenters [28,29], Nienow [13] on 20 m^3 mammalian cell cultures, Kelly and Humphrey [30] and Moilanen *et al.* [31] provide complementary information on the large bioreactor scale.

16.3
Scale-Down

Downscaling, that is, mimicking the full-scale industrial process in the pilot or lab, would be perfect if the cells face an identical environment at all scales. This environment is not constant, but a dynamic exposure to varying oxygen and substrate concentrations, shear rates, temperatures, and pH. Perfect downscaling of the complete environment of the cells, in all its details, to the lab is not only impossible but also not needed. Downscaling research is successful if the right approximations are made, concentrating on the aspects to which the cells are most sensitive. For proper downscaling, there are three key guidelines:

1) Keep the process within the same operating regimes under the same rate-limiting mechanism.
2) Geometric similarity is not necessarily maintained. Instead, experimental setups in the lab can be designed within a wider space with the purpose to create similarity in the environmental conditions for the cells.
3) Any scale-down work should be guided by a mathematical framework. It is emphasized that this is not only important in order to quantify the items of study but also a mathematical model for the process helps to systematically organize know-how and data. This improves experimental design (beyond trial-and-error), judgment of experimental results beyond intuition, and learning via comparison of predicted and measured data. Here as elsewhere, the final objective of a mathematical model is to arrive faster to the desired result than without the support of such mathematical model.

The first attempt to give a quantitative estimation of the relative importance of the various transport rates is the *regime analysis* [32–34]. Here, all the relevant characteristic times for the physical and biological mechanisms are estimated and mutually compared. It is the primary task of the downscaling researcher to stay within the same physiological regimes of the cells as on the large scale. For example:

- If the liquid circulation or mixing time is larger than the consumption time of the limiting substrate, then cells will face substrate concentration gradients.
- If the mass transfer time is smaller than the gas residence time, then the gas phase will get depleted.
- If the mass transfer time is larger than the oxygen consumption time, then oxygen concentration gradients will develop.

Granted that this preliminary insight has been gained, then in a further assessment the magnitude of the rates will be found, with concentration and shear rate gradients being similar at different scales. But for high precision, more advanced mathematical models are needed, and they can be ranked according to their level of sophistication:

- Single parameter models, that is, the axial flow with dispersion model, or the tanks-in-series model. Here, with one parameter the deviation from ideal plug flow or well-mixed flow is defined. Although simple, such models can be very valuable to assess the primary flow characteristics in scale-down equipment.
- Compartment models [19] capture the more intricate flow and reaction dynamics in virtual compartments, in which different mechanisms occur, or in which rates grossly differ. For instance, in one part of the large bioreactor, close to the impeller, mass transfer is determined mainly by the impeller power input, while in more remote zones the bubble velocity is dominant. The model may then be composed of compartments in which different $k_l a$ correlations apply (e.g., Eqs. (16.8) and (16.10)). Another example is concerned with cells that may move between zones where the metabolism is different, for example, between overflow metabolism close to the sugar inlet point, and reabsorption of remaining sugar and the overflow metabolite elsewhere. There can even be a third zone where only glucose is consumed and not the metabolite [24] or where the cells are starved for substrates, as happens in poorly mixed regions.
- Computational models, linking together fluid and reaction dynamics in finite element simulations. Enabled by massive computation power, these models have now developed to levels that make detailed descriptions of the local state possible, in millions of finite volumes describing together the full reactor [22,23,35,36].

These models usually start with a description and characterization of the liquid flow, while aspects of gas flow, mass transfer, shear rates, and reaction kinetics are included later, superimposed on a "frozen" fluid flow. More recently, also gas–liquid fluid dynamics has been computed from the start, accounting for the mutual interactions between gas and liquid.

Essential in successful use of models is that the key parameters are determined by independent experiments. For large bioreactor processes, such validation is not easy due to the limited amount of published material available and lack of test facilities. The computational models do, however, have a tremendous potential to reconstruct missing process data. An ideal design approach would be the following:

1) A regime analysis is executed based on coarse estimation and comparison of subprocesses that make up the real process (mixing, mass transfer, reaction, etc.).
2) Based on a combination of fundamental rules and (scarce) experimental data, a computational model is set up, which is able to describe known characteristics of a process, in a given geometry.
3) The computed results are checked against the primary outcome of the regime analysis. Inconsistencies should be addressed and removed.
4) The large bioreactor is modeled as two or more compartments, in which the main regimes are different or rates have largely different values [37].
5) This information is used to define pilot or lab-scale simulators [38], where the same compartments are recognized and where the critical conditions are as far as possible mimicking the large bioreactor.
6) After execution of the experiment, fine-tuning of previous parts may be needed, and some iterative cycles made.

In the literature, there are many recent reports of detailed experimental scale-down studies [38,39]. For the down-scale simulators, we can recognize two different approaches: the one-compartment systems and two-compartment systems.

16.3.1 One-Compartment Systems

In the one-compartment model, varying conditions are created via intermittent feeding of limiting substrate, on–off control of aeration, cycling of stirrer speed, pressure, temperature, inlet gas composition, and pH. This will give insight into the main impact of single, repeated disturbances on the bioprocess performance. The disturbances are usually regular, but can also be variable, for example, via a Monte Carlo simulation or randomized dynamics that better mimics the variable circulation times in a large bioreactor. Yegneswaran, Gray, and Thompson [40] have thus varied the dissolved oxygen concentration in *S. clavuligerus* cultures *with cycle times* between 8 and 44 s. In extended studies, the impact of multiple variations can be studied as well.

By keeping all the cells exposed to the same conditions at all times, the variation in exposure time as in real reactors is not taken into account. Another point of concern is that the average time of cycles is often chosen to be longer than needed. For example, De Jonge *et al.* [41] studied the impact of 360 s cycles with on–off glucose feed on *Penicillium chrysogenum* cultures. And, in an earlier study, Vardar and Lilly [42] varied the dissolved oxygen concentration with *P. chrysogenum*, with a regular cycle time of 2 min. From our previous assessment (Table 16.2), it was found that even in 1000 m^3 vessels the mean circulation time would not exceed 60 s if the total power input is at a reasonable value of 2 kW m^{-3}. Of course, the reported investigations could be judged as a worst case, but shorter cycle times in principle would be preferred. A particular note is made on a recent report applying repeated cycles of high and low stirrer speeds at 2 l scale to mimic the shear rate gradients of a 14 m^3 industrial reactor with mammalian (CHO) cells producing therapeutic proteins [43]. A cycle time of 36 s was applied, with power input varying between 0.01 and 0.4 W kg^{-1}, which are values estimated for the low and high turbulence zones in the large reactor, respectively. Nienow [44] made clear that this scale-down protocol is not appropriate as the cells are already going through such cycles in a bench bioreactor with constant stirrer speed. The cycling of the average power input is not necessary and subjects the cells to a 40× wider range of shear rates than needed.

An advantage of single-compartment models is that they also are amenable to high-throughput screening. Microbioreactors have been proposed, downsized to shake-flask [45], microtiter plate [46], or even smaller. It should be understood that at this scale, one moves further away from the large scale, and regime changes will certainly happen, thus increasing the risks that the process is not representative for the real bioprocess. Table 16.7 shows some key characteristics of such small bioreactors.

Table 16.7 Characteristics of microbioreactors.

Reactor volume (ml)	Height (mm)	Diameter (mm)	Wall area/volume (m^2 m^{-3})	Surface area/volume (m^2 m^{-3})	N (s^{-1})	Reynolds number (–)	N_p (–)
100 000[a)]	1050	350	11	1	12.4	24 300	2
100	105	35	114	10	42	800	5[b)]
10	48	16	247	21	70	300	5[b)]
1	22	7.5	533	44	118	100	5[b)]

Aspect ratio H/T is 3. Agitation is assumed via a small Rushton turbine impeller with $d_s/T = 0.4$, and $P/V = 2$ kW m^{-3}. Aeration is done only via the gas–liquid surface. Interfacial effects will become much more dominant than in larger reactors, as well as the loss of turbulent flow (reflected by the Reynolds number $Re_{mix} = \rho\ N\ d_s^2/\eta$, with $\eta = 10$ mPa·s), and can result in artifacts such as wall growth, cell settling, morphology changes, influence of the vessel material, diffusion limitation, and so on.

a) Reference for comparison.

b) Unaerated.

It should also be understood that many studies have been executed in one-compartment bioreactors with single disturbances, for example, a pulse of glucose. Such studies give insight into the initial response of microorganisms to changes in their environment, but have little relevance for downscaling of a large-scale fermentation where the disturbances are repeated many times. If the fermentation lasts for 3 days, then with an average cycle time of 1 minute the cells initially present undergo more than 4000 cycles. A further disadvantage is that the culture is disturbed after the change and may not be useful for further research and elucidation of the *in vivo* kinetics and needs to be stopped. To overcome the latter, solutions have been presented such as the Bioscope [47] and the sensor reactor [48] that operate in a master–slave mode and do not disturb the main culture.

16.3.2
Two-Compartment Systems

Two-compartment setups have been applied in other studies [32,38,49–53]. In the early days, two interconnected stirred tanks were used, between which the broth was recirculated, with feeding or aeration in only one compartment. The size ratio of the compartments was somewhat arbitrarily set as, for example, 1 : 10, as was also the power input in each compartment, and the recirculation rate was varied from seconds to minutes. The shear fields generated by the pump and the flow in the circulation loops, however, were not accounted for. Although qualitatively good comparison with large-scale data was obtained, precise matching appeared difficult. Later on, based on the reasoning that outside the impeller region the broth circulates more as a plug flow with dispersion than as well mixed, a tube or static mixer was applied as the second compartment. Although intuitively better, the complex flow patterns in large fermenters are still not fully captured. And in all circumstances, the fast fluctuations as measured by Larsson *et al.* [9] and calculated by Enfors *et al.* [8] were not accounted for. The stochastic nature of such micromixing variations, on the scale of seconds, and with amplitude of more than twice the average value, could also have a large impact on microbial performance.

For scale-down studies with emphasis on the significance of a one- or two-compartment model, the use of well-characterized biological systems is an important point of consideration. The occurrence of overflow metabolism, the morphology, and the cell viability are often quantitatively related to the (lack of) control of limiting substrate and oxygen concentrations, or shear rate, or pH and temperature. Amanullah *et al.* [50] published important results on the oxygen-sensitive acetoin and 2,3-butanediol formation with *Bacillus subtilis*. Hewitt *et al.* [21] explained the loss of viability of *E. coli* cultures as a result of high glucose and low oxygen levels, but only on smaller scales where the variations are smaller and more defined and not on the large scale where the concentrations vary in a more severe and chaotic manner.

From this analysis, it can be summarized that there are five degrees of freedom for the proper design and operation of scale-down simulators: the number of compartments, the compartment volume ratio, the recycle flow rate, the flow pattern per compartment, and the relevance of fast feed fluctuations superimposed on the feed or bioreactor control. From literature analysis, these five requirements for precise scale-down to our best knowledge have probably never been all fulfilled within a single study.

16.4 Investigations at Lab Scale

In this section, four examples of lab-scale down-scale studies will be presented in somewhat more detail, to illustrate what has been accomplished and what are still issues and needs for further development. The examples concern the fermentative production of

- gluconic acid, a small metabolite, with a robust bacterium, *G. oxydans*;
- lipase, an enzyme used in food application, with a sensitive bacterium, *Pseudomonas alcaligenes*;
- baker's yeast, used as living biomass in dough, with yeast *S. cerevisiae*; and
- penicillin, a secondary metabolite with antibiotic activity, using a filamentous fungus, *P. chrysogenum*.

There are more good examples available in the literature, but most of the typical aspects encountered are covered in this limited set.

16.4.1 Gluconic Acid

One of the first reports on fermentation downscaling was by Oosterhuis *et al.* [32]. Observations in a 25 m^3 stirred tank reactor with *Gluconobacter oxydans* converting glucose into gluconic acid showed that local oxygen concentrations varied along the vertical axis. The maximum oxygen concentration was found at the level of the two Rushton turbine impellers in the tank, and close to the headspace it dropped to nearly zero. In between the impellers and closer to the bottom, intermediate oxygen concentrations were measured. A more detailed experiment revealed also a radial decrease of the oxygen concentration toward the tank wall. Downscaling of these observations covered three topics:

- Mathematical modeling, describing several virtual compartments, to cover local mass transfer and mixing phenomena.
- Lab-scale experiments with varying oxygen concentrations in one reactor. Gas blending was used to vary the dissolved oxygen concentration

independent of the stirrer speed. The cycle time was 20 s, derived from large-scale regime analysis, but other cycle times were also tested.
- Lab-scale experiments using two interconnected stirred tanks, with aeration or pure oxygen gassing of one unit and nitrogen gas sparged to the other. The relative residence time in the compartments was varied, so was the recycle rate.

It was found that a mathematical model with stacked stirred and bubbled compartments, using the related $k_l a$ correlations (16.8)–(16.10) and the impeller pumping flow rate, described well the axial dissolved oxygen distribution. As a refinement, a more detailed five-compartment model also covered the radial contours. In agreement with the experimental results, the model showed that from the impeller discharge to the tank wall, the dissolved oxygen concentration dropped from 15 to 3% of saturation.

The downscaled gluconic acid fermentations showed differences in the product formation rate for the one- and two-compartment tests, even when the same average times of aeration and oxygen depletion were used. In the single compartment, exposure times to low dissolved oxygen concentrations above 180 s caused a lower maximum oxygen consumption rate. In the two-compartment simulator, cells did not suffer from the variations in oxygen tension. Short exposure of cells to high dissolved oxygen levels, however, reduced the cell growth rate and the ability of the cells to make gluconic acid. This demonstrates that results are sensitive to the mean residence time in the recycle loop and the magnitude of dispersion or back mixing, as well as to the absolute concentrations of dissolved oxygen. These variables should be mimicked as precisely as possible.

16.4.2
Lipase

An industrial study of the scale-up of a lipase enzyme production, with *P. alcaligenes* growing on soybean oil in fed-batch mode, demonstrated an empirical scale-down approach to resolve a scale-up problem [52,54]. The studies revealed a gradual decline in enzyme productivity from 10 l, 100 l, 4 m^3 to 100 m^3. At 100 m^3 the productivity was only one-third of that in 10 l. The researchers embarked on a root-cause analysis, and identified five possible reasons for the decline (Table 16.8). In a parallel study, each of the causes was independently verified.

It was concluded that, upon scale-up, the aeration rate needed to be reduced and the pressure was increased. This had resulted in an increase of the CO_2 concentration and inhibition of product formation. Other causes contributed only marginally or not at all to the production losses, and could be ruled out. In a final validation of the modeling work, increasing the aeration rate (together with more precise foam control), minimizing the headspace pressure to 1.1 bar, and reducing pH (from 7.0 to 6.7) solved the problem. The scale-up loss was reduced to a satisfactory 10%.

Table 16.8 Root-cause analysis and down-scaled results helping to solve the lipase scale-up problem (based on [52,54]).

Possible cause	Downscale action executed	Result obtained
Concentration gradients of the feed soy bean oil, the titrant ammonia, and the pH	One-compartment scale-down reactor, with on–off control of soy bean oil (fast with 2 s on and 20 s off, and slow with 10 s on and 200 s off), with NH_3 as N-source, and with up-down control of pH, using amplitudes of 0.3 and 0.5	No effect found
The use of technical-grade compounds in the factory, while pure in the lab, the inoculum preparation from flask in the lab and from a fermenter in the plant, and medium steam sterilization in the large scale compared to autoclaving in the lab	From pilot plant scale, a sample was taken immediately after inoculation of the main fermenter, and transferred to a lab fermenter for a master–slave comparative study	No effect found
Dissolved oxygen gradients in the 100 m^3 reactor, or even oxygen depletion, which was not the case on lab scale	Two-compartment setup, with two stirred tanks having volume ratio 1 : 10. The larger tank was aerated and the smaller not. A recycle time of 200 s was used	Under these conditions, the lipase production was 25% lower compared to the reference, which is significant but not enough to explain the full effect
Higher shear stress on the large scale, from impeller action or gas–liquid interfaces	In the lab, a worst-case test was done with higher stirrer speed and aeration rate, such that P/V increased by a factor 3. Measurement of lipase degradation	For this study, 20% less lipase was found, again not enough to explain the losses observed. In addition, lipase appeared to be stable and not degraded by high shear stresses
The combination of lower gas flow rate and higher pressure in the large reactor would result in higher CO_2 levels	On lab scale, CO_2 gas was added to the air inlet, such that 5 and 10% CO_2 was supplied	*10% CO_2 alone could explain a lipase reduction of 70%* – the most important conclusion!

This study presents a comprehensive overview of what can be achieved in industry to address a scale-up problem in the most practical way, and it also underlines the importance of CO_2 ventilation control. However, it is tempting to state that a good regime analysis could have presented the solution perhaps in a more direct way.

16.4.3
Baker's Yeast

Sweere *et al.* [26,49,55] operated both a single-stirred tank and a two-stirred tank combination to study the impact of glucose and dissolved oxygen concentration gradients on the aerobic growth of *S. cerevisiae*, the baker's yeast. They first applied a regime analysis based on a hypothetical large-scale operation. This was used to design scale-down simulator experiments. Different volume ratios and recirculation rates were applied, to span the full window of possible large-scale operations. With glucose feeding in only one reactor, they found that less biomass and more ethanol was produced. The physiological mechanism was related to a lower maximum specific oxygen consumption rate. For the by-products glycerol and acetate, no clear effect of the recycle rate was observed.

In a one-compartment system, cycling oxygen levels already showed biomass reductions of 15% (for cycling 30 s) and 35% (for cycling 60 s), and the formation of ethanol. The yield of biomass got progressively lower, while the yield of ethanol increased at longer cycle times. Similar results were observed in the two-compartment simulator for circulation times of 30 s and 60 s. For slower variations, the single reactor performed just a little better. A difference was found for glycerol and acetate: in the single compartment 2–3× more glycerol was produced at slow cycling (>1 min), and 2–4× more acetate throughout (see Figure 16.3).

This information illustrates that by-products can be important signaling sensors for good scale-down. No real scale-up was included in this work. A later study by Abel, Hübner, and Schugerl [56] confirmed these results when another *S. cerevisiae* strain was used, although they reported more ethanol formation when the cycling was extremely fast. This was attributed to mechanical stress by the pumping action.

However, George *et al.* [27] did a combined effort, scaling down from a 215 m^3 industrial bubble column producing baker's yeast, to a two-compartment lab setup with a stirred tank and a static mixer, providing plug flow with some dispersion in an external loop. Glucose was fed at the beginning of the loop and this resulted in a temporary exposure of the cells to concentrations of $0.5–1.9\,g\,l^{-1}$. The volume of the mixer was 10% and the residence time 60 s, as derived from coarse fluid flow estimations and fluid dynamics computations for the large tank. In parallel, a reference fermentation was run in the lab without the external loop, keeping everything else unchanged.

As a conclusion of this study, a final biomass yield reduction of 6–7% was observed at the end of the process in both the scale-down setup and in the industrial reactor compared to the lab-scale reference. This was accompanied by a temporary higher peak of the ethanol concentration, revealing first ethanol overflow and subsequently reabsorption, a process with a net ATP consumption that can well explain the yield loss. The biomass quality was also measured in terms of dough gassing power. For this work, the quality under varying conditions was significantly better, and the increase in the quality parameter may well mean more than the decrease of the biomass yield. Unfortunately, no report of

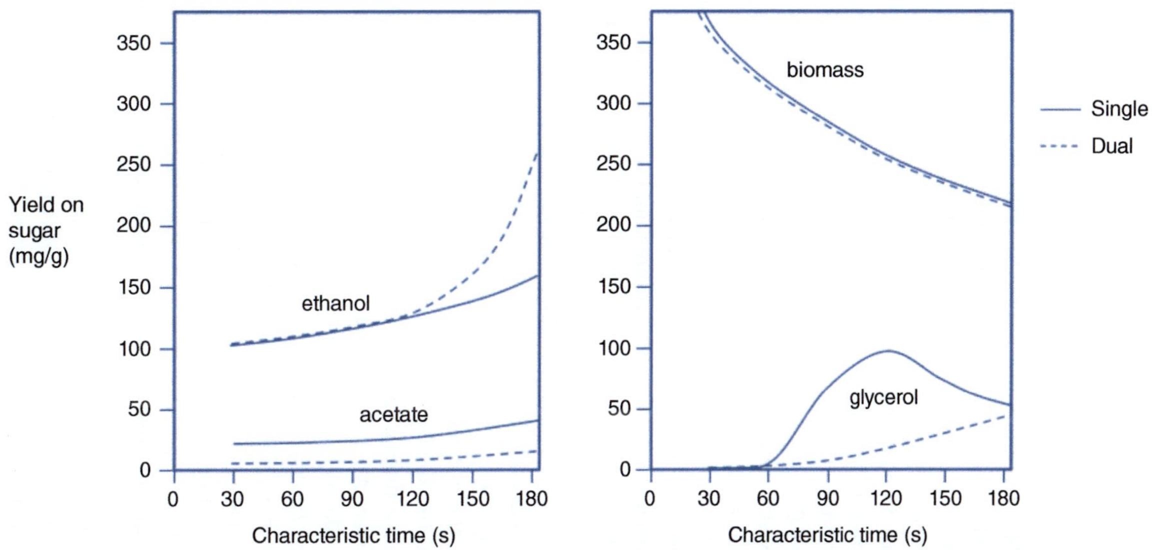

Figure 16.3 Comparison of product yields on sugar, experimentally achieved in a single stirred tank reactor with continuous cycling of the dissolved oxygen concentration ("single") or in a two-compartment setup in which one stirred tank was aerated and the other sparged with nitrogen gas ("dual"). Yields are expressed as a function of the average cycling time, denoted here as the characteristic time for the cycle (based on [49,54).

the acetate and glycerol levels was made, and it was not discussed why the Sweere studies resulted in bigger losses under similar conditions (it could be because another strain was used). Finally, no leads for large-scale process improvement were provided.

Still, despite these unresolved questions, the extensive studies of baker's yeast production present a key example of a proper scale-down approach, reconfirming the power of the method, provided that the scale-down design is quantitatively coupled to the full-scale operation.

16.4.4 Penicillin

Vardar and Lilly [42] were one of the first to verify the impact of anticipated large-scale dissolved oxygen gradients on the production of penicillin, with the filamentous fungus *P. chrysogenum*. A lab-scale fermenter was operated with dissolved oxygen concentrations cycling by ±10% around the average 0.08 mol/m^3. Just below this value, in controlled reference fermentations, it was found that the specific penicillin production rate decreased sharply. The cycling experiment also resulted in a lower productivity, and the effect was reversible. With oxygen levels further reduced to 0.02 mol/m^3, an irreversible impairment of the penicillin production was observed, so was a reduced oxygen uptake rate. Larsson and Enfors [57] observed similar biological damage via an oxygen starvation experiment.

Larsson and Enfors [58] used a two-compartment setup, with an aerated stirred tank and an anaerobic plug-flow unit through which broth was recycled. If the plug-flow reactor was 6% of the volume, and at residence times higher than 5 min, the respiration was irreversibly reduced. However, with a smaller volume, and residence times of 2 min and less, no such effect was found.

De Jonge *et al.* [41] considered cycling glucose concentrations, via on–off feeding in a single lab-scale bioreactor. With a feeding time of 36 s and subsequent nonfeeding for 324 s, they operated a 6 min cycle. After an initial peak = 0.4 mmol/l, the glucose concentration decreased to virtually zero after 3 min. As a result, the specific penicillin formation rate was about 60% lower than in the reference during the first 100 h, and after 200 h it had gradually moved to 40% lower. An increased rate of storage metabolism and energy consumption in futile cycling was held responsible.

The sensitivity of *P. chrysogenum* for shear damage was investigated via the energy dissipation–circulation function [15]. Mycelial damage could be well described at different scales and with different impellers, and it was strongly correlated with the loss of penicillin productivity. Thus, on a smaller scale there is more fragmentation at equal P/V and d_s/T. Interestingly, impeller tip speed commonly used as an indicator of mycelial damage did not appear to correlate well with damage to the mycelia.

The loss of productivity as a result of cycling concentrations applies to *P. chrysogenum*, but cannot be generalized to other filamentous fungi. For example, *A. oryzae* showed a similar correlation between the mycelial damage and the energy dissipation–circulation function [59]. However, enzyme production was *not* impaired.

Such observations, although providing valuable information, illustrate that even for important industrial fermentations like production of penicillin, the scale-down approach has been merely applied in an *ad hoc* fashion, without a careful analysis of the large-scale situation and precise, systematic scale-down of the key environmental conditions. Wang *et al.* [39] have recognized this and proposed to select *P. chrysogenum* as a good model for an exhaustive study, integrating computational fluid and reaction dynamics in a multiscale approach.

All these examples have provided good insights, but appear to be of limited scope since usually only one or two scale items are studied. Also, the design and operation of scale-down simulators is not always based on careful large-scale analysis, and simultaneous application of the five quoted degrees of freedom in scale-down design has never been applied in a single study.

16.5 Scale-Up

This last section is an attempt to describe how the large-scale process can be improved based on good down-scale studies. Of course, such improvements are in the field of industrial application and usually not in the public domain. Still, there are some good published examples and in the remainder of the chapter some more general guidelines can be given.

From the comparison of characteristic times, rate-limiting steps can be identified. It is proposed to relieve the (transport) limitation or, alternatively, reduce the consumption rate of the limiting nutrient. Characteristic transport times can be decreased in many different ways, and the same is true for the characteristic substrate uptake times. It is not possible to be exhaustive here, but some examples can be provided:

- *Retrofitting of the large-scale bioreactor.* Consider, for example, the production of baker's yeast from sugar. A single glucose inlet point, at the top of the vessel, may be suboptimal and result in local ethanol overflow. Improvement may be gained by feeding close to the bottom impeller, or a dual glucose inlet point or an inlet in the middle section of the reactor can be used. In this way, concentration gradients can develop only over half the distance [7]. Another retrofit could be the use of different impeller designs to achieve a better axial mixing. Buckland *et al.* [60] demonstrated this for *Streptomyces* and *Nocardia* cultures. The use of new mixer designs can also be a key to reduction of mycelial damage in fermentations with filamentous fungi as was found by Kold (2010), cited in Villadsen *et al.* [61], p. 539: an *A. oryzae* fed-batch fermentation was extended for many hours without degradation of the mycelium when rotating jet heads were used for mixing rather than Rushton turbines.

- *Tuning of the experimental protocols*. It was reported for *A. oryzae* that a less filamentous morphology could be induced via intermittent feeding of glucose in fed-batch culture. As a result, the broth viscosity was lower and this significantly increased the oxygen transfer capacity, at the same power input. When the extra oxygen transfer capacity was used to add extra sugar, this resulted in up to 75% higher product formation on industrial scale [62].
- *Changing the quality or nature of the feedstock*. If cells are sensitive to external glucose concentrations, one may select a lower quality sirup and feed oligosaccharides, instead. The cells are now likely to excrete enzymes to hydrolyze the feedstock, and glucose will be slowly released into the medium. In this way, variations in glucose concentrations as a result of poor mixing will be dampened out. Figure 16.4 shows an illustrative example of replacing glucose by maltose.

Another means is to choose another feedstock, or use mixed feedstocks, and by rebalancing of the stoichiometry the sensitivity for the limiting substrate could be reduced. For example, if in the PDO case 50% of the sugar would be replaced by glycerol, then we will get:

$$0.5\, C_6H_{12}O_6 + C_3H_8O_3 + 2.5\, O_2 \rightarrow C_3H_8O_2 + 3\, CO_2 + 3\, H_2O. \qquad (16.16)$$

Under these conditions, depending on the affinity for glycerol compared to glucose, it could be that the carbon feeding into the metabolism is more even. However, for this example the oxygen demand is higher compared to the stoichiometry of (16.13), which might introduce a more severe oxygen stress. It is noted that the same would be true for other reduced cosubstrates, such as ethanol and vegetable oil. Neither of the stoichiometries (16.13) and (16.16) is, however, optimal. Except for the ATP used to produce biomass (and this has to be synthesized through glucose oxidation to CO_2 in the TCA cycle and respiration), the process is best run anaerobically with a carbon yield of $^3/_4$ when glucose is used as substrate, and $^7/_8$ when glycerol as substrate. The fundamental change in the process – from aerobic to anaerobic – is discussed below.

Adjustment of the properties of the microorganism. A general direction could be to reduce the formation rate of biomass (if this is not the main product) or the energy required for cell maintenance as this will, as discussed above, also decrease the oxygen and carbon substrate demand. In parallel, if the metabolic network could be made more efficient toward the product, then usually less oxygen is required. We may, for example, distinguish a low and high biomass yield on glucose:

$$CH_2O + 0.69\, O_2 + 0.06\, NH_3 \rightarrow 0.3\, CH_{1.8}O_{0.5}N_{0.2} + 0.7\, CO_2 + 0.82\, H_2O. \qquad (16.17)$$

$$CH_2O + 0.58\, O_2 + 0.08\, NH_3 \rightarrow 0.4\, CH_{1.8}O_{0.5}N_{0.2} + 0.6\, CO_2 + 0.76\, H_2O. \qquad (16.18)$$

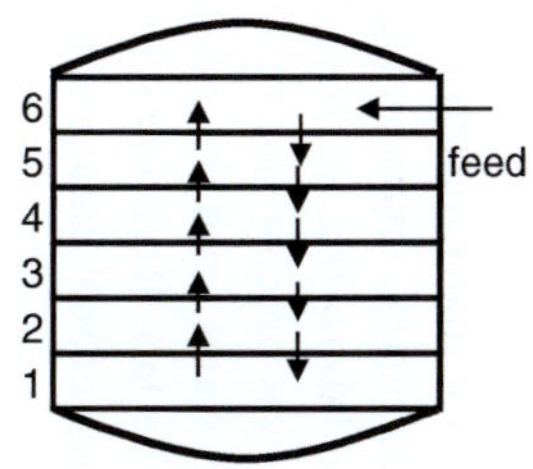

Maltose ⇢ Glucose ⇢ Consumption

maltose degradation kinetics: rM = c×M/(M+m)
glucose uptake kinetics: rG = c×G/(G+g)×Cx

local concentrations in mg/l:

Feed: glucose

maltose	glucose
0	34
0	14
0	4.8
0	1.5
0	0.4
0	0.2

Feed: maltose

maltose	glucose
29	18
13	11
6.1	5.5
2.8	2.3
1.4	1.1
0.8	0.5

affinity const. maltose: m=1 g/l

Feed: maltose

maltose	glucose
130	6.9
110	5.8
88	4.9
75	3.8
66	3.4
62	3.0

affinity const. maltose: m=10 g/l

Figure 16.4 Simple linear compartment model for a large bioreactor. With poor mixing, a steep glucose concentration gradient may develop. The sections depicted in red numbers are below a critical glucose limit of 0.5 g/l. Replacing glucose by maltose, and assuming slow-release kinetics for maltose hydrolysis, the glucose gradients will be less severe, and concentrations are more likely to stay within critical limits.

It is found that for a given glucose consumption rate, a higher biomass yield also results in a lower oxygen demand. Then also oxygen concentration gradients will be less severe.

Taking again the example of PDO, the ultimate microorganism will be able to produce the product with a carbon yield of 3/4 without any involvement of oxygen:

$$C_6H_{12}O_6 \rightarrow 1.5\, C_3H_8O_2 + 1.5\, CO_2. \tag{16.19}$$

Another change may be in the affinity for limiting substrates. If this can be decreased (higher Monod constant, K_s) by changing the uptake transporter nature or properties, then the intermediate substrate concentration will be higher, which gives less tendency for local depletion.

The third target for strain improvement may be the ability to buffer external variations via an internal storage pool, such as trehalose or glycogen. Then, the cell may be better capable of controlling intracellular intermediate pool levels.

Further, in the lipase example the scale-up problem could also have been addressed by development of a strain that is more tolerant to CO_2. Then, upon upscaling the process would have been less sensitive to the CO_2 increase.

In all cases, careful judgment is needed on the economic impact of all the changes, and the outcome will determine the preferred way to go. Feedstock changes result in different variable costs, and sometimes more oxygen brings more utility costs, but these may be compensated by producing more product with lower fixed costs per amount of product formed. Biological solutions usually require higher upfront R&D costs, and there is some delay before implementation can be done. Retrofitting and other hardware adjustments need investment costs, and during construction the fermenter is out of operation, maybe for a considerable period. And finally, the availability and intimate use of a good modeling framework will give higher precision and a faster route to results.

16.6 Outlook

It is expected that the number of bioprocess factories and bio-refineries to serve the growing need for more sustainable bio-products will further increase in coming years, and the expansion may very well last for many decades.

Commercialization of innovations in biochemistry, engineering, and biology will require a fast and reliable path from the lab to the factory. This brings a strong need to further evolve fast and simple, yet precise and sufficiently representative methods for scale-down. Scale-down starts from the perspective of the full-scale industrial process, and not from the lab. Reconstruction of process details in a factory may be hampered by the lack of real data and limited accessibility to new data.

However, the further implementation of computational fluid and reaction dynamics is expected to become complementary in filling gaps in the detailed representation. Integration of metabolic and fluid dynamics, also including intracellular dynamics, will be one of the key bridging challenges [36,63,64]. Biological sensing systems that are built into living cells and return recordings on the experiences of the cell traveling through the large bioreactor are also expected to become a very useful novel instrument to gain information from the large scale [65].

On the laboratory experiment side, design and tests with good scale-down simulators will require further developments in the art. For high-precision scale-down, it is recommended to consider the 5 degrees of freedom for design and operation, as explained in Section 16.2.

For a faster approach, practical suggestions are given in Sections 16.3 and 16.4. Under all circumstances, care should be taken that downscaled work in the lab remains quantitatively representative for the real conditions in the large-scale bioreactor. This is not a trivial point, and it appears that in the literature this essential rule has generally been loosely applied.

References

1 Nienow, A.W., Warmoeskerken, M.M.C.G., Smith, J.M., and Konno, M. (1985) On the flooding/loading transition and the complete dispersal condition in aerated vessels agitated by a Rushton-turbine. Proceedings of 5th European Mixing Conference, Germany, BRHA, Cranfield, pp. 143–154.

2 Nienow, A.W., Konno, M., and Bujalski, W. (1985) Studies on three phase mixing; a review and recent results. Proc. 5th European Mixing Conference, Germany, BRHA, Cranfield, pp. 1–13.

3 Joshi, J.B. (1980) Axial mixing in multiphase contactors: a unified correlation. *Trans. Inst. Chem. Engrs.*, **58**, 155–165.

4 Heijnen, J.J. and Van't Riet, K. (1984) Mass transfer, mixing and heat transfer phenomena in low viscosity bubble column reactors. *Chem. Eng. J.*, **28**, B21–B42.

5 Heijnen, J.J., Hols, J., Van der Lans, R.G.J.M., Van Leeuwen, H.L.J.M., Mulder, A., and Weltevrede, R. (1996) A simple hydrodynamic model for the liquid circulation velocity in a full-scale two- and three-phase internal airlift reactor operating in the gas recirculation regime. *Chem. Eng. Sci.*, **52**, 2527–2540.

6 Van't Riet, K. and Tramper, J. (1991) *Basic Bioreactor Design*, Marcel Dekker, New York.

7 Van't Riet, K. and Van der Lans, R.G.J.M. (2011) Mixing in bioreactor vessels, in *Comprehensive Biotechnology*, 2nd edn, vol. **2** (ed. M. Moo-Young) Elsevier, Amsterdam, pp. 63–80.

8 Enfors, S.O., Jahic, M., Rozkov, A., Xu, B. *et al.* (2001) Physiological responses to mixing in large scale bioreactors. *J. Biotechnol.*, **85**, 175–185.

9 Larsson, G., Tornkvist, M., Stahl-Wernersson, E., Tragardh, C., Noorman, H., and Enfors, S.O. (1996) Substrate gradients in fed-batch bioreactors: origin and consequences. *Bioproc. Eng.*, **14**, 281–289.

10 Van't Riet, K. (1979) Review of measuring methods and results in nonviscous gas–liquid mass transfer in stirred vessels. *Ind. Eng. Chem. Proc. Des.*, **18**, 357–364.

11 Feijen, J. and Hofmeester, J.J.M. (1991) Gradients in production scale bioreactors. Proceedings of the 7th European Conference on Mixing, Antwerp, Belgium, vol. **1**, 269–276.

12 Metzner, A.B. and Otto, R.E. (1957) Agitation of non-Newtonian fluids. *AIChE J.*, **3**, 3–10.

13 Nienow, A.W. (2006) Reactor engineering in large scale animal cell culture. *Cytotechnology*, **50**, 9–33.

14 Sanchez Perez, J.A., Rodriguez Porcel, E.M., Casas Lopez, J.L., Fernandez Sevilla, J.M., and Chisti, Y. (2006) Shear rate in stirred tank and bubble column bioreactors. *Chem. Eng. J.*, **124**, 1–5.

15 Jüsten, P., Paul, G.C., Nienow, A.W., and Thomas, C.R. (1996) Dependence of mycelial morphology on impeller type and agitation intensity. *Biotechnol. Bioeng.*, **20**, 672–684.

16 Godoy-Silva, R., Chalmers, J.J., Casnocha, S.A., Bass, L.A., and Ma, N. (2009) Physiological responses of CHO cells to repetitive hydrodynamic stress. *Biotechnol. Bioeng.*, **103**, 1103–1117.

17 Bylund, F., Collet, E., Enfors, S.O., and Larsson, G. (1998) Substrate gradient formation in the large-scale bioreactor lowers cell yield and increases by-product formation. *Bioproc. Eng.*, **18**, 171–180.

18 Cui, Y.Q., Van der Lans, R.G.J.M., Noorman, H.J., and Luyben, K.C.A.M. (1996) Compartment mixing model for stirred reactors with multiple impellers. *T. I. Chem. Eng. Lond.*, **74A**, 261–271.

19 Vrábel, P., Van der Lans, R.G.J.M., Cui, Y.Q., and Luyben, K.C.A.M. (1999) Compartment model approach: mixing in large scale aerated reactors with multiple impellers. *Chem. Eng. Res. Des.*, **77**, 291–302.

20 Vrábel, P., Van der Lans, R.G.J.M., Luyben, K.C.A.M., Boon, L., and Nienow, A.W. (2000) Mixing in large-scale vessels stirred with multiple radial or radial and axial up-pumping impellers: modelling and measurements. *Chem. Eng. Sci.*, **55**, 5881–5896.

21 Hewitt, C.J., Nebe-Von Caron, G., Axelsson, B., McFarlane, C.M., and Nienow, A.W. (2000) Studies related to the scale-up of high-cell-density *E. coli* fed-batch fermentations using multiparameter flow cytometry: effect of a changing microenvironment with respect to glucose and dissolved oxygen concentration. *Biotechnol. Bioeng.*, **20**, 381–390.

22 Gunyol, O. and Mudde, R.F. (2009) Computational study of hydrodynamics of a standard stirred tank reactor and a large-scale multi-impeller fermenter. *Int. J. Multiscale Com. Eng.*, **7**, 559–576.

23 Gunyol, O., Noorman, H.J., and Mudde, R.F. (2009) CFD simulations of a large-scale fermenter with muliple impellers, in *Proceedings of the 9th International Conference on Gas–Liquid Solid Reactor Engineering* (eds J. Chaouki and P. Tanguy), University of Montreal, Montreal, Canada, pp. 1–4.

24 Noorman, H. (2011) An industrial perspective on bioreactor scale-down: what we can learn from combined large-scale bioprocess and model fluid studies. *Biotechnol. J.*, **6**, 934–943.

25 Oosterhuis, N.M.G. and Kossen, N.W.F. (1984) Dissolved-oxygen concentration profiles in a production-scale bioreactor. *Biotechnol. Bioeng.*, **26**, 546–550.

26 Sweere, A.P.J., Matla, Y.A., Zandvliet, J., Luyben, K.Ch.A.M., and Kossen, N.W.F. (1988) The influence of continually changing glucose concentrations on the fed-batch bakers' yeast production. *Appl. Microbiol. Biotechnol.*, **28**, 109–115.

27 George, S., Larsson, G., Olsson, K., and Enfors, S.O. (1998) Comparison of the baker's yeast process performance in laboratory and production scale. *Bioproc. Eng.*, **18**, 135–142.

28 Li, Z.J., Shukla, V., Wenger, K., Fordyce, A., Pedersen, A.G., and Marten, M. (2002) Estimation of hyphal tensile strength in production-scale *Aspergillus oryzae* fungal fermentations. *Biotechnol. Bioeng.*, **77**, 601–613.

29 Marten, M.R., Bhargava, S., and Wenger, K.S. (2003) Pulsed addition of limiting-carbon during *Aspergillus oryzae* fermentation leads to improved productivity of a recombinant enzyme. *Biotechnol. Bioeng.*, **82**, 111–117.

30 Kelly, W.J. and Humphrey, A.E. (1998) Computational fluid dynamics model for predicting flow of viscous fluids in a large fermentor with hydrofoil flow impellers and internal cooling coils. *Biotechnol. Prog.*, **14**, 248–258.

31 Moilanen, P., Laakkonen, M., and Aittamaa, J. (2006) Modeling aerated fermenters with computational fluid

dynamics. *Ind. Eng. Chem. Res.*, **45**, 8656–8663.

32 Oosterhuis, N.M.G., Kossen, N.W.F., Olivier, A.P., and Schenk, E.S. (1985) Scale-down and optimization studies of the gluconic acid fermentation by *Gluconobacter oxydans*. *Biotechnol. Bioeng.*, **27**, 711–720.

33 Sweere, A.P.J., Luyben, K.C.A.M., and Kossen, N.W.F. (1987) Regime analysis and scale-down: tools to investigate the performance of bioreactors. *Enzyme Microb. Technol.*, **9**, 386–398.

34 Langheinrich, C., Nienow, A.W., Eddleston, T., Stevenson, N.C., Emery, A.N., Clayton, T.M., and Slater, N.K.H. (1998) Liquid homogenisation studies in animal cell bioreactors up to 8 m^3 in volume. *Food Bioprod. Proc.*, **76**, 107–116.

35 Kelly, W.J. (2008) Using computational fluid dynamics to characterize and improve bioreactor performance. *Biotechnol. Appl. Biochem.*, **49**, 225–238.

36 Lapin, A., Klann, M., and Reuss, M. (2010) Multi-scale spatio-temporal modeling: lifelines of microorganisms in bioreactors and tracking molecules in cells, in *Biosystems Engineering II*, Springer, pp. 23–43.

37 Delafosse, A., Collignon, M.L., CAlvo, S., Delvigne, F., Crine, M., Thonart, P., and Toye, D. (2014) CFD-based compartment model for description of mixing in bioreactors. *Chem. Eng. Sci.*, **106**, 76–85.

38 Neubauer, P. and Junne, S. (2010) Scale-down simulators for metabolic analysis of large-scale bioprocesses. *Curr. Opin. Biotechnol.*, **21**, 114–121.

39 Wang, G., Chu, J., Noorman, H., Xia, J., Tang, W., Zhuang, Y., and Zhang, S. (2014) Prelude to rational scale-up of penicillin production: a scale-down study. *Appl. Microbiol. Biotechnol.* doi: 10.1007/s00253-013-5497-2

40 Yegneswaran, P.K., Gray, M.R., and Thompson, B.G. (1991) Experimental simulation of dissolved oxygen fluctuations in large fermenters: effect on *Streptomyces clavuligerus*. *Biotech. Bioeng.*, **38**, 1203–1209.

41 De Jonge, L.P., Buijs, N.A., ten Pierick, A., Deshmukh, A., Zhao, Z., Kiel, J.A., Heijnen, J.J., and van Gulik, W.M. (2011) Scale-down of penicillin production in *Penicillium chrysogenum*. *Biotechnol. J.*, **6**, 944–958.

42 Vardar, F. and Lilly, M.D. (1982) Effect of cycling dissolved oxygen concentrations on product formation in penicillin fermentations. *Eur. J. Appl. Microbiol. Biotechnol.*, **14**, 203–211.

43 Sieck, J.B., Cordesa, T., Budacha, W.E., Rhiela, M.H., Suemeghya, Z., Leist, C., Villiger, T.K., Morbidelli, M., and Soos, M. (2013) Development of a scale-down model of hydrodynamic stress to study the performance of an industrial CHO cell line under simulated production scale bioreactor conditions. *J. Biotechnol.*, **64**, 41–49.

44 Nienow, A.W. (2014) Re "Development of a scale-down model of hydrodynamic stress to study the performance of an industrial CHO cell line under simulated production scale bioreactor conditions" [Sieck, J.B., Cordes, T., Budach, W.E., Rhiel, M.H., Suemeghy, Z., Leist, C., Villiger, T.K., Morbidelli, M., Soos, M., 2013 . . . . (2013). *J. Biotechnol.* doi: http://dx.doi.org/10.1016/j.jbiotec.2013.12.002

45 Klöckner, W. and Büchs, J. (2012) Advances in shaking technologies. *Trends Biotechnol.*, **30**, 307–314.

46 Marques, M.P.C., Cabral, J.M.S., and Fernandes, P. (2010) Bioprocess scale-up: quest for the parameters to be used as criterion to move from microreactors to lab-scale. *J. Chem. Technol. Biotechnol.*, **85**, 1184–1198.

47 Mashego, M.R., van Gulik, W.M., Vinke, J.L., Visser, D., and Heijnen, J.J. (2006) *In vivo* kinetics with rapid perturbation experiments in *Saccharomyces cerevisiae* using a second-generation BioScope. *Metab. Eng.*, **8**, 370–383.

48 El Massaoudi, M., Spelthahn, J., Drysch, A., de Graaf, A., and Takors, R. (2003) Production process monitoring by serial mapping of microbial carbon flux distributions using a novel sensor reactor approach: I. Sensor reactor system. *Metab. Eng.*, **5**, 86–95.

49 Sweere, A.P., Janse, L., Luyben, K.C., and Kossen, N.W. (1988) Experimental simulation of oxygen profiles and their influence on baker's yeast production: II. Two-fermentor system. *Biotechnol. Bioeng.*, **20**, 579–586.
50 Amanullah, A., McFarlane, C.M., Emery, A.N., and Nienow, A.W. (2001) Scale-down model to simulate spatial pH variations in large-scale bioreactors. *Biotechnol. Bioeng.*, **73**, 390–399.
51 George, S., Larsson, G., and Enfors, S.O. (1993) A scale-down two-compartment reactor with controlled substrate oscillations: metabolic response of *Saccharomyces cerevisiae. Bioproc. Eng.*, **9**, 249–257.
52 Gerritse, G., Hommes, R.W.J., and Quax, W.J. (1998) Development of a lipase fermentation process that uses a recombinant *Pseudomonas alcaligenes* strain. *Appl. Environ. Microbiol.*, **64**, 2644–2451.
53 Hewitt, C.J., Onyeaka, H., Lewis, G., Taylor, I.W., and Nienow, A.W. (2007) A comparison of high cell density fed-batch fermentations involving both induced and non-induced recombinant *Escherichia coli* under well-mixed small-scale and simulated poorly mixed large-scale conditions. *Biotechnol. Bioeng.*, **15**, 495–505.
54 Geraats, S.M.G. (1994) Scaling-up of a lipase fermentation process: A practical approach, in *Advances in Bioprocess Engineering*, Springer pp. 41–46.
55 Sweere, A.P., Mesters, J.R., Janse, L., Luyben, K.C., and Kossen, N.W. (1988) Experimental simulation of oxygen profiles and their influence on baker's yeast production: I. One-fermentor system. *Biotechnol. Bioeng.*, **20**, 567–578.
56 Abel, C., Hübner, U., and Schugerl, K. (1994) Transient behaviour of baker's yeast during enforced periodical variation of dissolved oxygen concentration. *J. Biotechnol.*, **32**, 45–57.
57 Larsson, G. and Enfors, S.O. (1985) Influence of oxygen starvation on the respiratory capacity of *Penicillium chrysogenum. Appl. Microbiol. Biotechnol.*, **21**, 228–233.
58 Larsson, G. and Enfors, S.O. (1988) Studies of insufficient mixing in bioreactors: effects of limiting oxygen concentrations and short-term oxygen starvation on *Penicillium chrysogenum. Bioprocess Eng.*, **3**, 123–127.
59 Amanullah, A., Christensen, C.H., Hansen, K., Nienow, A.W., and Thomas, C.R. (2002) Dependence of morphology on agitation intensity in fed-batch cultures of *Aspergillus oryzae* and its implications for recombinant protein production. *Biotech. Bioeng.*, **77**, 815–826.
60 Buckland, B.C., Gwewonyo, K., DiMasi, D., Hunt, G., Westerfield, G., and Nienow, A.W. (1988) Improved performance in viscous mycelial fermentations by agitator retrofitting. *Biotechnol. Bioeng.*, **31**, 737–742.
61 Villadsen, J., Nielsen, J., and Lidén, G. (2011) *Bioreaction Engineering Principles*, 3rd edn, Springer, New York.
62 Bhargava, S., Nandakumar, M.P., Roy, A., Wenger, K.S., and Marten, M.R. (2003) Pulsed feeding during fed-batch fungal fermentation leads to reduced viscosity without detrimentally affecting protein expression. *Biotechnol. Bioeng.*, **81**, 341–347.
63 Bajpai, R.K. and Reuss, M. (1982) Coupling of mixing and microbial kinetics for evaluating the performance of bioreactors. *Can. J. Chem. Eng.*, **60**, 384–392.
64 Schmalzriedt, S., Jenne, M., Mauch, K., and Reuss, M. (2003) Integration of physiology and fluid dynamics. *Adv. Biochem. Eng. Biotechnol.*, **80**, 19–68.
65 Garcia, J.R., Cha, H.J., Rao, G., Marten, M.R., and Bentle, W.E. (2009) Microbial nar-GFP cell sensors reveal oxygen limitations in highly agitated and aerated laboratory-scale fermentors. *Microb. Cell Fact.*, **8**, 6–12.

17
Commercial Development of Fermentation Processes

Thomas Grotkjær

Summary

This chapter links fermentation technology with economics and business development on the track from research to commercial production. After the introduction Section 17.2 describes the overall challenges of developing fermentation processes and how to deal with the high degree of uncertainty in the initial phase. The techno-economic analysis of fermentation processes is described in Section 17.3. It is shown how the key fermentation parameters, namely, carbon yield, productivity, and product titer impact production costs. Five fermentation-based chemicals are used as examples in the subsequent analysis (ethanol, *n*-butanol, succinic acid, adipic acid, and 1,3-propanediol). In each case, a corresponding petrochemical route is used as benchmark and the concept of an "indifference curve" is introduced.

Succinic acid is an emerging building block in the chemical industry and hence several technical and commercial aspects of this product are analyzed in detail. In Section 17.4, basic concepts of market and value chain analysis are introduced. The value chain analysis is used to illustrate the importance of feedstock and how brand owners can have an impact on how renewable products are commercialized and marketed. Fermentation technology can be used to produce new chemicals and it is discussed how renewable resources and biotechnology bring innovation to the industry. It is also shown how new products can replace existing chemicals, for example, how succinic acid can replace adipic acid. Sections 17.5–17.6 conclude with an analysis of the new succinic acid industry and exemplify how companies (e.g., fermentation technology providers and chemical companies) find market opportunities.

17.1
Introduction

Chapter 17 deals with the fundamental aspects of developing and commercializing products based on fermentation technology. Fermentation development

Fundamental Bioengineering, First Edition. Edited by John Villadsen.

projects based on intensive metabolic engineering are resource demanding, and it often takes more than 10 years to advance ideas from proof-of-concept in R&D to commercial-scale production and successful market entry. The long time to market means that the initial assumptions on technology, market, and so on are likely to change over time and therefore project plans continuously need adjustment. The commercial and organizational challenges are to navigate in this environment and find solutions to advance the project. A prerequisite of success is that both technical and commercial milestones in the project plan are met, and hence new business development must pay close attention to multiple factors at the same time, and right from the start of new projects.

There has been a vivid and intensive research in academia on how companies should mature and commercialize business ideas based on new, advanced technology [1,2]. Today, biotechnology is a key technology enabler in a wide range of industries, but it would be wrong to conclude that these industries are similar. In fact, many industries are fundamentally different. For instance, it is easy to imagine that producing ethanol in Brazil or the US Midwest with large equipment under nonsterile conditions requires different competence from what is needed to produce proteins in pharmaceutical labs that are likely to be in big cities.

The purpose of this chapter is to bring the technical and commercial aspects of early business development closer to each other and demonstrate their strong interdependence. The main focus is the business of large-scale fermentation processes, and in that sense ethanol production is a better industry benchmark than protein production. However, commercialization of new technology for chemical intermediates presents several challenges that are different from the ethanol industry. The successful companies understand the differences in detail and will design the development and commercialization plans to maximize the business potential while reducing risk. In most cases, fermentation processes also include a downstream process where the chemical of interest is recovered from the fermentation broth and purified to meet a set of product specifications. Downstream processing is definitely vital in the development of new fermentation processes but also a very broad topic. The reason is that new fermentation processes require a specific arrangement of multiple unit operations in order to meet the product specification. This chapter will focus on one particular unit operation, namely, fermentation, and not analyze general concepts of downstream processing in detail. Readers are referred to the specialized literature that deals with this topic, including specific unit operations (see e.g., Harrison *et al.* [3], Heinzle, Biwer, and Cooney [4], and Forciniti [5]).

During the last decade, the companies BioAmber, Myriant, Succinity (joint venture between Corbion and BASF), and Reverdia (joint venture between Roquette and DSM) have invested heavily in developing fermentation technology for production of succinic acid. As of 2014, the companies have all scaled up their production processes and have sold their first batches of succinic acid in commercial quantities. The technical and commercial development of succinic acid will be a key case study throughout the chapter. Furthermore, technical and commercial aspects of ethanol, *n*-butanol, adipic acid, and 1,3-propanediol production will be discussed in order to get a broader perspective of the development process. A common denominator of the chemicals mentioned is that they all have attracted considerable industrial attention in recent years.

In Section 17.2, the basic aspects of a development project plan is discussed. The reason for devoting some time to project development and management is that the project plan is a guide to which aspects of new business ideas should be analyzed – and at what time during the project. In Section 17.3, the primary scope is to link metabolic engineering and fermentation physiology to economics through the techno-economic analysis. It is important to answer questions on how long it takes to develop a new process and on the likelihood of meeting the technical targets. Every single commodity product must meet a cost target in order to be commercially attractive. The output of the techno-economic analysis defines the boundaries of the market analysis and is linked to the commercial development track. It may come as a surprise to many people how clear some of the commercial boundaries of fermentation technology can be drawn before going to the laboratory. The commercial analysis will shed more light on the value created by developing a new fermentation process. A common trait of the case studies is that there are several companies involved in the series of production steps (supply chain) to convert raw material to end product via a new fermentation process. Who is interested in the new products and how much are they willing to pay for renewable products?

Finally, it should be stressed that the technical and commercial analyses presented in this chapter *illustrate* different topics of planning and analysis, but the analysis is definitely not exhaustive in a commercial setting. In each development project, the analyses form a necessary basis in a decision process, but the analysis must continue to be based on new data and learning throughout the development phase. There are several technical and commercial issues that are beyond the scope of this chapter. For instance, one important area is the internal analysis where the company evaluates the strategic fit between a new business area and the internal competences of the company. This "simple question" is extremely important to address, especially in a competitive field where several companies have the same ambitions. It is a catastrophe to invest significantly in a new area that is flooded by perhaps stronger competitors a few years later.

17.2 Basic Principles of Developing New Fermentation Processes

Biotechnology is used for numerous products in different industries and the annual, global production of different molecules ranges from small-scale pharmaceutical production ($<1\ kg\ year^{-1}$) to very large scale commodity production, for example, ethanol production (>10 000 000 metric tons a year). In Figure 17.1, it is shown that for numerous chemicals, from complex biological molecules to cement, there is a close relationship between global market volume and market price. The power law relationship in Figure 17.1 shows that the global market

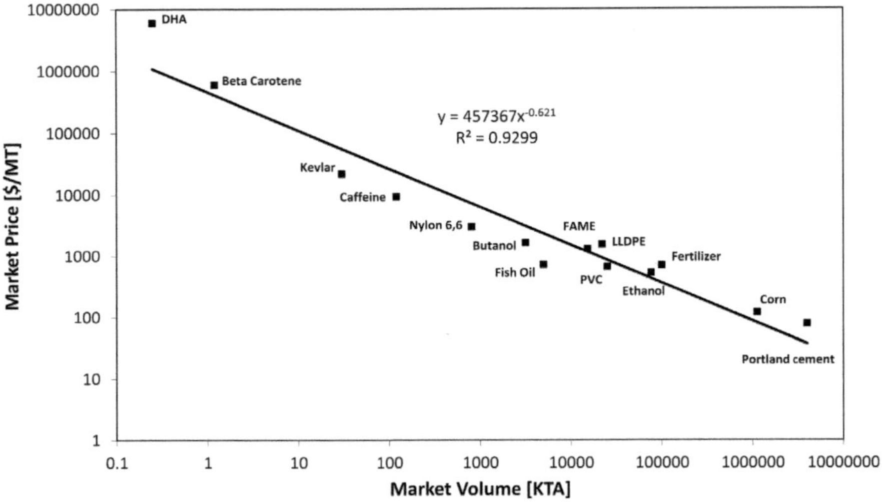

Figure 17.1 The power law relationship between market price ($/MT) and volume (kilo metric tons per annum = KTA) for a series of chemicals and commodity products [6]. The relationship is independent of feedstock and production technology, for example, fermentation and catalysis. From a fermentation development perspective, it is tempting to target high-value added products, but the "penalty" is a drastically reduced market volume. Specialized products sold at higher market prices are often marketed with a larger degree of documentation, application development support, and so on. Commodity products are typically sold with less support. Abbreviations: DHA, docosahexaenoic acid is an omega-3 fatty acid; FAME, fatty acid methyl ester is a biodiesel, commonly made from fats/fatty acids and methanol; LLDPE, linear low-density polyethylene commonly made by copolymerization of ethylene with longer-chain alkenes.

price of a molecule/product typically decreases by 35% if the market volume is doubled and by 58% when the market volume triples. The strong relationship between market price and volume suggests that an anticipated market price will set an upper limit for market penetration (volume). Cascone and Burke [6] call the phenomenon the "cost exclusion curve" since you cannot decide on both volume and price at the same time.

The price/volume nature of a fermentation product gives immediate guidance on the development plan and what is important for commercial success. In Table 17.1, it is stated that the use of biotechnology is marginal in the production of low-cost commodity chemicals, and it is crucial for commercial production that a commodity chemical can be recovered with simple means and with low cost. Thus, a project plan must be designed such that the feasibility of a simple and cheap recovery process can be tested sooner rather than later. If a project is infeasible because of a technical hurdle (e.g., the cost of the recovery process), it should be stopped as fast as possible ("fail fast philosophy") and not years down the road when money has been spent in vain. The prioritization of tasks is important and can lead to great savings when there is a time span of

Table 17.1 Production processes based on biotechnology account only for a small market share within most industries. The exception is the pharmaceutical industry where biotechnology plays a very important role [4].

Scale	Industry	Products	Biotech Market share	Biocatalyst	Downstream Complexity
Very large	Commodity chemicals	Small molecules	Very low	MO, EN	Low
Very large	Food/feed	Protein, sugars, and so on	Medium	MO, EN	Low-medium
Very large	Mining	Metals	Very low	MO	Low
Very large	Waste treatment	Purified water, air, soil	High	MO	Low
Large	Detergents	Enzymes	Medium	MO	Low
Medium	Fine chemicals	Small molecules	Low	MO, EN	Medium
Small-medium	Health care/ cosmetics	Small molecules, proteins	Medium	MO, EN, MC	Medium-high
Medium	Conventional Pharmaceutical	Small molecules	Low-medium	MO	Low-medium
Small	Biopharma	Proteins	High	MO, MC	High

Abbreviations: MO, microorganism; EN, enzyme; MC, mammalian cell culture.

10 years between the first and last tasks in the project plan. In pharmaceutical production the recovery processes are complex, which suggests that the value drivers are different from projects concerned with production of commodity chemicals [4].

In Figure 17.2, the technical and commercial timelines are given for a typical staged project plan where the purpose is to both develop and commercialise fermentation technology in large scale. In many large companies, it is common practice to include additional dimensions in the analysis such as strategic fit with an existing product portfolio, technical and manufacturing synergies, and financial risk/return, but these concerns are outside the scope of the present analysis. Smaller companies are often established based on a couple of specific ideas and hence there is no product portfolio to analyze. This is one of the reasons why smaller companies can move faster than larger companies.

The project is evaluated at each gate of the project plan and it is decided whether the project should continue or be stopped. Milestones are used to determine whether a project period has been successful or not. They give direction to a project plan and form a future basis for decision making. One often talks about "SMART" milestones, that is, milestones that are *s*pecific, *m*easurable, *a*ctionable (or *a*mbitious!), *r*ealistic, and *t*ime-bound. For a project based on fermentation technology, a logical milestone could be to achieve a specific fermentation performance in terms of carbon yield, productivity, and titer. The decision makers do not always have the time to acquire a deep understanding of

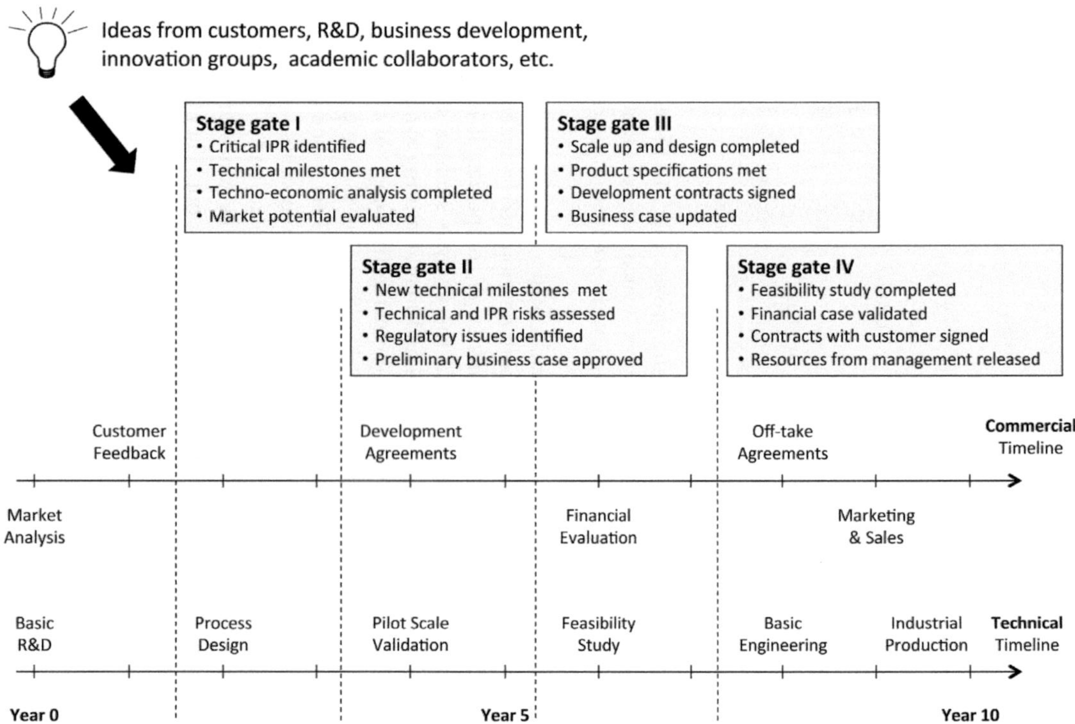

Figure 17.2 The timeline from basic R&D in biotechnology to commercial production is long and can easily exceed 10 years. Most companies employ a stage gate approach to continuously evaluate the technical and commercial feasibility of projects. In this example, there are four major decision points from basic R&D to commercialization. In order to monitor the progress of the project, it is common practice to include more milestones than shown in the figure.

complex projects and hence they only follow milestones. This is an important reason for carefully deciding on milestones that reflect project progress. It can be difficult for outsiders to understand how creativity and innovation have accumulated over time while a missed milestone is fully understood by everyone in less than a minute!

In a classical context, the first period of a project (stage gate I) is used to scope the project both in terms of technical proof-of-concept and in terms of market analysis. Ideas can be generated from multiple sources, for example, customers, R&D, business development, innovation groups, academic collaborators, and so on (Figure 17.2). In biotechnology, the intellectual property rights (IPRs) are often an important decision factor since the lack of rights to produce and sell the product can be an immediate showstopper. Conversely, a strong IPR position can be the foundation of a strong business case, leaving the company in a situation with no/few competitors and therefore in a market position where it is possible to gain additional profit. If case the company does not have the necessary IPR then the first business task could be to acquire the rights.

In the next phase (stage gate II), resource allocation typically is higher and milestones become quantitative and more difficult to reach (Figure 17.2). In recent years, there has been great progress in metabolic engineering and fermentation technology [7,8], but this part of the development phase remains long and technically difficult. In the future, it is believed that strain development will be faster, but it should be kept in mind that strain development only is one out of many time consuming tasks. In the early phase of strain development, it is important to ensure that the selected host can be used in an industrial setting, and hence it is advisable to initiate the necessary regulatory approval. The first commercial milestones could be to get approval of the business case and get further certainty on IPR.

The technology is ready for scale-up to commercial scale production after completion of basic R&D and testing in pilot scale (Figure 17.2). Alternatively, companies may choose to build a demonstration-scale facility to prove technology as well as testing product quality and business potential in a larger scale. Although a demonstration-scale production facility does not take full advantage of economy of scale, it serves as a tool to manage capital risk. The investment in a commercial large-scale production facility is often much higher than the cost of entire R&D phase, and consequently plant construction is also a staged process with numerous checkpoints, for example, completion of a feasibility study and completion of an engineering package (stage gate III).

In the last phase, the technology is complete – or at least good enough – and the project is now in a phase where most critical issues are related to business, for example, financing the production plant and obtaining a firm commitment of customers through contracts. A take-or-pay agreement is an example of a commercial contract where a customer agrees to buy a certain amount of product from a plant, for example, Vinmar and BioAmber's agreement on 210 KTA succinic acid [9]. The take-or-pay agreements give the chemical producer added certainty on commercial feasibility, and the major investment in a new plant will appear less risky for the investors (stage gate IV). For a chemical company, the entire investment (R&D, pilot plant, production plant, samples, marketing, etc.) must be recovered through sales of product, that is, there can be a very long development phase where the project does not generate any value.

The short introduction to project management illustrates the interdependence of technical and commercial issues. In general terms, the technical project members are responsible for technical feasibility, a safe/robust production process and an attractive cost structure, while the commercial project members quantify the value of the technology product and make the commercial agreements. With the right people and competences in place, the challenge in project management and development is to make sure that commercial assumptions are based on technical insight and vice versa. The techno-economic analysis discussed in next section is a good example of a task that depends on several input parameters.

17.3 Techno-economic Analysis: the Link Between Science, Engineering, and Economy

The first techno-economic analysis is a fundamental step in bridging science and engineering with economy. The analysis may be a simple "back-of-the-envelope calculation" to estimate a minimum cost of production, but it should continuously be updated as the project progresses. During the course of the project, critical issues should be given attention and validated with experimental work and more advanced simulation tools. In the initial phase, a simple spreadsheet may be sufficient to account for the overall cost structure, while subsequent analyses may be done with advanced simulation software such as SuperPro Designer, Aspen, ChemCad, and so on, Advanced software packages can be used to simulate mass/energy balances, separation processes in detail, sizing of equipment, and so on. The detailed capital investment analysis is typically addressed only at the later stages of the project when there is more certainty on unit operations, location of the plant, and equipment vendors. Consultants specialized in the chemical industry can also be used to get an overview of competitiveness of different processes. For instance, the company Nexant offers process evaluation research planning (PERP) reports on specific topics, for example, evaluation of technologies for *n*-butanol and adipic acid production [10] as well as industry trends Biorenewable Insights [11]. Experience from similar projects is extremely useful in the early phase since the open literature offers little help. As will be shown in the next sections, there are still several issues that can be analyzed and evaluated at a very early stage without detailed process knowledge.

17.3.1 Value Drivers and Production Costs of Fermentation Processes

The efficiency of a fermentation process is typically evaluated by a set of parameters that are important for the economic feasibility of the process (see Chapters 4 and 10). It is common practice to describe fermentation performance by three parameters that are directly related to microbial physiology:

- *Product yield* (e.g., ethanol, *n*-butanol, or succinic acid) on carbon source (e.g., glucose, sucrose, or xylose), measured in g g^{-1}.
- *Productivity*, also known as space time yield, measured in g l^{-1} h^{-1}.
- *Concentration* (or titer) of product, measured in g l^{-1}.

The economic impact of improving the three performance parameters for succinic acid production will be discussed in further details. It is, however, important to look for other value drivers in the development phase. The three parameters above are always important in a fermentation process, but there could be equally important process specific parameters. In the production of succinic acid, one can add another three parameters that play a role:

- *Medium costs* and utility chemicals, measured in US dollars per metric tons product ($/MT) in the fermentation broth at the end of fermentation. For instance, the cost of adding expensive vitamins to the medium and maintaining a constant pH can add considerable costs to the fermentation and recovery processes (see Example 17.1).
- *By-product formation* during fermentation, measured as purity of product (%) in the broth, typically excluding biomass. Formation of a by-product that is structurally similar to the product can cause drastically increased costs while CO_2 is a volatile gas that leaves the fermentor and in most cases has no impact on recovery.
- *Stability of microorganism*, measured in successful fermentations (%). A robust microorganism gives stable production, and may even allow for recycling the spent biomass. Strain stability and robustness are often overlooked topics in research.

In general, the first task in a project plan is to show that production is technically feasible (first three parameters) and later to refine the process by reducing costs (last three parameters). Without detailed information about physiology and reactor design, the simplest approach is to investigate whether a given cost target can be met through metabolic engineering and fermentation optimization, that is, improving product yield, productivity, and concentration. In most cases, there exist several combinations of fermentation performance that lead to the same production cost (see Example 17.2 for further details). Any project team should define the milestones such that R&D has the flexibility and the best chance to meet the fermentation performance needed. The techno-economic analysis should guide R&D on which parameters that play an important role for success. The succinic acid case offers some interesting insights.

Example 17.1 Organic acid production and impact of pH

The metabolism of many microorganisms is ideally suited to produce oxidized products such as organic acids. A consequence of organic acid production is that the pH of the fermentation medium rapidly drops to a level where the low pH inhibits both growth and product formation. For instance, without base addition it can easily be shown that the pH of the fermentation medium with succinic acid drops to $pH < 2.27$ for a titer of $50\,g\,l^{-1}$, that is, uncontrolled succinic acid formation will lead to strong product inhibition and low titer. In the production of many organic acids, it is therefore a common approach to control pH at a constant level by adding an alkaline neutralizing agent during the fermentation. The amount of neutralizing agent added depends on the dissociation constant(s) of the organic acid, its concentration. and the pH set point.

Succinic acid, in this example written as H_2A, is a divalent, weak organic acid with $pK_{A,1} = 4.16$ and $pK_{A,2} = 5.61$. The Henderson–Hasselbalch equation describes the relationship between pH, pK_A, and the ratio between acid and the corresponding base:

$$pH = pK_{A,1} + \log\frac{[HA^-]}{[H_2A]}.$$

$$pH = pK_{A,2} + \log\frac{[A^{2-}]}{[HA^-]}.$$

The succinic acid concentration at the end of the fermentation, $[H_2A]_f$, is the sum of the undissociated, H_2A, and the two dissociated forms, HA^- and A^{2-}:

$$[H_2A]_f = [H_2A] + [HA^-] + [A^{2-}].$$

It can be shown by insertion that the relative concentration of each of the three forms of succinic acid is a function of pH and the two pK_A values:

$$\frac{[H_2A]}{[H_2A]_f} = \frac{1}{1 + 10^{(pH-pK_{A,1})} + 10^{(pH-pK_{A,1})}10^{(pH-pK_{A,2})}}.$$

$$\frac{[HA^-]}{[H_2A]_f} = \frac{10^{(pH-pK_{A,1})}}{1 + 10^{(pH-pK_{A,1})} + 10^{(pH-pK_{A,1})}10^{(pH-pK_{A,2})}}.$$

$$\frac{[A^{2-}]}{[H_2A]_f} = \frac{10^{(pH-pK_{A,1})}10^{(pH-pK_{A,2})}}{1 + 10^{(pH-pK_{A,1})} + 10^{(pH-pK_{A,1})}10^{(pH-pK_{A,2})}}.$$

If $pH < pK_{A,1}$, then most of the succinic acid will be undissociated while a higher pH can be attained only if a larger fraction of the succinic acid is in the dissociated forms. For $pH = 5.5$, the fractions of H_2A, HA^- and A^{2-} are 0.025, 0.549, and 0.426, respectively. In order to obtain this distribution of fractions, one must add 0.70 mol of calcium carbonate per mol succinic acid based on the following equation:

$$H_2A + CaCO_3 \rightarrow CaA + CO_2 + H_2O.$$

Succinic acid is primarily sold in the undissociated form, H_2A, and hence an equimolar amount of sulfuric acid must be added in order to convert the salt into the free succinic acid form.

$$H_2SO_4 + CaA \rightarrow H_2A + CaSO_4.$$

Figure 17.3 shows the associated variable costs of chemicals and gypsum disposal for different pH values. In this example, it is assumed that calcium carbonate and sulfuric acid cost 250 \$/MT and 50 \$/MT, respectively, while calcium sulfate is disposed at a cost of 100 \$/MT (value highly dependent on location). If it is assumed that the reactions are 95% efficient, then the additional variable costs of chemicals and gypsum disposal are 294 \$/MT compared to a low pH fermentation process. There are alternatives to calcium carbonate as neutralizing agents, for example, calcium hydroxide and ammonia sulfate. The preferred choice is a trade-off between cost of raw materials, downstream processing, and the cost of waste disposal.

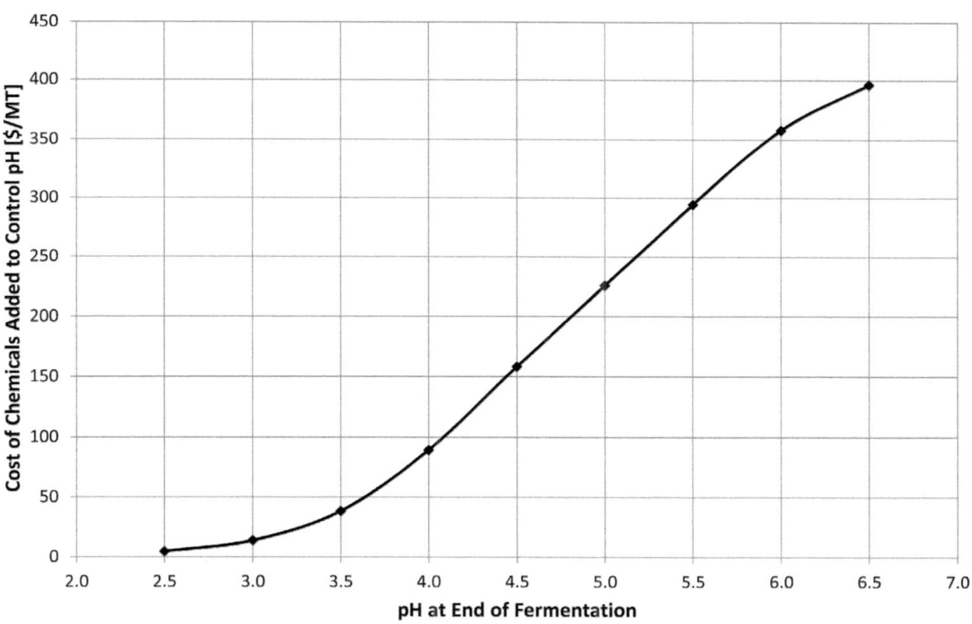

Figure 17.3 Cost of chemicals when succinic acid is produced at a predefined, constant pH between 2.5 and 6.5. In this example, the fermentation medium is continuously neutralized with base (calcium carbonate) to maintain the specified pH. After fermentation, the resulting calcium succinate is converted to the product of interest, free succinic acid, through addition of sulfuric acid. Gypsum, calcium sulfate dihydrate, is formed as a by-product with negative (or neutral/little) commercial value. Microorganisms have different tolerance to acidity: *Saccharomyces cerevisiae* can produce organic acids at pH < 5.0 without seriously stressing the organism while a typical bacterial fermentation (e.g., with *Actinobacillus succinogenes*) is inhibited at pH < 6.0.

The example illustrates that addition of chemicals to the fermentation medium can be associated with significant costs – not only in terms of chemicals usage but also in terms of disposal. It is evident that the traditional gypsum process adds significant cost to a commodity production process and hence several academic institutions and companies have worked on alternative fermentation and recovery process solutions [12]. Succinity has developed a new recovery process where the ammonia-based neutralizing agent is regenerated, while Reverdia and BioAmber have developed low pH microorganisms [13–15]. An attempt to substitute the cost of extra chemicals with higher capital costs (electro-enhanced membrane filtration) is discussed in Chapter 10, Example 10.8.

The analyzed succinic acid production process is illustrated in Figure 17.4. The overall process has been simplified and divided into two major sections in order to investigate the influence of improving the fermentation process: fermentation including primary recovery (removal of biomass and water) and

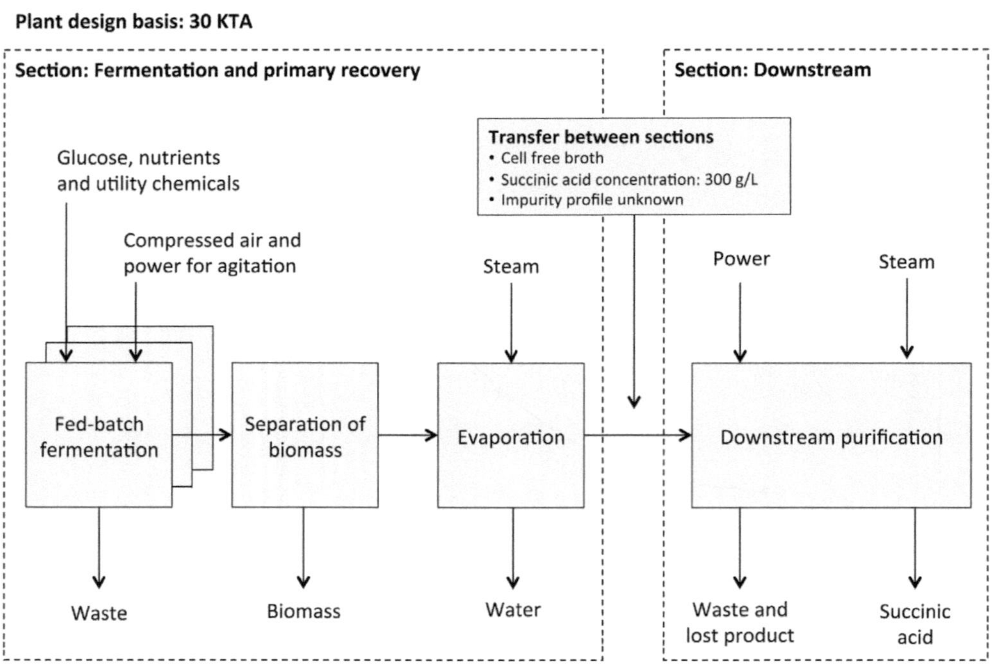

Figure 17.4 Simplified overview of the production process from glucose to succinic acid (nameplate capacity of 30 KTA). The process is divided into two major sections: fermentation including primary recovery and downstream processing to deliver a polymer-grade succinic acid. For simplicity, the cost of chemicals is all assumed to be used in the fermentation and primary recovery section. It is assumed that fermentation and primary recovery should produce a crude succinic acid with a concentration of 300 g l^{-1}.

downstream processing to deliver a polymer-grade succinic acid. It is assumed that the recovery process requires a feed stream of cell-free broth with a concentration equivalent to $300\,g\,l^{-1}$ of free succinic acid. The impact of fermentation performance and primary recovery on production costs can now be studied for constant downstream processing costs. The techno-economic model of succinic acid production is adapted from Novozymes A/S and based on the principles described in Hermann and Patel [16] and Peters, Timmerhaus, and West [17].

The capital investment needed for constructing a 30 KTA succinic acid plant depends on the productivity of the microorganism and the turnaround time, that is, the time needed to prepare tanks in a production cycle (e.g., sterilization, preparation of medium, actual fermentation time, pumping the product out of the fermentor and cleaning). In the reference case, succinic acid is produced by fed-batch fermentation in 48 h and 6 h are needed for turnaround. Thus, the total cycle time of 54 h defines the cycle time and thereby the fermentation volume needed to meet the required production capacity. In this example, succinic acid is produced by fermentation at constant pH = 5.5 (see Example 17.1), and calcium carbonate is added to keep constant pH and a high productivity throughout the production phase. In the reference scenario, it is assumed that $120\,g\,l^{-1}$ of succinic acid is produced in 48 h using $5\,g\,l^{-1}$ of biomass corresponding to an average productivity of $2.5\,g\,l^{-1}\,h^{-1}$ (Table 17.2). Fermentation performance targets should be realistic and therefore it is always prudent to benchmark with existing and similar processes (Section 17.2). It is known that succinic acid can be produced with titers above $100\,g\,l^{-1}$ [18] and recently Novozymes A/S demonstrated that the similar L-malic acid can reach more than $300\,g\,l^{-1}$ [19]. The precedence for high titers in organic acid production suggests that the titer is realistic and it can be improved if needed.

The next question is to estimate what can be expected in terms of carbon yield when biomass and by-product formation must be taken into account. The analysis is based on the principles of Villadsen, Nielsen, and Lidén [20] where chemicals and biomass in the reaction scheme are normalized to one carbon atom (C-mol), that is, glucose ($C_6H_{12}O_6$) is written as CH_2O. The maximum theoretical yield of succinic acid on glucose on a C-mol basis is

$$CH_2O + 1/7\,CO_2 \rightarrow 8/7\,CH_{3/2}O + 1/3\,ATP.$$

The balanced reaction from glucose to succinic acid is ATP positive and if CO_2 is available as substrate, then the maximum theoretical yield of succinic acid on glucose is $1.12\,g\,g^{-1}$. In an industrial process, it is not realistic to achieve a 100% efficient conversion of glucose to succinic acid, and hence it is assumed that only 85% of glucose is converted to succinic acid while the remaining glucose is converted into other products such as CO_2, organic acids, additional biomass, and so on. Biomass formation under aerobic conditions on a C-mol basis is assumed to be

$$CH_2O + 0.398\,O_2 \rightarrow 0.574\,X(CH_{1.8}O_{0.5}N_{0.2}) + 0.426\,CO_2.$$

Table 17.2 Scenarios of succinic acid production costs (30 KTA). The costs can be reduced by addressing several technical performance parameters. The table is divided into four sections: (1) assumptions on physiology, (2) dimensioning of fermentation vessels, (3) investment in major equipment for fermentation, and (4) cost of production for each scenario. The table illustrates how an improved fermentation process can lead to reduced costs (scenarios A–F).

Performance		**Reference (REF.) and improved performance (A–F)**						
Fermentation assumptions		**REF.**	**A**	**B**	**C**	**D**	**E**	**F**
Fermentation time	h	48	96	96	84	58	48	38
Turnaround time	h	6	6	6	6	6	6	6
Cycle time	h	54	102	102	90	64	54	44
Biomass concentration	$g\,l^{-1}$	5.0	5.0	5.0	5.0	5.0	5.0	3.0
Product titer	$g\,l^{-1}$	120	80	100	100	120	120	150
pH	—	5.5	7.0	6.5	5.5	5.5	4.5	2.0
Aeration	vvm	1.0	1.0	1.0	1.0	1.0	0.5	0.3
Agitation	$kW\,m^{-3}$	0.30	0.30	0.30	0.30	0.30	0.20	0.10
Product throughput								
Yield after fermentation	$g\,g^{-1}$	0.882	0.615	0.625	0.748	0.758	0.882	0.917
Yield after recovery	$g\,g^{-1}$	0.837	0.584	0.593	0.711	0.720	0.837	0.871
Throughtput per fermentor	MT/batch	48	32	40	40	48	48	60
Batches per year	#/yr	658	987	789	789	658	658	526
Number of fermentors	#	4.1	11.5	9.2	8.1	4.8	4.1	2.7
Investment								
Fermentation, $n = 0.90$	MM$	11	29	24	21	13	11	8
Compression, $n = 0.75$	MM$	3	6	5	4	3	2	1
Evaporation, $n = 0.80$	MM$	1	2	2	2	1	1	1
Downstream, $n = 0.60$	MM$	28	28	28	28	28	28	28
Inside battery limits (ISBL)	MM$	44	65	59	56	46	43	38
Outside battery limits (OSBL)	MM$	26	39	35	33	27	25	22
Total plant capital	MM$	70	104	93	89	73	68	60
Other project costs (OPC)	MM$	17	26	23	22	18	17	15
Total project investment (TPI)	MM$	87	130	117	111	91	85	76

Production costs								
Glucose	\$/MT	478	685	674	563	556	478	459
Nutrients	\$/MT	21	31	25	25	21	21	10
Chemicals	\$/MT	225	412	396	294	294	158	1
Aeration	\$/MT	81	244	195	171	97	41	16
Agitation	\$/MT	2	8	5	4	2	1	0
Evaporation	\$/MT	39	72	53	53	39	39	26
Downstream	\$/MT	201	201	201	201	201	201	201
Variable cost	\$/MT	1048	1654	1549	1311	1211	939	714
Total direct costs	\$/MT	1238	1872	1759	1517	1404	1127	896
Total cash cost	\$/MT	1400	2067	1944	1697	1569	1288	1050
Depreciation	\$/MT	318	475	427	406	334	311	276
Plant gate cost	\$/MT	1718	2542	2371	2103	1903	1599	1326

The fermentation vessels are assumed to be 500 m^3 (height/diameter ratio of 3) with a working volume of 80%. It is assumed that 95% of the succinic acid produced by fermentation is recovered. The analysis has been adapted from Novozymes A/S and modified to illustrate the impact of fermentation performance.

Here, X is a standard biomass with 5% ash and an elemental composition as given above. The reaction scheme corresponds to a biomass yield on glucose of 0.50 $g\,g^{-1}$ (see Table 17.5 for further details). When combining succinic acid yield of (85%) and biomass production, the overall reaction scheme on a C-mol basis is

$$0.862\,CH_2O + 0.032\,O_2 + 0.078\,CO_2 \rightarrow 0.043\,X + 0.897\,CH_{3/2}O + 0.271\,ATP.$$

In this reaction scheme, the stoichiometric coefficient of glucose (CH_2O) is less than 1 since the by-product profile is unknown. At a later stage of the development, it may be discovered that the remaining glucose (0.138 C-mol/C-mol glucose fed) is converted to a by-product (e.g., fumaric acid, citric acid, lactic acid, etc.) or CO_2 due to excessive respiration. The estimated yield of succinic acid on glucose is 29.5 g C-mol^{-1} 30 g C-mol^{-1} $\cdot$ 0.897 $g\,g^{-1}$ = 120/136 $g\,g^{-1}$ = 0.882 $g\,g^{-1}$ glucose fed since the loss of glucose to biomass and cellular maintenance energy must be taken into account (see Table 17.5 for further details).

The biomass itself does not bring any positive value to the overall production economy besides the crucial ability to produce the product. As shown in Chapter 2, the productivity (volumetric production rate), q_P, is given as

$$q_P = r_P x,$$

where q_P is measured in g product produced per liter per hour. The specific production rate, r_P, is measured in gram product per gram dry cell mass per hour, and the biomass concentration, x, in gram biomass per liter. The relationship between productivity and biomass concentration suggests that there is a trade-off between fermentation time and overall yield: a high productivity can be obtained by high biomass concentration but a high biomass concentration also means less sugar for product formation (see Example 17.2 for an example of trade-off). There is no apparent limit for productivity, but as a guiding principle the productivity should exceed 1.0–2.5 $g\,l^{-1}\,h^{-1}$ in batch or fed-batch fermentation mode. Novozymes A/S has demonstrated that L-malic acid can be produced with a productivity of 2.5 $g\,l^{-1}\,h^{-1}$ in the filamentous fungi *Aspergillus oryzae* [19,21], and hence this productivity can be used as a target for a new succinic acid project. In some cases, it may be possible to increase productivity through process engineering solutions that include continuous fermentation mode, *in situ* removal of product, and so on [22]. In general, more advanced fermentation systems are associated with increased technical risk and therefore it is common practice to aim for simple fermentation systems (batch or fed-batch fermentation).

In the reference scenario, it is assumed that the fermentor is sparged with air (aeration of 1.0 vvm) and agitated (specific power input of 0.30 $kW\,m^{-3}$). Aeration is a significant cost driver for long aerobic fermentation processes (244 $/MT in scenario A compared to 81 $/MT in the reference scenario, Table 17.2), but in the case of succinic acid production there is room for optimization. In scenario F the cost of aeration is heavily reduced due to lower aeration (0.3 vvm) and shorter fermentation time (44 h). Agitation has normally a minor impact on the overall cost, and succinic acid production is no exception. Highly aerobic fermentation systems are typically more expensive to build than anaerobic systems due to more expensive equipment and additional utility costs. In this simple example, the investment of the fermentation vessels is entirely a function of productivity. In Table 17.2, it is shown that the cost of evaporation is 72 $/MT in scenario A (80 $g\,l^{-1}$) and 26 $/MT in scenario F (150 $g\,l^{-1}$). The conclusion is that the *direct* effect of product titer is relatively small. It should be noted that in practice it is difficult to obtain high productivity and high carbon yield if the product titer remains low. The cost of production (defined as plant gate cost) of the reference scenario is estimated to 1718 $/MT and the total investments for the plant 87 million US dollars (MM$). The cost of glucose is the major cost driver followed by the costs of chemicals and downstream processing (Table 17.2).

In the early stages of project development, one must evaluate if the reference scenario is competitive and if the target reflects a conservative or an ambitious R&D target. A conservative target ("easy target to reach") would mean that the new product could be commercialized fast (highly valued by the marketing department), while the downside is that sales margin is lower than for a more ambitious target (highly valued by the finance department as long as R&D investment is unchanged). In scenarios A–F in Table 17.2, it can be observed that plant gate costs can be as high as 2542 $/MT (scenario A) and as low as 1326 $/MT (scenario F) for a very ambitious target that requires optimization of several parameters. In reality, the fermentation targets will be a function of the required return on investment (also known as ROI) of the company (see Crundwell [23] for a thorough introduction to the finance of investment projects). If a business case requires that cost of production is 1326 $/MT (scenario F), then the R&D hurdle seems very high and the project is risky to start. The fundamental problem of developing fermentation processes is that a prolonged R&D period often gives only a marginal return (the theoretical yield is an upper limit) and not an additional upside. If the cost target is 1718 \$/MT (reference scenario), then the project may seem feasible and there is an upside to improve the process further when the plant has been built and the product is on the market. In industrial biotechnology, it is the norm to improve processes over the lifetime of the project to stay competitive. Commodity chemical production processes are normally always highly efficient due to the large incentives to improve the production processes. For instance, in the case of succinic acid production, a cost reduction of 50 \$/MT results in annual savings of 1.5 MM\$ for a 30 KTA succinic acid plant. The annual savings increase with plant size and hence it is possible to justify R&D expenses to reduce cost even when the cost reduction potential decreases.

The total investment of a plant, typically known as capital expenses or just CapEx, will change as a function of plant size (capacity). The CapEx (typically measured in MM\$) for scaling up or down can be approximated by the following empirical relationship:

$$\mathrm{CapEx} = \mathrm{CapEx}_0 \left(\frac{S}{S_0}\right)^n,$$

where CapEx_0 is the capital investments of the reference case, n is a scaling factor (exponent), while S and S_0 are the plant capacity of the approximated and reference case (e.g., measured in KTA), respectively. The relationship can also be used to estimate the investment cost of one or more unit operations at different scale in the plant, for example, fermentation and unit operations of downstream processing. Tank volume (measured in m^3) is the natural reference point in the case of fermentation. In Table 17.2, it is observed that the pessimistic scenario A requires 11.5 fermentors, while scenario F only requires 2.7 fermentors. Clearly, a plant is designed with an integer number of fermentors, but in a sensitivity analysis this additional constraint only complicates the analysis. A scaling factor of approximately $n = 0.90$ is typical for the investment in fermentors, while the values of n for compressors, evaporators, and the downstream recovery equipment is lower (Table 17.2, [17]). The scaling factor for the fermentors assumes that the vessels have reached a maximum size (e.g., 500 m^3 as in this example) and further capacity increase is obtained by investing in additional, identical fermentors. The CapEx in scenario A is estimated to 130 MM\$ but only 76 MM\$ for a highly optimized process (scenario F), that is, fermentation optimization can have a very important impact on an investment case.

The variable costs of chemicals for the gypsum process (Example 17.1), aeration and fixed costs can be reduced significantly but glucose remains an important cost driver no matter how much the process is optimized. This conclusion suggests that one should pay close attention to carbon yield and the cost of glucose. Tables 17.3 and 17.4 shed further light on the financial implications of the succinic acid case. The analysis is on a consolidated basis and includes operating expenses and the total investment for the complete plant. Consumption of glucose for the fermentation section is 1.194 MT MT^{-1} succinic acid produced accounting for 478 \$/MT.

Table 17.3 Production economics of polymer-grade succinic acid produced by the gypsum process based on the reference scenario: variable cost and investment. The table summarizes the total investments for three scenarios (15, 30, and 60 KTA).

	Unit	Cost	Consumption/		$/MT
Raw materials					
Calcium carbonate	250	$/MT	0.478	MT	120
Calcium sulfate disposal	100	$/MT	0.823	MT	82
Corn steep liquor, solids	282	$/MT	0.047	MT	13
Glucose	400	$/MT	1.194	MT	478
Sulfuric acid	50	$/MT	0.469	MT	23
Misc. chemicals	840	$/MT	0.063	MT	53
Gross raw material cost					778
Utilities					
Cooling water	4	¢/m^3	442.3	m^3	18
Electricity	8	¢/kWh	75.4	kWh	6
Process water	35	¢/m^3	3.2	m^3	1
Steam	25	$/MT	9.8	MT	245
Total variable cost					270
Capacity, KTA	15		30		60
Investment, MM$					
Inside battery limits (ISBL)	27		44		71
Outside battery limits (OSBL)	16		26		42
Total plant capital	43	$n = 0.70$	70	$n = 0.70$	113
Other project costs (OPC)	11		17		28
Total project investment (TPI)	54		87		141

The values $n = 0.70$ and $n = 0.70$ are plant scaling factors (investment section). The analysis has been adapted from Novozymes A/S.

Cost related to the chemicals of the gypsum process accounts for (120 + 82 + 23) $/MT = 225 $/MT while the remaining part is related to fermentation medium and utility chemicals. Utility costs account for 270 $/MT.

Economy of scale is an important competitive factor in the chemical industry. The evaluation of succinic acid is based on a plant of 30 KTA. In Table 17.3, it is shown that doubling the capacity from 30 KTA to 60 KTA results in an investment increase of 62% (less than 100%), that is, decreased investment per unit mass succinic acid produced. In the reference scenario, it is assumed that 65% of the capital investment is related to the downstream process that is assumed to have a lower scaling factor ($n = 0.60$) than fermentation (Table 17.2). Thus,

Table 17.4 Economics of polymer-grade succinic acid production by the gypsum process based on the reference scenario: overall cost structure and a summary addressing product value. The table summarizes in the three columns the total production economics for three scenarios (15, 30, and 60 KTA).

Cost items, $/MT			
Raw materials	778	778	778
By-product credits	0	0	0
Utilities	270	270	270
Variable costs	1048	1048	1048
Maintenance materials	43	35	28
Operating supplies	20	10	5
Operating labor, 8.0/shift	202	101	51
Maintenance labor	29	23	19
Control laboratory	40	20	10
Total direct costs	1383	1238	1161
Plant overhead	217	116	64
Taxes and insurance	57	46	38
Total cash cost	1657	1400	1262
Depreciation	392	318	258
Plant gate cost	2049	1718	1521
G + A, sales, res., 4.0%	97	81	71
Production costs, $/MT			
At 120% capacity	1963	1674	1501
At 100% capacity	2146	1799	1592
At 80% capacity	2421	1987	1728
Product value (cost + 10%/yr ROI before taxes), $/MT			
At 120% capacity	2202	1868	1658
At 100% capacity	2432	2031	1780
At 80% capacity	2778	2277	1963

Variable costs are transferred from Table 17.3. The total direct costs include fixed costs for maintenance, labor, and laboratory control that are independent of production volume. Depreciation consists of two components: ISBL is depreciated over 10 years, OSBL over 20 years (both straight line). For each case the production economics is investigated for a plant operating at 80%, 100%, and 120% of the nameplate capacity (30 KTA). A market price higher than the product value indicates that chemical producer has an instantaneous return on investment that exceeds 10%. Abbreviation: G + A, sales, res., general expenses and administration, sales, research, that is, cost not directly related to production but important for an ongoing business. G + A, sales, res. is estimated to 4.0% of the product value. The analysis has been adapted from Novozymes A/S.

economy of scale is much more pronounced for the downstream process compared to fermentation. The large weight of the downstream processing section results in a plant scaling factor of $n = 0.70$ (Table 17.3). A second benefit of a large-scale producer is that all cost items except for the variable costs also decrease as a function of scale. In the current analysis, the plant gate costs decrease from 1718 \$/MT to 1521 \$/MT when capacity is expanded to 60 KTA (Table 17.4). Note that this observation is also in agreement with the general observation in Figure 17.1 where it was shown that the market price decreases with increased production volume.

In Table 17.4, it is sown that operating a plant at low capacity will result in drastically increased production costs. The example of Tables 17.2–17.4 clearly shows that economy of scale must be considered, but only if you can take advantage of the scale. The product value, listed in Table 17.4, is a concept often used by the chemical and energy consultant company IHS to investigate the sensitivity of plant throughput [24]. Product value includes production costs and returns on investment before tax. If the market price is higher than the product value, then the analysis indicates that the project will meet the investment criteria.

Example 17.2 The trade-off between fermentation performance parameters

Productivity and carbon yield are important key fermentation performance parameters that must be optimized in every research project. In the reference scenario, the productivity is 2.5 $g\,l^{-1}\,h^{-1}$, the succinic acid yield on glucose is 0.882 $g\,g^{-1}$, and the estimated plant gate cost is 1718 \$/MT. Let us assume that after a couple of years of research it is discovered that the productivity target is too ambitious. Based on experimental evidence, it is estimated that the fermentation process is more likely to reach a maximum of 1.7 $g\,l^{-1}\,h^{-1}$. The question is whether the project should be stopped due to the inferior productivity or could a higher yield of succinic acid on glucose compensate for the inferior productivity?

In Figure 17.5, the productivity is plotted as a function of plant gate cost and the succinic acid yield on glucose is plotted as a function of plant gate cost. The function is hyperbolic, that is, total costs are very high for very low productivity, while the costs decrease to an almost constant level for high productivity (approximately >5 $g\,l^{-1}\,h^{-1}$). It is also observed that there is an almost linear relationship between plant gate cost and succinic acid yield. The better utilization of feedstock (glucose) always has a positive impact on cost, while the marginal benefits of higher productivity decreases. In this example, the lower productivity (1.7 $g\,l^{-1}\,h^{-1}$ versus the target 2.5 $g\,l^{-1}\,h^{-1}$) results in a plant gate cost increase of approximately 40 \$/MT that can be compensated by increasing the succinic acid yield on glucose from approximately 0.88 $g\,g^{-1}$ to 0.97 $g\,g^{-1}$. The two curves in Figure 17.5 can be used to guide R&D on finding the best alternatives to optimize the strains in terms of cost rather than intuition.

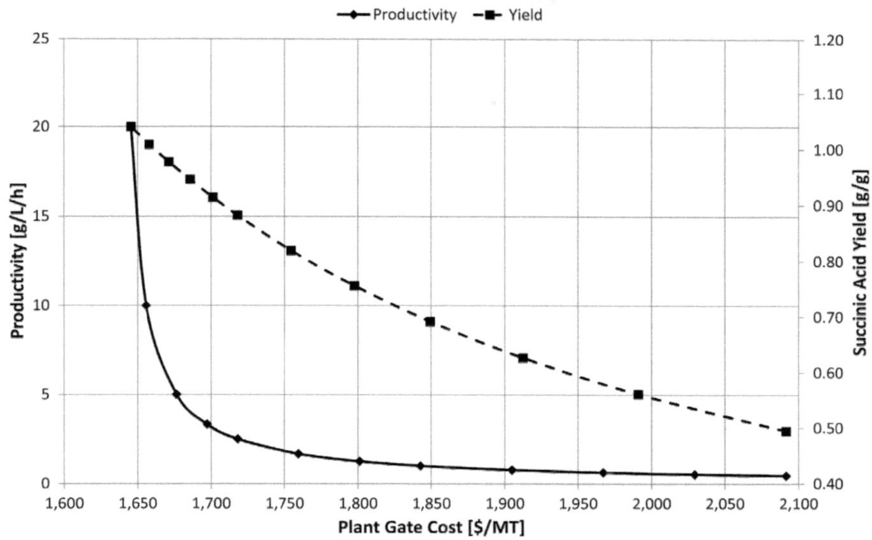

Figure 17.5 Sensitivity analysis of the reference scenario (Table 17.2): productivity plotted as a function of plant gate cost (left axis) and succinic acid yield on glucose plotted as a function of plant gate cost (right axis). For a plant gate cost of 1718 \$/MT (reference scenario), the productivity is 2.5 g l^{-1} h^{-1} and the succinic acid yield on glucose is 0.882 g g^{-1}. The productivity can be improved through optimization of the specific production rate, r_P, while a higher succinic acid yield, Y_{SP}, requires a more efficient fermentation process, less glucose used for cellular maintenance energy, and/or elimination of coproducts or by-products.

As a final comment, it should be mentioned that the key assumption in this example is that the succinic acid plant will be designed based on the fermentation performance of the reference scenario. As mentioned earlier, in many cases strain optimization is done on a strain already operating in an existing plant. If there is overcapacity in the recovery section, then fermentation is a production bottleneck, and consequently even smaller improvements in fermentation productivity will lead to increased production capacity. Modification of an existing plant with the purpose of expanding capacity is known as "debottlenecking" in the chemical industry. Capacity expansion for a small investment can be extremely useful and attractive if an increased market demand must be met in a short period of time.

17.3.2 Assessment of New Fermentation Technologies

As shown in Table 17.1, renewable feedstock and biotechnology still play a limited role in the production of commodity chemicals. The brutal fact is that in many cases the production route from renewable sources is not competitive. It is, therefore, of utmost importance to evaluate whether a new fermentation process is competitive or there are more attractive production methods. In most cases, the competing technology would be based on a major petrochemical feedstock, for example, ethylene and propylene, which is converted to the desired product by chemical catalysis. In order to give a broader perspective on competing technologies, the analysis is now expanded to include an additional four chemicals that can be produced by both fermentation and chemical catalysis, namely, ethanol, *n*-butanol, adipic acid, and 1,3-propanediol. The analysis in this section is primarily based on techno-economic data from Novozymes A/S that has been adapted to the assumptions on fermentation performance and conversion yields given in Tables 17.5 and 17.6.

Example 17.3 A new route from renewable resources to adipic acid via *cis,cis*-muconic acid

In a study it was shown that adipic acid (1,6-hexanedicarboxylic acid) can be produced from renewable resources [25]. The idea is based on a two-step production process where glucose is converted to *cis,cis*-muconic acid (MA) by fermentation and subsequently hydrogenated to adipic acid by chemical catalysis. In this example, the technical feasibility (risk profile) of developing a fermentation process is discussed.

The first dedicated step toward MA is the condensation of the 3-carbon molecule phosphoenolpyruvate (PEP) and the 4-carbon molecule D-erythrose 4-phosphate (E4P) to the 7-carbon molecule 3-deoxy-D-arabinoheptulosonate-7-phosphate (DAHP):

$$\text{PEP} + \text{E4P} \rightarrow \text{DAHP}.$$

PEP is a key intermediate in glycolysis, while E4P is an intermediate of the pentose-phosphate pathway (see Chapter 4). The flux to PEP is high in most microorganisms, while balancing an equal molar flux to E4P is a challenge that most likely requires significant metabolic engineering efforts. The most natural route to E4P is through the oxidative branch of the pentose-phosphate pathway but that will lead to a lower yield due to loss of CO_2 [26]. From DAHP to MA there are 4 enzymatic steps of which the last step involves incorporation of molecular oxygen and ring opening of catechol to yield the final product, MA. Based on the above assumptions. The overall stoichiometry for conversion of glucose to MA is

$$7\,\text{Glucose} + 6\,\text{ATP} + 6\,\text{O}_2 \rightarrow 6\,\text{MA} + 6\,\text{NADH} + 6\,\text{CO}_2.$$

The incorporation of molecular oxygen suggests that the aeration in an industrial production process must be carefully controlled in order to reach an efficient conversion of glucose to MA. The most important parameter for efficient ATP generation in aerobic fermentation processes is the P/O ratio that is defined as the ATP produced (P) by reduction of an oxygen atom (O) during oxidative phosphorylation. The parameter is

Table 17.5 Fermentation performance targets based on simple microbial physiology. For ethanol (ETOH), *n*-butanol (BUOH), succinic acid (SUCC), adipic acid (ADIPIC) and 1,3-propanediol (PDO).

Chemical		ETOH	BUOH	SUCC	ADIPIC	PDO
Physiology						
Max. biomass yield on glucose (growth)	$g\,g^{-1}$	0.10	0.10	0.50	0.50	0.50
Max. product yield on glucose (growth)	$g\,g^{-1}$	0.43	0.34	0	0	0
Max. product yield on glucose (nongrowth)	$g\,g^{-1}$	0.51	0.41	1.12	0.70	0.63
ATP yield on glucose (nongrowth)	$mol\,mol^{-1}$	2	2	2	0.43	0
Biomass concentration	$g\,l^{-1}$	5	3	5	8	5
Fermentation efficiency (nongrowth)	%	95%	55%	85%	85%	85%
Fermentation time	h	30	36	48	50	50
Estimation of product yield						
ATP Maintenance	$mol\,l^{-1}$	0.90	0.65	1.44	2.40	1.50
ATP generation	$mol\,l^{-1}$	2.26	0.72	1.39	0.40	0.00
Additional ATP supply	$mol\,l^{-1}$	0.00	0.00	0.05	2.00	1.50
Product from glucose (growth)	g l	22	10	0	0	0
Product from glucose (nongrowth)	$g\,l^{-1}$	98	10	120	100	150
Product titer	$g\,l^{-1}$	120	20	120	100	150
Biomass from glucose (growth)	$g\,l^{-1}$	52	29	10	16	10
Product from glucose (nongrowth)	$g\,l^{-1}$	191	25	107	144	237
Coproducts/waste	$g\,l^{-1}$	12	40	19	25	42
Maintenance	$g\,l^{-1}$	0	0	0	17	13
Glucose consumed	$g\,l^{-1}$	255	93	136	202	301
Product yield (product titer/glucose cons.)	$g\,g^{-1}$	0.47	0.21	0.88	0.49	0.50
Efficiency (product yield/max. yield)	%	92%	52%	78%	71%	79%
Productivity (product titer/ferm. time)	$g\,l^{-1}\,h^{-1}$	4.0	0.6	2.5	2.0	3.0
Feedstock						
Corn	MT/MT	2.59	5.67			
Coproduct credits						
Acetone	MT/MT	0	0.33			
DDGS	MT/MT	0.78	2.41			
Ethanol	MT/MT	0	0.17			

Ethanol and *n*-butanol production (anaerobic processes) are based on corn (maize) as feedstock, while succinic acid, adipic acid (via *cis,cis*-muconic acid), and 1,3-propanediol are based on glucose (aerobic or microaerobic processes). The calculated product yield, Y_{SP}, is based on metabolic pathway analysis where losses of glucose to coproducts, by-products, biomass, and cellular maintenance energy have taken into account. It is conservatively assumed that biomass is formed at the start of the fermentation and 6 mmol ATP g^{-1} DW h^{-1} is needed for cellular maintenance energy. Ethanol and *n*-butanol production is associated with net ATP generation that satisfies the overall need for ATP.

Table 17.6 Overview of petrochemical (PETRO) and fermentation (BIO)-based production routes to five chemicals.

Chemical		Ethanol		*n*-Butanol		Succinic acid		Adipic acid		1,3-Propanediol	
Process		**PETRO**	**BIO**	**PETRO**	**BIO**	**PETRO**	**BIO**	**PETRO**	**BIO**	**PETRO**	**BIO**
Capacity	KTA	150	150	100	100	30	30	300	75	50	50
Investment/capacity	MM$/KTA	0.98	0.63	1.25	1.10	2.10	2.90	1.94	3.72	2.00	2.27
Major raw material		Ethylene	Corn	Propylene	Corn	Maleic anhydride	Glucose	Cyclo-hexane	Glucose	Ethylene oxide	Glucose
Max. conversion	$g\,g^{-1}$	1.64	0.51	1.76	0.41	1.20	1.12	1.74	0.70	2.71	0.63
Conversion yield	%	98%	92%	98%	52%	98%	78%	98%	71%	70%	79%
Recovery yield	%	98%	98%	90%	97%	95%	95%	95%	95%	97%	97%
Consumption	MT/MT	0.634	2.169	0.644	4.807	0.892	1.194	0.619	2.131	0.543	2.072
Major raw material											
Unit cost	$/MT	1050	193	1200	193	1250	400	1125	400	1150	400
Total cost	$/MT	666	499	772	928	1115	478	696	853	624	829
Major coproduct			DDGS	IBA	DDGS, ACT, ETOH			Fuel	Fuel		
Value of major coproduct	$/MT		−78	−77	−300			−40	−25		

Capacity indicates the plant size of the analysis. For each production process, the major raw material is listed: the consumption factors indicate the mass of major feedstock that is needed to produce one unit of product and how much feedstock contributes to the overall production cost. Abbreviations: DDGS, Distillers Dried Grains with Solubles; IBA, isobutyraldehyde; ACT, acetone; ETOH, ethanol. The analysis has been adapted from Nexant [10] and Novozymes A/S.

specific to the host microorganism used and is therefore an important parameter to consider in strain design. Under aerobic conditions and for a typical P/O ratio of 1.5 the overall stoichiometry is

$$7\,\text{Glucose} + 9\,O_2 \rightarrow 6\,\text{MA} + 6\,CO_2 + 3\,\text{ATP}.$$

The NADH has been used to generate ATP through oxidative phosphorylation. The yield of *cis,cis*-muconic acid on glucose is $0.68\,\text{g}\,\text{g}^{-1}$. On a C-mol basis, the overall reaction scheme is

$$CH_2O + 3/14\,O_2 \rightarrow 6/7\,CHO_{2/3} + 1/7\,CO_2 + 1/14\,\text{ATP}.$$

The long pathway from DAHP to MA in addition to the requirement for significant engineering of the central carbon metabolism, in particular to obtain a high flux of E4P, suggests that it is a long-term R&D project. Moreover, the ATP gain (1/14) in connection with product formation is lower than for a typical anaerobic process (anaerobic process typically gives 1/3), which means that additional carbon is needed for cellular maintenance energy. An R&D project that requires a carbon yield close to the theoretical limit is, therefore, considered risky.

The competitiveness of a fermentation process is to a great extent dependent on the resources spent on research and optimization, though feedstock cost is an equally important factor (Table 17.2). In order to achieve a realistic goal (milestone) in a timeline manner (cf. SMART targets), it is important to make some assumption on the fermentation performance that does not require perfection in the development project. In Table 17.5, examples of fermentation performance targets have been estimated for five fermentation technologies. Each fermentation technology targets one of the following five chemicals: ethanol [27], *n*-butanol [22,28,29], succinic acid [18], adipic acid [25], and 1,3-propanediol [30,31]. The targets estimated in Table 17.5 could serve as an important milestone (stage gate II in Figure 17.2). It is expected that the research projects of the five technologies will be significantly different in terms of metabolic engineering, fermentation, and recovery developments. The metabolic pathways to each of the products are different and so are the physical properties.

The theoretical yield of the five chemicals on glucose gives the upper limit (max. product yield on glucose (nongrowth) in Table 17.5), but the practical yield in an industrial process will be lower. From Table 17.5, it is assumed that glucose is needed for biomass formation, coproduct/by-product formation, coproducts/waste, and possibly also cellular maintenance energy. In this example, it is assumed that $3–8\,\text{g}\,\text{l}^{-1}$ of biomass is needed for the five different fermentation processes. For the anaerobic conditions (ethanol and *n*-butanol), biomass production must be coupled to product formation, while biomass is uncoupled from production formation under aerobic conditions (succinic acid, adipic acid, and 1,3-propanediol).

Anaerobic fermentation conditions offer the best opportunities for low-cost production, but anaerobic conditions are not always possible due to high ATP requirements in connection with product formation [32]. As commonly known, the yeast *S. cerevisiae* is an ideal host for ethanol production: ethanol yields are high (close to the theoretical limit), productivity is high due to a very high glycolytic flux, and high titers can be obtained. Furthermore, in many ethanol production plants biomass can be recycled after fermentation, which leads to an even more efficient fermentation process. *S. cerevisiae* for ethanol production is "designed by nature" and probably the most efficient fermentation process one can imagine. The overall reaction of glucose to ethanol on a C-mol basis is

$$\begin{aligned} 0.952\ \mathrm{CH_2O} \rightarrow\ & 0.023\mathrm{X} + 0.613\ \mathrm{ETOH(CH_3O_{1/2})} + 0.007\ \mathrm{GLYC(CH_{8/3}O)} \\ & + 0.309\ \mathrm{CO_2} + 0.250\ \mathrm{ATP}. \end{aligned}$$

Here, ETOH is ethanol and GLYC is glycerol. From the reaction scheme, it can be seen that glycerol is the only by-product and it is formed only in small quantities in connection with biomass formation. Furthermore, years of operation and optimization experience has resulted in additional cost savings [33]. It is important that one does not require the same performance of new fermentation processes. As discussed previously, if the targets of a new project are too optimistic, the research project will inevitably be delayed and closed down when the targets are not met!

The traditional fermentation route to *n*-butanol is the so-called ABE fermentation process, where ABE is short for the key fermentation products acetone, *n*-butanol, and ethanol. In this process, different species of *Clostridium*, for example, *C. beijerinckii*, convert sugars to a mixture of acetone, *n*-butanol, and ethanol in a mass ratio of 3 : 6 : 1 [28]. The reaction of glucose to *n*-butanol on a C-mol basis is

$$\begin{aligned} 0.799\ \mathrm{CH_2O} \rightarrow\ & 0.037\ \mathrm{X} + 0.318\ \mathrm{BUOH(CH_{1/2}O_{1/2})} \\ & + 0.047\ \mathrm{ETOH(CH_3O_{1/2})}\ + 0.111\ \mathrm{ACT(CH_2O_{1/3})} \\ & + 0.041\ \mathrm{FA(CH_3O_{1/2})} + 0.216\ \mathrm{CO_2} + 0.271\ \mathrm{ATP}. \end{aligned}$$

Here, BUOH is *n*-butanol, ACT is acetone, and FA is formic acid. The fermentation process has a high carbon yield of combined products on glucose, but suffers from a major drawback: the fermentation products are very toxic and lead to strong inhibition even at low titer ($<20\,\mathrm{g\,l^{-1}}$). A century of research has shown that the barrier for increasing toxicity is very high, and therefore companies and academic institutions have developed technology for removing the solvents during fermentation as part of an integrated fermentation production process [22,34].

In this study, it is assumed that ethanol and *n*-butanol are produced from the agricultural commodity corn (maize). Corn contains approximately 74% starch (variation depends on location, season, etc.), while the remaining part is a non-fermentable ingredient that can be used as a valuable feed known as "distillers dried grains with solubles" (DDGS). Thus, the use of corn as feedstock has the interesting consequence that ethanol and *n*-butanol production require that a second business is developed for the coproducts. In the case of *n*-butanol, there are two additional by-products, namely, acetone and ethanol. For a succinic acid producer, another way of generating value would be to develop a market for the gypsum produced, that is, converting a disposal problem into a value generating

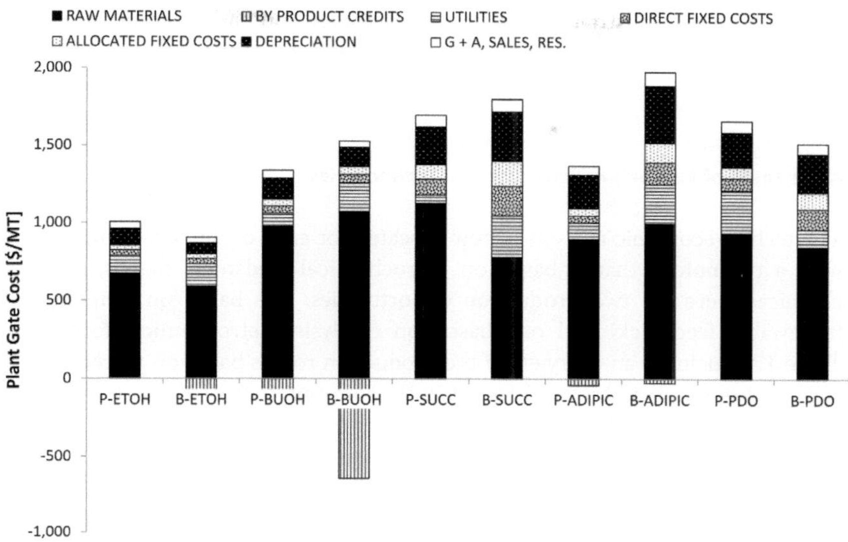

Figure 17.6 Plant gate cost of ethanol (ETOH), *n*-butanol (BUOH), succinic acid (SUCC), adipic acid (ADIPIC), and 1,3-propanediol (PDO). The prefixes "P" and "B" indicate whether the production process is based on petrochemical or renewable feedstock. All production processes are highly dependent on the cost of raw materials, utility, and depreciation. The *n*-butanol production route gives by-product credits in the form of acetone, ethanol, and DDGS. The analysis has been adapted from Nexant [10] and Novozymes A/S.

coproduct. In case the customer takes the gypsum for free, then the succinic acid plant gate costs are reduced by 82 $/MT (Table 17.3). In most cases, coproducts have an important influence on the overall feasibility of a project, and hence the market aspects of all coproducts must be carefully evaluated before an investment decision (see also Figure 17.6). In many ways, bio-refineries are not of the future – they exist today and are necessary for economic feasibility of most projects. Every Joule and carbon atom is important in commodity business!

Succinic acid, adipic acid, and 1,3-propanediol are all produced from glucose and in practice under microaerobic or aerobic conditions. The oxygen uptake rate is kept as low as possible to avoid an unnecessary loss of carbon to CO_2 while still obtaining sufficient ATP to sustain biomass formation. In all cases, it is expected that some glucose must be used for respiration and ATP generation since product formation is associated with little/no ATP formation. Furthermore, ATP is also needed for cellular maintenance energy of the probably stressed culture. The efficiency of the succinic acid and 1,3-propanediol production processes is estimated to 78 and 79% of the theoretical yield, respectively, while the target of adipic acid is 71%. The metabolic pathway to adipic acid is long and complicated to engineer suggesting that it is difficult to reach close to theoretical performance (Example 17.3). Optimization close to the limits of microbial physiology can require years of R&D and large investments. In the best business cases, the microorganisms do not need to be highly optimized and hence R&D can be completed fast. Industrial validation of technology and early commercialization are important parameters for most companies, and as mentioned earlier fermentation processes can be improved in parallel or at a later stage.

17.3.3 Assessment of Competing Petrochemical Technologies

The techno-economic analysis is now repeated for each of the five chemicals but with a technology that is based on petrochemical feedstock, that is, for each chemical there are two production opportunities: one based on fermentation (renewable feedstock) and one based on catalysis (petrochemical feedstock). Table 17.6 includes an overview of the production routes based on petrochemical (PETRO) and renewable feedstock (BIO). The unit cost of the petrochemical feedstock is higher (>1000 \$/MT) than sugars and corn (<400 \$/MT), but in general more renewable feedstock is needed compared to petrochemical feedstock.

In the traditional chemical industry, ethanol can be formed by simple catalytic hydration of the commodity chemical ethylene. The process is considered competitive and is used in the chemical industry, but it is of course not based on renewable resources. Today, most ethanol is produced by fermentation as a consequence of political support for renewable ethanol. Approximately, 60% of all organic chemicals are derived from ethylene, and hence it is the prime building block of the petrochemical industry [35]. Production of *n*-butanol is almost exclusively done by catalytic hydroformylation of propylene to butyraldehyde (the so called "oxo process"). In a subsequent step butyraldehyde is hydrogenated to *n*-butanol. The interesting feature of this process is that the feedstock, propylene, is a 3-carbon (C3) that is extended by a formyl group, CHO (C1). A prerequisite for cheap *n*-butanol production is the availability of cheap propylene and syngas. Hence it is not surprising that major integrated petrochemical companies like BASF and Dow Chemical have some of the most cost-competitive production sites.

Succinic acid can be produced by hydrogenation of maleic anhydride followed by hydrolysis and in fact maleic anhydride is also used for the structurally similar molecules, malic acid and fumaric acid. Maleic anhydride is also a commodity chemical but more expensive and produced in much smaller scale than ethylene and propylene. Adipic acid is formed from cyclohexane in a number of oxidation steps that includes ring opening. The process has been optimized for many years and in today's market it requires scale to be competitive. A typical adipic acid plant has a capacity of 300 KTA, which is considerably larger than plants using renewable feedstock. A large adipic acid plant (75 KTA) based on a renewable route therefore has a significant investment disadvantage compared to the petrochemical route: 3.72 MM\$/KTA compared to 1.94 MM\$/KTA. The petrochemical production route of 1,3-propanediol has similarities to *n*-butanol production: ethylene oxide (a C2 produced from ethylene) is converted to 3-hydroxypropionaldehyde (C3) and hydrogenated to 1,3-propanediol [36].

Figure 17.6 displays the plant gate cost of production for all production routes investigated. For all processes and regardless of technology there are three major cost drivers: feedstock, utilities, and depreciation. The cost of feedstock is crucial for all processes, which justifies the focus on this parameter both in the research phase (product carbon yield) and in industrial production (competition and sourcing of feedstock).

Example 17.4 The indifference curve

For all production processes covered in this chapter, the major raw material accounts for a very large proportion of the plant gate cost. This means that changes in raw material prices can make a significant impact on competitiveness. It is, therefore, of great commercial interest to investigate how feedstock price volatility impacts different, competing processes. In the simplest case, two competing processes are compared.

Let the succinic acid cost of production (plant gate cost), based on petrochemical feedstock and renewable feedstock, be a simple linear function of the major raw materials:

$$\mathrm{COP_P} = 0.892\,P_{\mathrm{MA}} + 261.$$

$$\mathrm{COP_B} = 1.194\,P_{\mathrm{G}} + 922.$$

All other cost components are constant. P_{MA} is the market price of maleic anhydride and P_{G} is the market price of glucose. The slopes of the linear expressions are the consumption factors, that is, how much feedstock in MT is needed to produce 1 MT of succinic acid (Table 17.6). Expressing P_{G} as a function of P_{MA} for $\mathrm{COP_P} = \mathrm{COP_B}$, one obtains the indifference curve (here a linear relationship) for the production of succinic acid:

$$P_{\mathrm{G}} = 0.747\,P_{\mathrm{MA}} - 554.$$

The slope and intercept of the indifference curve provide immediate and valuable information about the competitiveness of the two processes: the slope (<1) indicates that more glucose than maleic anhydride is consumed to produce the same amount of succinic acid. The intercept (<0) indicates that the petrochemical route has lower nonfeedstock-related cost. In a situation with equal production costs (e.g., maleic anhydride at 1,250 $/MT and glucose at 380 $/MT), a constant percentage increase of price for both feedstocks will lead to a more competitive fermentation process. The analysis of future competitiveness is extremely important for new processes especially when they take several years to develop.

The indifference curves in Figure 17.7 demonstrate that production of succinic acid and 1,3-propanediol by fermentation is competitive on production costs (market price points close to indifference curve), while production of adipic acid is noncompetitive (market price way above the indifference curve). In this analysis, it is assumed that the market price of glucose is 400 $/MT, but competitiveness for adipic acid requires that the market price drops to approximately 200 $/MT. It is beyond the scope of this analysis to discuss feedstock price dynamics in detail, but a 50% cost reduction is drastic, especially for a

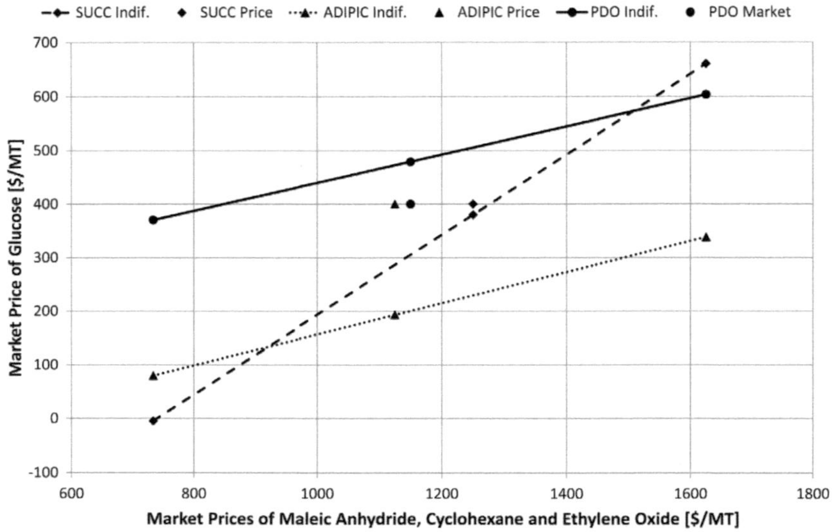

Figure 17.7 The indifference curve (here a straight line) illustrates when two production processes based on different feedstock (e.g., glucose and maleic anhydride) have identical production costs for a given product (e.g., succinic acid). In this example, the indifference curve and market price (single-price point) are illustrated for three chemicals: succinic acid, adipic acid, and 1,3-propanediol. The price point of cyclohexane (1125 $/MT) and glucose (400 $/MT) is above the indifference curve that indicates that the petrochemical route is favorable unless the market price of glucose drops to less than 200 $/MT. The indifference curve is an important tool to assess the competitiveness and feedstock sensitivity of several technologies.

commodity product. The indifference curve for succinic acid has a much steeper slope than the corresponding curves for adipic acid and 1,3-propanediol, that is, more glucose is needed for production of adipic acid and 1,3-propanediol compared to succinic acid and therefore competitiveness for these processes is more sensitive to glucose price volatility. In the chemical market, it is not unusual to see price volatility and hence it is advisable to base the analysis on a longer time period rather than using a horizon of a few months where supply/demand imbalances have created unnatural low or high prices. In Figure 17.7, only a single price point has been added to the chart for each of the three chemicals, and hence the analysis could be improved significantly by populating the figure with additional price points (e.g., a 5 year time window).

17.4
From Fermentation Process Development to the Market

It is an interesting exercise to imagine a dream scenario when introducing new technology and new chemicals to the market. Dreams are normally quite far away from reality, but they can give guidance on whether a product is likely to be successful and where the technology and the end product suffer from commercial short-comings. In a commercial setting, this is often referred to as whether the product has a strong "value proposition." What would a "dream scenario" look like when introducing fermentation-based chemicals to the market?

First, technology development should be done fast in order to market the product before the competitors and to minimize R&D costs (low investment in general). Second, the new technology and products should be covered by several broad patents that will give the company a unique market position and a product that is difficult to copy. Third, the new technology should lead to low production costs. Fourth and maybe more surprisingly, it is desirable to produce chemicals that already are marketed. It is a lengthy process to develop chemical markets and hence market penetration is expected to be much faster than creating a new chemical market. However, in addition the renewable chemicals should be recognized and valued by the end consumer as unique due to their renewable character, the environmental footprint, and so on.

It is definitely not obvious how an already existing chemical can be perceived differently by an end consumer and it is also not evident that the cost of production will be favorable. In fact, it was concluded in Section 17.3.3 that cost competitiveness could be a challenge for products based on renewable feedstock. In this section, we will look further into the value of the chemicals and how that value is distributed between companies that take part in bringing the product to market.

As can been seen in Table 17.7, the five chemicals in the analysis were intentionally chosen to have very different market characteristics. Ethanol is not only

Table 17.7 Estimated market volume and size of the five chemicals analyzed (2014).

Chemical	Market			Production		Major Players	Applications
	Price $/MT	Vol. Growth %/YR	Market Vol. KTA	Capacity KTA	Utilization %		
Ethanol	600	4.0%	85 000	100 000	85%	ADM, POET	Biofuel
n-Butanol	1500	2.5%	3500	3800	92%	BASF, DOW	Coatings and paints
Succinic acid	2400	>15%	50	65	77%	Bioamber, Myriant	Polymers and solvents
Adipic acid	1900	1.5%	2600	3500	74%	Invista, Solvay	Nylons and polyesters
1,3-Propanediol	~2200	>10%	~70	~70	—	Dupont	Polyester fibers

The capacity refers to the total volume of chemicals that can be produced on an annual basis (total nameplate capacity). The plant utilization rate (actual production) in the selected cases is between 74 and 92%, which illustrates that it is not given that a plant will be operated at its nameplate capacity. The analysis has been adapted from Novozymes A/S.

a major biofuel but also a building block in the chemical industry, *n*-butanol and adipic acid are both commodity chemicals with many applications, while succinic acid and 1,3-propanediol are emerging building blocks for polymer production (small volume with high growth rates). Table 17.7 confirms the findings of Figure 17.1 where it was shown that market prices and volume are negatively correlated. In this chapter, it is assumed that ethanol sells for 600 \$/MT, while dedicated low-cost commodity chemicals such as *n*-butanol and adipic acid sell for approximately three times the value of ethanol (1500–1900 \$/MT).

17.4.1
The Value Chain of the Chemical Industry

The value chain analysis is a very important tool in assessing the market opportunity of new products. Most companies in the chemical industry are part of a larger supply chain and hence these companies have to be analyzed in a framework that includes both upstream suppliers and downstream customers. Products based on fermentation processes are typically placed upstream in the value chain and several steps from the end consumer (Figure 17.8). In order to have a successful market introduction and profitable business, it is crucial that the value of new products is recognized throughout the value chain.

It is important to recognize that there is a tremendous difference between products (chemicals) and technology (conversion of intermediates) participating in the value chain. Some people might even rightfully argue that the (fermentation) technology does not participate in the value chain and is irrelevant to the

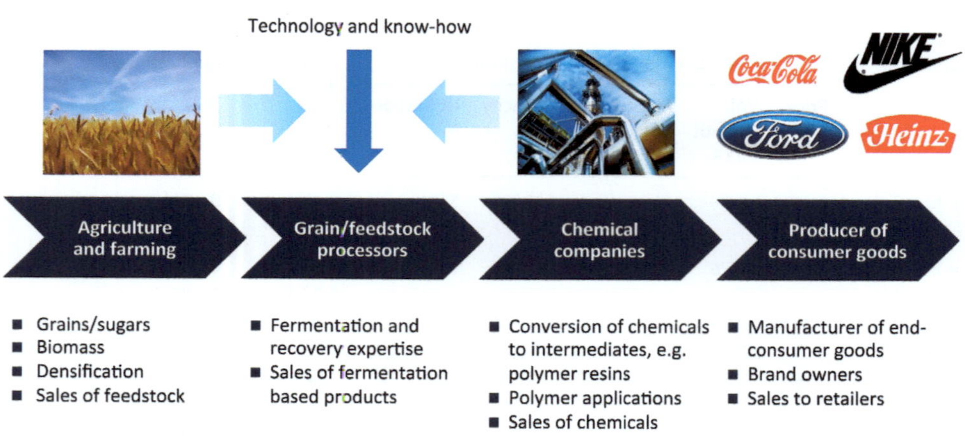

Figure 17.8 The value chain from renewable feedstock to chemicals and end products owned by brand owners is long and includes complicated production processes. It is crucial for technology developers to have a deep understanding of all the players and their needs. Technology is needed to enable new value chains, but technology is typically not visible to brand owners and consumers.

value chain analysis. This might sound as bad news for scientists, engineers, and technology developers, but it should be understood in its right context. New fermentation technology is undoubtedly recognized as valuable by a chemical company, but it is probably of little value or completely irrelevant to brand owners at the end of the value chain. Brand owners and end consumers simply do not "see" the technology in the final product.

The chemical 1,3-propanediol from DuPont is an example of a chemical that is part of a long value chain. In the next production step after fermentation, 1,3-propanediol and terephthalic acid are combined to produce the polyester polytrimethylene terephthalate (PTT) that is branded under the name Sorona®. Polyesters have many applications but PTT is predominately used for production of fibers that serve as a key raw material in carpets. It is highly unlikely that end consumers will care more about the fermentation process (if they know what it is) than the price and properties of the carpet, for example, its softness and stain resistance. However, the fermentation process may indirectly be recognized as a valuable driver to every single company in the value chain if the product is based on renewable feedstock, and the technology improves the environmental footprint compared to existing technology. Similarly, a brand owner may reject to market and sell the product if the product has a negative environmental footprint that will create negative consumer awareness.

The purpose of the remaining part of the section is to investigate the interest of the players in the value chain. In Table 17.3, it was shown that the cost of feedstock (both glucose and corn) has a major impact on the economic feasibility of a fermentation process. A highly relevant question is whether the feedstock can be sourced at the same cost in the future? It is impossible to predict future commodity prices (or stock market prices for that matter), but one can and one should definitely look into the underlying market drivers of feedstock production. In Table 17.8, it is observed that the United States produces considerably

Table 17.8 Historical and forecasted corn yields (metric tons per hectare) in different countries.

	Corn yield trends (MT/hectare)				Growth %
	1990	2000	2010	2015E	
Global average	3.7	4.4	5.1	5.6	2.8%
USA	7.1	8.6	9.6	10.7	2.8%
Argentina	3.8	5.8	7.8	7.8	5.0%
China	4.6	4.9	5.3	5.8	1.5%
Brazil	2.1	3.0	4.1	4.3	5.0%
India	1.4	1.8	2.4	2.3	3.2%

The table indicates that the agricultural output can be improved significantly worldwide over the next few years. Growth is calculated as compound annual growth rate between 1990 and 2015E. The table is adapted from Mackey *et al.* [37].

more corn per hectare than the global average. It is not without reason that United States is a major country for the production of fermentation-based products. Furthermore and most importantly, introduction of new and improved crops are expected to boost the yield on a global scale even further. Yield is not the same as production costs, but the analysis suggests that crop output will increase.

Today, cellulosic sugars are produced in commercial scale and this may soon lead to a second wave of feedstock development and possibly cheaper sugars in certain geographies. At the same time, the emergence of shale gas has provided the petrochemical industry with a new, cheap raw material that has improved cost competitiveness of processes based on light feedstock such as ethylene (e.g., feedstock for chemical production of 1,3-propanediol). New production methods will emerge based on feedstock availability and consequently make competitor analysis a moving target. The recent development clearly demonstrates that the need for cheap and abundant feedstock is important to commodity production regardless of whether the feedstock is petrochemical or renewable. Furthermore, petrochemical and renewable feedstocks are fundamentally different, (availability, composition, transportability, etc.) which will have a large impact on future supply chains [38,39].

As mentioned earlier, at the other end of the value chain we have consumers who typically see the chemicals and polymers only as part of a finished product, for example, packaging material and toys made of polymer resins, clothes made of polymer fibers, and so on. The management consultant company McKinsey & Co. conducted a study on consumers' willingness to pay a premium on green products [40]. In Figure 17.9, the share of consumers picking green is plotted as a function of green premium (i.e., the extra cost of buying "green" compared to traditional products). Packaging material visible to the consumers can drive a larger premium than materials used for production of automobiles, furniture, electronics, and buildings. As expected, the addressable market drops as a function of increased green premium. The conclusion of the study was that only very few consumers are willing to pay a premium exceeding approximately 25%. In the chemical industry, a premium of 25% is substantial and therefore a company marketing renewable products should carefully investigate whether such a premium is realistic and design their market segmentation and pricing strategy accordingly [40]. The optimal pricing strategy is dependent on a number of factors, including sales margins, economy of scale, competitors, and so on. A good stress test of a business case is whether the business can compete in a market where the premium has been eroded or does not exist at all. It is not unlikely that many business cases have been rejected due to the lack of cost competitiveness in a market without premium (cf. indifference curves in Figure 17.7). The years to come will demonstrate whether a premium can be sustained, but there are reasons to believe that the green trend is strong [40].

The value chain shown in Figure 17.8 is to some extent simplified. Between the chemical producer and the brand owner there are often additional companies

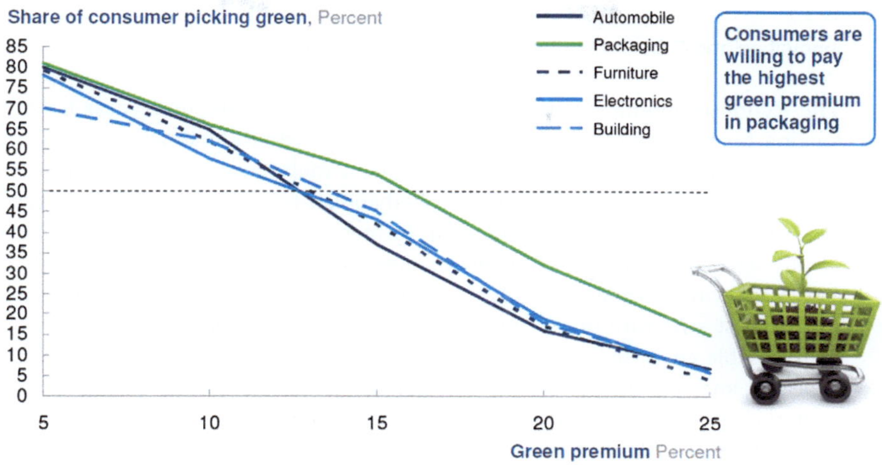

Figure 17.9 Consumers are willing to pay a limited premium for green products. There is an almost linear relationship between the share of consumers picking green and the maximum achievable premium. The analysis suggests that pricing is crucial for a company marketing renewable products: there is a clear trade-off between market price (premium) and market share. Material used for packaging is believed to get a higher premium than material used for automobiles, furniture, electronics, and buildings. (The figure is from Kaiser *et al.* [40].)

involved, for example, compounders and original equipment manufacturers (OEM) who manufacture the actual products or components that are sold in retail under the name of the brand owner. A longer value chain will undoubtedly have even bigger impact on how the premium of the end products will be distributed between the players.

It is definitely not necessary and advisable to involve all companies in a value chain at the same time, but upon completion of development work the value chain must be enabled through one or multiple commercial agreements that span the whole value chain. The companies in the value chain have to agree on who should invest and who should benefit from higher market prices of chemicals sold with a green premium (if any). Chemical companies would argue that they need higher prices due to a new and more valuable product, while brand owners would argue that they have invested heavily in marketing in order to make sure that the value of green products is visible and understood by end consumers. Research has shown that several players expect to benefit from a premium and that will undoubtedly lead to tough negotiations [40–42]. It requires a lot of commercial effort and time to bridge the interest of all parties and draft the necessary commercial contracts to take an idea to commercial production. It is not only science that is time consuming.

17.4.2
Innovation and Substitution Patterns in the Chemical Industry

The techno-economic analysis demonstrated that cost competitiveness is challenging for the commodity chemical adipic acid, while the emerging building blocks succinic acid and 1,3-propanediol have a more favorable and competitive cost structure. The low product yields of ethanol, *n*-butanol, and 1,3-propanediol ($0.41–0.63\,g\,l^{-1}$) is not a question about fermentation optimization but the laws of physics! Conversion of relatively oxidized sugars into reduced chemicals is associated with considerable CO_2 generation regardless of production process. If today's cost structure of a fermentation based chemical is unfavorable due to high variable costs, then one of the most important value propositions has disappeared and one has to analyze whether the remaining drivers in the "dream scenario" are sufficient to justify a development project.

In Figure 17.10, it is shown that polymer innovation based on petrochemical building blocks has decreased dramatically since 1960. The decrease suggests that the traditional seven building blocks are insufficient to generate product differentiation, or there is no market driver for engaging in further

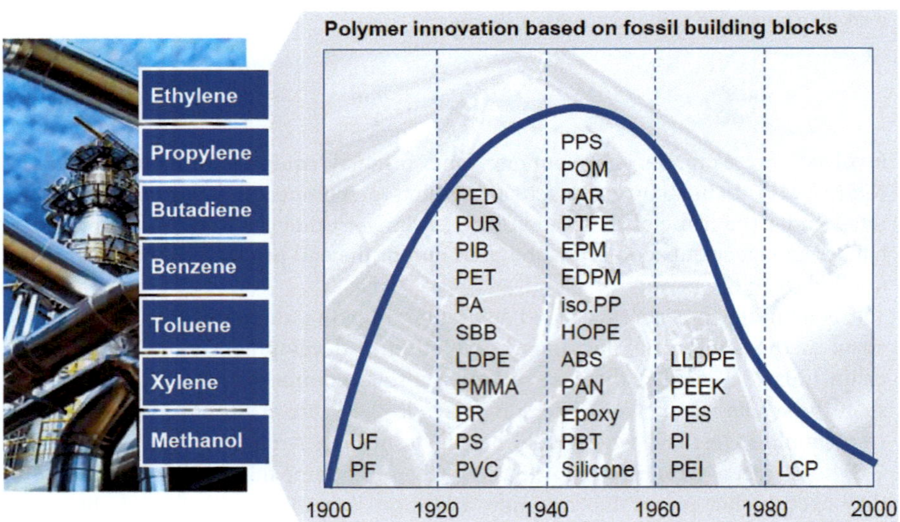

Figure 17.10 The petrochemical industry has built a large portfolio of products based on only seven key building blocks. Several polymers were developed and marketed between 1940 and 1960 but then innovation ceased. New building blocks based on renewable feedstock and biotechnology could boost innovation and create a new era of product development in the chemical industry. Succinic acid and 1,3-propanediol are examples of chemicals with new functionality that have been marketed after year 2000. (The figure is reproduced from Ref. [43].)

innovation, in the chemical industry. If it is assumed that there is a need for innovation, then another market approach for the chemical industry would be to develop new chemicals rather than producing existing petrochemicals from renewable feedstock.[1)] This approach is fundamentally different from the "dream scenario," but it addresses one of the major key concerns of renewable products.

A new building block offers new product differentiation opportunities, but the development and introduction of new products are also associated with considerable investments. As already discussed, it requires testing and validation of existing products when the chemical industry changes from petrochemical to renewable feedstock. However, introduction of a new chemical also requires fundamental research and process development throughout the value chain. Every single step in the production process from sugar to chemical must be rethought and tested in great detail before a new product can be introduced to the market. Furthermore, when a new chemical is introduced to the market there is limited economy of scale and no experience in large-scale production. The upside of the investment is that new chemicals and products can have a game changing effect on the chemical industry and bring long-term growth opportunities to a company.

In order to estimate the market potential of emerging building blocks such as succinic acid and 1,3-propanediol, it is also necessary to study to what degree these emerging building blocks can substitute existing chemicals (Figure 17.11). Traditionally, substitution of chemicals boils down to two key parameters: cost and performance. New chemicals are per definition different from existing chemicals, but the beauty of chemistry is that the chemical structure immediately points to a hypothesis about application opportunities. Companies in position of the necessary chemical know-how are in a good position to qualify new products and evaluate application opportunities.

Succinic acid (C4 dicarboxylic acid) has the opportunity to substitute adipic acid (C6 dicarboxylic acid) in many applications [48,49]. Furthermore, succinic acid could also replace phthalic anhydride and find applications within plasticizers [47]. Many plasticizers have a bad reputation (rightfully or not), and there is political pressure on the chemical industry to find alternative solutions. The oxalic acid (C2 dicarboxylic acid) market is much smaller than the adipic acid market and is expected to offer much fewer market opportunities. Similarly, 1,3-propanediol (C3 diol) is closely related to 1,2-ethanediol (C2 diol, also known as ethylene glycol) and 1,4-butanediol (C4 diol) that all are used in production of polyesters, typically with the aromatic compound terephthalic acid. As discussed in Chapter 3, DuPont also markets the Sorona® polyester based on 1,3-propanediol [50].

1) The expression "drop in chemical" is often used to indicate that an existing petrochemical is produced from renewable resources. The expression is a bit unfortunate because ethanol is also considered a "drop in fuel" though it is fundamentally different from traditional petrol used for transportation.

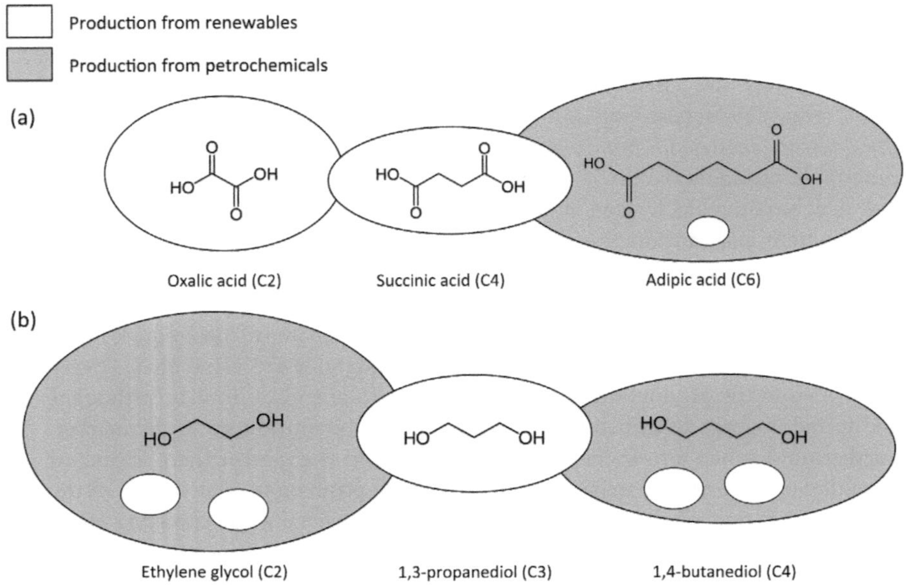

Figure 17.11 Example of chemical substitution: similar chemicals can to some extent substitute each other based on price and performance. (a) Succinic acid (C4) is structurally similar to adipic acid (C6) and is expected to be a replacement in several applications [44,45]. The oxalic acid (C2) market is much smaller than the adipic acid market and is expected to offer fewer, opportunities. (b) 1,3-propanediol (C3) is structurally similar to 1,2-ethanediol (C2) and 1,4-butanediol (C4) that are produced in very large quantities. In all markets mentioned, there are initiatives to produce products from renewables for example, 1,4-butanediol [46,47], though the vast majority of products are currently produced from petrochemical resources.

The advantage of substituting commodity chemicals with new chemicals is that even a small market share is attractive for succinic acid and 1,3-propanediol. Today, the succinic acid market is less than 2% of the total adipic acid market (Table 17.7), and hence there is room for market expansion without a major reaction from existing adipic acid producers. The same holds true for the small market of 1,3-propanediol compared to the markets of 1,2-ethanediol and 1,4-butanediol. If the succinic acid and 1,3-propanediol markets keep growing then it is expected that market penetration will become more difficult unless the market price drops. As shown in Figure 17.1, market price will sooner or later be a limiting factor for market penetration. The difference in cost of production between adipic acid (petrochemical feedstock) and succinic acid (renewable feedstock) is estimated to approximately 450 \$/MT, which is a significant barrier (Figure 17.6). Many chemicals will not be produced cheaply, but a small niche chemical market can definitely be very profitable and attractive (Table 17.1). As already mentioned, the relationship between price reduction and market expansion is an interesting topic that will not be discussed further. It is, however, clear that a price reduction in the hunt for market expansion is dangerous without a concomitant reduction of production cost.

17.5
The Industrial Angle and Opportunities in the Chemical Industry

Fermentation as a production process has impact on the new succinic acid industry, but it would be incorrect to say that feedstock and the production process define the industry. For instance, succinic acid can be produced from multiple sources (e.g., glucose and maleic anhydride) and in the previous sections it was shown that the nature of the feedstock plays a larger role than the technology itself. In order to enter the succinic acid industry, it is necessary to develop (Figure 17.2) or acquire production technology that meets product specifications [51]. Technology is often an entrance barrier in the chemical industry, and it is also a barrier that will prevent companies to enter the succinic acid industry. Currently, the succinic acid industry has many traits from the commodity chemical industry – it is research and capital intensive with only a few specialized players (oligopoly). However, as an emerging building block with high growth rates, the succinic acid industry has many unique features that must be analyzed separately. Commodity chemicals have growth rates tightly connected to the gross domestic product (GDP) of the major markets, and annual growth rates below 5% should be expected (Table 17.7; [35]).

In order to understand the attractiveness of the future succinic acid business, it is important to conclude on the direction of the industry rather than current status. The analysis of the succinic acid industry in this chapter is based on Porter's five forces model where it is assumed that the state of an industry can be described by five competitive forces [52]. Improved agricultural methods (Table 17.8) and cellulosic feedstock could derisk the exposure to feedstock volatility and hence the bargaining power of suppliers is expected to decrease over time. However, feedstock costs do play an important role (Table 17.6) and hence location and feedstock security will continue to play an important role. It is expected that all players in the industry will find solutions to deal with feedstock costs. Maleic anhydride could be a future, alternative petrochemical feedstock, but it will most likely not lead to reduced cost (Figures 17.6 and 17.7).

Fast market penetration is a key success factor for an emerging building block, and it is expected that for the next few years the relationship with downstream customers in the value chain will be the key focus area for all players in the succinic acid industry (Figure 17.12). Smaller companies like BioAmber and Myriant are expected to form partnerships with downstream partners to develop products [9], while the larger companies Succinity and Reverdia can use their parent companies for product development, that is, succinic acid is used as an internal intermediate and will not necessarily be marketed. In the chemical industry, vertical integration (companies own several steps in the supply chain)

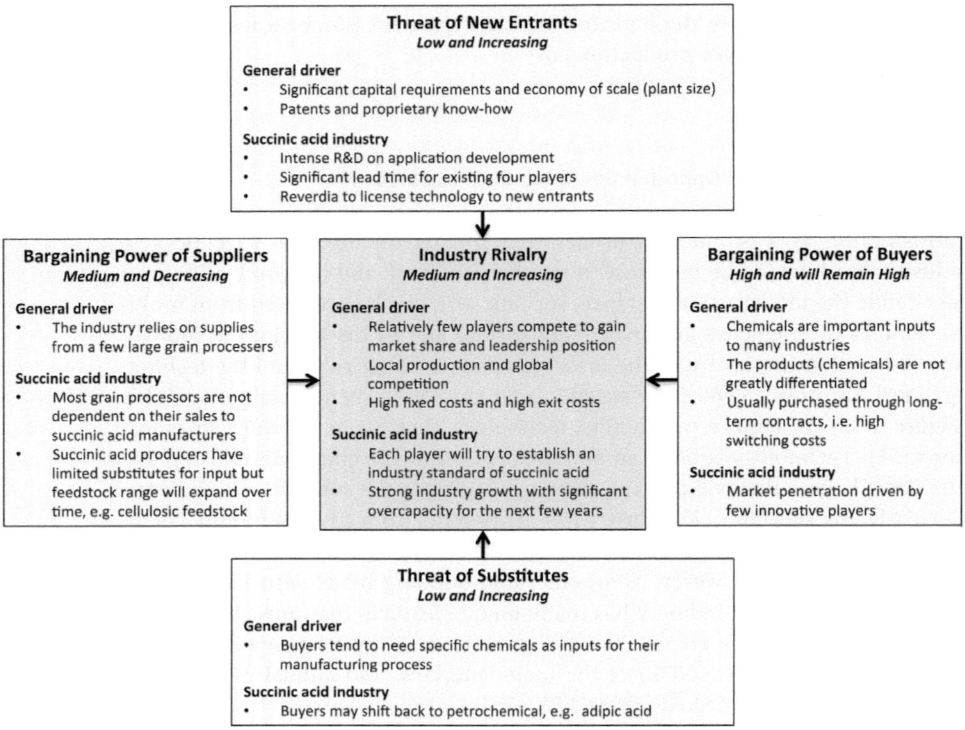

Figure 17.12 Competition in the succinic acid market based on Porter's five forces model [52]. The assumption is that the state of an industry depends on five basic competitive forces. The strength of the forces determines the profitability of the industry. For each force, the expected time-dependent change of the force is indicated; that is, is it expected that the force will play an increasing or decreasing role in the future. Drivers have been classified as general drivers in the chemical industry (top) and drivers specifically related to the new succinic acid industry (bottom). The analysis illustrates that the bargaining power of buyers is the main barrier in the industry.

is normal. Vertical integration is not a guarantee for success and it should be chosen for the right reasons [52,53]. The four companies mentioned in this chapter all produce succinic acid based on renewable feedstock. It remains to be seen if a competing company will enter the market based on large-scale production of succinic acid from maleic anhydride. If customers have a strong preference for a renewable product, then succinic acid from maleic anhydride will have a hard time to enter the market.

As shown in Section 17.4.3, succinic acid and adipic acid can substitute each other. If succinic acid is successful, then succinic acid may start to impact the adipic acid market and a retaliation response is expected [52]. Adipic acid has a strong market position and attractive cost structure due to economy of scale (Table 17.6), and it is expected that the threat of substitution will increase in the future. Thus, in the years to come, it is not expected that adipic acid producers will consider succinic acid as a strong threat to their business.

Perhaps, the most interesting question is to see how the rivalry in the succinic acid business will change in the years to come. Each of the four companies mentioned in this chapter has ramped up production and is ready to aggressively develop the market based on their production capacities [44,45]. Capacity expansion and economy of scale is the name of the game in commodity (chemical) business [52], and it is not believed that succinic acid will be any different (Tables 17.3 and 17.4). Economy of scale can lead to a strong market position and hence there is a strong industry incentive to build capacity that exceeds demand (Table 17.7). In case all companies can sell their succinic acid, that is, production equals nameplate capacity, then there is a chance for building a profitable industry. If demand does not meet supply then a second option could lead to fierce price competition and decreased industry profits. It will be interesting to follow the future market development and the strategic steps of the companies.

The succinic acid industry is new and it remains to be seen whether the competing companies will get a payback on the initial investments, but there are definitely important drivers that could pave the way for success: the threats of new entrants are low but could increase since Reverdia has announced that they license their technology [51], the bargaining power of suppliers are medium, and the threats of substitutes low (Figure 17.2). The success of succinic acid will depend on the capability of the players to build a strong, growing market in collaboration with partners (drive demand) rather than too aggressive capacity expansion (driving supply). R&D on application development will still play an important role (not metabolic engineering); the players will look for further process improvements and cheaper feedstock opportunities.

The industry analysis of succinic acid can be repeated for the other four cases but with different conclusions that will impact a decision process. Below are a few main conclusions:

- *Ethanol* is a commodity biofuel that is sold at spot market price like crude oil, and hence there is a tremendous pressure on cheap raw materials and production excellence. The good news is that the ethanol market is highly developed and all ethanol can be sold as long as it meets the market price. Fermentation technology is necessary but not a differentiating factor in the industry.
- *n-Butanol* is primarily used for applications within paints and coatings followed by a number of smaller markets such as adhesives, plastics, and textiles. BASF and Dow are major players due to their internal demand in the acrylate business. Smaller market segments are covered by a number of smaller regional players where fermentation could be an alternative production method. Several studies have shown that *n*-butanol is a good biofuel and hence some of the attractiveness of *n*-butanol has been attributed to its future potential to penetrate the biofuel market.

- *Adipic acid* is primarily used for production of nylon 6,6 fibers, which is the starting material for carpet yarn, home furnishing, and a number of applications in the automotive sector. Market growth is, therefore, tightly coupled to the state of the market. The recent recession has created overcapacity and lower margins in the industry (Table 17.7). New production capacity is, therefore, not needed in a foreseeable future, but the industry does struggle with pollution due to the harsh chemical conditions of existing production methods. A new and environment-friendly fermentation-based route to adipic acid seems very challenging, but it could be an alternative to existing technology.
- *1,3-propanediol* has many similarities to succinic acid in terms of market development, but the notable difference is that most of the 1,3-propanediol produced is consumed by a single company (captive production), namely, DuPont. It is, therefore, expected that market development will be very challenging for entrants who only want to market 1,3-propanediol.

The analysis of the different chemical markets demonstrates that each market has different opportunities and challenges. Market opportunities require different competences and hence the attractiveness of a market segment also depends on what a specific company can bring to the industry.

17.6
Evaluation of Business Opportunities

The analysis presented in this chapter has for pedagogical reasons focused on a business development plan that includes a series of chronological events: basic research, scale-up of fermentation technology, industrial-scale production followed by market introduction and market expansion. Furthermore, it is implicitly assumed that the same company is responsible for all tasks related to the long development plan and production (Figure 17.2). This development track is far from the "normal behavior" in biotechnology/chemistry since the development plan requires a long series of competences that only few companies possess. It is also time-consuming to develop all technologies in-house and in many cases larger companies prefer to acquire base technology to shorten time to market. The technical risk profile of a business case is a critical evaluation criterion for most companies, and in some cases the risk of technical failure is a good reason for not starting a project. Instead of embarking on an internal R&D project, it is a common approach (especially for larger companies) to explore additional opportunities such as acquiring IPR and technology that has already been developed by other companies.

One of the more aggressive development plans is to acquire the entire company that excels in a certain area and then expand the business. In business development, it is the task to investigate these opportunities and establish the necessary commercial agreements that will maximize the likelihood of success and shorten time to market.

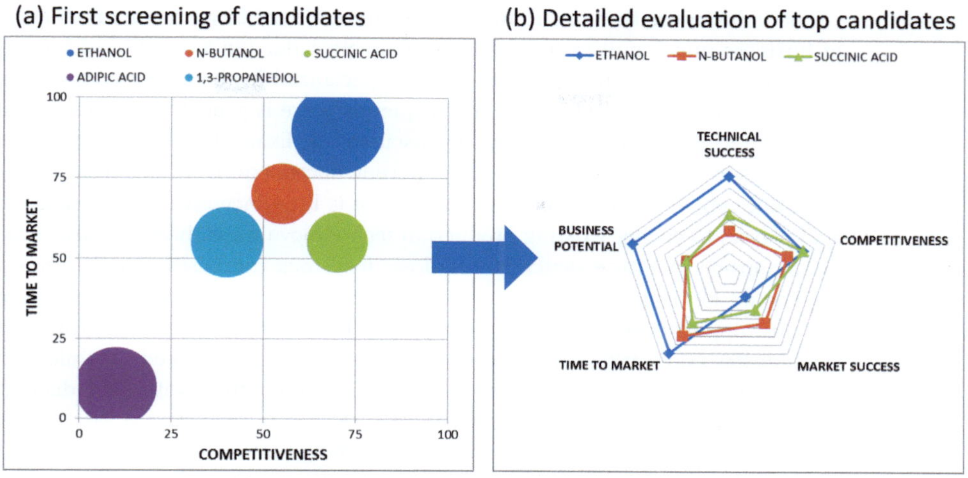

Figure 17.13 Example of a product candidate selection process. (a) In the first screening, only chemicals with competitiveness scores and time to market scores of at least 50 are considered. A score of 0 indicates low competitiveness and long time to market (e.g., >15 years), while a score of 100 indicates very competitive and short time to market (e.g., <2 years). (b) In the next round of analysis, a more detailed evaluation is applied, including business potential (annual sales), likelihood of successful R&D, and likelihood of successful market introduction. In the example, ethanol, *n*-butanol, and succinic acid show different strengths and weaknesses.

The development of fermentation technology for chemical production could also be evaluated by a company that has interest in developing and licensing/selling technology. Novozymes A/S is an example of a technology provider that develops enzymes and microorganisms but does not produce commodity chemicals. Based on a preliminary analysis, the technology provider may have shortlisted the five chemicals analyzed in this chapter, that is, ethanol, *n*-butanol, succinic acid, adipic acid, and 1,3-propanediol.

In Figure 17.13, it is exemplified how a company can use a screening process to select the best chemical candidates for developing fermentation technology. In the first round of screening, the working hypothesis is that the technology must be developed from scratch, and only competitive processes with a short to medium time to market will be considered. The first screening criteria are not neutral because each company will have their own interpretation of competitiveness and time to market. The important message is that a screening process is needed in order to reduce the number of business opportunities to a manageable size that can be further scrutinized. Additional parameters will be considered in a more detailed analysis, for example, business potential (revenue potential, internal rate of return, etc.), technical success, and market success. In the second round, the five candidate technologies are reduced to three candidates (ethanol, *n*-butanol, and succinic acid). The remaining candidate technologies are analyzed in further details and it is revealed that each technology has distinct pros and cons (Figure 17.13).

The industry analysis of succinic acid is an important input for the evaluation of business opportunities for the technology provider (Sections 17.5–17.6): all companies discussed in this chapter have specialized in metabolic engineering and fermentation technology, and consequently there is probably no direct need for new fermentation technology. As discussed in Example 17.1, it is believed that the succinic acid producers use different microorganisms and have developed different recovery processes. Thus, there is little incentive for the succinic acid players to change existing production technology unless there is a large economic upside. A future switch to cellulosic feedstock (next-generation technology) could be of interest to multiple succinic acid producers and thereby be a business opportunity for a technology player. Furthermore, sustained growth in the industry could trigger interest of new players that might need succinic acid technology. In this case, the technology player has to compete with Reverdia that offers a license to their own technology [51].

The analysis presented here has a focus on fermentation technology developers, and the analysis cannot be extrapolated to other companies that may have another position in the value chain. The industry analysis in Section 17.5 is needed to understand the overall dynamics of an industry sector, but each company will have a very different perception of an industry. Classical catalyst companies may draw a conclusion on the industry that is different from a fermentation technology provider (Figure 17.13).

The present analysis sheds light on some of the fundamental aspects of analyzing technology for fermentation-based chemicals in a competitive world. Key focus areas include the competitiveness of technology and industry analysis. In addition there are corporate questions that must be answered. It is beyond the scope of the present review to discuss all implications, but it is an organizational challenge to allocate resources and nurture basic research projects over longer time periods, especially in larger companies when new business opportunities do not generate significant revenue. As a consequence many companies have developed new organizational structures to sustain breakthrough innovation [54].

17.7
Concluding Remarks and Outlook

Biotechnology is a key technology in multiple industries and the applications range from pharmaceuticals to commodity chemicals and biofuels. This chapter analyzed fermentation processes for production of commodity chemicals in a commercial framework. Five fermentation products were used as examples: ethanol, *n*-butanol, succinic acid, adipic acid, and 1,3-propanediol. It was shown that developing commercially relevant fermentation processes is challenging and in all cases cost competitiveness is extremely important for commercial success. In most cases, it is necessary to achieve a carbon yield of the product on the feedstock (glucose, corn) that is close to the theoretical maximum (Section 17.3.2).

The risk of not reaching the target must be taken into account throughout the R&D project. It is, therefore, crucial that a project team allocates the necessary time and resources to set up realistic milestones before embarking on the long R&D journey (Section 17.2).

The value chain analysis is a necessary and basic tool in the commercial evaluation of new technology for conversion of renewable feedstock to products (Section 17.4.2). Renewable feedstock plays an important role in the value chain, much more than the fermentation technology itself and therefore it is important to understand how all downstream customers perceive a new product. In some cases, products produced from renewable resources are sold with a premium compared to existing products, but it is not always the case (Section 17.4.2). Failure to understand how each company reacts to new products can result in a dysfunctional value chain and failed market introduction.

The succinic acid case presented in this chapter is an example of an emerging chemical that is based on renewable feedstock and fermentation (Section 17.5). It is believed that the lessons from this case, as well as other emerging chemicals, will pave the way for further expansion of renewable products. It is important to stress that successful market introduction is based not only on technical success but also on commercial success where multiple players in the value chain have endorsed renewable products (Section 17.6).

Acknowledgment

Ronald Cascone, Alexander Coker, and Marcos Nogueira Cesar from Nexant are acknowledged for their great contribution to the cost exclusion curve and the techno-economic evaluation of adipic acid production. Mark L. Morgan from IHS is acknowledged for his insight into the chemical industry and his contribution to the market analysis. Alan Berry and Ejner Bech Jensen from Novozymes A/S are acknowledged for their valuable comments on the final manuscript. The author would like to thank Novozymes A/S in Denmark for providing data for the techno-economic analysis and the editor, John Villadsen, for guidance and comments on the manuscript.

This chapter is dedicated to the memory of my wonderful parents, Inga Christine Grotkjær (1933–2012) and Heinrich Grotkjær (1930–2014).

References

1 Rothwell, R. (1992) Successful industrial innovation: critical factors for the 1990s. *R&D Manag.*, **22**, 221–239.

2 Daubenfeld, T., Bergmann, T., Frank, N., and Weidenfeld, L. (2014) On the current practice of new business development in the German chemical industry. *J. Business Chem.*, **11**, 87–97.

3 Harrison, R.G., Todd, P.W., Rudge, S.R., and Petrides, D.P. (2002) *Bioseparations Science and Engineering*, 1st edn, Oxford University Press, New York.

4 Heinzle, E., Biwer, A., and Cooney, C. (2006) *Development of Sustainable Bioprocesses*, John Wiley & Sons, Ltd., Chichester, England.

5 Forciniti, D. (2008) *Industrial Bioseparations*, 1st edn, Blackwell Publishing, Ames, USA.
6 Cascone, R. and Burke, B. (2010) Biorenewables update: what is beyond ethanol and biodiesel? *Hydrocarbon Proc.*, **89** (9), 51–55.
7 Hong, K.K. and Nielsen, J. (2012) Metabolic engineering of *Saccharomyces cerevisiae*: a key cell factory platform for future biorefineries. *Cell. Mol. Life Sci.* doi: 10.1007/s00018-012-0945-1.
8 Zhang, J., Babtie, A., and Stephanopoulos, G. (2012) Metabolic engineering: enabling technology of a bio-based economy. *Curr. Opin. Chem. Eng.*, **1**, 355–362.
9 BioAmber (2014) Press Release from 4 February 2014, investor.bio-amber.com (accessed 7 July 2014).
10 Nexant (2010) *PERP Program 08/09-2 Adipic Acid*, White Plains, New York, USA.
11 Nexant (2014) Biorenewable Insights, http://thinking.nexant.com/program/biorenewable-insights (accessed 9 December 2014).
12 Kurzrock, T. and Weuster-Botz, D. (2010) Recovery of succinic acid from fermentation broth. *Biotechnol. Lett.*, **32**, 331–339.
13 Succinity (2014) Technology Overview, http://www.succinity.com/biobased_succinic_acid/technology (accessed 6 July 2014).
14 Reverdia (2014) Technology Overview, http://www.reverdia.com/technology/technology-overview (accessed 6 July 2014).
15 BioAmber (2013) Press release, 1 October 2013, investor.bio-amber.com.
16 Hermann, B.G. and Patel, M. (2007) Today's and tomorrow's bio-based bulk. *Appl. Biochem. Biotechnol.*, **136**, 361–388.
17 Peters, M.S., Timmerhaus, K.D., and West, R.E. (2004) *Plant Design and Economics for Chemical Engineers*, 5th edn, McGraw Hill, Singapore.
18 Songa, H. and Lee, S.Y. (2006) Production of succinic acid by bacterial fermentation. *Enzyme Microb. Technol.*, **39**, 352–361.
19 Berry, A. (2014) Bio-based bulk chemicals: technical advances and commercial realities. Presented at RTP Chemical Biology & Biotechnology Symposium. Raleigh, USA, May 2014.
20 Villadsen, J., Nielsen, J., and Lidén, G. (2011) *Bioreaction Engineering Principles*, 3rd edn, Springer, New York.
21 Brown, S.H., Bashkirova, L., Berka, R., Chandler, T., Doty, T., McCall, K., McCulloch, M., McFarland, S., Thompson, S., Yaver, D., and Berry, A. (2013) Metabolic engineering of *Aspergillus oryzae* NRRL 3488 for increased production of L-malic acid. *Appl. Microbiol. Biot.*, **97**, 8903–8912.
22 Tao, L., He, X., Tan, E.C.D., Zhang, M., and Aden, A. (2014) Comparative techno-economic analysis and reviews of *n*-butanol production from corn grain and corn stover. *Biofuel. Bioprod. Bioref.*, **8**, 342–361.
23 Crundwell, F.K. (2008) *Finance for Engineers*, 1st edn, Springer, London, UK.
24 HIS (2010) Process Economics Program (PEP), http://www.ihs.com/products/chemical/technology/pep/index.aspx. (accessed 9 December 2014).
25 Niu, W., Draths, K.M., and Frost, J.W. (2002) Benzene-free synthesis of adipic acid. *Biotechnol. Prog.*, **18**, 201–211.
26 Curran, K.A., Leavitt, J.M., Karim, A.S., and Alper, H.S. (2013) Metabolic engineering of muconic acid production in *Saccharomyces cerevisiae*. *Metab. Eng.*, **15**, 55–66.
27 Kwiatkowski, J.R., McAloon, A.J., Taylor, F., and Johnston, D.B. (2006) Modeling the process and costs of fuel ethanol production by the corn dry-grind process. *Ind. Crop. Prod.*, **23**, 288–296.
28 Jones, D.T. and Woods, D.R. (1986) Acetone–butanol fermentation revisited. *Microbiol. Rev.*, **50**, 484–524.
29 Lee, S.Y., Park, J.H., Jang, S.H., Nielsen, L.K., Kim, J., and Jung, K.S. (2008) Fermentative butanol production by *Clostridia*. *Biotechnol. Bioeng.*, **101**, 209–224.
30 Nakamura, C.E. and Whitedy, G.M. (2003) Metabolic engineering for the microbial production of 1,3-propanediol. *Curr. Opin. Biotechnol.*, **14**, 454–459.
31 Antoniewicz, M.R., Kraynie, D.F., Laffend, L.A., González-Lergier, J., Kelleher, J.K., and Stephanopoulos, G. (2007) Metabolic flux analysis in a nonstationary system:

fed-batch fermentation of a high yielding strain of *E. coli* producing 1,3-propanediol. *Metab. Eng.*, **9**, 277–292.

32 Weusthuis, R.A., Lamot, I., van der Oost, J., and Sanders, J.P.M. (2011) Microbial production of bulk chemicals: development of anaerobic processes. *Trends Biotechnol.*, **29**, 153–158.

33 Hettinga, W.G., Junginger, H.M., Dekker, S.C., Hoogwijk, M., McAloon, A.J., and Hicks, K.B. (2009) Understanding the reductions in US corn ethanol production costs: an experience curve approach. *Energ. Policy*, **37**, 190–203.

34 Green, E.M. (2011) Fermentative production of butanol: the industrial perspective. *Curr. Opin. Biotechnol.*, **22**, 337–343.

35 Jones, T., Dunwoodie, M., Boucher-Ferte, V., and Reiff, O. (2011) *Chemicals for Beginners*, 5th edn, Deutsche Bank AG, London, UK.

36 Lawrence, R. (2014) Davy Process Technology Develops a New Process for 1,3 Propanediol, www.davyprotech .com. (accessed 7 July 2014).

37 Mackey, P.J., Walsh, P., Srethapramote, S., Wells, A.D., Maheshwari, M., Gupta, L., Hwang, H.H., and Andrews, V. (2012) Green is good: the potential of bioplastics. *Morgan Stanley Research*.

38 Kircher, M. (2012) The transition to a bio-economy: emerging from the oil age. *Biofuels Bioprod. Bioref.*, **6**, 369–375.

39 Kircher, M. (2014) The emerging bioeconomy: industrial drivers, global impact, and international strategies. *Indus. Biotechnol.*, **10**, 11–18.

40 Kaiser, H.M., Miremadi, M., Musso, C., and Weihe, U. (2012) The growing demand for green. McKinsey on Chemicals, Number 4, Spring edition.

41 Mirabal, A.S., Scholz, L., and Carus, M. (2013) *Market Study on Bio-based Polymers in the World*, Nova Institute, Hürth, Germany.

42 Carus, M., Eder, A., and Beckmann, J. (2014) GreenPremium prices along the value chain of biobased products. *Indus. Biotechnol.*, **10**, 83–89.

43 Riese, J. (2009) McKinsey & Co. on White Biotechnology, www.dsm.com (accessed 7 July 2014).

44 Kockelkoren, J. (2014) Biosuccinium accelerate bio-based market adoption. Presented at World Congress on Industrial Biotechnology. Philadelphia, USA, May 2014.

45 Walter, P. (2014) Is renewable succinic acid a commercial reality? Presented at World Congress on Industrial Biotechnology. Philadelphia, USA, May 2014.

46 Yim, H., Haselbeck, R., Niu, W., Pujol-Baxley, C., Burgard, A., Boldt, J., Khandurina, J., Trawick, J.D., Osterhout, R.E., Stephen, R. *et al.* (2011) Metabolic engineering of *Escherichia coli* for direct production of 1,4-butanediol. *Nat. Chem. Biol.*, **7**, 445–452.

47 Weastra (2013) Determination of market potential for selected platform chemicals: itaconic acid, succinic acid, 2,5-furandicarboxylic acid. EU funded (FP7) project BioConSepT, www.bioconcept.eu (accessed 13 July 2014).

48 Lynd, L.R., Wyman, C., Laser, M., Johnson, D., and Landucci, R. (2005) Strategic Biorefinery Analysis: Analysis of Biorefineries. Subcontract report NREL/SR-510-35578.

49 Bechthold, I., Bretz, K., Kabasci, S., Kopitzky, R., and Springer, A. (2008) Succinic acid: a new platform chemical for biobased polymers from renewable resources. *Chem. Eng. Technol.*, **31**, 647–654.

50 Kurian, J.V. (2005) A new polymer platform for the future: Sorona® from corn derived 1,3-propanediol. *J. Polym. Environ.*, **13**, 159–167.

51 Reverdia (2014) Press release, 1 October 2014, http://www.reverdia.com/press-releases/reverdia-adds-licensing-biosuccinium-portfolio (accessed 10 December 2014).

52 Porter, M.E. (1980) *Competitive Strategy*, 1st edn, Free Press, New York.

53 Stuckley, J. and White, D. (1993) When and when not to vertically integrate. The McKinsey Quarterly, August edition.

54 O'Connor, G.C., Leifer, R., Paulson, A.S., and Peters, L.S. (2008) *Grabbing Lightning: Building a Capability for Breakthrough Innovation*, 1st edn, John Wiley & Sons, Inc., San Francisco, USA.

Index

Fundamental Bioengineering, First Edition. Edited by John Villadsen.

c